HANDBOOK OF LOGIC
IN COMPUTER SCIENCE

Editors

S. Abramsky, Dov M. Gabbay, and T. S. E. Maibaum

HANDBOOKS OF LOGIC IN COMPUTER SCIENCE
and
ARTIFICIAL INTELLIGENCE AND LOGIC PROGRAMMING

Executive Editor
Dov M. Gabbay

Administrator
Jane Spurr

Handbook of Logic in Computer Science

Handbook of Logic in Artificial Intelligence and Logic Programming

Handbook of Logic in Computer Science

Volume 1
Background: Mathematical Structures

Edited by

S. ABRAMSKY
Professor of Computing Science

DOV M. GABBAY
Professor of Computing Science

and

T. S. E. MAIBAUM
*Professor of Foundations of
Software Engineering*

*Imperial College of Science, Technology and Medicine
London*

Volume Co-ordinator

DOV M. GABBAY

CLARENDON PRESS · OXFORD
1992

Oxford University Press, Walton Street, Oxford OX2 6DP

Oxford New York Toronto
Delhi Bombay Calcutta Madras Karachi
Kuala Lumpur Singapore Hong Kong Tokyo
Nairobi Dar es Salaam Cape Town
Melbourne Auckland Madrid
and associated companies in
Berlin Ibadan

Oxford is a trade mark of Oxford University Press

Published in the United States
by Oxford University Press Inc., New York

© Chapter authors, 1992

A catalogue record for this book is available from the British Library

Library of Congress Cataloging in Publication Data
(Data available on request)

ISBN 0–19–853735–2

Printed in Great Britain by
Biddles Ltd,
Guildford and King's Lynn

Preface

We are happy to present the first volumes of the *Handbook of Logic in Computer Science*. Logic is now widely recognized to be one of the foundational disciplines of computing and has found applications in virtually all aspects of the subject, from software engineering and hardware to programming language and artificial intelligence. There is a growing need for an in-depth survey of the application of logic in computer science and AI. The *Handbook of Logic in Computer Science* and its companion, the *Handbook of Logic in Artificial Intelligence and Logic Programming* have been created in response to this need.

We see the creation of the Handbook as a combination of authoritative exposition, comprehensive survey, and fundamental research exploring the underlying unifying themes in the various areas. The intended audience is graduate students and researchers in the areas of computing and logic, as well as other people interested in the subject. We assume as background some mathematical sophistication. Much of the material will also be of interest to logicians and mathematicians.

The tables of contents of the volumes were finalized after extensive discussions between handbook authors and second readers. The first two volumes present the background logic and mathematics extensively used in computer science. The point of view is application oriented. The other four volumes present major areas in which the methods are used. These include Volume 3 — Semantic Structures; Volume 4 — Semantic Modelling; Volume 5 — Specification and Verification; and Volume 6 — Logical Methods.

The chapters, which in many cases are of monographic length and scope, are written with emphasis on possible unifying themes. The chapters have an overview, introduction, and a main body. A final part is dedicated to more specialized topics.

Chapters are written by internationally renowned researchers in the respective areas. The chapters are co-ordinated and their contents were discussed in joint meetings.

Each chapter has been written using the following procedures:

1. A very detailed table of contents was discussed and co-ordinated at several meetings between authors and editors of related chapters. The discussion was in the form of a series of lectures by the authors to everyone present. Once an agreement was reached on the detailed table of contents the authors wrote a draft and sent it to the editors and to other related authors. For each chapter there is a second reader (the first reader is the author) whose job it has been to scrutinize the

chapter together with the editors. The second reader's role is very important and has required effort and serious involvement with the authors.

Second readers for this volume are:

Chapter 1: Valuation Systems and Consequence Relations — M. Fourman
Chapter 2: Recursion Theory — M. Hennessy and J. Tucker
Chapter 3: Universal Algebra — A. Poigne and M. Fourman
Chapter 4: Category Theory — E. Wagner and M. Fourman
Chapter 5: Topology — H. Barendregt
Chapter 6: Model Theory — I. Hodkinson and D. Gabbay.

2. Once this process was completed (i.e. drafts seen and read by a large enough group of authors), there were other meetings on several chapters in which authors lectured on their chapters and faced the criticism of the editors and audience. The final drafts were prepared after these meetings.

3. We attached great importance to group effort and co-ordination in the writing of chapters. The first two parts of each chapter, namely the Introduction–Overview and Main Body, are not completely under the discretion of the author, as he/she had to face the general criticism of all the other authors. Only the third part of the chapter is entirely for the authors' own tastes and preferences.

The Handbook meetings were generously financed by OUP, by SERC contract SO/809/86, by the Department of Computing at Imperial College, and by several anonymous private donations.

We would like to thank our colleagues, authors, second readers, and students for their effort and professionalism in producing the manuscripts for the Handbook. We would particularly like to thank the staff of OUP for their continued and enthusiastic support, and Mrs Jane Spurr, our OUP Administrator for her dedication and efficiency.

London   The Editors
June 1992

Contents

Universal algebra

Valuation Systems and Consequence Relations

Mark Ryan and Martin Sadler

Contents

1   Introduction

This chapter is an introduction to some of the basic concepts and machinery of logic, concentrating on the contemporary uses of logic in computer science. It is not comprehensive, since many topics (such as equational reasoning, algebras, categories and computability) are introduced in other chapters. In particular, we concentrate on the applications of logics as reasoning systems about computing rather than on the foundational aspects of computer science which logic addresses. This is not a standard introduction to classical first order logic and its model theory, such treatments being widely available elsewhere [Hamilton, 1978; Hodges, 1983; Makowsky,]. Rather, we attempt to bring together some topics from the logics currently used for reasoning about programs and about computer systems, and see them within a unifying framework at an appropriate level of abstraction.

Prerequisites. Most of the constructions in this chapter are based on 'naïve set theory', whose use we take for granted. We also assume some exposure to the use and manipulation of the usual propositional operators $(\wedge, \rightarrow, \vee, \neg)$, and to predicates. Some familiarity with discrete mathematics, including the use of λ-notation, and the idea of proofs both *within* and *about* a formal system would be helpful.

1.1   Logics and computer science

It has been recognized for a long time that there is an intimate connection between logic and computer science:

> 'It is reasonable to hope that the relationship between computation and mathematical logic will be as fruitful in the next century as that between analysis and physics in the last. The development of this relationship demands a concern for both applications and mathematical elegance.' (J. McCarthy, 1965)

We begin this chapter by describing some aspects of the interconnections between logic and computer science. We will introduce the machinery of logic later, here we rely on the reader's intuitions about *truth, proofs* and *propositions.*

First we can distinguish between a 'foundational' aspect and a 'reasoning system' aspect of the connection. At the foundational aspect, logic is used to provide a model of the phenomenon of computation. For example: the *Curry–Howard isomorphism* is a description of the parallel between computation and proof transformation in intuitionistic logic. An idealized view of computation is adopted: it is taken to mean reducing typed lambda terms to normal form. We can justify this idealization by appealing to the formal power of the typed λ-calculus and of Turing machines; see for example [Barendregt,] and [Phillips,] in this volume.

The key idea is the identification of *propositions* with *types*, and *proofs* with *terms*. A proposition is a type whose terms are the proofs of the proposition. This programme is called *intuitionistic type theory*, and started with P. Martin-Löf [Martin-Löf, 1975]. Reducing lambda terms to normal form (which is our model of computation) is the same as eliminating redundant steps in proofs. We return to this topic briefly later on.

Girard's *Linear logic* is a further step in this foundational direction. It abandons certain structural properties present in both classical and intuitionistic logic and by doing so provides a new approach to many issues in functional programming, such as lazy evaluation, side-effects, memory allocation and parallelism.

This is not, however, the level of interconnections between logic and computer science with which we are primarily concerned in this chapter; the interested reader should see other chapters in this Handbook. Instead, we concentrate on *logics as reasoning systems*—on their claim to be systems for making deductions from premises. Logic is used for describing and implementing systems which reason about a particular domain (e.g. in *specification theory* and *artificial intelligence*). Temporal logic is used to reason about domains in which time plays a key role, deontic logic for domains involving normative behaviour, default logic when only partial information is available, and so on. Specification theory provides many examples of this relationship between logic and computer science. The expressions in specification languages usually denote logical theories, often based on *multi-modal logics* (see [Stirling,] in this volume). Multi-modal logics have also been used to reason about concurrent systems and non-deterministic systems.

Already in this chapter we have referred to several different 'logics': intuitionistic logic, temporal logic, linear logic, deontic logic and others. There is a recurrent debate between *absolutism* and *pluralism*, whether there is 'one true logic' for reasoning about all domains or whether different logics should be studied for different domains. We attempt in this chapter to argue that computer science demands a pluralist view.

It is often stated that classical logic (the logic which is usually taken to be the 'one true logic' by absolutists) is capable of expressing arguments phrased in any 'non-standard' logic, and so removes the need for studying them. Indeed, we shall see in the last section that many logics can be embedded in classical logic in a precise sense. But to argue along these lines misses an important point. As V. Pratt says,

> 'There is a tradition in logic, carried over into computer science,
> to think of pure first order logic as a universal language. In fact

first order logic is about as useful in verification as a Turing
machine is in software engineering: cute to watch but not very
useful. The sense in which first order logic is universal is that
knowledge about domains can be embedded in it in the form
of axioms, from which arguments may proceed. Unfortunately,
axioms without methods tend not to be very useful in practice.'

It is not the formal power of classical logic that is at issue, but rather
how useful it is as a tool for representing arguments. Similarly in the choice
of which computer language to use to program a particular application,
expressive power is not the whole issue either. What seems important in
both cases is the suitability of the expressive medium for the application.
Let us see an example of this.

Example 1.1.1. *Temporal reasoning.* Sometimes we want to reason about
propositions which are not true or false once and for all, but which change
their truth values through time, as in the following.

1   John will never become a lecturer unless he has previously
    completed his Ph.D.
2   John has not yet completed his Ph.D.

---

3   Either John will complete his Ph.D., or he will never become
    a lecturer.

Sentences 1 and 2 (above the line) are the premises of the argument and 3 is
the conclusion we want to draw. This example really does involve temporal
reasoning: you cannot just have propositions saying 'John is a lecturer', or
'John has completed his Ph.D.' because the argument involves these things
being true and false at different moments of time.

We might consider expressing it in first-order logic as follows: let $L(t)$
and $P(t)$ respectively mean that John becomes a lecturer at time t and
completes his Ph.D. at time t. Let n be a constant (to be interpreted as
'now'), and $t_1 < t_2$ mean t_1 is earlier than t_2. The argument above can
then be coded as follows:

1   $\forall t\,(n{<}t \wedge L(t) \rightarrow \exists t'\,(t'{<}t \wedge P(t')))$
2   $\neg\exists t\,((t{<}n \vee t{=}n) \wedge P(t))$

---

3   $\exists t\,(n{<}t \wedge P(t)) \vee \neg\exists t'\,(n{<}t' \wedge L(t'))$

Already one might object that this is not a very perspicuous representation of a simple argument. Worse still, the argument is not even valid

as it stands. To make it valid we need to add the following premise which expresses the transitivity of the 'earlier than' relation:

4    $\forall t_1, t_2, t_3 \ (t_1 < t_2 \wedge t_2 < t_3) \rightarrow t_1 < t_3$

Other axioms, expressing the irreflexivity of the relation and the linearity of time may also be needed in temporal arguments:

5    $\forall t \ \neg(t < t)$
6    $\forall t_1, t_2 \ (t_1 = t_2 \vee (t_1 < t_2 \vee t_2 < t_1))$

So we should add these too. The problem is that our premises are cluttered with complications to do with the flow of time which we really want to be handled 'automatically' by the logic. This is what we achieve with temporal logic.

We will define unary operators $\square$ and $\boxdot$ with the following interpretations: $\square A$ means that 'A will be forever true in the future' and $\boxdot A$ means that 'A was always true in the past'. They are called temporal operators and can be applied to any formulas in the logic, including formulas which already have temporal operators. We can define their duals: $\lozenge A$ means $\neg\square\neg A$, that is, A will be true at some point in the future; and $\diamondsuit A$ is $\neg\boxdot\neg A$, which means that A was true at least once in the past.

Then let the propositions l and p mean that John is a lecturer and that John completes his Ph.D. Time is now 'part of the logic' rather than being explicitly represented. The argument goes as follows:

1    $\square(l \rightarrow \lozenge p)$
2    $\neg(\lozenge p \vee p)$
_____

3    $\lozenge p \vee \neg\lozenge l$

This argument is rather simpler than the one in predicate logic. Moreover, it is valid in the appropriate temporal logic. Temporal logics are examples of a class of logics called 'modal logics', which we will look at in this chapter.

The example above is evidence that it is appropriate to use logics which are designed with a particular domain in mind, in this case the temporal domain. We accepted that we could have represented the argument in classical logic but showed that it would be clumsier to do so. The next example illustrates an argument which would be even harder to represent in classical logic.

Example 1.1.2. The folklore of modal logic begins with Aristotle's sea battle argument. It is a paradox about what can be said to be *necessarily* true. Something is necessarily true if it is not possible for it to be false; but exactly what this amounts to varies from context to context. Sometimes it means *physically* necessary, as in 'bodies of like charge repel'; sometimes it means *logically* necessary, as in 'bachelors are unmarried'. Certainly, we can imagine a context in which a sea-captain's orders are necessarily carried out.

1 If the captain gives the order to attack, then necessarily there will be a sea battle tomorrow.
2 If he does not give the order, then necessarily there will not be one.
3 Either he gives the order, or he does not.

---

4 Either it is necessary that there is a sea battle, or it is necessary that none occurs.

The conclusion is paradoxical because of the fact that the battle's taking place depends on whether the captain gives the order or not. Neither it nor its failure to take place can be necessary; the outcome is contingent precisely on whether the order is given.

In fact, the solution is not very hard to find. What is necessary is that the captain's orders, if given, are followed, but the orders themselves are not necessary. Suppose p is the sentence 'the captain gives the order' and q is 'there will be a sea battle'. The unary operator $\Box$ is used to mean 'necessarily'. Here are two ways of coding the argument:

1	$p \to \Box q$	1 $\Box(p \to q)$
2	$\neg p \to \Box \neg q$	2 $\Box(\neg p \to \neg q)$
3	$p \vee \neg p$	3 $p \vee \neg p$
4	$\Box q \vee \Box \neg q$	4 $\Box q \vee \Box \neg q$

The difference between these two is the scope of the of the necessity operator in lines 1 and 2. Only the first version (the one on the left) is valid. The second version, which more correctly translates the intuition of the necessity of carrying out the orders, is not valid so the paradox disappears.

The notation makes the distinction between the two readings transparent. We have just the right level of expressibility; not so much that the notation is cluttered and confusing, but enough to separate the two readings precisely.

This example involves the simplest modal language, in which the unary operators $\Box$ and $\Diamond$ (box and diamond) are introduced. In the previous example there were two forms of box and two of diamond, but in a more general setting there can be infinitely many modalities, each corresponding to an action or a program, as the next example shows.

Example 1.1.3. *Multi-modal logic.* For each action or program a, there are modal operators $[a]$ and $\langle a \rangle$. If A is a sentence, $[a]A$ means: after a

occurs, A *must* be true. $\langle a \rangle A$ means that A *might* be true after a occurs. The distinction is important only if a is non-deterministic; $[a]A$ asserts A after all possible executions, while $\langle a \rangle A$ asserts that A holds after at least one. Examples:

1. $\neg \text{in}(x) \rightarrow [\text{add}(x)]\text{in}(x)$. Adding x to a database results in x's presence.

2. $([\text{push}(x)][\text{pop}]A) \leftrightarrow A$. The state of a stack is unchanged by a push followed by a pop. Notice the order in which actions are sequenced; $[a][b]A$ means 'after a, then after b, A holds'.

A full account of multi-modal logic is given elsewhere in this volume [Stirling,].

The next example has more the flavour of the foundational aspect rather than the reasoning-system aspect. It shows that intuitionistic logic and type theory have the same 'structure'. This is called the Curry–Howard isomorphism.

Example 1.1.4. *Type theory.* From a collection of atomic types we can generate a bigger collection of types by combining them in various ways. Given types A and B we have the following compound types:

1. A term of type $A \times B$ (Cartesian product) consists of a pair (a, b), where a is of type A and b is of type B.

2. A term of type $A + B$ (disjoint union) consists of a term of type A or a term of type B, together with an indication of which of A or B it is.

3. A term of type $A \rightarrow B$ (function space) consists of a functional term which, applied to a term of type A, returns a term of type B.

4. A term of type $\prod_{i \in I} A_i$ (Cartesian product indexed by I) consists of a function which takes each element i in I to a term of type A_i.

5. A term of type $\sum_{i \in I} A_i$ (disjoint union indexed by I) consists of an element i of I together with a term a which is of type A_i.

It turns out that there is a precise parallel between this table and the interpretation given to the logical operators $\wedge$, $\vee$, $\rightarrow$, $\forall$ and $\exists$ in *intuitionistic logic*. Intuitionistic logic is described later in this chapter (Section 4), but for now we will introduce the idea of constructive proof, which is its historical motivation.

Intuitionistic logic arose from the view that only proofs of a 'constructive' nature should be allowed in mathematics. This idea is due to Brouwer.

Intuitionictic logic, due to Heyting, It is not the only formalisation of this view, but is the most influential. All intuitionistic proofs count as ordinary (classical) proofs, but not conversely. There is a proof of $p \vee \neg p$ in classical logic, but it is not a constructive proof, so it is not valid in intuitionistic logic. An intuitionistic proof of the sentence $A \vee B$ must consist of either a proof of A or a proof of B, and since neither p nor $\neg p$ can be proved (from no premises), so $p \vee \neg p$ cannot be proved. Likewise, an intuitionistic proof of a sentence of the form $\exists x\, A(x)$, say one in which the variable x ranges over the natural numbers, is a proof of a specific instance $A(n)$ for some specific natural number n or an effective means of finding such a proof.

The following table summarizes the definition of intuitionistic proof.

1. A proof of $A \wedge B$ consists of a pair (a, b), where a is a proof of A and b is a proof of B.

2. A proof of $A \vee B$ consists of a proof of A or a proof of B.

3. A proof of $A \rightarrow B$ consists of a construction (i.e. a function) which transforms any proof of A into a proof of B.

4. A proof of $\forall x\, A$ consists of a construction which, given an individual c, returns a proof of $A[c/x]$ (this notation means A with all free occurrences of x replaced by c).

5. A proof of $\exists x\, A$ consists of an instance c and a proof of $A[c/x]$.

This interpretation of the logical operators was introduced in the 1930s, and is due to Heyting. These tables show the Curry–Howard isomorphism between constructive proof on one hand and type theory on the other. This is further explained in [Girard, 1989].

1.2 Summary

The examples given show a tiny number of the diverse logics available. We made a distinction between the 'foundational' connection between logic and computer science and the 'reasoning system' connection. It is the latter that we concentrate on in this chapter.

The remainder of this chapter is structured as follows. Section 2 describes the satisfaction-based view of logic, in which the basic structure is a *valuation system* in which sentences denote a *truth value*. Section 3 is about the consequence-based view of logic, which concerns which conclusions can be drawn from which premises. We show that this view arises naturally from the satisfaction-based view. In Section 4, we look at syntactic approaches to defining consequence; this topic is called *proof theory*.

We give the proof theory for the three main logics discussed in the chapter, namely classical logic, intuitionistic logic, and the modal logic S4. In Section 5 some further topics are discussed, including maps between logics, and more about valuation systems.

2 Valuation systems

There are two fundamental and well established views as to what logic is about. We will call them the *satisfaction-based* and *proof-based* views. The first one focuses on the manipulation of *valuation systems*—reasoning by truth tables is an example of this. We will begin this section by reviewing and then developing this view. The second view is in many ways more basic, focusing on the manipulation of syntactic expressions to formulate proofs of arguments. This is the proof-based view—the premises and the conclusion are linked by a *proof.*

The two views of logic differ in what they think is fundamental. In this chapter we will remain largely agnostic as to what is the right view of logic, if indeed there is such a thing! What is important is that both of these views gives rise to a basic mathematical structure called a consequence relation.

Valuation systems, or variations on them, appear again and again in the interplay between logic and computer science. It is with them that we begin rather than logical systems *per se*. In section 3 we generalise this to consequence relations, and section 4 considers the other important special case of consequence relations, namely proofs.

2.1 Satisfaction

In both views of logic we start with some basic linguistic entities, called *sentences.* These sentences are usually given by *inductive definitions* (such as the definition which follows) and constitute a *formal language.* The satisfaction-based view is concerned with whether these sentences are *true* or *false* (or have some other *truth value*), and this question is decided with respect to an *interpretation* (sometimes also called an *algebra* or a *structure,* depending on the context). These concepts are illustrated by the following examples.

Example 2.1.1. Consider a language in which sentences are made using the symbols: a, $'$ (prime), $\oplus$, and $=$. These symbols can be combined to make expressions called 'terms'. The rules for making terms are:

1. a is a term.

2. If t is a term then t' is a term.

3. If t_1 and t_2 are terms then $t_1 \oplus t_2$ is a term.

With any such set of rules there is an implicit rider: anything which cannot be formed using these rules is not a term. Examples: a' and $a' \oplus a''$ are terms; but $a' \oplus$ is not, for rule 3 only allows $\oplus$ to be introduced between two terms. The notation is ambiguous; for example, there are two distinct terms both of which are noted $a \oplus a' \oplus a''$. There is the term formed by combining the terms a and a' according to rule 3, and combining the result with a'' according to rule 3 again. But also, there is the term formed by combining a with the combination of a' and a''. We can use brackets to disambiguate terms, so these two are more perspicuously noted $(a \oplus a') \oplus a''$ and $a \oplus (a' \oplus a'')$ respectively. Also, we use brackets to distinguish $a \oplus (a')$ and $(a \oplus a)'$. We will take the unbracketed expression $a \oplus a'$ to mean the former of these two, and insist on using brackets when the latter is meant.

The reader may be more familiar with the convention that brackets are a formal part of the syntax, which clearly they are not here. There are advantages and disadvantages to both conventions. The formal-syntax convention leads to a proliferation of brackets.

Terms may be composed to form bigger linguistic entities; 'sentences' are made by the following rule.

If t_1 and t_2 are terms then $t_1 = t_2$ is a sentence.

Examples: $a = a'$, $a \oplus a' = a''' \oplus a'$ and $(a \oplus a')' = a$ are all sentences. $(a = a') \oplus a''$ is not because it is not formed according to the rules.

So far we have specified a language (syntax), but we have not assigned any meaning (semantics) to it. Here is an interpretation of the language we have described:

- terms are interpreted as natural numbers;

- the symbol a is interpreted as the number 0;

- the symbol $'$ is interpreted as the successor function on numbers (the successor of n is $n + 1$);

- the symbol $\oplus$ is interpreted as addition;

- the symbol $=$ is interpreted as equality between numbers.

Now we can decide which sentences in the language are true in the interpretation we have given, and which are false. The sentence $a'' \oplus a'' = (a \oplus a)''''$ is true (both terms are interpreted as 4), but the sentence $a'' = ((a \oplus a) \oplus a)$ is false ($2 \neq 0$).

Remember that this is just one interpretation—there are many others. Consider, for example, the interpretation of the language which is like the one above, so that terms are interpreted as integers, except that the symbol $=$ is interpreted as 'greater-than'. The fact that it was written $=$ suggests that it was *intended* to be interpreted as equality, but the constraint that the syntax imposes is merely that it sits between two terms, and interpreting it as greater-than meets this constraint. The interpretations of the sentences $a'' \oplus a'' = (a \oplus a)''''$ and $a'' = ((a \oplus a) \oplus a)$ are then '4 is greater than 4' and '2 is greater than 0', which are false and true respectively.

Example 2.1.2. We can extend the language described in Example 2.1.1 with the following symbols: $\wedge$, $\vee$ and $\neg$. The symbols $\wedge$ and $\vee$ are binary operators between sentences and are called *conjunction* and *disjunction*; $\neg$ is a unary operator on sentences called *negation*. Thus, sentences are defined as follows.

1. As before, if t_1 and t_2 are terms then $t_1 = t_2$ is a sentence.

2. If A is a sentence so is $\neg A$.

3. If A and B are sentences so are $A \wedge B$, $A \vee B$ and $A \rightarrow B$.

Again, brackets will be used to disambiguate where necessary. In $A \wedge B$, A and B are called *conjuncts*; in $A \vee B$ they are called *disjuncts*. Our intended interpretation is one in which: $\wedge$ is interpreted as 'and', $\vee$ is 'or', $\rightarrow$ is 'implies' and $\neg$ is 'not'. In the simplest interpretation $A \wedge B$ is true if both A and B are true; otherwise it is false. $A \vee B$ is true if either A or B is true, and false if they are both false. $\neg A$ is true if A is false, and false otherwise. $A \rightarrow B$ is true if B is true whenever A is true. We say 'the simplest interpretation' because we will look at some other interpretations of the operators later in this section.

Notice that in the above examples, we happened to have a three-tiered language: first 'terms' like $a' \oplus a$, then what we might call 'atomic sentences' which consist of the $=$ symbol placed between two terms, and then 'compound sentences' like $(a'' \oplus a'' = (a \oplus a)'''') \wedge \neg(a'' = (a \oplus a) \oplus a)$. Our interpretations were given in three stages as well: first we interpreted the terms (as integers) by giving the meanings of a, $'$ and $\oplus$. Then we interpreted the atomic sentences (as truth values) by stating that $=$ meant equality between integers, and the (compound) sentences by providing the meanings of $\wedge$, $\vee$ and $\neg$. The important thing to remember is that we have sentences in a language which are true or false *in an interpretation*.

The notion of sentences being true or false in interpretations is pervasive. Here is a small example which shows a way in which the idea occurs in computer science. We will keep the sentence operators $\neg$, $\wedge$ and $\vee$ of the last example (Example 2.1.2), but our atomic sentences will look quite different.

Example 2.1.3. *Straight-line programs.* Let $S_1; S_2; \ldots; S_n$ be a sequence of possibly guarded assignments in an assignment-based programming language. An assignment is said to be 'guarded' if it is executed only if a certain condition (the 'guard') is true. Let $s_0, s_1, \ldots, s_n$ be the points in the program text between the statements (s_0 and s_n being the points in the text before and after the program). For each such point we will specify an interpretation of the language. Atomic sentences will express arithmetic properties of the values of variables. For example, in the following program

$$
\begin{array}{l}
\bullet s_0 \\
X := 1 \\
\quad \bullet s_1 \\
Y := 3 \\
\quad \bullet s_2 \\
X := X + 1 \\
\quad \bullet s_3 \\
Y := Y \times 2 \\
\quad \bullet s_4 \\
\text{if } Z > 0 \text{ then } X := 1 \\
\quad \bullet s_5
\end{array}
$$

the sentence $Y = 3 \times X$ is true in the interpretation s_2 and s_4, and false in s_3. It is undefined in s_0, s_1 and s_5, because we do not know the values of the variables when this piece of code is entered. For some entry values it is true and for some it is false. The sentence $(Z \leqslant 0) \vee (X = 1)$ is true in s_1, s_2 and s_5. In s_1 and s_2 it is true because the second disjunct is true. In s_5 one of the disjuncts must be true (which one depends on the initial value of Z). At the remaining points the sentence depends on the initial values of X and Z, and so is undefined here.

Notice that we are not insisting that every sentence be true or false in an interpretation, and have allowed some to be undefined. We can handle this formally by making 'undefined' a third truth value. The restriction to assignments (disallowing looping constructs) in this example is because we wanted to make the interpretations points in the text. Looping programs traverse points several times, so it would have been necessary to make interpretations be 'execution paths' of the program rather than just points.

2.1.1 Notation

We will use the symbols A, B, etc. to range over sentences, and Γ, Δ, etc. for sets of sentences. If A is true of an interpretation m, we write $m \Vdash A$, and say m *satisfies* A. We also say that A *holds in* m, and that

m is a *model of A*. We call ⊩ a *satisfaction* relation. Given a collection of sentences and a collection of interpretations, the satisfaction-based view of logic is concerned with identifying this relation.

We can equally well think of *evaluating* a sentence in an interpretation and obtaining a truth value. $\llbracket A \rrbracket_m$ means the truth value of A in m. In the simple case of having just two truth values **t** and **f** (true and false), we will write $m \Vdash A$ iff $\llbracket A \rrbracket_m = \mathbf{t}$. The reader acquainted with *denotational semantics* of programming languages will be familiar with these ideas. Here, a sentence denotes a truth value.

These two styles, the relational style (⊩) and the functional ($\llbracket \cdots \rrbracket$), are two sides of the same coin. But the functional style is more expressive, because instead of just saying that an interpretation satisfies a sentence, we can use different truth values to distinguish between degrees of satisfaction. We have only talked about two truth values up to now, but shortly we will see examples of several and even infinitely many truth values.

2.2 Valuation systems

We have seen in Examples 2.1.1 and 2.1.2 how the languages we use can have bigger syntactic categories[1] defined in terms of simpler ones. From a logical point of view we are most interested in the behaviour of the operators between sentences (so far we have seen the operators ∧, ∨ and ¬), and not so much in how atomic sentences are constructed. For the remainder of this section we will leave atomic sentences unanalysed, and write them as p, q, r, etc. There is an analogy between the languages we are considering here and programming languages in computer science. Our concentrating on the logical operators and leaving the atomic sentences unanalysed is analogous to concentrating on the constructs in a programming language which form bigger statements from smaller ones (e.g. **while**, **if** and so on) and ignoring the exact nature of the atomic statments.

A *propositional language* is a collection of atomic sentences together with a collection of operators for putting them together. Let us assume for now the atomic sentences $\{p, q, r, \ldots\}$, and the operators ∧, ∨, →, ¬ and ⊥ (and, or, implies, not and falsity). As usual, ∧, ∨ and → are binary operators, ¬ is unary; ⊥ is a nullary operator. Sentences in the language are the atomic sentences together with sentences made of atomic sentences using the operators: they include p, $(p \wedge q) \rightarrow r$, $\neg p \rightarrow (q \rightarrow r)$, $p \rightarrow \bot$ and so on. Formally, the following rules define the language, which we will call L:

1. If A is an atomic sentence then A is in L.

---

[1]A syntactic category is simply a class of expressions in a language. This is not to be confused with categories in Category Theory, studied elsewhere in this volume!

2. $\perp$ is in L.

3. If A is in L then so is $\neg A$.

4. If A and B are in L, then so are $A \wedge B$, $A \vee B$ and $A \to B$.

The interpretation we have in mind for $\perp$ is that it stands for a contradiction.

We now give several examples of interpretations for L. The first one is the standard interpretation which yields classical logic.

Example 2.2.1. *Two-valued truth tables.* These specify a class of interpretations for L which fixes the intended meaning of the operators, while leaving the atomic sentences undefined. For example, it prevents us having $m \Vdash p \wedge q$ ($p \wedge q$ is true in m) without also having $m \Vdash p$, because the intended meaning of $\wedge$ includes that if $p \wedge q$ is true in the interpretation m then so must p be. We specify this by means of the following truth tables.

$\wedge$	t	f		$\vee$	t	f		$\to$	t	f		$\neg$			$\perp$	
t	t	f		t	t	t		t	t	f		t	f			f
f	f	f		f	t	f		f	t	t		f	t			

If A and B both evaluate to **t** in a particular interpretation m (i.e. $[\![A]\!]_m = \mathbf{t}$ and $[\![B]\!]_m = \mathbf{t}$) then $A \wedge B$ must evaluate to **t** as well. If $[\![A]\!]_m = \mathbf{t}$ then $[\![\neg A]\!]_m = \mathbf{f}$, and vice versa. Also, $[\![\perp]\!]_m = \mathbf{f}$ for any m.

Having given these truth tables, we don't have to give the truth value of every sentence to specify an interpretation, because the truth tables act as constraints on what interpretations can be. We can specify interpretations simply by stating what the atomic sentences evaluate to. Let m be an interpretation such that $[\![p]\!]_m = \mathbf{t}$, $[\![q]\!]_m = \mathbf{t}$ and $[\![r]\!]_m = \mathbf{f}$. Then since $[\![\neg p \to (q \to r)]\!]_m = \mathbf{t}$ and $[\![(p \wedge q) \to r]\!]_m = \mathbf{f}$, we write $m \Vdash \neg p \to (q \to r)$ and $m \nVdash (p \wedge q) \to r$.

Truth tables are a functional way of describing valuation systems. A corresponding relational description for classical logic is

$$m \Vdash A \wedge B \quad \text{iff} \quad m \Vdash A \text{ and } m \Vdash B$$
$$m \Vdash A \vee B \quad \text{iff} \quad m \Vdash A \text{ or } m \Vdash B$$
$$m \Vdash A \to B \quad \text{iff} \quad m \nVdash A \text{ or } m \Vdash B$$
$$m \Vdash \neg A \quad \text{iff} \quad m \nVdash A$$
$$m \nVdash A$$

Here is another class of interpretation for L.

Example 2.2.2. *Underdetermination.* Here we distinguish a third possibility, that the truth value of a sentence is currently *undefined*. There

are many ways of doing this, depending on the intuitions we are trying to
capture. The most common way of doing this is the following:

∧	t	u	f
t	t	u	f
u	u	u	f
f	f	f	f

∨	t	u	f
t	t	t	t
u	t	u	u
f	t	u	f

→	t	u	f
t	t	u	f
u	t	u	u
f	t	t	t

¬	
t	f
u	u
f	t

⊥
f

This system is due to Kleene. For example, if $[\![A]\!]_m = \mathbf{f}$ and $[\![B]\!]_m = \mathbf{u}$,
then $[\![A \vee B]\!]_m = \mathbf{u}$. We cannot say that it is **t** or **f** because it depends on
what B turns out to be. But if $[\![A]\!]_m = \mathbf{f}$ and $[\![B]\!]_m = \mathbf{u}$ then $[\![A \wedge B]\!]_m = \mathbf{f}$
because it doesn't matter what B evaluates to. One way of thinking about
this is that we allow **u** to 'grow' into **t** or **f** as information arrives.

Now that we have more than two truth values, when should we write
$m \Vdash A$ in the relational style and when not? To answer this we can split the
truth values between those which are *designated* and those which are not,
and write $m \Vdash A$ iff $[\![A]\!]_m$ is designated. Designated truth values are the
ones which 'count as true' when one is not interested in the fine structure
of the truth value set. Non-designated ones count as false. Typically,
only **t** will be designated in a three valued system. But if our interest is in
potential truth then **t** and **u** will be designated; only **f** lacks potential truth.
If our interest is in knowledge, then just **t** and **f** will be designated. On that
view, the decision about designated values should reflect two orderings on
the space of truth values: a truth ordering and a knowledge ordering.

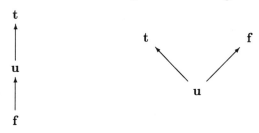

truth ordering　　　　　　　　knowledge ordering

The designated values are the higher ones according to an ordering—which
one depends on the application.

We can now abstract out the essential aspects of the examples so far.

Definition 2.2.3. A *propositional language* L is a pair $\langle P, O \rangle$ consisting
of a countable set P of atomic sentences together with a set O of operators

(sometimes called connectives). Operators come with an *arity*, which is the number of sentences they take to form another sentence. Operators ∧, ∨ and → have arity 2, ¬ has arity 1 and ⊥ has arity 0 (it is already a sentence).

The set of *sentences* of a propositional language L is the smallest set containing P closed under the operators O. L is ambiguously used to mean the language and the set of sentences of the language. Brackets (and sometimes precedence orderings) are used to disambiguate where necessary.

Definition 2.2.4. A *valuation system* **M** for a propositional language L is a triple $\langle M, D, F \rangle$ where:

1. M is a set with at least two elements, the set of truth values;

2. D is a non-empty proper subset of M, the set of designated truth values;

3. $F = \{f_{o_1}, f_{o_2}, \ldots f_{o_n}\}$ is a set of functions, one corresponding to each operator in $O = \{o_1, o_2, \ldots, o_n\}$, such that $f_o : M^{n_o} \to M$ (where n_o is the arity of the operator o). We say that f_o *interprets* o.

The set F, which is the interpretation of the operators in O, was represented by truth tables in the examples above. The notion of valuation systems with designated truth values is due to M. Dummett [Dummett, 1977].

It is easy to see how our earlier examples fit into the valuation system mould we have introduced. Example 2.2.1 is the most important class of interpretations for propositional languages so we will repeat it in the form of a definition using valuation systems.

Definition 2.2.5. *Classical propositional logic.* Let L be a propositional language with the operators $\{\land, \lor, \to, \neg, \bot\}$. The classical valuation system for L is $\langle M, D, F \rangle$ where $M = \{\mathbf{t}, \mathbf{f}\}$, $D = \{\mathbf{t}\}$ and F is given as follows:

$f_\land$	t	f
t	t	f
f	f	f

$f_\lor$	t	f
t	t	t
f	t	f

$f_\to$	t	f
t	t	f
f	t	t

$f_\neg$	
t	f
f	t

$f_\bot$	
	f

These tables are the same as the truth tables of Example 2.2.1, but written in the new notation we have introduced.

We might prefer to make the truth values mathematically more 'structured'.

Example 2.2.6. The valuation system for classical propositional logic can also be given as follows:

$$
\begin{aligned}
M' &= \{0,1\} \\
D' &= \{1\} \\
f'_\wedge(x,y) &= x \times y \\
f'_\vee(x,y) &= x + y \ \text{(but with } 1+1=1) \\
f'_\neg(x) &= 1 - x \\
f'_\rightarrow(x,y) &= (1-x) + y \\
f'_\perp &= 0
\end{aligned}
$$

This valuation system is easily seen to be equivalent to classical propositional logic (Definition 2.2.5) in the following sense: let $\langle M, D, F \rangle$ be the valuation system of Definition 2.2.5. Then there is a bijection $* : M \to M'$ such that D is mapped onto D' ($*$ preserves designated values) and for all operators i, $(f_i(t_1, \ldots, t_n))^* = f'_i(t_1^*, \ldots, t_n^*)$; $*$ preserves the interpretation of the operators. The topic of structure-preserving maps between algebras is taken up in [Tucker and Meinke,] (this volume).

A valuation system gives us a *compositional account* of the operators of our language, but it does not tell us how to evaluate atomic sentences. To get off the ground we need to assign truth values to the atomic sentences.

Definition 2.2.7. An *assignment* a relative to a valuation system $\mathbf{M} = \langle M, D, F \rangle$ for a language $L = \langle P, O \rangle$ is a function $a : P \to M$.

Given such an assignment we can now compute the truth value of any sentence of L.

Definition 2.2.8. Each assignment a relative to a valuation system $\mathbf{M}$ induces an *interpretation* (or *valuation*) v_a given by:

1. $v_a(p) = a(p)$, for $p \in P$.

2. $v_a(o(A_1, \ldots A_n)) = f_o(v_a(A_1), \ldots v_a(A_n))$, where n is the arity of o and f_o interprets o in $\mathbf{M}$.

An interpretation is a valuation system plus an assignment. The valuation system part guarantees a compositional evaluation of the operators, while the assignment part specifies the truth values of the atomic sentences. The reason for splitting interpretations into two components is to allow the truth value of the atomic sentences (the assignment) to vary, while the treatment of the operators remains fixed. So $\mathbf{M}$ is a collection of interpretations each of which treats the operators in the language in the same way.

Definition 2.2.9. A sentence A is a *tautology* in a valuation system $\mathbf{M}$ if for every assignment a relative to $\mathbf{M}$, $v_a(A) \in D$.

Example 2.2.10. The *law of excluded middle*, $p \vee \neg p$, is a tautology in classical propositional logic. For let a be an assignment for the valuation system in Definition 2.2.5. Either $a(p) = \mathbf{t}$ or $a(p) = \mathbf{f}$. Suppose $a(p) = \mathbf{t}$. Then

$$
\begin{aligned}
v_a(p \vee \neg p) &= f_\vee(v_a(p), v_a(\neg p)) \\
&= f_\vee(\mathbf{t}, f_\neg(\mathbf{t})) \\
&= f_\vee(\mathbf{t}, \mathbf{f}) \\
&= \mathbf{t} \\
&\in D
\end{aligned}
$$

Similarly, if $a(p) = \mathbf{f}$ then $v_a(p \vee \neg p) = \mathbf{t} \in D$, so $p \vee \neg p$ is a classical tautology. It is not, however, a tautology in the three-valued valuation system of Example 2.2.2, as the reader should check.

Definition 2.2.11. Sentences A and B are *equivalent* in a valuation system $\mathbf{M}$ if for every assignment a relative to $\mathbf{M}$, $v_a(A) = v_a(B)$.

Example 2.2.12. (Classical logic continued.) In classical logic, the following pairs of sentences are equivalent.

$$
\begin{aligned}
A \wedge B \quad &\text{and} \quad \neg(A \rightarrow \neg B) \\
A \vee B \quad &\text{and} \quad \neg A \rightarrow B \\
\neg(A \wedge B) \quad &\text{and} \quad \neg A \vee \neg B \\
\neg(A \vee B) \quad &\text{and} \quad \neg A \wedge \neg B \\
\neg\neg A \quad &\text{and} \quad A \\
\neg(A \rightarrow A) \quad &\text{and} \quad \bot
\end{aligned}
$$

The proofs are by case analysis; we will show it for the first pair. Suppose $v_a(A) = \mathbf{t}$. If $v_a(B) = \mathbf{t}$ then $v_a(\neg(A \rightarrow \neg B)) = f_\neg(f_\rightarrow(\mathbf{t}, f_\neg(\mathbf{t})))$ and $v_a(A \wedge B) = f_\wedge(\mathbf{t}, \mathbf{t})$, both of which are $\mathbf{t}$ according to the tables of Definition 2.2.5. If, on the other hand, $v_a(B) = \mathbf{f}$, then $v_a(\neg(A \rightarrow \neg B))$ and $v_a(A \wedge B)$ both come out to be $\mathbf{f}$. The argument for the cases that $v_a(A) = \mathbf{f}$ is similar.

Note that these equivalences do not hold for all logics considered in this chapter.

The following examples illustrate the power of the valuation system framework.

Example 2.2.13. *Sequential evaluation.* Let L be the language given by $P = \{p, q, r, \ldots\}$ and $O = \{\wedge, \vee, \rightarrow, \neg\}$ as before, and let $M = \{\mathbf{t}, \mathbf{f}, \mathbf{u}\}$ and

$D = \{\mathbf{t}\}$ as in Example 2.2.2. But let $f_\wedge$ and $f_\to$ be given by the following truth tables:

$f_\wedge$	t	u	f
t	t	u	f
u	u	u	u
f	f	f	f

$f_\to$	t	u	f
t	t	u	f
u	u	u	u
f	t	t	t

$f_\neg$ and $f_\perp$ are as before (Example 2.2.2). The idea behind these definitions is as follows: sentences can evaluate to **t** or to **f**, or their evaluation can fail to terminate (because of lack of information, perhaps, or because of the nature of the algorithm). If the sentence $p \wedge q$ is evaluated sequentially, and p fails to terminate, then the whole sentence fails to terminate (and so its denotation is **u**). This is in spite of the fact that evaluating q may yield **f**, which, under *parallel* evaluation (Example 2.2.2) would make the denotation of $p \wedge q$ be **f**.

Example 2.2.14. Another variation on the three-valued theme (due to Goddard and Routley) forms the basis for modelling *insignificance*. This time **u** can be thought of as 'non-significant' or 'meaningless'. $M = \{\mathbf{t}, \mathbf{u}, \mathbf{f}\}$ and $D = \{\mathbf{t}\}$ as before, but

$f_\wedge$	t	u	f
t	t	u	f
u	u	u	u
f	f	u	f

$f_\to$	t	u	f
t	t	u	f
u	u	u	u
f	t	u	t

	$f_\neg$
t	f
u	u
f	t

Any sentence involving **u** evaluates to **u**, while sentences not involving **u** evaluate to their standard propositional truth value.

Examples 2.2.2, 2.2.13 and 2.2.14 have in common that they are about a third truth value whose intuitive meaning is 'undefined'. The next example introduces a fourth value, *overdefined* or *ambiguous*.

Example 2.2.15. The simplest case of a bilattice consists of the four truth values in $M = \{\mathbf{u}, \mathbf{t}, \mathbf{f}, \mathbf{o}\}$. Bilattices were introduced by Ginsberg [Ginsberg, 1990], and by Fitting [Fitting, 1989]. As before we consider both the truth ordering and the knowledge (or evidence) ordering. Just as undefined (**u**) is neither true nor false, overdefined (**o**) can be considered both true and false at once. Thus one might have written the truth value set as $M = \{\emptyset, \{\mathbf{t}\}, \{\mathbf{f}\}, \{\mathbf{t}, \mathbf{f}\}\}$, as Ginsberg [Ginsberg, 1990] does, making the knowledge ordering more explicit.

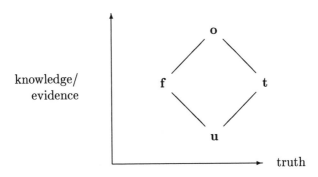

knowledge/
evidence

truth

Ginsberg defines the operators ∧, ∨ and ¬ as follows:

$f_\wedge$	t	u	f	o
t	t	u	f	o
u	u	u	f	f
f	f	f	f	f
o	o	f	f	o

$f_\vee$	t	u	f	o
t	t	t	t	t
u	t	u	u	t
f	t	u	f	o
o	t	t	o	o

	$f_\neg$
t	f
u	u
f	t
o	o

In lattice-theoretic terms, $f_\wedge$ and $f_\vee$ are meet and join respectively on the truth ordering, and $f_\neg$ inverts the truth ordering. With this number of truth values, there is more choice of how → should be defined, depending on the application. A natural choice is to make $f_\to$ be ⩽ on the truth ordering:

$f_\to$	t	u	f	o
t	t	f	f	f
u	t	t	f	?
f	t	t	t	t
o	t	?	f	t

This is easy in all but two cases, noted ?, which again must be determined by the application in mind.

More generally, a bilattice is defined to be any set of truth values with two partial orderings, $\leqslant_t$ and $\leqslant_k$, satisfying the condition that the meet and join operation of each ordering is monotonic with respect to the other ordering (see [Fitting, 1989] for more details).

This last example shows the beginnings of how we can structure the set of truth values. The next example shows how the space of truth values need not be finite or even countable. Both these ideas lead to modal logic, as we will see in the next section.

Example 2.2.16. *Venn diagrams.* Let M be the set of all subsets of a bounded region, R, of the plane $\mathbb{R}^2$ (in fact any set will do here but a bounded region of the plane suggests Venn diagrams drawn on the page);

and let D be a subset of M which is closed under union ($X, Y \in D$ implies $X \cup Y \in D$). Then let:

$$
\begin{aligned}
f_\wedge(X, Y) &= X \cap Y \\
f_\vee(X, Y) &= X \cup Y \\
f_\neg(X) &= R - X \\
f_\to(X, Y) &= (R - X) \cup Y \\
f_\perp &= \emptyset
\end{aligned}
$$

There is a structure preserving map from this valuation system to the standard valuation system for classical logic (Definition 2.2.5) which respects the meanings of the operators. (Such maps are called homomorphisms.) In this case, every proper subset of R is mapped to **f**, and R is mapped to **t**. This means that the extra machinery—the use of the region on the plane—has not bought us anything new, but it will in the next section when we exploit the fact that the truth values are sets.

Here the truth values are objects made up of smaller objects, in this case the points of the region of the plane we are considering. When this is so, we can define a new notation which will be useful in the next section: given an assignment, a, we say A is *true* at $(x, y) \in R$ iff $(x, y) \in v_a(A)$. In this case, we also say that (x, y) forces A and write $(x, y) \Vdash_a A$.

2.3 Modal logic and possible worlds

We have seen how valuation systems may be made up of smaller units (points in the example above), which themselves behave like interpretations of the language in the sense that they respect the meanings of the operators. In this section we will extend this idea by adding structure to the set of points in an appropriate way, and introduce operators to exploit the structure. We start by giving the language (which is the same as the language we have been using up to now but with two more operators called modal operators) and then we give the valuation system. The valuation system is called 'S4' and is one of a family of valuation systems for languages with the new operators.

Let L be the language given by $P = \{p, q, r, \ldots\}$ and $O = \{\wedge, \vee, \to, \neg, \square, \lozenge\}$ (see Definition 2.2.3). Operators $\square$ and $\lozenge$ are unary and are usually pronounced 'box' and 'diamond'. They are called *modal* operators, for reasons which will become clear. The following family of interpretations for L is called S4, which is the name given to the equivalent system in the classification of modal logics in 1932 by one of the founders of the subject, C. I. Lewis.

Definition 2.3.1. *Valuation systems for modal logic S4.* Let $(W, \leqslant)$ be a pre-order (that is, W is a set and $\leqslant$ is a reflexive and transitive order on it), and let w_0 be a distinguished element of W. W might again be a bounded region of the plane, with $\leqslant$ a pointwise or lexicographic ordering; we only require that W is a set pre-ordered in some way. Let $\mathbf{M}$ be the valuation system given by:

1. $M = \mathcal{P}(W)$ (the power set of W);

2. $D = \{X \subseteq W \mid w_0 \in X\}$;

3. F is the following collection of functions:

$$\begin{aligned}
f_\wedge(X, Y) &= X \cap Y \\
f_\vee(X, Y) &= X \cup Y \\
f_\neg(X) &= W - X \\
f_\rightarrow(X, Y) &= (W - X) \cup Y \\
f_\perp &= \emptyset \\
f_\square(X) &= \{x \in W \mid \forall y\, x \leqslant y \text{ implies } y \in X\} \\
f_\diamond(X) &= \{x \in W \mid \exists y\, x \leqslant y \text{ and } y \in X\}
\end{aligned}$$

This specifies a family of valuation systems, one for each triple $(W, \leqslant, w_0)$. In the literature, $(W, \leqslant)$ is known as a modal frame; W is a set of 'worlds', and $\leqslant$ is known as the accessibility relation. The term w_0 is the distinguished 'actual' world.

Notice that the operators $\square$ and $\diamond$ (unlike the others) involve the ordering on W: if the sentence A contains a $\square$ or a $\diamond$, its truth value relative to an assignment a depends not only on the truth values of the atomic sentences in A, but also on how the points in W are connected together. This is more perspicuously seen by the definition of each $f \in F$ in terms of the forcing relation $\Vdash$ we gave in Example 2.2.16, for we have the following result.

Proposition 2.3.2. Let $x \Vdash_a A$ iff $x \in v_a(A)$, for a given assignment a. Then we have:

$$\begin{aligned}
x \Vdash_a p \quad &\text{iff} \quad x \in a(p) \\
x \Vdash_a A \wedge B \quad &\text{iff} \quad x \Vdash_a A \text{ and } x \Vdash_a B \\
x \Vdash_a A \vee B \quad &\text{iff} \quad x \Vdash_a A \text{ or } x \Vdash_a B \\
x \Vdash_a A \rightarrow B \quad &\text{iff} \quad x \nVdash_a A \text{ or } x \Vdash_a B \\
x \Vdash_a \neg A \quad &\text{iff} \quad x \nVdash_a A \\
x \Vdash_a \square A \quad &\text{iff} \quad y \Vdash_a A \text{ for every } y \text{ such that } x \leqslant y \\
x \Vdash_a \diamond A \quad &\text{iff} \quad y \Vdash_a A \text{ for some } y \text{ such that } x \leqslant y
\end{aligned}$$

Proof. We will prove lines 1, 2 and 6 to give the idea.

1. $x \Vdash_a p$ iff $x \in v_a(p)$ (definition of $\Vdash_a$); and $x \in v_a(p)$ iff $x \in a(p)$ (Def. 2.2.8).

2. $x \Vdash_a A \wedge B$ iff $x \in v_a(A \wedge B)$. But, $v_a(A \wedge B) = v_a(A) \cap v_a(B)$ (Defs. 2.2.8 and 2.3.1), so the condition is equivalent to $x \in v_a(A)$ and $x \in v_a(B)$, i.e. $x \Vdash_a A$ and $x \Vdash_a B$.

6. $x \Vdash_a \Box A$ iff $x \in v_a(\Box A)$ iff $x \in \{x \in W \mid \forall y\, x \leqslant y \text{ implies } y \in v_a(A)\}$, i.e. iff $x \leqslant y$ implies $y \Vdash_a A$. ∎

The idea of this is that we can now give 'simple' truth values (true or false) to sentences at points in W. For example, the first line of this table says: $A \wedge B$ is 'true at x' iff both A and B are true there. Compare this with the table given after example 2.2.1. We have from the definitions that

$$v_a(A) \in D \quad \text{iff} \quad w_0 \Vdash_a A$$

Thus, having the designated truth value in the whole system simply means having the value 'true' in the distinguished world w_0. What is special about $\Box$ and $\Diamond$ is that the truth value of $\Box A$ at x does not depend only on the truth value of A at x, but on the truth value of A at a lot of other points as well. The same goes for $\Diamond$.

What this shows is that each point in W acts like a smaller valuation system which respects the meanings of the non-modal operators ($\wedge$, $\vee$, $\neg$, $\to$) in the sense of the standard valuation system for classical logic (Definition 2.2.5). The modal operators $\Box$ and $\Diamond$ are non-local in this sense: the truth value of $\Box A$ or $\Diamond A$ at x depends on the truth value of A at other points as well as x. This interpretation of the modal operators was invented by S. Kripke in a seminal paper [Kripke, 1963]. For a comprehensive account of modal logic, as well as a guide to the literature, see [Hughes and Cresswell, 1968] and [Hughes and Creswell, 1984]. More recently, J. van Benthem has surveyed this idea of 'gluing' classical interpretations together [van Benthem, 1985].

Example 2.3.3. Let L be the language given by $P = \{p, q\}$ and with the same operators as above. Let $W = \{w_0, w_1, w_2\}$ with $w_0 \leqslant w_1$ and $w_0 \leqslant w_2$ (and also $w_i \leqslant w_i$ for each $i = 0, 1, 2$; see the figure below). **M** is the valuation system given by the triple $(W, \leqslant, w_0)$ in the way described above. Let a be the assignment defined as follows: $a(p) = \{w_1, w_2\}$ and $a(q) = \{w_0, w_1\}$.

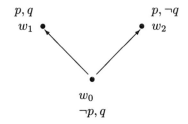

1. We can compute $v_a(\Diamond(p \to \Box q))$ directly from the definitions of $f_\Diamond$, $f_\to$ and $f_\Box$ as follows.

$$
\begin{aligned}
v_a(\Diamond(p \to \Box q)) &= \{x \mid \exists y \, x \leqslant y \text{ and } y \in v_a(p \to \Box q)\} \\
&= \{x \mid \exists y \, x \leqslant y \text{ and } y \in (W - v_a(p)) \cup v_a(\Box q)\} \\
&= \{x \mid \exists y \, x \leqslant y \text{ and } y \in \{w_0\} \cup \{w_1\}\} \\
&= \{w_0, w_1\}
\end{aligned}
$$

We can paraphrase this as: the formula $\Diamond(p \to \Box q)$ is true at the worlds w_0 and w_1.

2. Equivalently, we can use Proposition 2.3.2 to check that $w_0 \Vdash_a \Diamond(p \to \Box q)$ and $w_1 \Vdash_a \Diamond(p \to \Box q)$, while $w_2 \nVdash_a \Diamond(p \to \Box q)$. For example,

$$w_0 \Vdash_a \Diamond(p \to \Box q) \text{ iff } x \Vdash_a p \to \Box q \text{ for some } x \text{ in } \{w_0, w_1, w_2\};$$

so now we check the right-hand side:

$$
\begin{aligned}
w_1 \Vdash_a p \to \Box q \quad &\text{if} \quad w_1 \Vdash_a \Box q \\
&\text{if} \quad w_1 \Vdash_a q
\end{aligned}
$$

(since $\{w \mid w_1 \leqslant w\} = \{w_1\}$), which is correct.

So, we can talk either of the truth value of a sentence in the interpretation, in which case the truth value is a subset of W; or we can talk of a sentence being true or false at a point in W, which is simply saying that the point is in or is not in the truth value subset.

Elements of W are called possible worlds because of the history of modal logic. As mentioned in Example 1.1.2 in the introduction, it began as an attempt to elucidate the meanings of *necessarily* ($\Box$) and *possibly* ($\Diamond$). These words qualify the sense in which something is true or false — it may be possibly true, or necessarily true, or possibly necessarily possibly true, etc. The qualifications are *modes* of truth, hence the term modal logic. A sentence is *necessarily true* at a point $w \in W$ if it is true in every way in which the world might develop from w, i.e. at every point $\geqslant w$. Likewise, it is possibly true if there is some world $\geqslant w$ at which it is true.

In computer science, elements of W are most often the state of a system, or of the execution of a program. We hinted at this when we used $\Box$ to mean 'always in the future' or 'henceforth' (Example 1.1.1) and in the programs Examples (1.1.3 and 2.1.3). The idea of $\Box$ meaning 'in the future' can be made more precise, yielding *temporal logic* as a branch of modal logic. $\langle W, \leqslant \rangle$ is interpreted as a set of moments in time, with $\leqslant$ to mean 'precedes'. The truth value of a sentence A is the set of points of time at which A holds. The reader should check that the definitions of $f_\wedge$, $f_\vee$, $f_\rightarrow$ and $f_\neg$ are the right ones for this interpretation of the operators. For example, $f_\neg(X) = W - X$ means that $\neg A$ holds precisely at those points at which A does not hold. We do not consider temporal logic in this chapter; the interested reader should see a short book by R. Goldblatt [Goldblatt, 1987].

2.4 Predicate languages

So far in Section 2 we have analysed sentences by breaking them down into constituent atomic sentences, regarding these atomic sentences as the building blocks. We showed how one can introduce operators in the language to capture arguments in several specialized domains, but still retain the general framework of the *valuation system*. That was because the interpretation of the operators remained fixed. In this section we will consider examples with new operators (the quantifiers) whose interpretations vary with the assignment. We have to generalize the notion of valuation system to incorporate them. For example, consider the argument:

1   Most people are happy
2   All happy people will go to heaven

---

3   Most people will go to heaven

Intuitively, it is clear that this argument is correct. But if we try to symbolise the sentences in the argument by propositional letters we just get something like: p, q therefore r. This is because it is not possible to break down the sentences using the propositional operators, nor even with the modal ones. Since we can easily find an interpretation in which p and q are true and r is false, we have failed to represent the argument.

It should be clear that although the future tense is mentioned in the argument, it is not temporal in the sense of Example 1.1.1, where the argument depended on the properties of time. This argument involves dealing with the phrases 'all' and 'most'.

What we need are two operators, $\forall$ (for all) and **most**, which act on variables in sentences.

1   **most** x (happy(x))
2   $\forall x$ (happy$(x) \rightarrow$ heaven(x))

---

3   **most** x (heaven(x))

(We assume we are talking only of *people* in the argument.)

Operators like $\forall$ are called quantifiers. Because of the special way they are used in sentences, they cannot be made to fit into the framework of Definition 2.2.3. This is therefore generalized as follows:

Definition 2.4.1. A *predicate language* L is a tuple $\langle P, T, V, O, Q \rangle$ where:

1. P is a set of *predicate symbols*, each associated with a non-negative integer, its *arity*.

2. T is a set of *terms*.

3. V is a set of *variables*.

4. O is a set of *operators*, each with a specified arity.

5. Q is a set of *quantifiers*.

Compare this definition with the propositional version, Definition 2.2.3. P and T correspond to P in that definition, for they are extra-logical symbols which make up the atomic formulas. O, V and Q correspond to the O of the propositional definition, because they are the logical symbols which make complex formulas out of atomic ones.

We have not given any information about the construction of terms. Typically (and in an example to follow) they are formed from variables and a further syntactic category, *function symbols*. At this stage, all that counts is that they are formed somehow and that their formation rules allow them to contain variables. In particular, any variable is a term; thus, $V \subseteq T$. We will write var(t) for the set of variables in a term t.

An atomic formula is formed from a predicate symbol with an appropriate number of terms:

If $t_1, \ldots, t_n$ are in T and p is in P with arity n, then $p(t_1, \ldots, t_n)$ is an atomic formula.

Again, constant atomic formulas are available; they are formed from zero-arity predicate symbols and no terms.

Formulas are either atomic formulas, or are built from other formulas by the operators in O (as in the propositional case), or the quantifiers in Q.

1. If A is an atomic formula then A is a formula.

2. If $A_1, \ldots, A_n$ are formulas and o is in O with arity n, then $o(A_1, \ldots, A_n)$ is a formula.

3. If A is a formula, q is in Q and x is in V then $q\,x\,A$ is a formula.

An important question which will arise is whether variables are in the scope of a quantifier or not. Such variables are said to be *bound*. Unbound variables are said to be *free*. For example, suppose $\{\exists, \forall\} \subseteq Q$ and $\{\wedge, \vee, \neg\} \subseteq O$, $\{x, y, z\} \subseteq V$ and $\{p, q\} \subseteq P$. Then, in the formula

$$\exists x \, (p(y, z) \wedge \forall y \, (\neg q(\underline{y}, \underline{x}) \vee p(\underline{y}, z)))$$

the variables underlined are bound. Note that there is both a free occurrence and a bound occurrence of y. The set of free variables of A is written free(A). We will define this by induction.

Definition 2.4.2. Let $p \in P$ with arity n, $o \in O$ with arity m, $t_1, \ldots, t_n \in T$, $q \in Q$ and let $A_1, \ldots, A_m$ be formulas. Then

1. free$(p(t_1, \ldots, t_n)) = \mathrm{var}(t_1) \cup \ldots \cup \mathrm{var}(t_n)$.

2. free$(o(A_1, \ldots, A_m)) = \mathrm{free}(A_1) \cup \ldots \cup \mathrm{free}(A_m)$.

3. free$(q\,x\,A) = \mathrm{free}(A) - \{x\}$.

For example:

$$\begin{aligned}
\mathrm{free}(\exists x \, &(p(y, z) \wedge \forall y \, (\neg qy, x \vee py, z))) \\
&= \mathrm{free}(p(y, z) \wedge \forall y \, (\neg q(y, x) \vee p(y, z))) - \{x\} \\
&= (\mathrm{free}(p(y, z)) \cup ((\mathrm{free}(q(y, x)) \cup \mathrm{free}(p(y, z))) - \{y\})) - \{x\} \\
&= (\{y, z\} \cup \{x, z\}) - \{x\} \\
&= \{y, z\}
\end{aligned}$$

Notice that we are using the word 'formula' where previously we used 'sentence'. That is because the formulas we have defined are not necessarily fully fledged sentences. A *sentence* is a formula which has no free variables.

The above definitions are quite general in the sense that they allow us to introduce a variety of quantifiers. However, it should be noted that any language satisfying the definitions is first-order, in the sense that quantification is allowed only over terms, not over sets of terms or properties of terms. Such languages are called higher-order. See [van Benthem and Doets, 1983] for a readable survey.

In particular, this means that it is not possible to capture the most general use of the 'most' quantifier by these definitions, because it is known that **most** is a second-order quantifier: its arguments are predicates, not variables. One cannot express a fact like 'most people are happy' in a general situation in which there are non-people as well as people in the domain, without resorting to a second-order **most**.

This is not the case for the quantifiers 'all' and 'some'. One can express 'all people are happy' and 'some people are happy' as $\forall x \, (\text{person}(x) \rightarrow \text{happy}(x))$ and $\exists x \, (\text{person}(x) \wedge \text{happy}(x))$ respectively.

Now we turn to the predicate analogues of the other definitions which were given in Section 2.2 for the propositional case. Let $L = \langle P, T, V, O, Q \rangle$ be a predicate language.

Definition 2.4.3. A valuation system **M** for L is a tuple $\mathbf{M} = \langle M, D, F, G \rangle$ such that:

1. M is a set with at least two elements and D is a non-empty proper subset of M.

2. F is a set of functions from M to M, one corresponding to each operator in O. The function f_o corresponds to the operator o.

3. G is a set of functions from $\mathcal{P}(M)$ to M, one corresponding to each quantifier in Q. The function g_q corresponds to the quantifier q. These functions map possibly infinite subsets of M onto an element of M. Such functions are called infinitary functions from M to M.

This should be compared with Definition 2.2.4. M, D and F have the same role here as there. The new component, G, is used to interpret quantifiers.

In the propositional case, an assignment just gave truth values (elements of M) to each of the atomic sentences. In this case an assignment must also give some kind of denotation to the terms in T. Therefore, it comes with a *domain*[2] *of individuals* which interpret T and thereby gives truth values to formulas.

Let $L = \langle P, T, V, O, Q \rangle$ be a predicate language and $\mathbf{M} = \langle M, D, F, G \rangle$ be a valuation system for L.

---

[2]The word 'domain' is often used in computer science to mean a partially ordered set satisfying certain properties; such domains are used to define the semantics of programming language constructs. That is *not* the sense in which we use it here. Here it is simply the 'carrier' set on which functions and relations are defined to interpret the function symbols and predicate symbols of the language.

Definition 2.4.4. An *assignment* for **M** is a pair $\langle a, I \rangle$ consisting of a function a together with a non-empty set I (called the domain of individuals) such that:

1. If $t \in T$ then $a(t)$ is an element of I.

2. If $x \in V$ then $a(x) \in I$.

3. If $p \in P$ has arity n then $a(p)$ is a function from the set I^n to the set M (i.e. $a(p) : I^n \to M$).

Thus, an assignment gives us the basis for defining an interpretation (i.e. a map from formulas in the language to M). The assignment interprets terms as actual elements of I, and predicate symbols as actual predicates on the domain I. It also associates an element of I with each variable. This may seem counterintuitive at first, because we want to think of variables as representing arbitrary elements of the domain. But in a moment we will see that the quantifiers ignore how the variables they quantify are assigned. The assignment $a(x)$ of a variable x only makes a difference when the variable is free.

Before we can define the interpretation induced by the assignment, we need a piece of notation: given a variable x and two assignments a and a' with the same domain of individuals I, we will write $a \sim_x a'$ to mean that a and a' agree on everything except possibly the element of I assigned to x. That is to say, $a \sim_x a'$ iff:

1. If $t \in T$ is such that $x \notin \text{var}(t)$, then $a(t) = a'(t)$; and

2. If $y \in V$ is such that $x \neq y$, then $a(y) = a'(y)$; and

3. If $p \in P$ then $a(p) = a'(p)$.

Definition 2.4.5. Given a valuation system $\mathbf{M} = \langle M, D, F, G \rangle$ for a predicate language L and an assignment a for **M**, a *valuation* is a function v_a from sentences in L to M defined as follows:

1. If $p \in P$ has arity n and $t_1, \ldots, t_n \in T$ then $v_a(p(t_1, \ldots, t_n)) = a(p)(a(t_1), \ldots, a(t_n))$

2. $v_a(o(A_1, \ldots, A_n)) = f_o(v_a(A_1), \ldots, v_a(A_n))$, as before, and

3. $v_a(q\, x\, A) = g_q(\{v_{a'}(A) \mid a \sim_x a'\})$.

The compositional flavour of these definitions should by now be clear. Here, atomic sentences are interpreted by interpreting the predicate symbol and applying the result to the interpretation of the terms. Interpreting sentences with operators proceeds by interpreting the operator and applying the result to the interpretations of its arguments, and so on.

We have given general definitions to allow other kinds of quantifiers to be used, but the most usual quantifiers are of course $\forall$ and $\exists$. *Classical predicate logic* or *first-order* logic has the language given by

1. $O = \{\wedge, \vee, \rightarrow, \neg, \bot\}$.

2. $V = \{x, y, z, x_1, x_2, \ldots\}$.

3. $Q = \{\forall, \exists\}$.

Its valuation system is given by the following definition.

Definition 2.4.6. *Classical first-order logic.* The valuation system for such predicate languages which gives classical logic is $\langle M, D, F, G \rangle$:

1. $M = \{\mathbf{t}, \mathbf{f}\}$ and $D = \{\mathbf{t}\}$.

2. F is defined as in the propositional case (Definition 2.2.5).

3. $G = \{g_\forall, g_\exists\}$; they are defined by:

$$g_\forall(X) = \begin{cases} \mathbf{f} & \text{if } \mathbf{f} \in X \\ \mathbf{t} & \text{otherwise;} \end{cases} \quad \text{and} \quad g_\exists(X) = \begin{cases} \mathbf{t} & \text{if } \mathbf{t} \in X \\ \mathbf{f} & \text{otherwise.} \end{cases}$$

2.5   Summary

Our goal in this section (and indeed throughout the chapter) has been to illustrate how much freedom there is in defining logics for computer science. We have seen a variety of languages and valuation systems, and how the truth values in a valuation system may be structured in a way meaningful to the application. One especially important example of this is modal logic. We saw how the possible worlds in modal logic are just like small valuation systems glued together by the accessibility relation (Proposition 2.3.2). Finally, the definitions of languages and valuation systems were extended to allow for predicate logic.

3   Consequence relations and entailment relations

We started Section 2 by discussing the two views of logic which we called *satisfaction-based* and *proof-based*. So far we have only considered the satisfaction-based view, which focuses on satisfaction of sentences in interpretations. We now generalize this satisfaction view to a *consequence* view and consider *validity of arguments* rather than the *truth of sentences*. This is done by considering a relation, called a *consequence relation*, between the premises of an argument and its conclusion. The proof-based view of logic (to be described in Section 4) also yields a consequence relation. Thus, the consequence view is a generalisation both of the satisfaction view and the proof view. In the following diagram, we show the three views, the primary objects in each one and the sections in which they are described.

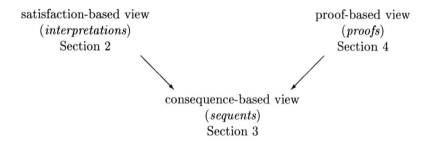

satisfaction-based view   proof-based view
(*interpretations*)   (*proofs*)
Section 2   Section 4

consequence-based view
(*sequents*)
Section 3

    Valuation systems automatically give us one kind of consequence relation, which we call an entailment relation. In this section, we first consider consequence relations in the abstract, and then we look at the particular example of entailment relations.

    There is a large amount written on consequence relations. For a good introduction, see Avron [Avron, 1987].

3.1 Consequence relations

Let L be a language. As before, L viewed as a set is the set of sentences of L.

Definition 3.1.1. A *consequence relation* over L is a relation, written $\vdash$, between $\mathcal{P}(L)$ and L which satisfies the following properties:

- Inclusion: if $A \in \Gamma$ then $\Gamma \vdash A$.

- Monotonicity: $\dfrac{\Gamma \vdash A}{\Gamma, \Delta \vdash A}$.

- Cut: $\dfrac{\Gamma \vdash C \quad \Delta, C \vdash A}{\Delta, \Gamma \vdash A}$.

    The horizontal bar is read, 'if...then...'. The commas mean set-union; thus Γ, Δ means $\Gamma \cup \Delta$ and Δ, C means $\Delta \cup \{C\}$.

$\Gamma \vdash A$ means that the sentence A follows from, or is a consequence of, the sentences in Γ. The expression $\Gamma \vdash A$ is a *sequent*, and expresses the argument with *premises* Γ and *conclusion* A. Thus:

- Inclusion states that if A is mentioned explicitly as a premise of an argument, then it is a valid consequence.

- Monotonicity says that if A follows from Γ, then it follows from any superset of Γ. In other words, adding more premises should not upset any conclusion. The term monotonicity comes from the fact that the set of consequences of Γ cannot decrease as Γ increases. It is also sometimes called *weakening* or *thinning*, because making unnecessary assumptions (extra premises) in an argument weakens it or thins it out.

- Cut says that if A can be derived from some premises Δ and a sentence C, then it can be derived from those premises plus some other premises which yield C. It is called cut because we can cut out the formula C and go straight to A.

These three properties are often thought of as the essential properties of a logical system. Although there are alternatives to definition 3.1.1 to be found in the literature (see eg. [Avron, 1987]), they usually have at least these properties.

Certainly these properties hold for formal arguments, like those in mathematics. We don't expect that new facts which emerge about mathematics will invalidate old theorems (unless they were wrong in the first place, of course)—this is monotonicity. And the use of lemmas in mathematical argument corresponds to cut, for which reason cut is sometimes called lemma generation. If I can prove a lemma (C) from one set of premises, and a theorem (A) from another set together with the lemma, then I should be able to prove the theorem from both sets of premises together.

It is debatable whether one should accept these properties for informal arguments, however; consider the following 'counterexample' to monotonicity, due to van Benthem [van Benthem, 1985]:

I put sugar in my coffee $\vdash$ it tastes fine;

I put sugar and I put diesel oil $\nvdash$ it tastes fine.

But, we are idealizing.

Two more properties are of interest:

Definition 3.1.2. *Compactness.* The consequence relation $\vdash$ is *compact* if, whenever $\Gamma \vdash A$, there is a finite subset Γ_0 of Γ such that $\Gamma_0 \vdash A$.

There's something rather suspicious about an argument that relies on an infinite number of premises, so compactness is a property we should hope for. Moreover, if we define consequence relations by means of *proofs*, as we will in Section 4, then they must be compact because proofs are finite objects.

Another property we should hope for can be thought of as 'independence of the underlying language'. If L is a propositional language we can state this as follows. If $\Gamma \vdash A$ and Γ^* and A^* are obtained by substituting one of the atomic sentences in the language by some other sentence, then we should expect that $\Gamma^* \vdash A^*$. For if not, then $\Gamma \vdash A$ because of some property of the atomic sentence which is not represented in the argument. Since these sequents are supposed to be generic arguments, all the information needed in the argument should be explicit in the premises.

Definition 3.1.3. Let $L = \langle P, O \rangle$ be a propositional language. A *substitution* $*$ on L is a map $P \to L$

A substitution $*$ on L can be extended to a map $L \to L$. Given a formula A, the formula A^* is obtained by replacing all occurrences of atomic sentences in A by their image under the substitution: if $o \in O$ and $A_1, \ldots, A_n \in L$ then

$$(o(A_1, \ldots, A_n))^* = o(A_1^*, \ldots, A_n^*)$$

For example, suppose $P = \{p, q\}$ and $p^* = p \wedge q$ and $q^* = q$. Then, if A is $(p \vee q) \to p$, A^* is $((p \wedge q) \vee q) \to (p \wedge q)$. If Γ is a set of sentences, then Γ^* means $\{B^* \mid B \in \Gamma\}$.

Definition 3.1.4. A consequence relation $\vdash$ has the *substitution property* if, for any substitution $*$, $\Gamma \vdash A$ implies $\Gamma^* \vdash A^*$.

Definition 3.1.5. A consequence relation $\vdash$ is *standard* if it has the compactness and substitution properties.

Another important question about consequence is whether it can be internalized in the language:

Definition 3.1.6. The consequence relation $\vdash$ has the *deduction property* if there is a binary operator $\to$ in L such that for all sentences A and B and sets of sentences Γ,

$$\Gamma \vdash A \to B \quad \text{iff} \quad \Gamma, A \vdash B$$

This is obviously an important property, because it links the 'meta-level' implication of the consequence relation with the 'object-level' implication of the language.

3.2   Entailment relations

Now we turn to a special case of consequence relations, called entailment relations. Entailment relations are defined in terms of a valuation system. They are written ⊨, possibly subscripted by the valuation system in question. We suppose again we are working in some fixed language L, and that Γ is a set of sentences and A a sentence in L. Let **M** be a valuation system for L with truth values M and designated values D.

Definition 3.2.1. Γ *entails* A in **M** , written $\Gamma \vDash_M A$, if for every assignment a such that $v_a(B) \in D$ for each $B \in \Gamma$, then $v_a(A) \in D$ also.

This says that if the premises are all designated then so is the conclusion. $\Gamma \vDash_M A$ means that (according to the valuation system **M**) the argument with premises Γ and conclusion A is a valid one.

We have to check that entailment relations really are consequence relations:

Proposition 3.2.2. *Structural properties.* An entailment relation $\vDash_M$ has the following structural properties:

1. Inclusion: if $A \in \Gamma$ then $\Gamma \vDash_M A$.

2. Monotonicity: if $\Gamma \vDash_M A$ then $\Gamma, \Delta \vDash_M A$.

3. Cut: if $\Gamma \vDash_M C$ and $\Delta, C \vDash_M A$ then $\Delta, \Gamma \vDash_M A$.

Proof. Let a be any assignment relative to **M**.

1. If $v_a(B) \in D$ for all $B \in \Gamma$ then certainly $v_a(A) \in D$, since $A \in \Gamma$.

2. If $v_a(B) \in D$ for all $B \in \Gamma \cup \Delta$ then $v_a(B') \in D$ for all $B' \in \Gamma$, since $\Gamma \subseteq \Gamma \cup \Delta$. But $\Gamma \vDash_M A$, so $v_a(A) \in D$.

3. If $v_a(B) \in D$ for all $B \in \Gamma \cup \Delta$ then, as in the previous cases, we reason that the following must be in D: $v_a(B')$ for each $B' \in \Gamma$, since $\Gamma \subseteq \Gamma \cup \Delta$; $v_a(C)$, since $\Gamma \vDash_M C$; $v_a(B'')$ for each $B'' \in \Delta \cup \{C\}$; and finally, $v_a(A)$.   ∎

Suppose L is a propositional language. Then we have the substitution property described for consequence relations.

Lemma 3.2.3. Let $*$ be a substitution for the propositional language L and let **M** be a valuation system for L. If assignments a and a' are such that $a'(p) = a(p^*)$ for all $p \in P$, then $v_{a'}(A) = v_a(A^*)$ for all sentences A in L:

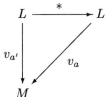

Proof. By induction on the structure of A. If A is atomic, then $v_{a'}(A) = a'(A) = a(A^*) = v_a(A^*)$. Now suppose A is $o(A_1, A_2, \ldots, A_n)$ where n is the arity of $o \in O$, and that the result holds for $A_1, A_2, \ldots$ and A_n. Then

$$
\begin{aligned}
v_{a'}(A) &= v_{a'}(o(A_1, \ldots, A_n)) \\
&= f_o(v_{a'}(A_1), \ldots, v_{a'}(A_n)) & \text{definition of } f_o \\
&= f_o(v_a(A_1^*), \ldots, v_a(A_n^*)) & \text{inductive hypothesis} \\
&= v_a(o(A_1^*, A_2^*, \ldots, A_n^*)) & \text{definition of } f_o \\
&= v_a(A^*).
\end{aligned}
$$

∎

Proposition 3.2.4. *Substitution.* Given a propositional language L and a valuation system **M** for L and a substitution $*$ on L,

$$\Gamma \vDash_M A \text{ implies } \Gamma^* \vDash_M A^*$$

Proof. Suppose $\Gamma \vDash_M A$, and let a be an assignment such that $v_a(B^*) \in D$ for each $B \in \Gamma$. Let a' be an assignment defined by $a'(p) = a(p^*)$. Then for all $B \in \Gamma$, $v_{a'}(B) = v_a(B^*)$ by the lemma. Therefore $v_{a'}(B) \in D$ for any $B \in \Gamma$, so $v_{a'}(A) \in D$. But $v_{a'}(A) = v_a(A^*)$, so $v_a(A^*) \in D$. ∎

The notion of substitution can be extended to predicate languages too, and analogues of the results follow.

Neither compactness nor the deduction property can be established for entailment relations based on valuation systems in the abstract. We need to look at particular examples.

3.3   The systems $\mathcal{C}$ and $\mathcal{S}4$

In Section 2 we saw many examples of valuation systems, including valuation systems for classical logic and the modal logic S4. For classical logic we had a valuation system for both propositional and predicate languages

(Defs. 2.2.5 and 2.4.6 respectively). The corresponding entailment relations come straight from Definition 3.2.1. Suppose Γ and A are respectively a set of sentences and a sentence of L.

Definition 3.3.1. *Propositional $\mathcal{C}$.* If L is a propositional language and **M** is the valuation system of Def. 2.2.5, then $\Gamma \vDash_{\mathcal{C}} A$ if $\Gamma \vDash_{\mathbf{M}} A$.

Definition 3.3.2. *Predicate $\mathcal{C}$.* If L is a predicate language and **M** is the valuation system of def. 2.4.6, then $\Gamma \vDash_{\mathcal{C}} A$ if $\Gamma \vDash_{\mathbf{M}} A$.

The situation for modal logic S4 is a little different, because in Definition 2.3.1 we defined a family of valuation systems.

Definition 3.3.3. *$\mathcal{S}4$.* If L is a propositional modal language then $\Gamma \vDash_{\mathcal{S}4} A$ if for *each* valuation system **M** satisfying Definition 2.3.1, we have that $\Gamma \vDash_{\mathbf{M}} A$.

It should be noted that, while only one valuation system is needed for the entailment relation for classical logic, an infinite number are needed to define the entailment relation in the case of modal logic.

These systems, together with another one called $\mathcal{I}$ (intuitionistic logic), are studied for the remainder of this chapter. In Section 4 we look at their *proof theory*, and the system $\mathcal{I}$ is also defined there. In Section 5, we see valuation systems for $\mathcal{I}$ as well as the relations between the three systems $\mathcal{C}$, $\mathcal{S}4$ and $\mathcal{I}$.

All three of the systems so far defined, propositional and predicate $\mathcal{C}$ and (propositional) $\mathcal{S}4$, have the compactness property and the deduction property. We will only prove compactness for propositional $\mathcal{C}$. It follows easily from the following lemma.

Lemma 3.3.4. Let L be a propositional language, **M** the valuation system of Def. 2.2.5, and Γ a set of sentences of L. If for every finite subset Γ_0 of Γ there is an assignment a' relative to **M** such that $v_{a'}(B) = \mathbf{t}$ for each $B \in \Gamma_0$, then there is also an assignment a such that $v_a(B) = \mathbf{t}$ for each $B \in \Gamma$.

Proof. Let $X \subseteq_f Y$ mean X is finite and $X \subseteq Y$. Let $\{p_1, p_2, \ldots\}$ be the atomic sentences of L. The proof is simultaneously a definition by induction and a proof by induction. We will define $a(p_i)$ by induction on i, and then show that $v_a(B) = \mathbf{t}$ for each $B \in \Gamma$. Notation: given two assignments a and a', let $a \equiv_i a'$ mean $\forall j\, 1 \leqslant j < i$ implies $(a(p_j) = \mathbf{t}$ iff $a'(p_j) = \mathbf{t})$.

Suppose $a(p_j)$ is defined for all $1 \leqslant j < i$. At each step we will show the following property of i: for every $\Gamma_0 \subseteq_f \Gamma$ there is an assignment a' such that $v_{a'}(B) = \mathbf{t}$ for each $B \in \Gamma_0$, and $a \equiv_i a'$. (Notice that this requirement holds trivially for the base case $i = 1$.) Now assume that the property holds for i, that is, for each $\Gamma_0 \subseteq_f \Gamma$ there is an assignment as described. The induction step is as follows.

1. *If for each $\Gamma_0 \subseteq_f \Gamma$ there is such an assignment a' with the additional property that $a'(p_i) = \mathbf{t}$, then let $a(p_i) = \mathbf{t}$. In this case the inductive condition holds easily for $i + 1$.*

2. *Otherwise*, let $a(p_i) = \mathbf{f}$. We need to show that the inductive condition holds for $i + 1$. First, notice that there is a $\Gamma_1 \subseteq_f \Gamma$ such that: for every a' for which $v_{a'}(B) = \mathbf{t}$ (each $B \in \Gamma_1$) and $a \equiv_i a'$, it is the case that $a'(p_i) = \mathbf{f}$; otherwise we would be in case 1. Now let $\Gamma_0 \subseteq_f \Gamma$. $\Gamma_1 \cup \Gamma_0$ is finite, so there is an a' such that $v_{a'}(B) = \mathbf{t}$ for each $B \in \Gamma_1 \cup \Gamma_0$ and $a'(p_j) = \mathbf{t}$ iff $a(p_j) = \mathbf{t}$ for all $j < i$. But $v_{a'}(B) = \mathbf{t}$ for each $B \in \Gamma_1$, so $a'(p_i) = \mathbf{f}$ as required.

Now to show that $v_a(B) = \mathbf{t}$ for each $B \in \Gamma$. let $B \in \Gamma$. Let i be the index of the highest atomic sentence in B. There is an assignment a' such that $a \equiv_i a'$ and $v_{a'}(B) = \mathbf{t}$. Therefore $v_a(B) = \mathbf{t}$. ∎

Proposition 3.3.5. $\vDash_C$ is compact.

Proof. Suppose $\Gamma \vDash_C A$. Let $\mathbf{M}$ be the valuation system for classical logic, as before. Then there is no assignment a such that $v_a(B) = \mathbf{t}$ for each $B \in \Gamma \cup \{\neg A\}$; this follows from the properties of $\neg$. By the lemma, there is a finite subset Γ_0 of Γ such that there is no assignment a for which $v_a(B) = \mathbf{t}$ for each $B \in \Gamma_0 \cup \{A\}$. Hence, again using the property of $\neg$, we obtain $\Gamma_0 \vDash_C A$. ∎

Proposition 3.3.6. The entailment relations propositional $\vDash_C$, predicate $\vDash_C$ and (propositional) $\vDash_{S4}$ have the deduction property.

Proof. For propositional and predicate C: Let $\mathbf{M}$ be the relevant valuation system. Suppose $\Gamma \vDash_C A \to B$. We prove $\Gamma, A \vDash_C B$. Let a be such that $v_a(C) = \mathbf{t}$ for each $C \in \Gamma \cup \{A\}$. Then $v_a(A \to B) = \mathbf{t}$ and $v_a(A) = \mathbf{t}$. By inspection of $f_\to$, it follows that $v_a(B) = \mathbf{t}$.

Conversely, suppose $\Gamma, A \vDash_C B$. We prove $\Gamma \vDash_C A \to B$. Suppose $v_a(C) = \mathbf{t}$ for each $C \in \Gamma$. If $v_a(A) = \mathbf{t}$ then $v_a(C) = \mathbf{t}$ for each $C \in \Gamma \cup \{A\}$, so $v_a(B) = \mathbf{t}$, and (inspecting again) $v_a(A \to B) = \mathbf{t}$. On the other hand, if $v_a(A) = \mathbf{f}$ then $v_a(A \to B) = \mathbf{t}$. Therefore, $\Gamma \vDash_C A \to B$.

Now for $S4$. Let $\mathbf{M}$ be a valuation system satisfying Definition 2.3.1. Suppose $\Gamma \vDash_M A \to B$. We prove $\Gamma, A \vDash_M B$. Let a be such that $w_0 \in v_a(C)$ (i.e. $v_a(C) \in D$) for each $C \in \Gamma \cup \{A\}$. Then $w_0 \in v_a(A)$ and $w_0 \in v_a(A \to B)$ (since $\Gamma \vDash_M A \to B$). By inspection of $f_\to$ in Def. 2.3.1, we have that either $w_0 \notin v_a(A)$ (a contradiction), or $w_0 \in v_a(B)$. Therefore, $v_a(B) \in D$ and so $\Gamma, A \vDash_M B$.

Conversely, suppose $\Gamma, A \vDash_M B$. We prove $\Gamma \vDash_M A \to B$. Suppose $w_0 \in v_a(C)$ for each $C \in \Gamma$. Either $w_0 \in v_a(A)$ or $w_0 \notin v_a(A)$. If $w_0 \in v_a(A)$ then $w_0 \in v_a(C)$ for each $C \in \Gamma \cup \{A\}$, so $w_0 \in v_a(B)$. Since $v_a(Y) \subseteq v_a(X \to Y)$ for any X, Y, it follows that $w_0 \in v_a(A \to B)$ as required. On the other hand, if $w_0 \notin v_a(A)$ then $w_0 \in v_a(A \to B)$. Therefore, $\Gamma \vDash_M A \to B$. ∎

3.4 Levels of implication

At this stage it might be useful to consider the various levels of *implication* ('if... then...') we have introduced, and their relationships. First we looked at the operator $\to$ between sentences. The sentence $A \to B$ is true (in classical logic, say) if: A is true implies B is true. Then we introduced the relation $\vDash_M$ and generalized it to $\vdash$. $\Gamma \vDash_M A$ means: if Γ is designated in **M**, then so is A. $\Gamma \vdash A$ means A is a consequence of Γ. Now we have expressions like

$$\frac{\Gamma \vdash A}{\Delta \vdash B}$$

with the meaning: if $\Gamma \vdash A$ then $\Delta \vdash B$. The horizontal bar is thus an implication between linguistic entities containing $\vdash$. Why do we have these levels of implication, and what are the relations between them?

We have them both for historical and logical reasons. Historically, material implication ($\to$) and other forms of 'object-level' implication (see the examples in Section 2.2) were studied before the idea of deriving consequences from theories was introduced by Tarski. The horizontal bar was introduced by Gentzen soon after this to talk about relationships between statements about consequence. Once we formalize one level of implication, we inevitably discuss its properties at the next level up. So now that we have rules to express relations between sequents, we are led to questions like: if this and this rule are valid in a system, must that one be valid too? That introduces yet another level of implication.

Concerning relationships, suppose $\to$ and $\Rightarrow$ are two adjacent levels of implication; $\Rightarrow$ relates expressions formed from $\to$. One can consider the relationship between $\Rightarrow (A \to B)$ and $(\to A) \Rightarrow (\to B)$. For example,

- In the case of $\to$ and $\vdash$, the comparison is between (i) $\vdash A \to B$ and (ii) $A \vdash B$. The equivalence between these two is precisely the deduction property (Definition 3.1.6).

- In the case of $\vdash$ and the horizontal bar between sequents, we compare (i) $\Gamma \vdash A$, and (ii) $\vdash \Gamma / \vdash A$. For finite Γ, (i) implies (ii) by the cut rule. If (ii) implies (i), the logic is said to be *smooth*. Classical logic is smooth, but intuitionistic and modal logic are rough. For more details of smoothness, see [Dummett, 1977].

3.5 Consequence operator

Rather than model intuitions about valid arguments in terms of a binary relation ($\vdash$), we can use a unary *operator* C (for consequence). In fact consequence was first formalized by Tarski in the 1930s using such an operator (see for example the collected papers in [Tarski, 1956]). Given a set of sentences Γ, $\mathsf{C}(\Gamma)$ is to denote the set of consequences of Γ.

C and $\vdash$ are interchangeable in the following way:

$$\mathsf{C}(\Gamma) = \{A \mid \Gamma \vdash A\} \quad \text{and} \quad \Gamma \vdash A \text{ iff } A \in \mathsf{C}(\Gamma)$$

The key properties of consequence relations given in Definitions 3.1.1 and 3.1.5 are just as neatly expressed using this operator:

Proposition 3.5.1. $\vdash$ satisfies:

1. Inclusion, iff $\Gamma \subseteq \mathsf{C}(\Gamma)$ for all sets of sentences Γ.

2. Monotonicity, iff $\Gamma \subseteq \Delta$ implies $\mathsf{C}(\Gamma) \subseteq \mathsf{C}(\Delta)$ for all sets Γ and Δ.

3. Cut, iff $\Sigma \subseteq \mathsf{C}(\Gamma)$ implies $\mathsf{C}(\Delta \cup \Sigma) \subseteq \mathsf{C}(\Delta \cup \Gamma)$, all Γ, Δ and *finite* Σ.

4. Compactness, iff $\mathsf{C}(\Gamma) = \bigcup_{\Gamma_0 \subseteq_f \Gamma} \mathsf{C}(\Gamma_0)$. ($\Gamma_0 \subseteq_f \Gamma$ means $\Gamma_0 \subseteq \Gamma$ and Γ_0 is finite.)

5. Substitution, iff if $\Delta \subseteq \mathsf{C}(\Gamma)$ then $\Delta^* \subseteq \mathsf{C}(\Gamma^*)$, all Γ and Δ.

Proof. In each case we prove (a) 'if', and (b) 'only if'.

1. (a) Suppose $A \in \Gamma$. We want to show that $\Gamma \vdash A$. By hypothesis, $A \in \mathsf{C}(\Gamma)$. Then by definition, $\Gamma \vdash A$. (b) We want to show that $\Gamma \subseteq \mathsf{C}(\Gamma)$. Suppose $A \in \Gamma$. By hypothesis, $\Gamma \vdash A$, and by definition, $A \in \mathsf{C}(\Gamma)$. But A was arbitrary, so $\Gamma \subseteq \mathsf{C}(\Gamma)$.

2. (a) Suppose $\Gamma \vdash A$. Then $A \in \mathsf{C}(\Gamma)$, so $A \in \mathsf{C}(\Gamma \cup \Delta)$ by hypothesis, so $\Gamma, \Delta \vdash A$. (b) Suppose $\Gamma \subseteq \Delta$ and $A \in \mathsf{C}(\Gamma)$. Then $\Gamma \vdash A$, so $\Delta \vdash A$, so $A \in \mathsf{C}(\Delta)$.

3. (a) Suppose $\Gamma \vdash C$ and $\Delta, C \vdash A$. We want to show that $\Delta, \Gamma \vdash A$. $\{C\} \subseteq \mathsf{C}(\Gamma)$, so by hypothesis $\mathsf{C}(\Delta \cup \{C\}) \subseteq \mathsf{C}(\Delta \cup \Gamma)$. But, $A \in \mathsf{C}(\Delta \cup \{C\})$, so $A \in \mathsf{C}(\Delta \cup \Gamma)$, so $\Delta, \Gamma \vdash A$.

(b) Suppose $\Sigma \subseteq C(\Gamma)$ and $A \in C(\Delta \cup \Sigma)$. We want to show that $A \in C(\Delta \cup \Gamma)$. Since Σ is finite, let $\Sigma = \{C_1, C_2, \ldots, C_n\}$.

$$\cfrac{\Gamma \vdash C_n \qquad \cfrac{\Gamma \vdash C_2 \qquad \cfrac{\Gamma \vdash C_1 \quad \Delta, \{C_1, C_2, \ldots, C_n\} \vdash A}{\Delta, \Gamma, \{C_2, \ldots, C_n\} \vdash A}}{\vdots} \qquad \Delta, \Gamma, \{C_n\} \vdash A}{\Delta, \Gamma \vdash A}$$

Thus, the result is achieved by $|\Sigma|$ applications of cut. (Recall that Σ is finite.)

4. and 5. Easy. ∎

3.6 Summary

Consequence relations stand between the premises of an argument and its conclusion, and satisfy certain natural properties. Entailment relations are examples of consequence relations; they are those which arise from valuation systems.

4 Proof theory and presentations

In the last section we saw how consequence relations could be defined from valuation systems (we wrote them as ⊨) and how they could be considered as the primitives of a logical system (when we wrote them as ⊢). In this section we are concerned with *syntactic characterizations* of consequence relations (however they are defined). Consequence relations are specified by defining a means of constructing *proofs*. A proof is a syntactic demonstration that a conclusion follows from a set of premises. By 'syntactic', we mean that there are rules, based on the syntactic form of the premises and the conclusion, for constructing the proof. The study of syntactic characterizations of consequence relations is called proof theory. We will call a proof-theoretic description of a logic a *presentation*. A presentation specifies a consequence relation in the following way: $\Gamma \vdash A$ if there is a *proof of A from* Γ; that is, a proof with *conclusion A* and *premises* Γ.

One reason why syntactic characterizations of consequence relations are important is that they can potentially yield *decision procedures* which will enable us to determine, for a particular Γ and A, whether $\Gamma \vdash A$ or not.

Here, and for much of the remainder of the chapter, we will concentrate on three logics which we will refer to as $\mathcal{C}$, $\mathcal{I}$ and $\mathcal{S}4$-they are classical logic, intuitionistic logic and the modal logic S4. We looked at the entailment relations for C and $S4$ in Section 3.3. We have considered both propositional and predicate versions of C, but we only consider propositional versions of $\mathcal{I}$ and $S4$.

In the logic literature there are three styles of presentation in widespread use: Hilbert presentations, after D. Hilbert; natural deduction presentations; and sequent presentations. We will present C and $S4$ in each of the styles, and show how $\mathcal{I}$ arises naturally by making an adjustment to the natural deduction presentation of C. Along the way, we will briefly sketch equivalences between the styles; for the full details the reader is advised to consult [Dummett, 1977].

4.1   Hilbert presentations

In the *Hilbert* (sometimes called *axiomatic*) style, we characterize a consequence relation by giving a set of axioms together with a finite set of rules of inference. This style of proof originates with Hilbert (using earlier work of Frege), and is often called the Hilbert style or the Hilbert-Frege style. The axioms represent 'basic truths' and the idea is to generate further truths by applying the rules of inference. A *proof* of A from the premises Γ is a finite sequence of formulas ending with A such that each formula is either

- one of the axioms, or

- a member of Γ, or

- derivable from previous formulas in the sequence by means of a rule of inference.

As already stated, we write $\Gamma \vdash A$ if there is such a proof of A from Γ.

A sentence A is a *theorem* of the presentation if there is a proof of A without any premises (i.e. if $\emptyset \vdash A$).

Propositional Classical Logic
Propositional C may be characterized by the following axiom schemes:

A1. $A \rightarrow (B \rightarrow A)$
A2. $(A \rightarrow (B \rightarrow C)) \rightarrow ((A \rightarrow B) \rightarrow (A \rightarrow C))$
A3. $(\neg A \rightarrow \neg B) \rightarrow (B \rightarrow A)$

and by the following rule of inference, called *modus ponens*:

$$\frac{A \quad A \rightarrow B}{B}$$

A1–3 are called axiom schemes because each one represents the family of axioms given by all instantiations of A, B and C to actual sentences of the language. For example, $(p \land q) \to (p \to (p \land q))$ is an axiom in the A1 scheme (A is $p \land q$ and B is q). The rule of inference *modus ponens* (M.P. for short) means: from A and $A \to B$, derive B. That is to say, B can be introduced in a proof if A and $A \to B$ are already in it.

We will write this proof presentation as $\vdash_C$. Here is an example of a proof of $\emptyset \vdash_C \neg B \to (B \to A)$.

1.	$\neg B \to (\neg A \to \neg B)$	A1
2.	$(\neg A \to \neg B) \to (B \to A)$	A3
3.	$((\neg A \to \neg B) \to (B \to A)) \to$	
	$\quad\quad (\neg B \to ((\neg A \to \neg B) \to (B \to A)))$	A1
4.	$\neg B \to ((\neg A \to \neg B) \to (B \to A))$	M.P. 2,3
5.	$(\neg B \to ((\neg A \to \neg B) \to (B \to A))) \to$	
	$\quad\quad ((\neg B \to (\neg A \to \neg B)) \to (\neg B \to (B \to A)))$	A2
6.	$(\neg B \to (\neg A \to \neg B)) \to (\neg B \to (B \to A))$	M.P. 4,5
7.	$\neg B \to (B \to A)$	M.P. 1,6

This presentation of classical logic only refers to the operators $\to$ and $\neg$. The operators $\land$, $\lor$ and $\perp$ are defined as abbreviations in terms of these as follows:

$$A \land B \;\equiv\; \neg(A \to \neg B)$$
$$A \lor B \;\equiv\; \neg A \to B$$
$$\perp \;\equiv\; \neg(A \to A)$$

We have already observed these equivalences in Example 2.2.12.

We have yet to show that $\vdash_C$ is a consequence relation in the sense of Definition 3.1.1. In fact we can show that any Hilbert relation is a consequence relation. The properties of inclusion and monotonicity are instantly apparent; in the case of inclusion, if $A \in \Gamma$ then the proof of A from Γ is simply the sequence consisting of one formula, namely A. In the case of monotonicity, the proof of A from $\Gamma \cup \Delta$ is identical to the proof of A from Γ. To verify the cut property, we need to show that a proof of A from Γ, Δ can be found, given proofs of some C from Γ and of A from $\Delta \cup \{C\}$. This can be achieved by replacing the C in the proof $\Delta, C, \ldots, A$ with the proof $\Gamma, \ldots, C$, yielding the proof $\Delta, \Gamma, \ldots, C, \ldots, A$.

Indeed, $\vdash_C$ is also *standard* in the sense of Definition 3.1.5. These properties are also straightforward to check. Compactness follows from the fact that proofs are *finite* sequences; and substitution is guaranteed because the axioms are expressed in terms of axiom schemes, as described above.

We can also check that $\vdash_C$ has the deduction property, viz. that $\Gamma, A \vdash_C B$ implies $\Gamma \vdash_C A \to B$. This verification is a more sophisticated example of surgery on proofs than the verification of the consequence relation properties. The basic idea is to transform the proof of $\Gamma, A \vdash_C B$ into one of $\Gamma \vdash_C A \to B$, and this is done by induction on the length of the proof of

$\Gamma, A \vdash_C B$. The details, which are straightforward but instructive to work out, are left to the reader. They can also be found in [Hamilton, 1978].

Of course the most important aspect of $\vdash_C$ in which we are interested is its relation to $\vDash_C$. The point is that they should be equivalent:

$$\Gamma \vdash_C A \quad \text{iff} \quad \Gamma \vDash_C A$$

and indeed they are. We discuss this in Section 4.7.

Classical Predicate Logic

The presentation above can be extended to the predicate case, thus defining predicate $\vdash_C$. To do this we have to use the syntactic properties *free* and *free for*. A variable in a formula is said to be *free* in the formula if it is not in the scope of any quantifier; this has already been defined in Definition 2.4.2. The other concept, free for, is a little more complicated.

The intuition is the following: a term t is free for x in A if substituting every free occurrence of x in A does not introduce any interactions with the quantifiers in A. For example, z is free for y in $\forall x\, p(y)$, because substituting every free occurrence of y in $\forall x\, p(y)$ with z yields $\forall x\, p(z)$, and no new quantifier interactions have been introduced. On the other hand, x is not free for y in that formula, because the corresponding substitution is $\forall x\, p(x)$, and a new quantifier interaction has appeared. So for this particular case, t is free for y in $\forall x\, p(y)$ iff $x \notin \text{var}(t)$.

Definition 4.1.1. Let A be a formula and x a variable. A term t is free for x in A if x does not occur free in A within the scope of a quantifier of a variable in $\text{var}(t)$.

Note that in particular, x is free for x in A, whether or not x occurs in A, free or otherwise.

Starting from propositional $\vdash_C$, we arrive at predicate $\vdash_C$ by adding the axioms

A4. $(\forall x\, A) \to A$
A5. $(\forall x\, A(x)) \to A(t)$, if t is free for x in $A(x)$
A6. $\forall x\, (A \to B) \to (A \to \forall x\, B)$,
         if A contains no free occurrences of x

and the following rule of inference, called *generalization*:

$$\frac{A}{\forall x\, A}.$$

In A5, the expression $A(x)$ means a formula possibly with a free occurrence of x in it; and $A(t)$ means the same formula with the term t

substituted for x. Thus, A is a context into which x or t may be placed. Some authors write this axiom in the following way:

A5′. $(\forall x\, A) \to A[t/x]$,   if t is free for x in A

which more clearly shows the substitution but makes referring to the substitution instances less easy. The expression $A[t/x]$ means the result of substituting all free occurrences of x in A with t.

Again concerning A5, it should be clear why the restriction that t be free for x in A applies to the scheme A5. If t were not free for x in A then the substitution would introduce unintended bindings between quantifiers and variables. The restriction in A6 is necessary because otherwise it would be possible to prove $\forall x\, B$ from $\forall x\, (A \to B)$ and a particular instance of A, which is not wanted.

The rule of inference, called generalization, means that at any point in a proof which contains the formula A we can introduce $\forall x\, A$.

Propositional $S4$

The presentation for classical propositional logic can be extended by axioms for manipulating the modal operators $\Box$ and $\Diamond$, yielding a proof system for $S4$. The axioms for $\vdash_{S4}$ are the propositional axioms A1–3 above together with

A7. $\Box(A \to B) \to (\Box A \to \Box B)$
A8. $\Box A \to A$
A9. $\Box A \to \Box\Box A$

and the rules are *modus ponens* (above), and the following rule called *necessitation*:

$$\frac{\vdash A}{\vdash \Box A}$$

This rule says that if A is a *theorem* (can be derived from no premises) then so is $\Box A$. The operator $\Diamond$ is defined to be $\neg\Box\neg$. In Section 5.3 we discuss A7–A9, and prove that A8 and A9 correspond respectively to the reflexivity and transitivity conditions on the ordering $\leqslant$ of Section 2.3.

Hilbert presentations are concise, a fact which makes them attractive when it comes to proving properties of logics. But they are unpopular as presentations of logics for a number of reasons. For example, Hilbert presentations are not perspicuous about the meanings of the operators. It is impossible to match the definitions of $\to$ and $\neg$ given by the axioms and rules with our intuition, whereas this was easy to do in the case of the valuations system (Definition 2.2.5). The problem is that the presentation is too packaged-up to be recognizable as a correct presentation of the system. This means that proofs are hard to find. A related problem is that the proofs are rather impenetrable, as the example above of the proof of $\neg B \rightarrow (B \rightarrow A)$ shows.

The other systems of proof presentation do not suffer from these drawbacks.

4.2 Natural deduction presentations

The *natural deduction* style has no axioms, there are only rules of inference. To compensate for the lack of axioms, at each stage of a proof it is permitted to introduce any formula as hypothesis. Such hypotheses may be discharged later in the proof by other rules. There are many ways of representing the introduction and discharge of hypotheses on the page. The one we present here employs boxes to control the scope of assumptions. When an assumption is introduced, a box is opened. Discharging assumptions is achieved by rules which have boxes in their premises.

Natural deduction was invented by G. Gentzen. His original paper, Chapter 3 of [Gentzen, 1969], is readable and contains many insights. Further work on natural deduction was done by D. Prawitz [Prawitz, 1965], who describes the box notation we use here and attributes it to S. Jaśkowski.

A box contains a fragment of a proof which can use the assumption which opened it. Some rules discharge hypotheses by closing boxes. Typically there are two types of rules of inference for each operator, one which shows how to *introduce* the operator in an argument and one how to *eliminate* it.

Consequence in natural deduction is defined as follows: $\Gamma \vdash A$ if there is a natural deduction proof of A with premises in Γ.

Classical Logic

For the operator $\wedge$ in classical logic $\mathcal{C}$ we have the following rules: the introduction rule

$$\frac{A \quad B}{A \wedge B} \wedge\text{i}$$

and the elimination rules

$$\frac{A \wedge B}{A} \wedge\text{e}_1 \quad \text{and} \quad \frac{A \wedge B}{B} \wedge\text{e}_2$$

Notice the labelling: 'i' means introduction, 'e' means elimination. Given A and B as provable, we can prove $A \wedge B$ by means of the rule $\wedge$i; conversely, given $A \wedge B$, we can prove A using $\wedge$e$_1$, or B using $\wedge$e$_2$—hence the need for two rules which eliminate $\wedge$.

The rules for $\wedge$ do not involve boxes and assumptions. Here are the introduction and elimination rules for $\rightarrow$:

$$\frac{\boxed{\begin{array}{c} A \\ \vdots \\ B \end{array}}}{A \rightarrow B} \rightarrow\text{i} \qquad\qquad \frac{A \quad A \rightarrow B}{B} \rightarrow\text{e}$$

The rule $\rightarrow$i says: if you can prove B by assuming A as a hypothesis, then you can prove $A \rightarrow B$. Moreover, the proof of $A \rightarrow B$ doesn't depend on having made the assumption A; we get to discharge that hypothesis. The box notation is used to show that the rule allows us to discharge the A. Procedurally, we can describe this as follows: to prove $A \rightarrow B$, open a box and write A into it. Using the A, together with every other formula which is available for the proof of $A \rightarrow B$, prove B. If you can do this, then you have proved $A \rightarrow B$ from those other formulas alone. In a more declarative way we can state this as follows: $A \rightarrow B$ holds if assuming A makes B hold.

The other rules for natural deduction for $\mathcal{C}$ are shown in Fig. 1. For the operators $\wedge$, $\vee$, and $\rightarrow$, there are introduction and elimination rules. We have already described the rules for $\wedge$ and $\rightarrow$, and the rules $\vee$i$_1$ and $\vee$i$_2$ are straightforward; if you can prove A then of course you can prove $A \vee B$ ($\vee$i$_1$). The rule $\vee$e works as follows. If you can prove $A \vee B$, and you can prove C by assuming A, and separately by assuming B, then it should be possible to prove C without assuming either A or B explicitly, but just by having $A \vee B$. We know that one of A or B is provable, and whichever it is we know that C is provable; therefore, C must be provable.

The rule $\neg$i says that one can prove $\neg A$ if assuming A leads to a contradiction. The rule $\neg$e has the effect of identifying all contradictions. It should be noted that if one identifies $\neg A$ with $A \rightarrow \perp$, then the rules $\neg$i and $\neg$e are just special cases of $\rightarrow$i and $\rightarrow$e, namely, the cases in which $B = \perp$.

There is no rule for introducing $\perp$; in practice it is often introduced by $\neg$e. The rule $\perp$e says that everything can be derived from a contradiction.

The rules for $\forall$ and $\exists$ have the flavour of infinitary variations on the rules for $\wedge$ and $\vee$ respectively. For example, $\forall$i says that if one introduces a new variable y and proves a formula $A(y)$ containing y, then one is entitled to deduce $\forall x\, A(x)$. (Recall that if $A(y)$ is a formula containing zero or more free occurrences of y then $A(x)$ is the formula obtained by substituting x for those free occurrences.) The rationale for this rule is the following: if one can prove $A(y)$ without any assumptions about what y is, then $A(x)$ is provable for any x, hence $\forall x\, A(x)$. The box in this rule guarantees that y is entirely new; nothing is known about it; y is not allowed to occur anywhere outside the box.

$$\frac{A \quad B}{A \wedge B} \wedge i \qquad\qquad \frac{A \wedge B}{A} \wedge e_1 \qquad \frac{A \wedge B}{B} \wedge e_2$$

$$\frac{A}{A \vee B} \vee i_1 \qquad \frac{B}{A \vee B} \vee i_2$$

$$A \vee B \quad \boxed{\begin{array}{c} A \\ \vdots \\ C \end{array}} \quad \boxed{\begin{array}{c} B \\ \vdots \\ C \end{array}}$$
$$\overline{\qquad\qquad C \qquad\qquad} \vee e$$

$$\boxed{\begin{array}{c} A \\ \vdots \\ B \end{array}}$$
$$\frac{}{A \rightarrow B} \rightarrow i \qquad\qquad \frac{A \quad A \rightarrow B}{B} \rightarrow e$$

$$\boxed{\begin{array}{c} A \\ \vdots \\ \bot \end{array}}$$
$$\frac{}{\neg A} \neg i \qquad\qquad \frac{A \quad \neg A}{\bot} \neg e$$

(no introduction rule for $\bot$) $\qquad\qquad \dfrac{\bot}{A} \bot e$

$$\boxed{\begin{array}{c} y \\ \vdots \\ A(y) \end{array}}$$
$$\frac{}{\forall x\, A(x)} \forall i \qquad\qquad \frac{\forall x\, A(x)}{A(t)} \forall e$$

$$\frac{A(t)}{\exists x\, A(x)} \exists i \qquad\qquad \exists x\, A(x) \quad \boxed{\begin{array}{cc} y & A(y) \\ & \vdots \\ & B \end{array}}$$
$$\overline{\qquad\qquad B \qquad\qquad} \exists e$$

$$\frac{\neg\neg A}{A} \neg\neg$$

In $\forall i$ and $\exists e$, y must not occur outside the boxes.

Fig. 1. Natural deduction rules for classical logic

The rules ∀e and ∃i are straightforward, but ∃e deserves a comment. Firstly, the purpose of the box is the same; it circumscribes the occurrences of y. It also introduces the assumption $A(y)$, in the same way that boxes in other rules introduce assumptions. The rule works as follows. We have $\exists x\, A(x)$; and we are capable of proving B from our current assumptions together with the further assumption of $A(y)$ for some new y which doesn't occur anywhere except in this subproof. In particular, y doesn't occur in B. Thus, we know $A(x)$ is provable for some x; call this y. But B follows from $A(y)$ and makes no reference to what y actually is; so, whatever y is, B must be provable.

This proof system is called *natural* deduction because the rules embody patterns familiar in ordinary reasoning, or at least in mathematical proof. For example, a proof of a statement about the natural numbers typically proceeds as follows: 'Let n be a natural number. Then..., thus the assertion holds for n. But n was an arbitrary natural number, so the assertion holds for *all* natural numbers.' This is the rule ∀i. Another example is proofs by contradiction, or *reductio ad absurdum*. To prove A, we suppose $\neg A$, and show that this leads to a contradiction. According to the rule ¬i, this entitles us to conclude $\neg\neg A$. But then, the rule ¬¬ allows us to deduce A. So, proofs by contradiction involve the two rules ¬i and ¬¬

Now to see some example proofs. The following is a proof of $\neg B \to (B \to A)$:

1	$\neg B$	Assumption
2	B	Assumption
3	$\bot$	¬e 1, 2
4	A	$\bot$e 3
5	$B \to A$	→i 2–4
6	$\neg B \to (B \to A)$	→i 1–5

Since the main operator in the formula we want to prove is →, it is a reasonable guess that →i is the last rule to be applied. Thus we try to construct a proof with a box, beginning with the assumption $\neg B$ and ending with the conclusion $B \to A$. Now, to prove $B \to A$ we do the same thing: open a box with B, and try to obtain A. At this point we have both $\neg B$ and B as current assumptions, so proving $\bot$ is easy with the rule ¬e (line 3), and A follows from this by $\bot$e (line 4). From now, we will not bother to write 'Assumption' each time a box is opened. If a box begins with a formula, that formula is an assumption. (Every box except the box in the rule ∀i begins with a formula.)

The following proofs show the equivalence of $\neg A \lor B$ and $A \to B$. First, we prove $\neg A \lor B \vdash A \to B$.

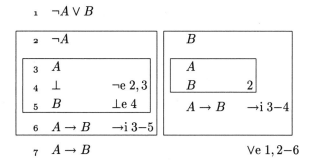

The formula $\neg A \lor B$ is outside all the boxes in line 1 because it is the premise of our argument. Unlike the assumptions made in lines 2 and 7, it does not have to be discharged by a rule. (The assumptions made in lines 2 and 7 are discharged by the $\lor$e rule in line 11.) Notice how the outermost structure of the proof is the elimination of the $\lor$ in the premise. We have $\neg A \lor B$, and prove $A \to B$ on the assumption first of $\neg A$ and then of B. These computations are carried out 'in parallel', as the rule $\lor$e allows.

Here is the proof of the converse, $A \to B \vdash \neg A \lor B$.

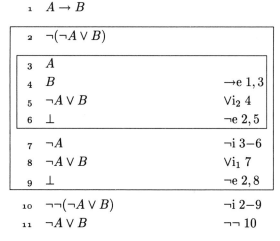

This proof seems less natural than the previous one. One might wonder if it was really necessary to prove $\neg\neg(\neg A \lor B)$ and use the $\neg\neg$ rule; could we have proceeded more directly? In fact, as we will see shortly, the $\neg\neg$ rule is essential in this proof, in the sense that one cannot find a proof which does not use it.

To illustrate the rules for quantifiers, we show

$$\forall x\,(Fx \to Hx), \forall x\,(Gx \to Hx), \forall x\,(Fx \lor Gx) \vdash \forall x\,Hx$$

We are omitting brackets; Fx is short for $F(x)$, etc.

1	$\forall x\,(Fx \to Hx)$	
2	$\forall x\,(Gx \to Hx)$	
3	$\forall x\,(Fx \lor Gx)$	

y 4

5	$Fy \lor Gy$	$\forall$e 3

6	Fy		Gy	
7	$Fy \to Hy$	$\forall$e 1	$Gy \to Hy$	$\forall$e 2
8	Hy	$\to$e 6,7	Hy	$\to$e 6,7

9	Hy	$\lor$e 6–8

10	$\forall x\,Hx$	$\forall$i 4–9

To illustrate the necessity of the condition that the variable which introduces a box in the rules $\forall$i and $\exists$e must not occur outside the box, consider the following 'proof' of $\exists x\,Fx, \exists x\,Gx \vdash \exists x\,(Fx \land Gx)$.

1	$\exists x\,Fx$
2	$\exists x\,Gx$

y 3 Fy

y 4 Gy

5	$Fy \land Gy$	$\land$i 3,4
6	$\exists x\,(Fx \land Gx)$	$\exists$i 5

7	$\exists x\,(Fx \land Gx)$	$\exists$e 2,4–6

8	$\exists x\,(Fx \land Gx)$	$\exists$e 1,3–7

In the inner box, y should have been something new. If it had been, line 5 would read $Fy \land Gz$ (say), and the application of $\exists$i at line 6 would be blocked.

Natural deduction is a considerable improvement on the Hilbert style in terms of giving us a perspicuous presentation. But the method of handling assumptions is not well suited to being handled by computer, nor to mathematical reasoning about proofs. In the next section, we recast these rules into a form in which the manipulation of assumptions is clearer; even though the rules are just the same.

4.3 Natural deduction in sequent style

The sequent calculus is similar to natural deduction except that the discharging of assumptions is clearer. Instead of manipulating formulas as in natural deduction, we manipulate whole sequents; the set on the left of the turnstile contains the set of hypotheses and current assumptions. The full set of rules for C is given in Fig. 2. Notice how the discharging of assumptions is represented. In the rule $\to$i, for example, the discharging of A is made explicit by removing it from the left of the turnstile. The same conditions for the quantifier rules apply:

1. y is a variable and t is a term and both are free for x in $A(x)$;

2. $A(y)$ and $A(t)$ result from $A(x)$ by replacing every free occurrence of x by y and t respectively;

3. In $\forall$i, y does not occur free in Γ or $\forall x\, A(x)$, and in $\exists$e, y does not occur free in Γ, $\exists x\, A(x)$, or in B.

A proof in the sequent calculus is a tree, whose root is the sequent which is being proved and whose leaves are 'basic sequents' of the form $A \vdash A$. Here, then, is our former proof of $A \to B \vdash_C \neg A \vee B$:

$$
\cfrac{
\cfrac{
\cfrac{
\cfrac{
\cfrac{A \vdash A \quad A \to B \vdash A \to B}{A \to B, A \vdash B}\to\text{e}
}{A \to B, A \vdash \neg A \vee B}\vee\text{i}_2 \quad \neg(\neg A \vee B) \vdash \neg(\neg A \vee B)
}{
\cfrac{
\cfrac{A \to B, \neg(\neg A \vee B), A \vdash \bot}{A \to B, \neg(\neg A \vee B) \vdash \neg A}\neg\text{i}
}{A \to B, \neg(\neg A \vee B) \vdash \neg A \vee B}\vee\text{i}_1 \quad \neg(\neg A \vee B) \vdash \neg(\neg A \vee B)
}\neg\text{e}
}{
\cfrac{A \to B, \neg(\neg A \vee B) \vdash \bot}{A \to B \vdash \neg\neg(\neg A \vee B)}\neg\text{i}
}\neg\text{e}
}{A \to B \vdash \neg A \vee B}\neg\neg
$$

One thing that should be clear about natural deduction proofs in sequent style is that they involve a lot of writing!

There are many variations on the presentation of natural deduction in sequent form. For example, if one admits a more general form of basic sequent, namely $\Gamma, A \vdash A$ instead of $A \vdash A$, then it is possible to write the

$$\frac{\Gamma \vdash A \quad \Delta \vdash B}{\Gamma, \Delta \vdash A \wedge B} \wedge i$$

$$\frac{\Gamma \vdash A \wedge B}{\Gamma \vdash A} \wedge e_1 \qquad \frac{\Gamma \vdash A \wedge B}{\Gamma \vdash B} \wedge e_2$$

$$\frac{\Gamma \vdash A}{\Gamma \vdash A \vee B} \vee i_1 \qquad \frac{\Gamma \vdash B}{\Gamma \vdash A \vee B} \vee i_2$$

$$\frac{\Gamma \vdash A \vee B \quad \Delta, A \vdash C \quad \Theta, B \vdash C}{\Gamma, \Delta, \Theta \vdash C} \vee e$$

$$\frac{\Gamma, A \vdash B}{\Gamma \vdash A \to B} \to i$$

$$\frac{\Gamma \vdash A \quad \Delta \vdash A \to B}{\Gamma, \Delta \vdash B} \to e$$

$$\frac{\Gamma, A \vdash \bot}{\Gamma \vdash \neg A} \neg i$$

$$\frac{\Gamma \vdash A \quad \Delta \vdash \neg A}{\Gamma, \Delta \vdash \bot} \neg e$$

(no introduction rule for $\bot$)

$$\frac{\Gamma \vdash \bot}{\Gamma \vdash A} \bot e$$

$$\frac{\Gamma \vdash A(y)}{\Gamma \vdash \forall x\, A(x)} \forall i$$

$$\frac{\Gamma \vdash \forall x\, A(x)}{\Gamma \vdash A(t)} \forall e$$

$$\frac{\Gamma \vdash A(t)}{\Gamma \vdash \exists x\, A(x)} \exists i$$

$$\frac{\Gamma \vdash \exists x\, A(x) \quad \Delta, A(y) \vdash B}{\Gamma, \Delta \vdash B} \exists e$$

$$\frac{\Gamma \vdash \neg\neg A}{\Gamma \vdash A} \neg\neg$$

In $\forall i$, y must not occur free in Γ or $A(x)$.
In $\exists e$, y must not occur free in Γ, $A(x)$ or B.

Fig. 2. Natural deduction rules for $\mathcal{C}$ in sequent style.

rules with more than one premise with Γ in place of Δ and Θ, so that, for instance, $\wedge$i would appear as

$$\frac{\Gamma \vdash A \quad \Gamma \vdash B}{\Gamma \vdash A \wedge B}$$

But, this leads to more unwieldy proofs; for example, our previous proof now becomes:

$$\frac{\cfrac{A{\to}B, \neg(\neg A{\vee}B), A \vdash A \quad A{\to}B, \neg(\neg A{\vee}B), A \vdash A{\to}B}{\cfrac{A{\to}B, \neg(\neg A{\vee}B), A \vdash B}{A{\to}B, \neg(\neg A{\vee}B), A \vdash \neg A{\vee}B} \vee ib \quad A{\to}B, \neg(\neg A{\vee}B), A \vdash \neg(\neg A{\vee}B)}}{\cfrac{\cfrac{A{\to}B, \neg(\neg A{\vee}B), A \vdash \bot}{\cfrac{A{\to}B, \neg(\neg A{\vee}B) \vdash \neg A}{A{\to}B, \neg(\neg A{\vee}B) \vdash \neg A{\vee}B} \vee ia} \neg i \quad A{\to}B, \neg(\neg A{\vee}B) \vdash \neg(\neg A{\vee}B)}{\cfrac{A{\to}B, \neg(\neg A{\vee}B) \vdash \bot}{\cfrac{A{\to}B \vdash \neg\neg(\neg A{\vee}B)}{A{\to}B \vdash \neg A{\vee}B} \neg\neg} \neg i} \neg e} \neg e$$

Another variation is whether the rule of monotonicity

$$\frac{\Gamma \vdash A}{\Gamma, B \vdash A}$$

is an explicit rule in the system or not. As things stand, it is derivable:

$$\frac{\cfrac{\Gamma \vdash A \quad B \vdash B}{\Gamma, B \vdash A \wedge B} \wedge i}{\Gamma, B \vdash A} \wedge e_1$$

but introducing it explicitly can yield more elegant proofs.

Propositional $\mathcal{S}4$

The rules for propositional $\mathcal{S}4$ are the rules of *propositional $\mathcal{C}$* together with the following rules for manipulating $\Box$.

$$\frac{\Gamma \vdash A}{\Box\Gamma \vdash \Box A} \Box_N \qquad\qquad \frac{\Box\Gamma \vdash A}{\Box\Gamma \vdash \Box A} \Box_T$$

We also have the sequent scheme $\Box A \vdash A$, called $\Box_R$. The labels here respectively stand for Necessitation, Transitivity and Reflexivity. The rule $\Box_N$ corresponds to our earlier rule for necessitation, and $\Box_R$ and $\Box_T$ correspond to the axioms A8 and A9 respectively. (The Appendix to [Hughes and Cresswell, 1968] contains a discussion on modal logic in natural deduction.)

Equivalence of natural deduction and Hilbert systems

Given a proof of a sequent $\Gamma \vdash_C A$ in the Hilbert style, it is possible to transform it into a proof of the same sequent in natural deduction. To show this, we have first to show that each of the axioms is provable in natural deduction from no premises. For example, for the axiom A3, which was $(\neg A \to \neg B) \to (B \to A)$, one has the following proof:

$$
\cfrac{
\cfrac{
\cfrac{\neg A \vdash \neg A \quad \neg A \to \neg B \vdash \neg A \to \neg B}{\neg A \to \neg B, \neg A \vdash \neg B} \to e \quad B \vdash B
}{
\cfrac{
\cfrac{
\cfrac{
\cfrac{\neg A \to \neg B, B, \neg A \vdash \bot}{\neg A \to \neg B, B \vdash \neg\neg A} \neg i
}{\neg A \to \neg B, B \vdash A} \neg\neg
}{\neg A \to \neg B \vdash B \to A} \to i
}{\vdash \neg A \to \neg B \to (B \to A)} \to i
} \neg e
}
$$

It is much more interesting to work out such proofs than to read them, so the reader is recommended to find proofs for the other axioms and for the definitional equivalences for $\wedge$ and $\vee$. The next stage of the proof transformation is to justify the use of the rules *modus ponens* and generalization. *Modus ponens* is essentially no more than the rule $\to e$, as the reader can check. Fuller details of this proof can be found in [Dummett, 1977].

The situation for the modal operators is similar. For example, it is easy to prove the axiom A9 from the rule $\Box_T$:

$$
\cfrac{
\cfrac{\Box A \vdash \Box A}{\Box A \vdash \Box\Box A} \Box_T
}{\vdash \Box A \to \Box\Box A} \to i
$$

Conversely, we sketch how to transform a natural deduction proof of a sequent $\Gamma \vdash A$ into a Hilbert-style axiomatic proof. Essentially, we justify each of the natural deduction rules in terms of the Hilbert system. We illustrate this with the rule $\neg\neg$. It says, if $\Gamma \vdash \neg\neg A$ then $\Gamma \vdash A$, so we have to show how to transform an Hilbert-style proof of $\neg\neg A$ from Γ to a proof of A from Γ. The way to do this is to append the following Hilbert-style proof of A from $\neg\neg A$ (of course, some renumbering of the lines will be required):

1.	$\neg\neg A$	hypothesis
2.	$\neg\neg A \to (\neg\neg\neg\neg A \to \neg\neg A)$	A1
3.	$\neg\neg\neg\neg A \to \neg\neg A$	M.P. 1,2
4.	$(\neg\neg\neg\neg A \to \neg\neg A) \to (\neg A \to \neg\neg\neg A)$	A3
5.	$\neg A \to \neg\neg\neg A$	M.P. 3,4
6.	$(\neg A \to \neg\neg\neg A) \to (\neg\neg A \to A)$	A3
7.	$\neg\neg A \to A$	M.P. 5,6
8.	A	M.P. 1,7

Another example to show how difficult Hilbert-style proofs are!

We will also prove the rule $\Box_T$ from the axiom A9. The first thing to observe is that if the sequence $A_1, A_2, \ldots, A_n$ is a Hilbert-style proof, then so is $\Box A_1, \Box A_2, \ldots, \Box A_n$. To check this it is necessary only to observe that the following 'boxed' form of the rule *modus ponens* is guaranteed by the axiom $\Box(A \to B) \to (\Box A \to \Box B)$:

$$\frac{\Box A \quad \Box(A \to B)}{\Box B}$$

We prove this as follows:

$$\cfrac{\Box A \quad \cfrac{\Box(A \to B) \quad \Box(A \to B) \to (\Box A \to \Box B)}{\Box A \to \Box B}\ \text{M.P.}}{\Box B}\ \text{M.P.}$$

Now suppose we have an Hilbert-style proof of A from $\Box B_1, \ldots, \Box B_m$. As described, we can convert this into a proof of $\Box A$ from $\Box\Box B_1, \ldots, \Box\Box B_m$. But, we have the axiom scheme $\Box B \to \Box\Box B$, so by instantiating B to each of the B_i and applying M.P. m times, we arrive at the following proof of $\Box A$ from $\Box B_1, \ldots, \Box B_m$:

$\Box B_1$	hypotheses
$\vdots$	
$\Box B_m$	
$\Box B_1 \to \Box\Box B_1$	A9
$\Box\Box B_1$	M.P. 1, $m+1$
$\vdots$	repeat for $B_2, \ldots, B_m$
$\vdots$	insert proof of $\Box A$ from
$\vdots$	$\Box\Box B_1, \ldots, \Box\Box B_m$
$\Box A$	

4.4 Intuitionistic logic

We mentioned in Section 1 that intuitionistic logic ($\mathcal{I}$) is a logic which rejects the basic assumption of classical logic that every sentence is true or false regardless of our ability to demonstrate its truth value. This section digresses from the subject of proof theory to describe intuitionistic logic in more detail, because it arises very naturally as a subsystem of the natural deduction presentation of classical logic (whether in proof-box style or in sequent style).

The rules given in Figs. 1 and 2 for classical logic displayed a pleasing symmetry, because with one exception, each of the rules either introduces or eliminates an operator. The exception is the rule $\neg\neg$. Intuitionistic logic has all the rules given in those figures except this rule.

Intuitionistic logic requires that if we say that something is true then we have shown that it is true by constructing a proof or disproof. We have a proof of $\neg\neg A$ when we can show that we shall never have a proof of $\neg A$, that is, when we can show that we will never have a proof that A will never be proved. Clearly, however, this does not amount to a proof of A itself, and hence on this view one cannot infer A from $\neg\neg A$. On the other hand, a proof of A does count as a proof that A will never be disproved; so, inferring $\neg\neg A$ from A is valid. We can check that it holds, given the remaining rules:

$$\frac{\dfrac{A \vdash A \quad \neg A \vdash \neg A}{A, \neg A \vdash \bot} \neg e}{A \vdash \neg\neg A} \neg i$$

All the remaining rules are valid intuitionistically, because they do not have the non-constructive character of the rule $\neg\neg$. For example, Scott gives the following argument for retaining the rule $\vee e$ [Scott,]:

> Suppose we have derivations of C from A and also from B. Then... the assertability of A yields the assertability of C, and the assertability of B yields the assertability of C. Suppose in addition we have $A \vee B$ assertable, which is to say we have A assertable or we have B assertable. Whichever obtains we have grounds for asserting C.

The loss of the $\neg\neg$ rule has a big impact on the set of provable sequents; here are some examples of some classically valid sequents and whether they are rejected or accepted intuitionistically:

rejected	accepted
$\vdash A \vee \neg A$	$\vdash \neg\neg(A \vee \neg A)$
$\neg\neg A \vdash A$	$A \vdash \neg\neg A$
$\neg A \rightarrow A \vdash A$	$A \rightarrow \neg A \vdash \neg A$
$\neg A \rightarrow B \vdash \neg B \rightarrow A$	$A \rightarrow \neg B \vdash B \rightarrow \neg A$
$\neg A \rightarrow \neg B \vdash B \rightarrow A$	$A \rightarrow B \vdash \neg B \rightarrow \neg A$
$\neg A \rightarrow B, \neg A \rightarrow \neg B \vdash A$	$A \rightarrow B, A \rightarrow \neg B \vdash \neg A$

But notice that intuitionistic logic does not remove all of the argument forms which one might consider objectionable in classical logic. The following are all valid intuitionistically:

$$A, \neg A \quad \vdash \quad B$$

$$\neg A \quad \vdash \quad A \rightarrow B$$

$$B \quad \vdash \quad A \rightarrow B$$

Notice also that the classical equivalence between $A \rightarrow B$ and $\neg A \vee B$ fails for intuitionistic logic. We have $\neg A \vee B \vdash_{\mathcal{I}} A \rightarrow B$, but $A \rightarrow B \nvdash_{\mathcal{I}} \neg A \vee B$. Recall that in the proof of the latter in $\mathcal{C}$ we had to use the $\neg\neg$ rule.

4.5 Gentzen sequent calculus for $\mathcal{I}$

Natural deduction is an improvement on Hilbert-style deduction from the point of view of finding proofs. But, even so, some of the natural deduction proofs given so far in this section have been less than obvious. One might ask, can we use them as an effective procedure for determining the validity of an argument? Some of the rules do lend themselves to providing proof procedures. For example, the rule $\wedge$i:

$$\frac{\Gamma \vdash A \quad \Gamma \vdash B}{\Gamma \vdash A \wedge B}$$

can be read backwards with the meaning: to prove something of the form $\Gamma \vdash A \wedge B$, prove $\Gamma \vdash A$ and $\Gamma \vdash B$ separately. But the rule $\wedge e_1$:

$$\frac{\Gamma \vdash A \wedge B}{\Gamma \vdash A}$$

is useless in this respect, because it makes the task of proving $\Gamma \vdash A$ more difficult (proving $\Gamma \vdash A \wedge B$ is harder than proving $\Gamma \vdash A$). Nor can the rule ¬e.

$$\frac{\Gamma \vdash A \quad \Gamma \vdash \neg A}{\Gamma \vdash \bot}$$

be used backwards to prove $\Gamma \vdash \bot$ for some Γ, because we have no way of knowing what a suitable A might be. These observations lead to the following definitions.

Definition 4.5.1. The *subformula relation* is the smallest transitive relation between formulas in the language such that

- $\bot$ is a subformula of everything.

- A is a subformula of itself and of $\neg A$.

- A and B are subformulas of $A \wedge B$, $A \vee B$ or $A \to B$.

- $A(t)$ is a subformula of $\forall x \, A(x)$ and $\exists x \, A(x)$, for any term t.

Definition 4.5.2. The proof rule

$$\frac{\Gamma \vdash A}{\Gamma' \vdash A'}$$

has the *subformula property* if every formula in $\Gamma \cup \{A\}$ is a subformula of a formula in $\Gamma' \cup \{A'\}$.

The idea is that rules with the subformula property can be used backwards to provide a decision procedure. All of the introduction rules seen so far have the subformula property. So do the rules $\Box_N$, $\Box_R$ and $\Box_T$ in $\mathcal{S}4$. But none of the elimination rules do. For example, in $\wedge e_2$ the formula A is not a subformula of any formula in the bottom. Nor does the rule ¬¬, since ¬¬A doesn't appear in the bottom (A does but that's not enough).

It turns out that for $\mathcal{I}$ we can rewrite these rules so that they do have the subformula property. The full set of rules is given in Fig. 3.

The left hand column of the figure shows the introduction rules of $\mathcal{I}$ in natural deduction sequent-style; compare Fig. 2. (Recall that $\mathcal{I}$ is $\mathcal{C}$ without the rule ¬¬.)

The right hand column also shows introduction rules, but they introduce the operators on the left. Notice the labelling convention; for example, $\wedge r$ means $\wedge$ is introduced on the *r*ight, as before, while $\wedge \ell$ means $\wedge$ is introduced on the *ℓ*eft.

MONO$_r$ is our former $\bot$e; $\bot$ has now taken on a structural role rather than a logical role, as it had as an operator. There is no great significance to

Structural rules

$\Gamma, A \vdash A$ (inclusion)

$$\dfrac{\Gamma \vdash C \quad \Delta, C \vdash A}{\Delta, \Gamma \vdash A} \text{ CUT}$$

$$\dfrac{\Gamma \vdash \bot}{\Gamma \vdash A} \text{ MONO}_r$$

$$\dfrac{\Gamma \vdash A}{\Gamma, B \vdash A} \text{ MONO}_\ell$$

Logical rules

$$\dfrac{\Gamma \vdash A \quad \Delta \vdash B}{\Gamma, \Delta \vdash A \wedge B} \wedge r$$

$$\dfrac{\Gamma, A, B \vdash C}{\Gamma, A \wedge B \vdash C} \wedge \ell$$

$$\dfrac{\Gamma \vdash A}{\Gamma \vdash A \vee B} \vee r_1 \qquad \dfrac{\Gamma \vdash B}{\Gamma \vdash A \vee B} \vee r_2$$

$$\dfrac{\Gamma, A \vdash C \quad \Delta, B \vdash C}{\Gamma, \Delta, A \vee B \vdash C} \vee \ell$$

$$\dfrac{\Gamma, A \vdash B}{\Gamma \vdash A \to B} \to r$$

$$\dfrac{\Gamma \vdash A \quad \Delta, B \vdash C}{\Gamma, \Delta, A \to B \vdash C} \to \ell$$

$$\dfrac{\Gamma, A \vdash \bot}{\Gamma \vdash \neg A} \neg r$$

$$\dfrac{\Gamma \vdash A}{\Gamma, \neg A \vdash \bot} \neg \ell$$

$$\dfrac{\Gamma \vdash A(y)}{\Gamma \vdash \forall x\, A(x)} \forall r$$

$$\dfrac{\Gamma, A(t) \vdash C}{\Gamma, \forall x\, A(x) \vdash C} \forall \ell$$

$$\dfrac{\Gamma \vdash A(t)}{\Gamma \vdash \exists x\, A(x)} \exists r$$

$$\dfrac{\Gamma, A(y) \vdash C}{\Gamma, \exists x\, A(x) \vdash C} \exists \ell$$

In $\forall r$, y must not occur in Γ or $A(x)$.
In $\exists \ell$, y must not occur in Γ, $A(x)$ or C.

Fig. 3. Gentzen rules for $\mathcal{I}$, and the rule CUT

this, but the point is that it denotes an empty right-hand side of a sequent. Thus, sequents in this style can have zero or one formula on the right. This structural restriction is what makes it intuitionistic logic; for, as the next section shows, removing the restriction (sequents there can have zero or more formulas on the right) gives classical logic.

Equivalence of natural deduction and Hilbert systems

In the presence of the structural rules, it is not difficult to prove that the left-introduction rules are equivalent to their elimination counterparts. Dealing first with the structural rules, MONO_ℓ is our former Monotonicity rule and, as mentioned, MONO_r is our former $\bot e$, while the inclusion and cut rules are just the same. The proofs are straightforward for the logical rules. For example, for the operator $\to$, first assume $\to \ell$ and prove $\to e$:

$$\cfrac{\cfrac{\Gamma \vdash A \quad \Gamma, B \vdash B}{\Gamma, A \to B \vdash B} \to \ell \qquad \Gamma \vdash A \to B}{\Gamma \vdash B} \text{ CUT}$$

Conversely, assume the rule $\to e$ and derive $\to \ell$:

$$\cfrac{\cfrac{\cfrac{\Gamma \vdash A}{\Gamma, A \to B \vdash A} \text{MONO}_\ell \quad \cfrac{A \to B \vdash A \to B}{\Gamma, A \to B \vdash A \to B} \text{MONO}_\ell}{\Gamma, A \to B \vdash B} \to e \qquad \cfrac{\Gamma, B \vdash C}{\Gamma, A \to B, B \vdash C} \text{MONO}_\ell}{\Gamma, A \to B \vdash C} \text{ CUT}$$

The proofs for the other operators are similar. What this proves, then, is that a sequent can be proved by the natural deduction rules for $\mathcal{I}$ iff it can be proved by Gentzen's rules for $\mathcal{I}$ together with the cut rule. But, the cut rule does not have the subformula property, so this still does not give us a decision procedure. However, in 1934 Gentzen proved a famous theorem called the 'cut elimination theorem'. (Again, see Chapter 3 of [Gentzen, 1969].)

Proposition 4.5.3. *Gentzen's cut elimination theorem.* Every proof in this system which employs the CUT rule can be replaced by one which does not.

Proof. The proof is quite lengthy. The idea is to show how proofs which use cut can be transformed into proofs which have the cut pushed higher up the proof tree. Repeated application of the transformation removes the cut rule altogether. The full details are given in [Dummett, 1977]. ∎

What this means is that we have a decision procedure for $\mathcal{I}$, by employing the rules of Fig. 3 without the cut rule. To see how this works, consider trying to prove the sequent $\neg A \vee B \vdash_{\mathcal{I}} A \to B$. Look in Fig. 3 for a rule whose bottom sequent matches this: if we find one, then all we have to do is prove its top sequent. The rules MONO$_\ell$, MONO$_r$, $\vee\ell$ and $\to r$ all match, but the first two leave us trying to prove $\vdash A \to B$ and $\neg A \vee B \vdash \bot$ respectively. The rule $\vee\ell$ is more hopeful, giving us $\neg A \vdash A \to B$ and $B \vdash A \to B$. Thus, we have the partial proof:

$$\frac{\vdots \qquad\qquad \vdots}{\neg A \vdash A \to B \qquad B \vdash A \to B} \vee\ell$$
$$\overline{\neg A \vee B \vdash A \to B}$$

Continuing to match against the bottom sequent of the rules in this way we arrive at the full proof:

$$\frac{\dfrac{\dfrac{\dfrac{A \vdash A}{\neg A, A \vdash \bot} \neg\ell}{\dfrac{\neg A, A \vdash B}{\neg A \vdash A \to B} \to r} \text{MONO}_\ell \qquad \dfrac{B, A \vdash B}{B \vdash A \to B} \to r}{\neg A \vee B \vdash A \to B}} \vee\ell$$

Now let us try to prove $\vdash_{\mathcal{I}} p \vee \neg p$. We know from Section 4.4 that this is not a theorem of $\mathcal{I}$. Again, we could use MONO$_r$ and try to prove $\vdash \bot$, which will fail, or we can use $\vee r_1$ or $\vee r_2$ which give us $\vdash p$ and $\vdash \neg p$. In the first case all we can do is use MONO$_r$ again, leaving us trying to prove $\vdash \bot$ once more; in the second, we could use MONO$_r$ or $\neg$i, but we quickly get stuck again:

$$\frac{\dfrac{\vdots}{\dfrac{\vdash \bot}{\vdash p} \text{MONO}_r}}{\vdash p \vee \neg p} \vee r_1 \qquad\qquad \frac{\dfrac{\vdots}{\dfrac{\vdash \bot}{\vdash \neg p} \text{MONO}_r}}{\vdash p \vee \neg p} \vee r_2 \qquad\qquad \frac{\dfrac{\vdots}{\dfrac{p \vdash \bot}{\vdash \neg p} \neg i}}{\vdash p \vee \neg p} \vee r_2$$

None of these 'proofs' succeeds, so we conclude $\nvdash_{\mathcal{I}} p \vee \neg p$.

This technique is not as simple as one might at first think, however. Consider trying to prove $\vdash_{\mathcal{I}} \neg\neg(p \vee \neg p)$. We know from Section 4.4 that this is a theorem of $\mathcal{I}$. The only rule which matches (besides MONO$_r$) is $\neg r$, which gives us $\neg(p \vee \neg p) \vdash \bot$. Again, MONO$_\ell$ doesn't help so we use $\neg\ell$ and find ourselves trying to prove $\vdash p \vee \neg p$. We might be tempted to conclude that it is not valid intuitionistically. However, here is the required proof:

$$
\cfrac{
  \cfrac{
    \cfrac{
      \cfrac{
        \cfrac{
          \cfrac{
            \cfrac{p \vdash p}{p \vdash p \vee \neg p} \vee \ell
          }{\neg(p \vee \neg p), p \vdash \bot} \neg \ell
        }{\neg(p \vee \neg p) \vdash \neg p} \neg r
      }{\neg(p \vee \neg p) \vdash p \vee \neg p} \vee r_2
    }{\neg(p \vee \neg p), \neg(p \vee \neg p) \vdash \bot} \neg \ell
  }{\neg(p \vee \neg p) \vdash \bot} *
}{\vdash \neg\neg(p \vee \neg p)} \neg r
$$

The trick is to exploit the fact that we have *sets* on the left at the point marked $*$.

4.6 Gentzen sequent calculus for $\mathcal{C}$ and $\mathcal{S}4$

4.6.1 Classical logic

Unfortunately we cannot find a left-introduction rule with the subformula property for $\neg\neg$. If we could, then we could obtain $\mathcal{C}$ by adding this new rule to the rules of Fig. 3. However, we can obtain the rules of Fig. 4.

This allows us to have 'anonymous disjunctions' on the right. We can prove $\vdash_{\mathcal{C}} p \vee \neg p$ in the following way:

$$
\cfrac{
  \cfrac{
    \cfrac{p \vdash p}{\vdash p, \neg p} \neg r
  }{\vdash p \vee \neg p} \vee r
}{}
$$

This was not available to us intuitionistically, because of the strong requirement of the intuitionistic rule $\vee r$.

Notice also that the fact that $\mathcal{I}$ is a subsystem of $\mathcal{C}$ is clear at this level also. The Gentzen rules for $\mathcal{I}$ are just those for $\mathcal{C}$ with the restriction that one can have at most one formula on the right.

4.6.2 Modal logic

Rules with the subformula property can also be found for $\mathcal{S}4$. They are given in Fig. 5. Again, showing that these rules are equivalent to the ones previously given is straightforward.

Structural rules

$$\Gamma, A \vdash A, \Delta \ \text{(inclusion)}$$

$$\frac{\Gamma \vdash \Theta, C \quad C, \Delta \vdash \Lambda}{\Gamma, D \vdash \Theta, \Lambda} \ \text{CUT}$$

$$\frac{\Gamma \vdash \Delta}{\Gamma \vdash A, \Delta} \ \text{MONO}_r$$

$$\frac{\Gamma \vdash \Delta}{\Gamma, A \vdash \Delta} \ \text{MONO}_\ell$$

Logical rules

$$\frac{\Gamma \vdash A, \Delta \quad \Gamma \vdash B, \Delta}{\Gamma \vdash A \wedge B, \Delta} \ \wedge r$$

$$\frac{\Gamma, A, B \vdash \Delta}{\Gamma, A \wedge B \vdash \Delta} \ \wedge \ell$$

$$\frac{\Gamma \vdash A, B, \Delta}{\Gamma \vdash A \vee B, \Delta} \ \vee r$$

$$\frac{\Gamma, A \vdash \Delta \quad \Gamma, B \vdash \Delta}{\Gamma, A \vee B \vdash \Delta} \ \vee \ell$$

$$\frac{\Gamma, A \vdash B, \Delta}{\Gamma \vdash A \rightarrow B, \Delta} \ \rightarrow r$$

$$\frac{\Gamma \vdash A, \Delta \quad \Gamma, B \vdash \Delta}{\Gamma, A \rightarrow B \vdash \Delta} \ \rightarrow \ell$$

$$\frac{\Gamma, A \vdash \Delta}{\Gamma \vdash \neg A, \Delta} \ \neg r$$

$$\frac{\Gamma \vdash A, \Delta}{\Gamma, \neg A \vdash \Delta} \ \neg \ell$$

$$\frac{\Gamma \vdash A(y), \Delta}{\Gamma \vdash \forall x \, A(x), \Delta} \ \forall r$$

$$\frac{\Gamma, A(t) \vdash \Delta}{\Gamma, \forall x \, A(x) \vdash \Delta} \ \forall \ell$$

$$\frac{\Gamma \vdash A(t), \Delta}{\Gamma \vdash \exists x \, A(x), \Delta} \ \exists r$$

$$\frac{\Gamma, A(y) \vdash \Delta}{\Gamma, \exists x \, A(x) \vdash \Delta} \ \exists \ell$$

In $\forall r$ and $\exists \ell$, y must not occur in Γ, Δ or $A(x)$.

Fig. 4. Gentzen rules for $\mathcal{C}$, and the rule CUT.

$$\frac{\Gamma \vdash A, \Delta}{\Box\Gamma \vdash \Box A, \Diamond\Delta}\ \Box_N \qquad\qquad \frac{\Gamma, A \vdash \Delta}{\Box\Gamma, \Diamond A \vdash \Diamond\Delta}\ \Diamond_N$$

$$\frac{\Gamma, A \vdash \Delta}{\Gamma, \Box A \vdash \Delta}\ \Box_R \qquad\qquad \frac{\Gamma \vdash A, \Delta}{\Gamma \vdash \Diamond A, \Delta}\ \Diamond_R$$

$$\frac{\Box\Gamma \vdash A, \Diamond\Delta}{\Box\Gamma \vdash \Box A, \Diamond\Delta}\ \Box_T \qquad\qquad \frac{\Box\Gamma, A \vdash \Diamond\Delta}{\Box\Gamma, \Diamond A \vdash \Diamond\Delta}\ \Diamond_T$$

Fig. 5. Additional Gentzen rules for $\mathcal{S}4$.

4.7 Properties of presentations

We have given several proof-theoretic definitions of $\vdash_C$ and $\vdash_{S4}$, and we have shown their equivalence in each case. In this section we state that they are equivalent to $\vDash_C$ and $\vDash_{S4}$ of Section 3. Proving that $\Gamma \vdash A$ implies $\Gamma \vDash A$ is called *soundness*—it shows that the proof system is 'sound' with respect to the valuation system in the sense of not proving anything which is not valid. The other direction, that $\Gamma \vDash A$ implies $\Gamma \vdash A$, is called *completeness*, because it shows that the proof system is strong enough to prove everything that is valid.

Definition 4.7.1. Let $\vdash$ be a consequence relation defined proof-theoretically and $\vDash$ an entailment relation.

- $\vdash$ is *sound* with respect to $\vDash$ if

$$\Gamma \vdash A \quad \text{implies} \quad \Gamma \vDash A.$$

- $\vdash$ is *complete* with respect to $\vDash$ if

$$\Gamma \vDash A \quad \text{implies} \quad \Gamma \vdash A.$$

Showing soundness is generally much easier than showing completeness. To show soundness, it is sufficient to show that the rules for constructing proofs preserve entailment. In the case of Hilbert presentations, this amounts to induction on the length of the proof. For the base case (proofs of length 1) it is sufficient to show that the axioms are tautologies. For the inductive case, it is necessary to show that the rules of inference preserve entailment; so for M.P. we show: if the proofs of A and $A \to B$ are valid from certain premises then the proof of B is valid from the same premises.

The strategy for proving soundness is to show that application of the rules preserve entailment according to the valuation systems. This is a proof by structural induction. The base case is just the inclusion rule. The inductive step consists in showing, for each rule in the presentation, that the sequent on the bottom of the rule is valid in the valuation system whenever the top ones are. Proposition 3.2.2 establishes it for the structural rules, so all we need to check are the logical rules in the figures of the preceding section.

Showing completeness, on the other hand, is much harder. One way of proceeding is to show that the proof system can be extended by adding as an axiom any sentence for which neither it nor its negation is a theorem, in such a way that there is an assignment for the valuation system which makes all the theorems true. If A is a tautology which is not a theorem, then extending the proof system by $\neg A$ gives an assignment in which both A and $\neg A$ are true. Since this is impossible, there are no tautologies which are not theorems; therefore $\vDash A$ implies $\vdash A$. If we can then show that both the valuation system and the proof system have the deduction property, then this proves that $\Gamma \vDash A$ implies $\Gamma \vdash A$ at least for finite Γ. But the task of showing that the proof system can always be extended in the way described is hard. The full details for Hilbert systems can be found in [Hamilton, 1978].

Proposition 4.7.2. The proof systems given for systems $\mathcal{C}$ and $\mathcal{S}4$ are sound with respect to their entailment relations.

Proof. As we have said, it is just a matter of checking each rule. A selection of four rules is given.

1. The rule $\rightarrow$i in $\mathcal{C}$. It is necessary to show that this rule preserves entailment, that is, that $\Gamma, A \vDash_{\mathcal{C}} B$ implies $\Gamma \vDash_{\mathcal{C}} A \rightarrow B$. Suppose $\Gamma, A \vDash_{\mathcal{C}} B$ and a is an assignment such that $v_a(C) = \mathbf{t}$ for each $C \in \Gamma$. We want to show that $v_a(A \rightarrow B) = \mathbf{t}$. Since $M = \{\mathbf{t}, \mathbf{f}\}$, $v_a(A)$ is either $\mathbf{t}$ or $\mathbf{f}$. Suppose $v_a(A) = \mathbf{t}$. Then, since $\Gamma, A \vDash_{\mathcal{C}} B$, $v_a(B) = \mathbf{t}$. Then $v_a(A \rightarrow B)$ is $f_{\rightarrow}(\mathbf{t}, \mathbf{t})$, which is $\mathbf{t}$. Suppose now that $v_a(A) = \mathbf{f}$. Then $v_a(A \rightarrow B) = f_{\rightarrow}(\mathbf{f}, v_a(B))$, which is $\mathbf{t}$ no matter what $v_a(B)$ is.

2. The rule $\forall$e in $\mathcal{C}$. We need to show that if $v_a(\forall x\, A(x)) = \mathbf{t}$ for an assignment $\langle a, I \rangle$, then $v_a(A(t)) = \mathbf{t}$ for any term t. Suppose for some t we have that $v_a(A(t)) = \mathbf{f}$. Let a' be the assignment identical to a except that $a'(x) = a_T(t)$; it assigns to x the member of I which a assigns to the vagrant term t. We have $a \sim_x a'$. Then $v_{a'}(\forall x\, A(x)) = \mathbf{f}$, and since $v_a(\forall x\, A(x)) = \bigwedge\{v_{a'}(A(x)) \mid a \sim_x a'\}$, it follows that $v_a(\forall x\, A(x)) = \mathbf{f}$.

3. The rule →i in $\mathcal{S}4$. Suppose $\Gamma, A \vDash_M B$, where $\mathbf{M}$ is a valuation system meeting the conditions of Def. 2.3.1, with worlds W. Suppose $w \Vdash_a \Gamma$ for some assignment a and $w \in W$. We have to show that $w \Vdash_a A \to B$. If $w \Vdash_a A$, then $w \Vdash_a B$ and so $w \Vdash_a A \to B$. Otherwise, $w \Vdash_a B$ immediately.

4. The rule $\Box_T$ in $\mathcal{S}4$. Suppose $\Box\Gamma \vDash A$. We need to prove $\Box\Gamma \vDash \Box A$. Suppose $w \Vdash_a \Box\Gamma$ for some assignment a. We need to show $w \Vdash_a \Box A$. Pick $x, y \in W$ such that $w \leqslant x \leqslant y$. Since $w \leqslant y$ and $w \Vdash_a \Box\Gamma$, $y \Vdash_a \Gamma$. Since y is arbitrary, $w \Vdash_a \Box A$. ∎

The proofs of the soundness of the other rules are left to the reader. Notice that the proof of the rule $\Box_T$ relies on the transitivity of $\leqslant$. Similarly, $\Box_R$ relies on its reflexivity. We explore this in Section 5.3.

Completeness is the converse of soundness, but this simple observation hides the fact that completeness is really a deeper property than soundness, and is generally much harder to prove. It is in general easier to show assertions of the form $\Gamma \vdash A$ ('there is a proof') than to show that $\Gamma \nvdash A$ ('there is no proof')—showing the former is just exhibiting the proof, while the latter involves quantifying over all potential proofs. On the other hand, it is easier to show $\Gamma \nvDash A$ ('there is a countermodel') than $\Gamma \vDash A$ ('there is no countermodel') because a single assignment is needed for the former, while we have to consider possibly an infinite number of assignments to show the latter. Here are the definitions of soundness and completeness recast in view of the observation that of $\vdash$ and $\nvdash$ are 'basic':

- Soundness: there is not both a proof and a countermodel.

- Completeness: either there is a proof or there is a countermodel.

To show soundness we can suppose we have a proof and a countermodel, and use them to derive a contradiction. Showing completeness, on the other hand, is much harder. We could show that if there is no proof then there must be a countermodel, but this is difficult because the absence of a proof gives us nothing to *use* to construct the countermodel. On the other hand, showing that the absence of a countermodel implies the existence of a proof is no easier; again, there is nothing to use to help construct the proof.

5 Some further topics

This is only the beginning of the story. In this section we look briefly at a number of extensions and further ideas.

5.1 Valuation systems for $\mathcal{I}$

$\mathcal{I}$ (intuitionistic logic) was defined in Section 4.4 as having the proof rules of $\mathcal{C}$ (classical logic) but without the rule $\neg\neg$ for eliminating double-negations:

$$\frac{\Gamma \vdash \neg\neg A}{\Gamma \vdash A}$$

It was argued that this rule failed the criterion of *constructiveness*, which is the hallmark of intuitionistic logic, because a proof of $\neg\neg A$, which is merely a proof that there is no proof for $\neg A$, cannot be translated into a proof of A. This in turn means that the 'principle of excluded middle' $(p \vee \neg p)$ is not valid in $\mathcal{I}$, because a proof of $p \vee \neg p$ must be translatable either into a proof of p, or a proof of $\neg p$, neither of which are wanted as theorems. This strong reading of $\vee$ is another characteristic of $\mathcal{I}$; for in $\mathcal{C}$, one can (indeed, one does) have a proof of $p \vee \neg p$ without the ability to translate it into a proof of p or of $\neg p$.

This move towards partiality is already present in the many-valued valuation systems of Section 2.2. One might therefore ask the question, will one of these valuation systems provide a semantics for $\mathcal{I}$? A likely contender is the valuation system of Example 2.2.2 in Section 2.

However, one can show that this is not so. Indeed, we can show a much stronger result, due to Gödel, that there is no *finite* valuation system for $\mathcal{I}$.

Proposition 5.1.1. There is no finite valuation system for which $\mathcal{I}$ is sound and complete.

Proof. Suppose $\mathbf{M} = \langle M, D, F \rangle$ is a valuation system in which M has n elements, and suppose $\mathcal{I}$ is sound and complete with respect to $\mathbf{M}$. Consider the following formula:

$$A = \bigvee_{1 \leqslant i < j \leqslant n+1} (p_i \leftrightarrow p_j)$$

We prove (1) $\vDash_M A$; and (2) $\nvdash_{\mathcal{I}} A$.

(1) Let a be an assignment for $\mathbf{M}$. Since A has $n+1$ atomic sentences in it and there are n truth values, there must be i and j such that $a(p_i) = a(p_j)$. Since $\vdash_{\mathcal{I}} p \leftrightarrow p$, we have $\vDash_M p \leftrightarrow p$, and so $v_a(p_i \leftrightarrow p_j) \in D$. But also, $p_i \leftrightarrow p_j \vdash_{\mathcal{I}} A$, and so $p_i \leftrightarrow p_j \vDash_M A$, so $v_a(A) \in D$. But a was arbitrary, and so $v_a(A) \in D$ for every a, i.e. $\vDash_M A$. (2) This follows by inspecting the Gentzen rules of Section 4.5. ■

We can, however, find a class of finite valuation systems. The following one, due to Kripke, combines the idea of partiality with the modal semantics of Section 2.3.

Let $(W, \leqslant)$ be a finite partial order, that is, W a finite set and $\leqslant$ a reflexive, antisymmetric and transitive relation on W; and let w_0 be a distinguished element of W. Let $M = \langle M, D, F \rangle$ be the valuation system given by:

1. $M = \mathcal{P}(W)$ (the power set of W);

2. $D = \{X \subseteq W \mid w_0 \in X\}$. So far this is the same as the definition of valuation systems for $\mathcal{S}4$ (Def. 2.3.1); but now,

3. F is the following collection of functions:

$$
\begin{aligned}
f_\wedge(X, Y) &= X \cap Y \\
f_\vee(X, Y) &= X \cup Y \\
f_\perp &= \emptyset, \text{ so far as before} \\
f_\neg(X) &= \{x \in W \mid \forall y \, x \leqslant y \text{ implies } y \notin X\} \\
f_\rightarrow(X, Y) &= \{x \in W \mid \forall y \, x \leqslant y \text{ implies } y \in ((W - X) \cup Y)\}.
\end{aligned}
$$

The 'modal operators'—those whose interpretation depends on the ordering $\leqslant$—are $\neg$ and $\rightarrow$ in this definition. The apparent similarity between valuation systems for $\mathcal{I}$ and for $\mathcal{S}4$ will be clarified in the next section.

An assignment a relative to this valuation system is a map $P \rightarrow M$, as before, with the added condition that for each $p \in P$, $a(p)$ is $\leqslant$-closed. That is, if $x \in a(p)$ and $x \leqslant y$ then $y \in a(p)$.

Again it is more perspicuous to translate this definition into the forcing relation of Section 2.3. Given a valuation system $\mathbf{M}$ and an assignment a as described, we write $x \Vdash A$ if $x \in v_a(A)$. Then:

$$
\begin{aligned}
x \Vdash p \quad &\text{iff} \quad x \in a(p) \\
x \Vdash A \wedge B \quad &\text{iff} \quad x \Vdash A \text{ and } x \Vdash B \\
x \Vdash A \vee B \quad &\text{iff} \quad x \Vdash A \text{ or } x \Vdash B \\
x \Vdash A \rightarrow B \quad &\text{iff} \quad \forall y \, x \leqslant y \text{ implies } y \nVdash A \text{ or } y \Vdash B \\
x \Vdash \neg A \quad &\text{iff} \quad \forall y \, x \leqslant y \text{ implies } y \nVdash A
\end{aligned}
$$

The entailment relation for intuitionistic logic is defined exactly as the entailment relation for $\mathcal{S}4$. Thus: $\Gamma \vDash_{\mathcal{I}} A$ if, for each valuation system given by $(W, \leqslant, w_0)$ and assignment a, we have that

If $w_0 \Vdash B$ for each $B \in \Gamma$ then $w_0 \Vdash A$.

5.2   Maps between logics

Given two systems, we are often interested in how they compare with each other. Is one 'stronger' than the other? In this section we consider the propositional versions of the systems $\mathcal{I}$, $\mathcal{C}$ and $\mathcal{S}4$.

Example 5.2.1. $\mathcal{I}$ is a 'subsystem' of $\mathcal{C}$; and $\mathcal{C}$ is a 'subsystem' of $\mathcal{S}4$. This is apparent simply by examining the rules given in Section 4. The rules of $\mathcal{I}$ are a subset of those for $\mathcal{C}$, which in turn form a subset of the $\mathcal{S}4$ rules. Therefore, we have

$$\Gamma \vdash_{\mathcal{I}} A \text{ implies } \Gamma \vdash_{\mathcal{C}} A \text{ implies } \Gamma \vdash_{\mathcal{S}4} A$$

However, this is far from the full story.

Example 5.2.2. $\mathcal{C}$ can be viewed as a weaker subsystem of $\mathcal{I}$, in the following sense:

$$\Gamma \vdash_{\mathcal{C}} A \quad \text{iff} \quad \neg\neg\Gamma \vdash_{\mathcal{I}} \neg\neg A$$

$\neg\neg\Gamma$ is an abbreviation for the set $\{\neg\neg B \mid B \in \Gamma\}$.

The direction right to left follows from Example 5.2.1 and the fact that $\neg\neg A$ and A are equivalent in classical logic. The proof of the converse is by consideration of the proof of $\Gamma \vdash_{\mathcal{C}} A$. It is most convenient to take the proof in sequent-style natural deduction. Take this proof, and prefix every formula with $\neg\neg$. We have to show that each step of the resulting structure is valid in $\mathcal{I}$. For example, for uses of the rule $\wedge$i in the original proof we need to show that $\neg\neg\Gamma \vdash_{\mathcal{I}} \neg\neg A$ and $\neg\neg\Gamma \vdash_{\mathcal{I}} \neg\neg B$ imply $\neg\neg\Gamma \vdash_{\mathcal{I}} \neg\neg(A \wedge B)$. Of course this really only involves showing that $\neg\neg A, \neg\neg B \vdash_{\mathcal{I}} \neg\neg(A \wedge B)$, for we have

$$\cfrac{\neg\neg\Gamma \vdash \neg\neg B \qquad \cfrac{\neg\neg\Gamma \vdash \neg\neg A \qquad \neg\neg A, \neg\neg B \vdash \neg\neg(A \wedge B)}{\neg\neg\Gamma, \neg\neg B \vdash \neg\neg(A \wedge B)} \text{ CUT}}{\neg\neg\Gamma \vdash \neg\neg(A \wedge B)} \text{ CUT}$$

The proof, then, of $\neg\neg A, \neg\neg B \vdash_{\mathcal{I}} \neg\neg(A \wedge B)$ is the following:

$$\cfrac{\cfrac{A \vdash A \quad B \vdash B}{A, B \vdash A \land B} \land i \qquad \neg(A \land B) \vdash \neg(A \land B)}{\cfrac{\cfrac{\cfrac{\neg(A \land B), A, B \vdash \bot}{\neg(A \land B), A \vdash \neg B} \neg i \qquad \neg\neg B \vdash \neg\neg B}{\cfrac{\neg\neg B, \neg(A \land B), A \vdash \bot}{\neg\neg B, \neg(A \land B) \vdash \neg A} \neg i \qquad \neg\neg A \vdash \neg\neg A}}{\cfrac{\neg\neg A, \neg\neg B, \neg(A \land B) \vdash \bot}{\neg\neg A, \neg\neg B \vdash \neg\neg(A \land B)} \neg i} \neg e} \neg e} \neg e$$

We adopt the same approach for the other rules. So, use of the rule $\neg i$ in the original proof involves justifying passing from $\neg\neg\Gamma, \neg\neg A \vdash \neg\neg\bot$ to $\neg\neg\Gamma \vdash \neg\neg\neg A$. This is justified if one can prove $\neg\neg\bot \vdash_{\mathcal{I}} \bot$, which can be done as follows:

$$\cfrac{\cfrac{\bot \vdash \bot}{\vdash \neg\bot} \neg i \qquad \neg\neg\bot \vdash \neg\neg\bot}{\neg\neg\bot \vdash \bot} \neg e$$

One might have expected the rule $\neg\neg$ to present difficulties, since it is the one which is not valid intuitionistically; but the proof goes through. We must justify passing from $\neg\neg\Gamma \vdash \neg\neg\neg\neg A$ to $\neg\neg\Gamma \vdash \neg\neg A$, so again it suffices to show that $\neg\neg\neg\neg A \vdash_{\mathcal{I}} \neg\neg A$:

$$\cfrac{\cfrac{\cfrac{\neg A \vdash \neg A \quad \neg\neg A \vdash \neg\neg A}{\neg A, \neg\neg A \vdash \bot} \neg e}{\neg A \vdash \neg\neg\neg A} \neg i \qquad \neg\neg\neg\neg A \vdash \neg\neg\neg\neg A}{\cfrac{\neg\neg\neg\neg A, \neg A \vdash \bot}{\neg\neg\neg\neg A \vdash \neg\neg A} \neg i} \neg e$$

The remaining cases are left as an exercise.

Example 5.2.3. We saw in Example 5.2.1 that $\mathcal{I}$ is a subsystem of $\mathcal{S}4$, but it may be embedded in $\mathcal{S}4$ in another, more satisfying, way, which is also due to Gödel. $*$ maps $\mathcal{I}$ formulas to $\mathcal{S}4$ formulas in the following way:

$$
\begin{aligned}
p^* &= \Box p && \text{for atomic } p\\
(A \land B)^* &= A^* \land B^*\\
(A \lor B)^* &= A^* \lor B^*\\
(A \to B)^* &= \Box(A^* \to B^*)\\
(\neg A)^* &= \Box\neg(A^*)
\end{aligned}
$$

Thus, for example, the intuitionistic formula $\neg p \to (p \to q)$ becomes the modal formula $\Box(\Box\neg\Box p \to \Box(\Box p \to \Box q))$.

We have that

$$\Gamma \vdash_{\mathcal{I}} A \quad \text{iff} \quad \Gamma^* \vdash_{\mathcal{S}4} A^*$$

This shows that the strong readings of $\to$ and $\neg$ seen above (Section 5.1) can be coded up in terms of $\square$. One might have expected this, given the correspondence between the valuation systems for $\mathcal{S}4$ and $\mathcal{I}$.

5.3 Correspondence theory

Certain modal formulas are satisfied in the valuation systems for $\mathcal{S}4$ given by Definition 2.3.1 by virtue of the properties of the relation $\leqslant$ on W. In this section we start by looking at examples of such tautologies, and then we see what happens if $\leqslant$ is generalized to an arbitrary relation R. It turns out that some properties of R correspond exactly to the validity of formulas.

Proposition 5.3.1. Let L be the language given by $P = \{p, q, r, \ldots\}$ and $O = \{\wedge, \vee, \to, \neg, \square, \diamond\}$ (see Definition 2.2.3), and let $(W, \leqslant, w_0)$ specify a valuation system $\mathbf{M}$ for L in the way described in Definition 2.3.1. Let a be an assignment relative to $\mathbf{M}$. Then for any sentences A and B of L,

1. $v_a(\square A \to A) = W$

2. $v_a(\square A \to \square\square A) = W$

3. $v_a(\square(A \to B) \to (\square A \to \square B)) = W$

4. $v_a(\square A) = v_a(\neg\diamond\neg A)$

5. if $v_a(A) = W$ then $v_a(\square A) = W$.

Paraphrasing: the sentences $\square A \to A$, $\square A \to \square\square A$ and $\square(A \to B) \to (\square A \to \square B)$ are all tautologies in $\mathcal{S}4$ (since they have designated truth values). Also, $\square A$ and $\neg\diamond\neg A$ are equivalent; and if A is a tautology then so is $\square A$.

Proof. 1. Since $v_a(\square A \to A) = f_\to(v_a(\square A), v_a(A)) = (W - f_\square(v_a(A))) \cup v_a(A)$, it is sufficient to show that $f_\square(X) \subseteq X$ for all $X \subseteq W$. Let $x \in f_\square(X)$. Then by definition of $f_\square$, $y \geqslant x$ implies $y \in X$. In particular, $x \in X$.

2. $v_a(\square A \to \square\square A) = (W - f_\square(v_a(A))) \cup f_\square(f_\square(v_a(A)))$, so it's sufficient to show that $f_\square(X) \subseteq f_\square(f_\square(X))$ for any $X \subseteq W$. Let $x \in f_\square(X)$. To show $x \in f_\square(f_\square(X))$, we need to show that $y_1 \geqslant x$ implies $y_1 \in f_\square(X)$, that is, $y_1 \geqslant x$ implies $(y_2 \geqslant y_1$ implies $y_2 \in X)$. Suppose y_1

and y_2 are such that $y_2 \geqslant y_1 \geqslant x$. Then $y_2 \geqslant x$ by transitivity, and so $y_2 \in X$ as required.

3. Again, it is sufficient to show that, for all $X, Y \subseteq W$, $f_\Box((W - X) \cup Y) \subseteq (W - f_\Box(X)) \cup f_\Box(Y)$. Let $x \in f_\Box((W - X) \cup Y)$, and suppose $x \in f_\Box(X)$. We need to show that $x \in f_\Box(Y)$, i.e. that $\forall y\, x \leqslant y$ implies $y \in Y$. Suppose $x \leqslant y$; then $y \in X$ since $x \in f_\Box(X)$, and $y \in (W - X) \cup Y$ since $x \in f_\Box((W - X) \cup Y)$. Therefore, $y \in Y$ as required.

4. It is sufficient to show that $f_\Box(X) = f_\neg(f_\Diamond(f_\neg(X)))$.

$$
\begin{aligned}
x \in f_\neg(f_\Diamond(f_\neg(X))) \quad &\text{iff} \quad x \in (W - f_\Diamond(W - X)) \\
&\text{iff} \quad x \notin f_\Diamond(W - X) \\
&\text{iff} \quad x \leqslant y \text{ for no } y \in W - X \\
&\text{iff} \quad x \leqslant y \text{ implies } y \notin W - X \\
&\text{iff} \quad x \leqslant y \text{ implies } y \in X \\
&\text{iff} \quad x \in f_\Box(X)
\end{aligned}
$$

5. This follows from the fact that $f_\Box(W) = W$, which is immediate. ∎

These proofs are much easier if we use the equivalent formulation of Proposition 2.3.2. We will use that for the remainder of this section.

In more general applications of modal logic, valuation systems are generated by a set of worlds W equipped with an arbitrary relation R instead of $\leqslant$. The relation R is called an *accessibility relation*. The functions $f_\Box$ and $f_\Diamond$ are then given by:

$$
\begin{aligned}
f_\Box(X) &= \{x \in W \mid \forall y\, Rxy \text{ implies } y \in X\} \\
f_\Diamond(X) &= \{x \in W \mid \exists y\, Rxy \text{ and } y \in X\}
\end{aligned}
$$

In this section, we look at the correspondences between properties of the accessibility relation and validity of sentences. The proofs of Proposition 5.3.1 are easily generalized to show that

1. If R is reflexive, $\Box A \rightarrow A$ is true at every point of W, and

2. If R is transitive, $\Box A \rightarrow \Box\Box A$ is true at every point of W.

We can also show that the converses are true:

Proposition 5.3.2. Let $(W, \leqslant, w_0)$ be a modal frame and A a sentence.

1. R is reflexive if, for all assignments a and $x \in W$, $x \Vdash_a \Box A \rightarrow A$.

2. R is transitive if, for all assignments a and $x \in W$, $x \Vdash_a \Box A \rightarrow \Box\Box A$.

Proof. 1. Let $x \in W$ and p be any atomic sentence. Pick an assignment a such that $a(p) = \{y \in W \mid Rxy\}$. Then $x \Vdash_a \Box p$. Since $\Box p \rightarrow p$ is true at all points, $x \Vdash_a \Box p \to p$, so $x \Vdash_a p$. Hence $x \in a(p)$, therefore Rxx.

2. Let $x, y, z \in W$ be such that Rxy and Ryz. Let a be an assignment such that $a(p) = \{w \in W \mid Rxw\}$. Then $x \Vdash_a \Box p$. To show Rxz we need to show that $z \Vdash_a p$. Since $x \Vdash_a \Box p \to \Box\Box p$, $x \Vdash_a \Box\Box p$. Since Rxy, $y \Vdash_a \Box p$. And since Ryz, $z \Vdash_a p$ as required. Hence Rxz. ∎

Another correspondence example is the following: the accessibility relation R is said to be *directed* if whenever Rxy and Rxz there is a $w \in W$ such that Ryw and Rzw.

Proposition 5.3.3. *R is directed iff, for all a and x, $x \Vdash_a \Diamond\Box A \to \Box\Diamond A$.*

Proof. (If) Let $x, y, z \in W$ be such that Rxy and Rxz. We need to establish the existence of an appropriate w. Let a be an assignment such that $a(p) = \{s \in W \mid Rys\}$. Then $y \Vdash_a \Box p$, so $x \Vdash_a \Diamond\Box p$, so $x \Vdash_a \Box\Diamond p$ (using $x \Vdash_a \Diamond\Box A \to \Box\Diamond A$), so $z \Vdash_a \Diamond p$. Hence there is a w such that Rzw and $w \Vdash_a p$. So $w \in a(p)$, and Ryw.

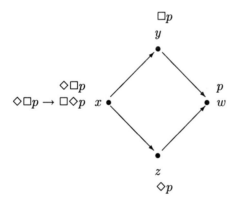

(Only if) Suppose R is directed and $x \Vdash_a \Diamond\Box A$ for some assignment a. We will show that $x \Vdash_a \Box\Diamond A$. Since $x \Vdash_a \Diamond\Box A$ there is a y such that Rxy and $y \Vdash_a \Box A$. Consider z such that Rxz. Since R is directed there is a w such that Ryw and Rzw. Now $w \Vdash_a A$, so $z \Vdash_a \Diamond A$ and, since z was arbitrary, $x \Vdash_a \Box\Diamond A$. ∎

In general we can construct systems of modal logic in two ways. We can either stipulate properties of the accessibility relation R, or we can insist on the validity of certain axioms or rules. We end this section by mentioning a negative result whose proof is quite hard. First notice that each of the conditions of reflexivity, transitivity and directedness are first-order conditions on R:

1. reflexivity is $\forall x\, Rxx$

2. transitivity is $\forall x\, \forall y\, \forall z\, ((Rxy \wedge Ryz) \to Rxz)$

3. directedness is $\forall x\, \forall y\, \forall z\, \exists w\, (Rxy \wedge Rxz) \to (Ryw \wedge Rzw)$.

We might think that every modal sentence of an appropriate form corresponds to a first-order constraint on R. This may be true, but the 'appropriate form' is not as simple as one might think:

Proposition 5.3.4. There is no sentence φ of first-order predicate logic such that $\langle W, R \rangle$ satisfies φ iff for all assignments a and worlds $x \in W$, $x \Vdash_a \Box \Diamond A \to \Diamond \Box A$.

For an excellent survey of other results in correspondence theory, see [van Benthem, 1984].

5.4 Consistency

Much is often made of consistency, and the mainstream of logical development has avoided treating inconsistencies seriously. However, tolerating or even courting inconsistencies has become increasingly important, particularly in applications in AI. As one might expect there is not a single concept that captures exactly what we mean by consistency in the general setting. Here we will examine a number of simple concepts and the relationships between them.

Definition 5.4.1. *Consistency.* Given a consequence relation, $\vdash$, we say that *a set of sentences* Γ, is:

- *negation consistent* if there is no A such that both $\Gamma \vdash A$ and $\Gamma \vdash \neg A$

- *absolutely consistent* if there exists an A such that $\Gamma \nvdash A$

- *maximally negation consistent* if

  1. Γ is negation consistent, and

  2. if $\Gamma \nvdash A$ then $\Gamma \cup \{A\}$ is not negation consistent

- *complete* if for all A either $\Gamma \vdash A$ or $\Gamma \vdash \neg A$.

Notice that *complete* as used here has nothing to do with the use of the word in Definition 4.7.1. Complete in the present sense could be called 'fully opinionated': on any issue A, either Γ is in favour of A or it is against A.

Negation consistency captures much of our intuitions but it is clearly inappropriate if there is no suitable negation symbol ($\neg$) in the language. We will now look at the relationships between the various notions.

Proposition 5.4.2.

1. If Γ is negation consistent then Γ is absolutely consistent.

2. If Γ is absolutely consistent, and for all A and B we have that $A, \neg A \vdash B$ then Γ is negation consistent.

3. If Γ is negation consistent and complete then it is maximally negation consistent.

4. If Γ is maximally negation consistent and $\vdash$ satisfies the rule

$$\frac{\Gamma, A \vdash B \quad \Gamma, A \vdash \neg B}{\Gamma \vdash \neg A}$$

then Γ is complete. (The rule, called *reductio ad absurdum*, is derivable from $\neg$e and $\neg$i in Fig. 2.)

5. If Γ is not absolutely consistent then Γ is complete.

The proofs are left as an exercise.
In the logics $\mathcal{C}$, $\mathcal{I}$ and $\mathcal{S}4$ we have

negation consistent	iff	absolutely consistent;
maximally negation consistent	iff	negation consistent and complete.

Absolute consistency is a minimal requirement on a set of sentences in order that it can be regarded a *theory*, or a coherent 'body of knowledge'. A set of sentences is useless if every sentence is derivable from it. But the same is not true of negation consistency, since we can tolerate inconsistency on a single issue if we have consistency for others. On this view, their equivalence in the standard logics is a serious drawback.

Inconsistent sets of sentences often arise when one wants to add new information to a theory which conflicts with what is there already. This has led to a new field concerned with the 'dynamics' of theories: how they can be updated with contradicting information in such a way as to preserve (negation) consistency. For an introduction, see [Gärdenfors, 1988; Ryan, 1992].

The notions of negation consistency and absolute consistency can be applied to logics as well as to sets of sentences within a logic; but that is outside the scope of this chapter.

Acknowledgements

The authors are grateful to the following people for helpful comments and useful discussions: Mark Dawson, José Fiadeiro, Lex Holt, Ian Mackie, Ruy de Queiroz and Alex Simpson. We are particularly grateful to Lex Holt who has read through two early drafts and provided extensive feedback. Paul Taylor's TEX macros were used.

References

[Avron, 1987] Arnon Avron. Simple consequence relations. Technical Report ECS-LFCS-87-30, LFCS, Department of Computer Science University of Edinburgh, 1987.

[Barendregt, 1992] H. Barendregt. Lambda calculi with types. *Handbook of Logic in Computer Science*, volume 2, 1992.

[Dummett, 1977] M. Dummett. *Intuitionism.* Oxford University Press, 1977.

[Enderton, 1972] H. B. Enderton. *A Mathematical Introduction to Logic.* Academic Press, New York, 1972.

[Fitting, 1989] M. Fitting. Bilattices. *Journal of Philosophical Logic*, 18:225–256, 1989.

[Gärdenfors, 1988] P. Gärdenfors. *Knowledge in Flux: Modelling the Dynamics of Epistemic States.* MIT Press, 1988.

[Gentzen, 1969] G. Gentzen. Investigations into logical deduction. In M. E. Szabo, editor, *The Collected Papers of Gerhard Gentzen*, chapter 3, pages 68–129. North-Holland Publishing Company, 1969.

[Ginsberg, 1990] M. Ginsberg. Bilattices and modal operators. *Journal of Logic and Computation*, 1(1), 1990.

[Girard, 1989] J.-Y. Girard. *Proofs and Types.* Cambridge University Press, 1989. Translated and with appendices by P. Taylor and Y. Lafont.

[Goldblatt, 1987] R. Goldblatt. *Logics of Time and Computation.* CSLI Lecture Notes, 1987.

[Haack, 1978] S. Haack. *Philosophy of Logics*. Cambridge University Press, 1978.

[Hacking, 1979] Ian Hacking. What is logic. *Journal of Philosophy*, 76:285–318, 1979.

[Hamilton, 1978] A. G. Hamilton. *Logic for Mathematicians*. Cambridge University Press, 1978.

[Hodges, 1983] W. Hodges. Elementary predicate logic. In D. M. Gabbay and F. Guenthner, editors, *Handbook of Philosophical Logic*, volume 1. Dordrecht: D. Reidel, 1983.

[Hughes and Cresswell, 1968] G. E. Hughes and M. J. Cresswell. *An Introduction to Modal Logic*. Methuen, London, 1968. Reprinted with corrections, 1972.

[Hughes and Creswell, 1984] G. Hughes and M. Creswell. *A Companion to Modal Logic*. Methuen, 1984.

[Kripke, 1963] S. Kripke. Semantical considerations on modal logic. *Acta Philosophica Fennica*, 16:83–94, 1963. Reprinted in L. Linsky, editor, *Reference and Modality*, Oxford University Press, 1971.

[Makowsky, 1992] J. A. Makowsky. Model theory. In this volume, chapter 6, 1992.

[Martin-Löf, 1975] P. Martin-Löf. An intuitionistic theory of types: Predicative past. In H. E. Rose and J. C. Shepherdson, editors, *Logic Colloquium '73*, pages 73–118. Amsterdam and Oxford, 1975.

[Phillips, 1992] I. C. C. Phillips. Recursion theory. In this volume, chapter 2, 1992.

[Prawitz, 1965] D. Prawitz. *Natural Deduction: A Proof-Theoretical Study*. Almqvist & Wiksell, 1965.

[Ryan, 1992] M. Ryan. *Ordered Presentations of Theories*. PhD thesis, Department of Computing, Imperial College, 1992.

[Scott, 1981] D. S. Scott. Notes on the formalization of logic. Department of Philosophy, University of Oxford, 1981.

[Stirling, 1992] C. Stirling. Modal and temporal logic. *Handbook of Logic in Computer Science*, volume 2, 1992.

[Tarski, 1956] A. Tarski. *Logic, Semantics, Metamathematics*. Oxford University Press, 1956.

[Tucker and Meinke, 1992] J. Tucker and K. Meinke. Universal algebra. In this volume, chapter 3, 1992.

[van Benthem and Doets, 1983] J. van Benthem and K. Doets. Higher order logic. In D. Gabbay and F. Guenthner, editors, *Handbook of Philosophical Logic*, volume 1. Dordrecht: D. Reidel, 1983.

[van Benthem, 1984] J. van Benthem. Correspondence theory. In D. Gabbay and F. Guenthner, editors, *Handbook of Philosophical Logic*, volume 2. Dordrecht: D. Reidel, 1984.

[van Benthem, 1985] J. van Benthem. *A Manual of Intensional Logic*. CSLI Lecture Notes, 1985.

Notation index

Notation	Meaning	Definition
$\langle P, O \rangle$	propositional language	def. 2.2.3
$\langle P, T, V, O, Q \rangle$	predicate language	def. 2.4.1
$\mathbf{M}$	interpretation or model	defs. 2.2.4, 2.4.3
a	assignment	defs. 2.2.7, 2.4.4
$A, B, \ldots$	sentences	page 12
$\Gamma, \Delta, \ldots$	sets of sentences	page 12
$v_a(A)$	truth value of A	def. 2.2.8
$(W, \leqslant, w_0)$	modal frame	def. 2.3.1
$f_\wedge, f_\vee, \ldots$	interpretation of $\wedge, \vee, \ldots$	def. 2.2.4
$\vdash$	consequence relation	def. 3.1.1
$\vDash_\mathbf{M}$	entailment relation from valuation system $\mathbf{M}$	def. 3.2.1
$\mathcal{C}, \mathcal{S}4$	classical and modal logics	defs. 3.3.1–3.3.3
$\mathcal{I}$	intuitionistic logic	sect. 4.4
$\mathsf{C}(\Gamma)$	consequence operator	sect. 3.5
$A(x)$	formula with a free variable	page 43
$A[t/x]$	substitution of t for x in A	page 43
$\wedge\text{i}, \vee\text{i}_1, \ldots$	natural deduction intro. rules	sect. 4.2
$\wedge\text{e}_1, \vee\text{e}, \ldots$	natural deduction elim. rules	sect. 4.2
$\wedge\ell, \vee\ell, \ldots$	Gentzen rules for left intro.	sect. 4.5 and 4.6
$\wedge r, \vee r, \vee r_1, \ldots$	Gentzen rules for right intro.	sect. 4.5 and 4.6

Recursion Theory

I. C. C. Phillips

Contents

0   Introduction

0.1   Opening remarks

Computability theory might be a better name for recursion theory, since
it is the mathematical study of what is computable. However the name
is traditional. It arose because recursion theory started as the study of
the *recursive functions*, a mathematical characterization of the functions
which are computable (by some program). 'Recursive' because, just as in
most programming languages, recursion is an important and very powerful
mechanism for constructing these functions.

We shall study the question of how powerful various programming languages are – in principle, that is, without limitations of time and space.
We shall examine various important computational principles, including
recursion, and investigate their sometimes surprising power. This will aid
our theoretical understanding of programming languages. We shall see that
there is a well-defined notion of what is computable, and there are programming languages which are *universal*, in the sense that if something can be
computed then it can be computed in that language. Furthermore there
are certain problems which in a precise sense are too hard to be computed.
It is clearly of more than purely theoretical interest to know what these
are, so that we do not waste time searching for what cannot exist.

Recursion theory seems alien to many computing scientists today, perhaps because the basic results of the subject pre-date the first computers,
having been discovered in the early 1930s. Initially programming languages
developed along quite different lines, being very closely tied to the computer architecture at a low level. With the advent of higher level languages,
particularly the declarative ones, the relevance of recursion theory to programming languages has become much more manifest.

Moreover, some features of recursion theory to which objections are often raised by computing scientists need never have caused difficulty. A case
in point is coding (or Gödel numbering). This was introduced in Gödel's
famous 1931 paper [Gödel, 1931] as a way of showing how a formal language could express 'meta' statements about provability in formal systems.
It was then employed to show how e.g. Turing machine programs could be
encoded as data so that a 'universal' Turing machine program could interpret them. Along then came the first stored program digital computers,
and we find that instructions are stored in ultimately the same way as the
data to which they are to be applied, namely as bits. This is of course
very far from the 'Industrial Revolution' view of machines manufacturing
objects (though the machines also have to be manufactured); it is really
closer to the much longer-standing use of pen and paper as a universal vehicle of description. So regarding programs as data objects should not seem

strange. Neither should the concept of a formal language, since what else is a programming language? One might wonder how differently recursion theory might be viewed if it had arisen out of practical developments in computing, instead of pre-dating them.

Most of the results reported in this chapter are over fifty years old. The reader should not get the impression that recursion theory stopped years ago. This chapter is introductory in nature and we have concentrated on the material which we feel is of most relevance for today's computing scientists.

The term recursion theory may conjure up Turing machines for many people. Important though these are, we shall argue in this chapter that other approaches to computability have much to offer. We shall aim to show how they reflect some of the underlying structure of modern programming languages.

Recursion theory aims to provide *part* of the foundation for computing science, by addressing basic issues about the power of programming languages and the nature of some of their structuring mechanisms. Awareness of these issues and of the common basis for many diverse programming languages is valuable for the computing scientist. However it would in my view be quite wrong to see the subject as providing the sole foundation. This *Handbook* provides an opportunity to demonstrate how recursion theory can play its part along with more recent approaches to the foundations of computing.

0.2 A taster

We give some idea of the flavour of our subject by discussing informally an important limitation on the power of programming languages. Let P be a program whose intended behaviour is to input a member of $\mathbf{N}$, the set of natural numbers, and then output another natural number. The meaning of P may be regarded as a function $f : \mathbf{N} \to \mathbf{N}$. In general P may not terminate on some inputs, and so f will be a *partial* function, which we write $f : \mathbf{N} \to_p \mathbf{N}$. As we would prefer such simple programs always to give a result, it would clearly be desirable to have a language L, all of whose programs were guaranteed to terminate on all inputs. As already mentioned there are universal languages which can compute everything which can be computed. Now for the limitation: recursion theory tells us that if L is universal then it must have non-total programs. We are forced to choose between totality and universality. Let us outline the argument.

The first idea is that we can list all the programs of L as an enumeration $P_0, P_1, \ldots, P_n, \ldots$ There are many ways in which this can be done. One systematic procedure would be to list programs in order of length. Since L will only have finitely many symbols, there can only be finitely many programs of a given length. These could be ordered lexicographically.

Now suppose that L is universal and that all its programs are total. Let g_n be the total function computed by P_n.

Now we employ a *diagonal argument*. Define $f : \mathbf{N} \to \mathbf{N}$ by $f(n) = g_n(n) + 1$. Then f is certainly a total function. Moreover f is computable by the following procedure. Given input n, generate the list of L-programs far enough to find P_n. Then run P_n on input n, and add 1 to the result. But now since L is universal, f must be computable by some L-program and must therefore be g_m (some $m \in \mathbf{N}$). Then $f(m) = g_m(m)$. But $f(m) = g_m(m) + 1$. Contradiction.

Some explanation is called for of such a rapid (though fully stated) argument. In defining f we expressly chose it to differ from each g_m by ensuring $f(m) \neq g_m(m)$. Imagine the functions g_m together with their values on all arguments listed as an infinite array:

$$
\begin{array}{cccc}
g_0(0) & g_0(1) & g_0(2) & \cdots \\
g_1(0) & g_1(1) & g_1(2) & \cdots \\
g_2(0) & g_2(1) & g_2(2) & \cdots \\
\cdots & \cdots & \cdots & \cdots
\end{array}
$$

f differs on every input from the function enumerated by the principal diagonal of the array (hence the term diagonal argument). It cannot appear as some row of the array since it would then differ from itself where the row met the diagonal. Hence our enumeration of total computable functions is not complete.

On first encountering such an argument the reader may be reduced to bafflement, awe or frustration or sheer disbelief. A commonly raised objection is that we can simply extend the language L to include a program for f and thereby evade the contradiction. But this is misconceived, since the argument given is that given any terminating language we can enumerate its programs and then diagonalize to produce a total computable function not computable within the language, and this argument can equally well be applied to the extended language. A more telling objection is that f is pathological, that is, it is not a function which we have any reason to wish to compute.

We therefore have two options:

1. Use a language which can compute all total computable functions but which necessarily admits non-terminating programs.

2. Use a language whose programs always terminate but which cannot compute some total computable functions.

In fact option (1) is usually chosen, but (2) is possible. For instance the primitive recursive functions described in Section 1.4 provide such a language.

The argument we have just given can be adapted to show another important limitation on languages – not all functions are computable.

Let L be a universal language whose programs are enumerated as $P_1, \ldots$ as before. Let $g_1, \ldots$ be as before. Define f by

$$f(n) = \begin{cases} g_n(n) + 1 & \text{if } g_n \text{ defined} \\ 0 & \text{otherwise} \end{cases}$$

Suppose that f is computable. Even though f is not total, since L is universal, $f = g_m$, some m. Consider $f(m)$. If $g_m(m)$ is defined then $f(m) = g_m(m) + 1$ by definition of f, which contradicts $f = g_m$. On the other hand, if $g_m(m)$ is undefined then $f(m) = 0$, and this again contradicts $f = g_m$. Either way we have a contradiction, and so we conclude that f is not computable.

Of course f is not a particularly useful function, but we shall see that there are non-computable functions of great practical importance. These are associated with so-called *decision problems*. Two of the most important decision problems will serve as examples:

- Does program P terminate when assigned input n? (Halting Problem)

- Is formula ϕ a theorem of the predicate calculus?

Of course particular examples of these problems can be and have been decided. But what we mean by a *solution* of the Halting Problem is an algorithm which would take as inputs P, n and return yes or no depending on whether P terminated on n. We shall see that such an algorithm is not to be had – in the jargon, the Halting Problem is *unsolvable* or *undecidable*. The second problem is also undecidable; whatever theorem-proving and model-theoretic procedures we have at our disposal there will always be formulas whose validity cannot be decided by those methods. On the positive side there are problems which can be decided, for instance the 'bounded' Halting Problem:

- Does program P terminate within k steps when assigned input n?

The algorithm to decide this takes inputs P, k, n, and just consists of running P on n for k steps. Unlike the case of the Halting Problem proper, we can return a definite 'no' if P has not terminated after k steps, instead of having to wait an unspecified (and possibly infinite) period to see whether it terminates.

0.3 Contents of the chapter

In Section 1 we discuss simple programming languages exemplifying imperative and functional paradigms. Of course we miss out a lot of important features of modern structured languages. Our intention is to make the link between recursion theory and programming languages explicit while omitting complexities which will simply get in the way.

In Section 2 we cover the main elementary results about the decidability of computational problems such as the famous Halting Problem. We treat the $S-m-n$ Theorem and Rice's Theorem, and also discuss recursive and recursively enumerable sets.

In Section 3 we take an apparently quite different tack, with an introduction to inductive definitions. These are of widespread use in computer science as they enable us to deal with recursion in programming languages and logics. We show that they are related to fixed points of operators on sets and functions. The material forms an introduction to domain theory and denotational semantics, as well as being used in the remainder of this chapter. As a case study we provide a denotational semantics for a simple functional language and relate this to the operational semantics discussed in Section 1. The material in Sections 3.3, 3.4 on ordinals and fixed points of non-continuous operators is more advanced, and may be omitted if desired.

In Section 4 we bring together strands from Sections 2 and 3 by considering fixed point results in recursion theory. We show how functions on programs have fixed points (the Second Recursion Theorem, related to the paradoxical combinator of lambda calculus). We also show how computable operators on functions and sets have computable fixed points (the First Recursion Theorem). This enables us to show directly that our denotational semantics of Section 3 gives us computable functions. We also discuss how recursion theory can be approached in an abstract manner using the notion of acceptable programming system. We show that any two codings of the computable functions are isomorphic.

1 Languages and notions of computability

In this section we examine various programming languages which differ quite radically in style. We shall see that nevertheless they all have the same computing power, the so-called Fundamental Result. We then discuss Church's Thesis, which states that these languages are universal – if a function is computable in some informal sense then they can compute it.

1.1   Data types and coding

Before discussing programming languages it is useful to consider the objects
which will be the data manipulated by those languages. We shall be dealing with theoretical foundations, as distinct from an actual computational
problem, and so we wish to be as economical as possible, and consider only
those data types which are most simple and basic. In fact we would like to
adopt a single most basic type from which all other types can be built up.

Let us now consider three candidates for our most basic type:

- lists

- trees (the *S*-expressions of Lisp)

- natural numbers

We shall in fact choose the third of these possibilities. First we say a
little about the first two options:

Lists over a set A are defined by

$$l ::= \text{nil} \mid [a : l] \qquad\qquad a \in A$$

Thus $[a_1 : [a_2 : [a_3 : \text{nil}]]]$ is the list of three elements, which may more
conveniently be written $[a_1, a_2, a_3]$. We would like more structure, and so
we allow lists of lists, etc. starting from some set of atoms. This allows
us to represent expressions and thereby the syntax of any programming
language. For instance the command

$$\text{while } x > 0 \text{ do } x := x - 2 \text{ od}$$

has the syntax tree of Fig. 1.1, which we can represent by the list

$$[\text{while}, [>, x, 0], [\text{assign}, x, [-, x, 2]]]$$

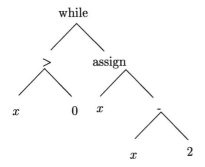

Fig. 1.1

To be more canonical, we could avoid having an underlying set A and build up all lists from the empty list nil. For instance, [nil:[nil:nil]], which may be written [nil, nil] in the usual list notation. We must now reserve certain lists to represent the atoms of whatever data type we are modelling (such as 'while' in the above example). This gives us the type defined by

$$l ::= \text{ nil } | [l : l].$$

Our second candidate base type is McCarthy's *S-expressions* (*S*-exp for short)[McCarthy, 1960]. Here one has a stock of atomic symbols A. Pairing is taken as fundamental:

$$s ::= a | (s, s)$$

Thus an *S*-exp is a binary branching tree with atoms at the leaves. In Lisp the operators car, cdr act as unpairing functions. They are partial, being undefined on atoms. One can do a surprising amount with *S*-exps. For instance lists of *S*-exps can be defined:

$$(s_1, (s_2, (s_3, \ldots (s_n, \text{nil}) \ldots)))$$

represents the list $[s_1, s_2, s_3, \ldots, s_n]$. So clearly any list can be represented by an *S*-exp. Given this ability to form 'tuples' of any length one can represent any syntax as *S*-exps.

As with lists we can decide to make A empty to be more canonical; this can be done by having the single unanalysed *S*-exp nil:

$$s ::= \text{nil} | (s, s)$$

and reserving certain *S*-exps to provide enough atoms. In this case as can be seen from the BNF definitions there is really no difference from the 'atomless' lists even if we think of them in different ways, so that it does not matter whether we work with lists or *S*-exps.

If we are to use the natural numbers (denoted $\mathbf{N}$) as our basic type we need to know that we can 'simulate' the type of *S*-exps. So we now indicate how *S*-exps can be 'coded' as natural numbers. First we need a pairing function on $\mathbf{N}$. We can get this by arranging the members of $\mathbf{N} \times \mathbf{N}$ as a table:

	0	1	2	3	...
0	1	3	6	10	...
1	2	5	9	...	
2	4	8	...		
3	7	...			
4	11	...			

and counting up the diagonals as indicated. We start at 1 instead of 0 as we wish to reserve 0 for the nil S-exp. It is easy to calculate that the function $p : \mathbf{N} \times \mathbf{N} \to \mathbf{N}$ described above is given arithmetically by

$$p(m, n) = (m + n)(m + n + 1)/2 + n + 1$$

Notice that p is one–one, that every member of $\mathbf{N}$ other than 0 has a pre-image under p, and that if $p(m, n) = k$ then $m, n < k$. Now we can use p to create a one–one map $f : S\text{-exp} \to \mathbf{N}$ as follows:

$$f(nil) = 0$$
$$f((s_1, s_2)) = p(f(s_1), f(s_2))$$

In fact this map is also onto in view of the above-mentioned properties of p. We can now regard natural numbers as S-exps and apply the operations of pairing and car, cdr (left and right unpairing) to them. For instance car(8)=2, and car(cdr(9))=1.

The upshot of this is that if we have access to the natural numbers and some simple arithmetic operations we can simulate a data type such as S-exp. Of course the converse is also true: S-exp is rich enough to simulate $\mathbf{N}$.

We shall adopt $\mathbf{N}$ as our fundamental data type — there is certainly no simpler choice, it is well-understood (Fermat's Last Theorem notwithstanding), and it is traditional for recursion theory. Our 'coding' of S-exp in $\mathbf{N}$ serves as a first example of how we can think of ourselves as working with S-expressions while having the theoretical advantage of working with $\mathbf{N}$. If we do not take this dual view then we shall either have the considerable theoretical overhead of carrying around many different data types, or else we shall have to manipulate a lot of hard-to-remember and unsuggestive arithmetical operations.

1.2 The imperative paradigm

We understand an imperative language to be one which is state-based. Computations proceed from state to state until they reach a final state. The state of a computation is typically the values currently stored in the various registers, together with the point where control resides in the program, normally described by the label of the next instruction due to be executed.

It might seem that all programming languages are of this kind, but in fact mathematicians were working with algorithmic languages for describing functions some time before imperative languages were devised. The latter came into being to exploit the architecture of the new computers. Nowadays there is also substantial interest in using non-imperative languages. The hope is that these will be easier to use and analyse since they are more mathematical, and that they will allow more flexibility in implementation since they are not tied to a particular view of how a computer functions.

In this subsection we examine

1. Turing machines

2. Register machines

3. While programs

Our treatment owes much to [Bird, 1976]. Our languages will be rather stripped-down and theoretical in character, but we hope that the reader will feel that they capture the essence of more practical languages, such as BASIC and Pascal.

1.2.1   Turing machines

Before defining Turing machines we discuss Turing's analysis of computation, which will be our main 'direct' evidence for Church's Thesis.

Turing set out to answer the question: what can be computed by a human being following a set of definite instructions? In his analysis of this question he identified a number of ideas which will be important for our understanding of what a computable function is. He imagined a person and a sheet of paper. On this paper can appear various symbols from some finite set of symbols Σ. The person can write further symbols or erase ones which are already there. Computation proceeds by this writing and deletion happening in the following manner:

> The paper is divided into cells, each of which may contain nothing or precisely one symbol. The person's attention is focussed on a single cell at any one time, and each stage of the computation consists of writing a new symbol in the cell, erasing the symbol in the cell, or moving to an adjacent cell. The person performs each stage of the computation according to a fixed finite set of rules, which specify what may happen. Possible actions are governed entirely by this set of rules together with the particular symbol being scanned.

The set of rules is of course the program, while the sheet of paper is the store. The program is fixed and finite but the paper while finite can be extended as necessary — Turing was interested in what can be computed regardless of limitations of time and space. It does not really matter whether we think of the computation being performed by a human agent or by a machine.

Notice how computation is *sequential* (only one action at a time) and *local* - one cannot read more than one cell at a time, and one cannot move around arbitrarily without computational cost. Thus for instance the addition of two numbers cannot be performed 'in one fell swoop', even though this is regarded as a computable operation — as the numbers may be arbitrarily large they will in general occupy more than one cell, so that even reading the numbers takes several steps. The usual pen-and paper method for adding two numbers serves as a good illustration of what Turing envisaged.

The above gives a number of insights into the nature of computation, and gives us the means to take a given scenario and say whether it is a genuine computation, or whether non-computable means are being employed. However it does not allow us to determine what the class of computations is, and thereby what the class of computable objects is. For this we need the precise mathematical description of the paper and the program embodied in the *Turing Machine*. This concept is meant to be a convincing precise realization of the necessarily imprecise concepts described above.

A Turing machine is actually a machine plus a program. It consists of a two-way infinite tape marked out into squares together with a read–write head which can move up and down the tape.

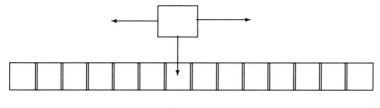

Fig. 1.2

Each square contains a symbol from a finite set $\Sigma = \{\sigma_0, \ldots, \sigma_n\}$. Symbol σ_0 is a blank, which we also denote B. Initially all squares of the tape outside a finite segment are blank (contain B), reflecting the requirement that the tape is not truly infinite, but of finite size, though capable of extension in either direction. The head scans exactly one square at any moment. It can replace the symbol being scanned with another symbol, or move either left or right by one square. Additionally the head is in one of a finite number of states $Q = \{q_0, \ldots, q_m\}$ at any stage. In performing a computation it follows a set of instructions. These tell it what action to carry out on the basis of

- the symbol being scanned

- the state of the head

— again the local nature of computation is made clear. Thus the instructions are of the form

$$q_i \sigma_j A q$$

in other words, when in state q_i scanning a square with symbol σ_j, perform A and change to state q. Here A is one of σ_k (replace σ_j by σ_k), L (move one square to the left), or R (move one square to the right). Initially the head starts in state q_0 scanning the leftmost non-blank square.

Thus formally, a Turing machine may be defined to be a quadruple (Σ, Q, q_0, I) where Σ, Q, I are finite, $B \in \Sigma, q_0 \in Q$ and $I \subseteq Q \times \Sigma \times (\{L, R\} \cup \Sigma) \times Q$. Notice that we do not impose an order on I; at each stage the 'enabled' instruction to be executed is given by the state of the head and the symbol being scanned. It can happen that more than one instruction is enabled; in this case the machine can choose any enabled instruction. If the set I contains overlapping but non-identical instructions $q_i \sigma_j A q, q_i \sigma_j A' q'$, then the machine is said to be *non-deterministic*; otherwise it is *deterministic*. We shall assume here that we are always dealing with deterministic machines — it turns out that allowing non-determinism does not give any extra computing power, though it can be useful, for instance in the study of formal languages.

The *state* or *configuration* of a Turing machine computation is given by the symbols on the tape, the state of the head and the square being scanned. We can omit all leading and trailing blanks, restricting attention to a finite portion of the tape, and we can also ignore the absolute position of the squares on the tape — there is no need for them to be indexed by the integers. So a handy representation of the configuration of a computation is given by $w_1 q w_2$, where $w_1, w_2 \in \Sigma^*$ and $q \in Q$. This means that the contents of the tape are $w_1 w_2$ (omitting leading and trailing blanks) and that the head is scanning the first symbol of w_2.

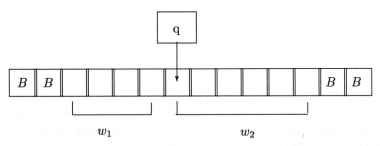

Fig.1.3

The initial configuration is $\Lambda q_0 w_0$ where w_0 is the starting string of non-blank symbols (Λ denotes the empty string).

Clearly a TM computation may carry on for ever, or it may halt. The latter can only be because the computation arrives in a configuration $w_{1q}w_2$ where no instruction is enabled. It is of course a simple matter to check whether a configuration is final by reference to the program of the TM. We shall see that it is a different matter to tell whether a TM set to work on a given string on a tape will eventually halt — this is the famous Halting Problem, which will turn out to be undecidable.

In this chapter we shall compare programming languages according to what functions they can compute. We therefore need notions of input and output for TMs. This is most naturally expressed in terms of strings from Σ. We say that w is the *input* to a computation with Turing machine M if M is started in configuration $\Lambda q_0 w$. We say that a configuration of M is *halting* if it is of the form $\Lambda q w$ and no instruction of M is enabled. Note that this is stronger than merely insisting that no instruction is enabled. If a computation terminates in a halting configuration $\Lambda q w$ then its *output* is w; otherwise it does not output (output is undefined). With this definition it is plain that each TM defines a partial function $f : \Sigma^* \to_p \Sigma^*$. However as we discussed in the previous subsection we prefer to work with the natural numbers, and so we show how TMs compute over $\mathbf{N}$. The idea is to use the unary representation of $n \in \mathbf{N}$ as a string of 1s. So take $\Sigma = \{B, 1\}$ and let M be a TM with alphabet Σ. M computes the k-ary partial function $f : \mathbf{N}^k \to_p \mathbf{N}$ defined as follows: for any $m_1, \ldots, m_k, f(m_1, \ldots, m_k) = n$ if when started in initial configuration $\Lambda q_0 \mathbf{m_1} B \mathbf{m_2} \ldots B \mathbf{m_k}, M$ eventually halts in configuration $\Lambda q_0 \mathbf{n}$; otherwise $f(m_1, \ldots, m_k)$ is undefined. Here $\mathbf{m}$ denotes a string of $m + 1$ occurrences of the symbol 1 — we use $m + 1$ instead of m so that 0 can be differentiated from the empty string.

Definition 1.2.1. The partial function $f : \mathbf{N}^k \to_p \mathbf{N}$ is TM-*computable* if there is some TM M over the alphabet $\{B, 1\}$ such that M computes f.

Computing over $\mathbf{N}$ rather than words over some alphabet does not involve any loss of computational power since we can 'code up' words as natural numbers in a computable manner. Indeed we shall see later that we can code up a TM and its computations so that all questions concerning it are reduced to questions about computing over $\mathbf{N}$.

Turing machines are not meant to be used for practical programming, as they exhibit no useful structure. We shall explore the class of TM-computable functions via alternative characterizations. The importance of Turing machines lies in their ability to articulate the nature of computation, albeit at the lowest level. They are also fundamental to the study of computational complexity (outside the scope of this chapter), since they provide the standard definitions of the time and space required for computations.

Oracles. It can be useful to imagine that some object X (typically a set of data values) is known and to ask what can be computed given X. X may be regarded as available during a computation by means of an *oracle* which when presented with the question 'does $x \in X$?' will answer yes or no. We can imagine this formalized in the case of Turing machines by defining an oracle TM to be a TM with extra instructions of the form $q_i \sigma_j q q'$. This is to be interpreted as: if the head is in state q_i, scanning symbol σ_j, then change to state q if the oracle answers yes, and to q' if the oracle answers no. Of course we then have to stipulate what the question is. It might be: does the string to the right of the head belong to a set $X \subseteq \Sigma^*$? Or it might be: does the number of 1s on the tape belong to $Y \subseteq \mathbf{N}$? These are ways of giving the computation access to information about the oracle set. Of course such sets will be infinite to be of any interest, and a terminating computation cannot use more than a finite portion of such information; it may only consult the oracle a finite number of times. Naturally the same oracle TM can be run with different questions.

1.2.2   Flow charts and register machines

The reader is probably familiar with the use of flow charts to describe programs. They consist of boxes with lines connecting them to show the flow of control. Boxes contain *commands* to be executed, or *tests* to decide which branch should be followed at a fork. We first give a formal definition of flow charts, and then specialize this to consider register machines.

Let a set Cmd of commands and a set Tst of tests be given (ranged over by C, T respectively). Let $l, \dots$ range over a set of labels for statements.

A *flow chart program* is a set of instructions of the form

l: C goto l'
l: if T goto l else goto l''

A *start* label l_0 is specified, and the program is executed in the obvious fashion until a label is reached which has no instruction. Usually such programs are written as a list of statements, and then it is understood that control passes to the next statement in textual order unless otherwise specified, so that some labels can be omitted. A label which occurs after a goto in a program, but which has no corresponding instruction, is said to be an *exit* label. One can easily give a precise definition for execution by positing a store with values in a set V, and associating to each command a function from V to V, and to each test a predicate over V.

As a particular example we consider register machine programs. Let $\mathbf{A} = (A, f_1, \dots, f_j, P_1, \dots, P_k)$ be an algebra with functions f_i and predicates P_i. A *register machine* (RM) over $\mathbf{A}$ consists of an infinite set of registers $R_0, R_1, \dots$, each of which can store an element of A. Denote the contents of R_n by r_n. For our set of commands $Cmd_\mathbf{A}$ we choose

$$r_n := f_i(\mathbf{r})$$

and for $Tst_{\mathbf{A}}$

$$P_i(\mathbf{r})$$

Here $\mathbf{r}$ stands for some list of r_m's, of the appropriate length for the arity of the function or predicate, and possibly including repeats. Executing the command $r_n := f_i(\mathbf{r})$ consists of replacing the contents of R_n by $f_i(\mathbf{r})$. As an example of the kind of algebra which might be considered we mention groups (with multiplication). We here restrict ourselves to a case of particular interest, namely the natural numbers $\mathbf{N}$ and refer the reader to [Meinke *et al.*, 1992] for analysis of register machines over general algebras. We choose a generous set of functions and predicates for our algebra so that $Cmd_{\mathbf{N}}$ is

1. $r_n := r_n + 1$

2. $r_n := r_n - 1$

3. $r_n := r_n + r_m$

4. $r_n := r_n \times r_m$

and $Tst_{\mathbf{N}}$ is

5. $r_n = 0$

6. $r_n = r_m$

We let $0 - 1 = 0$ to keep the predecessor function total.

Definition 1.2.2.

1. Let P be a *RM* program. For any natural number k, we may define the k-ary partial function $f : \mathbf{N}^k \to_p \mathbf{N}$ computed by P:

   Let $n_1, \ldots, n_k$ be natural numbers. Then $f(n_1, \ldots, n_k)$ is obtained as follows: let the register machine be started at instruction l_0 of P with $n_1, \ldots, n_k$ in registers $R_0, \ldots, R_{k-1}$ respectively. If and when the *RM* terminates, read the contents of R_0. This is $f(n_1, \ldots, n_k)$. If the *RM* does not terminate then $f(n_1, \ldots, n_k)$ is undefined.

2. The partial function $f : \mathbf{N}^k \to_p \mathbf{N}$ is *RM-computable* if there is some *RM* program P such that P computes f.

Remark 1.2.3. Of course allowing (3), (4) and (6) violates our locality condition (showing that we are dealing with computation at a higher level than that of the Turing machine), but there is no harm in including them since they can be simulated using (1), (2) and (5). Here for instance is a 'macro' for (3):

$$l_0 : \text{ if } r_m = 0 \text{ goto } l \text{ else goto } l_1$$
$$l_1 : r_m := r_m - 1 \text{ goto } l_2$$
$$l_2 : r_n := r_n + 1 \text{ goto } l_0$$

The main program resumes at instruction l. Notice that this destroys r_m by setting its contents to 0. If r_m is required further it can be retained by storing it in an auxiliary register not used in the main program. The reader is invited to write the appropriate macro. Notice that any flow chart RM program uses only finitely many registers, so that there is always a new one available. The reader may like to provide a similar program for (4) and (6).

1.2.3 While programs

Unlike flow charts, while programs are structured, so that there are no jumps and no need for labels. Let Cmd, Tst be as before. Let P, Q range over while programs. Programs are of one of the following forms:

$$C$$
$$P; Q$$
$$\text{if } T \text{ then } P \text{ else } Q \text{ fi}$$
$$\text{while } T \text{ do } P \text{ od}$$

They are executed in the obvious way (';' is sequential composition), which as in the case of flow charts can be made precise in terms of its effect on a store with values V.

It is not hard to see that whatever Cmd and Tst are, while programs are no more powerful than flow charts. To demonstrate this we show how to construct a flow chart program P' with a single exit label corresponding to each while program P. The proof is by induction on while programs.

Base case: C becomes $l_0 : C$ goto l. Here l is the single exit label.

For '$P; Q$', suppose that P', Q' are constructed by induction hypothesis. Ensure that P', Q' have no labels in common (by relabelling if necessary). Now get P'' by making the exit label of P' the same as the start label of Q'. Combining P', Q' gives a flow chart program for $P; Q$.

For 'if T then P else Q fi', suppose that P', Q' are as before, with start labels l_0, l_1 respectively. Get P'' by relabelling the exit label of P' to be the same as that of Q'. The required flow chart program is

$$l : \text{ if } T \text{ goto } l_0 \text{ else goto } l_1$$
$$P''$$
$$Q'$$

The only case where there is any real difference between the languages is 'while T do P od'. Assume P' as before, with start label l_0 and exit label l_1. The required flow chart program is

$$l_1 : \text{ if } T \text{ goto } l_0 \text{ else goto } l$$
$$P'$$

(the exit label of this program is of course l). The proof is complete.

In general while programs are weaker than flow charts [Bird, 1976] (p. 15). However we can get the full power of flow charts using while programs if we add a very simple family of tests and commands, as we shall now see.

Our aim is to translate every flow chart into a corresponding while program. The first step is to convert each flow chart into mutually recursive procedures.

To this end define the set Trm of terms:

$$t ::= \text{ skip } \mid I \mid C; t \mid \text{ if } T \text{ then } t \text{ else } t \text{ fi}$$

where C, T range over Cmd, Tst as before and I ranges over a new set Ide of procedure identifiers. Skip is a command whose meaning is 'do nothing'. Each term represents a procedure. A procedure program is of the form

$$I_i \text{ where}$$
$$I_1 = t_1(I_1, \ldots, I_n)$$
$$\ldots$$
$$I_n = t_n(I_1, \ldots, I_n)$$

Here I_i is the identifier of the main procedure; execution of the program consists of calling I_i, which executes t_i, calling other procedures as specified in the list of equations. Recursive calls are of course allowed. Notice that there is no looping or infinite behaviour through use of goto's or explicit loop constructs such as while. Such behaviour will come about purely through mutual recursion.

We now describe how any flow chart can be interpreted by recursive procedures. As an example consider the flow chart of Fig. 1.4.

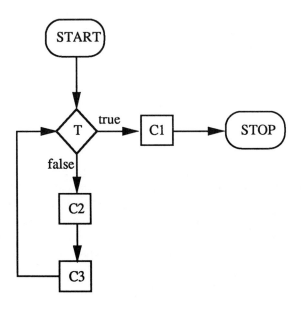

Fig. 1.4

This may be written more formally as:

$$l_0 : \text{ if } T \text{ goto } l_1 \text{ else goto } l_2$$
$$l_1 : C_1 \text{ goto } l_4$$
$$l_2 : C_2 \text{ goto } l_3$$
$$l_3 : C_3 \text{ goto } l_0$$

Let us assign a procedure identifier to each label of the flow chart and view a goto as calling the corresponding procedure. Thus:

$$I_0 \text{ where}$$
$$I_0 = \text{ if } T \text{ then } I_1 \text{ else } I_2 \text{ fi}$$
$$I_1 = C_1; I_4$$
$$I_2 = C_2; I_3$$
$$I_3 = C_3; I_0$$
$$I_4 = \text{ skip}$$

The example could be rendered more succinctly by

$$I \text{ where}$$
$$I = \text{ if } T \text{ then } C_1 \text{ else } C_2; C_3; I \text{ fi}$$

but of course we are interested in illustrating the general method of translation.

It should be clear that this translation will work for any flow chart, translating one with n labels into a program with n procedures which has the same effect. We have therefore shown informally that our procedures are at least as powerful as flow charts. We apologize to the reader for not providing a rigorous semantics to allow the validity of the translation to be formally established.

We show how to convert a procedure program P of the form

$$I_i \text{ where}$$
$$I_1 = t_1(I_1, \ldots, I_n)$$
$$\ldots$$
$$I_n = t_n(I_1, \ldots, I_n)$$

into an equivalent while program. The idea for what follows is from [Smyth, 1975].

The first stage is to represent P by a single recursive term. The recursive terms $RTrm$ are got by adding a variable-binding operation rec $I._$ to the definition of Trm:

$$t ::= \text{ skip } \mid I \mid C; t \mid \text{ if } T \text{ then } t \text{ else } t \text{ fi } \mid \text{ rec } I.t$$

We illustrate the representation of P by a term t in the case $n = 3, i = 2$:

$$t = \quad \text{rec } I_2.t_2(\text{rec } I_1.t_1(I_1, I_2, \text{ rec } I_3.t_3(I_1, I_2, I_3)),$$
$$I_2,$$
$$\text{rec } I_3.t_3(\text{rec } I_1.t_1(I_1, I_2, I_3), I_2, I_3)))$$

From this it should be clear how to get such a term in general. Such a term will be closed (all variables bound).

We now show how to represent each recursive term as a while program. This is where we need an assumption about the tests and commands. Let $T_0, T_1, \ldots$ be an infinite family of mutually exclusive tests (in other words at most one T_i can be true at a time in any interpretation we give to them). These T_i are to be distinct from existing members of Tst. Let $C_0, C_1, \ldots$ be new commands, distinct from existing members of Cmd. The effect of C_i is to make T_i true (and therefore every $T_j, j \neq i$, false) and to leave everything else unchanged. Using the augmented sets of tests Tst' and commands Cmd' we can realize every recursive term. The idea is roughly that we assign T_i to I_i for each i. Whenever we encounter I_i we set T_i to true using C_i. If we are dealing with a closed term we first encounter I_i bound by rec, in which case we also enter a while loop with the test T_i. We now give the translation $[\![\_]\!]$ by induction on $RTrm$:

$$[\![\text{skip}]\!] = C_0$$
$$[\![I_i]\!] = C_i \qquad\qquad i \ge 1$$
$$[\![C;t]\!] = [\![C]\!]; [\![t]\!]$$
$$[\![\text{if } T \text{ then } t \text{ else } t' \text{ fi}]\!] = \text{if } T \text{ then } [\![t]\!] \text{ else } [\![t']\!]$$
$$[\![\text{rec } I_i.t]\!] = C_i; \text{ while } T_i \text{ do } [\![t]\!] \qquad\qquad i \ge 1$$

As an example we translate the flow chart of Fig.1.5. It is taken from [Bird, 1976] (p. 15), where it is established that it does not have an equivalent while program.

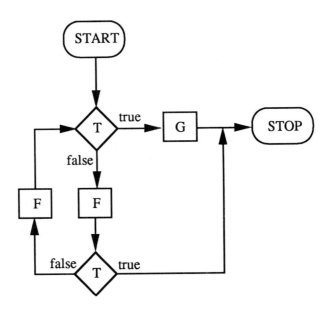

Fig. 1.5

We may translate it into the procedure program

I_1 where
$\quad I_1 =$ if T then $(G;$ skip$)$
$\qquad\qquad$ else $F;$ (if T then skip else $F; I_1$ fi)
$\qquad\qquad$ fi

As a recursive term this is:

$\quad$ rec I_1. if T then $(G;$ skip$)$
$\qquad\qquad$ else $F;$ (if T then skip else $F; I_1$ fi)
$\qquad\qquad$ fi

The corresponding while program is:

> C_1;
> while T_1 do
>   if T then $G; C_0$
>   else $(F;$ if T then C_0 else $F; C_1$ fi)
>   fi
> od

(In fact the final C_1 is redundant.) Notice how C_0 is used to exit from the program.

The translations we have just given are entirely natural ones. We pay a certain price in that the while program may be longer textually owing to the repetition of terms involved in moving from procedures to a recursive term. Clearly the T_i and C_i are to be independent of Tst, Cmd, requiring a division of the store for data values. The effect can be easily obtained on a register machine with a register R_k not used otherwise storing a 'flag' natural number variable. Then T_i is '$r_k = i$' and C_i is '$r_k := i$'. We can specialize while programs to RMs using Cmd_N, Tst_N as before, and we can define the k-ary function corresponding to a while RM program in the same way. From the discussion above it is clear that while programs and flow charts are of equal power on register machines.

Remark 1.2.4. [Bird, 1976] defines a more general syntax for procedure programs incorporating full sequential composition. We simply used procedures as a technical tool in our translation of flow charts into while programs. Our methods would not apply to a more general definition of procedure programs.

1.3   The functional paradigm

A functional language is declarative. A program consists of a list of statements defining the result of applying various function symbols to various arguments. The program does not tell one what to do and there is no notion of state. To give meaning to such programs we can define an evaluation strategy. The aim of a strategy is as follows. Suppose we are interested in the value of the function associated with function symbol f applied to arguments $x_1, \ldots, x_k$. We apply the strategy to the term $f(x_1, \ldots, x_k)$ and attempt to reduce this to a normal, that is, fully computed form. If $f(x_1, \ldots, x_k)$ is of natural number type Nat, then a normal form will just be a natural number. More generally one can think of reducing $f(x_1, \ldots, x_k)$ to simplest possible terms, as in school algebra. After introducing a simple functional language in Section 1.3.1, we discuss how its programs may be evaluated — its operational semantics — in Section 1.3.2. There are various reduction strategies, and they can give different results. We shall define two of the most common ones, namely call-by-value and call-by-name. We shall formulate these strategies as syntax-directed proof systems. Then in Section 1.3.3 we describe informally how our language may be given a different semantics using fixed points. Again we are indebted to [Bird, 1976].

1.3.1 A simple functional language

We use a toy functional language, which we shall call FL for short. For simplicity FL will not have more sophisticated features of functional languages such as pattern-matching and higher-order functions. We shall restrict ourselves to two base types, namely

   Nat   natural numbers
   Bool   booleans, namely *tt, ff*

and the types
   $\text{Nat}^k \to \text{Nat}, \text{Nat}^k \to \text{Bool},$

the types of k-ary functions from Nat to Nat, Bool respectively. Functions of type $\text{Nat}^k \to \text{Bool}$ will sometimes be referred to as *predicates.*

We shall define sets of *terms* of the various types. Programs will then be recursive definitions constructed from terms. It is important to realize that programs do not mention natural numbers directly. Instead they use numerals $\text{num}(n)$, one for each $n \in \mathbf{N}$. This will aid understanding greatly when we come to regard programs as data objects later. Here num is a constructor. We shall also require an infinite supply of variables for numerals, functions and predicates, and we use the same device. We shall allow some sugaring of the formal syntax.

We give first the *pre-defined* terms together with their types. As well as
   $\text{num}(n) : \text{Nat}$ all $n \in \mathbf{N}$
   tt, ff: Bool

there will be various functions of type $\text{Nat}^k \to \text{Nat}, \text{Nat}^k \to \text{Bool}$. (To be truly rigorous we should refer to these as function *symbols*, as they are syntactic objects which will be interpreted by functions.) In particular we shall include

   iszero: Nat $\to$ Bool
   equals: $\text{Nat}^2 \to$ Bool
   succ, pred: Nat $\to$ Nat
   plus, times: $\text{Nat}^2 \to$ Nat

We shall write $\text{num}(n)$ as n and $\text{iszero}(x)$ as $x = 0$, etc. when speaking less formally. We have used somewhat more cumbersome notation to make explicit the form of the various constructs.

Next the *variables*:

var(n): Nat all $n \in \mathbf{N}$
fun(k, n): $\text{Nat}^k \to \text{Nat}$ all $k, n \in \mathbf{N}$
pred(k, n): $\text{Nat}^k \to \text{Bool}$ all $k, n \in \mathbf{N}$

We shall allow the use of metavariables both for ease of writing programs and to refer to arbitrary variables. So let

$x, y, \ldots$ range over numeric variables
$f, g, \ldots$ range over function (and predicate) variables

Terms of type $\text{Nat}^k \to \text{Nat}$ or $\text{Nat}^k \to \text{Bool}$ are said to have arity k.

Let Tpe be either Nat or Bool. As well as the pre-defined terms of type Tpe we have

$f(t_1, \ldots, t_k)$: Tpe   where $t_1, \ldots, t_k$: Nat and f: $\text{Nat}^k \to$ Tpe is
pre-defined or variable

if b then t_1 else t_2 : Tpe where b: Bool and t_1, t_2: Tpe

This completes our definition of the terms and their types, which is of course mutually recursive.

An *FL* program P is a list of recursive equations

$$f_1(x_1^1, \ldots, x_{n_1}^1) = t_1$$
$$\ldots$$
$$f_k(x_1^k, \ldots, x_{n_k}^k) = t_k$$

where $f_1, \ldots, f_k$ are function variables and $t_1, \ldots t_k$ are terms such that variables appearing in t_i are drawn from $x_1^i, \ldots, x_{n_i}^i, f_1, \ldots f_k$. We naturally insist that for each i, $f_i(x_1^i, \ldots, x_{n_i}^i)$ and t_i are of the same basic type. P is to be viewed as a mutually recursive definition of a family of functions. Sometimes we regard f_1 as the principal function, $f_2, \ldots, f_k$ being auxiliary functions whose purpose is to assist in the definition of f_1.

It is convenient to abbreviate lists $x_1, \ldots, x_k$ by $\mathbf{x}$, so that we may write $f(\mathbf{x})$ instead of $f(x_1, \ldots, x_k)$.

1.3.2 Methods of evaluation

We now turn to how our functional programs are to be interpreted. We first consider a few example programs. [Bird, 1976] gives the following program to compute the integer part of x/y: div(x, y)= if $x < y$ then 0 else div($x - y, y$) + 1 (here we assume that programs for $<$ and $-$ are given).

We can evaluate for instance div(3, 2) by rewriting:

div(3, 2) $\to$ if $3 < 2$ then 0 else div($3 - 2, 2$) + 1
                    $\to$ div(1, 2) + 1
                    $\to$ (if $1 < 2$ then 0 else div($1 - 2, 2$) + 1) + 1
                    $\to 0 + 1$
                    $\to 1$

Let us try div(3, 0):

$$\begin{aligned}
\text{div}(3,0) \quad &\rightarrow \text{ if } 3 < 0 \text{ then } 0 \text{ else div}(3 - 0, 0) + 1 \\
&\rightarrow \text{div}(3,0) + 1 \\
&\rightarrow (\text{if } 3 < 0 \text{ then } 0 \text{ else div}(3 - 0, 0) + 1) + 1 \\
&\rightarrow (\text{div}(3,0) + 1) + 1 \\
&\rightarrow \ldots
\end{aligned}$$

Our evaluation will plainly not terminate. This corresponds of course to division by 0 being undefined. In a rewriting $t \rightarrow t'$ we intend that t and t' denote the same values. Plainly this is not the case with div(3, 0)and div(3, 0) + 1 unless both sides are undefined.

Now consider the following program for multiplication:

$$\text{mult}(x, y) = \text{ if } y = 0 \text{ then } 0 \text{ else } x + \text{mult}(x, y - 1)$$

At first sight mult is the same as the pre-defined function times, but they may behave differently as we shall now see. Consider mult(div(3, 0), 0). Informally we are not sure whether to say that this is 0 because multiplication by 0 always gives 0, or to say that it is undefined because division by 0 is always undefined. This uncertainty is reflected in the various ways that we can rewrite the term:

$$\begin{aligned}
\text{mult}(\text{div}(3,0), 0) \quad &\rightarrow \text{ if } 0 = 0 \text{ then } 0 \text{ else } x + \text{mult}(x, 0 - 1) \\
&\rightarrow 0
\end{aligned}$$

if we take the former approach, and

$$\begin{aligned}
\text{mult}(\text{div}(3,0), 0) \quad &\rightarrow \text{mult}(\text{if } 3 < 0 \text{ then } 0 \text{ else div}(3 - 0, 0) + 1, 0) \\
&\rightarrow \text{mult}(\text{div}(3,0) + 1, 0) \\
&\rightarrow \ldots
\end{aligned}$$

if we take the latter. Notice that the first approach involves rewriting the outermost function symbol, while in the second we rewrite the innermost function symbol first. If we need to evaluate a particular argument to a function in order to obtain a result we say that the function is *strict* in that argument (a precise definition of strictness is given in Section 3.2.1). We have just seen that mult may or may not be strict in its first argument (it is definitely strict in its second). The latter is more in accord with usual mathematical practice, where arguments are evaluated first before applying a function. We definitely expect times(div(3,0),0) to be undefined, in other words times should be strict in both arguments.

Remark 1.3.1. Evaluation of the conditional. This is of course a function of three arguments. When evaluating div(1, 2) we get

$$\text{if } 1 < 2 \text{ then } 0 \text{ else div}(1 - 2, 2)$$

If we immediately evaluate the third argument of the conditional, we clearly get an unintended negative number (remember we are working with

Nat). Obviously we should evaluate the Boolean first argument before the other two arguments, so that we can ignore the one which does not apply. We shall insist on this order of evaluation.

As indicated at the start of this subsection, we shall regard the meaning of an *FL* program as being the results we get when we supply numerical arguments to the principal function symbol and evaluate the resulting term according to certain rules. The discussion above indicates that there is more than one plausible strategy. In fact there are many in use. This means that a program does not have a single intended meaning as in imperative programming — we have to know what strategy is to be employed. In what follows we shall single out two strategies. This will be enough to illustrate what is going on, and in particular to make the point that it does not matter which strategy is employed from the point of view of what functions can be computed, our guiding interest in this section.

In general an *FL* program is of the form

$$f_1(\mathbf{x}^1) = t_1$$
$$\ldots$$
$$f_k(\mathbf{x}^k) = t_k$$

For simplicity we shall only define evaluation rules for the case of a single equation $f(\mathbf{x}) = t$ (where t contains variables $\mathbf{x}$ and possibly f) though this can be generalized.

In what follows let u vary over terms of type Nat or Bool with no function variables other than f. Let $c, \ldots$ range over pre-defined terms of type Nat or Bool. These are called normal forms, in other words they are fully reduced.

Define an evaluation relation $\Rightarrow$ from terms to normal forms by structural induction on the formation of u as follows:

1. $c \Rightarrow c$

2. if g is a pre-defined function of arity m then

$$\frac{u_i \Rightarrow c_i \quad (i \leq m) \quad g(\mathbf{c}) = c}{g(\mathbf{u}) \Rightarrow c}$$

3.

$$\frac{u \Rightarrow tt \quad u_1 \Rightarrow c}{\text{if } u \text{ then } u_1 \text{ else } u_2 \Rightarrow c} \qquad \frac{u \Rightarrow ff \quad u_2 \Rightarrow c}{\text{if } u \text{ then } u_1 \text{ else } u_2 \Rightarrow c}$$

$4V$.

$$\frac{u_i \Rightarrow c_i \quad (i \leq n) \quad t[\mathbf{c}/\mathbf{x}] \Rightarrow c}{f(\mathbf{u}) \Rightarrow c}$$

$4N$.

$$\frac{t[\mathbf{u}/\mathbf{x}] \Rightarrow c}{f(\mathbf{u}) \Rightarrow c}$$

In fact we define two relations, $\Rightarrow_V$ and $\Rightarrow_N$, depending on which version of rule (4) we adopt. Relation $\Rightarrow$ on its own will mean either $\Rightarrow_V$ or $\Rightarrow_N$. The intended meaning of $u \Rightarrow c$ is that u evaluates to c. Our definition is obviously inductive, and is given by rules in the style of a logical system (cf [Ryan and Sadler, 1992]). Notice that the rules are *syntax directed*, or *compositional*, in that they follow the syntactic structure of the term. This style of defining how to evaluate terms is known as *structured operational semantics* ([Plotkin, 1981; Abramsky, 1983]). It is clear that if $u \Rightarrow c$ then we can find a proof of it by systematically working 'backwards'. At each stage there is only one rule which could have been applied. Notice also that the rules reflect the semi-strictness of the conditional: considering if 'u then u_1 else u_2', if u evaluates to tt then we don't need u_2 to be defined (i.e. evaluate to something).

Lemma 1.3.2. *For any u there is at most one c such that $u \Rightarrow c$.*

Proof. By induction on proofs.   ∎

The two semantics emphasize the point that in the functional style the semantics is not dictated by the syntax. The V system corresponds to call-by-value, in which arguments to a function variable fare fully evaluated before f is evaluated at all. The N system corresponds to call-by-name, in which if f occurs outermost in a term then it is evaluated before starting on its arguments. We shall see that the two systems can yield different results.

Although our two systems are such that for any term u there is at most one c such that $u \Rightarrow c$, they do not prescribe exactly how u should be found. Working backwards from the conclusion on rule $(4V)$ for example we see that the u_i must be evaluated to c_i before $t[\mathbf{c}/\mathbf{x}]$ can be evaluated, but the u_i can be evaluated in any order, or simultaneously. The systems may be realized by rewriting terms, and the difference between the systems is reflected in the different *strategies* which may be employed, in other words in what order the various subterms may be rewritten. Our systems may seem a perverse way of expressing the evaluation of terms, since it appears that we have to build up the whole proof tree before we can make a single move. However the advantage is that we have a clear structural definition which does not impose a particular order of attack, allowing concurrency, reuse of proofs etc. Each system corresponds to a *family* of strategies.

Before continuing we give a simple example to illustrate the application of the rules, using the well-known (and well-worn) factorial function.

Example 1.3.3. $\mathrm{fac}(n)=$if $n = 0$ then 1 else $n\times \mathrm{fac}(n - 1)$
Here $=, \times, n - 1$ are pre-defined functions. Let us see how the proof of $\mathrm{fac}(\mathrm{fac}(3))=720$ looks in the two systems:

With $(4V)$ the proof finishes

$$\frac{\mathrm{fac}\ (3) \Rightarrow 6 \quad \text{if } 6 = 0 \text{ then } 1 \text{ else } 6\times \mathrm{fac}(6 - 1) \Rightarrow 720}{\mathrm{fac}(\mathrm{fac}(3))\Rightarrow 720}$$

With $(4N)$ the proof finishes

$$\frac{\text{if } \mathrm{fac}(3) = 0 \text{ then } 1 \text{ else } \mathrm{fac}\ (3)\times \mathrm{fac}(\mathrm{fac}(3) - 1) \Rightarrow 720}{\mathrm{fac}(\mathrm{fac}(3))\Rightarrow 720}$$

Again taking the example of $\mathrm{fac}(\mathrm{fac}(3))$ we may rewrite it as follows:

$$
\begin{aligned}
\mathrm{fac}(\mathrm{fac}(3)) \quad &\to \mathrm{fac}(\text{if } 3 = 0 \text{ then } 1 \text{ else } 3\times \mathrm{fac}(2 - 1)) \\
&\to \mathrm{fac}(\text{if } \mathrm{ff} \text{ then } 1 \text{ else } 3\times \mathrm{fac}(2 - 1)) \\
&\to \mathrm{fac}(3\times \mathrm{fac}(2 - 1)) \\
&\to \mathrm{fac}(3\times \mathrm{fac}(1)) \\
&\to \mathrm{fac}(3\times (\text{if } 1 = 0 \text{ then } 1 \text{ else } 1\times \mathrm{fac}(1 - 1))) \\
&\to \ldots
\end{aligned}
$$

We always rewrote the *innermost* occurrence of a function symbol, the exception being that we evaluated the *outermost* conditional. This corresponds to system V. As an alternative we might have proceeded

$$
\begin{aligned}
\mathrm{fac}(\mathrm{fac}(3)) \quad \to \quad & \text{if } \mathrm{fac}(3) = 0 \text{ then } 1 \text{ else } \mathrm{fac}(3)\times \mathrm{fac}(\mathrm{fac}(3) - 1)) \\
\to \quad & \text{if}(\text{if } 3 = 0 \text{ then } 1 \text{ else } 3\times \mathrm{fac}(2 - 1)) = 0 \text{ then } 1 \\
& \text{else } \mathrm{fac}(3)\times \mathrm{fac}(\mathrm{fac}(3) - 1)) \\
\to \ldots &
\end{aligned}
$$

Here we rewrote the outermost occurrence of fac. This corresponds to the system N.

Remark 1.3.4. For readers familiar with term-rewriting [Klop, 1992] we indicate how evaluation could be carried out using a term-rewriting system

(TRS), making the above informal remarks rather more precise. We use the following set of reduction rules:

2. $g(\mathbf{c}) \to c$ if $g(\mathbf{c}) = c$

3. if tt then x else $y \to x$
   if $f\!f$ then x else $y \to y$

4. $f(\mathbf{x}) \to t$

Items (2) and (4) are of course classes of reduction rules (cf. recursive program schemes [Klop, 1992], though our special treatment of the conditional means that we do not have a recursive program scheme in his sense). The TRS is regular (and left-normalizing). Our proof systems are more general than reduction strategies since they may be implemented by a variety of the latter. System N can be implemented by outermost strategies. One can prove by structural induction on proofs in N that $u \Rightarrow_N c$ iff $u \to_R^* c$ where R is either O (outermost) or PO (parallel outermost). System V does not exactly correspond to any standard strategy, since it requires innermost treatment of $f(\mathbf{x}) \to t$ redexes and this will not work for the conditional, which would prefer outermost treatment. The reader may care to formulate a satisfactory strategy and prove that it is faithful to system V.

Consider an FL program P of the form $f(\mathbf{x}) = t$. Again we restrict ourselves to one equation for simplicity. Associate with P two functions defined as follows:

Definition 1.3.5. Let $f_V : \mathbf{N}^k \to_p \mathbf{T}$ (where $\mathbf{T}$ is $\mathbf{N}$ or $\mathbf{B}$ (the Booleans), whichever is appropriate) be given by

$$f_V(\mathbf{n}) = \begin{cases} \text{the unique } c \text{ such that } f(\mathbf{n}) \Rightarrow_V c & \text{if one exists} \\ \text{undefined} & \text{otherwise} \end{cases}$$

f_N is defined similarly.

We shall also refer to f_V, f_N as $O_V(P), O_N(P)$ respectively, since they are the operational semantics of P according to our two evaluation disciplines. It is intuitively reasonable that $O_N(P)$ is more defined than $O_V(P)$ since with system V we can only evaluate an $f(\mathbf{u})$ if all its arguments have been evaluated, whereas with system N we rewrite it without waiting for the arguments and hence may achieve a result despite non-termination of some argument.

Theorem 1.3.6. *For any* $P, O_V(P) \subseteq O_N(P)$.

Proof. We show that for any term u, if $u \Rightarrow_V c$ then $u \Rightarrow_N c$ by induction on the proof of $u \Rightarrow_V c$. We can then apply this to $f(\mathbf{n})$ to get the desired result. There are cases, according to which rule is used, to derive $u \Rightarrow_V c$:

(1), (2), (3) are trivial since they are common to both V and N systems.

(4V). $f(\mathbf{u}) \Rightarrow_V c$. Then $u_i \Rightarrow_V c_i$ and $t(\mathbf{c}/\mathbf{x}) \Rightarrow_V c$. By induction, $u_i \Rightarrow_N$ c_i and $t(\mathbf{c}/\mathbf{x}) \Rightarrow_N c$. We want $t(\mathbf{u}/\mathbf{x}) \Rightarrow_N c$. This will follow from the next result.

Lemma 1.3.7. *For any terms* u, u_i, *if* $u_i \Rightarrow_N c_i$ *and* $u[\mathbf{c}/\mathbf{y}] \Rightarrow_N c$ *then* $u[\mathbf{u}/\mathbf{y}] \Rightarrow_N c$.

Proof. By induction on the proof of $u[\mathbf{c}/\mathbf{y}] \Rightarrow_N c$.

1. Straightforward

2. $u = g(\mathbf{u}')$ and $u[\mathbf{c}/\mathbf{y}] = g(\mathbf{u}'[\mathbf{c}/\mathbf{y}])$. We have $u_i'[\mathbf{c}/\mathbf{y}] \Rightarrow_N c_i'$ and $g(\mathbf{c}') = c$. By induction $u_i'[\mathbf{u}/\mathbf{y}] \Rightarrow_N c_i'$. Hence $u[\mathbf{u}/\mathbf{y}] = g(\mathbf{u}'[\mathbf{u}/\mathbf{y}]) \Rightarrow_N c$ by rule (2).

3. Straightforward

(4N). $u = f(\mathbf{u}')$. Then $f(\mathbf{u}'[\mathbf{c}/\mathbf{y}]) \Rightarrow_N c$ and so $t[\mathbf{u}'[\mathbf{c}/\mathbf{y}]/x] \Rightarrow_N c$. But $t[\mathbf{u}'[\mathbf{c}/\mathbf{y}]/x] = t[\mathbf{u}'/\mathbf{x}][\mathbf{c}/\mathbf{y}]$. Hence by induction $t[\mathbf{u}'/\mathbf{x}][\mathbf{u}/\mathbf{y}] = t[\mathbf{u}'[\mathbf{u}/\mathbf{y}]/\mathbf{x}] \Rightarrow_N c$. Now by (4N) $f(\mathbf{u}'[\mathbf{u}/\mathbf{y}]) = u[\mathbf{u}/\mathbf{y}] \Rightarrow_N c$ and the proof is complete. Notice that the proof is genuinely an induction on proofs in system N and not on term formation, since in this last case $t[\mathbf{u}'/\mathbf{x}]$ is in general more complex than $f(\mathbf{u}')$. ∎

Definition 1.3.8. A partial function is *FL-computable* if there is an *FL* program P such that $f = O_V(P)$.

Apparently if we had chosen $O_N(P)$ instead of $O_V(P)$ in the above definition we should have got a different definition of *FL*-computable, but it turns out that this is not the case.

1.3.3 Least fixed point semantics

We now introduce a radically different way of assigning meaning to functional programs. Instead of saying how they are to be evaluated, as hitherto, we can regard them as inductive definitions, and then say that their meaning is the least fixed point. Such a semantics is often referred to as *denotational*, by which is meant that we give each program a *denotation* in a semantic domain, and this denotation is derived from the denotations of the constituent parts of the program. The denotational semantics is more 'mathematical', and makes it easier to analyse programs and prove things about the programs and the functions they compute.

In this subsection we provide an informal introduction. The definitions and the equivalence of operational and denotational semantics are treated in Section 3.2 since they depend on an understanding of inductive definitions, the subject of Section 3.

Let us start with a simple example. Consider the program

$$f(n) = \text{if } n = 0 \text{ then } 1 \text{ else } 2 * f(n-1)$$

This should compute the total function $\exp:\mathbf{N} \to \mathbf{N}$ given by $\exp(n) = 2^n$.

Let $\mathbf{N} \to_p \mathbf{N}$ denote the set of partial functions from $\mathbf{N}$ to $\mathbf{N}$ ordered by inclusion. This is a partial order, with least element $\uparrow$, the totally undefined function. Define the operator $\Phi : (\mathbf{N} \to_p \mathbf{N}) \to (\mathbf{N} \to_p \mathbf{N})$ by

$$\Phi(h)(n) = \text{if } n = 0 \text{ then } 1 \text{ else } 2 * h(n-1)$$

As h may not be total, then of course $\Phi(h)$ may not be either. We call Φ a *functional*, since it is an operator on functions. Notice that Φ is total. Clearly the function defined by our program should satisfy

$$f = \Phi(f)$$

A function with this property is said to be a *fixed point* of Φ. In this case we should have $\exp = \Phi(\exp)$, which is indeed easily seen to be the case. But how can we construct the fixed point if we don't know what it is?

Let us see how to build up the total function exp from successively larger partial functions lying below exp in the inclusion ordering. These will be finite (in other words their domain of definition is finite) and they are to be thought of as approximations to f, which is the 'limit' of its approximations. We start with $\uparrow$, which we know approximates any function, and build up larger approximations by successively applying Φ. We present the results in a table:

	0	1	2	3	$\ldots$
$\uparrow$	$\uparrow$	$\uparrow$	$\uparrow$	$\uparrow$	$\ldots$
$\Phi(\uparrow)$	1	$\uparrow$	$\uparrow$	$\uparrow$	$\ldots$
$\Phi^2(\uparrow)$	1	2	$\uparrow$	$\uparrow$	$\ldots$
$\Phi^3(\uparrow)$	1	2	4	$\uparrow$	$\ldots$
$\ldots$					

Plainly the limit or least upper bound of this increasing sequence is the intended function exp and we have constructed our desired fixed point. We shall see that in general, a functional corresponding to a program may have more than one fixed point, and that the procedure we have outlined actually constructs its *least* fixed point (lfp), which may be a partial function unlike exp. We can still take this to be the meaning of the program, since any larger fixed point will depend on introducing arbitrary values which are not forced on us by the program text. This leads us to adopt the following.

Thesis 1.3.9. The (denotational) meaning of any *FL* program is the lfp of the functional associated with it.

Remark 1.3.10. It may not be clear at this stage why we use the word 'denotational'. It refers to the way in which the functional, whose lfp we want, is built up from simpler functionals according to the construction of the program. In our simple example above we took this to be entirely obvious.

1.4 Recursive functions

So far we have considered imperative and functional languages. Both styles give the same set of computable functions as we shall see in Section 1.4. We now give an alternative presentation which gives more insight into the structure of this set of computable functions.

We start by defining the class of *primitive recursive functions*, which we shall call PR. We first do this in an informal mathematical style. PR is derived by taking a set of base functions and 'closing up' this set by applying various function-forming operations to it. All functions are of type $\mathbf{N}^n \to \mathbf{N}$ for some n. The base functions are as follows:

- The *zero* function $0 : \mathbf{N} \to \mathbf{N}$ defined by
  $$0(x) = 0 \qquad \text{all } x \in \mathbf{N}$$
  (Using a zero function instead of the number 0 is just a technical device to ensure that everything discussed is a function.)

- The *successor* function $S : \mathbf{N} \to \mathbf{N}$ defined by
  $$S(x) = x + 1 \qquad \text{all } x \in \mathbf{N}$$

- The *projection* functions $\pi_m^n : \mathbf{N}^n \to \mathbf{N} (n \geq m \geq 1)$ defined by
  $$\pi_m^n(x_1, \ldots, x_m, \ldots, x_n) = x_m$$

This completes the base functions. Plainly zero and successor give us access to the natural numbers. The projection functions are useful for removing, adding and reordering variables.

The two function-forming operations are

- *Composition.* Given $f : \mathbf{N}^n \to \mathbf{N}$ and $g_i : \mathbf{N}^m \to \mathbf{N} (i = 1, \ldots, n)$ their composition is the function $h : \mathbf{N}^m \to \mathbf{N}$ defined by

$$h(\mathbf{x}) = f(g_1(\mathbf{x}), \ldots, g_n(\mathbf{x}))$$

If desired we can give h the name $\text{comp}_m^n(f, g_1, \ldots, g_n)$, making the composition operator explicit, though the usual applicative notation is more perspicuous.

- *Primitive recursion.* Given $f : \mathbf{N}^n \to \mathbf{N}$ and $g : \mathbf{N}^{n+2} \to \mathbf{N}$ we form new function $h : \mathbf{N}^{m+1} \to \mathbf{N}$ by primitive recursion as follows:

$$h(\mathbf{x}, 0) = f(\mathbf{x})$$

$$h(\mathbf{x}, y+1) = g(\mathbf{x}, y, h(\mathbf{x}, y))$$

Again we can make the operation explicit by giving h the name $\mathrm{prim}^n(f, g)$. Function h is defined recursively, but the definition is clearly valid since it only uses smaller values of y, and we have a base case from which to start given by f.

Definition 1.4.1. The *primitive recursive functions* (PR) are the least set including the zero, successor and projection functions and closed under the operations of composition and primitive recursion.

All primitive recursive functions are computable in an informal sense since the base functions are clearly computable and if we can compute f, g, etc. then we can plainly compute $\mathrm{comp}^n_m(f, g_1, \ldots, g_n)$, $\mathrm{prim}^n(f, g)$. Similarly all primitive recursive functions are total.

Examples 1.4.2.

1. Addition. The usual recursive definition

$$f(x, 0) = x$$
$$f(x, y+1) = f(x, y) + 1$$

To see $f \in \mathrm{PR}$ we rewrite this as

$$f(x, 0) = \pi^1_1(x)$$
$$f(x, y+1) = g(x, y, f(x, y))$$

where $g(x, y, z) = S(\pi^3_1(x, y, z))$. Thus

$$f = \mathrm{prim}^1(\pi^1_1, \ \mathrm{comp}^1_3(S, \pi^3_1))$$

2. Multiplication. The usual recursive definition

$$g(x, 0) = 0$$
$$g(x, y+1) = g(x, y) + x$$

Thus $g \in \mathrm{PR}$ since it is derived from 0 and a form of addition by primitive recursion. The reader is invited to supply the details as for addition.

Of course we shall normally use the standard infix notations $x+y, x.y$ etc.

It turns out that PR is actually an enormous set of functions. It contains all (total) functions that we could conceivably want to compute for any practical purpose, as well as many more whose computational complexity is too high for them to be feasibly computable. But it does not contain all the computable functions, by the argument outlined in Section 0.2 - since all functions in PR are total, we can diagonalize to get a new computable function not in PR.

For a more concrete example consider *Ackermann's function* $A : \mathbf{N}^2 \to \mathbf{N}$ defined by

$$A(0, y) = y + 1$$
$$A(x + 1, 0) = A(x, 1)$$
$$A(x + 1, y + 1) = A(x, A(x + 1, y))$$

This is clearly an FL program and so A is FL-computable. It is not obvious that it terminates, since $A(x+1, y+1)$ depends on $A(x, z)$ where $z = A(x + 1, y)$, and $A(x + 1, y)$ will be much larger than y. However we can prove termination for all x, y using the following double induction hypothesis:

$P(x, y)$ iff $A(x', y')$ terminates for all (x', y') such that $x' < x$ or
$$x' = x, y' < y$$

It is plain that

$$P(x, y) \Rightarrow A(x, y) \text{ terminates}$$

by inspection of the defining equations of A. Ackermann's function grows too fast to be primitive recursive. In fact one may show the following:

Proposition 1.4.3. *If $f : \mathbf{N} \to \mathbf{N}$ is PR then there is n such that for all $x, f(x) < A(n, x)$.*

Proof. Omitted. ∎

The reader may care to calculate $A(n, x)$ for the first few values of n.

It turns out that we can get all computable functions by just adding one more function-forming operation.

Notation 1.4.4. Let $f : \mathbf{N}^n \to_p \mathbf{N}$. We write $f(\mathbf{x}) \downarrow$ to indicate that $f(\mathbf{x})$ is defined, $f(\mathbf{x}) \uparrow$ to indicate that $f(\mathbf{x})$ is undefined. We have to be careful when dealing with equality. When we write say $f(\mathbf{x}) = 3$ we mean $f(\mathbf{x}) \downarrow$ and $f(\mathbf{x}) = 3$. If also $g : \mathbf{N}^n \to_p \mathbf{N}$ then when we write $f(\mathbf{x}) = g(\mathbf{x})$ we mean $f(\mathbf{x}) \downarrow$ iff $g(\mathbf{x}) \downarrow$, and if $f(\mathbf{x}) \downarrow$ then $f(\mathbf{x}) = g(\mathbf{x})$. The equality $f = g$ means that for all $\mathbf{x}$, $f(\mathbf{x}) = g(\mathbf{x})$.

Definition 1.4.5. *Minimization.* Let $f : \mathbf{N}^{n+1} \to_p \mathbf{N}$. We form a new function $g : \mathbf{N}^n \to_p \mathbf{N}$ by minimization as follows:

$$g(\mathbf{x}) = \text{ least } y \text{ s.t. } f(\mathbf{x}, y) = 0$$

We may write $g = min^n(f)$.

We have to be careful in interpreting this. Notice that we have let f be partial. This is because minimization can introduce partiality; if there is no y for which $f(\mathbf{x}, y) = 0$ then $g(\mathbf{x}) \uparrow$. But what if say $f(\mathbf{x}, 0) \uparrow$ and $f(\mathbf{x}, 1) = 0$? In general we cannot tell whether $f(\mathbf{x}, 0) \uparrow$ (due to the undecidability of the Halting Problem) and so even if we discover that $f(\mathbf{x}, 1) = 0$ we cannot let $g(\mathbf{x}) = 1$ as we cannot be confident that 1 is the least y s.t. $f(\mathbf{x}, y) = 0$. Thus computational considerations lead us to say that $g(\mathbf{x}) \uparrow$. Generalizing we define

$$min^n(f)(\mathbf{x}) = z \quad \text{if } f(\mathbf{x}, z) = 0 \text{ and } \forall y < z(f(\mathbf{x}, y) \downarrow \,\&\, f(\mathbf{x}, y) \neq 0)$$
$$min^n(f)(\mathbf{x}) \uparrow \quad \text{if no such } z \text{ exists}$$

Plainly this defines a unique partial function. It is also clear that if f is computable then so is $min^n(f)$; we simply compute $f(\mathbf{x}, 0), f(\mathbf{x}, 1), \ldots$ successively until we get the value 0.

Definition 1.4.6. The *recursive functions* (Rec) are the least set including the zero, successor and projection functions and closed under the operations of composition, primitive recursion and minimization. A *total recursive* function is a member of Rec which is a total function.

Notice that we are now allowing composition and primitive recursion to apply to partial functions. The recursive functions are often called the *partial* recursive functions, as recursive is sometimes taken to mean total recursive. They are also sometimes called μ-recursive, where μ refers to the minimisation operator.

All recursive functions are computable in an informal sense since the base functions are computable and the operations preserve computability. We shall see that the recursive functions in fact comprise all computable functions.

It is of interest to see how our language of terms for the recursive functions may be translated into our functional language FL. This is a straightforward matter for $0, S, \pi$, comp and prim. If $g = min^n(f)$ we can write $g(\mathbf{x}) = h(\mathbf{x}, 0)$ where

$$h(\mathbf{x}, y) = \text{ if } f(\mathbf{x}, y) = 0 \text{ then } y \text{ else } h(\mathbf{x}, y + 1)$$

adding equations for the definition of f of course. We see that there is a non-well-founded recursion, possibly leading to non-termination.

Our translation raises the question as to which evaluation strategy we intend to be employed. In mathematics it is usual to use the V-semantics, where if an argument to a function is undefined then so is the result. By adopting this, our translation gives us a precise meaning for our language for the recursive functions.

Definition 1.4.7. Let $R(\mathbf{x})$ be a predicate on $\mathbf{N}^k$. R is *recursive (primitive recursive)* if the (total) characteristic function $\mathrm{ch}_R : \mathbf{N}^k \to \mathbf{N}$ defined by

$$\mathrm{ch}_R(\mathbf{x}) = \left\{ \begin{array}{ll} 0 & \text{if } R(\mathbf{x}) \\ 1 & \text{if not } R(\mathbf{x}) \end{array} \right.$$

is recursive (primitive recursive).

Remark 1.4.8. In imperative programming terms, the primitive recursive functions may be computed using 'for' loops. Consider the primitive recursion

$$h(\mathbf{x}, 0) = f(\mathbf{x})$$
$$h(\mathbf{x}, y+1) = g(\mathbf{x}, y, h(\mathbf{x}, y))$$

h can be computed by the following program:

```
input x, y
z := f(x)
if y = 0 then return z else
for i = 0 to y − 1
    z := g(x, i, z)
return z.
```

Minimization requires the power of unbounded iteration, such as 'while' loops.

1.5 Universality

Let us say that a function is *effectively computable* if it can be computed by some means that we intuitively recognize to be valid. Let us also say that a programming language (and accompanying semantics) is *universal* if it can compute every effectively computable function. In this subsection we explain the remarkable fact that all the languages we have considered (and many others) are considered to be universal. First we show that the various computing formalisms we have examined all give the *same* class of computable functions, the so-called Fundamental Result. On the basis of this we are led to Church's Thesis, the statement that this class of computable functions is precisely the set of those functions which are effectively computable.

1.5.1 Coding and the Fundamental Result

In the preceding section we have defined a variety of notions of computability: via Turing machines (TM), register machines (RM), while programs, functional programs (FL), and finally the (partial) recursive functions. We saw in Section 1.1.3 that under a simple assumption while programs have the same computational power as register machines. Our aim now is to show that in fact all the different notions of computability coincide.

Theorem 1.5.1 (Fundamental Result). *Let $f : \mathbf{N}^n \to_p \mathbf{N}$. The following are equivalent:*

1. *f is TM-computable*

2. *f is RM-computable (flowchart)*

3. *f is FL-computable*

4. *f is recursive*

We had better admit right away that we are not going to prove this. It would take a long time, and there are many texts where variants of the above are proved. Instead we aim to give the reader some idea of the concepts involved in the proof. Of course the most economical way to prove that four statements are equivalent is to prove a circle of four implications. However in the present circumstance this would require some ingenuity in relating, for instance, functional programs and Turing machines. Instead our strategy will be to regard the recursive functions as the lynch-pin, and prove all the other statements equivalent to (4). Our method then has the advantage that when confronted with yet another notion of computability, we can hope to adapt our methods and prove that it in its turn is equivalent to (4).

Proving that (4) implies the others is relatively straightforward. We have already in effect shown (4) $\Rightarrow$ (3) in giving our translation of the recursive functions into FL (with the V-semantics) at the end of Section 1.3. In the case of RMs, we proceed by induction on the definition of the recursive functions to show that each of them can be programmed on a RM. This simply requires a little care to make sure that computations do not overlap, moving the contents of registers around when necessary. Turing machines are somewhat lower level, and it is probably advisable to show that they can simulate RMs rather than attacking the recursive functions directly. Of course one can save some work here by taking some minimal set of RM commands (e.g. addition can be programmed in terms of successor, so that we don't have to write a TM version of the addition command). The reader may find the details for (4) $\Rightarrow$ (2)$\Rightarrow$ (1) in [Boolos and Jeffrey, 1989] (Ch. 6, 7].

So we must now consider the harder problem of the reverse implications. With each of our programming languages we had a notion of what it means to input a number (or tuple of numbers) and of what the result of the computation would be. The computation was rather different depending on whether we were dealing with an imperative or a functional language. In the imperative case we have the idea of a computation passing through successive *configurations*, each described by the state of the store (for instance the contents of the registers or of the TM tape), and where the control is in the program (for instance the label of the next instruction to be executed). At each state there is a way of calculating the next configuration. The computation is the sequence of configurations. In the functional case we defined the result of a computation by means of a proof system. Computing the result consists of building up a proof using the syntax-directed rules. The computation is the proof.

So these two paradigms are very different. But what they have in common is that in each case, given program P and input tuple $\mathbf{x}$, we can say that the result of the computation *if there is a result* is z and that it is *witnessed* by a completed computation y. Furthermore the result z is actually recorded in y (for instance z is the contents of a particular register when the computation reaches a halting configuration, or $f(\mathbf{x}) = z$ is the conclusion of a proof). For each of the languages we have considered we can therefore define a predicate T as follows:

$T(P, \mathbf{x}, y)$ iff y is a completed computation of program P on input $\mathbf{x}$

This is *Kleene's T-predicate*. We furthermore can define a function U such that for any $P, \mathbf{x}$, if for some y, $T(P, \mathbf{x}, y)$, then $U(y)$ is the result of P on input $\mathbf{x}$. The function U will probably be very simple to define, and can well be total. Of course the function associated with P is not necessarily total. If it is undefined on input x, then there will be no y such that $T(P, \mathbf{x}, y)$.

One way to look at the foregoing is that we have a very general paradigm for computation. If we wish to compute the result of P on input x, we can search through all computations y, checking in each case whether $T(P, \mathbf{x}, y)$. This will be a purely mechanical task requiring no insight and we can program a computer to do it. If we find such a y then the result is defined and equal to $U(y)$. Of course this is not a sensible way of carrying out the computation, since it means trying out a lot of wrong y's, and we have ways of generating the correct y (if such exists) using the operational semantics of the language. However it is important for the matter in hand, which is to show that all notions of computability can be interpreted by the recursive functions.

Given a language L we want to show that if f is computable by program P then f is recursive. Our strategy is as follows: convert the T-predicate into a predicate on *numbers* by coding up programs and computations as numbers. Show that T is now recursive. Convert U into a function on numbers and show that U is recursive. Then in the light of the previous paragraph, for any $\mathbf{x}$

$$f(\mathbf{x}) = U(\ \min y.\ T(e, \mathbf{x}, y))$$

Here e is the number coding P, and $\min y.\ T(e, \mathbf{x}, y)$ finds the least y such that $T(e, \mathbf{x}, y)$. Since T is recursive and Rec is closed under minimization, $\min y.\ T(e, \mathbf{x}, y)$ is recursive. Hence f is recursive.

It remains to fill in the details of how the above is to be done for our particular example languages, in particular for TMs, and for FL — this is of course where all the work is! Let us first deal with TMs.

Our first task is to show how to code programs as numbers. We shall often refer to codes for programs as *indices*. Here the problem is not so much to assign a number index to each program, but to do this in such a way that we can manipulate the indices to extract information about the program. For instance we may wish to know what the first instruction of program M is. Let us say that it is (q, σ, L, q'). Then we should be able to extract a code for this from the code for M. In other words there is a function first: $\mathbf{N} \to_p \mathbf{N}$ such that

first $(\ulcorner M \urcorner) = \ulcorner (q, \sigma, L, q') \urcorner$

Here the 'ears' are the traditional notation for codes or Gödel numbers — a sort of quotation mark. If we can show that all such syntactic manipulations of programs are recursive when applied to the codes, then we are on our way to showing that TMs can be interpreted using the recursive functions. So our technical requirement is to show that operations on programs translate under coding into recursive functions on the codes. The way that we achieve this is to require our coding to be *effective*, that is, computable in an informal sense. Here is a general definition of a coding:

Definition 1.5.2. A *coding* of a set A is a 1–1 function $g : A \to \mathbf{N}$ such that

1. g is effective

2. range (g) is decidable

3. if $n \in$ range (g) then $g^{-1}(n)$ can be found effectively

We think of g^{-1} as an enumeration of A. Note that 'effective' here must remain informal.

If we have such a coding of TMs then the function first given above will indeed be effective: one can calculate it by taking the code of M, recovering M from it, finding its first instruction (clearly effective) and coding that.

Our task is still to prove that the first is recursive, since we do not at this stage want to appeal to Church's Thesis.

Let us now indicate how TMs can be coded. First we need to know how to code tuples or lists of numbers. Recall from Section 1.0 the pairing function $p : \mathbf{N} \times \mathbf{N} \to \mathbf{N}$. Much as we did there, we can use p to code up lists. For instance $(3, 5, 4)$ will be coded as $p(3, p(5, p(4, 0)))$. Number 0 is the code for the empty list. It can be shown that p is a recursive function (in fact primitive recursive), as are the unpairing functions left, right: $\mathbf{N} \to \mathbf{N}$ (conventionally we let left(0)=right(0)=0). We thus have an effective coding of tuples from $\mathbf{N}$. We denote the code of $(n_1, \ldots, n_k)$ by $\langle n_1, \ldots, n_k \rangle$. We shall need the function of two arguments which finds the members of a list. We use the notation $(n)_i$ for the ith member of the list coded by n, so that $(\langle n_1, \ldots, n_k \rangle)_i = n_i$ if $1 \leq i \leq k$. This may be defined as follows:

$$(n)_0 = 0$$
$$(n)_{i+1} = \text{left}(\text{right}^i(n)) \quad (i \geq 0)$$

To see that this is primitive recursive, notice that $\text{right}^i(n)$ is defined by the following primitive recursion:

$\text{right}^0(n) = n$
$\text{right}^{i+1}(n) = \text{right}(\text{right}^i(n))$

We shall also need the function which finds the length of a list. The length of a list is given by

length(nil) = 0
length([a : l]) = 1+ length(l)

The corresponding operation lth: $\mathbf{N} \to \mathbf{N}$ on codes is clearly defined by

lth(0)=0
lth(n + 1) = 1+ lth(right(n + 1))

Clearly this satisfies $\text{lth}(\langle n_1, \ldots, n_k \rangle) = k$ as we wanted. It is not defined by a true primitive recursion as it stands. However it is a valid recursion, because $\text{right}(n + 1) \leq n$ for all n, so that we only use previously defined values of lth. We need a more powerful form of definition by recursion, which we now give:

Definition 1.5.3. Let $f : \mathbf{N}^{n+2} \to \mathbf{N}$. The function $g : \mathbf{N}^{m+1} \to \mathbf{N}$ is said to be defined by *course-of-values recursion* from f if it is given by:

$$g(\mathbf{x}, y) = f(\mathbf{x}, y, \overline{g}(\mathbf{x}, y))$$

where $\overline{g}(\mathbf{x}, y) = \langle g(\mathbf{x}, y - 1), \ldots, g(\mathbf{x}, 0) \rangle$.

We see that g is allowed to appeal to *all* previous values, rather than just the single preceding value, as in primitive recursion. We write the list in reverse order for irritating technical reasons.

Proposition 1.5.4. *Let f be PR, and let g be defined by course-of-values recursion from f. Then g is PR.*

Proof. We show first that $\bar{g}$ is PR.

$$\bar{g}(\mathbf{x}, 0) = 0$$
$$\bar{g}(\mathbf{x}, y + 1) = p(f(\mathbf{x}, y), \bar{g}(\mathbf{x}, y))$$

Now g is got from f and $\bar{g}$ by composition.   ∎

Let us apply this to lth. Recall that $\overline{\mathrm{lth}}(n) = \langle \mathrm{lth}(n-1), \dots, \mathrm{lth}(0) \rangle$ so that if we wish to pick out $\mathrm{lth}(\mathrm{right}(n))$ we must take the $(n-\mathrm{right}(n))$th entry. Thus

$$\mathrm{lth}(n) = \text{if } n = 0 \text{ then } 0 \text{ else } 1 + (\overline{\mathrm{lth}}(n))_{n-\mathrm{right}(n)}.$$

It is a simple matter to show that the use of the conditional does not take us out of PR, so that we see that lth is got by course-of-values recursion from PR functions, and is therefore in PR as we wanted.

This completes our discussion of how lists are coded. Let us now apply this to the task of coding Turing machines. We consider machines over the alphabet $\{B, 1\}$, as this is what is used in the definition of TM-computability (Definition 1.2.1). Let us also assume that TMs have states described by natural numbers, so that $Q = \mathbf{N}$, and that the start state is always 0. A particular TM M is described as a list of instructions: $M \in$ List $(\mathbf{N} \times \{B, 1\} \times \{B, 1, L, R\} \times \mathbf{N})$. Notice that we have changed to talking about lists rather than sets, although the order in which the instructions are written does not matter. Let us assign numbers to $B, 1, L, R$ in a purely arbitrary way, say $\ulcorner B \urcorner = 0, \ulcorner 1 \urcorner = 1, \ulcorner L \urcorner = 2, \ulcorner R \urcorner = 3$. Now M may be regarded as a list of 4-tuples, and may be coded in $\mathbf{N}$ – call it $\ulcorner M \urcorner$ – using our coding of lists above. For instance, the TM given by the following instructions

(0, 1, R, 1)
(0, B, R, 0)
(1, 1, B, 0)

(whose behaviour is not terribly sensible) is coded by

$$\langle \langle 0, \ulcorner 1 \urcorner, \ulcorner R \urcorner, 1 \rangle, \langle 0, \ulcorner B \urcorner, \ulcorner R \urcorner, 0 \rangle, \langle 1, \ulcorner 1 \urcorner, \ulcorner B \urcorner, 0 \rangle \rangle$$

Our notation is designed to show that we do not have to be troubled with what number this actually is, a matter of no interest.

Let us check that we have a valid coding in terms of Definition 1.5.2. Certainly it is effective, and we can effectively uncode a number into the corresponding machine. But we need to deal with the problem that not every number is the code of a valid TM. We must show that we can decide whether or not a given number n is the code of a TM. This means exhibiting a total recursive function $g : \mathbf{N} \to \mathbf{N}$ such that

$g(n) = 0$ if n is the code of a TM
$g(n) = 1$ otherwise.

We first describe how the matter is decided informally. Take the given number n. Decode it into the corresponding sequence of numbers $\langle n_1, \ldots, n_k \rangle$. Check that each n_i is the code of a 4-tuple $\langle m_1, \ldots, m_4 \rangle$ where $m_2 \leq 1, m_3 \leq 3$. We must also check that the machine is deterministic. This means that we do not have both $\langle m_1, m_2, m_3, m_4 \rangle$ and $\langle m_1, m_2, m'_3, m'_4 \rangle$ where $m_3 \neq m'_3$ or $m_4 \neq m'_4$. Let us write this more formally:

n is the code of a TM iff
$\forall_i < \text{lth}(n)[\text{lth}((n)_{i+1}) = 4 \wedge ((n)_{i+1})_2 \leq 1 \wedge ((n)_{i+1})_3 \leq 3]$
$\wedge \forall i < \text{lth}(n) \forall j < \text{lth}(n)$
$[(n)_{i+1} \neq (n)_{j+1} \to (((n)_{i+1})_3 \neq ((n)_{j+1})_3 \vee ((n)_{i+1})_4 \neq ((n)_{j+1})_4)]$

The benefit of writing it out in this fashion is that we can now see that this predicate is decidable if we can show that the logical operations of conjunction, disjunction, implication, negation, and *bounded* quantification preserve decidability. For apart from these the definition simply uses functions (and predicates in the case of $=$) which we already know to be recursive.

Proposition 1.5.5.

1. Let R, S be decidable predicates. Then the following predicates are also decidable:

    $R \wedge S, R \vee S, R \to S, \neg R$

2. Let $R(\mathbf{x}, y)$ be decidable. Then so are $R_1(\mathbf{x}, y), R_2(\mathbf{x}, y)$ defined by

    $R_1(\mathbf{x}, y)$ iff $\forall z < y R(\mathbf{x}, z)$
    $R_2(\mathbf{x}, y)$ iff $\exists z < y R(\mathbf{x}, z)$

Proof. 1. Left to the reader.

2. R_1 may be defined by the following primitive recursion:

$R_1(\mathbf{x}, 0) = 0$
$R_1(\mathbf{x}, y + 1) = R_1(\mathbf{x}, y) \wedge R(\mathbf{x}, y)$

For R_2 we can use the well-known equivalence $\exists z < y R(\mathbf{x}, z)$ iff $\neg \forall z < y \neg R(\mathbf{x}, z)$ ∎

It is not the case in general that if $R(\mathbf{x}, y)$ is decidable then so are $\forall y R(\mathbf{x}, y)$ and $\exists y R(\mathbf{x}, y)$.

With the help of the proposition we have now established that we have a proper coding of TMs. Now we must turn to coding up computations. Recall that a configuration c is a triple (w_1, n, w_2) where $w_1 w_2$ is the contents of the tape, and the head is in state n and scanning the first symbol of w_2. It is convenient to assume that both w_1 and w_2 are non-empty, which can always be achieved by adding blanks. Words w in $\{B, 1\}^*$ can be coded as $\ulcorner w \urcorner$ using our coding of lists. Thus $\ulcorner B11 \urcorner = \langle \ulcorner B \urcorner, \ulcorner 1 \urcorner, \ulcorner 1 \urcorner \rangle$. Further define $\ulcorner c \urcorner = \langle \text{rev}(w_1) \urcorner, n, \ulcorner w_2 \urcorner \rangle$. Here $\text{rev}(w_1)$ denotes that we write w_1 in reverse order, for technical convenience. A computation consists of a sequence of configurations. We need a relation $\text{next}(\ulcorner M \urcorner, \ulcorner c_1, \urcorner, \ulcorner c_2 \urcorner)$ which holds iff c_2 is the next step after c_1 in a computation with machine M. Informally this says that there is an instruction in M which is enabled in c_1, and that c_2 is the result of applying this instruction. It is defined as follows:

$\text{next}(e, m_1, m_2)$ iff $\exists i < \text{lth}(e)$.
$((e)_{i+1})_1 = (m_1)_2 \land ((e)_{i+1})_2 = ((m_1)_3)_1 \land (m_2)_2 = ((e)_{i+1})_4 \land$
$[((e)_{i+1})_3 = \ulcorner B \urcorner \rightarrow ((m_2)_1 = (m_1)_1 \land (m_2)_3 = p(\ulcorner B \urcorner, \text{right}((m_1)_3)))] \land$
$[((e)_{i+1})_3 = \ulcorner 1 \urcorner \rightarrow ((m_2)_1 = (m_1)_1 \land (m_2)_3 = p(\ulcorner 1 \urcorner, \text{right}((m_1)_3)))] \land$
$[((e)_{i+1})_3 = \ulcorner R \urcorner \rightarrow ((m_2)_1 = p(\text{left}((m_1)_3), (m_1)_1) \land$
$\quad (m_2)_3 \quad = \text{right}((m_1)_3) \quad$ if $\text{lth}((m_1)_3)) > 1$
$\quad \quad \quad = p(\ulcorner B \urcorner, 0) \quad \quad$ otherwise $)] \land$
$[((e)_{i+1})_3 = \ulcorner L \urcorner \rightarrow ((m_2)_3 = p(\text{left}((m_1)_1), (m_1)_3) \land$
$\quad (m_2)_1 \quad = \text{right}((m_1)_1) \quad$ if $\text{lth}((m_1)_1)) > 1$
$\quad \quad \quad = p(\ulcorner B \urcorner, 0) \quad \quad$ otherwise $)]$

We apologize for including this unreadable formula. It could obviously be made more palatable in various ways. The point we wish to make is that using logic and our coding of lists, we are able to describe quite complex properties - the next relation is in fact where the main 'nuts and bolts' of Turing machines reside. It is now clear from our formula that the next relation is recursive.

We can now obtain the Kleene T-predicate: $T_m(e, \mathbf{x}, y)$ says that y codes a sequence $y_1, \ldots, y_k$ of configurations such that

y_1 is the initial configuration for input $\mathbf{x} = x_1, \ldots, x_m$

$\forall i < k. \, i \geq 1 \rightarrow \text{next}\,(e, y_i, y_{i+1})$

y_k is a halting configuration with the head scanning the leftmost of a string of 1s, the tape otherwise being blank

The reader should now be happy that this predicate is indeed recursive without us spelling out further details. Notice that only bounded quantification is used. In fact T_m is primitive recursive, since we did not need minimization in defining it. It remains to define the function U which extracts the result from y. Plainly it just counts the 1s left on the tape in configuration y_k and subtracts 1. We omit the detailed definition, which shows that U is also primitive recursive.

We have shown that there are primitive recursive T, U such that if f is TM-computable by M, then

$f(\mathbf{x})$ is defined iff there exists a unique y such that $T_m(e, \mathbf{x}, y)$

$$f(\mathbf{x}) = U(\min y . T_m(e, \mathbf{x}, y))$$

We can conclude that f is recursive. This completes our proof that (1) $\Rightarrow$ (4).

Our final task is to show (3) $\Rightarrow$ (4), in other words if a function is FL-computable then it is recursive. Our approach must be similar to that for (1) $\Rightarrow$ (4), but of course the notion of computation is different. Recall the proof rules for the evaluation relation $\Rightarrow$ in Section 1.2.2. First we need to define codes for terms, ranged over by u, etc. This is done as follows. First assign codes to the constructors. There is an important point when it comes to the numerals num(n). If we simply code these by n, then we have used up all the available codes. Instead we code them as $\langle \ulcorner \text{num} \urcorner, n \rangle$ where $\ulcorner \text{num} \urcorner$ is a particular number. The rest of the coding is as follows:

$\ulcorner \text{var}(n) \urcorner = \langle \ulcorner \text{var} \urcorner, n \rangle$
$\ulcorner \text{plus}\,(u_1, u_2) \urcorner = \langle \ulcorner \text{plus} \urcorner, \ulcorner u_1 \urcorner, \ulcorner u_2 \urcorner \rangle$

(and similarly for the other pre-defined functions)

$\ulcorner \text{fun}(k, n)(u_1, \ldots, u_k) \urcorner = \langle \ulcorner \text{fun} \urcorner, k, n, \ulcorner u_1 \urcorner, \ldots, \ulcorner u_k \urcorner \rangle$
$\ulcorner \text{pred}(k, n)(u_1, \ldots, u_k) \urcorner = \langle \ulcorner \text{pred} \urcorner, k, n, \ulcorner u_1 \urcorner, \ldots, \ulcorner u_k \urcorner \rangle$
$\ulcorner \text{ if } b \text{ then } u_1 \text{ else } u_2 \urcorner = \langle \ulcorner \text{cond} \urcorner, \ulcorner b \urcorner, \ulcorner u_1 \urcorner, \ulcorner u_2 \urcorner \rangle$

It does not matter which numbers we choose for the first component – $\ulcorner \text{num} \urcorner, \ulcorner \text{var} \urcorner, \ulcorner \text{plus} \urcorner$, etc. – so long as they are all different. Many different codings could be employed, but notice that a convenient feature of this one is that if u' is a subterm of u then $\ulcorner u' \urcorner < \ulcorner u \urcorner$. Now we can define the predicate

Term(m) iff m is the code of an FL term

by course-of-values recursion and show that Term is decidable. We define the relation

Proof(e, y) iff e codes an FL program of the form $f(\mathbf{x}) = t$ and y codes a proof (of some $u \Rightarrow c$) using the rules (1), (2), (3), (4V)

How do we code a proof? For instance if the last rule to be applied is

$$(3)\frac{u \Rightarrow tt \qquad u_1 \Rightarrow c}{\text{if } u \text{ then } u_1 \text{ else } u_2 \Rightarrow c}$$

and y_1, y_2 code the proofs of $u \Rightarrow tt, u_1 \Rightarrow c$ respectively, then the code of the whole proof can be $\langle 3, \ulcorner \text{if } u \text{ then } u_1 \text{ else } u_2 \urcorner, \ulcorner c \urcorner, y_1, y_2 \rangle$.
Now we can define the T-predicate $T_m(e, x_1, \ldots x_m, y)$ iff e codes an FL
program $f(z_1, \ldots, z_m) = t$
and Proof(e, y)
and y proves $f(\text{num}(x_1), \ldots, \text{num}(x_m)) \Rightarrow c$ (some c) Note that two different kinds of variable are used in the above. The z_i's are formal variables in FL — var(n)'s for some n. The x_i's vary over numbers, and appear as constants num(x_i) in FL.

Now just as with TMs we can show that each FL-computable function is recursive. We apologize for only providing a sketch, but hope that the reader is by now feeling comfortable with the techniques, and appreciates that the particular details are of little importance. This completes our proof of the Fundamental Result.

A noteworthy feature of the proof that FL-computable implies recursive is that it makes no essential difference whether we choose rule $(4V)$ or $(4N)$ – everything is still recursive. Indeed one main reason why we chose to introduce two different methods of evaluation was to point out that one can always get a (primitive) recursive T-predicate no matter which of the available methods of evaluation one considers. An immediate corollary of our proof is that FL-N-computable implies FL-V-computable, since we have shown the latter is equivalent to recursive. In fact the reverse implication also holds, since we can force call-by-value even when using the N-semantics – simply rewrite the program $f(x_1, \ldots, x_m) = t$ as

if $x_1 = x_1$ then($\ldots$ (if $x_m = x_m$ then t else t)$\ldots$)else t)

In view of the Fundamental Result we can use the word computable to refer to any of the notions of computability that we have encountered, since they are all equivalent. The following very useful result emerged from our proof:

Theorem 1.5.6 (Kleene Normal Form Theorem). *There is a PR predicate T_m, and a PR function U such that if $f : \mathbf{N}^m \to_p \mathbf{N}$ is computable then there is e such that for all* $\mathbf{x}$,
  $f(\mathbf{x})$ *is defined iff* $\exists y \, T_m(e, \mathbf{x}, y)$
  and
  $f(\mathbf{x}) = U(\text{min} y. T_m(e, \mathbf{x}, y))$ ∎

An immediate consequence of the Normal Form Theorem is that every recursive function can be defined using minimization just once. For definiteness we can assume we are using the T-predicate associated with our coding of TMs, but it does not matter which we choose. Notice that even if the number e does not code a TM, then $T_m(e, \mathbf{x}, y)$ is still defined (and of course false), and $U(\min y.T_m(e, \mathbf{x}, y))$ is still a partial recursive function — in fact the totally undefined function.

Notation 1.5.7. Let $\{e\}_m(\mathbf{x})$ denote the partial recursive m-ary function
$U(\min y.T_m(e, \mathbf{x}, y))$

It is clear that as e ranges over N, $\{e\}_m$ ranges over all partial recursive m-ary functions. We shall supress the subscript m when the arity is clear.

1.5.2 Church's Thesis

Recall that we use the word 'effective' to mean 'computable by some means which we intuitively recognize to be valid'. Let us use 'computable' to mean recursive or computable by Turing machines, register machines, or FL programs. This makes sense because by the Fundamental Result all these are equivalent.

Thesis 1.5.8 (Church's Thesis). A partial function is effective precisely if it is computable

It is of course clear that computable implies effective, and so the content of the Thesis is that any effective function is computable. It is called a Thesis (rather than a theorem) because by its very nature it cannot be proved — the concept of effectiveness has to remain somewhat nebulous. If we give it a precise meaning we can hope to prove the Thesis, but it would have lost its force, since it would no longer encompass all possible methods of computation. Of course we can *disprove* it simply by exhibiting a function which is not computable in the ways we have discussed, but for which we can provide some other means of computing it which are recognized by us and others to be valid.

There are three main reasons why computer scientists universally accept Church's Thesis:

1. We discussed earlier Turing's analysis of what is involved in computation. This seems so general that it is hard to imagine some other method which falls outside the scope of his description. This being granted, we can further argue that the Turing machine provides a good formalization of his informal description, so that anything which can be computed can be computed by a Turing machine. Thus there are 'philosophical' grounds for believing the Thesis.

2. Every precise notion of computability which has yet been proposed (and there are a number which we have not discussed) gives the same set of computable functions. We of course devoted considerable effort to establishing a version of this Fundamental Result. It is plausible to explain this correspondence by postulating that there is a well-defined

notion of effective computability and that this has been captured by the various notions put forward.

3. It has not been disproved, despite having been proposed over 50 years ago.

The Thesis has something of the nature of a scientific hypothesis — we can provide supporting evidence and we know what it would mean to disprove it.

Church's Thesis is important for two reasons. Firstly it gives significance to the work we do within particular computational formalisms, since we know that they have universal computing power. Secondly it allows us to shorten and clarify some of our arguments. Suppose we wish to show that some function is computable. We could devise a Turing machine to compute it, or a recursive function definition, but this might well get very complex and tedious. So instead we shall often argue informally (but precisely) that it is effective, and then appeal to Church's Thesis to assert that it is computable. Arguments which depend on this sort of procedure are called 'proofs by Church's Thesis'. They are not true proofs, but of course a full proof with no appeal to Church's Thesis could be supplied if required.

2 Computability and non-computability

In this Section we cover the basics of computability theory. We start in Section 2.1 with limitative results showing that certain matters, such as the famous Halting Problem, are non-computable (by any means). In Section 2.2 we turn to positive results, such as the existence of a universal program. Section 2.3 deals with recursive and recursively enumerable sets. Section 2.4 looks at a technical tool of recursion theory connected with partial evaluation. This enables us to establish further undecidability results in Section 2.5, notably Rice's Theorem. Finally Section 2.6 deals with the comparison of computational problems (showing that one problem is in some sense no harder than another).

2.1 Non-computability

The main purpose of Section 1 was to show how the notion, of computability could be made precise in a number of ways, corresponding to different programming paradigms. It was then shown in the Fundamental Result (1.5.1) that all of these notions had the same power. By Church's Thesis we can identify our notions of computability with effective computability.

We are now in a position to show that certain matters must remain in principle non-computable by any means, since we can show that they are not computable according to some precise notion and then appeal to Church's Thesis. Our main tool for showing that certain objects are non-computable will be diagonalization. We employed this in Section 0.2 where we showed that a programming language whose programs were all guaranteed to terminate could not compute all computable functions. Clearly our first task should be to answer the question of whether all functions are computable in the negative. This we can do in quite a strong way by giving an explicit definition of a non-computable function. Recall from the Normal Form Theorem 1.5.6 that $\{e\}_1$ ranges over all (partial) unary computable functions as the natural number e ranges over (codes for) programs. Our strategy is to define a function $f : \mathbf{N} \rightarrow \mathbf{N}$ such that $f(n)$ is different from $\{n\}(n)$(suppressing the subscripted 1). We have to bear in mind that $\{n\}(n)$ may not be defined, but in this case we can just choose any value for $f(n)$. So let

$$f(n) = \begin{cases} \{n\}(n) + 1 & \text{if } \{n\}(n) \downarrow \\ 0 & \text{otherwise} \end{cases}$$

Then by construction $f(n) \neq \{n\}(n)$ for any n. So $f \neq \{n\}$ for any n. It is therefore clear that f is not computable. Notice that f is total, showing that there are non-computable total functions. Why is f not computable? In fact the function taking n to $\{n\}(n) + 1$ is computable; this will be discussed further in Section 2.2. The problem lies with testing whether $\{n\}(n) \downarrow$. This is the famous *self-halting problem*, and is an example of a predicate which is not decidable. We have just proved this, since we have claimed that if it were decidable then f would be computable, which is false. However the reader may be more convinced by a direct argument, which we now give.

First a remark about the use of the word 'problem'. With each predicate $P(x)$ of a natural number variable x we associate a *decision problem* $?P(x)?$ This is simply the problem of determining for each x whether $P(x)$ holds. The problem is said to be *solvable* if the predicate is decidable. Of course this does not just mean that we can solve each individual instance of the problem; we require an algorithm for solving any instance in a uniform manner.

Theorem 2.1.1 (Turing). *The self-halting problem $?n(n) \downarrow?$ is unsolvable.*

Proof. Suppose for a contradiction that the problem is solvable. Then there is a Turing machine M which, when given n as input, outputs 1 if $\{n\}(n) \downarrow$ and 0 otherwise. Let us modify M to produce a new machine M' which behaves exactly as M except that if M halts with 1 on the tape then

I. C. C. Phillips

M' has further instructions which then send the computation into a loop so that it fails to terminate. Let M' have index e. Note that for any x

$$\{e\}(x) \downarrow \text{ iff } M' \text{ halts when given } x \text{ as input.}$$

Now suppose we give M' the number e as input. From the above

$$\{e\}(e) \downarrow \text{ iff } M' \text{ halts when given } e \text{ as input.} \qquad (*)$$

But if $\{e\}(e) \downarrow$ then M outputs 1 when given e as input, and so M' loops and does not halt. Also if $\{e\}(e) \uparrow$ then M outputs 0 when given e as input, and M' does the same and so halts. So by construction of M'

$$\{e\}(e) \downarrow \text{ iff } M' \text{ does not halt when given } e \text{ as input.}$$

But this last statement is a clear contradiction of $(*)$. ∎

Remark 2.1.2. Even though we carried out the above proof in the context of Turing machines, it is clear that it would go through for any programming language providing for a non-terminating program, conditional branching and sequential composition of programs.

We have now shown rather rapidly that not every problem is solvable. However the reader may object that giving a program itself as an input is not something one would normally wish to do. Are there any *significant* problems which are unsolvable? We can immediately answer in the affirmative: we are surely interested in the problem of whether a program will terminate on a given input. If we had a program to decide this, we could use it to improve our compilers by having them issue an error message before carrying a computation out if it was going to fail to terminate, and thereby save time and trap errors. So we would like to solve the problem $?\{e\}(x)?$ for e ranging over all (codes of) programs, and x over all inputs. But we can easily deduce that this is unsolvable from the unsolvability of the self-halting problem. The point is that the latter is actually an *easier* problem. Suppose that we can solve the general halting problem. Then we have a program P into which we can feed inputs e, x, and which tells us whether $\{x\}(y) \downarrow$. If we want to know whether $\{x\}(x) \downarrow$ all we have to do is plug x into P twice, once as program and once as value. So we have solved the self-halting problem. Since the easier problem is unsolvable, the harder one must be also.

Corollary 2.1.3. *The halting problem* $?\{x\}(y) \downarrow?$ *is unsolvable.*

We have just seen an example of *problem reduction*. If we have two problems $?P(x)?$ and $?P'(x)?$ and we show that P is easier than P', then we say P is *reduced* to P'. This will be formalized in the notion of *many-one reduction* (Section 2.6). In our example we reduced the self-halting problem to the general halting problem. A common strategy for demonstrating that a problem is undecidable is to reduce one or other version of the halting problem to it.

Remark 2.1.4. Although the statements of unsolvability given above are both precise and proven, they gain significance from Church's Thesis, which tells us that if a problem is unsolvable by Turing machines then it cannot be solved by *any* effective means.

Example 2.1.5. Here are three further unsolvable problems: $?\{x\}(y) = z?$, $?\{x\}(y) = 0?$,$?\{x\}(x) = 0?$ We leave it to the reader to show that the last is undecidable by a direct diagonal argument as in Theorem 2.1.1. This problem can then be reduced to each of the other two to show them unsolvable.

2.2 Computability

In contrast to the previous subsection, in this subsection we collect two fundamental positive results on computabiity; the existence of self-interpreters and the solvability of the bounded halting problem. We conclude with the useful technique of dovetailing. The reader is probably familiar with the idea of an *interpreter*— a program I which takes any program P in a given language L and 'runs' P to produce the same effect as P. Suppose that L is universal. Then we can think of I as being a sort of 'universal program'. If I is in a language L' different from L then I provides a means of compiling L-programs P into L'-programs $I(P)$, though this may be very inefficient; a proper compiler will likely perform various optimizations. Now since L is universal, then by Church's Thesis we can write an interpreter in L for itself — a self-interpreter. This is the content of the next result:

Theorem 2.2.1 (Universal Program Theorem). *Universal Program Theorem For each m there is a computable function Ψ_m such that $\{e\}_m(\mathbf{x}) = \Psi_m(e, \mathbf{x})$.*

Proof. We have essentially already proved this in the Kleene Normal Form Theorem (1.5.6). The T-predicate $T_m(e, \mathbf{x}, y)$ is a computable predicate in e as well as $\mathbf{x}, y$. From this we immediately get that $\{e\}_m(\mathbf{x})$ may be regarded as a computable function of e as well as $\mathbf{x}$. Notice that we don't actually need to appeal to Church's Thesis to get that the universal program exists and is computable. ∎

In contrast to the usual halting problem, the *bounded* halting problem $?\{x\}(y) \downarrow$ in $\leq t$ steps $?$ is solvable. 'Step' here means 'step in a Turing machine computation'. Informally we solve this problem by running program x on input y for t steps, and returning 1 if the computation has terminated and 0 otherwise. This does not solve the full halting problem, because if the computation is still running after t steps, we do not know whether it will eventually terminate. If we define a predicate

$B(x, y, t)$ iff $\{x\}(y) \downarrow$ in $\leq t$ steps

corresponding to the bounded halting problem, it is not hard to see that B is in fact primitive recursive. Define a primitive recursive function next: $\mathbf{N}^2 \to \mathbf{N}$ such that if c is a configuration of a computation with a Turing machine M then next $(\ulcorner M \urcorner, \ulcorner c \urcorner)$ is the code of the next configuration (unless c is halting, in which case it doesn't matter what value we assign). Using the function next we can then define a function config: $\mathbf{N}^3 \to \mathbf{N}$ by primitive recursion such that config($\ulcorner M \urcorner, n, i$) is the ith configuration of the computation of M on input n (again taking an arbitrary value if this is undefined). Now we have

$B(x, y, t)$ iff $\exists i \leq t$ [config(x, y, i) is halting $\land \forall j < i$ config(x, y, j) is not halting]

This shows that B is PR since PR predicates are closed under bounded quantification by (the proof of) Proposition 1.5.5.

Example 2.2.2 (Dovetailing). Suppose we are given a partial recursive function $f : \mathbf{N} \to_p \mathbf{N}$ and we wish to search for a value such that $f(n) \downarrow$. The obvious way to do this is to use a program for f to calculate $f(0), f(1), f(2), \ldots$ successively until we find n such that $f(n) \downarrow$. However if $f(0) \uparrow$ then we will never finish computing it, even if $f(1)$ is defined. So plainly this is the wrong approach. Notice that in this case applying the minimization operator (Definition 1.4.4) to f to find the least x such that $f(x) = 0$ will yield undefined. To avoid the problem of getting trapped by a non-terminating computation we adopt the following approach, known as *dovetailing*. Perform one step of the computation of $f(0)$. Store the new configuration and perform one step of the computation of $f(1)$, storing the new configuration. Return to $f(0)$ and perform another step. Continue in this fashion, bringing in new computation $f(2), f(3) \ldots$ until one of them terminates (if this ever happens), in which case we output the corresponding value of n. We describe a possible discipline for dovetailing the computations of $f(0), f(1), \ldots$ by the following table. The numbers on the columns represent steps in the computation, and the numbers in the table represent steps in the dovetailed computation.

	1	2	3	4	...
$f(0)$	1	3	6	10	...
$f(1)$	2	5	9	...	
$f(2)$	4	8	...		
$f(3)$	7	...			
$f(4)$	11	...			

Notice the similarity with the pairing function p of Section 1.1. Suppose $f(n)$ terminates. Then we have certainly found an n such that $f(n) \downarrow$. However it may not be the least, since it may be that $f(n-1) \downarrow$, only the computation takes more steps than for $f(n)$. On the other hand suppose that $f(n) \downarrow$ for some n, and that the computation takes t steps. Then the dovetailing computation must terminate, because if it does not, then eventually it will perform t steps of the computation of $f(n)$ which must force it to terminate. A more abstract approach to our search problem is to look for a pair n, y such that $T(e, n, y)$, where e is the code of a program for f. This is because we know that

$$f(n) \downarrow \text{ iff } \exists y. T(e, n, y)$$

(by Theorem 1.5.6). We can certainly find the least value of $p(m, n)$ such that $T(e, n, y)$, since T is decidable. Of course as before this may not be the least value of n such that $f(n) \downarrow$.

2.3 Recursive and recursively enumerable sets

In this subsection we examine the theory of computable sets. Of course the essence of set theory is the membership relation, and so the essential computational problem posed by a set is the problem of membership: given a set $A \subseteq \mathbf{N}$ and $n \in \mathbf{N}$, is n a member of A? Unlike functions, where there is a single notion of computability, for sets there are two forms of computability, corresponding to strong and weak solutions to the membership problem. We have already defined the strong version, namely recursiveness or decidability (Definition 1.4.6). For A to be decidable there must be a program which when given an input n will always terminate, and will output 'yes' if $n \in A$ and 'no' if $n \notin A$. The weaker version is called recursive enumerability, or semi-decidability. For A to be semi-decidable, there must be a program which given an input n will output 'yes' if $n \in A$ and will not terminate if $n \notin A$. So in the case $n \notin A$ we get no information from the program. The reason for the name 'semi-decidable' should be obvious from our informal definition. If the answer is 'yes' then we know we will eventually receive it, but if the answer is 'no' then we may have to wait for ever in a state of uncertainty. Why is this also called 'recursively enumerable' (usually abbreviated to r.e.)? This name reflects a different but equivalent definition. A set A is r.e. if its members can be generated (enumerated) as

$$a_0, a_1, a_2, \ldots, a_m, \ldots$$

by some program. Of course if A is infinite it is never generated fully, but we can be sure that any member of A will eventually be output by the program. So if $n \in A$ then we will eventually discover this. On the other hand, if $n \notin A$ then it will never appear in the sequence of outputs, but we never discover this and have to wait for ever. We hope that this explanation indicates why these two definitions are in fact equivalent. We now give precise definitions and proofs, first observing that many important computational problems are semi-decidable, such as the halting problem (Section 2.1) and theoremhood in first-order logic (see Section 4.4 below).

Definition 2.3.1. A set $A \subseteq \mathbf{N}$ is *recursively enumerable (r.e.) or semi-decidable* if its semi-characteristic function

$$s_A(n) = \begin{cases} 1 & n \in A \\ \uparrow & n \notin A \end{cases}$$

is (partial) recursive.

Proposition 2.3.2. *If a set is recursive then it is r.e.* ∎

Theorem 2.3.3. *The following are equivalent for $A \subseteq \mathbf{N}$:*

1. *A is r.e.*

2. *A is the domain of definition of a partial recursive function.*

Proof. (1) ⇒ (2) Obvious.
(2) ⇒ (1) Let f be partial recursive, and suppose that A is the domain of f. We can compute s_A by the following informal procedure: For any n, compute $f(n)$. If the computation terminates return 1. By Church's Thesis s_A is recursive. This is an example of proof by Church's Thesis, where its use is not essential. If we wished to take the trouble we could write down an explicit definition of s_A as a recursive function derived from f. ∎

Notation 2.3.4. Denote the domain of definition of the partial recursive unary function $\{e\}$ by W_e. In view of Theorem 2.3.3 any r.e. set is W_e for some e.

Theorem 2.3.5. *Let $A \subseteq \mathbf{N}$ be non-empty. A is r.e. iff A can be enumerated as the range of a total recursive function.*

Proof. ⇒ Suppose $A = W_e$. Then $x \in A$ iff $\exists y T(e, x, y)$. Pick any $k \in A$. Define $f : \mathbf{N} \to \mathbf{N}$ by

$$f(z) = \begin{cases} \text{left}(z) & \text{if } T(e, \text{ left}(z), \text{ right}(z)) \\ k & \text{otherwise} \end{cases}$$

This function is recursive (indeed primitive recursive). It is perhaps easier to understand if we write it as

$$f(p(x,y)) = \begin{cases} (x) & \text{if } T(e, x, y) \\ k & \text{otherwise} \end{cases}$$

(with a special case for $f(0)$). It is clear that if x is in the image set of f then $\exists y T(e, x, y)$ so that $x \in A$. Conversely, if $x \in A$ then there is some y such that $T(e, x, y)$, and then $f(p(x,y)) = x$, so that x is in the image set of f.

⇐ Let $f : \mathbf{N} \to \mathbf{N}$ be recursive and have image set A. Define $g(x) = \min z(f(z) = x)$. Then plainly $g(x) \downarrow$ iff xA. Also g is recursive since the recursive functions are closed under minimization. Hence A is the domain of a partial recursive function and so is r.e. by Theorem 2.3.3. ∎

Notation 2.3.6. For $A \subseteq \mathbf{N}$, let $\overline{A}$ denote the complement $\mathbf{N} - A$.

Theorem 2.3.7 (Post's Theorem). *A set $A \subseteq \mathbf{N}$ is recursive iff both A and $\overline{A}$ are r.e.*

Proof. ⇒ Assume A is recursive. Let P be a program computing the characteristic function ch_A, which outputs 1 if $n \in A$ and 0 if $n \notin A$. By modifying P so that instead of outputting 0 it goes into a loop we construct a program which computes s_A. By modifying P so that instead of outputting 1 it loops and instead of outputting 0 it outputs 1, we construct a program which computes $s_{\overline{A}}$.

⇐ Suppose both A and $\overline{A}$ are r.e. with their semi-characteristic functions computed by P, Q respectively. To compute ch_A we run both P and Q on input n *in parallel*. If P terminates ($n \in A$) then we output 1, and if Q terminates ($n \notin A$) then we output 0. By the definitions of s_A and $s_{\overline{A}}$, exactly one of P, Q must terminate. We have thus given an effective procedure for computing ch_A, which *by Church's Thesis* can be realised by a program in any of the programming formalisms we considered in Section 1. Again notice that we use Church's Thesis simply to shorten and clarify the proof. We could with more effort give an explicit construction of a Turing machine, recursive function, etc. ∎

Using Post's Theorem we can immediately show the existence of sets which are r.e. but not recursive, and sets which are not even r.e.

Notation 2.3.8. Let $K = \{x \mid \{x\}(x) \downarrow\}$.

By the unsolvability of the Halting Problem (Theorem 3.1.1) K is not recursive. However it is r.e., since it is the domain of the function $f(x) = \min y\, T(x, x, y)$. Now consider $\overline{K}$. If it is r.e. then by Post's Theorem K is recursive, which is a contradiction. So $\overline{K}$ is an example of a non-r.e. set.

As we have discussed, if an infinite set A is enumerated by some program, then we can know finitely that a certain number n is in A, but not that it is not in A. In other words the membership problem is semi-decidable. However if A is enumerated in increasing order we can also determine finitely that n does not belong to A. We examine the numbers output until either n appears or a number greater than n appears. One of these must eventually happen. In the former case $n \in A$ of course, whereas in the latter $n \notin A$. So A is decidable. This is the content of the next Theorem. The foregoing may remind the reader of the difference between searching ordered and unordered lists.

Theorem 2.3.9. *The range of a strictly increasing total recursive function is recursive.*

Proof. Let $f : \mathbf{N} \to \mathbf{N}$ be recursive with image set A. Define

$$h(n) = \min z.f(z) \geq n$$

h is defined by minimization over a recursive predicate, and is therefore recursive. It is also total, since f is total and the range of f is not bounded by any n. Plainly $n \in A$ iff $f(h(n)) = n$, which is clearly recursive. ∎

Exercise 2.3.10. Show that the intersection and union of any two recursive sets is recursive. Similarly for r.e. sets.

The definition of r.e. may be extended to sets of tuples of natural numbers by coding the tuples as natural numbers.

Definition 2.3.11.

$x \subseteq \mathbf{N}^k$ is r.e. iff $\{< x_1, \ldots, x_k >\mid (x_1, \ldots, x_k) \in X\}$ is r.e.

Exercise 2.3.12. Let $f : \mathbf{N}^k \to_p \mathbf{N}$. The *graph* of f is the set

$$X = \{(x_1, \ldots, x_k, y) \mid f(x_1, \ldots, x_k) = y\}$$

Show that f is recursive iff the graph of f is r.e.
(Hint: You may wish to use dovetailing (Example 2.2.2) for the $\Leftarrow$ direction.)

Theorem 2.3.13. *Let $R \subseteq \mathbf{N}^k$ be r.e. and let $S \subseteq \mathbf{N}^k$ be defined by*

$$S(\mathbf{x}) \text{ iff } \exists y R(\mathbf{x}, y)$$

Then S is r.e.

Proof. R is the domain of a partial recursive function, and so we have

$$R(\mathbf{x}, y) \text{ iff } \exists z T(e, < \mathbf{x}, y >, z)$$

for some index e. Let

$$f(\mathbf{x}) = \ \text{min} w.T(e, < \mathbf{x}, \ \text{left}(w) >, \ \text{right}(w))$$

Then f is (partial) recursive and it is not hard to see that

$$S(\mathbf{x}) \text{ iff } f(\mathbf{x}) \downarrow$$

Hence S is r.e. ∎

2.4 The S–m–n theorem and partial evaluation

In this subsection we discuss the S–m–n theorem. This is an extremely simple technical result concerning multiple inputs to programs. We shall need it for establishing further undecidability results (Section 2.5) and for the fixed point theorems (Section 4.1). It has recently acquired a greater significance for computing science as a simple form of partial evaluation of functional programs. Imagine that we wish to run a program P which expects two inputs, I_1 and I_2. It may be that having received the first input, a say, we can make some progress on the computation before receiving the second input, b say. Once we know a we may be able to transform or optimize the program so that it will run more efficiently whatever the value of b. One way to think about this process of *partial evaluation* is to regard us as having transformed the original program P into a new program Q, which depends on P and a. Of course the real interest in partial evaluation lies in getting Q such that for any b, Q on b runs faster than P on inputs a and b. But there is always a 'trivial' way in which Q can be produced. Informally Q will be the following program which expects a single input:

Generate a;
Input generated value for I_1;
Input I_2;
P

The following diagram may make this clearer:

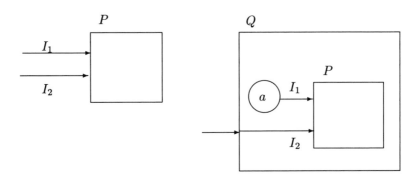

Fig. 2.1

This transformation of P into Q is quite easy to achieve with Turing machines. P expects an input string of the form $1^{a+1}B1^{b+1}$ (representing the input a, b). Q is the same as P except that before running P it changes an input 1^{b+1} (any b) into $1^{a+1}B1^{b+1}$, leaving the head pointing to the leftmost 1. Essentially all Q has to do is to move leftwards leaving one blank and then printing $a + 1$ 1's. Obviously Q depends on a.

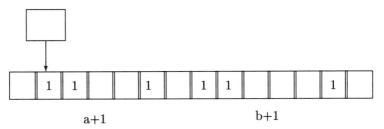

Fig. 2.2

(We note in passing that things are equally easy with register machines. Suppose that P is run with a, b in registers $0, 1$, and with all other registers set to 0. Q is run with b in register 0 and all other registers set to 0. First b is moved from register 0 to 1, and then a is entered in register 0 by simply adding 1 to it a times).

It should be plain that (a code for) Q can be obtained effectively from (a code for) P and a. In fact this function — call it S — is primitive recursive, as are all functions which relate to syntactic and coding matters, though we shall not take the time and effort to show this. The upshot of this is that for any program index e and any inputs x, y

$$\{e\}_2(x, y) = \{S(e, x)\}_1(y)$$

This can be generalized to arbitrary numbers of inputs in an obvious way:

Theorem 2.4.1 (S–m–n Theorem). *Let $m, n \geq 1$. There is a primitive recursive function S_n^m such that for any index e and m–tuple $\mathbf{x}$, $n - -$tuple $\mathbf{y}$,*

$$\{e\}_{m+n}(\mathbf{x}, \mathbf{y}) = \{S_n^m(e, \mathbf{x})\}_n(\mathbf{y})$$

2.5 More undecidable problems

Many problems associated with programs are really about the functions which they calculate. For instance we might ask whether two programs compute the same function, or whether a given program computes a total function. Are such problems solvable? We can give a rather sweeping negative answer to this question using a powerful theorem we shall now give: any non-trivial property of computable functions is undecidable. More precisely:

Theorem 2.5.1 (Rice's Theorem). *Let $B \subseteq C_1$ (the class of unary partial computable functions) and suppose $B \neq \varnothing, C_1$. Then $\{x \in \mathbf{N} \mid \{x\} \in B\}$ is not recursive (in other words $?\{x\} \in B?$ is unsolvable).*

Proof. (Based on [Rogers, 1967; Cutland, 1980] We are told that both B and $C_1 - B$ are non-empty. The totally undefined function $\uparrow$ must be in one of $B, C_1 - B$. Let h be some function not in the same set as $\uparrow$. Assume without loss of generality that $h \in B, \uparrow \notin B$; if in fact $\uparrow \in B, h \notin B$ the proof is essentially the same. Define

$$f(x, y) = \begin{cases} h(y) & \{x\}(x) \downarrow \\ \uparrow & \text{otherwise} \end{cases}$$

Then f is computable. Notice that this does not entail solving the halting problem, since f is left undefined if $\{x\}(x) \uparrow$. By the S–m–n theorem $f(x, y) = \{g(x)\}(y)$ for some primitive recursive function g. Now if $\{x\}(x) \downarrow$ then $\{g(x)\}(y) = h(y)$ for all y, and if $\{x\}(x) \uparrow$ then $\{g(x)\}(y) = \uparrow$ for all y. Hence

$$\{x\}(x) \downarrow \text{ iff } \{g(x)\} \in B$$

and we have reduced the halting problem to $?\{x\} \in B?$ It follows that $?\{x\} \in B?$ is unsolvable. ∎

Rice's Theorem can be applied to such examples as the following:

Example 2.5.2. $?\{x\}(0) \downarrow?$ is unsolvable.

I. C. C. Phillips

This is essentially the same problem as that of deciding whether a given Turing machine will halt when started on an empty tape.

Example 2.5.3. $?\{x\}$ is a total function ? is not even semi-decidable. It is immediate from Rice's Theorem that the problem is unsolvable but we wish to show a stronger result. So we use the S–m–n Theorem to give a direct reduction of the complement of the self-halting problem (which we know is not r.e.) to the total function problem. Define

$$f(x,t) = \begin{cases} 1 & \text{if } \{x\}(x) \text{ does not halt in } \leq t \text{ steps} \\ \uparrow & \text{otherwise} \end{cases}$$

Then f is computable. By the S–m–n Theorem there is a computable g such that $f(x,t) = \{g(x)\}(t)$. Notice that if $\{x\}(x) \uparrow$ then $\{g(x)\}$ is total, whereas if $\{x\}(x) \downarrow$ then $\{g(x)\}$ is undefined above a certain value. Therefore if the total function problem is r.e. then so is the problem $?\{x\}(x) \uparrow$? We have the required reduction, and can now deduce that the total function problem is not r.e.

Exercise 2.5.4. Deduce from Example 2.5.3 that the problem of deciding whether two programs compute the same function, $?\{x\} = \{y\}$? is not semi-decidable.

Example 2.5.5 (Hilbert's Tenth Problem). A *diophantine equation* is an equation of the form $p(\mathbf{x}) = 0$, where $p(\mathbf{x})$ is a polynomial in variables $\mathbf{x}$ with integer coefficients, for which we are interested in solutions in integers. For instance $x^2 - x - 6 = 0$ is soluble (put $x = 3$), but $x^2 - 5$ is not. *Hilbert's Tenth Problem* is the problem of determining for a given diophantine equation whether it has a solution. The word 'problem' is being used in a different sense here from that of decision problem. When Hilbert proposed his famous set of problems in 1900, the only solution that could be imagined was to provide an algorithm [Davis, 1958]. It was only after the work of Turing and others that mathematicians could envisage showing the problem to be unsolvable, so that no such algorithm existed. After substantial effort by a number of mathematicians, the problem was eventually shown to be undecidable by Matiyasevich in 1970. See [Bell and Machover, 1977] for the proof.

2.6   Problem reduction and r.e. completeness

We have seen a number of examples of one problem being reduced to another. For instance the self-halting problem was reduced to the general halting problem (Corollary 2.1.3), and the complement of the self-halting problem was reduced to the total function problem (Example 2.5.3). All the examples we have considered are captured by the following definition:

Definition 2.6.1. Let $A, B \subseteq \mathbf{N}$. A is *many–one reducible* to B, written $A \leq_m B$ iff there is a total recursive function $f : \mathbf{N} \to \mathbf{N}$ such that for all x,

$$x \in A \text{ iff } f(x) \in B$$

If both $A \leq_m B$ and $B \leq_m A$ we write $A \equiv_m B$.

The interpretation is that A is no harder to compute than B, since given a program to compute B one can compute A using the computable function f. The name 'many–one' refers to the reduction function f, and is to distinguish this notion from Turing reduction, which we shall define below.

Remark 2.6.2. A problem such as the the halting problem is more naturally associated with a subset of $\mathbf{N}^2$ rather than $\mathbf{N}$, but we can use our pairing function and sequence coding to coerce a subset A of $\mathbf{N}^k$ into a subset A' of $\mathbf{N}$. Thus

$$(x_1, \ldots, x_k) \in A \text{ iff } x_1, \ldots, x_k \in A'$$

It is easy to check that $\leq_m$ is a pre-ordering. It is not a partial ordering in view of the following.

Proposition 2.6.3. *Let $A, B \subseteq \mathbf{N}$ be such that A is recursive and B is non-trivial, that is $\neq \varnothing, \mathbf{N}$. Then $A \leq_m B$.*

Proof. Exercise. ∎

Corollary 2.6.4. *If A and B are recursive and non-trivial, then $A \equiv_m B$.*

Proposition 2.6.5. *If $A \leq_m B$ and B is recursive then A is recursive.*

We see that if A is recursive and B is r.e. but not recursive then $A <_m B$. This is because $A \leq_m B$ by Proposition 2.6.3, but not $B \leq_m A$, or else B would be recursive by Proposition 2.6.5. We have an example of such a set B in the shape of K, the set associated with the self-halting problem. It turns out that K is maximal among r.e. sets for $\leq_m$, in other words any r.e. problem can be reduced to K.

Definition 2.6.6. Let A be an r.e. set. A is *r.e. complete* iff for any r.e. set B, $B \leq_m A$.

Theorem 2.6.7. *K is r.e. complete.*

Proof. Similar to that of Rice's Theorem 2.5.1. Let A be r.e. Then $x \in A$ iff $\{e\}(x) \downarrow$ for some e. Define

$$f(x,y) = \begin{cases} 1 & \{e\}(x) \downarrow \\ \uparrow & \text{otherwise} \end{cases}$$

Then f is computable and by the S–m–n theorem $f(x,y) = \{g(x)\}(y)$ for some recursive function g. Now if $\{e\}(x) \downarrow$ then $\{g(x)\}(y) = 1$ for all y, and if $\{e\}(x) \uparrow$ then $\{g(x)\}(y) = \uparrow$ for all y. Hence

$$\cdot \; x \in A \quad \text{iff} \; \{g(x)\}(g(x)) \downarrow$$
$$\text{iff} \; g(x) \in K$$

∎

Post (1944) investigated the question of whether there are r.e. sets which are neither recursive nor r.e. complete. He was able to show that there are. We refer the reader to [Rogers, 1967, Chapter 8] for the details. It remains true that all the r.e. sets we are likely to encounter in connection with computational problems will be either recursive or r.e.-complete.

The picture we have presented so far is very simplified. In fact many notions of reducibility have been studied. The most important of these is *Turing reducibility*, which we now briefly describe. Intuitively A is Turing reducible to B, written $A \leq_T B$ if A can be computed relative to B, that is, if we are given access to B then we can compute A. What does 'access to B' mean? The reader is reminded of the discussion of oracle Turing machines at the end of Section 1.2.1. We are to imagine that B is an oracle to the computation, so that in the course of computing whether a given n is a member of A, we are supplied with the answer to any question of the form 'does $x \in B$?' without any computational effort. Turing reducibility has all the properties of many–one reducibility which we described above. In particular Proposition 2.6.5 holds with $\leq_m$ replaced by $\leq_T$, so that recursive oracles do not allow us to compute any new sets. The relationship between our two forms of reducibility is easy to establish:

Proposition 2.6.8. *For any A, B, if $A \leq_m B$ then $A \leq_T B$.*

Proof. Suppose that there is a total recursive function $f : \mathbf{N} \to \mathbf{N}$ such that $x \in A$ iff $f(x) \in B$. Then to decide whether $x \in A$, we compute $f(x)$ and consult the oracle as to whether $f(x) \in B$. It follows that A is decidable relative to an oracle for B. ∎

For a trivial example of how the notions differ, consider the sets $\varnothing, \mathbf{N}$. It is easy to see that neither can be m-reduced to the other. However since an oracle for any set A becomes an oracle for its complement $\overline{A}$ by swapping 'yes' and 'no', we have the following:

Proposition 2.6.9. *For any A, $A \equiv_T \overline{A}$.*

Example 2.6.10. For a more interesting distinction consider K and $\overline{K}$. These are T-equivalent in view of the preceding, but m-incomparable. This follows easily from the following, since we know that $\overline{K}$ is not r.e.

Proposition 2.6.11. *If $A \leq_m B$ and B is r.e. then A is r.e.*

The result is false for T-reducibility in view of our example. Post also investigated the question of whether there are r.e. sets which are neither recursive nor r.e. T-complete. This proved much harder than the corresponding question for m-reducibility, and became known as Post's Problem. It was eventually solved independently in 1956 by Friedberg and Muchnik, who showed the existence of two r.e. sets which are T-incomparable. We refer the reader to [Rogers, 1967] for further details on all the matters discussed in this section.

3 Inductive definitions

Infinite objects occur widely in computing science. Some obvious examples are data types such as the integers, or functions on such types. If these are to be computable then we must be able to grasp them in a systematic way by finite means. This will mean writing a program to define them (with 'program' possibly interpreted in a wider sense than usual), which will in many cases involve the use of recursion. This section outlines a general theory into which such definitions can be fitted which shows how they can be given a precise meaning. This should serve to sharpen intuition about some subtleties concerning the use of recursions, both in programming and perhaps even more so in specification. It will also be a necessary preliminary to understanding any semantics of a programming language, in which programs are mapped in a structured way into some semantic domain of mathematical objects. We shall have a simple example of this in Section 3.2. Sections 3.3 and 3.4 contain more advanced material on ordinals and fixed points of non-continuous operators. This may be omitted if desired.

3.1   Operators and fixed points

3.1.1   An example

Here is the syntax of the simple language of while programs from Section 1.2.2, given in Backus–Naur form (BNF):

$$P ::= C \mid P; P \mid \text{if } T \text{ then } P \text{ else } P \text{ fi} \mid \text{while } B \text{ do } P \text{ od}$$

Here C ranges over Cmd (commands) and T over Tst (tests). Both of these are assumed to have been previously defined. In the terminology of data types our BNF definition gives us *constructors* for programs — nullary constructors (each $C \in$ Cmd) which create commands 'out of nothing' and constructors of higher arity which create new commands out of existing ones (sequential composition, conditional, while).

The point of the example is that the set of programs Prog, ranged over by P, is defined in terms of itself. Notice also that we are defining an infinite set by finite means.

In formal language theory our simple BNF definition is a grammar for a context-free language. We are using it as a simple example to motivate a much more general concept — that of an *inductive definition*. Such definitions crop up frequently in computing science. They are definitions which refer to the thing which they define, in recursive fashion. They are typically needed when defining infinite sets constructively (and therefore by finite means). Even when an infinite set is defined explicitly, as for example

$$\text{Even} = \{n \in \text{Nat} \mid n \text{ is divisible by } 2\}$$

then it will tend to depend on a pre-defined infinite set (in this case Nat) which needed an inductive definition.

It is therefore worthwhile for us to examine what the meaning of such definitions should be, and also to what extent they define computable objects. We shall arrive at a quite abstract formulation to take advantage of the fact that many of the ideas are of very wide application.

Let us return to our simple BNF example. The set we are defining is the one got by repeatedly applying the constructors until we get nothing new. We therefore want Prog to be closed under the constructors. We could write out the definition in more detail as a series of clauses

$$\left.\begin{array}{l} \text{Cmd} \subseteq \text{Prog} \\ \text{if } T \in \text{Tst}, P_1, P_2 \in \text{Prog then} \\ \quad \text{if } T \text{ then } P_1 \text{ else } P_2 \text{ fi} \in \text{Prog} \\ \quad \text{while } B \text{ do } P_1 \text{ od} \in \text{Prog} \\ \quad P_1; P_2 \in \text{Prog} \end{array}\right\} (*)$$

This certainly puts any term built up from the constructors into Prog. However what about a term such as 'while skip do true', which we certainly don't want? Our rewritten definition (∗) does not exclude it. We therefore add the extra condition that nothing gets into Prog except by virtue of one of the clauses of (∗) (this was implicit in the BNF definition). Another way of putting this is that Prog is the *least* set defined by (∗).

Let us abstract away somewhat from the particular details of the definition to see what is happening in general. Define a single command-building operation:

$$\Gamma(X) = \quad \text{Cmd} \cup$$
$$\{\text{if } T \text{ then } x \text{ else } y \text{ fi, while } T \text{ do } x \text{ od}, x; y \mid T \in \text{ Tst}, x, y \in X\}$$

Γ acts on sets unlike the constructors. We can clearly rephrase (∗) as $\Gamma \,(\text{Prog}) \subseteq \text{Prog}$, in other words Prog is closed under Γ . So our definition can be expressed as:

Prog is the least set X such that $\Gamma(X) \subseteq X$

There is an alternative way of obtaining Prog from Γ . W can build up Prog in stages as follows. Define sets $X_n(n \geq 0)$ by

$$X_0 = \varnothing$$
$$X_{n+1} = \Gamma(X_n)$$

It should be intuitively clear that $C_n \subseteq \text{Prog}$ (each n), and moreover that any member of Prog is got by some finite number of applications of Γ to the empty set. Hence $\text{Prog} = \bigcup\{X_n \mid n \geq 0\}$.

It is no accident that these two different ways of arriving at Prog give the same results, as we shall now see. It is useful to be very general in our approach and introduce the notions of complete partial orders and fixed points of monotonic functions.

3.1.2 Complete partial orders and least fixed points

First recall that a *partial order* (po) is a set A together with a binary relation $\sqsubseteq$ on A which is reflexive, transitive and anti-symmetric (viz. for all $a, b \in A$, if $a \sqsubseteq b$ and $b \sqsubseteq a$ then $a = b$). We denote such an ordering by $(A, \sqsubseteq)$. A *linear* (or total) order is a po $(A, \sqsubseteq)$ which has the additional property that for all $a, b \in A, a \sqsubseteq b$ or $b \sqsubseteq a$.

Definition 3.1.1. Let $(A, \sqsubseteq)$ be a po and let $B \subseteq A$. An element $a \in A$ is an *upper bound* for B iff for all $b \in B, b \sqsubseteq a$. Term a is a *least* upper bound (lub) or *supremum* for B if it is an upper bound and for all a', if a' is an upper bound for B then $a \sqsubseteq a'$. We write $a = \bigsqcup B$. It is plain using the anti-symmetry of $\sqsubseteq$ that lubs *if they exist* are unique. A set $C \subseteq A$ is a *chain* if it is linearly ordered by $\sqsubseteq$. Furthermore $(A, \sqsubseteq)$ is *complete* if it has a least element and every chain $C \subseteq A$ has a lub. Complete partial order is usually abbreviated to cpo. Least elements are often denoted by $\bot$. Notice that $\bot = \bigsqcup \varnothing$ since every member of A is trivially an upper bound for $\varnothing$, so that we can infer the existence of a least element. Of course any *finite* chain has a top element which will automatically be the supremum, so that completeness is only of significance with infinite chains.

Remark 3.1.2. There are variant definitions of completeness involving chains indexed by the natural numbers and directed sets.

A commonly arising example of a cpo is $(\mathbf{P}S, \subseteq)$ where S is any set. Here the supremum $\bigsqcup$ is just the set-theoretic union $\bigcup$.

Definition 3.1.3. Let $(A, \sqsubseteq)$ and $(B, \sqsubseteq)$ be cpo's. A function $f : A \to B$ is *monotonic* if for all $a, a' \in A$, if $a \sqsubseteq a'$ then $f(a) \sqsubseteq f(a')$. Function f is *continuous* if for all chains $C \subseteq A$,

$$\bigsqcup f(C) \text{ exists and } f(\bigsqcup C) = \bigsqcup f(C)$$

where of course $f(C)$ abbreviates $\{f(a) \mid a \in C\}$.

Remark 3.1.4. It is not hard to check that if f is monotonic then $\bigsqcup f(C)$ exists and $\bigsqcup f(C) \sqsubseteq f(\bigsqcup C)$. Furthermore it is an easy exercise to check that if f is continuous then it is also monotonic.

Let us see how this applies to our example: choose some large enough set T of terms — for instance all finite strings which can be typed at the keyboard (the particular choice is irrelevant). Then we are attempting to define Prog as a subset of T. Now $(\mathbf{P}T, \subseteq)$ is a cpo and Γ may be viewed as a function from T to itself which it is easy to see is monotonic.

We claim that Γ is also continuous: take a chain $C \subseteq \mathbf{P}T$. By the Remark we must show $\Gamma(\bigcup C) \subseteq \bigcup \Gamma(C)$. So take $t \in \Gamma(\bigcup C)$. Observe that t is formed using a single constructor from at most two members of $\bigcup C$. Hence there is a finite set $X \subseteq \bigcup C$ such that $t \in \Gamma(X)$. Now for each member x of X pick some $Z_x \in C$ such that $x \in Z_x$. Then $\{Z_x \mid x \in X\}$ forms a finite chain which must have a supremum $Y \in C$. Plainly $X \subseteq Y$ and so $t \in \Gamma(X)$ and $t \in \bigcup \Gamma(C)$ as required.

This argument can of course be used quite generally:

Criterion 3.1.5. Let $(\mathbf{P}B, \subseteq)$ be a cpo and let $f : \mathbf{P}B \to \mathbf{P}B$ be a monotonic function. Then f is continuous provided that whenever $b \in f(Y)$ then there is a finite $X \subseteq Y$ such that $b \in f(X)$.

In the light of the criterion we see that we naturally expect that a computable operation on sets should be continuous — an element should be included on the basis of only a finite amount of information.

Definition 3.1.6. Let $(A, \sqsubseteq)$ be a cpo and let $f : A \to A$. Element $a \in A$ is a *fixed point* (fp) of f if $f(a) = a$. Term a is a *least* fixed point (lfp) of f if it is a fixed point and if b is any other fixed point then $a \sqsubseteq b$. The lfp of f is sometimes denoted μf. Element $a \in A$ is a *pre-fixed point* of f if $f(a) \sqsubseteq a$.

Of course least fixed points if they exist are unique.

Theorem 3.1.7 (Knaster–Tarski Fixed Point Theorem). *Let $(A, \sqsubseteq)$ be a cpo and let $f : A \to A$ be continuous. Then f has a lfp which is also the least pre-fixed point.*

Proof. We shall construct μf as the lub of a chain. Define $x_n (n \geq 0)$ by

$$x_0 = \bot$$
$$x_{n+1} = f(x_n)$$

Claim. $x_n \sqsubseteq x_{n+1}$.

By induction. Certainly $x_0 \sqsubseteq x_1$ since $\bot$ is least. Assume that $x_n \sqsubseteq x_{n+1}$. Then $f(x_n) \sqsubseteq f(x_{n+1})$ since f is monotonic. So $x_{n+1} \sqsubseteq x_{n+2}$ as required.

By the claim, $C = \{x_n \mid n \geq 0\}$ is a chain and has a lub $\bigsqcup C$.

Claim. $\bigsqcup C = \bigsqcup f(C)$.

By the previous claim $f(C) = \{x_n \mid n \geq 1\}$. This is also a chain and clearly has the same lub as C.

Now by continuity of f we have $f(\bigsqcup C) = \bigsqcup f(C)$. Combining this with the claim gives $f(\bigsqcup C) = \bigsqcup C$.

It remains to show that $\bigsqcup C$ is the least pre-fixed point, which will entail that it is the lfp. So take any y such that $f(y) \sqsubseteq y$. We show by induction that $x_n \sqsubseteq y$ for all n. Clearly $x_0 \sqsubseteq y$. Assume $x_n \sqsubseteq y$. Then by monotonicity $f(x_n) \sqsubseteq f(y)$. But $x_{n+1} = f(x_n)$ and $f(y) \sqsubseteq y$. Hence $x_{n+1} \sqsubseteq y$ and the induction is complete. So y is an upper bound for C and therefore $\bigsqcup C \sqsubseteq y$ by definition. ∎

Returning to the BNF example we note that we defined Prog as the least pre-fixed point of Γ. Theorem 3.1.7 tells us that such a point exists and moreover the proof shows that it is equal to the set derived by building up Prog 'in stages'.

Inductive definitions arise in many areas of computing science. As well as BNF syntax definitions we may mention transitive closure of a relation, theoremhood in a logic and structured operational semantics of programming languages.

The BNF example can be connected with logic in a suggestive way. We can think of the nullary constructors as corresponding to axioms — thus $C \in$ Prog can be deduced from no assumptions if $C \in$ Cmd. The non-nullary constructors correspond to rules. Thus we may write

$$\frac{P_1 \in \text{Prog} \qquad P_2 \in \text{Prog}}{\text{if } B \text{ then } P_1 \text{ else } P_2 \in \text{Prog}}$$

It should be clear from this example how to view a logic presented in terms of axioms and rules as an inductive definition. The introduction rules of a natural deduction system correspond loosely to constructors. However logics may of course contain other types of rule (such as eliminations and cuts) so that we need our more general framework to view them as inductive definitions.

With any inductive definition in constructor form is associated a principle of *structural induction*. This allows us to prove an assertion that all members of the lfp satisfy some property by finite means, following the structure of the definition. To take a well-known example, the natural numbers Nat may be defined by

$$0 \in \text{Nat}$$
$$\text{if } n \in \text{Nat then } S(n) \in \text{Nat}$$

(where S is successor — $S(n) = n + 1$). Here the constructors are $0, S$. The principle of induction over Nat states

If $P(0)$
and $\forall n(P(n) \rightarrow P(S(n)))$
then $\forall n P(n)$

where P is some property of natural numbers. Notice that there is one 'proof-obligation' for each constructor. The reader is invited to formulate the structural induction principle associated with our definition of Prog.

How may this be justified in terms of our understanding of inductive definitions as lfp's? Structural induction obviously requires us to prove that the property P is preserved by the constructors — for instance, if $P(n)$ then $P(S(n))$. In terms of the general theory with inductive operator Γ this is simply $\Gamma(P) \sqsubseteq P$. We arrive at:

Principle 3.1.8 (Park Induction). Let $(A, \sqsubseteq)$ be a cpo and let $\Gamma : A \rightarrow A$ be continuous. For any $P \in A$, if $\Gamma(P) \sqsubseteq P$ then $\mu\Gamma \sqsubseteq P$.

Proof. P is a pre-fixed point and so this is clear by Theorem 3.1.7. ∎

The topic of structural induction will be treated more rigorously in the next chapter [Meinke and Tucker, 1992].

Remark 3.1.9. Constructor-form inductive definitions are important not only because they are particularly common and clear examples of inductive definitions. With any inductive definition Γ, a natural question to ask is, how can one tell whether a given object is a member of $\mu\Gamma$. If Γ can be presented in constructor form there is an obvious terminating algorithm for determining membership. We illustrate it using our BNF example. To discover whether some term t is a member of Prog, we see whether we can match t with one of the patterns given. If for instance t is 'while T do t'' then we have partially succeeded and can go on to test whether t' is in Prog. This procedure is of course redeemed from circularity because t' is a simpler term than t and we shall eventually push back beyond the constructors of arity > 0 to the nullary constructors, at which point no further recursive call of our procedure is necessary. The existence of such an algorithm, which we may call a decision procedure for the inductive definition, is of course highly relevant to the writing of parsers for programming languages. We shall later address the issue of deciding membership for inductive definitions in general, and we shall see that we cannot expect a terminating algorithm to exist (in particular we cannot 'decide' theoremhood for a number of important logics).

3.1.3 Greatest fixed points

Usually the set being defined is the least fp of the associated inductive operator, and some authors take this to be 'the' meaning of an inductive definition (e.g. [Moschovakis, 1974]). However there are important examples in computing where *greatest* fixed points are called for. We give one such now concerning specification of non-deterministic processes (our example uses in effect a simplified recursive Hennessy-Milner logic – see e.g. [Bradfield and Stirling, 1990]; a logician might see it in terms of Kripke frames and modal logic – [Stirling, 1992]). We shall demonstrate that certain natural properties can be specified as least fp's and others as greatest fp's.

For our example we shall view a process as something which can move between states. Mathematically we shall have a set P of states and a binary transition relation $\rightarrow$ on P. We identify a process-in-a-state with a process, so that when we write $p \rightarrow q$ we think of process p evolving into process q. We are interested in formulating *properties* of processes and we shall illustrate how inductive definitions play their part in this. Note that we shall not distinguish between properties and the sets they define (their extensions). Formally we shall build up a logic, with formulas ranged over by φ and a satisfaction relation $\models$ on processes and formulas with $p \models \varphi$ meaning that φ is true of p. Thus

$$p \models A \text{ iff } A(p) \qquad (A \text{ a primitive predicate})$$
$$p \models \varphi \wedge \psi \text{ iff } p \models \varphi \text{ and } p \models \psi$$

and similarly for the other propositional connectives. We also use a 'next stage' operator next:

$$p \vDash \text{next}(\varphi) \text{ iff } q \vDash \varphi \text{ for all } q \text{ such that } p \to q$$

For the moment we assume that P is finite branching, namely for any p there are only finitely many q such that $p \to q$.

We hope that the reader will not be too confused if we note that we have defined the satisfaction relation $\vDash$ inductively. The reader may care to work out what operator $\vDash$ is the least pre-fixed point of. What set does the operator act on?

Consider the formula next(ff) where ff means falsity — for no p does $p \vDash$ ff. Plainly $p \vDash \text{next}(\text{ff})$ iff for no q does $p \to q$, in other words p is a terminal process that can do nothing. We would like to express the weaker property that a process will *eventually* terminate, in other words it cannot perform an infinite series of transitions. A little thought shows that next(ff) $\lor$ next(next(ff)) expresses 'must terminate immediately or on the next step'. We are therefore looking for something like

$$\text{next}(\text{ff}) \lor \text{next}(\text{next}(\text{ff}) \lor \text{next}(\text{next}(\text{ff}) \lor \ldots))$$

Unfortunately we are not allowed to form infinite formulas and it is not clear what they should mean. But the formula does remind us of the chain of approximations to a least fixed point which we constructed. Let us try another tack. We want to define a set E of processes which satisfies

if p terminates immediately (viz. $p \vDash \text{next}(\text{ff})$) then $p \in E$
if for every q s.t. $p \to q, q \in E$ then $p \in E$

and we want E to be the *least* such set. We would like a way of expressing this as a logical formula. The property E we want is the *strongest* property X such that

$$\text{next}(\text{ff}) \to X$$
$$\text{next}(X) \to X$$

We may rewrite this as

$$X \leftarrow (\text{next}(\text{ff}) \lor \text{next}(X)) \tag{$\dagger$}$$

In other words we want the least pre-fixed point of the operator Γ on properties defined by $\Gamma(X) = \text{next}(\text{ff}) \lor \text{next}(X)$. It is easy to see that next and $\lor$ are monotonic. Hence Γ is monotonic since the composition of monotonic functions is monotonic. We remark that not all logical operators are monotonic, negation being an obvious example. Let us check that Γ is continuous. Suppose $p \in \Gamma(X)$. Then either $p \vDash \text{next}(\text{ff})$ in which case $p \in \Gamma(\varnothing)$ or else for all q s.t. $p \to q, q \in X$. In this case let $Q = \{q \mid p \to q\}$. We have assumed that Q is finite, and clearly $Q \subseteq X, p \in \Gamma(Q)$. Hence Γ is continuous by Criterion 3.1.5.

Thus by Theorem 3.1.7 we know that the lfp $\mu\Gamma$, which we wanted to express eventual termination, will exist. If we don't wish to make the operator Γ explicit there are a couple of notations in common use: we may write $\mu X.\,\text{next}(\text{ff}) \vee \text{next}(X)$, where X is a bound variable, or if we wish to name the lfp we can write

$$E \Leftarrow \text{next}(\text{ff}) \vee \text{next}(E)$$

The backwards arrow is explained by formula (†).

It is worth noting that E can in fact be defined more simply. Let F be given by

$$F \Leftarrow \text{next}(F)$$

In fact $E = F$. To prove this it is enough to show that E, F satisfy each other's defining implication. It is clear that $\text{next}(E) \to E$, showing that F is at least as strong as E since F is the *least* x such that $\text{next}(x) \to x$. For the converse, we know $\text{ff} \to F$. Since next is monotonic we have $\text{next}(\text{ff}) \to \text{next}(F)$. But $\text{next}(F) \to F$. Hence $(\text{next}(\text{ff}) \vee \text{next}(F)) \to F$ as required.

Remark 3.1.10. The proof of Theorem 3.1.7 shows that E is equivalent to the infinite disjunction

$$\text{ff} \vee \text{next}(\text{ff}) \vee \text{next}(\text{next}(\text{ff})) \vee \ldots$$

Now suppose that we wish to specify that a process should always satisfy some property R, no matter what state it reaches. This could be for instance that a certain register is ≥ 0, or that a certain component does not overheat, or a host of other things. Informally we could attempt to express this by the infinite conjunction

$$R \wedge \text{next}(R) \wedge \text{next}(\text{next}(R)) \wedge \ldots$$

Let us try to see this as an inductive definition. The property I we are defining should satisfy

if $p \vDash I$ then $p \vDash R$
if $p \vDash I$ then $q \vDash I$ for all q s.t. $p \to q$

This may be written more briefly as

$$I \to (R \wedge \text{next}(I))$$

Consider the operator $\Phi(X) = R \wedge \text{next}(X)$. This is monotonic and continuous just like Γ above. But unlike above, where we were looking for E such that $\Gamma(E) \to E$, we want $I \to \Phi(I)$. The strongest property for which this holds is ff, in other words the empty set of processes, which is plainly not what we want. Suppose that A is any set of processes which is included in R and satisfies $A \subseteq$ next(A). Take $p \in A$. Then any process which may evolve from p is also in A, using $A \subseteq$ next(A) as many times as needed. So all these processes satisfy R and p satisfies I. We conclude from this that if $X \to \Phi(X)$ then $X \to I$. In other words I is the weakest property satisfying $X \to \Phi(X)$. Let us examine under what circumstances such 'greatest post-fixed points' exist.

We can simply mimic the approach we took for least fp's by "turning the po upside-down". Instead of requiring lub's of ascending chains we require greatest lower bounds (glb's) of descending chains.

Theorem 3.1.11 (cf Theorem 3.1.7). *Let $(A, \sqsubseteq)$ be a po with a greatest element $\top$ and such that glb's of descending chains exist (in other words an upside-down cpo). Let $f : A \to A$ be monotonic and for all descending chains $C \subseteq A$, $f(\sqcap C) = \sqcap f(C)$ (in other words f is upside-down continuous). Then f has a gfp (sometimes denoted νf) which is also the greatest post-fixed point.*

Proof. Same as for Theorem 3.1.7 *mutatis mutandis*.   ∎

There is an alternative approach to the existence of gfp's which will be explored in Section 3.4.

Example 3.1.12. Now let us apply Theorem 3.1.11 to our example. We wanted the gfp of the operator $\Phi : \mathbf{P}P \to \mathbf{P}P$ given by $\Phi(X) = R \wedge$ next(X). The po $(\mathbf{P}P, \subseteq)$ has P as top element arbitrary meets (intersections) and so satisfies the conditions of the Theorem. It is straightforward to check that Φ preserves meets of descending chains; we conclude that $\nu\Phi$ exists and is equal to

$$R \wedge \text{ next}(R) \wedge \text{ next(next}(R)) \wedge \ldots$$

Note that we did not require $\to$ to be finite branching.

A property such as E is sometimes referred to as a *liveness* property (Lamport), namely one that states that something good must eventually happen. These are lfp's. On the other hand I is a *safety* property, since it states that something bad (in this case the failure of R) never happens. Such properties are gfp's.

What happens if Γ is not continuous? Then it is not necessarily the case that $\bigsqcup X_n$ is fixed by Γ. However we can try forming $\bigsqcup X_n$ anyway and apply Γ to it again. It turns out that using this process of repeatedly applying Γ together with forming unions will always construct the least fixed point of Γ. However before we discuss this further we need to have a language for describing precisely what we are doing. We shall index the approximations to the lfp of Γ which we are building up by the *ordinals*, which we introduce in the next subsection.

To sum up this subsection, we have seen how infinite sets which arise naturally in computing can be described in a finite clear and compact way as fixed points of inductive definitions, and we have discovered conditions under which such fixed points exist.

3.2 The denotational semantics of the functional language *FL*

As an application of the theory of inductive operators we give a denotational semantics of the language *FL* we introduced in Section 1.3.1. Our intention is to assign to each *FL* program in a syntax-directed way, an inductive operator, and to take the lfp of the operator to be the meaning, or denotation of the program. In particular our operator will be a monotonic and continuous function on a cpo.

As motivation consider the following example FL program:

$$f(x) = \text{if } x = 0 \text{ then } 1 \text{ else (if } x \geq 2 \text{ then } x * f(x - 2))$$

We expect this program to compute the function $g : \mathbf{N} \rightarrow_p \mathbf{N}$ defined by:

$$g(x) = \begin{cases} \text{fac}(x/2) & x \text{ even} \\ \uparrow & x \text{ odd} \end{cases}$$

As in Section 1.3.3 we can associate an operator ϕ: $(\mathbf{N} \rightarrow_P \mathbf{N}) \rightarrow (\mathbf{N} \rightarrow_p \mathbf{N})$ with the program:

$$\Phi(h)(x) = \text{if } x = 0 \text{ then } 1 \text{ else (if } x \geq 2 \text{ then } x * h(x - 2))$$

This operator is monotonic (if h becomes more defined then so does $\Phi(h)$) and continuous (because $\Phi(h)(x)$ can be calculated using only finitely many vaues of h). We can check that g is indeed the lfp of Φ. Notice that Φ has other fixed points, for example the total function $\text{fac}(\llcorner x/2 \lrcorner)$. We are free to assign any value to $f(2)$, after which f is determined on all inputs. As the value assigned to $f(1)$ is purely arbitrary, it is preferable to leave it undefined rather than introduce spurious information. This justifies our choice of the *least* fixed point as the meaning of the program. Before setting up the denotational semantics in Section 3.2.2 we need some further definitions and results about cpos and functions over them.

We have found the approaches of [Bird, 1976] and [Loeckx and Sieber, 1987] helpful, and the reader is referred to them for a more detailed account than we can provide here.

150   I. C. C. Phillips

3.2.1   Partial values and strict functions

To deal with arbitrary FL programs we have to change from working with
partial functions to a more sophisticated approach to undefined values. As
an indication of why we do this, take the conditional and the function
symbol $+$. We want to regard these both as functions, but we shall need
to reflect the difference between them that the conditional does not need
all its arguments to be defined in order to be defined, whereas $+$ does.
It is therefore natural to allow undefined elements in the domains of our
functions, so that we can be explicit about what values they should yield.
So for $\mathbf{T} = \mathbf{N}$ or $\mathbf{B}$, let $\mathbf{T}_\perp$ be $\mathbf{T}\sup\{\perp_\mathbf{T}\}$ where $\perp_\mathbf{T}$ represents undefined
(we shall omit the subscript). Notice that by including $\perp$ in the range of
our functions we make them total.

When we add $\perp$ to a set such as $\mathbf{N}$ we make it into a partial ordering
$\mathbf{N}_\perp$, with $\perp$ as least element (Fig. 3.1). In fact it is a cpo.

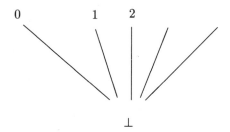

Fig. 3.1

Definition 3.2.1. Let $(D,\sqsubseteq), (E,\sqsubseteq)$ be cpo's. Define the *product* ordering
$(D \times E, \sqsubseteq)$ by

$$(d,e) \sqsubseteq (d',e') \text{ iff } d \sqsubseteq d' \& e \sqsubseteq e'$$

Define the *function space* ordering $([D \to E], \sqsubseteq)$ by letting $[D \to E]$ be
the set of all (total) continuous functions from D to E, and letting

$$f \sqsubseteq g \text{ iff for all } d \in D, f(d) \sqsubseteq g(d)$$

We shall tend to suppress the orderings and speak of $D \times E, [D \to E]$.

Proposition 3.2.2. *If D, E are cpos then so are $D \times E, [D \to E]$.*

As a natural extension, for any k, D^k is a cpo when equipped with the
ordering

$$(d_1,\ldots,d_k) \sqsubseteq (d'_1,\ldots,d'_k) \text{ iff } d_i \sqsubseteq d'_i, \text{ for all } i \leq k$$

With this ordering we know what it means for a function $f : D^k \to E$ to be monotonic. Function f is said to be *monotonic in the ith argument* if

whenever $d_i \sqsubseteq d'_i$ then $f(d_1, \ldots, d_i, \ldots, d_k) \sqsubseteq f(d_1, \ldots, d'_i, \ldots, d_k)$.

It is easy to check that that f is monotonic precisely if it is monotonic in each argument.

Consider as an example a monotonic function $f : \mathbf{N}_\perp \to \mathbf{N}_\perp$. Either $f(\perp) = n$ (some $n \in \mathbf{N}$) in which case we know that $f(x) = n$, all $n \in \mathbf{N}$ and so $f(x)$ can be calculated without evaluating the argument x, or else $f(\perp) = \perp$, in which case we cannot get any result from $f(x)$ before x is evaluated. In the latter case we say that f is *strict*. This operational discussion motivates the following definition.

Definition 3.2.3. Let $f : D^1 \times \ldots D^k \to D$, where $D_1, \ldots, D_k, D$ are either $\mathbf{N}_\perp$ or $\mathbf{B}_\perp$. Function f is said to be *strict in its ith argument* if whenever arguments $(x_1, \ldots, x_k)$ are supplied to f and $x_i = \perp$ then the result $f(x_1, \ldots, x_k) = \perp$. Function f is said to be *strict* if it is strict in all its arguments.

The intuition is that if a function is strict in a particular argument then that argument is definitely required to compute f. The conditional is strict in its first argument (the test), but not in its other two.

Definition 3.2.4. Let $f : \mathbf{N}^k \to \mathbf{T}$. The *strict extension* of f is the function $f^+ : \mathbf{N}_\perp^k \to \mathbf{T}_\perp$ defined by

$$f^+(x_1, \ldots, x_k) = \begin{cases} f(x_1, \ldots, x_k) & \text{if all } x_i \neq \perp \\ \perp & \text{if some } x_i = \perp \end{cases}$$

This is clearly the unique strict function extending f. We shall work with the strict extensions of basic functions such as $+$.

Notice that strict functions are monotonic, because $\mathbf{N}_\perp$ is 'flat' (it consists of a bottom element and a set of maximal elements). In fact they are continuous, in view of the following:

Proposition 3.2.5. *Let D, E be cpo's, and suppose that D has no infinite increasing chain. If $f : D \to E$ is monotonic then it is continuous.* ∎

We shall require some simple results on continuous functions.

Proposition 3.2.6. *Let $f : D \to D'$, $g : D' \to D''$ be continuous. Then $g \circ f : D \to D''$ is continuous.* ∎

Proposition 3.2.7. *Let $f : D \to E_1$, $g : D \to E_2$ be continuous. Then $f \times g : D \to E_1 \times E_2$ is continuous.* ∎

3.2.2 Denotational semantics of FL

Having disposed of these preliminaries we can set up the denotational semantics. Our strategy will be to assign meanings to the symbols of FL as members of appropriate 'domains'. Terms of type Nat or Bool will be interpreted as elements of $\mathbf{N}_\perp$ or $\mathbf{B}_\perp$ respectively. Function symbols of type $\mathrm{Nat}^k \to \mathrm{Nat}$, $\mathrm{Nat}^k \to \mathrm{Bool}$ will be interpreted as members of $[\mathbf{N}_\perp^k \to \mathbf{N}_\perp], [\mathbf{N}_\perp^k \to \mathbf{B}_\perp]$ respectively. Since terms will in general have variables, we assign this meaning relative to an environment which gives the meaning of these variables. We associate a continuous functional to each FL program in the manner of the example given at the beginning of Section 3.2, and define the meaning of a program to be the lfp of its functional.

An *environment* is an assignment to individual and function variables. Environments are ranged over by ρ. Formally we have

$$\rho(\mathrm{var}(n)) \in \mathbf{N}_\perp$$
$$\rho(\mathrm{fun}(k,n)) \in [\mathbf{N}_\perp^k \to \mathbf{N}_\perp]$$
$$\rho(\mathrm{pred}(k,n)) \in [\mathbf{N}_\perp^k \to \mathbf{B}_\perp]$$

Definition 3.2.8. The *meaning* $[\![u]\!]\rho$ of a term u in an environment ρ is defined by structural induction on terms:

$$[\![x]\!]\rho = \rho(x)$$
$$[\![c]\!]\rho = c$$
$$[\![g(\mathbf{u})]\!]\rho = g([\![u_1]\!]\rho, \ldots, [\![u_k]\!]\rho)$$
$$[\![f(\mathbf{u})]\!]\rho = \rho(f)([\![u_1]\!]\rho, \ldots, [\![u_k]\!]\rho)$$

Constants c, g are interpreted by strict members of the appropriate domains. They will be continuous in view of Proposition 3.2.5. We use the standard interpretation under which they have their expected meaning.

Recall from Section 1.3.1 that an FL program P has the form

$$f_1(x_1^1, \ldots, x_{n_1}^1) = t_1$$
$$\ldots$$
$$f_k(x_1^k, \ldots, x_{n_k}^k) = t_k$$

For convenience let $D^* = [D^{n_1} \to D_1] \times \ldots \times [D^{n_k} \to D_k]$, where the D's are of the appropriate type for P. An environment must interpret f_i as a member of $[D^{n_i} \to D_i]$. If we follow the procedure of the example at the beginning of Section 3.2, we shall define functionals Φ_i by

$$\Phi_i(f_1, \ldots, f_k)(x_1^i, \ldots, x_{n_i}^i) = t_i$$

However this is too naive, as it contains purely formal objects with no meaning. So take any $(h_1, \ldots, h_k) \in D^*$ and any $(d_1, \ldots, d_{n_i}) \in D^{n_i}$

and let ρ be the environment which assigns $\rho(x_1^i) = d_1, \ldots, \rho(x_{n_i}^i) = d_{n_i}$, $\rho(f_1) = h_1, \ldots, \rho(f_k) = h_k$. Define $\Phi_i : D^* \to (D^{n_i} \to D)$ by

$$\Phi_i(h_1, \ldots, h_k)(d_1, \ldots, d_{n_i}) = [\![t_i]\!]\rho$$

Note For those familiar with λ-notation [Barendregt, 1992] this may be more succinctly expressed as

$$\Phi_i = [\![\lambda \mathbf{h}.\lambda \mathbf{x}^i.t_i]\!]$$

Since the term in brackets contains no variables (either over functions or individuals) no environment is needed to interpret it.

As we are interested in finding fixed points we want to know that Φ_i is continuous. But it follows that $[\![t_i]\!]\rho$ is continuous as a function of $h_1, \ldots, k_k, d_1, \ldots, d_{n_i}$ from Propositions 3.2.6, 3.2.7 (we omit the details). Hence $\Phi_i : [D^* \to [D^{n_i} \to D]]$.

Now let $\Phi : [D^* \to D^*]$ be $\Phi_1 \times \ldots \times \Phi_k$, which is also continuous by Proposition 3.2.7. It therefore has a lfp $\mu\Phi$. Define the denotation of P as follows. $M(P) : D^{n_1} \to \mathbf{T}$ is given by

$$M(P)(\mathbf{d}) = \begin{cases} \pi_1(\mu\Phi)(\mathbf{d}) & \text{if this is } \neq \perp \\ \uparrow & \text{otherwise} \end{cases}$$

Here π_1 is projection on the first coordinate and $\mathbf{T}$ is $\mathbf{N}$ or $\mathbf{B}$ as appropriate. Notice that we make this a partial function.

3.2.3 Equivalence of operational and denotational semantics

We wish to show that the operational semantics $O_N(P)$ (Definition 1.3.5) and the denotational semantics $M(P)$ coincide. We first show that our operational semantics is *safe*, in the sense that it gives more information than $M(P)$.

Lemma 3.2.9 (Substitution Lemma). *For any terms $u, \mathbf{u}$ and any environment ρ,*

$$[\![u]\!]\rho([\![\mathbf{u}]\!]\rho/\mathbf{x}) = [\![u[\mathbf{u}/\mathbf{x}]]\!]\rho.$$

Proof. This is a completely general result of denotational semantics. The proof proceeds by induction on u and is left to the reader. ∎

Proposition 3.2.10. *The evaluation strategies N, V are safe (viz. $f_N, f_V \subseteq \mu\Phi$).*

Proof. The successive approximations to $M(P)$ are

$$\pi_1(\Phi^0(\bot)), \ldots, \pi_1(\Phi^k(\bot)), \ldots$$

We wish to show that if $f_1(\mathbf{d}) \Rightarrow c$ then $\exists k$ such that $\pi_1(\Phi k(\bot))(\mathbf{d}) = c$, since then clearly $M(P)(\mathbf{d}) = \pi_1(\mu\Phi)(\mathbf{d}) = c$. As often in these situations, we cannot prove this directly. Instead we prove a more general result by structural induction on proofs in system N. We shall show that for any term u with no free individual variables,

$$\text{if } u \Rightarrow_N c \text{ then } \exists k \text{ such that } [\![u]\!](\Phi^k(\bot)) = c. \tag{$\dagger$}$$

Here we abuse notation by letting $(\Phi^k(\bot))$ denote the environment which assigns $\Phi^k(\bot)$ to $f_1, \ldots, f_k$. Suppose that this has been done. To get our desired result, set $u = f_1(\mathbf{d})$ and assume $f_1(\mathbf{d}) \Rightarrow c$. Then by $(\dagger)$ we have k such that $[\![f_1(\mathbf{d})]\!](\Phi^k(\bot)) = c$. But $[\![f_1(\mathbf{d})]\!](\Phi^k(\bot)) = \pi_1(\Phi^k(\bot))(\mathbf{d})$. Hence result.

Now we prove $(\dagger)$ by induction on the proof of $u \Rightarrow_N c$ using the rules of Section 1.3. We drop the N suffix. There are cases depending on which rule is used to derive $u \Rightarrow_N c$.

1. Here u must be c and $[\![c]\!]\rho = c$ for any environment ρ .

2. $u = g(\mathbf{u})$. We have $u_i \Rightarrow c_i$ and $g(\mathbf{c}) = c$. By induction $\exists k_i$ such that $[\![u_i]\!](\Phi^{k_i}(\bot)) = c_i$. Let $k = \max\{k_i\}$. Then $[\![u_i]\!](\Phi^k(\bot)) = c_i$ using $\Phi^{k_i}(\bot) \sqsubseteq \Phi^k(\bot)$. Now

$$[\![u]\!](\Phi^k(\bot)) = g([\![u_1]\!](\Phi^k(\bot)), \ldots) = g(\mathbf{c}) = c$$

3. $u = \text{if } u_1 \text{ then } u_2 \text{ else } u_3$. We have $u_1 \Rightarrow tt$ (the case for ff is similar) and $u_2 \Rightarrow c$. By induction $\exists k_1$ such that $[\![u_1]\!](\Phi^{k_1}(\bot)) = tt$ and $\exists k_2$ such that $[\![u_2]\!](\Phi^{k_2}(\bot)) = c$. Let $k = \max\{k_1, k_2\}$. Then

$$\begin{aligned} [\![u]\!](\Phi^k(\bot)) &= [\![u_2]\!](\Phi^k(\bot)) \quad \text{since } [\![u_1]\!](\Phi^k(\bot)) = tt \\ &= c \end{aligned}$$

4N. $u = f_i(\mathbf{u})$. We have $t_i[\mathbf{u}/\mathbf{x}] \Rightarrow c$. By induction $\exists k$ such that $[\![t_i[\mathbf{u}/\mathbf{x}]]\!](\Phi^k(\bot)) = c$. Now

$$\begin{aligned} [\![f_i(\mathbf{u})]\!](\Phi^{k+1}(\bot)) &= (\pi_i \circ \Phi \circ \Phi^k(\bot))([\![u_1]\!](\Phi^{k+1}(\bot))\ldots) \\ &= \Phi_i(\Phi^k(\bot))([\![u_1]\!](\Phi^{k+1}(\bot)), \ldots) \\ &= [\![t_i]\!](\Phi^k(\bot))([\![u_1]\!](\Phi^{k+1}(\bot))/x_1, \ldots) \\ &\qquad \text{by definition of } \pi \circ \Phi \\ &\sqsupseteq [\![t_i]\!](\Phi^k(\bot))([\![u_1]\!](\Phi^k(\bot))/x_1, \ldots) \\ &= [\![t_i[\mathbf{u}/\mathbf{x}]]\!](\Phi^k(\bot)) \\ &\qquad \text{by Lemma 3.2.9} \\ &= c \end{aligned}$$

Hence $[\![u]\!](\Phi^{k+1}(\perp)) = c$ as required. ∎

Theorem 1.3.6 stated that $O_V(P) \subseteq O_N(P)$, and so it is clear that the V strategy is also safe.

Theorem 3.2.11. $O_N(P) = M(P)$.

Proof. It is enough to show $M(P) \subseteq O_N(P)$. We prove that for all n, and all u with no free individual variables,

if $[\![u]\!](\Phi^n(\perp)) = c$ then $u \Rightarrow c$ $\qquad$ (‡)

Applying this to the term $f_1(\mathbf{c})$ as in the previous proposition then gives the result.

Proof of (‡) By induction on n and a subsidiary induction on u. Suppose result true for n. We show it for $n + 1$. There are cases according to the construction of u:

1. $u = c$. Immediate.

2. $u = g(\mathbf{u})$. Suppose $[\![u]\!](\Phi^{n+1}(\perp)) = c$. Now

   $$[\![g(\mathbf{u})]\!](\Phi^{n+1}(\perp)) = g([\![u_1]\!](\Phi^{n+1}(\perp)), \ldots)$$

   Since g is strict, there must be c_i such that $[\![u_i]\!](\Phi^{n+1}(\perp)) = c_i$. But then by induction hypothesis $u_i \Rightarrow c_i$ and $g(\mathbf{c}) = c$. Hence $u \Rightarrow c$ by rule (2).

3. $u = $ if u_1 then u_2 else u_3. No particular difficulty.

4. $u = f_i(\mathbf{u})$. This is the heart of the proof and the reason why we need induction on n as well as u. Suppose $[\![u]\!](\Phi^{n+1}(\perp)) = c$. Now

   $$\begin{aligned}[\![f_i(\mathbf{u})]\!](\Phi^{n+1}(\perp)) &= \Phi_i(\Phi^n(\perp))([\![u_1]\!](\Phi^{n+1}(\perp)), \ldots) \\ &= [\![t_i]\!](\Phi^n(\perp))([\![u_1]\!](\Phi^{n+1}(\perp))/x_1, \ldots) \\ &\qquad \text{by definition of } \pi \circ \Phi. \end{aligned}$$

   Put

   $$u'_j = \begin{cases} c_j & \text{if } [\![u_j]\!](\Phi^{n+1}(\perp)) = c_i \\ u_j & \text{if } [\![u_j]\!](\Phi^{n+1}(\perp)) = \perp \end{cases}$$

   Then for each j,

   $$[\![u'_j]\!](\Phi^n(\perp)) = [\![u_j]\!](\Phi^{n+1}(\perp))$$

   Hence

   $$\begin{aligned}[\![u]\!](\Phi^{n+1}(\perp)) &= [\![t_i]\!](\Phi^n(\perp))([\![u'_1]\!](\Phi^n(\perp)) \ldots) \\ &= [\![t_i[\mathbf{u'}/\mathbf{x}]]\!](\Phi^n(\perp)) \text{ by Lemma 3.2.9} \end{aligned}$$

   Now by induction $t_i[\mathbf{u'x}] \Rightarrow c$. Moreover $u'_j = u_j$ or $u_j \Rightarrow u'_j$ (again by induction). So we obtain $t_i[\mathbf{u}/\mathbf{x}] \Rightarrow c$ using Lemma 1.3.7 from this we infer $f(\mathbf{u}) \Rightarrow c$ by rule (4). ∎

3.3   Ordinals

Lists and sequences occur frequently in mathematics and computing science. They have *length*, which may be finite or infinite, but of course *order* matters as well. It is customary to regard them as 'indexed' by natural numbers, thus

$$a_1, a_2, \ldots, a_k$$

or in the case of an infinite sequence

$$a_1, a_2, \ldots, a_n, \ldots$$

In this subsection we shall see how to index longer sequences by ordinals in a way which is still finitely based. In Section 3.4 we shall use ordinals to extend our study of inductive definitions.

Imagine taking linear orderings of size n (each $n \in \mathbf{N}$) and forming their limit as suggested by the following diagram (with the orderings increasing from left to right):

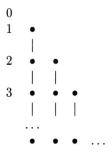

We embedded each ordering into an initial segment of the next and formed a new infinite ordering consisting of the vertical 'threads' connecting the orderings. Instead of making the embeddings explicit we could 'squash' the diagram down vertically by identifying the points which are connected by the threads. We then have a diagram like the following:

$$\bullet \quad \bullet \quad \bullet \quad \ldots$$
$$0 \quad 1 \quad 2 \quad 3 \quad \ldots \quad \omega$$

where we adopt the convention that the ordering corresponding to n is all the dots which are strictly to the left of n. It is usual to call this new ordering ω, and we have written this to the right of all the dots. It is plain that we form the successor ordering labelled by $n + 1$ from the ordering labelled by n, by simply placing an extra dot at the top (right-hand) end of the ordering. Let us do this with our new infinite ordering:

$$\begin{array}{cccccccc} \bullet & \bullet & \bullet & \bullet & \ldots & \bullet & \\ 0 & 1 & 2 & 3 & \ldots & \omega & \omega+1 \end{array}$$

We denote the new ordering by $\omega + 1$.

We should be clear that by ω, $\omega + 1$ we don't have particular representations of these orderings in mind (hence the dots). They are to stand for order types, that is equivalence classes of orderings under isomorphism and they are called *ordinals*. Though it will be convenient for some purposes to have a 'canonical' representation of the ordinals as sets, we shall be free to choose other 'notations' for them depending on the task in hand.

It is important to realize that although ω is an infinite object and $\omega+1$ involves placing a new object 'beyond infinity', there is no difficulty in constructing them out of familiar mathematical objects, such as the natural numbers. Thus ω is plainly just the order type of $\mathbf{N}$ itself, and $\omega + 1$ is the type of the ordering defined by using the usual ordering on the positive numbers and placing 0 at the top:

$$n <_{\omega+1} m \text{ iff } 0 < n < m \text{ or } m = 0$$

$$1 \quad 2 \quad 3 \quad \ldots \quad 0$$

Now that we have formed $\omega + 1$ it is clearly open to us to form $\omega + 2$

$$\begin{array}{cccccc} \bullet & \bullet & \bullet & \cdots & \bullet & \bullet \end{array}$$

and $\omega + 3$

$$\begin{array}{ccccccc} \bullet & \bullet & \bullet & \cdots & \bullet & \bullet & \bullet \end{array}$$

We can form $\omega + n$ for any $n \in \mathbf{N}$ and then take the limit as before to get

$$\begin{array}{ccccccc} \bullet & \bullet & \bullet & \cdots & \bullet & \bullet & \bullet & \ldots \end{array}$$

which we naturally denote $\omega + \omega$. Once again we emphasize that there is no difficulty in constructing a set with this order type out of the natural numbers. For example we can take $\mathbf{N}$ with the ordering

$$n <_{\omega+\omega} m \quad \text{iff } n, m \text{ are even and } n < m$$
$$\text{or } n, m \text{ are odd and } n < m$$
$$\text{or } n \text{ is even and } m \text{ is odd}$$

Needless to say there are many different orderings on $\mathbf{N}$ with the same order type.

Given $\omega + \omega$ we can again add a new top element 'beyond infinity' to form $\omega + \omega + 1$. The reader is invited to draw an appropriate diagram.

Since $\omega + \omega$ is 2 copies of ω, one placed after the other, we can also call it $\omega.2$ (the reason for the notation is that '.' is actually a multiplication on ordinals just as $+$ is addition); $\omega + \omega + 1$ is then $\omega.2 + 1$. We can now produce ordinals $\omega.m + n$ for any $m, n \in \mathbf{N}$. We can think of $\omega.m + n$ as m copies of ω followed by n. These may be produced successively by using the successor constructor ($+1$ in other words) to get from $\omega.m + n$ to $\omega.m + n + 1$ and then taking the limit of

$$\omega.m, \omega.m + 1, \omega.m + 2, \ldots, \omega.m + n, \ldots$$

to get $\omega.(m+1)$. Notice that even though each $\omega.m+n$ is in a sense infinite, having infinitely many ordinals below it, it has a finite description.

Even now we have not exhausted the ordinals, for it is open to us to form the limit of all the $\omega.m + n$'s . This may be thought of as ω copies of ω :

$$\bullet \; \bullet \; \bullet \; \ldots \; \bullet \bullet \; \bullet \; \ldots \; \bullet \; \bullet \; \bullet \; \ldots \ldots$$

It is natural to call it $\omega.\omega$. It too, can be represented as an ordering in $\mathbf{N}$. First note that $\omega.\omega$ is the order type of the following 'lexicographic' ordering on $\mathbf{N} \times \mathbf{N}$:

$$(m, n) <_L (m', n') \text{ iff } \begin{aligned} &m < m' \text{ or} \\ &m = m' \text{ and } n < n' \end{aligned}$$

Here (m, n) corresponds to $\omega.m + n$ of course. We shall refer to it as a *notation* for $\omega.m + n$. We now need a 1-1 mapping of $\mathbf{N} \times \mathbf{N}$ into $\mathbf{N}$ in order to represent $<_L$ as an ordering $<_{\omega.\omega}$ on $\mathbf{N}$; the pairing function p of Section 1.1 will do, though of course there are many others.

We can continue to form new ordinals using successor and taking limits. Let us abbreviate $\omega.\omega$ to ω^2 . Having taken ω copies of ω to get ω^2, we can take ω copies of ω^2 to get ω^3 . This process can be continued to give ω^n for any n. We can then take the limit of this ascending sequence to get a new ordinal which it is natural to call ω^ω . We shall discuss how to obtain a finite representation for ordinals below ω^ω later.

Let us pause for breath and develop the ordinals more formally. Here is an inductive definition of the ordinals:

Definition 3.3.1.

    0 is an ordinal

    if α is an ordinal then so is $S(\alpha)$

    if X is a downwards-closed segment of the ordinals with no greatest element then $\bigsqcup X$ is an ordinal.

It is conventional to let $\alpha, \beta, \gamma \ldots$ range over the ordinals. By downwards-closed we mean that if $\alpha \in X$ and $\beta < \alpha$ then $\beta \in X$. Our definition has the merit of presenting the ordinals as a data type with constructors $0, S, \bigsqcup$. This makes it clear that every ordinal is either 0 or a successor or a limit. Furthermore each ordinal has a unique construction. Our definition can be the basis for the more sophisticated descriptions of ordinals in terms of multiplication, etc. initiated above.

However Definition 3.3.1 has certain flaws. In order to understand 'downwards-closed' we have to define the ordering on ordinals. This will have to be defined simultaneously with the ordinals. Furthermore the inductive clause for limits uses an infinite set X. So the definition is not finitary. We shall use descriptions of limit ordinals as limits of a variety of ascending chains of ordinals (partly in order to obtain finitary descriptions or *notations*). These considerations make us turn to an alternative definition of the ordinals.

First a definition:

Definition 3.3.2. A *well-ordering* (wo) is a partial ordering $(X, <)$ such that any subset $Y \subseteq X$ has a $<$ −least element.

(Apologies are in order for the curious grammar — the terminology is standard and comes from the German.) One easily sees that any well-ordering is a linear (total) ordering.

Characterization 3.3.3. Let $(X, <)$ be a linear ordering. Then it is a well-ordering iff it contains no descending ω-sequence (viz. a sequence $x_0 > x_1 > \ldots > x_n > \ldots$ of members of X).

Proof. $\Rightarrow$ Assume $(X, <)$ is a well-ordering and suppose for a contradiction that we have an ω -sequence $x_0 > x_1 > \ldots$ Let $Y = \{x_0, x_1, \ldots\}$. Then Y has a least element xn, some n. But this is plainly impossible.

$\Leftarrow$ Assume there are no descending ω-sequences. Take $Y \subseteq X$. Pick any $y_0 \in Y$. If y_0 is least then we are done; otherwise we can pick $y_1 \in Y, y_1 < y_0$. We can continue to pick out y_i's to form a descending sequence until we reach a least element. This we must eventually do since we have assumed that such a sequence cannot be infinite. Hence Y has a least element as required. (The cognoscenti will recognize a use of the Axiom of Choice here.) ∎

Our guiding idea is that an ordinal is the isomorphism type of a well-ordering. One can check that all the ordinals we have described so far are indeed well-ordered. First of all ω is wo. It of course contains infinite ascending sequences; however any descending sequence in ω must start at some finite number n, and there are only finitely many numbers below n, so that we can use Characterization 3.3.3. As a further example take ω^2. Suppose $\alpha_0, \alpha_1, \ldots$ is a descending sequence in ω^2. Let $\alpha_i = \omega.m_i + n_i (i = 0, 1, \ldots)$. Then $m_0, m_1, \ldots$ must form a non-increasing sequence;

it is therefore eventually constant (since ω is wo). Say m_i is constant, all $i > k$. Then the $n_i(i > k)$ must form a strictly decreasing sequence. But this is impossible, again since ω is wo. We conclude that ω^2 is wo.

Instead of defining an ordinal to be the order type of a well-ordering, we shall pick on a particular canonical representative of each isomorphism class. An ordinal will be the set of all ordinals below it. This representation is due to von Neumann.

Definition 3.3.4. An *ordinal* is a transitive set well-ordered by $\in$. If α, β are ordinals then let $\alpha < \beta$ iff $\alpha \in \beta$.

(A set S is *transitive* if $\in$ is transitive on the members of S, i.e. $\forall x, y, z \in S.\ x \in y \in z \to x \in z$). This has the advantage of definiteness, though we really think of an ordinal as being or representing an order type.

We shall denote the class of all ordinals by On. We say *class* since the collection of ordinals is too large to form a set. How many ordinals there are depends on our set theory and is a question that need not concern us here.

Proposition 3.3.5. *If $\alpha \in On$ and $\beta < \alpha$ then $\beta \in On$.*

Proof. Exercise.   ∎

Thus every ordinal is the set of its predecessors. The least ordinal 0 is plainly the empty set $\emptyset$. If $\alpha \in On$ then $S(\alpha)$ should contain α and all ordinals below α. Hence $S(\alpha) = \alpha \cup \{\alpha\}$. Thus $1 = \{\emptyset\}, 2 = \{0, 1\} = \{\emptyset, \{\emptyset\}\}, \ldots, \omega = \{0, 1, \ldots\}$. One can check that $S(\alpha)$ is an ordinal; it is the least ordinal $> \alpha$.

Lemma 3.3.6. *A proper initial segment I of On is an ordinal.*

Proof. First we check that I is transitive: suppose $\alpha \in \beta \in I$. Then $\alpha < \beta$ and so $\alpha \in I$ since I is an initial segment. Now suppose $A \subseteq I$ is non-empty. We wish to show that A has a least element. Take $\beta \in A$. Either β is least or else $A \cap \beta$ is non-empty. In the latter case $A \cap \beta$ has a least element α since β is an ordinal. Ordinal α will also be the least element of A.   ∎

Remark 3.3.7. The proof of the lemma also clearly shows that On is transitive and wo by $\in$. However if we then deem it to be an ordinal we have $On \in On$, which leads to paradox. This is why On is not a set, unlike I above.

Proposition 3.3.8. *If X is any set of ordinals then $\bigcup X$ is an ordinal.*

Proof. First check that $\bigcup X$ is an initial segment of On: clearly if $x \in \bigcup X$ then x is an ordinal by Proposition 3.3.5. If $\alpha < \beta \in \bigcup X$ then $\alpha < \beta \in$

$\gamma \in X$ and so $\alpha \in \gamma \in X$, in other words $\alpha \in \bigcup X$. (Of course $<$ and $\in$ mean exactly the same thing for ordinals.) Now apply Lemma 3.3.6. ∎

What about the limit-forming operation in general? If X is any set of ordinals then $\bigcup X$ is an ordinal and is the least upper bound $\bigsqcup X$. If X has a greatest element α then $\bigcup X = \alpha$, but if X has no greatest element then $\bigcup X > \alpha$, every $\alpha \in X$, and $\bigcup X$ is a limit ordinal.

Definition 3.3.9. An ordinal λ is a *limit ordinal* if $\lambda \neq 0$ and $\forall \alpha < \lambda$. $S(\alpha) < \lambda$ (it is usual to let λ range over limit ordinals). An ordinal α is a *successor ordinal* if $\alpha = S(\beta)$, some β.

It is easy to see that any ordinal must be either 0 (the empty set), a successor or a limit ordinal.

With any well-ordering there is associated an induction principle:

Principle 3.3.10 (Induction over a Well-ordering). Let $(X, <)$ be a well-ordering and let P be some property of members of X. If $\forall x \in X(\forall y < xP(y) \rightarrow P(x))$ then $\forall x \in XP(x)$.

Proof. Assume for a contradiction that $\forall x \in X(\forall y < xP(y) \rightarrow P(x))$ but not $\forall x \in XP(x)$. Let $Z = \{x \in X \mid \neg P(x)\}$. Then $Z \neq \varnothing$. Hence Z has a least element x. But then $\forall y < x.y \notin Z$, in other words $\forall y < xP(y)$. Hence $P(x)$, which is a contradiction. ∎

This principle can be of course be used for the ordinals, since they are wo. In this context it is known as the *Principle of Transfinite Induction*. Ordinals $\geq \omega$ are called transfinite. Transfinite induction is a generalization of finite induction, which is just ordinary induction on the natural numbers. Finite induction comes in two brands:

1. If $\forall n \in \mathbf{N} \forall m < n(P(m) \rightarrow P(n))$ then $\forall n \in \mathbf{N}P(n)$.

2. If $P(0)$ and $\forall n \in \mathbf{N}(P(n) \rightarrow P(S(n)))$ then $\forall n \in \mathbf{N}P(n)$.

In the same way there are two forms of transfinite induction:

Principle 3.3.11 (Transfinite Induction). Let P be some property of ordinals.

1. If $\forall \alpha \in On \ \forall \beta < \alpha(P(\beta) \rightarrow P(\alpha))$ then $\forall \alpha \in OnP(\alpha)$.

2. If

$$P(O)$$
$$P(\alpha) \rightarrow P(S(\alpha)) \qquad \text{all } \alpha \in On$$
$$\forall \beta < \lambda P(\beta) \rightarrow P(\lambda) \qquad \text{all limit } \lambda \in On$$

then $\forall \alpha \in On P(\alpha)$.

Proof. Form (1) is clearly just 3.3.9 specialized to On.

To show (2), assume the hypotheses of (2) and suppose for a contradiction that P is not true of all ordinals. Let β be least such that $\neg P(\beta) - \beta$ exists since On is wo. Then it is easy to see that β cannot be 0, a successor or a limit without contradicting the appropriate hypothesis. Contradiction. ∎

Remarks 3.3.12. The second form of transfinite induction is clearly the appropriate structural induction for our constructor-form definition of the ordinals (Definition 3.3.1). This is the version that we shall find most useful. Note also that instead of proving a property of *all* ordinals we can also use transfinite induction to prove that all ordinals below a certain ordinal have a property; a simple example of this is induction on **N**, which we may now see as induction up to ω .

It is well-known that functions on data-types can be defined by structural recursion, in other words recursive definitions which pattern-match on the structure of the data-type. As a simple example consider our BNF definition of Prog in Section 3.1.1. A function f: Prog $\rightarrow$ **N** which measures the number of assignment statements on members of Prog may be defined by

$$f(C) = 1 (C \in \text{Cmd})$$
$$f \text{ (if } T \text{ then } P_1 \text{ else } P_2) = f(P_1) + f(P_2)$$
$$f \text{ (while } B \text{ do } P) = f(P)$$

Another example is addition on Nat

$$m + 0 = m$$
$$m + (S(n)) = S(m + n)$$

We have written $S(n)$ instead of the more usual $n + 1$ to emphasize the way that our definition follows the structure of the BNF definition of Nat:

$$n ::= 0 \mid S(n)$$

To each inductively defined structure we associate an induction principle and a pattern of structural recursion. We can extend this definition method to the ordinals, where it is known as transfinite recursion.

As a simple example, let us define addition on ordinals. Our intuition will be that to form the sum of two ordinals we place them (or their representatives) one after the other. Thus $\alpha + \beta$ is as follows.

$$\alpha \qquad\qquad\qquad \beta$$

Fig. 3.2

where all elements of α are less than the least element of β (note that if α is a limit then it does not have a greatest element). This addition has the merit of corresponding to the existing definition of addition when restricted to finite ordinals. It is important to be aware however that this addition is no longer commutative when transfinite ordinals are involved. For instance $\omega + 1$ is as follows.

$$0 \quad 1 \quad 2 \quad \cdots \qquad\qquad 0$$

Fig. 3.3

Whereas $1 + \omega$ is as follows.

$$\begin{array}{cc} \bullet & \\ 0 & \quad 0 \quad 1 \quad 2 \quad \cdots \end{array}$$

Fig. 3.4

This is plainly isomorphic to ω. The 1 has been absorbed by the ω. An elegant recursive definition of addition can be given as follows:

$$\alpha + 0 = \alpha$$
$$\alpha + S(\beta) = S(\alpha + \beta)$$
$$\alpha + \lambda = \bigcup_{\beta < \lambda}(\alpha + \beta)$$

Notice how this follows version (2) of the Principle of Transfinite Induction in the way that it is assumed that the function is already defined for smaller values of the second argument.

As a further example we can define multiplication on ordinals:

$$\alpha.0 = 0$$
$$\alpha.S(\beta) = (\alpha.\beta) + \alpha$$
$$\alpha.\lambda = \bigcup_{\beta < \lambda}(\alpha.\beta)$$

Notice how, as before, this generalizes the usual recursive definition for finite ordinals. Again we perform recursion on the second argument, and at limit stages we again simply take the supremum of all the values obtained so far. Multiplication is also non-commutative. For instance by our definition

$$w.2 = w + w$$

whereas

$$2.w = \bigcup_{n<w} 2.n = w.$$

This explains why earlier we wrote $w.2$ instead of $2.w$. To picture ordinal multiplication, define the following ordering on $\alpha \times \beta$:

$$(\alpha_1, \beta_1) <_{\alpha\times\beta} (\alpha_2, \beta_2 \text{ iff } \beta_1\beta_2 \text{ or } (\beta_1 = \beta_2 \text{ and } \alpha_1 < \alpha_2)$$

(so-called reverse lexicographic ordering). Then $(\alpha\times\beta, <_{\alpha\times\beta})$ has the order type of $\alpha.\beta$.

We can prove by transfinite induction that addition and multiplication are associative.

Note. Having now formally defined addition we shall freely write $\beta + 1$ etc. instead of $S(\beta)$ etc.

We earlier saw some simple examples of ordinal exponentiation. Here is the definition by transfinite recursion:

$$\alpha^0 = 1$$
$$\alpha^{\beta+1} = \alpha^\beta.\alpha$$
$$\alpha^\lambda = \bigcup_{\beta<\lambda} \alpha^\beta$$

Compare this with the informal definition of w^w we gave earlier.

Proposition 3.3.13. *For any α, β, γ.*

$$\alpha^\beta.\alpha^\gamma = \alpha^{\beta+\gamma}$$
$$(\alpha^\beta)^\gamma = \alpha^{\beta.\gamma}$$

Proof. By induction. ∎

In order to justify the use of such transfinite recursions we need to prove that the functions we are defining are well-defined, that is, they have a unique value for each ordinal. We now give this justification and take the opportunity to fill a gap in our presentation so far, by showing that any well-ordering is isomorphic to a (von Neumann) ordinal.

Proposition 3.3.14 (Definition by transfinite recursion (1)). *Let $(X, <)$ be a well-ordering. Let A be a set and let $^{<X}A$ denote the set of*

functions from initial segments of X into A. Let $G : X \times^{<X} A \to A$. Then there is a unique function $f : X \to A$ defined by

$$f(x) = G(x, f \restriction x)$$

where $f \restriction x$ denotes f restricted to the initial segment $\{y \mid y < x\}$.

Proof. We need to check that $f(x)$ exists and is unique for each $x \in X$. Suppose this is not the case. Since X is well-ordered we can choose the least $x \in X$ such that $f(x)$ is not well-defined. But then $f(y)$ must be well-defined for each $y < x$ so that $f \restriction x$ is defined and unique. It is then clear that there is a unique value for $f(x)$, given by $G(x, f \restriction x)$. ∎

Just as with the induction principles, the above can be easily specialized to On, where there is also a second version as follows:

Proposition 3.3.15 (Definition by transfinite recursion (2)). *Let A be a set and let $^{<On}A$ denote the class of functions from ordinals into A. Let $a \in A, F : On \times A \to A, G : On \times^{<On} A \to A$. Then there is a unique function $f : On \to A$ defined by*

$$\begin{aligned} f(0) &= a \\ f(\beta + 1) &= F(\beta, f(\beta)) \\ f(\lambda) &= G(\beta, f \restriction \lambda) \end{aligned}$$

where $f \restriction \lambda$ denotes f restricted to values $< \lambda$.

Proof. Similar to that of Proposition 3.3.12 ∎

In order to justify the definitions of addition, etc. we need a parametrized version of the above which we omit.

As an immediate application we prove a result which shows that the ordinals can act as canonical representatives for the well-orderings:

Proposition 3.3.16. *Every well-ordering $(X, <)$ is isomorphic to a unique ordinal.*

Proof. Define $f :\to On$ by transfinite recursion as follows. For $x \in X, f(x) = \{f(y) \mid y < x\}$. This is valid by Proposition 2.2.12. Firstly f is order-preserving: if $x, y \in X$ and $x < y$ then $f(x) \in f(y)$. We check that f is 1–1: If $x, y \in X$ and $x \neq y$ then either $x < y$ in which case $f(x) \in f(y)$, or else $y < x$ in which case $f(y) \in f(x)$. In either case $f(x) \neq f(y)$ (we are actually using the set-theoretic axiom of foundation or regularity, which has the consequence that no set can be a member of itself). It is also easy to show that the range $f[X]$ of f is transitive: if

$a \in b \in f[X]$ then $b = f(x)$ some x, and so $a = f(y)$ some $y < x$, so that
clearly $a \in f[X]$. To see that $f[X]$ is well-ordered take a non-empty subset
$A \subseteq f[X]$. Plainly $A = f[Y]$, some $Y \subseteq X$ and if y is $<$ −least in Y then
$f(y)$ is $\in$ -least in A. Hence X is isomorphic under f to the ordinal $f[X]$.
It remains to show that this ordinal is unique. To show this, suppose that
g is any order-preserving bijection from X to some ordinal γ . Take any
$x \in X$. Then $\{g(y) \mid y < x\}$ is an initial segment of On since g is onto and
order-preserving. Now $\alpha = \{g(y) \mid y < x\}$ is the least ordinal $> g(y)$ for
each $y < x$. Plainly $g(x) \geq \alpha$ since g is order-preserving. Moreover since
g is onto we must have $g(x) = \alpha$. We have shown that g must satisfy the
equation

$$g(x) = \{g(y) \mid y < x\}$$

for any $x \in X$. But this is how f was defined, and we know that such a
transfinite recursion defines a unique function. Hence $g = f$ and $\gamma = f[X]$
as required. ∎

Using exponentiation we can produce an ordinal larger than any we
have seen so far, which is the next natural 'stopping-off point' after ω .
Define

$$c(0) = 0$$
$$c(n + 1) = \omega^{c(n)}$$
$$c(\omega) = \bigcup_{n<\omega} c(n)$$

$c(\omega)$ is usually called ε_0. One might expect that ω^α is always $> \alpha$ but in
fact $\omega^{\varepsilon_0} = \varepsilon_0$. We check

$$\omega^{c(\omega)} = \bigcup_{n<\omega} \omega^{c(n)} = \bigcup_{n<\omega} c(n+1) = c(\omega)$$

(Here we implicitly use the fact that ω^α is increasing as a function of α.)

Exercise 3.3.17. Show that ε_0 is the least ordinal α such that $\omega^\alpha \leq \alpha$.

One reason why ε_0 is important is that ordinals $< \varepsilon_0$ can be given a
canonical representation using plus, times and exponentiation.

Lemma 3.3.18. *Let $0 < \alpha < \varepsilon_0$. Then there are unique ordinals β, γ and
a unique $0 < n < \omega$ such that*

$$\alpha = \omega^\beta.n + \gamma \text{ where } \alpha > \beta \text{ and } \gamma < \omega^\beta.$$

Proof. Let δ be least such that $\alpha < \omega^\delta$. Then $\delta \leq \alpha$ since $\alpha < \omega^\alpha$. Also
$\delta > 0$. Now δ cannot be a limit since then $\omega^\delta = \bigcup_{\gamma<\delta} \alpha^\gamma$ and $\alpha < \omega^\gamma$ some

$\gamma < \delta$, which is a contradiction. Hence $\delta = \beta + 1$, some β , and we have $\omega^\beta \le \alpha < \omega^{\beta+1} = \omega^\beta.\omega$. Now let n be such that $\omega^\beta.n \le \alpha < \omega^\beta.(n+1)$. Clearly $n \ge 1$. Since $\alpha \ge \omega^\beta.n, \alpha = \omega^\beta.n + \gamma$, some γ . If $\gamma \ge \omega^\beta$ then $\alpha \ge \omega^\beta.n + \omega^\beta = \omega^\beta.(n+1)$ which is a contradiction. Hence $\gamma < \omega^\beta$. It remains to show that β, γ, n are unique. But if $\alpha = \omega^\beta.n + \gamma$ where $\gamma < \omega^\beta$ then clearly $\omega^\beta \le \alpha$. Moreover $\alpha < \omega^\beta.n + \omega^\beta = \omega^\beta.(n+1) < \omega^\beta.\omega = \omega^{\beta+1}$. Hence β is unique. Also $\omega^\beta.n \le \alpha < \omega^\beta.(n+1)$ and so n is unique. It follows that γ is unique, since one can show that for any ordinals α, β, γ , if $\alpha + \beta = \alpha + \gamma$ then $\beta = \gamma$ ∎

Remark 3.3.19. In the above proof it is plain that β is in fact the *greatest* ordinal such that $\omega^\beta \le \alpha$. However we have to be careful about appealing to a greatest ordinal with some property, as it need not exist in general. This explains our indirect approach.

Theorem 3.3.20 (Cantor Normal Form). *Let* $0 < \alpha < \varepsilon_0$. *Then there is a unique* $k > 0$ *and there are unique ordinals* $\alpha_0, >, \alpha_k$ *such that* $\alpha > \alpha_0 > ... > \alpha_k$, *and unique* $n_0, ..., n_k > 0$ *such that*

$$\alpha = \omega^{\alpha}0.n_0 + \omega^{\alpha}1.n_1 + ... + \omega^{\alpha}k.n_k$$

Proof. By induction on α . By Lemma 3.3.16, $\alpha = \omega^\beta.n + \gamma$ where $\alpha > \beta$ and $\gamma < \omega^\beta$. But by induction hypothesis, $\gamma = \omega^\gamma0.n_0 + ... + \omega^\gamma j.n_j$ where $\gamma > \gamma_0 > ... > \gamma_j$. Now $\omega^\beta > \gamma > \omega^\gamma0$. Hence $\beta > \gamma_0$. We therefore have $\alpha = \omega^\beta.n + \omega^\gamma0.n_0 + ... + \omega^\gamma j.n_j$ where $\beta > \gamma_0 > ... > \gamma_j$. For uniqueness we see that if $\alpha = \omega^\alpha0.n_0 + ... + \omega^\alpha k.n_k$ then α_0 and n_0 must be unique by Lemma 3.3.16. The uniqueness of α_1 and n_1, etc. follows by induction. ∎

As an easy application of the Normal Form Theorem we can get a canonical representation for ordinals up to ω^ω . Clearly if $\alpha < \omega^\omega$ then there are unique $m_0 > ... > m_k, n_0, ..., n_k$ such that

$$\alpha = \omega^m0.n_0 + ... + \omega^m k.n_k$$

It follows that each such α can be represented as a member of list($\mathbf{N} \times \mathbf{N}$).

One reason why for most purposes we can restrict attention to the finite ordinals is that ω is 'closed' under all commonly arising ordinal operations on its members. Thus

$$\text{if } m, n < \omega \text{ then } m + n < \omega$$

and ω is similarly closed under multiplication and exponentiation. When we turn to ordinals $> \omega$, we see that even being a limit ordinal does not guarantee closure under addition. For instance $\omega + \omega = \omega$, so that $\omega + \omega$

is not closed under $+$. It is not hard to see that ω^2 is the first ordinal $> \omega$ which is closed under $+$. This is not closed under multiplication — for that we must go up as far as ω^ω.

Proposition 3.3.21. ε_0 *is the least ordinal $> \omega$ which is closed under exponentiation.*

Before giving the proof we introduce a helpful concept.

Definition 3.3.22. Let λ be any limit ordinal. An increasing sequence $\alpha_0 < \alpha_1 < \ldots < \alpha_\gamma < \ldots (\gamma < \delta)$ of ordinals indexed by δ is called *cofinal* in λ if $\bigcup_{\gamma\ <\delta}\alpha_\gamma = \lambda$.

Lemma 3.3.23. *For any α, ω^α is closed under addition.*

Proof. By induction on α . Base case: $\omega^0 = 1$, which is certainly closed under $+$.

Successor case: $\omega^{\alpha+1} = \omega^\alpha.\omega$. If $\beta, \gamma < \omega^\alpha.\omega$ then there are $m, n < \omega$ such that $\beta < \omega^\alpha.m, \gamma < \omega^\alpha.n$. But then $\beta + \gamma < \omega^\alpha.m + \omega^\alpha.n = \omega^\alpha.(m + n) < \omega^\alpha.\omega$. Hence $\omega^{\alpha+1}$ is closed under $+$ (we did not use the induction hypothesis here).

Limit case: $\omega^\lambda = \bigcup_{\beta<\lambda} \omega^\beta$. The sequence $(\omega^\beta)_{\beta<\lambda}$ is cofinal in ω^λ and each of its membes is closed under $+$ by induction hypothesis. Take any $\gamma, \delta < \omega^\lambda$. Then there is $\beta < \lambda$ such that $\gamma, \delta < \omega^\beta$. But then $\gamma + \delta < \omega^\beta < \omega^\lambda$ and so ω^λ is closed under $+$. ∎

Proof. [of Proposition 3.3.21] First show that ε_0 is closed under $+$. Now the sequence $(c(n))_{n<\omega}$ is clearly cofinal in ε_0 and by Lemma 3.3.23 each $c(n)$ is closed under $+$. Hence ε_0 is closed under $+$. Next show ε_0 is closed under multiplication. Take any $\gamma, \delta < \varepsilon_0$. Then there is n such that $\gamma, \delta < c(n + 1)$. But then

$$\gamma.\delta < \omega^{c(n)}.\omega^{c(n)} = \omega^{c(n)+c(n)} < \omega^{c(n+1)}$$

since $c(n + 1)$ is closed under $+$. Hence $\gamma.\delta < \varepsilon_0$ as required. Finally show ε_0 is closed under exponentiation. Take any $\gamma, \delta < \varepsilon_0$. Then as before there is n such that $\gamma, \delta < c(n + 1)$. But then

$$\gamma^\delta < c(n + 1)^{c(n+1)} = (\omega^{c(n)})^{c(n+1)} = \omega^{c(n).c(n+1)} < \omega^{c(m)}$$

for some m since ε_0 is closed under multiplication. Hence $\gamma^\delta < \varepsilon_0$ as required.

Now suppose for a contradiction that $\omega < \beta < \varepsilon_0$ and β is closed under exponentiation. Let n be least such that $\beta \leq c(n+1)$. Then $c(n)$ and ω are both $< \beta$. Therefore $\omega^{c(n)} < \beta$. But $\omega^{c(n)} = c(n+1) \geq \beta$. Contradiction. ∎

Exercise 3.3.24. Characterize the ordinals $< \varepsilon_0$ which are closed under (i) addition, (ii) multiplication.

Recall that a set X is *countable* if it is finite or can be put in bijection with **N**. An ordinal is countable if it is countable as a set. It is clear that if α is countable and $\beta < \alpha$, then β is also countable. Hence the countable ordinals form an initial segment of *On*. We let ω_1 denote the least uncountable ordinal. Notice that the set of countable ordinals is uncountable, having order type ω_1. This has the important consequence that even though we can find a mapping into **N** of an arbitrarily large countable ordinal (and therefore a representation for it and all its predecessors), we cannot cope with all countable ordinals by a single mapping.

3.4 The general case

Armed with some basic ideas about ordinals, let us return in this subsection to our study of inductive definitions. In Section 3.1.2, we showed the existence of fixed points under the assumption that our operator Γ was monotonic and continuous. Now we show how the continuity condition can be dropped.

Theorem 3.4.1. *Let $(A, \sqsubseteq)$ be a cpo and let $\Gamma : A \to A$ be monotonic. Then Γ has a lfp which is also the least pre-fixed point.*

In the proof, as before, we construct $\mu\Gamma$ as the lub of a chain. The difference is that in the more general case the chain may have length $> \omega$. For any $x \in A$, define by transfinite recursion

$$\Gamma^0(x) = x$$
$$\Gamma^{\alpha+1}(x) = \Gamma(\Gamma^\alpha(x))$$
$$\Gamma^\lambda(x) = \bigsqcup_{\beta < \lambda} \Gamma^\beta(x)$$

This is well-defined since A is a cpo.

Lemma 3.4.2. *If $x \sqsubseteq \Gamma(x)$ then for any $\alpha < \beta, \Gamma^\alpha(x) \sqsubseteq \Gamma^\beta(x)$.*

Proof. Left to the reader. ∎

Proof. [of Theorem 3.4.1] Consider the chain $(\Gamma^\alpha(\bot)), \alpha \in$ On. It is increasing by Lemma 3.4.2 since $\bot \sqsubseteq \Gamma(\bot)$. If for all $\alpha, \Gamma^\alpha(\bot) \neq \Gamma^{\alpha+1}(\bot)$ then we would have as many members of the chain as there are ordinals, which is impossible, since A is a set and *On* is a class which is too big to be a set.

So let α be least such that $\Gamma^{\alpha+1}(\bot) = \Gamma^{\alpha}(\bot)$. Then $\Gamma^{\alpha}(\bot)$ is plainly a fixed point of Γ . Let x be any prefixed point of Γ . Then it is easy to show by induction that $\Gamma^{\beta}(\bot) \sqsubseteq x$ for all β . It follows that $\Gamma^{\alpha}(\bot)$ is the desired least prefixed point $\mu\Gamma$ of Γ.   ∎

Remarks 3.4.3.

1. Notice that once we have reached α such that $\Gamma^{\alpha+1}(\bot) = \Gamma^{\alpha}(\bot), \Gamma^{\beta}(\bot)$ remains constant for $\beta \geq \alpha$.

2. The proof can be modified to construct the least fixed point above *any* post-fixed point x, not just $\bot$. A special case here is an operator Γ which is *inclusive*, meaning that for all $x, x \sqsubseteq \Gamma(x)$, in other words $\Gamma(x)$ includes x. Then our proof will construct a lfp above any point x.

3. It is clear that the Park Induction Principle (3.1.8) still holds with a monotonic but not necessarily continuous operator.

Definition 3.4.4. Let $\Gamma : A \to A$ be monotonic. The least ordinal α such that $\Gamma^{\alpha+1}(\bot) = \Gamma^{\alpha}(\bot)$ is called the *closure ordinal* of Γ . It is written $|\Gamma|$.

If we added a continuity condition we would have $\Gamma^{\omega+1}(\bot) = \Gamma^{\omega}(\bot)$, so that $|\Gamma| \leq \omega$. So in Section 2.1 we were dealing with inductive definitions with closure ordinal $\leq \omega$, which are the most common ones.

If A has no uncountable increasing chains then the chain $(\Gamma^{\alpha}(\bot)), \alpha \in On$ must close off at some countable ordinal.

Proposition 3.4.5. *Let B be countable and let Γ be a monotonic operator on $(\mathbf{P}B, \subseteq)$. Then $|\Gamma|$ is countable.*

Proof. It is enough to show that $(\mathbf{P}B, \subseteq)$ has no uncountable increasing chains. Suppose that $(X_{\alpha})_{\alpha<\gamma}$ is a strictly increasing chain of subsets of B. Let b_{α} be some member of $X_{\alpha+1} - X_{\alpha}$. Then each b_{α} is different, and $\{b_{\alpha} \mid \alpha < \gamma\}$ must be countable. Hence γ is countable.   ∎

For an example of a property defined using an inductive operator with closure ordinal $> \omega$ we return to the process example of Section 3.1.3. We showed how

$$E(p) \text{ iff } p \text{ eventually terminates}$$

can be obtained as the lfp of the operator

$$\Gamma(X) = \text{next}(f\!f) \vee \text{next}(X)$$

In order to justify this we had to show that Γ is continuous, and this depended on $\{q \mid p \to q\}$ being finite for each p. Suppose we drop this assumption. Then Γ is still monotonic, and so Γ has a lfp $\mu\Gamma$ by Theorem 3.4.1. It is instructive to see why $\Gamma^\omega(\varnothing)$ may not be strong enough to capture E. Consider the process p_1 with paths of every finite length shown in the diagram

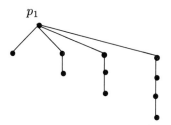

Fig. 3.6

$\Gamma^n(\varnothing)$ says that a process must terminate in $< n$ steps. So for each $n < \omega, p_1 \nvDash \Gamma^n(\varnothing)$. Hence $p_1 \nvDash \Gamma^\omega(\varnothing)$. However if $p_1 \to q$ then $q \vDash \Gamma^n(\varnothing)$, some n. Hence $p_1 \vDash \Gamma^{\omega+1}(\varnothing)$, so that $p \vDash E$. As a further example let p_2 be.

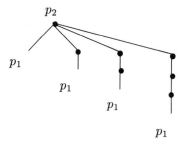

Fig. 3.7

Then one can check that $p_2 \vDash \Gamma^{\omega+\omega+1}(\varnothing)$ but not $\Gamma^{\omega+\omega}(\varnothing)$.

What bound can we put on $|\Gamma|$? Let us assign an ordinal to each process satisfying E, which we shall call its *rank*.

- If $\{q \mid p \to q\} = \varnothing$ then rank$(p) = 0$.

- If all q such that $p \to q$ have received a rank,
  then rank $(p) = (\bigcup_{p \to q}\text{rank}(q)) + 1$.

Then rank is a partial function from P to On, and those processes which receive a rank are precisely those which satisfy E. We can make a stronger statement, namely, if $\text{rank}(p) = \alpha$ then α is least such that $p \vDash \Gamma^{\alpha}(\varnothing)$. This may be checked on the example processes p_1, p_2. So $|\Gamma| = \bigcup_{p \in P} \text{rank}(p)$. This obviously depends on P.

If P is finite-branching then every member of E must have finite rank, and $|\Gamma| \leq \omega$. If P is countably-branching, that is, for every $p \in P, \{q \mid p \to q\}$ is countable, then every member of E must have countable rank. To see this, recall that a countable union of countable sets is countable, so that when we define rank $(p) = (\bigcup_{p \to q} \text{rank}(q)) + 1$, if we assume that $\text{rank}(q)$ is countable for each q, then so must be $\text{rank}(p)$. It follows that $|\Gamma| \leq \omega_1$.

Our reasoning has been somewhat informal in the last paragraph. We now indicate how it may be justified.

Definition 3.4.6. Let $<$ be a binary relation on a set X. Relation $<$ is *well-founded* if it has no infinite decreasing chain, viz. there is no sequence

$$x_0 > x_1 > \ldots > x_n > \ldots$$

of members of X.

Well-foundedness is a more general version of well-ordering (compare Characterization 3.3.3).

Principle 3.4.7 (Well-founded Induction). Let $(X, <)$ be well-founded. Let P be some property of members of X. Then

if $\forall x \in X (\forall y < x P(y) \to P(x))$ then $\forall x \in X P(x)$.

Proof. Exercise.   ∎

As ever there is an associated method of definition by recursion which is much as for a well-ordering (Proposition 3.3.14).

Now in our example clearly if we define $<$ on P by

$$q < p \text{ iff } p \to q$$

then $<$ is well-founded on E, and so we can justify the definition of rank as a well-founded recursion (clearly the definition of rank can be applied to any set with a well-founded relation).

4 Recursion theory

In this section we examine what is perhaps the heart of recursion theory, namely the fixed point or recursion theorems, which assert the existence of computable solutions to recursive definitions. We shall assume some slight acquaintance with λ-calculus in our explanatory material, as the connections are too close to ignore in a work of this kind, but the development will not depend on this. After examining the Second Recursion Theorem in Section 4.1, we see in Section 4.2 how we can approach recursion theory axiomatically via the notion of an acceptable programming system. In Section 4.3 we revisit operators on functions and show that if they are computable then they have a computable fixed point (the First Recursion Theorem). Finally in Section 4.4 we consider the computability of operators on sets and apply this to logic, concluding by discussing the undecidability of first-order logic.

4.1 Fixed point theorems

In this section let us assume some coding of programs such as that of Turing machines given in Section 1.5.1. Let the programs be enumerated as $P_0, P_1, \ldots$ and let $\{e\}_k$ denote the k-ary function computed by P_e. Recall the S–m–n Theorem 2.4.1 which stated that for any index e and m-tuple $\mathbf{x}$, n-tuple $\mathbf{y}$,

$$\{e\}_{m+n}(\mathbf{x}, \mathbf{y}) = \{S_n^m(e, \mathbf{x})\}_n(\mathbf{y})$$

This simple result gives us two profound fixed point theorems.

Theorem 4.1.1 (Rogers Fixed Point Theorem). *Let f be any computable unary function. Then there exists an index e which is a fixed point for f in the sense that for any x*

$$\{e\}(x) = \{f(e)\}(x)$$

This theorem has some remarkable consequences. For instance let f be the successor function. From the theorem we see that there is an index e such that P_e computes the same function as P_{e+1}. It does not matter what the details are of the particular coding we have chosen.

Before giving the proof, it is worth showing where the idea comes from, as otherwise the proof may seem baffling. In the (untyped) λ-calculus, one can have self-application, where a term can be applied to itself, serving as both function and argument. One can then form paradoxical terms such as $(\lambda x.xx)(\lambda x.xx)$ which β-reduces to itself. Furthermore each term f has a fixed point, that is, a term t such that $ft = t$ (where equality here means equivalence under β-reduction). The term t can be explicitly constructed from f as $(\lambda y.f(yy))(\lambda y.f(yy))$. Then

$$(\lambda y.f(yy))(\lambda y.f(yy)) \to f((\lambda y.f(yy))(\lambda y.f(yy)))$$

showing that $t = ft$.

Let us alter this a little. Let $g = \lambda y.\lambda x.f(yy)x, e = gg$. Then

$$
\begin{aligned}
ex &= ggx \\
&= f(gg)x \\
&= fex
\end{aligned}
$$

This shows e=fe with the use of the law of extensionality $(\lambda x.tx = t)$. The same law also gives $g = \lambda y.f(yy)$, so that $e = t$.

Let us apply these ideas to recursion theory. We cannot have true self-application of computable functions to themselves, as they must of course take natural number arguments. However we can apply a program to its own index. Thus $\{e\}(e)$ makes perfect sense for any $e \in \mathbf{N}$.

Proof. [of Theorem 4.1.1] Let $F(y, x) = \{f(S_1^1(y, y))\}(x)$. Then F is (partial) computable, using the existence of a universal program (Theorem 2.2.1) and the computability of S_1^1. Let g be an index for F, so that $\{g\}(y, x) = \{f(S_1^1(y, y))\}(x)$. Now let $e = S_1^1(g, g)$. We check that e is as required:

$$
\begin{aligned}
\{e\}(x) &= \{S_1^1(g, g)\}(x) \\
&= \{g\}(g, x) && \text{defn of } S_1^1 \\
&= F(g, x) && \text{defn of } g \\
&= \{f(S_1^1(g, g))\}(x) && \text{defn of } F \\
&= \{f(e)\}(x)
\end{aligned}
$$

∎

Theorem 4.1.2 (Second Recursion Theorem (Kleene)). *Let f be any (partial) computable 2-ary function. Then there is an index e such that for any x,*

$$\{e\}(x) = f(e, x).$$

Proof. This theorem also has a λ-calculus counterpart:
    Let $g = \lambda y.\lambda x.f(yyx), e = gg$. Then

$$
\begin{aligned}
ex &= ggx \\
&= f(ggx) \\
&= f(ex)
\end{aligned}
$$

So ex is another fixed point for f (cf. [Barendregt, 1984], Proposition 6.5.2(ii)).

Now for the proof proper: let $F(y,x) = f(S_1^1(y,y),x)$. Then F is computable, since f and S_1^1 are. Let g be an index for F. Now let $e = S_1^1(g,g)$. We check that e is as required:

$$\begin{aligned}
\{e\}(x) &= \{S_1^1(g,g)\}(x) \\
&= \{g\}(g,x) \\
&= f(S_1^1(g,g),x) \\
&= f(e,x)
\end{aligned}$$

∎

As an immediate consequence we have the existence of a self-reproducing program: let $f(y,x) = y$. Then Theorem 4.1.2 constructs for us an e such that for any x, $\{e\}(x) = e$, in other words a program P_e which on any input x outputs its own code e.

As noted in [Machtey and Young, 1978] we can use Theorem 4.1.1 to give a rapid proof of Rice's Theorem 2.5.1. Let B be a non-trivial set of unary partial recursive functions. Let e, e' be such that $\{e\} \in B, \{e'\} \notin B$. Define f by

$$f(n) = \begin{cases} e' & \text{if } \{n\} \in B, \\ e & \text{if } \{n\} \notin B \end{cases}$$

If $?\{x\} \in B$? is solvable, then f is total recursive. There must therefore be a fixed point n such that $\{f(n)\} = \{n\}$. But this is clearly impossible by construction of f. We conclude that $?\{x\} \in B$? is unsolvable.

Remark. We can deduce Theorem 4.1.1 from Theorem 4.1.2 as follows: let $F(e,x) = \{f(e)\}(x)$. Then F is computable using the existence of a universal program. So by Theorem 4.1.2 there is e such that $\{e\}(x) = \{f(e)\}(x)$ as required for Theorem 4.1.1.

The fixed point theorems are of very wide application in recursion theory, but are in a sense of less practical importance than the First Recursion Theorem, which we treat in Section 4.3 below. This is because they are typically used where we want to solve a recursive definition of a program in terms of itself. In fact this is not what we usually do when writing recursive definitions — normally these are extensional. For instance a divide-and-conquer list-sorting algorithm will be defined in terms of its own action (on smaller lists), but we do not pay any attention to the actual code of the program. All that matters is the input/output behaviour; the program is viewed as a black box. Examples such as the self-reproducing program, where the actual code really matters for the definition are rare in 'day-to-day' computing.

4.2   Acceptable programming systems

In Section 1 we described how a variety of computational formalisms gave
rise to the same notion of computability. We also described how the computable functions can be coded by natural numbers. The particular coding
we gave involved a great many arbitrary choices. For instance we might
have based it on register machines rather than Turing machines, and we
might have chosen a different pairing function, or different numbers to represent the individual instructions. We initially used the coding to prove
that every Turing-computable function is recursive. But then we saw that
it could be used to express the basic results of recursion theory, namely the
Universal Program Theorem (2.2.1) and the *S–m–n* Theorem (2.4.1). Once
these have been obtained, the reader will have observed that we did not
need to appeal to any specific features of our coding in proving the fixed
point theorems of Section 4.1. We would like to establish that it does not
matter which coding we chose. In this subsection we shall show that any
reasonable codings are computably isomorphic. A 'reasonable' coding will
simply be one that satisfies the two basic results mentioned above. This is
called an acceptable programming system.

Before giving the definition we must clear up a possible confusion. We
have used the word universal in two slightly different senses. We have
said that a computational formalism is universal if it can compute every
effectively computable function; thus Church's Thesis tells us that Turing
machines are universal. We have also said that a program or function is
universal if it allows us to interpret every program of some formalism (as
in the Universal Program Theorem). For this subsection let us be more
precise.

Definition 4.2.1 (as in [Kfoury *et al.*, 1982]). A *programming system*
is an assignment to each $n \in \mathbf{N}$ of a partial computable function $\{n\}_k$ (each
$k \geq 1$). It is *acceptable* if it satisfies the following:

1. It is *complete*; that is, it contains all the partial computable functions.

2. It has the *universal program property*; that is, for each k the $k+1$-ary
   universal function
   $$\Psi_k(e, x) = \{e\}_k(x)$$
   belongs to the system.

3. It satisfies the *S-m-n property*; that is, it contains a total $m + 1$-ary
   function S_n^m satisfying $\{S_n^m(e, \mathbf{x})\}_n(\mathbf{y}) = \{e\}_{m+n}(\mathbf{x}, \mathbf{y})$.

Of course the coding we gave in Section 1.5.1 is an acceptable programming
system. Also it is clear from their proofs that the fixed point Theorems
4.1.1, 4.1.2 are valid for any acceptable programming system.

Remarks 4.2.2.

1. Notice that we define completeness without reference to Church's Thesis, but simply in terms of Turing machines or any equivalent formalism.

2. There are other equivalent formulations of acceptability. For instance [Machtey and Young, 1978] have properties (1) and (2) but replace (3) by a requirement that the system should contain the composition of two functions. In some ways this is more satisfying, as composition is clearly a very basic property of any computational formalism.

Exercise 4.2.3. Show that for any acceptable programming system $\{\}$ there is a total computable function c such that for all $m, n, \{c(m,n)\} = \{m\} \circ \{n\}$, that is, for all x,

$$\{c(m,n)\}(x) = \{m\}(\{n\}(x))$$

It is clear that in any programming language, given a program P it is always possible to write a longer program which performs the same task. One simply 'pads' P with redundant instructions. This means that for any computable function there must be infinitely many programs which compute it. We shall show this for arbitrary codings, demonstrating that there can be no 'economical' coding of the computable functions - any function must be repeated infinitely often in the enumeration. The proof uses the Fixed Point Theorem:

Proposition 4.2.4. *Let $\{\}$ be an acceptable programming system. For any $e \in \mathbf{N}$ there exists $n > e$ such that $\{n\} = \{e\}$.*

Corollary 4.2.5. *For any $e \in \mathbf{N}$ there exist infinitely many n such that $\{n\} = \{e\}$.* ∎

Proof. [of Prop. 4.2.4] Take any $e \in \mathbf{N}$. Pick k such that $\{k\}$ is different from $\{0\}, \{1\}, \ldots, \{k\}$. Define

$$f(n) = \begin{cases} k & \text{if } n \le e \\ e & \text{if } n > e \end{cases}$$

Then f is certainly computable. So by Theorem 4.1.1 there is n such that $\{n\} = \{f(n)\}$. But if $n \le e$ then $\{n\} = \{k\}$, which is impossible. Hence $n > e$ and $\{n\} = \{e\}$. ∎

We remark that the above proof is non-constructive, in the sense that we do not say how k is to be found. We cannot expect k to be found recursively in e, since the problem of determining whether two numbers are codes for the same function is highly unsolvable. corollary 4.2.5 could also have been proved using Rice's Theorem 2.5.1, since any non-recursive set must be infinite. This still does not give us a way of generating other indices for programs equivalent to e. Although it is interesting to see that the Corollary can be proved quite abstractly for an arbitrary programming system, we would like to obtain a more effective padding result. Our strategy is to translate the padding of a particular programming system (the coding of Turing machines in Section 1.5.1) and translate it across to an arbitrary programming system.

Take any Turing machine M. Add some redundant instructions to get $pad(M)$, a new machine computing the same function. In terms of codes we clearly have a (primitive) recursive function pd such that for any $e, pd(e) > e$ and $\{pd(e)\} = \{e\}$. By iterating the padding we get a sequence of codes $pd^k(e)$.

Definition 4.2.6. A programming system has the *padding property* if there is a one–one total recursive function p such that for any e, k

$$\{e\} = \{p(e, k)\}$$

Proposition 4.2.7. *The coding given in Section 1.5.1 has the padding property.*

Proof. Set $p(e, k) = pd^k(e)$. ∎

Clearly similar arguments will show that any coding of a computational formalism will have the padding property. We would like to show that *any* acceptable programming system has the padding property. For this we need to be able to translate form one system to another. It is a simple matter to show that there are recursive translations between acceptable programming systems:

Proposition 4.2.8. *Let $\{\}, \{\}'$ be two acceptable programming systems. Then there is a total recursive function f such that for any $n, \{n\}' = \{f(n)\}$.*

Proof. Let Ψ' be the universal function associated with $\{\}'$ and let u be such that $\{u\}_2 = \Psi'$ (using the completeness of $\{\}$). Then for any n, x,

$$\begin{aligned}\{n\}'(x) &= \Psi'(n, x)\\ &= \{u\}_2(n, x)\\ &= \{S_1^1(u, n)\}(x)\end{aligned}$$

using the S–m–n property of $\{\}$. Now let $f(n) = S_1^1(u, n)$. ∎

However, in order to transfer the padding property from our particular coding to acceptable programming systems in general, we must prove a stronger version of Proposition 4.2.8 by making the translation function one–one.

Lemma 4.2.9. *Let* $f : \mathbf{N} \to \mathbf{N}$ *be total and computable and let* $\{\}$ *be an acceptable programming system. Then there is* i *such that* $\{f(x)\} = \{S_1^1(i, x)\}$ *for all* x *and such that* $S_1^1(i, x)$ *is one–one as a function of* x.

Proof. [as in [Machtey and Young, 1978; Kfoury *et al.*, 1982]] Our strategy is to find i such that $\{i\}_2(x, y) = \{f(x)\}(y)$ for all x, y. For then clearly $\{S_1^1(i, x)\}(y) = \{f(x)\}(y)$. But we have to impose extra conditions on i. We therefore define

$$
h(i, x, y) = \begin{cases} 0 & \text{if } \exists z < x[S_1^1(i, z) = S_1^1(i, x)] \\ 1 & \text{if } \forall z < x[S_1^1(i, z) \neq s_1^1(i, x)] \\ & \text{and } \exists z[x < z \leq y \text{ and } S_1^1(i, z) = S_1^1(i, x)] \\ \{f(x)\}(y) & \text{otherwise} \end{cases}
$$

By the Second Recursion Theorem 4.1.2 (generalized to arities greater than one) we can find i such that $\{i\}_2(x, y) = h(i, x, y)$, so that $\{S_1^1(i, x)\}(y) = h(i, x, y)$. We claim that $S_1^1(i, x)$ is one–one. For if not then we can take z, x least such that $z < x$ and $S_1^1(i, z) = S_1^1(i, x)$. But then $h(i, x, y) = 0$ for all y, and $h(i, z, y) = 1$ for all $y \geq z$. Hence $\{S_1^1(i, z)\} \neq \{S_1^1(i, x)\}$, and so $S_1^1(i, z) \neq S_1^1(i, x)$, which is a contradiction. Since $S_1^1(i, x)$ is one–one, it is clear from the definition of h that $h(i, x, y) = \{f(x)\}(y)$, and so we have $\{S_1^1(i, x)\}(y) = \{f(x)\}(y)$ for all x, y as required. ∎

Proposition 4.2.10. *Let* $\{\}, \{\}'$ *be two acceptable programming systems. Then there is a total recursive one–one function* g *such that for any* n, $\{n\}' = \{g(n)\}$.

Proof. Let f be a total recursive translation function as given by Proposition 4.2.8. Apply Lemma 4.2.9 to f to get i such that for all n, $\{S_1^1(i, n)\} = \{f(n)\}$. Then $S_1^1(i, n)$ is the required one–one translation function. ∎

Corollary 4.2.11. *Let* $\{\}, \{\}'$ *be two acceptable programming systems. If* $\{\}$ *has the padding property, then so does* $\{\}'$.

Proof. Suppose that f, g are one–one translation functions such that for all n

$$
\{n\}' = \{f(n)\}
$$
$$
\{n\} = \{g(n)\}'
$$

and p is a padding function for $\{\}$. Take any n. Then for all k

$$
\begin{aligned}
\{n\}' &= \{f(n)\} \\
&= \{p(f(n), k)\} \\
&= \{g(p(f(n), k))\}'
\end{aligned}
$$

Let $p' = g(p(f(n), k))$. Then p' is one–one, and is the required padding function for $\{\}'$. ∎

Now we are able to prove the promised isomorphism between acceptable programming systems.

Theorem 4.2.12 (Rogers Isomorphism Theorem). *Let* $\{\}, \{\}'$ *be two acceptable programming systems. Then there is a recursive bijection* f *such that for any* n, $\{n\}' = \{f(n)\}$.

Proof. We use a 'back-and-forth' argument similar to that used to prove the Schroeder-Bernstein Theorem, which states that given two sets A, B, if there are one–one maps from A to B and from B to A, then there is a bijection between A and B. We shall find it convenient to use the padding property. So suppose by Proposition 4.2.10 that we have one-one functions f, g such that for all n

$$\{n\}' = \{f(n)\}$$
$$\{n\} = \{g(n)\}'$$

and padding functions p, p' for $\{\}, \{\}'$. We build up the bijection in stages. At stage k we have a set A_k of k ordered pairs

$$\{(n_0, n_0'), (n_1, n_1'), \ldots, (n_{k-1}, n_{k-1}')\}$$

such that

$$\left. \begin{array}{l} \text{if } i < j < k \text{ then } n_i \leq n_j \text{ and } n_i' \leq n_j' \\ \text{for all } i < k \ \{n_i\} = \{n_i'\}' \end{array} \right\} \ *$$

We get A_{k+1} by adding a new pair (n_k, n_k') as follows. If k is even we choose n_k to be the least number not in $\{n_0, \ldots, n_{k-1}\}$. We know that $\{g(n_k)\}' = \{n_k\}$. So we could let n_k' be $g(n_k)$, except that possibly $g(n_k) \in \{n_0', \ldots, n_{k-1}'\}$. If this is the case then we use the padding function p' and choose n_k' to be $p'(g(n_k), m)$, where m is least such that $p'(g(n_k), m) \notin \{n_0', \ldots, n_{k-1}'\}$. We have now obtained A_{k+1} and it is easy to see that $(*)$ is satisfied. If k is odd then we proceed in similar fashion, but we take n_k' to be the least number not in $\{n_0', \ldots, n_{k-1}'\}$ and map it across to some new n_k using the padding function p if necessary. Now let $A = \bigcup_k A_k$. It is not hard to see that A defines a bijection from $\mathbf{N}$ to $\mathbf{N}$ such that if $(n, n') \in A$ then $\{n\} = \{n'\}'$. The back-and-forth procedure ensures that no number is omitted. Furthermore the procedure we followed is plainly effective, and so the bijection is recursive by Church's Thesis. Notice that we did not use the fact that f, g were one–one. However we did use the padding property (Corollary 4.2.11), and so we were relying indirectly on Proposition 4.2.10. ∎

Remark 4.2.13. The results from 4.2.4 to 4.2.12 can be generalized straightforwardly to arities greater than one.

4.3 Recursive operators

We have seen in Section 1.3.3 how functional programs may be given meaning as the least fixed point of a functional or operator on functions. We have looked at the general theory of monotonic and continuous operators in Section 3 and seen that they do have least fixed points. But are these fixed points computable, and under what circumstances? To be more particular, in Section 1.5 we showed that every FL-computable function is computable, by coding up the operational semantics using recursive predicates. In Section 3.2.3 we showed that the operational semantics coincides with the least fixed point semantics. Can we show directly that these lfps are computable? Clearly we must put some condition of computability on the operator. In this subsection we show how this may be done. For simplicity we shall not consider undefined values, and restrict ourselves to operators $\Phi : F_m \rightarrow F_n$, where F_k denotes the partial functions $\mathbf{N}^k \rightarrow_p \mathbf{N}$. Our treatment owes much to [Cutland, 1980; Rogers, 1967].

What properties do we expect a computable operator to have? It should certainly be *monotonic*. Given $\Phi : F_m \rightarrow F_n$ and $g \in F_m, \mathbf{x} \in \mathbf{N}^n$, we insist that calculating $\Phi(g)(\mathbf{x})$ only requires a finite amount of information about g, since we cannot manipulate infinite objects directly. Let θ range over partial functions with finite domains. We want

$$\Phi(g)(\mathbf{x}) = y \text{ implies } \Phi(\theta)(\mathbf{x}) = y, \text{ some finite approximant } \theta \sqsubseteq g \quad (*)$$

In fact F_k is a cpo for any k, with each $f \in F_k$ being the sup of its finite approximations, so that $(*)$ is equivalent to Φ being *continuous* as in Definition 3.1.3. We are saying that Φ is determined by its action on finite elements of F_m.

We must further impose the condition that $\Phi(\theta)(\mathbf{x})$ is calculated effectively. This must mean that there is a partial recursive function $\varphi : \mathbf{N}^{n+1} \rightarrow_p \mathbf{N}$ such that

$$\Phi(\theta)(\mathbf{x}) = \varphi(\theta, \mathbf{x})$$

Here we freely identify θ with some coding of it as a natural number.

Definition 4.3.1. Let $\Phi : F_m \rightarrow F_n$. Then Φ is a *recursive operator* if

1. Φ is monotonic

2. Φ is continuous

3. there is a partial recursive function $\varphi : \mathbf{N}^{n+1} \rightarrow \mathbf{N}$ such that for any finite $\theta \in F_m$ and any $\mathbf{x} \in \mathbf{N}^n, \Phi(\theta)(\mathbf{x}) = \varphi(\theta, \mathbf{x})$.

Examples 4.3.2.

1. Any FL program will give us a recursive operator. In particular the Ackermann function

$$A(0, y) = y + 1$$
$$A(x + 1, 0) = A(x, 1)$$
$$A(x + 1, y + 1) = A(x, A(x + 1, y))$$

(discussed in Section 1.4) gives us an operator $\Phi_A : F_2 \to F_2$. This is monotonic and continuous — in fact the calculation of $\Phi_A(g)(\mathbf{x})$ depends on at most two values of g. It is clear that, for any finite $\theta \in F_2$, $\Phi_A(\theta)(\mathbf{x})$ can be calculated effectively in θ and x. Hence by Church's Thesis it is computable by a recursive function φ. We conclude that Φ_A is recursive.

2. The minimization operator $\min^n : F_{n+1} \to F_n$ (Definition 1.4.5) is recursive.

We know that any monotonic and continuous operator has a lfp (Theorem 3.1.7). So any recursive operator Φ has a lfp f_Φ. We wish to show that f_Φ is recursive. First we show the following:

Theorem 4.3.3 (Myhill–Shepherdson Theorem). *Let $\Phi : F_m \to F_n$ be recursive. Then there is a total computable h such that $\Phi(\{e\}_m) = \{h(e)\}_n$.*

Proof. We first show that $g(e, \mathbf{x}) = \Phi(\{e\}_m)(\mathbf{x})$ is computable as a function of e and $\mathbf{x}$. Now

$$g(e, \mathbf{x}) = y \text{ iff } \exists \theta (\theta \subseteq \{e\}_m \wedge \varphi(\theta, \mathbf{x}) = y)$$

The predicate $\theta \subseteq \{e\}_m$ is semi-decidable by checking that for each of the finitely many $\mathbf{z}$ in the domain of θ, $\{e\}_m(\mathbf{z})$ is defined and equal to $\theta(\mathbf{z})$. So $(\theta \subseteq \{e\}_m \wedge \varphi(\theta, \mathbf{x}) = y)$ is r.e. by Exercise 2.3.10 and the graph of g is r.e. by Theorem 2.3.13. Hence g is recursive by Exercise 2.3.12. Now by the S–m–n Theorem 2.4.1 there is a total computable h such that $g(e, \mathbf{x}) = \{h(e)\}_n(\mathbf{x})$. ∎

It is worth remarking that a trivial consequence of Theorem 4.3.3 is that if $f \in F_m$ is computable then so is $\Phi(f)$.

Theorem 4.3.4 (First Recursion Theorem (Kleene)). Let $\Phi : F_m \to F_m$ be recursive. Then its lfp f_Φ exists and is computable.

Proof. We have already seen that f_Φ exists. Now we know from the proof of Theorem 3.1.7 that f_Φ is constructed as the lub of $\Phi^k(\uparrow)$. In other words

$$f_\Phi(\mathbf{x}) = y \text{ iff } \exists k \text{ such that } \Phi^k(\uparrow)(\mathbf{x}) = y$$

Let e_0 be an index for $\uparrow$. Then by Theorem 4.3.3 $\Phi^k(\uparrow)(\mathbf{x}) = \{h^k(e_0)\}(\mathbf{x})$, which is plainly computable in k and $\mathbf{x}$. Hence the graph of f_Φ is r.e., and by Exercise 2.3.12 f_Φ is computable. ∎

We can now conclude directly that the Ackermann function is computable (which we already knew from our proof in Section 1.5 that all *FL*-computable functions are recursive).

4.4   Inductive definitions and logics

In this subsection we consider the effectiveness of operators on sets (inductive definitions) and apply this to logics. We then discuss the undecidability of first-order logic.

Definition 4.4.1. An operator $\Gamma : \mathbf{PN} \to \mathbf{PN}$ is *recursive (r.e.)* if it is monotonic and continuous and the predicate

$$x \in \Gamma(X)$$

is recursive (r.e.) where X ranges over finite subsets of $\mathbf{N}$ (using some coding).

It turns out that recursive operators on sets are not particularly well-behaved. They can have lfp's which are not recursive, as the following example shows:

Example 4.4.2. Let start_x be the code of the starting configuration of Turing machine x applied to input x. For any x, y, z let $\text{next}(\langle x, y \rangle) = \langle x, z \rangle$ where z is the next configuration after y in the computation with TMx, if y is a possible configuration and not halting, $\text{next}(\langle x, y \rangle) = 0$ if y is halting, and y if y is not a possible configuration. Here we assume that 0 is not (the code of) a possible configuration. Let

$$\Gamma(X) = \{\langle x, \text{start}_x \rangle \mid x \in \mathbf{N}\} \cup \{\text{next}(\langle x, y \rangle) \mid \langle x, y \rangle \in X\}$$

Then Γ is easily seen to be recursive, and $\mu\Gamma$ is the set of all pairs $\langle x, y \rangle$ where y is a configuration of the computation of TM x on input x (or $y = 0$). Plainly

$$\langle x, 0 \rangle \in \mu\Gamma \text{ iff } \{x\}(x) \downarrow$$

It follows from the unsolvability of the halting problem that $\mu\Gamma$ is not recursive.

However the lfp of an r.e. operator is r.e. We have the following analogues of the Myhill–Shepherdson Theorem and the First Recursion Theorem:

Theorem 4.4.3. *If* $\Gamma : \mathbf{PN} \to \mathbf{PN}$ *is r.e. then there is a total computable* h *such that* $\Gamma(W_e) = W_{h(e)}$.

Proof. Similar to that of Theorem 4.3.3.   ∎

Theorem 4.4.4. *If* $\Gamma : \mathbf{PN} \to \mathbf{PN}$ *is r.e. then* $\mu\Gamma$ *is r.e.*

Proof. Similar to that of Theorem 4.3.4.   ∎

As an application of this, we note that we can associate an inductive operator with any logic L. Roughly speaking $\Gamma(X)$ is the set of formulas which
are immediate consequences of some set of formulas in X by a rule of L. Γ
is clearly monotonic, and will be continuous if the rules of L are all finitary.
It is reasonable to require that the rules of L are such that Γ is r.e.; in fact
it will probably be recursive. So we conclude by Theorem 4.4.4 that $\mu\Gamma$,
the set of all theorems of L, is r.e. We can argue this more informally by
simply enumerating in some systematic fashion all the proofs in L and all
theorems which they establish.

In particular the set of theorems of classical first-order logic is r.e. The
following famous theorem shows that we cannot do better:

Theorem 4.4.5 (Church's Theorem). *The set of theorems of first-order
logic is not recursive.*

This is often expressed as: first-order logic is not decidable. We sketch a
proof, the details of which are given in [Boolos and Jeffrey, 1989], Chapter
10. As the reader may expect, it involves expressing the halting of a Turing
machine computation as a theorem. Given a TM M we can effectively find
a finite set Δ of sentences expressing the program of M together with
the rules governing its operation, and a sentence H expressing the eventual
halting of a computation. These will be such that M eventually halts when
started on a blank tape iff $\Delta \vdash H$ (H is a consequence of Δ). But Δ is
finite, so that $\Delta \vdash H$ iff $\vdash \bigwedge \Delta \to H$. So if first-order logic is decidable we
can decide the problem of whether an arbitrary TM eventually halts when
started on a blank tape. But this is undecidable (Example 2.5.2), and so
we conclude that first-order logic is undecidable.

Further Reading

For the further development of recursion theory, including the study of degrees of reducibility (alluded to in section 3.6) the reader is referred to the
standard text [Rogers, 1967].

For the relationship of recursion theory with proof theory and Gödel's theorems on the incompleteness of formal systems, attractive expositions can be found in [Boolos and Jeffrey, 1989; Epstein and Carnielli, 1989]. The latter also covers the rôle of Church's Thesis within constructive mathematics.

For connections with typed λ-calculus, including Gödel's T (primitive recursion in higher types), see [Barendregt, 1992; Hindley and Seldin, 1986; Girard *et al.*, 1989].

For connections with formal language theory and complexity theory, see [Hopcroft and Ullman, 1979] and [Brairierd and Landweber, 1974].

For the various ways in which recursion theory has been generalized to sets and ordinals, as well as its place within mathematical logic, see [Barwise, 1977].

For computibility on data types other than the natural numbers, see [Meinke *et al.*, 1992].

Acknowledgements

My thanks go to the editors of this Handbook, and in particular Professor Dov Gabbay. Also to Professors John Tucker and Matthew Hennessy for reviewing drafts of this chapter, and to other contributors to this volume, including Professors Henk Barendregt and Jan-Willem Klop and Dr Mike Smyth. I am also indebted to Thomas Jensen, Bent Thomsen, Irek Ulidowski, Simon Thompson, Gillian Hill and Sarah Liebert for reading draft versions. Finally I wish not least to thank Mrs Jane Spurr for her able preparation of the manuscript.

References

[Abramsky, 1983] S. Abramsky. Experiments, powerdomains and fully abstract models for applicative multiprogramming. In *ICALP*, 1983.

[Barendregt, 1984] H.P. Barendregt. *The Lambda Calculus: Its Syntax and Semantics*. North-Holland, second edition, 1984.

[Barendregt, 1992] H.P. Barendregt. λ-calculus. In D. M. Gabbay S. Abramsky and T. S. E. Maibaum, editors, *Handbook of Logic in Computer Science, Volume 2*. Oxford University Press, 1992.

[Barwise, 1977] J. Barwise, editor. *Handbook of Mathematical Logic*. North-Holland, 1977.

[Bell and Machover, 1977] J.L. Bell and M. Machover. *A Course in Mathematical Logic*. North Holland, 1977.

[Bird, 1976] R. Bird. *Programs and Machines*. Wiley, 1976.

[Boolos and Jeffrey, 1989] G.S. Boolos and R.C. Jeffrey. *Computability and Logic*. Cambridge, third edition, 1989.

[Bradfield and Stirling, 1990] J. Bradfield and C. Stirling. Verifying temporal properties of processes. In J. C. M. Baeten and J-W. Klop, editors, *Concur '90*. Springer, Berlin, 1990. Lecture Notes in Computer Science, Vol 458.

[Brairierd and Landweber, 1974] W. S. Brairierd and L. H. Landweber. *Theory of Computation*. Wiley, 1974.

[Cutland, 1980] N. Cutland. *Computability: An Introduction to Recursive Function Theory*. Cambridge, 1980.

[Davis, 1958] M. Davis. *Computability and Unsolvability*. McGraw-Hill, London, 1958.

[Epstein and Carnielli, 1989] R. L. Epstein and W. A. Carnielli. *Computability: Computable Functions, Logic, and the Foundations of Mathematics*. Wadsworth and Brooks/Cole, 1989.

[Girard et al., 1989] J-Y. Girard, Y. Lafont, and P. Taylor. *Proofs and types*. Cambridge University Press, 1989.

[Gödel, 1931] K. Gödel. Über formal unentscheidbare sätze der principia mathematica und verwandte systeme, i. *Monatshefte für Mathematik und Physik*, 38:173–198, 1931. (Translated into English in [Davis, 1965]).

[Hindley and Seldin, 1986] J. R. Hindley and J. P. Seldin. *Introduction to combinators and λ-calculus*. Cambridge University Press, 1986.

[Hopcroft and Ullman, 1979] J. E. Hopcroft and J. D. Ullman. *Introduction to Automata Theory, Languages and Computation*. Addison-Wesley, 1979.

[Kfoury et al., 1982] A.J. Kfoury, R.N. Moll, and M.A. Arbib. *A programming approach to computability*. Springer, New York, 1982.

[Klop, 1992] J.-W. Klop. Term rewriting systems. In D. M. Gabbay S. Abramsky and T. S. E. Maibaum, editors, *Handbook of Logic in Computer Science, Volume 2*. Oxford University Press, 1992.

[Loeckx and Sieber, 1987] J. Loeckx and K. Sieber. *The Foundations of Program Verification*. Wiley-Teubner, second edition, 1987. First edition published in 1984.

[Machtey and Young, 1978] M. Machtey and P. Young. *An Introduction to the General Theory of Algorithms*. North-Holland, New York, 1978.

[McCarthy, 1960] J. McCarthy. Recursive functions of symbolic expressions and their computation by machine, part I. *Communciations of the ACM*, April 1960.

[Meinke and Tucker, 1992] K. Meinke and J. Tucker. Universal algebra. In D. M. Gabbay S. Abramsky and T. S. E. Maibaum, editors, *Handbook of Logic in Computer Science, Volume 1*. Oxford University Press, 1992.

[Meinke et al., 1992] K. Meinke, J. Tucker, and J. Zucker. Computability on abstract data types. In D. M. Gabbay S. Abramsky and T. S. E. Maibaum, editors, *Handbook of Logic in Computer Science, Volume 5*. Oxford University Press, 1992.

[Moschovakis, 1974] Y. Moschovakis. *Elementary Induction on Abstract Structures*. North-Holland, 1974.

[Plotkin, 1981] G. Plotkin. A structural approach to operational sematnics. Technical Report DAIMI FN-19, Computer Science Department, Aarhus University, 1981.

[Rogers, 1967] H. Rogers. *Theory of Recursive Functions and Effective Computability*. McGraw-Hill, New York, 1967.

[Ryan and Sadler, 1992] M. Ryan and M.R. Sadler. Valuation systems and consequence relations. In D. M. Gabbay S. Abramsky and T. S. E. Maibaum, editors, *Handbook of Logic in Computer Science, Volume 1*. Oxford University Press, 1992.

[Smyth, 1975] M. B. Smyth. Unique entry graphs, mu-terms and while programs. Unpublished manuscript, 1975.

[Stirling, 1992] C. Stirling. Modal and temporal logics. In D. M. Gabbay S. Abramsky and T. S. E. Maibaum, editors, *Handbook of Logic in Computer Science, Volume 2*. Oxford University Press, 1992.

Universal Algebra

K. Meinke and J. V. Tucker

Contents

> 'It has been said that 'the human mind has never invented
> a labor-saving machine equal to algebra.' If this be true, it is
> but natural and proper that an age like our own, characterised
> by the multiplication of labor-saving machinery, should be dis-
> tinguished by an unexampled development of this most refined
> and most beautiful of machines.'
>
> J. Willard Gibbs,
> Page 37 of *Multiple Algebra*,
> Proceedings of the American Association
> for the Advancement of Science,
> (35) (1887) pp. 37–66.

1   Introduction

This chapter is an introduction to the primary concepts, constructions and
results about algebras, their isomorphism and axiomatization. As with
other chapters in this *Handbook*, it is composed with the aims of teaching
the mathematical theory and explaining its relevance to problems in computer science. Among areas of application we will consider are: abstract
data types and their specification; syntax and semantics of programming
languages; and synchronous concurrent algorithms and architectures. Some
of these algebraic applications, and others such as asynchronous concurrent
processes, are developed fully as the subject matter of chapters in later volumes of the *Handbook*.

    We will begin with a short overview of the nature and origins of universal
algebra and its applications in computing; and we will discuss the contents
of the Chapter in detail.

1.1   What is universal algebra?

Algebra is about calculation. This is reflected in its ancient origins and
name; its historical development and contemporary abstract form; and in
its manifold applications, most recently in computer science.

Calculation can be formulated using a family

$$A = \langle\, A_s \mid s \in S \,\rangle$$

of sets of data, equipped with a collection F_0 of distinguished elements, each of the form

$$a \in A_s$$

for some $s \in S$; and with a collection F of basic functions, each of the form

$$f : A_{s_1} \times \ldots \times A_{s_n} \to A_s$$

for some $s_1, \ldots, s_n, \; s \in S$. These components are called *carriers, constants* and *operations*, respectively, and taken together these components constitute a *(many-sorted) algebraic structure* or, more simply, a *(many-sorted) algebra*.

On formulating an appropriate algebra, calculation can be performed directly by means of a sequence of applications of the operations of the algebra; or it can be specified indirectly by means of equations based on the operations. The formulation of algebras leads to methods for:

(i) the comparison and classification of algebras;

(ii) the construction of new algebras from existing algebras; and

(iii) the specification of algebras using axioms about their operations.

Universal algebra is a systematic theory concerned with these and other methods for the design and analysis of algebras *in general*.

1.2 Universal algebra in mathematics and computer science

In mathematics algebraists are involved mainly in the identification and detailed study of specific classes of algebras, defined axiomatically. The immense number of classes of algebras that are of interest and use to mathematicians have grown from the refinement of basic classes such as *groups* (important for symmetry); *rings and fields* (important for number systems and equations); *vector spaces* and *linear algebras* (important for linear analysis and geometry); and *Boolean algebras* and *lattices* (important for logic). The theories of many of these sub-classes of algebras are longstanding and very deep, and are widely applicable in mathematics and natural science.

Thus, in mathematics universal algebra is commonly considered to be the study of properties common to established algebraic systems; and to be useful because it organizes the important, but routine and expected, basic concepts and results for any specific class of algebras, prior to their analysis proper. However, the new classes of algebras that arise in mathematics are closely related to the familiar algebras, are few in number, and require natural adaptations of familiar ideas and results. Hence mathematicians and, indeed, algebraists often do not need to use universal algebra.

In computer science universal algebra is considered to be a general theory which is directly applicable to many problems. Whenever a specification or algorithm can be modelled by sets and functions there is an algebra. Universal algebra is needed to organize the great range of algebras that derive from the great range of data and functions that arise in computer science.

1.3 Overview of the chapter

The chapter is divided into four sections. The first two explain the basic concepts of algebra: Section 2 is a large collection of examples of algebras, chosen from computing to illustrate the concepts prior to their precise definition in Section 3.

The last two sections contain more advanced material and, for reasons of economy of space, are written in a more concise style. Section 4 is about constructions of algebras, including the subdirect product, direct and inverse limits, reduced product and ultraproduct. We emphasize ways in which these constructions formalize approximation methods in algebra.

Section 5 is about classes of algebras. It begins by explaining free, initial and final algebras, equational logic, and the proof of Birkhoff's Variety Theorem. It continues with characterization theorems for classes of algebras definable by conditional equations and equational Horn clauses.

Throughout the chapter, and especially in Section 2, we will refer to significant applications of the algebraic ideas in computer science. The material we have included has been used in abstract data type theory; concurrent process theory; hardware verification; programming language semantics; functional programming; and logic programming.

1.4 Historical notes

The history of the contemporary view of algebra and its applications in computer science is badly in need of research. It is not possible to write an accurate summary at this time; therefore we will make some remarks on just three aspects of the origins of our algebraic methods.

1.4.1 Origins of abstract algebra

An important step in the history of algebra was the development of Symbolic Algebra in the first half of the nineteenth century in Great Britain.

Symbolic Algebra transformed algebra from a science of arithmetical calculations, in which symbols stood for numbers, into a science of abstract calculations with uninterpreted symbols, using principles and rules that the symbolic expressions are assumed to obey. The symbolic calculations may, in appropriate circumstances, be interpreted as arithmetical calculations. A primary source for Symbolic Algebra is George Peacock's *A Treatise on Algebra* of 1830. For example, in Chapter 3, §78 we read the following:

> 'Algebra may be considered, in its most general form, as *the science which treats of the combinations of arbitrary signs and symbols by means of defined though arbitrary laws:* for we may *assume* any laws for the combination and incorporation of such symbols, so long as our assumptions are independent, and therefore not inconsistent with each other: in order, however, that such a science may not be one of useless and barren speculations, we choose some subordinate science as the guide merely, and not as the foundation of our assumptions, and frame them in such a manner that Algebra may become the most general form of that science, when the symbols denote the same quantities which are the objects of its operations: and as Arithmetic is the science of calculation, to the dominion of which all other sciences, in their application at least, are in a greater or less degree subject, it is the one which is usually, because most usefully, selected for this purpose.'

This *Treatise* is the first published account of Symbolic Algebra, and occupies a special place in the development of British Algebra and Logic by Peacock, D.L. Gregory, A. de Morgan, G. Boole and others.

Symbolic Algebra and the *Treatise* have much to interest the theoretical computer scientist. It is a remarkable fact that most of the key ideas about Symbolic Algebra discussed in the book had already been discussed in unpublished essays of 1822 by Charles Babbage. For details of Babbage's ideas and their relationship with those of Peacock see Dubbey [1977, 1978]. Further study of Babbage's work suggests that he should be considered as the father of abstract algebraic methods in computer science.

A second early landmark in the development of our subject is George Boole's *An Investigation of the Laws of Thought* of 1854. This book contains a detailed and polished account of an algebraic analysis of reasoning with propositions and classes. It expresses connectives algebraically, derives many algebraic laws for their behaviour, and shows how to calculate the validity of formalized arguments; it also treats probabilities. (See Kneale and Kneale [1962] and Smith [1982].)

A third landmark is Richard Dedekind's *What is a Number?* of 1888. In this book he considered the theory of different notation systems for representing natural numbers. In contemporary terms, he considered specific representations as algebras, and formulated the equivalence of representations as isomorphisms between algebras. Furthermore, Dedekind gave axioms to try to characterize the essential properties of any suitable number representation system and proved that any two algebras satisfying the axioms were isomorphic. Perhaps Dedekind should be considered as the father of the theory of abstract data types.

1.4.2 Universal algebra

The development of universal algebra must be seen in the context of the development of specific algebraic theories in the second half of the nineteenth century and the first half of the present century. A.N. Whitehead's *A Treatise on Universal Algebra* of 1898 sought a unification of algebraic theories through formal reasoning with equations. It was in G. Birkhoff's papers of 1933 and 1935 that the explicit concepts of universal algebra, subalgebra, congruence, free algebra, variety, and some fundamental results, were published.

The connections between universal algebra and lattice theory (made by Birkhoff, O. Øre, and others) and between universal algebra and model theory (made by A. Tarski, A.I. Malcev, and others), are important for our historical understanding of the subject. A useful short summary is contained in McKenzie *et al.* [1987].

Of special interest is the introduction of many-sorted universal algebras. Many-sorted relational systems appear in model theory, in A.I. Malcev's paper *Model Correspondences* published in 1959 (see Malcev [1971], Chapter 11). Many-sorted universal algebras were introduced in Higgins [1963] which generalized Birkhoff's Variety Theorem. Higgins' objective was to make available the algebra needed to formulate generators and relations for categories. In Birkhoff and Lipson [1970] many-sorted algebras are connected with automata theory.

The connections between universal algebra and category theory (made by F. Lawvere, J. Benabou and others) are important technically, and are useful for computer science.

1.4.3 Universal algebra and computer science

In the development of computer science since 1945, abstract algebra was first applied systematically in the theory of automata and formal languages. The use of power series and semigroups by M.P. Schützenberger, in papers of the period 1961–3, started an attractive and strong algebraic theory for automata that embraced earlier work of S.C. Kleene, J. Myhill, M. Rabin, D. Scott, A. Nerode and others, in the period 1956–9. This algebraic theory was further developed by specific results such as the theorems on machine synthesis in Krohn and Rhodes [1965], and general studies, such as Eilenberg and Wright [1967]. A convenient summary of some of these results can be found in Cohn [1981]; comprehensive accounts are available in Salomaa and Soittola [1978], Eilenberg [1974, 1976], Pin [1986] and Lallement [1979].

The routine use of universal algebra methods in computer science originates in the theory and practical development of programming languages in the 1960s. The problems of the period concerned machine computation with data other than numbers: symbolic computation with strings, natural language text, logical formulas etc. We are reminded of the problems addressed in Symbolic Algebra, more than a century earlier. The problem of specifying and implementing data types in programming languages, and in user's programs, led to the development of a comprehensive theory of data types, based on many-sorted universal algebra, in the 1970s.

In the theory of programs in the 1960s, computations on abstract structures had been considered by I.I. Ianov (as early as 1958), J. McCarthy, R.W. Floyd, E. Engeler, J.W. de Bakker and many others, mainly with the aim of proving mathematical results about either the equivalence or non-equivalence of programming constructs, or the correctness of programs. In the period several computer scientists were experimenting with 'algebraic ideas' in different contexts, including, D. Cooper, P. Landin, R. Burstall and R. Milner (see for example, Burstall and Landin [1969]).

The study of data types in the period has been conveniently surveyed in Gries [1978]. The use of many-sorted algebras and equations to model data types arose in unpublished work of S. Zilles (see Liskov and Zilles [1975]). The approach was developed by J.V. Guttag and J.J. Horning (in Guttag [1975] and Guttag and Horning [1978]); and especially by J.A. Goguen, J.W. Thatcher, E.G. Wagner and J.B. Wright in many influential papers, especially Goguen *et al.* [1978]. The work of this last group is discussed in Goguen [1989b].

1.5 Acknowledgements

We thank the editors of the *Handbook of Logic in Computer Science* for their invitation to write on universal algebra, and for their patient support over several years. We are indebted to the editors and our fellow authors for many stimulating discussions at Coseners House, Abingdon during the preparation of the *Handbook*. We also thank Jane Spurr for assistance in preparation of the final version of this chapter.

Our chapter has benefited considerably from discussions over many years with J.A. Bergstra, V. Stoltenberg-Hansen, B.C. Thompson and J.I. Zucker. Suggestions and encouragement at various stages of the project have been received from: J. Derrick, E. Engeler, J. Fauvel, J.A. Goguen, W. Hodges, J.W. Klop, M. Manzano, H. Simmons, M. Smyth and E. Wagner.

We are particularly indebted to K. Wicks for a very thorough review of the material which was invaluable for our work on later drafts. Several people have usefully commented on parts of the work including: B. McConnell, A. Poigné, P. Rodenburg, O. Schoett and K. Stephenson.

Parts of these notes have been used in short courses, funded by SERC, given at the Centre for Theoretical Computer Science at the University of Leeds, and at the Summer School on Algebraic Methods in Computer Science at the University College of Swansea. In addition the notes have been used in a postgraduate course on universal algebra given in the Department of Computer Systems at the University of Uppsala, and in second and third year undergraduate courses at the University College of Swansea. To the students of these courses we offer our thanks for their questions and comments.

1.6   Prerequisites

The prerequisites for the chapter are a little knowledge of set theory, algebra and logic, and a desire to learn more. The route we recommend through the chapter for a beginner is to study Sections 2 and 3 and then proceed to Section 5, studying subsections 5.1, 5.2 and 5.4. For a more experienced reader the material in Section 4 and Section 5.3 should hold more interest, though some of the examples in Section 2 may be new.

We have attempted to make this chapter self-contained except for an occasional use of Zorn's Lemma and König's Lemma.

2   Examples of algebras

As stated in the Introduction, a many-sorted algebra consists of a non-empty family of sets called carriers; a family of elements called constants; and a family of functions called operations. Whenever one meets some function $f : A \to B$ one meets an algebra with carriers A and B and operation f. The concept of an algebra is as general as that of a function.

Before developing the theory of algebras in the next section, we present many examples in order to familiarize the reader with some of the objectives of the theory. We will meet basic technical questions about subalgebras, homomorphisms and so forth in the context of algebras of Booleans; numbers; syntax; streams and arrays; programs and state transformations; concurrent algorithms and architectures; and software modules.

2.1 Some basic algebras

2.1.1 Algebras of Booleans

The set $\mathbf{B} = \{\ tt,\ f\!\!f\ \}$ of truth values or Booleans has associated with it many useful functions or connectives. For example, we assume the reader recognizes the maps

$$Not : \mathbf{B} \to \mathbf{B}$$

$$And : \mathbf{B} \times \mathbf{B} \to \mathbf{B}, \qquad Or : \mathbf{B} \times \mathbf{B} \to \mathbf{B},$$

$$Exor : \mathbf{B} \times \mathbf{B} \to \mathbf{B},$$

$$\Rightarrow : \mathbf{B} \times \mathbf{B} \to \mathbf{B}, \qquad \equiv\ : \mathbf{B} \times \mathbf{B} \to \mathbf{B},$$

$$Nand : \mathbf{B} \times \mathbf{B} \to \mathbf{B}, \qquad Nor : \mathbf{B} \times \mathbf{B} \to \mathbf{B}$$

which are normally given by truth tables. By choosing constants and a set of connectives we can make various useful algebras of Booleans such as

$$(\mathbf{B};\ tt,\ f\!\!f;\ Not,\ And)$$

$$(\mathbf{B};\ tt,\ f\!\!f;\ Not,\ Or)$$

$$(\mathbf{B};\ tt,\ f\!\!f;\ And,\ Or)$$

$$(\mathbf{B};\ tt,\ f\!\!f;\ And,\ \Rightarrow)$$

$$(\mathbf{B};\ tt,\ f\!\!f;\ Nand)$$

$$(\mathbf{B};\ tt,\ f\!\!f;\ Not,\ And,\ Or,\ \Rightarrow,\ \equiv)$$

$$(\mathbf{B};\ tt,\ f\!\!f;\ Not,\ And,\ Exor,\ Nand,\ Nor)$$

$$(\mathbf{B};\ tt,\ f\!\!f;\ Not,\ And,\ Or,\ Exor).$$

Now for X and Y non-empty sets, if $|X| = n$ and $|Y| = m$ then the number of maps $X \to Y$ is m^n. Thus, the number of k-ary functions $\mathbf{B}^k \to \mathbf{B}$ is 2^{2^k}. Furthermore, if $|X| = n$ then the number of subsets of X is 2^n. Thus, assuming operations are not duplicated, the number of algebras we can make from $\mathbf{B}$ with k-ary operations is $2^{2^{2^k}}$. In particular for $k = 2$ there are 16 binary connectives and 2^{16} algebras to be made from them.

Algebras of Booleans are used everywhere in computer science, usually to define and evaluate tests on data or to define and design digital hardware. In the case of work in hardware it is customary to use 1 for tt and 0 for $f\!\!f$, or 0 for tt and 1 for $f\!\!f$. Many designs can be developed from

$$(\{\ 0,\ 1\ \};\ 0,\ 1;\ Not,\ And,\ Exor,\ Nand,\ Nor).$$

This is clearly equivalent to the last but one algebra on $\mathbf{B}$ given above. Of course, if we use either the 1 for tt or the 0 for tt conventions then there is an obvious practical sense in which the choices result in equivalent algebras. In what precise theoretical sense can we formulate the equivalence of these different algebras? This leads us to the fundamental notion of *isomorphism* of algebras which we will study later.

We can usefully extend the design of these algebras of Booleans by introducing a special value u to model an unknown truth value in calculations. Let $\mathbf{B}^u = \mathbf{B} \cup \{\ u\ \}$. How do we incorporate the 'don't know' element in the definition of our logical connectives?

Consider the obvious principle that if one does not know the truth value of an input to a connective then one does not know the truth value of the output of the connective. This leads us to extend the definition of the connectives on $\mathbf{B}$ to connectives on $\mathbf{B}^u$ as follows. Let $f : \mathbf{B} \times \mathbf{B} \to \mathbf{B}$ be a binary connective. We define its extension $f^u : \mathbf{B}^u \times \mathbf{B}^u \to \mathbf{B}^u$ by

$$f^u(x, y) = \begin{cases} f(x, y), & \text{if } x, y \in \mathbf{B}; \\ u, & \text{if } x = u \text{ or } y = u. \end{cases}$$

By this method of definition the truth table of *And* is extended to

	tt	*ff*	*u*
tt	*tt*	*ff*	*u*
ff	*ff*	*ff*	*u*
u	*u*	*u*	*u*

This algebra for evaluating booleans may be used to raise errors and exceptions in programming constructs. However a case can be made for alternative methods. Consider the table for *And*. A reasonable alternative decision for the definition of this particular connective is given by the observation that in processing the conjunction of two tests, for any $y = tt, ff$ or u, the falsity of x can determine the output, i.e.

$$And(\mathit{ff}, y) = \mathit{ff}$$

and similarly for y. This results in a connective, well known to implementors of programming languages, called *parallel* or *concurrent and* or simply *Cand*. Here is its truth table:

	tt	*ff*	*u*
tt	*tt*	*ff*	*u*
ff	*ff*	*ff*	*ff*
u	*u*	*ff*	*u*

In the light of this remark about the concurrent calculation of a truth value, we may note that a *sequential calculation* that requires *both* inputs to be known for the output to be known would be described by the first truth table.

The addition of u affects all connectives such as *Or*, *And*, $\Rightarrow$, giving rise to new truth tables. What identities hold between the connectives? Is it the case that

$$\Rightarrow (x, y) = Not(Cand(x, Not(y)))?$$

2.1.2 Algebras of natural numbers

Many useful algebras are made by selecting a set of functions on the set $N = \{\ 0, 1, 2, \ldots\ \}$ of natural numbers. Consider the following functions.

$$Succ : \mathbf{N} \to \mathbf{N} \qquad Succ(x) = x + 1$$

$$Pred : \mathbf{N} \to \mathbf{N} \qquad Pred(x) = \begin{cases} x - 1, & \text{if } x \geq 1; \\ 0, & \text{otherwise.} \end{cases}$$

$$Add : \mathbf{N} \times \mathbf{N} \to \mathbf{N} \qquad Add(x, y) = x + y$$

$$Sub : \mathbf{N} \times \mathbf{N} \to \mathbf{N} \qquad Sub(x, y) = \begin{cases} x - y, & \text{if } x \geq y; \\ 0, & \text{otherwise.} \end{cases}$$

$$Mult : \mathbf{N} \times \mathbf{N} \to \mathbf{N} \qquad Mult(x, y) = x \cdot y$$

$$Quot : \mathbf{N} \times \mathbf{N} \to \mathbf{N} \qquad Quot(x, y) = \begin{cases} (\text{largest } k)(kx \leq y), & \text{if } x \neq 0 ; \\ 0, & \text{if } x = 0. \end{cases}$$

$$Rem : \mathbf{N} \times \mathbf{N} \to \mathbf{N} \qquad Rem(x, y) = y - Quot(x, y) \cdot x$$

$$Exp : \mathbf{N} \times \mathbf{N} \to \mathbf{N} \qquad Exp(x, y) = x^y$$

$$Log : \mathbf{N} \times \mathbf{N} \to \mathbf{N}$$

$$Log(x, y) = \begin{cases} (\text{largest } k)(x^k \leq y), & \text{if } x \neq 0 \text{ and } x \neq 1 \text{ and } y \neq 0 ; \\ 0, & \text{if } x = 0 \text{ or } y = 0 ; \\ 1, & \text{if } x = 1 . \end{cases}$$

$$Max : \mathbf{N} \times \mathbf{N} \to \mathbf{N} \qquad Max(x, y) = \begin{cases} x, & \text{if } x \geq y; \\ y, & \text{otherwise.} \end{cases}$$

$$Min : \mathbf{N} \times \mathbf{N} \to \mathbf{N} \qquad Min(x, y) = \begin{cases} x, & \text{if } x \leq y; \\ y, & \text{otherwise.} \end{cases}$$

$$Fact : \mathbf{N} \to \mathbf{N} \qquad Fact(x) = \begin{cases} x \cdot (x - 1) \cdot \ldots \cdot 2 \cdot 1, & \text{if } x \geq 1; \\ 1, & \text{if } x = 0. \end{cases}$$

Selected in various combinations these functions make, for example, the following algebras, each of which has interesting properties and applications:

$$(\mathbf{N};\ 0;\ Succ)$$

$$(\mathbf{N};\ 0;\ Pred)$$

$$(\mathbf{N};\ 0;\ Succ,\ Pred)$$

$$(\mathbf{N};\ 0;\ Succ,\ Add)$$

$$(\mathbf{N};\ 0;\ Succ,\ Pred,\ Add,\ Sub)$$

$$(\mathbf{N};\ 0;\ Succ,\ Add,\ Mult)$$

$$(\mathbf{N};\ 0;\ Succ,\ Pred,\ Add,\ Sub,\ Mult,\ Quot,\ Rem)$$

$$(\mathbf{N};\ 0;\ Succ,\ Add,\ Mult,\ Exp)$$

$$(\mathbf{N};\ 0;\ Succ,\ Pred,\ Add,\ Sub,\ Mult,\ Quot,\ Rem,\ Exp,\ Log)$$

To the above operations we may add the characteristic functions of relations such as:

$$Eq : \mathbf{N} \times \mathbf{N} \to \mathbf{B} \quad Eq(x,\ y) = \begin{cases} tt, & \text{if } x = y; \\ f\!f, & \text{otherwise.} \end{cases}$$

$$Lte : \mathbf{N} \times \mathbf{N} \to \mathbf{B} \quad Lte(x,\ y) = \begin{cases} tt, & \text{if } x \leq y; \\ f\!f, & \text{otherwise.} \end{cases}$$

$$Lt : \mathbf{N} \times \mathbf{N} \to \mathbf{B} \quad Lt(x,\ y) = \begin{cases} tt, & \text{if } x < y; \\ f\!f, & \text{otherwise.} \end{cases}$$

$$Case : \mathbf{B} \times \mathbf{N} \times \mathbf{N} \to \mathbf{N} \quad Case(x,\ y,\ z) = \begin{cases} y, & \text{if } x = tt; \\ z, & \text{otherwise.} \end{cases}$$

$$Prime : \mathbf{N} \to \mathbf{B} \quad Prime(x) = \begin{cases} tt, & \text{if } x \text{ is prime;} \\ f\!f, & \text{otherwise.} \end{cases}$$

These, and any other tests on $\mathbf{N}$, require us to add the Booleans to our algebras. For example, we will use the algebra called *Peano Arithmetic with Booleans*

$$PA(\mathbf{N},\ \mathbf{B}) = (\mathbf{N},\ \mathbf{B};\ 0,\ tt,\ f\!f;\ Succ,\ Add,\ Mult,\ And,$$
$$Not,\ Eq,\ Lt,\ Case).$$

However, if we replace $\{\ tt,\ f\!f\ \}$ by $\{\ 0,\ 1\ \}$ then tests can be defined as functions on $\mathbf{N}$ and the addition of a second sort can be avoided.

Here are some further types of functions that can be used in algebras of natural numbers.

Any bijection $Pair : \mathbf{N} \times \mathbf{N} \to \mathbf{N}$, together with its inverse $Unpair :$ $\mathbf{N} \to \mathbf{N} \times \mathbf{N}$ with coordinate functions $Unpair_1, Unpair_2 : \mathbf{N} \to \mathbf{N}$ such that

$$Unpair(x) = (Unpair_1(x),\ Unpair_2(x))$$

for $x \in \mathbf{N}$ can be added. For example $Pair(x,\ y) = 2^x(2y + 1) - 1$.

Many functions on **N**, and on the set **Z** of integers, are defined from functions on the set **Q** of rationals or the set **R** of real numbers. For example, we can make a function f on **N** from a function g on **R** using the *floor* and *ceiling maps*

$$\lfloor \ \rfloor : \mathbf{R} \to \mathbf{Z} \quad \lfloor x \rfloor = (\text{largest } k \in \mathbf{Z})[k \leq x]$$

$$\lceil \ \rceil : \mathbf{R} \to \mathbf{Z} \quad \lceil x \rceil = (\text{least } k \in \mathbf{Z})[x \leq k]$$

by defining

$$f(n) = \lfloor g(n) \rfloor \ \text{ or } \ f(n) = \lceil g(n) \rceil.$$

An important example is

$$\lambda : \mathbf{N} \to \mathbf{N} \quad \lambda(x) = \lfloor +\sqrt{x} \rfloor.$$

In these examples of algebras we have not considered the precise details of the representation of natural numbers in **N**. We have assumed a standard decimal representation. The functions and algebras described above can be developed for any number representation system, for example radix b systems for $b = 2$, 8 or 16. In particular the algebras obtained should be equivalent. Again this basic idea is formalized by the notion of *isomorphism*.

Further algebras arise from the attempt to make finite counting systems based on sets of the form

$$\{ \ 0, \ 1, \ 2, \ \ldots, \ m - 1 \ \}$$

for some $m \geq 1$. We will consider examples later in Section 3.1. (Exercise: devise ways of defining the functions listed above on finite sets of numbers of the form above.)

2.1.3   Algebras of integer, rational, real and complex numbers

The extension of the natural numbers by negative numbers to make the set **Z** of integers, facilitates calculation and leads to more interesting and useful algebras. The same remark is true of the extension of the integers to the set **Q** of rationals to accommodate division; the extension of the rationals to the set **R** of reals to accommodate measurements of line segments of irrational length; and the extension of the reals to the set **C** of complex numbers to accommodate the solution of polynomial equations.

However the study of algebras of these numbers has been dominated by the study of the algebras made from the basic operations of addition, subtraction, multiplication and, except in the case of the integers, division. These specific algebras have many properties in common and are best studied as examples of two general types of algebras called *rings* and *fields.* Other allied algebras made from matrices, polynomials and power series are also examples of rings and/or fields. The beauty and utility of the general theories of rings and fields, especially those parts that focus on these number algebras, is overwhelming and we will pay tribute to them by not attempting to discuss them in this chapter. The reader is recommended to study the elements of this theory independently. Elementary introductions are Birkhoff and MacLane [1965], Herstein [1964] and van der Waerden [1970]; and an advanced work is Cohn [1982].

2.1.4   Algebras of terms and expressions

We will consider the syntax and semantics of terms or expressions from an algebraic point of view.

Let s be a name for data called a *sort name.* Let F_0 be a non-empty set of constant symbols for distinguished data of sort s, and let F_n for $n \geq 1$ be a set of n-argument function or operator symbols; in particular, if $f \in F_n$ then f has domain type $s \times \ldots \times s$ (n times) and codomain type s. Collecting together the set $\{s\}$ of names and the family $\langle\, F_n \mid n \geq 0\,\rangle$ we form a single-sorted signature Σ.

Let X be a set of variable symbols of sort s. We assume X and the set F_0 of constant symbols are disjoint. The set $T(\Sigma, X)$ of all *terms* or *expressions* of sort s is inductively defined by:

(i)   each constant $c \in F_0$ is a term of sort s;

(ii)   each variable $x \in X$ is a term of sort s;

(iii)   if $t_1, \ldots, t_n$ are terms of sort s and $f \in F_n$ then $f(t_1, \ldots, t_n)$ is a term of sort s;

(iv)   nothing else is a term.

A short informal definition of $T(\Sigma, X)$ by means of a grammar is simply

$$t ::= c \mid x \mid f(t_1, \ldots, t_n),$$

for $c \in F_0$, $x \in X$, $t_1, \ldots, t_n \in T(\Sigma, X)$ and $f \in F_n$.

The set $T(\Sigma, X)$ (if non-empty) is the carrier for an important algebra of terms. The constants of this algebra are the constant symbols $c \in F_0$. The operations of this algebra are the mappings that apply function symbols to terms: for each $f \in F_n$ there is an operation

$$F : T(\Sigma, X)^n \to T(\Sigma, X)$$

defined by

$$F(t_1, \ldots, t_n) = f(t_1, \ldots, t_n)$$

for $t_1, \ldots, t_n \in T(\Sigma, X)$. We denote this algebra by $T(\Sigma, X)$.

Next we consider *term evaluation*. The semantics of terms is given by a set A and a map

$$\bar{v} : T(\Sigma, X) \to A,$$

where $\bar{v}(t)$ is the semantics or 'value' of term t. To calculate $\bar{v}$ we must interpret the constant symbols by elements of A and the operation symbols by maps on A; and we must assign elements of A to variables.

Let $c \in F_0$ be interpreted by an element $c_A \in A$. Let $f \in F_n$ be interpreted by a map $f_A : A^n \to A$. Clearly we have an algebra

$$(A; c_A \text{ for } c \in F_0; f_A \text{ for } f \in F_n, n \geq 1).$$

With respect to these fixed interpretations, term evaluation is given by the following.

Given an assignment $v : X \to A$ of an element $v(x)$ to each variable $x \in X$, we can define $\bar{v} : T(\Sigma, X) \to A$ by induction:

$$\bar{v}(c) = c_A;$$

$$\bar{v}(x) = v(x);$$

$$\bar{v}(\, f(t_1, \ldots, t_n)\,) = f_A(\, \bar{v}(t_1), \ldots, \bar{v}(t_n)\,).$$

Later we will see the construction of $\bar{v}$ from v as an important idea, generalizing the principle of induction to a general algebraic setting through the concepts of *freeness* and *initiality*. For the moment let us refer to the existence of $\bar{v}$, given v, as an *extension property* of $T(\Sigma, X)$.

Another operation on $T(\Sigma, X)$ worth noting is that of *term substitution*. Let t be a term, $\bar{x} = (x_1, \ldots, x_n)$ be a sequence of variable symbols and $\bar{t} = (t_1, \ldots, t_n)$ be a sequence of terms; we wish to define the term customarily denoted

$$t(\bar{x}/\bar{t}) \text{ or } t(x_1/t_1, \ldots, x_n/t_n)$$

obtained by substituting the term t_i for the variable symbol x_i, for each $i = 1, \ldots, n$, throughout t. This is done trivially by the extension property. Given $\bar{x} = (x_1, \ldots, x_n)$ and $\bar{t} = (t_1, \ldots, t_n)$ then we define an assignment $v = v(\bar{x}, \bar{t}) : X \to T(\Sigma, X)$ by

$$v(x) = \begin{cases} x, & \text{if } x \notin \{\, x_1, \ldots, x_n \,\}; \\ t_i, & \text{if } x = x_i. \end{cases}$$

Then by the method of extending v to $\bar{v}$ with the algebra A replaced by the algebra $T(\Sigma, X)$ we obtain

$$\bar{v} : T(\Sigma, X) \to T(\Sigma, X)$$

which carries out the required substitution of t_i for x_i, for $i = 1, \ldots, n$, in all terms.

This can be refined further by a new map

$$sub^n : T(\Sigma, X) \times X^n \times T(\Sigma, X)^n \to T(\Sigma, X)$$

which substitutes *any* n-tuple $\bar{t} = (t_1, \ldots, t_n)$ of terms for *any* n-tuple $\bar{x} = (x_1, \ldots, x_n)$ of variables into any term t. This is defined as follows: given $\bar{x} = (x_1, \ldots, x_n)$ and $\bar{t} = (t_1, \ldots, t_n)$ we define the obvious assignment $v = v(\bar{x}, \bar{t}) : X \to T(\Sigma, X)$, obtain $\bar{v}$, and uniformize $\bar{v}$ in defining

$$sub^n(t, \bar{x}, \bar{t}) = \bar{v}(t).$$

The extension property can also be used to discuss a change in variable names. Let X and Y be sets of variables of the same cardinality. We can define the effect on terms, of the transformation of variables from X to Y as follows. Consider the term algebras $T(\Sigma, X)$ and $T(\Sigma, Y)$. On choosing a variable transformation $v : X \to Y$ which is a bijection we obtain, by the extension property, a map $\bar{v} : T(\Sigma, X) \to T(\Sigma, Y)$ that transforms *all* the terms. It is possible to prove that $\bar{v}$ is a bijection.

Finally, let us note that the important equations that are used in the definition of the extension $\bar{v} : T(\Sigma, X) \to A$ of $v : X \to A$ can be seen to involve formally the algebraic structure of the algebra of terms:

$$\bar{v}(c) = c_A$$

$$\bar{v}(x) = v(x)$$

$$\bar{v}(f(t_1, \ldots, t_n)) = f_A(\bar{v}(t_1), \ldots, \bar{v}(t_n)).$$

We see that the map $\bar{v}$ preserves a relationship between the operations on the term algebra $T(\Sigma, X)$ and on the semantic algebra A. Such a structure-preserving mapping provides an example of a *homomorphism* between algebras; we will deal with these functions later.

2.1.5 Algebras and logic

In formalizing the syntax and semantics of logical systems we may form many algebras. Consider, for example, the inductive definition of a set P of propositional formulas from a set P_0 of propositional variables. We may conveniently define $p \in P$ by the grammar

$$p ::= T \mid F \mid p_0 \mid p_1 \wedge p_2 \mid p_1 \vee p_2 \mid \neg p_1,$$

where $p_0 \in P_0$ and $p_1, p_2 \in P$.

Let P denote an algebra with carrier P, constants T and F, and operations

$$\wedge : P \times P \to P, \quad \vee : P \times P \to P \text{ and } \neg : P \to P$$

for the connectives and, or and not. The algebra P has the same structure as an algebra of terms generated from $P_0 \cup \{ T, F \}$ by the application of connective symbols $\wedge$, $\vee$ and $\neg$.

One standard method of giving a semantics to propositions is to choose an assignment $v : P_0 \to \mathbf{B}$ of truth values, where $\mathbf{B} = \{ \ tt, \ f\!f \ \}$, and inductively define a map $\overline{v} : P \to \mathbf{B}$ which gives the truth value $\overline{v}(p)$ for $p \in P$ given the truth values of its variables. This semantics is best described algebraically as follows. First, we choose an algebra

$$B = (\mathbf{B}; \ tt, \ f\!f; \ And, \ Or, \ Not)$$

(using the functions of Section 2.1.1). Then we use $v : P_0 \to \mathbf{B}$ to construct $\overline{v} : P \to \mathbf{B}$ in the manner of Section 2.1.4. For any v there exists $\overline{v}$ which is a structure preserving map called a *homomorphism*.

In logic the algebras

$$(P; \ T, \ F; \ \wedge, \ \vee, \ \neg) \ \text{and} \ (P; \ T, \ F; \ \wedge, \ \vee)$$

are further refined by an equivalence relation $\equiv$ on P defined by $p \equiv q$ if, and only if, for all $v : P_0 \to \mathbf{B}$, $\overline{v}(p) = \overline{v}(q)$, which means that p and q have the same truth values in all circumstances (i.e. $p \leftrightarrow q$). We can turn the set $P/\equiv$ of equivalence classes into an algebra.

The set *Bool* of laws governing $\wedge$, $\vee$ and $\neg$ forms the set of axioms for *Boolean algebras*. A subset pertaining to $\wedge$ and $\vee$ form the set of axioms for *lattices*. Any algebra satisfying these axioms in *Bool* is called a Boolean algebra and is relevant for the semantics of P. Algebras of sets such as the power set algebra

$$(P(X); \ \emptyset, \ X; \ \cap, \ \cup, \ -)$$

satisfy these laws.

We will not define these sets of laws at this point. For an introduction to Boolean algebras and lattices we recommend Halmos [1963], Sikorski [1960], Birkhoff [1967], Grätzer [1978] and McKenzie *et al.* [1987]. The book Rasiowa and Sikorski [1970] is particularly interesting for its algebraic study of the semantics of logical calculi. Abstract characterizations of logics, using methods of universal algebra are given in Cleave [1992]. Some elementary material can be found in Birkhoff and MacLane [1965].

2.1.6 Algebras of functions

Algebras whose elements are functions occur everywhere. The simplest and most common have the following form: there is a set $S(X)$ of functions on a non-empty set X and the main operation of interest is function composition $\circ$. Two important examples are the set $F(X)$ of *all* functions $X \to X$ and the set $Sym(X)$ of all bijective functions $X \to X$. Elements of $Sym(X)$ are also called *permutations* or *symmetries* and, of course, have inverses, i.e. for each $\psi \in Sym(X)$ there is $\psi^{-1} \in Sym(X)$ such that for all $x \in X$

$$(\psi^{-1} \circ \psi)(x) = (\psi \circ \psi^{-1})(x) = x.$$

Let $i : X \to X$ be the identity function defined by $i(x) = x$ for all $x \in X$. Then the above equation becomes

$$\psi^{-1} \circ \psi = \psi \circ \psi^{-1} = i.$$

These observations lead us to the following algebras

$$(F(X); i; \circ) \quad \text{and} \quad (Sym(X); i; \circ, {}^{-1})$$

which are called the *semigroup of all functions with identity on X* and the *group of all permutations on X*.

The structure of the algebras of functions on X depends on the structure of X; important cases are when X is a finite set and when X has geometric structure. Through the extensive analysis and application of these particular function algebras in the 18th and 19th century their common properties were revealed and ultimately codified in the axiomatic theories of *semigroups* and *groups*. As with our notes on rings, fields and boolean algebras, we pay our respects to these subjects by not attempting to introduce the reader to them, though we will use groups occasionally.

A useful work on semigroups is Lallement [1979]. For introductions to group theory see for example Kargapolov and Merzljakov [1979] and Rotman [1973]. Some elementary material can be found in Birkhoff and MacLane [1965].

Exercises 2.1.7.

1. Using sort name *bool*, construct algebras of terms that may be evaluated in the algebras in Section 2.1.1.

2. Formalize further the details of the definition of the function $\bar{v}$ by justifying that the equations given for its definition have a unique solution.

3. Attempt to rework the above definitions in the case of two sorts s_1, s_2 and then in the general case of any non-empty set S of sorts.

4. Generalize the term substitution function sub^n of Section 2.1.4 to the term evaluation function

$$te^n : T(\Sigma, X) \times X^n \times A^n \to A$$

which substitutes $\bar{a} = (a_1, \ldots a_n) \in A^n$ for $\bar{x} = (x_1, \ldots, x_n)$ in $t \in T(\Sigma, X)$ and computes the value

$$te^n(t, \bar{a}, \bar{x}) = t(\bar{a}/\bar{x}).$$

(Warning: beware the case where t contains a variable not in the list $\bar{x}$.)

2.2 Some simple constructions

We will consider four examples of the construction of a new algebra from a given algebra.

2.2.1 Adding an algebra

Let A be an algebra. Suppose A does not contain an important data set such as the Booleans or natural numbers. Then, if it is required for some purpose, it is possible to add the missing data. For example, let B be any algebra. Then we can define the algebra $[A, B]$, called the *join* join of A and B, to be the algebra containing the carriers of A and B, and the constants and operations of A and B. Notice that there is no connection between A and B in this algebra. In the case of adding an algebra B of Booleans, a natural connection to include would be the operation of equality

$$Eq : A \times A \to B$$

of course.

Often in computation we find that the Booleans are a carrier in our algebras, and that tests in the form of Boolean valued functions are present. However, equally often we find that we add the natural numbers without connecting operations.

2.2.2 Algebras with an unspecified element

Let A be an algebra with one carrier A, and consider the effect of augmenting A with a special object $u \notin A$ to represent an *undefined* or *unspecified datum*. We will make a new algebra A^u with carrier

$$A^u = A \cup \{ u \}.$$

The constants of A^u are those of A together with

$$u$$

and the operations of A^u are derived from those of A as follows: let $f : A^n \to A$ be an operation of A, then define

$$f^u : (A \cup \{ u \})^n \to (A \cup \{ u \})$$

by

$$f^u(x_1, \ldots, x_n) = \begin{cases} f(x_1, \ldots, x_n), & \text{if } x_1, \ldots, x_n \in A; \\ u, & \text{otherwise.} \end{cases}$$

The condition on operations of A^u that makes them return an unspecified value if any of the input is unspecified, is often called the *strictness assumption*. Notice that we can extract A from A^u by forming the subalgebra of A^u generated by the elements of A.

2.2.3 Infinite streams

Consider augmenting a set A by infinite sequences of the elements of A. Let A be a non-empty set and let $\mathbf{N} = \{\ 0, 1, 2, \ldots\ \}$. An *infinite sequence* or *infinite stream* over A is a function

$$a : \mathbf{N} \to A.$$

For $i \in \mathbf{N}$, the ith element of the sequence is $a(i)$ and sequences are often written

$$a = a(0), a(1), \ldots \quad \text{or} \quad a = a_0, a_1, \ldots.$$

Let $[\mathbf{N} \to A]$ be the set of all infinite sequences over A.

Now given any algebra

$$A = (A; c_1, \ldots, c_p;\ f_1, \ldots, f_q)$$

we may make a new algebra $\overline{A}$, which augments A with the set of streams over the carrier set A. The *stream algebra* $\overline{A}$ has carriers A, $\mathbf{N}$ and $[\mathbf{N} \to A]$. Its constants are the constants of A and $0 \in \mathbf{N}$. Its operations are the operations of A together with the functions

$$Succ : \mathbf{N} \to \mathbf{N} \quad Succ(x) = x + 1,$$

$$Eval : [\mathbf{N} \to A] \times \mathbf{N} \to A \quad Eval(a, i) = a(i).$$

This algebra $\overline{A}$ simply involves infinite sequences as data, the elements of which may be read by means of the last two operations. There are many other operations that may be added to $\overline{A}$ to model other uses of infinite sequences.

To recover A from $\overline{A}$ we need only discard the new carriers and their constants and operations.

2.2.4 Infinite arrays

Consider augmenting a set A by infinite arrays of the elements of A. Let A be a non-empty set and let $\mathbf{N} = \{\ 0, 1, 2, \ldots\ \}$. An *infinite array* over A can be modelled by an infinite sequence $a : \mathbf{N} \to A$. The set $\mathbf{N}$ is used as a set of *addresses* and the value $a(i)$ is thought of as the element of A stored at address i. To read an element we again use

$$Eval : [\mathbf{N} \to A] \times \mathbf{N} \to A \quad Eval(a, i) = a(i).$$

To insert an element x at an address i we use a function

$$insert : A \times \mathbf{N} \times [\mathbf{N} \to A] \to [\mathbf{N} \to A]$$

$$insert(x, i, a)(j) = \begin{cases} a(j), & \text{if } j \neq i; \\ x, & \text{otherwise.} \end{cases}$$

for $i \in \mathbf{N}$, $x \in A$ and $a \in [\mathbf{N} \to A]$.

2.2.5 Finite arrays

Let us consider in detail a semantical model of finite arrays. Let A be a non-empty set of data and let $u \notin A$ be an object we may use to mark unspecified data. Let $A^u = A \cup \{ u \}$. Let $\mathbf{N} = \{ 0, 1, 2, \ldots \}$ be the set of addresses. We model a finite array by means of a pair

$$a^* = (a, l),$$

where $a : \mathbf{N} \to A^u$ and $l \in \mathbf{N}$ and, furthermore

$$a(i) = u$$

for all $i > l$. Thus a finite array is modelled as an infinite sequence a of data indexed by addresses, together with an upper bound l on the index of those addresses that have specified data. We define

$$A^* = \{ (a, l) \in [\mathbf{N} \to A^u] \times \mathbf{N} \mid a(i) = u \text{ for } i > l \}.$$

Given an algebra A with one carrier set, some constants and operations, we define the *algebra of finite arrays over* A as follows. First we form the algebra A^u which augments A with a special object u, adds u as a constant, and extends the operations of A to A^u by means of the strictness assumption (see Section 2.2.2). To the carrier of A^u we add new carriers $\mathbf{N}$ and A^*.

To the constants of A^u we add $0 \in \mathbf{N}$ and $null^* \in A^*$, defined by

$$null^* = (null, 0), \quad null(i) = u,$$

for all $i \in \mathbf{N}$.

To the operations of A^u we add $succ : \mathbf{N} \to \mathbf{N}$ and the following: for $a^* = (a, l) \in A^*$, $x \in A^u$ and $i, j \in \mathbf{N}$, a *read* operation

$$read : A^* \times \mathbf{N} \to A^u \quad read(a^*, i) = a(i);$$

an *array update* operation

$$aupdate : A^u \times \mathbf{N} \times A^* \to A^* \quad aupdate(x, i, a^*) = (b, l),$$

where

$$b(j) = \begin{cases} a(j), & \text{if } j < l \text{ and } j \neq i; \\ x, & \text{if } j < l \text{ and } j = i; \\ u, & \text{otherwise;} \end{cases}$$

a *bound update* operation

$$bupdate : A^* \times \mathbf{N} \to A^* \quad bupdate(a^*, i) = (b, i),$$

where

$$b(j) = \begin{cases} a(j), & \text{if } j \leq i; \\ u, & \text{otherwise}; \end{cases}$$

and an *upper bound* or *length* operation

$$ubound : A^* \to \mathbf{N} \quad ubound(a^*) = l.$$

To recover A from A^* we must discard the carriers $\mathbf{N}$ and A^* and consider the subalgebra of A^u generated by A.

2.3 Syntax and semantics of programs

We consider the algebraic structure of *while programs*.

2.3.1 Syntax of while programs

Consider the language $WP(P, T)$ consisting of programs made from a given set P of programs and a given set T of tests, by means of program forming constructs of *sequencing, conditional branching* and *iterating*. The class of program texts is given by a grammar which is defined for programs $S \in WP(P, T)$,

$$S ::= p \mid S_1 ; S_2 \mid if \, b \, then \, S_1 \, else \, S_2 \mid while \, b \, do \, S_0$$

for $p \in P$, $b \in T$, $S_1, S_2, S_0 \in WP(P, T)$.

The syntax can also be described as a certain term algebra of the type described in Section 2.1.4. The syntax of the language is an algebra $WP(P, T)$ with carrier sets $WP(P, T)$ and T. The constants of the algebra are all the elements of P and T. The operations of the algebra are defined by

$$seq : WP(P, T)^2 \to WP(P, T) \quad seq(S_1, S_2) = S_1 ; S_2$$

$$cond : T \times WP(P, T)^2 \to WP(P, T) \quad cond(b, S_1, S_2) = if \, b \, then \, S_1 \, else \, S_2$$

$$it : T \times WP(P, T) \to WP(P, T) \quad it(b, S_0) = while \, b \, do \, S_0.$$

Notice that each program can be constructed from P and T by applying the basic operations of the algebra. This construction is a special case of the term algebra construction, taking two sorts and sets of variables $X_1 = P$ and $X_2 = T$ and the above function symbols.

Suppose that the atomic programs are assignments of the form

$$x := e,$$

where x is a variable and e is an expression of the same type; and the tests are Boolean expressions. Formally the syntactic categories $Exp(\Sigma)$, $BExp(\Sigma)$ of expressions and Boolean expressions are constructed from a signature Σ as in Section 2.1.4, hence so is the set $Ass(\Sigma)$ of assignments. Thus, usually we work with the algebra of while programs over Σ

$$WP(Ass(\Sigma), BExp(\Sigma)).$$

We can extend this algebra to include not just $Ass(\Sigma)$ and $BExp(\Sigma)$, but the means of construction of $Ass(\Sigma)$. Let us add the carriers Var and Exp and the operation

$$ass : Var \times Exp \rightarrow WP(Ass(\Sigma), BExp(\Sigma)) \quad ass(x, e) = x := e.$$

As an exercise, experiment with further additions such as the means of constructing $Exp(\Sigma)$ and $BExp(\Sigma)$ from Var and Σ.

2.3.2 Algebras of state transformations

We will make an algebra of transformations and tests on states. Let S be any non-empty set and $\perp$ be an element of S. We call the elements of S *states* and $\perp$ the *undefined state*.

A *transformation* of S is a map $f : S \rightarrow S$ and a *test* on S is a map $b : S \rightarrow \{\ tt,\ ff\ \}$. Let $Trans(S)$ and $Test(S)$ be the set of all transformations and tests, respectively. We define an algebra $WT(Trans(S), Test(S))$ as follows.

The carriers of the algebra are:

$$Trans(S) \quad \text{and} \quad Test(S).$$

The constant of the algebra is: $\perp \in Trans(S)$ defined by $\perp(S) = \perp$.
The operations of the algebra are:

$$comp : Trans(S)^2 \rightarrow Trans(S) \quad comp(f_1, f_2)(s) = f_2(f_1(s)),$$

$$cond : Test(S) \times Trans(S)^2 \rightarrow Trans(S)$$

$$cond(b, f_1, f_2)(s) = \begin{cases} f_1(s), & \text{if } b(s) = tt; \\ f_2(s), & \text{if } b(s) = ff, \end{cases}$$

and

$$it : Test(S) \times Trans(S) \rightarrow Trans(S)$$

$$it(b,\ f)(s) = \begin{cases} f^k(s), & \text{if there is } k \in \mathbf{N} \text{ such that } b(f^i(s)) = tt, \\ & \text{for } 0 \le i \le k \text{ and } b(f^{k+1}(s)) = f\!\!f; \\ \bot, & \text{otherwise.} \end{cases}$$

Let us return to the constants. Let $F \subseteq Trans(S)$ and $B \subseteq Test(S)$. We denote by

$$WT(F,\ B)$$

the subalgebra consisting of all while transformations obtained from F and B by means of the operations.

2.3.3   Semantics of while programs

We will define the meaning or semantics of while programs by means of a function between sets

$$m : WP(P,\ T) \to Trans(S)$$

such that for each program $p \in WP(P,\ T)$

$$m(p) : S \to S$$

and for $s \in S$, $m(p)(s)$ is the final state after executing p on initial state s, if such a final state exists.

To define m we must also define the semantics of the tests by means of a function

$$w : T \to Tests(S).$$

Furthermore, we must define the semantics of the programs in the basic set P by means of a function

$$v : P \to Trans(S).$$

Suppose these maps w and v are given. Then the definition of m is by induction and can be written as follows:

$$m(p) = v(p)$$

$$m(seq(p_1,\ p_2)) = comp(m(p_1),\ m(p_2))$$

$$m(cond(b,\ p_1,\ p_2)) = cond(w(b),\ m(p_1),\ m(p_2))$$

$$m(it(b,\ p)) = it(w(b),\ m(p)).$$

These equations reveal that m preserves the operations of the algebras $WP(P,\ T)$ and $WT(Trans(S),\ Test(S))$. Formally the pair of maps m, w constitute a two-sorted homomorphism between the algebras.

We have noted that $WP(P, T)$ is an example of an algebra of terms of the form explained in Section 2.1.4. The maps m, w are obtainable by the extension property applied to v and w. In the notation of that section we may write the meaning function m for programs

$$m = \overline{v}$$

and since we assume that we have the semantics of all tests

$$w = \overline{w}.$$

2.3.4 General principles for semantics

Now let P be any non-empty set of program texts. A *semantics* for the programs of P is given by a non-empty set S and a meaning function

$$m : P \to S,$$

where $m(p)$ is the *meaning* or *semantics* of program p. We define *program equivalence* under such a semantics by the relation

$$p \equiv q \Leftrightarrow m(p) = m(q)$$

for any p, $q \in P$.

Suppose P is constructed from a set of basic programs A by the application of a set C of program forming constructs, each of the form

$$\sigma : P^n \to P.$$

Then we may say that the semantics $m : P \to S$ is *compositional with respect to the constructs in C and their semantics*, if for each construct $\sigma \in C$, $\sigma : P^n \to P$, there exists a function $\sigma_S : S^n \to S$ such that for all programs $p_1, \ldots, p_n \in P$

$$m(\sigma(p_1, \ldots, p_n)) = \sigma_S(m(p_1), \ldots, m(p_n)).$$

Thus the meaning of program $\sigma(p_1, \ldots, p_n)$ can be calculated or constructed from the meaning of programs $p_1, \ldots, p_n$. Clearly the sets P and S become algebraic structures with these functions, and compositionality of m relates the algebraic structure of P with that of S.

A second notion of compositionality can be defined based on program equivalence. We say that the semantics $m : P \to S$ is *compositional with respect to the constructs of C and program equivalence*, if for each construct

$\sigma \in C$, $\sigma : P^n \to P$ and any programs $p_1, \ldots, p_n, q_1, \ldots, q_n \in P$ we have that

$$p_1 \equiv q_1, \ldots, p_n \equiv q_n \;\Rightarrow\; \sigma(p_1, \ldots, p_n) \equiv \sigma(q_1, \ldots, q_n).$$

Thus equivalent programs can be composed by the constructs and result in equivalent programs.

In connection with the second notion above we can define a third notion. A *program context* t is any finite combination of constructs from C applied to a set X of variables for programs. In symbols,

$$t ::= x \mid \sigma(t_1, \ldots, t_n).$$

A context can be thought of as a program transformation. Then let $t(x_1, \ldots, x_n)$ be a context involving variables $x_1, \ldots, x_n$, and let $t(p_1, \ldots, p_n)$ denote the program in P obtained by substituting programs $p_1, \ldots, p_n$ for the variables $x_1, \ldots, x_n$ respectively. We define the semantics $m : P \to S$ to be *compositional with respect to contexts* if for any context $t(x_1, \ldots, x_n)$ and any $p_1, \ldots, p_n \in P$,

$$p_1 \equiv q_1, \ldots, p_n \equiv q_n \;\Rightarrow\; t(p_1, \ldots, p_n) \equiv t(q_1, \ldots, q_n).$$

Later (in Exercise 3.3.25.(7)) we will see that these three ideas are general algebraic ideas and we will deduce that:

Fact. *The three notions of compositionality are equivalent.*

The notion of program equivalence is used extensively in programming language theory.

Suppose a set P of programs has *two* semantics

$$m_1 : P \to S_1, \quad m_2 : P \to S_2$$

that must be compared. Let us say that m_1 is *more abstract* than m_2, and that m_2 is more *concrete* than m_1, if for any programs $p, q \in P$,

$$p \equiv_2 q \Rightarrow p \equiv_1 q,$$

where for $i = 1, 2$ we have $p \equiv_i q$ if, and only if, $m_i(p) = m_i(q)$ in S_i. This means that the equivalence classes of $\equiv_2$ in P are contained within those of $\equiv_1$ in P.

Different semantics arise from different tasks and, typically, an operational semantics, designed for the purposes of implementing programs, is more concrete than a mathematical semantics designed for verification.

Another semantic program equivalence relation can be derived from a specification method quite easily. A *specification system* for P consists of a non-empty set *Spec* and a satisfaction relation

$$sat \subseteq P \times Spec,$$

where program $p \in P$ satisfies specification $s \in Spec$ if, and only if, $sat(p, s)$ holds. We define a semantics $m : P \to P(Spec)$ by

$$m(p) = \{ \; s \in Spec \; | \; sat(p, s) \; \}$$

and note that the associated program equivalence notion is

$$p \equiv q \Leftrightarrow (\text{ for all } s \in Spec)[sat(p, s) \text{ if, and only if, } sat(q, s)].$$

That is, p and q are indistinguishable by the specification method. Usually a specification semantics is more abstract than an operational semantics. Sometimes, we may say that the specification method *uniquely determines* the operational semantics if their respective program equivalence relations coincide.

Examples of specification semantics are given in Bergstra *et al.* [1982], where $m(S) = \{ \; (p, q) \; | \; \{ \; p \; \}S\{ \; q \; \} \; \}$, and Fisher [1990] for temporal logic. Further general concepts concerning refinement are found in Back [1981].

2.4 Synchronous concurrent algorithms

A *synchronous concurrent algorithm* is an algorithm based on a network of modules and channels, computing and communicating data in parallel, and synchronized by a global clock. Synchronous algorithms process infinite streams of input data and return infinite streams of output data. Examples of synchronous concurrent algorithms include *clocked digital hardware*; *systolic algorithms* (Kung [1982]); *neural nets* (McCulloch and Pitts [1943], Minsky [1967]); and *cellular automata* (Codd [1968], Soulie *et al.* [1987] and Wolfram [1990]).

2.4.1 Informal definition

Consider a synchronous algorithm as a network of *sources*, *modules* and *channels* computing and communicating data as depicted in Figure 2.1.

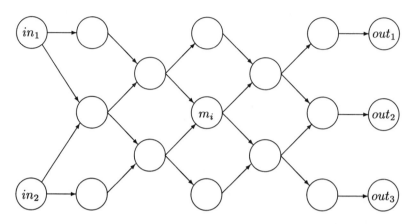

Figure 2.1

For simplicity we will assume that all the data is drawn from a single set A. The network has a global discrete clock $T = \{\ 0,\ 1,\ \dots\ \}$ to synchronize computations and the flow of data between modules. There are c *sources* labelled $s_1, \dots, s_c$ and k modules labelled $m_1, \dots, m_k$. The sources perform no computation; they are simply input ports where fresh data arrives at each clock cycle. Each module m_i has $p(i)$ inputs and a single output as depicted in Figure 2.2. The action of module m_i is specified by a function

$$f_i : A^{p(i)} \to A.$$

Results are read out from the channels of a subset of the modules $m_{\alpha_1}, \dots, m_{\alpha_d}$, which may be termed output modules or *sinks*.

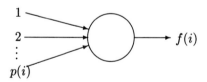

Figure 2.2

The interconnections between the sources and modules are represented by a pair of partial functions

$$\gamma : \mathbf{N}_k \times \mathbf{N} \to \{\ S,\ M\ \}$$

and

$$\beta : \mathbf{N}_k \times \mathbf{N} \to \mathbf{N},$$

where $\mathbf{N}_k = \{\ 1, 2, \dots, k\ \}$. Intuitively $\gamma(i,\ j) = S$ indicates that the jth input to module m_i comes from a source, whereas $\gamma(i,\ j) = M$ indicates that it comes from another module. The index of the source or module in question is given by $\beta(i, j)$.

We will assume that each module is initialized with a well defined output assigned to its output channel. Thus the initial state of the network is an element $b \in A^k$ and we will use the notation b_i to denote the value of the output of module m_i for $i = 1, \ldots, k$.

In terms of our intuitive picture, new data are available at each source and new results at each sink, at every tick of the global clock T. Thus the algorithm processes infinite sequences or streams of data. A stream $a(0), a(1), \ldots$ of data from A is represented by a map $a : T \to A$, and the set of all streams of data is represented by the set $[T \to A]$ of all such maps.

2.4.2 Formal definition

To specify formally a synchronous concurrent algorithm we first specify an algebra $\overline{A}$ which defines data, and the operations which may be performed on data by the modules.

The carriers of the algebra $\overline{A}$ are the sets

$$A, \quad T, \quad [T \to A] \text{ and } \mathbf{B}$$

of data, clock cycles, streams and Booleans.

The constants of the algebra are the constants $0 \in T$ and $tt, ff \in \mathbf{B}$.

The operations of the algebra are the maps

$$f_i : A^{p(i)} \to A$$

specifying the actions of the modules m_i for $i = 1, \ldots, k$; the successor function

$$t + 1$$

representing the tick of the clock; and the evaluation map

$$eval : [T \to A] \times T \to A$$

defined by

$$eval(a, t) = a(t)$$

representing the reading of a stream.

Some standard operations on $\mathbf{B}$ are also included. The algebra $\overline{A}$ is called a *stream algebra*. We met this construction in Section 2.2.3.

We represent the algorithm as the parallel evaluation of k functions, one for each module, recursively defined in a simple way.

We represent each module m_i by a *value function*

$$v_i : T \times [T \to A]^c \times A^k \to A.$$

Intuitively $v_i(t, a, b)$ gives the output of module m_i at time t, when the network is initialized with values b and is computing on the input streams a. The function is defined as follows

$$v_i(0, a, b) = b_i$$

$$v_i(t + 1, a, b) = f_i(arg_1, \ldots, arg_{p(i)}),$$

where for $j = 1, \ldots, p(i)$

$$arg_j = \begin{cases} eval(a_{\beta(i, j)}, t), & \text{if } \gamma(i, j) = S; \\ v_{\beta(i, j)}(t, a, b), & \text{if } \gamma(i, j) = M. \end{cases}$$

To simplify a little, if all of the channels to module i come from sources then

$$v_i(t + 1, a, b) = f_i(eval(a_{\beta(i, 1)}, t), \ldots, eval(a_{\beta(i, p(i))}, t))$$

and if all of the channels to module i come from other modules then

$$v_i(t + 1, a, b) = f_i(v_{\beta(i, 1)}(t, a, b), \ldots, v_{\beta(i, p(i))}(t, a, b)).$$

The algorithm is represented by

$$v_1, \ldots, v_k : T \times [T \to A]^c \times A^k \to A.$$

The input-output behaviour of the algorithm is determined by the sinks $m_{\alpha_1}, \ldots, m_{\alpha_d}$ and is represented by the maps

$$v_{\alpha_1}, \ldots, v_{\alpha_d} : T \times [T \to A]^c \times A^k \to A.$$

2.4.3 Algorithms as algebras

The description of algorithms allows us to use further algebraic methods. Consider the *design problem* for algorithms and architectures in hardware.

Given a specification S for a component that processes streams, select data and components C and design a concurrent algorithm or architecture A using C to implement S correctly. We will formulate this problem algebraically, based on the methods of Section 2.4.2.

A specification S for a component that processes streams is an algebra A_{task} with carriers

$$(A, T, [T \rightarrow A]; \ 0; \ t+1, \ eval, \ F_1, \ldots, F_d),$$

where

$$F_1, \ldots, F_d : T \times [T \rightarrow A]^n \rightarrow A$$

define elements on the output streams at the d output ports. Let us call this algebra the *task algebra*.

The building blocks for the design of the algorithm are selected and form a stream algebra $\bar{A}$; let us rename this A_{comp} and call it a *component algebra*.

The design of a concurrent algorithm or architecture leads to a synchronous concurrent algorithm represented by equations for the value functions

$$v_1, \ldots, v_k : T \times [T \rightarrow A]^n \times A^k \rightarrow A.$$

We add these functions to the component algebra A_{comp} together with constants $b_1, \ldots, b_k$ for initialization, to make an *algorithm algebra* A_{alg}.

The correctness of the design is formalized by the requirement that A_{alg}, with all operations removed except the operations $v_{\alpha_1}, \ldots, v_{\alpha_d}$ derived from sinks, is equivalent to A_{task}. This equivalence will be isomorphism and in symbols the requirement is

$$A_{task} \cong A_{alg}|_{\Sigma_{task}},$$

where Σ_{task} is the signature naming the operations of the sinks.

2.4.4 Conclusion

The functional definition of a synchronous concurrent algorithm by means of equations of the above form in Section 2.4.2 over stream algebras leads to a substantial mathematical theory, practical formal methods and software tools for these concurrent algorithms and architectures. See for example Thompson and Tucker [1985, 1991], Hobley *et al.* [1988], Eker and Tucker [1989] and McConnell and Tucker [1992], for the elements of the subject. The equations are primitive recursive equations and the theory connects with the theory of computability and specifiability on abstract algebras: see Tucker [1991]. Detailed studies of aspects of the theory are contained in: Thompson [1987], Meinke [1988], Meinke and Tucker [1988], Hobley *et al.* [1990]; and of its practical applications in: Martin [1989], Harman [1990] and Eker [1990]. A general survey of the theoretical foundations of hardware design is McEvoy and Tucker [1990].

2.5 Algebras and the modularisation of software

The discussion concerning the algebraic formulation of the design problem for devices, in Section 2.4.3, can be generalized to apply to the design problem for any class of algorithms expressed in terms of functions.

For example, suppose we want to compute a function F on a non-empty set A. Then the task is described by an algebra

$$A_{task} = (A; F).$$

Next suppose we are given A together with another non-empty set B; constants $c_1, \ldots, c_n$; and functions $g_1, \ldots, g_m$ on A and B; with which to design an algorithm for F. This collection of components is described by an algebra

$$A_{comp} = (A, B; c_1, \ldots, c_n; g_1, \ldots, g_m).$$

By means of algorithmic constructs we make various new functions $f_1, \ldots, f_l$ that perform subcomputations leading to the desired function F. The algorithm for F is described by an algebra with a distinguished output function

$$A_{alg} = (A, B; c_1, \ldots, c_n; g_1, \ldots, g_m, f_1, \ldots, f_l, F).$$

The correctness of the design is formalized by the requirement that A_{alg} with all its operations removed except F, is equivalent to A_{task}. This equivalence will be isomorphism and in symbols the requirement is

$$A_{task} = A_{alg}|_{\Sigma_{task}},$$

where Σ_{task} is the signature for A_{task} containing the output function symbol F.

3 Algebras and morphisms

In this section we give the definitions of many-sorted algebras, subalgebras, factor algebras, homomorphisms and direct products, together with many associated concepts and basic results. The material allows us to formalise and analyse in general terms specific algebras arising in computer science, such as those presented in Section 2; and to build a mathematical theory for many-sorted algebras in general, involving elaborate algebraic constructions (in Section 4) and axiomatic theories (in Section 5).

3.1 Signatures and algebras

A *many-sorted signature* Σ for a many-sorted algebra A fixes a formal notation for talking about the basic components of A by giving names for the sets, distinguished constants and algebraic operations occurring in A.

Definition 3.1.1. A *signature* Σ consists of:

(i) A non-empty set S, the elements of which we call *sorts*.

(ii) An $S^* \times S$-indexed family

$$\langle \Sigma_{w, s} \mid w \in S^*, s \in S \rangle$$

of sets, where for the empty word $\lambda \in S^*$ and any sort $s \in S$, each element

$$c \in \Sigma_{\lambda, s}$$

is called a *constant symbol* or *name* of sort s; and for each non-empty word $w = s(1) \ldots s(n) \in S^+$ and any sort $s \in S$, each element

$$\sigma \in \Sigma_{w, s}$$

is called an *operation* or *function symbol* or *name* of type (w, s); sometimes we term w the *domain type*, s the *codomain type* and n the *arity* of σ.

Thus we can define Σ to be the pair

$$(S, \langle \Sigma_{w, s} \mid w \in S^*, s \in S \rangle).$$

For emphasis, in certain circumstances we refer to Σ as an *S-sorted signature*.

Notice that we do not suppose that the sets of names for constants and operations are pairwise disjoint; thus we will allow so-called *overloading* of names in Σ.

When S is a singleton set we shall say that Σ is *single-sorted* and write Σ_n for the set $\Sigma_{s^n, s}$ of n-ary operations, $n = 0, 1, \ldots$.
    We say that Σ is *finite* if, and only if, only finitely many of the sets $\Sigma_{w, s}$ are non-empty and all non-empty sets $\Sigma_{w, s}$ are finite.

Examples 3.1.2.

(i) Consider the algebra $PA(\mathbf{N}, \mathbf{B})$ of natural numbers introduced in 2.1.2. We define a two-sorted signature $\Sigma = \Sigma^{PAB}$ for this algebra as follows. Let S be the set of sorts $S = \{\ nat,\ bool\ \}$ used to name the carriers of $PA(\mathbf{N},\ \mathbf{B})$. Let:

$$\Sigma_{\lambda,\ nat} = \{\ 0\ \}, \quad \Sigma_{\lambda,\ bool} = \{\ true,\ false\ \};$$

$$\Sigma_{nat,\ nat} = \{\ succ\ \}, \quad \Sigma_{bool,\ bool} = \{\ not\ \};$$

$$\Sigma_{nat\ nat,\ nat} = \{\ add,\ mult\ \}, \quad \Sigma_{bool\ bool,\ bool} = \{\ and\ \};$$

$$\Sigma_{nat\ nat,\ bool} = \{\ eq,\ lt\ \}, \quad \Sigma_{bool\ nat\ nat,\ nat} = \{\ case\ \}$$

and for all other $w \in S^*$ and $s \in S$ let $\Sigma_{w,\ s} = \emptyset$. Clearly this is a finite signature.

(ii) We can also give finite single-sorted signatures for semigroups with 1, groups, semirings, rings with 1 and fields using conventional notations, as follows.

Define the signature Σ^{SG} for semigroups with 1 by taking $S = \{\ sgroup\ \}$. Let:

$$\Sigma_0^{SG} = \{\ 1\ \}, \quad \Sigma_2^{SG} = \{\ \cdot\ \}$$

and $\Sigma_1^{SG} = \Sigma_n^{SG} = \emptyset$ for all $n \geq 3$.

Define the signature Σ^G for groups by taking $S = \{\ group\ \}$. Let:

$$\Sigma_0^G = \Sigma_0^{SG}, \quad \Sigma_1^G = \{\ ^{-1}\ \}, \quad \Sigma_2^G = \Sigma_2^{SG}$$

and $\Sigma_n^G = \emptyset$ for all $n \geq 3$.

Define the signature Σ^{SR} for semirings by taking $S = \{\ sring\ \}$. Let:

$$\Sigma_0^{SR} = \{\ 0,\ 1\ \}, \quad \Sigma_2^{SR} = \{\ +,\ \cdot\ \}$$

and $\Sigma_1^{SR} = \Sigma_n^{SR} = \emptyset$ for all $n \geq 3$.

Define the signature Σ^R for rings with 1 by taking $S = \{\ ring\ \}$. Let:

$$\Sigma_0^R = \Sigma_0^{SR}, \quad \Sigma_1^R = \{\ -\ \}, \quad \Sigma_2^R = \Sigma_2^{SR}$$

and $\Sigma_n^R = \emptyset$ for all $n \geq 3$.

Finally, define the signature Σ^F for fields by taking $S = \{\ field\ \}$. Let:

$$\Sigma_0^F = \Sigma_0^R, \quad \Sigma_1^F = \Sigma_1^R \cup \{\ ^{-1}\ \}, \quad \Sigma_2^F = \Sigma_2^R$$

and $\Sigma_n^F = \emptyset$ for all $n \geq 3$.

(iii) Define the single-sorted signature Σ^L for lattices by taking $S = \{\ latt\ \}$. Let:

$$\Sigma_2^L = \{\ \vee,\ \wedge\ \}$$

and $\Sigma_0^L = \Sigma_1^L = \Sigma_n^L = \emptyset$ for all $n \geq 3$.

Define the single-sorted signature Σ^B for Boolean algebras by taking $S = \{\ bool\ \}$. Let:

$$\Sigma_0^B = \{\ 0,\ 1\ \}, \quad \Sigma_1^B = \{\ \neg\ \}, \quad \Sigma_2^B = \Sigma_2^L$$

and $\Sigma_n^B = \emptyset$ for all $n \geq 3$.

In simple cases when the sort set and the domain type, codomain type and arities of constant and function symbols are familiar, it is more convenient to use a list notation for signatures: we may write

$$\Sigma = (\ s,\ s',\ldots;\ c,\ c',\ldots;\ \sigma,\ \sigma',\ldots\ ).$$

For example, the signature for fields can be written as

$$(\ field;\ 0,\ 1;\ -,\ ^{-1},\ +,\ \cdot\ ).$$

The fundamental object of interest in universal algebra is an (*S-sorted*) Σ algebra A, which fixes a particular interpretation of a signature Σ.

Definition 3.1.3. Let $\Sigma = (S, \langle \Sigma_{w,\,s} \mid w \in S^*,\ s \in S \rangle)$ be a signature. A Σ *algebra* A consists of:

(i) An S-indexed family $\langle A_s \mid s \in S \rangle$ of non-empty sets, where for each sort $s \in S$ the set A_s is called the *carrier* of sort s.

(ii) An $S^* \times S$-indexed family

$$\langle \Sigma_{w,\,s}^A \mid w \in S^*,\ s \in S \rangle$$

of sets of constants and sets of functions: for each sort $s \in S$

$$\Sigma_{\lambda,s}^A = \{\ c_A \mid c \in \Sigma_{\lambda,s}\ \},$$

where $c_A \in A_s$ is termed a *constant* of sort $s \in S$ which interprets the constant symbol $c \in \Sigma_{\lambda,s}$ in the algebra. For each non-empty word $w = s(1)\ldots s(n) \in S^+$ and each sort $s \in S$

$$\Sigma_{w,s}^A = \{\ \sigma_A \mid \sigma \in \Sigma_{w,s}\ \},$$

where $\sigma_A : A^w \to A_s$ is termed an *operation* or *function* with domain

$$A^w = A_{s(1)} \times \ldots \times A_{s(n)},$$

codomain A_s and arity n which interprets the function symbol σ in the algebra.

Thus we can define A to be the pair

$$(\langle\, A_s \mid s \in S \,\rangle,\ \langle\, \Sigma^A_{w,\,s} \mid w \in S^*,\, s \in S \,\rangle).$$

When no ambiguity arises we may use A to denote both an algebra and its S-indexed family of carrier sets.

Notice that in an algebra:

(i) every carrier set is a non-empty set; and,

(ii) every operation is a total function.

It is possible to develop extensions of the theory of many-sorted universal algebra (as presented in this chapter) that allow one or both conditions to be relaxed. (See for example Grätzer [1979], Ehrig and Mahr [1985], Burmeister [1986], Reichel [1987].)

For certain purposes, it is convenient to have a more suggestive notation for the constants in a Σ algebra A. Thus we let

$$Cons^A = \langle\, Cons^A_s \mid s \in S \,\rangle,$$

where

$$Cons^A_s = \Sigma^A_{\lambda,\,s}.$$

When a signature Σ is written as a list

$$(\ s,\, s',\ldots;\ \ c,\, c',\ldots;\ \ \sigma,\, \sigma',\ldots\ )$$

then we will also write a Σ algebra A as a list

$$(\ A_s,\, A_{s'},\ldots;\ \ c_A,\, c'_A,\ldots;\ \ \sigma_A,\, \sigma'_A,\ldots\ ).$$

Sometimes in the sequel we will assume for simplicity that signatures and algebras are single-sorted. Whenever appropriate (i.e. whenever it is not completely straightforward) we comment on the generalization to the many-sorted case.

Certain basic classifications of Σ algebras are immediate. If $|A_s| = 1$ for each sort $s \in S$ then A is termed a *unit algebra* or *trivial algebra*, otherwise A is termed *non-trivial*. We use $\mathbf{1}$ to denote a unit algebra. We say that A is *infinite* if, and only if, at least one carrier set A_s is infinite. We say that A is *countable* if, and only if, each carrier set A_s is a countable set, i.e. there exists an injection $\theta_s : A_s \to \mathbf{N}$.

Examples 3.1.4.

(i) The algebra of natural numbers $PA(\mathbf{N},\, \mathbf{B})$, introduced in Section 2.1.2, provides an obvious example of a Σ algebra $A = PA(\mathbf{N},\, \mathbf{B})$

for the signature $\Sigma = \Sigma^{PAB}$ defined in Example 3.1.2.(i). Thus we define $A_{nat} = \mathbf{N}$, $A_{bool} = \{\ tt,\ ff\ \}$. We define the constants $0_A = 0$, $true_A = tt$ and $false_A = ff$. The arithmetic operations are defined by

$$succ_A(x) = x + 1, \quad add_A(x,\ y) = x + y, \quad mult_A(x,\ y) = x \cdot y.$$

The logical operations are defined by

$$not_A(tt) = ff, \quad not_A(ff) = tt$$

$$and_A(x,\ y) = \begin{cases} tt, & \text{if } x = tt \text{ and } y = tt; \\ ff, & \text{otherwise.} \end{cases}$$

Also

$$eq_A(x,\ y) = \begin{cases} tt, & \text{if } x = y; \\ ff, & \text{otherwise.} \end{cases}$$

$$lt_A(x,\ y) = \begin{cases} tt, & \text{if } x < y; \\ ff, & \text{otherwise.} \end{cases}$$

and

$$case_A(x,\ y,\ z) = \begin{cases} y, & \text{if } x = tt; \\ z, & \text{otherwise.} \end{cases}$$

When we write algorithms on $\mathbf{N}$ it is convenient to have available several more basic operations such as division or exponentiation. Thus to Σ we might add the binary function symbols

$$div,\ mod,\ exp,\ log$$

of domain type *nat nat* and codomain type *nat*. Also the binary logical operation symbols

$$gt,\ gteq,\ lteq$$

of domain type *nat nat* and codomain type *bool*. To the algebra $PA(\mathbf{N},\ \mathbf{B})$ we adjoin the operations defined by

$$div(x,\ y) = (\text{greatest } n)(ny \le x), \quad mod(x,\ y) = x - (div(x,\ y) \cdot y),$$

$$exp(x,\ y) = x^y$$

$$log(x,\ y) = \begin{cases} (\text{largest } k)(x^k \le y), & \text{if } x \ne 0 \text{ and } x \ne 1 \text{ and } y \ne 0 ; \\ 0, & \text{if } x = 0 \text{ or } y = 0 ; \\ 1, & \text{if } x = 1 . \end{cases}$$

$$gt(x,\ y) = \begin{cases} tt, & \text{if } x > y; \\ ff, & \text{otherwise.} \end{cases}$$

$$gteq(x,\ y) = \begin{cases} tt, & \text{if } x \geq y; \\ ff, & \text{otherwise.} \end{cases}$$

$$lteq(x,\ y) = \begin{cases} tt, & \text{if } x \leq y; \\ ff, & \text{otherwise.} \end{cases}$$

(ii) We can construct another Σ algebra A' from the algebra $A = PA$ $(\mathbf{N},\ \mathbf{B})$ of the previous example by simply *reinterpreting* the function symbols *add* and *mult* so that

$$add_A(x,\ y) = x \cdot y, \quad mult_A(x,\ y) = x + y.$$

(iii) A number of structures based on the natural numbers are useful:

$$Succ(\mathbf{N}) = (\mathbf{N};\ 0;\ Succ),$$

$$Pred(\mathbf{N}) = (\mathbf{N};\ 0;\ Pred),$$

where

$$Pred(n) = \begin{cases} n - 1, & \text{if } n \geq 1; \\ 0, & \text{otherwise.} \end{cases}$$

Also

$$Count(\mathbf{N}) = (\mathbf{N};\ 0;\ Succ, Pred),$$

$$PA(\mathbf{N}) = (\mathbf{N};\ 0;\ Succ,\ Add,\ Mult),$$

$$SRing(\mathbf{N}) = (\mathbf{N};\ 0,\ 1;\ Add,\ Mult).$$

Using the constant and operation symbols of Σ^{PAB} we can define single-sorted signatures Σ^{Succ}, Σ^{Pred}, Σ^{Count}, Σ^{PA} and Σ^{SRing} for each of these algebras in the obvious way. Also of technical interest are the algebras:

$$(\mathbf{N};\ 0;\ Succ,\ Pred,\ Add);$$

$$PA_{pred}(\mathbf{N}) = (\mathbf{N};\ 0;\ Succ,\ Pred,\ Add,\ Mult).$$

(iv) In machine computation the range of numbers available is finite. This is often registered by the use of a notation for a maximum number M in programming languages (e.g. *Maxint* in PASCAL). To model this in algebras of natural numbers we proceed as follows.

Let $\Sigma^{count} = (nat; 0; succ, pred)$. Consider a Σ^{count} algebra

$$A = (\{\ 0, \ldots, M\ \}; 0_A; succ_A, pred_A)$$

with

$$succ_A(x) = x + 1$$

for $x < M$, and

$$pred_A(x) = x - 1$$

for $x > 0$. Two possibilities arise for the value of $succ_A(M)$:

(a) $succ_A(M) = M$, we call this the *linear option*;

(b) $succ_A(M) = 0$, we call this the *cyclic option*.

A third possibility arises if we attempt to model the case where a successor for M is 'not defined'. In this case we can make a new Σ^{count} algebra A' since we can adjoin a new element u to $\{\ 0, \ldots, M\ \}$. We can then define $succ_{A'}(u) = u$ to indicate that the operation does not return a number in the usual way. This matter of modelling unspecified elements was discussed earlier in Section 2.2.2. Another mathematical treatment would be to employ partial algebras.

Similarly replacing $succ_A$ by $pred_A$ we get the options:

(a) $pred_A(0) = 0$, the linear option;

(b) $pred_A(0) = M$, the cyclic option;

(c) $pred_{A'}(0) = u$, the unspecified option.

These options can be applied to other operations in the algebras of natural numbers of previous examples (see Exercise 3.1.8.(2)).

(v) Let us augment the natural numbers with infinite sequences of natural numbers in one of the above algebras, say $PA(\mathbf{N}, \mathbf{B})$. Thus we add to $\Sigma = \Sigma^{PAB}$ a new sort symbol $(nat \to nat)$ and a binary operation symbol *eval* of domain type $(nat \to nat)\ nat$ and codomain type nat to make a new signature $\bar{\Sigma}$. An *infinite sequence* over $\mathbf{N}$ is a function $a : \mathbf{N} \to \mathbf{N}$. Let $[\mathbf{N} \to \mathbf{N}]$ be the set of all such functions. We add to $PA(\mathbf{N}, \mathbf{B})$ the set $[\mathbf{N} \to \mathbf{N}]$ as a new carrier set of sort $(nat \to nat)$. We add the operation

$$eval_{\bar{A}} : [\mathbf{N} \to \mathbf{N}] \times \mathbf{N} \to \mathbf{N}$$

defined by

$$eval_{\bar{A}}(a,\ n) = a(n)$$

to make a $\overline{\Sigma}$ algebra $\overline{A} = \overline{PA}(\mathbf{N}, \mathbf{B})$.

(vi) Define the algebra A of finite and infinite sets of elements over a countable set X by

$$(X, P(X), \mathbf{N} \cup \{ \infty \}, \{ \text{tt}, \text{ff} \}; \emptyset; \cap, \cup, \neg, \textit{card}, \textit{mem}),$$

where $P(X)$ is the powerset of X, $\emptyset$ is the empty set and

$$\cup : P(X) \times P(X) \to P(X) \text{ and } \cap : P(X) \times P(X) \to P(X)$$

are the union and intersection operations on $P(X)$. The set complement operation $\neg : P(X) \to P(X)$ is defined by $\neg(s) = X - s$, the cardinality operation $\textit{card} : P(X) \to \mathbf{N} \cup \{ \infty \}$ is defined by $\textit{card}(s) = |s|$, and the set membership operation $\textit{mem} : X \times P(X) \to \{ \text{tt}, \text{ff} \}$ is defined by

$$\textit{mem}(x, s) = \begin{cases} \text{tt}, & \text{if } x \in s; \\ \text{ff}, & \text{otherwise.} \end{cases}$$

(Is the assumption that X is countable really necessary?)

(vii) Define the algebra A of finite sequences, strings or words over a non-empty countable set X by

$$A = (X, X^*, \mathbf{N}; \lambda, 0; \textit{inc}, \textit{head}, \textit{tail}, \textit{concat}, \textit{succ}, \textit{length}),$$

where X^* is the set of all finite sequences of members of X, including the empty sequence λ . The inclusion operation $\textit{inc} : X \to X^*$ is defined by

$$\textit{inc}(x) = x.$$

The head operation $\textit{head} : X^* \to X^*$ is defined by

$$\textit{head}(\lambda) = \lambda$$

and

$$\textit{head}(x_1 \ldots x_n) = x_1.$$

The tail operation $\textit{tail} : X^* \to X^*$ is defined by

$$\textit{tail}(x_1 \ldots x_n) = x_2 \ldots x_n$$

for $n > 1$ and

$$\textit{tail}(x_1) = \lambda, \quad \textit{tail}(\lambda) = \lambda.$$

The concatenation operation $\textit{concat} : X^* \times X^* \to X^*$ is defined by

$$\textit{concat}(x_1 \ldots x_n, x'_1 \ldots x'_m) = x_1 \ldots x_n \, x'_1 \ldots x'_m.$$

The length operation $length : X^* \to \mathbf{N}$ is defined by

$$length(\lambda) = 0$$

and for all $n \geq 1$

$$length(x_1 \ldots x_n) = n.$$

(viii) Let Σ be an S-sorted signature. We wish to define the algebra $T(\Sigma)$ of all *closed terms* or *closed expressions* over Σ. First for each sort $s \in S$, we define the set $T(\Sigma)_s$ of all closed terms over Σ of sort s by simultaneous induction over all sorts.

(a) For each sort $s \in S$ and each constant symbol $c \in \Sigma_{\lambda,s}$

$$c^s$$

is a closed term over Σ of sort s.

(b) For each sort $s \in S$, each non-empty word $w = s(1) \ldots s(n) \in S^+$, each function symbol $\sigma \in \Sigma_{w,s}$ and any terms $t_1, \ldots, t_n$ over Σ of sorts $s(1), \ldots, s(n)$ respectively,

$$\sigma^s(t_1, \ldots, t_n)$$

is a closed term over Σ of sort s.

(c) For each sort $s \in S$, nothing else is a closed term over Σ of sort s.

Recall that in the definition of a signature (3.1.1) we did not require the sets of constant and operation names to be pairwise disjoint. This allows us to reuse or overload names for constants and functions. For example the name 0 can denote a natural number, integer, real number etc. However, the purpose of a term of sort s is to denote a value in the carrier set A_s of an algebra A. The addition of a sort superscript to constant and function symbols ensures that the sets of terms are pairwise disjoint. Thus we may determine the sort of a term by inspection. Note we often omit sort superscripts on terms when no ambiguity will arise, e.g. in the case that Σ is single-sorted.

Note also that if there are no constant symbols of sort $s \in S$, and for each function symbol $\sigma \in \Sigma_{w,s}$ of domain type $w = s(1) \ldots s(n)$ and codomain type s there are no closed terms for at least one of the sorts $s(1), \ldots, s(n)$, then the carrier set $T(\Sigma)_s$ is empty. This means that $T(\Sigma)_s$ does not qualify as a carrier of sort s for an algebra of Σ terms.

Thus we suppose that Σ is *non-void* in the sense that for each sort $s \in S$

$$T(\Sigma)_s \neq \emptyset.$$

Then the constants and algebraic operations of our Σ algebra $T(\Sigma)$ are defined as follows. For each sort $s \in S$ and each constant symbol $c \in \Sigma_{\lambda, s}$

$$c_{T(\Sigma)} = c^s.$$

For each sort $s \in S$, each non-empty word $w = s(1)\ldots s(n) \in S^+$, each function symbol $\sigma \in \Sigma_{w,s}$ and any closed terms $t_1, \ldots, t_n$ over Σ of sorts $s(1), \ldots, s(n)$ respectively,

$$\sigma_{T(\Sigma)}(t_1, \ldots, t_n) = \sigma^s(t_1, \ldots, t_n).$$

(ix) Let Σ be an S-sorted signature. Let $X = \langle\, X_s \mid s \in S \,\rangle$ be an S-indexed family of sets X_s of *variable symbols* of sort $s \in S$. For each sort $s \in S$ we assume that the sets X_s and $\Sigma_{\lambda, s}$ are disjoint, i.e. no symbol is simultaneously a variable and a constant symbol of the same sort. We wish to define the algebra $T(\Sigma, X)$ of all *terms* or *expressions over Σ and X*. First for each sort $s \in S$ we define the set $T(\Sigma, X)_s$ of all terms over Σ and X of sort s, by simultaneous induction over all sorts.

(a) For each sort $s \in S$ and each constant symbol $c \in \Sigma_{\lambda,s}$

$$c^s$$

is a term over Σ and X of sort s.

(b) For each sort $s \in S$ and each variable symbol $x \in X_s$

$$x^s$$

is a term over Σ and X of sort s.

(c) For each sort $s \in S$, each non-empty word $w = s(1)\ldots s(n) \in S^+$, each function symbol $\sigma \in \Sigma_{w,s}$ and any terms $t_1, \ldots, t_n$ over Σ and X of sorts $s(1), \ldots, s(n)$ respectively

$$\sigma^s(t_1, \ldots, t_n)$$

is a term over Σ and X of sort $s \in S$.

(d) Nothing else is a term over Σ and X.

Once again, for $T(\Sigma, X)_s$ to qualify as the carrier set of sort s for a Σ algebra, $T(\Sigma, X)_s$ must be non-empty. This holds if, and only if,

Σ is non-void in sort s or else X_s is non-empty. Assuming each set $T(\Sigma, X)_s$ is non-empty, then the constants and algebraic operations of our Σ algebra $T(\Sigma, X)$ are defined as follows.

For each sort $s \in S$ and each constant symbol $c \in \Sigma_{\lambda, s}$

$$c_{T(\Sigma, X)} = c^s.$$

For each sort $s \in S$, each non-empty word $w = s(1) \ldots s(n) \in S^+$, each function symbol $\sigma \in \Sigma_{w,s}$ and any terms $t_1, \ldots, t_n$ over Σ and X of sorts $s(1), \ldots, s(n)$ respectively,

$$\sigma_{T(\Sigma, X)}(t_1, \ldots, t_n) = \sigma^s(t_1, \ldots, t_n).$$

When computing on an algebra A it may be convenient or necessary to involve new sets of data and appropriate operations, such as natural numbers; Booleans; finite and infinite sequences of elements from A. This leads to the construction of some new algebra B that is an *expansion* of A. For example, recalling Example 3.1.4.(iii), the algebra $PA(\mathbf{N})$ augments $Succ(\mathbf{N})$ by the operations of addition and multiplication; and the algebra $PA(\mathbf{N}, \mathbf{B})$ augments $PA(\mathbf{N})$ with the Booleans and the logical constants and definition-by-cases operation.

There is an extensive range of constructions for extending signatures and algebras. We will formulate one simple and fundamental definition of an expansion or augmentation of a signature and an algebra.

Definition 3.1.5. Let Σ^1 and Σ^2 be S_1 and S_2-sorted signatures respectively. We say that Σ^2 is an *expansion* of Σ^1 if, and only if,

$$S_1 \subseteq S_2,$$

and for each sort $s \in S_1$ and each word $w \in S_1^*$

$$\Sigma^1_{w,s} \subseteq \Sigma^2_{w,s}.$$

Equivalently we say that Σ^1 is a *subsignature* of Σ^2.

If Σ^2 is an expansion of Σ^1 and A and B are Σ^1 and Σ^2 algebras respectively then B is said to be a Σ^2-*expansion* of A if, and only if, for each sort $s \in S_1$

$$A_s = B_s,$$

for each sort $s \in S_1$ and each constant symbol $c \in \Sigma^1_{\lambda,s}$,

$$c_A = c_B$$

and for each sort $s \in S_1$, each non-empty word $w \in S_1^+$ and each function symbol $\sigma \in \Sigma^1_{w,s}$

$$\sigma_A = \sigma_B.$$

Equivalently we say that A is a Σ^1-*reduct* of B.

Clearly if Σ^2 is an expansion of Σ^1 then for each Σ^2 algebra B there is a unique Σ^1 algebra which is a Σ^1-reduct of B. This algebra is obtained by simply throwing away the carrier sets, constants and operations of B which are not named in Σ^1. We let $B|_{\Sigma^1}$ denote the unique Σ^1-reduct of B.

Examples 3.1.6.

(i) Consider augmenting an algebra by the natural numbers and Booleans. Recall the signature Σ^{PAB} of arithmetic given in Example 3.1.2.(i) and the Σ^{PAB} algebra $PA(\mathbf{N}, \mathbf{B})$ of arithmetic given in Example 3.1.4.(i). Let Σ be an S-sorted signature lacking arithmetic and Booleans. We can augment the signature Σ with the signature Σ^{PAB} of arithmetic giving an $S \cup \{\ nat,\ bool\ \}$-sorted signature Σ' as follows. Let $\Sigma'_{w,s} = \Sigma_{w,s}$ for each $w \in S^*$ and $s \in S$, and let $\Sigma'_{w,s} = \Sigma^{PAB}_{w,s}$ for each $w \in \{\ nat,\ bool\ \}^*$ and $s \in \{\ nat,\ bool\ \}$. In this example it is appropriate to augment the signature Σ' with operation symbols for definition-by-cases and equality on each sort $s \in S$. Thus for each sort $s \in S$, let $\Sigma'_{s\ s,\ bool} = \{\ eq\ \}$ and $\Sigma'_{bool\ s\ s,\ s} = \{\ case\ \}$. Otherwise let $\Sigma'_{w,s} = \emptyset$.

For any Σ algebra A we obtain a Σ' algebra B from A and $PA(\mathbf{N}, \mathbf{B})$ which has carrier sets $B_s = A_s$ for each sort $s \in S$, $B_{nat} = \mathbf{N}$ and $B_{bool} = \mathbf{B}$. The constants and operation symbols of Σ' are interpreted in the algebra B by the operations of A or $PA(\mathbf{N}, \mathbf{B})$ as appropriate. For each sort $s \in S$, the equality and definition-by-cases operations $eq_B : A_s \times A_s \to \mathbf{B}$ and $case_B : \mathbf{B} \times A_s \times A_s \to A_s$ of that sort $s \in S$ are defined in the usual way.

Clearly Σ' is finite if, and only if, Σ is finite. Also B is countable if, and only if, A is countable. Furthermore we have

$$B|_{\Sigma} = A$$

and

$$B|_{\Sigma^{PAB}} = PA(\mathbf{N}, \mathbf{B}).$$

(ii) Recall from Example 3.1.2.(ii) that Σ^F is the signature of a field and consider the field $\mathbf{Q}$ of rational numbers. Removing the division operation symbol $^{-1}$ from Σ^F results in the ring signature Σ^R and the ring of rational numbers $\mathbf{Q}|_{\Sigma^R}$. Removing subtraction $-$ from Σ^R results in the semiring signature Σ^{SR} and the semiring of rational numbers

$$(\mathbf{Q}|_{\Sigma^R})|_{\Sigma^{SR}} = \mathbf{Q}|_{\Sigma^{SR}}.$$

Alternatively, removing multiplication, $\cdot$ and 1 from Σ^R results in the group signature Σ^G and the additive group of rational numbers

$\mathbf{Q}|_{\Sigma^G}$. Finally, removing subtraction from Σ^G results in the semi-group signature Σ^{SG} of the additive semigroup of rationals $\mathbf{Q}|_{\Sigma^{SG}}$.

As we discussed in the Introduction, algebra studies calculation abstractly, in a way that seeks features common to many algebras. Thus, in the study of a significant property P of a particular algebra we invariably examine the family of *all* algebras satisfying P.

Definition 3.1.7. Let Σ be a signature. When considering a Σ algebra A satisfying a property P we may write $A \in Alg(\Sigma, P)$, and speak of the property P determining the *class* $Alg(\Sigma, P)$ of all Σ algebras satisfying P. In particular, if A is a Σ algebra we write $A \in Alg(\Sigma)$, and speak of the class $Alg(\Sigma)$ of all Σ algebras.

The problem with the informal concept of class is that it cannot be modelled on the informal concept of set. We may choose an appropriate signature Σ such that the class $Alg(\Sigma)$ contains all sets; hence if $Alg(\Sigma)$ were a set then we can construct Russell's Paradox. Does the concept of class lead to paradoxes?

Let us recall that the informal notion of a set given by Cantor was itself subject to paradoxes. Attempts to avoid these paradoxes led to the development of axioms for sets, designed to clarify the principles of set construction and avoid the paradoxes. These set theories, such as that of *Zermelo-Fraenkel set theory*, have been the subject of extensive logical investigation with the result that we believe that our informal notion of set, and informal principles of set construction, are safe from contradictions, providing that they conform to the axioms. Of course we have availed ourselves of this thesis without drawing attention to the existence of subtle foundational issues.

However, the informal concept of class that we wish to use does not conform to the axioms of set theory. Thus we need to turn to more general axiomatic frameworks to correct our faulty intuitions. Such a theory of classes defined by properties is provided by *von Neumann-Bernays-Gödel class theory*. In this theory some classes qualify as sets; satisfying essentially the same properties as sets as in Zermelo-Fraenkel set theory. Those classes which are not sets are called *proper classes*.

Thus, in addition to our thesis that our informal notions about sets are based on axiomatic set theory, we employ the thesis that our informal notions of class are based on axiomatic class theory.

Exercises 3.1.8.

1. Formulate signatures for the algebras of Section 2.1.

2. Consider the four possible designs for Σ^{count} algebras of the form

$$A = (\{\ 0, \ldots, M\ \};\ 0;\ succ_A,\ pred_A),$$

based on the linear and cyclic options for $succ_A$ and $pred_A$ in Example 3.1.4.(iv). Calculate the sets

$$\{\ x \in \{\ 0, \ldots, M\ \}\ : succ_A(pred_A(x)) = x\ \},$$

$$\{\ x \in \{\ 0, \ldots, M\ \}\ : pred_A(succ_A(x)) = x\ \},$$

in each of the cases. Hence deduce that A satisfies

$$succ_A(pred_A(x)) = x, \quad pred_A(succ_A(x)) = x$$

if, and only if, both $succ_A$ and $pred_A$ take the cyclic option on M and 0 respectively. Extend the cyclic and linear options for $succ_A$ and $pred_A$ to other operations in the algebras of natural numbers in 3.1.4.(ii).

3. Adapt the algebra of sets in Example 3.1.4.(vi) to create an algebra of all finite subsets of an arbitrary set X.

4. Recall the algebra of finite strings over a set X presented in Example 3.1.4.(vii). Generalize this algebra to an algebra of infinite strings.

5. Let Σ^1 be a subsignature of Σ^2. Is $T(\Sigma^2)|_{\Sigma^1} = T(\Sigma^1)$?

3.2   Subalgebras

An obvious construction on any S-sorted Σ algebra A arises by considering an S-indexed family of *subsets* $B = \langle\, B_s \subseteq A_s \mid s \in S\,\rangle$. If the subsets B_s contain the constants of A and are closed under the operations of A then we can make a structure that is again a Σ algebra.

Definition 3.2.1. Let A and B be S-sorted Σ algebras. Then B is said to be a Σ *subalgebra* of A if, and only if, for each sort $s \in S$,

$$B_s \subseteq A_s;$$

for each sort $s \in S$ and each constant symbol $c \in \Sigma_{\lambda,s}$

$$c_B = c_A;$$

and for each sort $s \in S$, each non-empty word $w = s(1) \ldots s(n) \in S^+$, each function symbol $\sigma \in \Sigma_{w,s}$ and any $(b_1, \ldots, b_n) \in B^w$,

$$\sigma_B(b_1, \ldots, b_n) = \sigma_A(b_1, \ldots, b_n).$$

We may also say that A is a Σ *extension* of B.

If B is a Σ subalgebra of A then we may omit reference to Σ and simply say that B is a subalgebra of A, writing $B \leq A$. If B is a subalgebra of A but $B \neq A$ then we say that B is a *proper subalgebra* of A (or A is a *proper extension* of B) and write $B < A$.

If $B = \langle\, B_s \subseteq A_s \mid s \in S \,\rangle$ is any S-indexed family of non-empty subsets of A then B is the family of carriers of a subalgebra of A if, and only if, B contains the constants of A and is closed under the operations of A. Furthermore, there is a unique subalgebra of A with an S-indexed family of carrier sets B, for if B_1 and B_2 are any subalgebras of A with identical carrier families then $B_1 = B_2$.

Examples 3.2.2.

(i) Consider the natural number algebra $PA(\mathbf{N})$ of Example 3.1.4.(iii). This algebra has no proper subalgebras because if $X \subseteq \mathbf{N}$ is the carrier of a subalgebra then the constant $0 \in X$ and for any $x \in X$, $Succ(x) \in X$. By the induction principle we deduce that all numbers $0, 1, 2, \ldots \in X$ and $X = \mathbf{N}$. Clearly this argument applies to all the algebras of Example 3.1.4.(iii) equipped with 0 and $Succ$; it also applies to the semiring example $SRing(\mathbf{N})$ of 3.1.4.(iii).

(ii) Recall that the fields $\mathbf{Q}, \mathbf{R}, \mathbf{C}$ of rational numbers, real numbers and complex numbers, respectively, have common signature Σ^F (of Example 3.1.2.(ii)). Clearly

$$\mathbf{Q} \leq \mathbf{R} \leq \mathbf{C}.$$

Notice that the statement that $\mathbf{Q}$ is a *subfield* of $\mathbf{R}$ is stronger than the statement that $\mathbf{Q}$ is a Σ^F subalgebra of $\mathbf{R}$ because it implies that $\mathbf{Q}$ also satisfies the field axioms.

If we consider $\mathbf{Q}, \mathbf{R}$ and $\mathbf{C}$ as semirings of signature Σ^{SR}, and consider the ring $\mathbf{Z}$ of integers as a semiring, then starting with the semiring $\mathbf{N}$ of natural numbers we have

$$\mathbf{N} < \mathbf{Z}|_{\Sigma^{SR}} < \mathbf{Q}|_{\Sigma^{SR}} < \mathbf{R}|_{\Sigma^{SR}} < \mathbf{C}|_{\Sigma^{SR}}.$$

Again these subalgebras are actually subsemirings. However, given a class K of Σ algebras that is defined axiomatically, it is common to find that Σ subalgebras of the algebras of K are not in K.

(iii) Consider the algebra $\overline{A} = \overline{PA}(\mathbf{N}, \mathbf{B})$ of numbers, Booleans and infinite sequences of numbers in Example 3.1.4.(v). The algebra $\overline{A}_S = \overline{PA}_S(\mathbf{N}, \mathbf{B})$ defined by simply replacing the carrier set $[\mathbf{N} \to \mathbf{N}]$ by any subset $S \subseteq [\mathbf{N} \to \mathbf{N}]$ qualifies as a subalgebra of $\overline{PA}(\mathbf{N}, \mathbf{B})$. This is clear because we have only to check the closure of S under evaluation

$$eval_{\overline{A}} : [\mathbf{N} \to \mathbf{N}] \times \mathbf{N} \to \mathbf{N}.$$

Thus we can make useful subalgebras by choosing S to be the set of *finite sequences* (i.e. infinite sequences that are zero almost everywhere), *primitive recursive sequences* or *recursive sequences*, for example.

Lemma 3.2.3. Let A, B and C be any S-sorted Σ algebras. If $B \leq A$, $C \leq A$ and $B \subseteq C$ then $B \leq C$.

Proof. For each sort $s \in S$ and each constant symbol $c \in \Sigma_{\lambda,s}$, since $B \leq A$ and $C \leq A$,

$$c_B = c_A = c_C.$$

For any word $w = s(1)\ldots s(n) \in S^+$ any sort $s \in S$, any function symbol $\sigma \in \Sigma_{w,s}$ and any elements $(b_1,\ldots,b_n) \in B^w$, since $B \leq A$ and $C \leq A$,

$$\sigma_B(b_1,\ldots,b_n) = \sigma_A(b_1,\ldots,b_n) = \sigma_C(b_1,\ldots,b_n).$$

Thus $B \leq C$.   ∎

When working with a particular class K of algebras it is often important that if $A \in K$ and B is a subalgebra of A then $B \in K$. We saw examples of this in the case of the classes of rings and fields.

Definition 3.2.4. A class K of Σ algebras is said to be *closed under the formation of subalgebras* if, and only if, whenever $A \in K$ and $B \leq A$ then $B \in K$.

Examples 3.2.5.

(i) The classes of semigroups, semirings, groups, rings and fields are closed under the formation of subalgebras. More generally, any class of algebras defined by equations is closed under the formation of subalgebras.

(ii) The class of all finite structures of any signature Σ is closed under the formation of subalgebras.

(iii) The class of all infinite structures of the signature $\Sigma^{Pred} = (nat; 0; pred)$ is not closed under subalgebras because subalgebras of $Pred(\mathbf{N})$ are finite. (Does this hold for all signatures?)

As we have seen, if $B = \langle\, B_s \subseteq A_s \mid s \in S \,\rangle$ does not contain each constant of A or is not closed under the operations of A then there is no Σ subalgebra of A which has B as its S-indexed family of carrier sets. Does there exist a unique 'smallest' subalgebra of A which contains the sets B_s?

For simplicity, we will answer this question in the single-sorted case.

Definition 3.2.6. Let A be any Σ algebra and $X \subseteq A$ be any subset of A. The subset $\langle X \rangle_A$ of A *generated by* X is the set

$$\langle X \rangle_A = \cap\{ \ B \mid X \subseteq B \text{ and } B \le A \ \}.$$

Thus $\langle X \rangle_A$ is the intersection of the family of carriers of *all* subalgebras of A.

When $X = \{ \ x_1, \ldots, x_n \ \}$ we will also write $\langle x_1, \ldots, x_n \rangle_A$ to denote $\langle X \rangle_A$. Notice if X is empty but Σ is non-void then $\langle X \rangle_A$ is not empty.

Lemma 3.2.7. *Let A be any Σ algebra and $X \subseteq A$ be any subset of A. If $\langle X \rangle_A$ is not empty then it is the carrier set of the smallest subalgebra of A containing X.*

Proof. First note that for every Σ subalgebra $B \le A$, the set of all constants $Cons^A \subseteq B$ and hence $Cons^A \subseteq \langle X \rangle_A$. Now consider any n-ary function symbol $\sigma \in \Sigma_n$ and $a_1, \ldots, a_n \in \langle X \rangle_A$. By definition $a_1, \ldots, a_n \in B$ for each subalgebra $B \le A$ such that $X \subseteq B$. For each such subalgebra B, $\sigma_A(a_1, \ldots, a_n) \in B$. Therefore

$$\sigma_A(a_1, \ldots, a_n) \in \cap\{ \ B \mid X \subseteq B \text{ and } B \le A \ \},$$

i.e. $\sigma_A(a_1, \ldots, a_n) \in \langle X \rangle_A$. Furthermore for any $x \in X$, $x \in B$ for each subalgebra $B \le A$ such that $X \subseteq B$. Hence

$$x \in \cap\{ \ B \mid X \subseteq B \text{ and } B \le A \ \},$$

i.e. $x \in \langle X \rangle_A$, so that $X \subseteq \langle X \rangle_A$. Thus if $\langle X \rangle_A \neq \emptyset$ then $\langle X \rangle_A$ is the carrier of a subalgebra of A containing X.

Suppose for a contradiction that there exists a subalgebra $C \le A$ containing X and that C is a proper subalgebra of $\langle X \rangle_A$. Then

$$C \in \{ \ B \mid X \subseteq B \text{ and } B \le A \ \}.$$

So $\langle X \rangle_A \subseteq C$ and hence $\langle X \rangle_A = C$ which contradicts C being a proper subalgebra of $\langle X \rangle_A$. ∎

The concept of *generating* an algebra by a set is a fundamental idea that will be encountered on many occasions.

Definition 3.2.8. Let A be any Σ algebra and $X \subseteq A$ be any subset. Then A is *generated* by X if, and only if,

$$\langle X \rangle_A = A.$$

We say that A is *finitely generated* if, and only if, A is generated by some finite set X.

Examples 3.2.9.

(i) Consider the natural number algebras of Example 3.1.4.(iii). As we noted earlier $PA(\mathbf{N})$ has no proper subalgebras and this means that

$$\langle \emptyset \rangle_{PA(\mathbf{N})} = PA(\mathbf{N}),$$

and so it is finitely generated. In the case of the algebra $Pred(\mathbf{N})$, for any finite set $\{ x_1, \ldots, x_n \} \subseteq \mathbf{N}$ we have

$$\langle x_1, \ldots, x_n \rangle_{Pred(\mathbf{N})} = (\{ 0, 1, \ldots, max(x_1, \ldots, x_n) \}; \ 0; \ Pred).$$

Thus $Pred(\mathbf{N})$ is not finitely generated.

(ii) In the field of real numbers $\mathbf{R}$ the subalgebra $\langle \emptyset \rangle_{\mathbf{R}}$ is the field of rational numbers.

To familiarize ourselves with this construction we prove some elementary properties.

Lemma 3.2.10. *Let A, B be any Σ algebras and let X, $Y \subseteq A$ be any subsets.*

(i) *If $B \leq A$ and $X \subseteq B$ then $\langle X \rangle_A \subseteq B$ and if $\langle X \rangle_A \neq \emptyset$ then*

$$\langle X \rangle_A \leq B.$$

(ii) *If $X \subseteq Y$ then $\langle X \rangle_A \subseteq \langle Y \rangle_A$ and if $\langle X \rangle_A \neq \emptyset$ then $\langle Y \rangle_A \neq \emptyset$ and*

$$\langle X \rangle_A \leq \langle Y \rangle_A.$$

Proof.   (i) By assumption $B \in \{ C \mid X \subseteq C \text{ and } C \leq A \}$. Thus

$$\langle X \rangle_A = \cap \{ C \mid X \subseteq C \text{ and } C \leq A \} \subseteq B.$$

By Lemma 3.2.7, if $\langle X \rangle_A \neq \emptyset$ then $\langle X \rangle_A \leq B$.

(ii) Suppose $X \subseteq Y$. If $C \leq A$ and $Y \subseteq C$ then $X \subseteq C$ and so, by definition, $X \subseteq \langle X \rangle_A \subseteq \langle Y \rangle_A$. Suppose $\langle X \rangle_A \neq \emptyset$. Then

$\langle Y \rangle_A \neq \emptyset$ and $\langle Y \rangle_A \leq A$. Therefore, since $X \subseteq \langle Y \rangle_A$, by (i), $\langle X \rangle_A \leq \langle Y \rangle_A$. ∎

We will consider the special case of the intersection set $\langle \emptyset \rangle_A$ of all subalgebras of A. If $\langle \emptyset \rangle_A \neq \emptyset$ then this set is the carrier of a subalgebra of A by Lemma 3.2.7. More generally, we note the following identity between sets:

Lemma 3.2.11. *Let A be any Σ algebra. Then $\langle \emptyset \rangle_A = \langle Cons^A \rangle_A$ and therefore A is generated by the empty set if, and only if, A is generated by the set $Cons^A$ of all constants of A.*

Proof. By definition

$$\langle \emptyset \rangle_A = \cap \{ B \mid \emptyset \subseteq B \text{ and } B \leq A \}$$
$$= \cap \{ B \mid Cons^A \subseteq B \text{ and } B \leq A \}$$
$$= \langle Cons^A \rangle.$$

∎

We are often concerned with *minimal* Σ algebras in applications of universal algebra to computer science.

Definition 3.2.12. A Σ algebra A is said to be *minimal* if, and only if, it has no proper subalgebra $B < A$.

Lemma 3.2.13. *Let Σ be a non-void signature and let A be a Σ algebra. The following are equivalent:*

(i) A is minimal;

(ii) A is generated by the empty set;

(iii) A is generated by the set $Cons^A$ of all constants of A.

Thus $\langle \emptyset \rangle_A$ is unique as the smallest subalgebra of A, i.e. if $B \leq A$ then $\langle \emptyset \rangle_A \leq B$.

Proof. (i) $\Rightarrow$ (ii). Suppose that A is minimal. Since Σ is non-void, $\langle \emptyset \rangle_A$ is the carrier of a subalgebra of A. Since A has no proper subalgebras, by definition $\langle \emptyset \rangle_A = A$.

The equivalence of (ii) and (iii) is given by Lemma 3.2.11.

(iii) $\Rightarrow$ (i). Suppose that $A = \langle Cons^A \rangle_A$. Let B be any subalgebra of A. Then $B \subseteq A$ and $Cons^A \subseteq B$. So by Lemma 3.2.10.(i), $A = \langle Cons^A \rangle_A$

$\subseteq B$. Hence $A = B$ so that A has no proper subalgebras, i.e. A is minimal. ∎

Examples 3.2.14.

(i) As we noted in Example 3.2.2.(i), the natural number algebras equipped with 0, $Succ$ are minimal, but $Pred(\mathbf{N})$ is not. The minimal subalgebra of $Pred(\mathbf{N})$ has carrier $\{\ 0\ \}$.

(ii) The integer ring $\mathbf{Z}$ is minimal as a ring. The rational field $\mathbf{Q}$ is minimal as a field.

(iii) The term algebra $T(\Sigma)$ is minimal for any non-void signature Σ.

(iv) Consider the algebra of integers

$$A = (\mathbf{Z};\, +,\, -)$$

with void signature Σ. We see that the set

$$\langle\, \emptyset\, \rangle_A = \{\ 0\ \}$$

because if B is any subalgebra of A then its carrier $B \neq 0$, and if $b \in B$ then $b - b = 0 \in B$; thus $\{\ 0\ \} \subseteq \langle\, \emptyset\, \rangle_A$. Conversely $\langle\, \emptyset\, \rangle_A \subseteq \{\ 0\ \}$ because $\{\ 0\ \}$ is the carrier of a subalgebra of A. Thus, A has a minimal subalgebra but no constants.

In the single-sorted case an inductive definition of the subset $\langle\, X\, \rangle_A$ generated by a subset X can be given as follows.

Definition 3.2.15. Let A be any Σ algebra and let $P(A)$ denote the power set of A. Define the map $E : P(A) \to P(A)$ by

$$E(X) = X \cup Cons^A \cup$$
$$\{\ \sigma_A(a_1,\ldots,a_n)\ |\ n \in \mathbf{N}^+,\, \sigma \in \Sigma_n\ \text{and}\ a_1,\ldots,a_n \in X\ \}.$$

Thus E applies the algebraic operations of A to the set X. Iterating E gives a sequence of maps $E^n : P(A) \to P(A)$ for each $n \in \mathbf{N}$, where

$$E^0(X) = X,\ E^{n+1}(X) = E(E^n(X)).$$

Clearly by definition of E, $E^n(X) \subseteq E^{n+1}(X)$ for all n. We define

$$E^\infty(X) = \cup_{n \in \mathbf{N}} E^n(X).$$

Lemma 3.2.16. *For any Σ algebra A and subset $X \subseteq A$, if $E^\infty(X)$ is non-empty then this set is a carrier set for a Σ subalgebra of A. If Σ is non-void then $E^\infty(X)$ is non-empty.*

Proof. By definition $Cons^A \subseteq E^1(X) \subseteq \cup_{n \in \mathbf{N}} E^n(X)$. Furthermore for each n-ary operation symbol $\sigma \in \Sigma_n$ and any $a_1, \ldots, a_n \in \cup_{n \in \mathbf{N}} E^n(X)$, since $E^m(X) \subseteq E^{m+1}(X)$, there exists $k \in \mathbf{N}$ such that $a_1, \ldots, a_n \in E^k(X)$. Therefore $\sigma_A(a_1, \ldots, a_n) \in E^{k+1}(X)$. So $\sigma_A(a_1, \ldots, a_n) \in \cup_{n \in \mathbf{N}} E^n(X)$, i.e. the set $E^\infty(X)$ is closed under the operations of A.

If Σ is non-void then Σ has constant symbols, hence $E^\infty(X)$ is non-empty. ∎

Theorem 3.2.17. *Let A be a Σ algebra and $X \subseteq A$ be any subset. Then:*

(i) $E^\infty(X) \subseteq \langle X \rangle_A$.

(ii) If $E^\infty(X)$ is non-empty then

$$E^\infty(X) = \langle X \rangle_A.$$

However $\langle \emptyset \rangle_A \neq \emptyset$ does not imply $E^\infty(X) \neq \emptyset$.

(iii) If Σ is non-void then for any Σ algebra A, A is minimal if, and only if, each element in the carrier of A can be obtained by a finite number of applications of operations of A to constants of A.

Proof. (i) Observe that $E^\infty(X) \subseteq \langle X \rangle_A$ if, and only if, for each $n \in \mathbf{N}$,

$$E^n(X) \subseteq \langle X \rangle_A.$$

The latter condition can be established by induction on n.

Basis. Suppose that $n = 0$. Then $E^0(X) = X \subseteq \langle X \rangle_A$, by Lemma 3.2.7.

Induction Step. Suppose that $n > 0$. Then $n = m + 1$, for some $m \in \mathbf{N}$. Now

$$E^n(X) = E^{m+1}(X)$$

$$= E^m(X) \cup Cons^A \cup$$

$$\{ \sigma_A(a_1, \ldots, a_n) \mid n \in \mathbf{N}^+, \sigma \in \Sigma_n \text{ and } a_1, \ldots, a_n \in E^m(X) \}.$$

By the induction hypothesis, $E^m(X) \subseteq \langle X \rangle_A$. Clearly $Cons^A \subseteq \langle X \rangle_A$. For any $a_1, \ldots, a_n \in E^m(X)$ we have $a_1, \ldots, a_n \in \langle X \rangle_A$ by the induction hypothesis. So for any Σ algebra B such that $X \subseteq B$ and $B \leq A$ we have

$a_1, \ldots, a_n \in B$. Hence for any n-ary operation symbol $\sigma \in \Sigma_n$ we have $\sigma_A(a_1, \ldots, a_n) \in B$. Therefore

$$\sigma_A(a_1, \ldots, a_n) \in \langle X \rangle_A.$$

Since σ and $a_1, \ldots, a_n$ were arbitrarily chosen,

$$E^n(X) \subseteq \langle X \rangle_A.$$

It follows from the induction principle that for all $n \in \mathbf{N}$, $E^n(X) \subseteq \langle X \rangle_A$, so $\cup_{n \in \mathbf{N}} E^n(X) \subseteq \langle X \rangle_A$, and therefore $E^\infty(X) \subseteq \langle X \rangle_A$.

(ii) We note that by definition

$$X \subseteq E^0(X) \subseteq \cup_{n \in \mathbf{N}} E^n(X)$$

and by Lemma 3.2.16, $E^\infty(X) \leq A$.
So, by Lemma 3.2.10.(i),

$$\langle X \rangle_A \subseteq E^\infty(X).$$

Finally note that Example 3.2.14.(iv) is an appropriate counterexample.

(iii) Follows from (ii) and Definition 3.2.15. ∎

Exercises 3.2.18.

1. Let Σ^1 be a subsignature of Σ^2. Is $T(\Sigma^1) \leq T(\Sigma^2)$?

2. Show that the subalgebra relation $\leq$ is a partial ordering relation on the set of all subalgebras of any Σ algebra A. Let A be a Σ algebra and K be a set of Σ subalgebras of A. Show that if $\cap K \neq \emptyset$ then this set is the carrier of a Σ subalgebra of A. Under what conditions on K is $\cup K$ a subalgebra of K? Suppose Σ is non-void. Define join and meet operations on the Σ subalgebras of A to show that this set forms a complete lattice under $\leq$.

3. Consider the algebra $Pred(\mathbf{N}) = (\ \mathbf{N};\ 0;\ Pred)$ of Example 3.1.4.(iii). Prove that the subalgebras of $Pred(\mathbf{N})$ are precisely the following

$$(\{\ 0, 1, \ldots, n\ \};\ 0;\ Pred), n = 0, 1, \ldots,$$

$$(\{\ 0, 1, \ldots\ \};\ 0;\ Pred).$$

Notice that every proper subalgebra is finite.

4. Check that the subalgebras of $\overline{PA}(\mathbf{N}, \mathbf{B})$ are precisely the algebras $\overline{PA}_S(\mathbf{N}, \mathbf{B})$ for non-empty $S \subseteq [\mathbf{N} \to \mathbf{N}]$.

5. Let Σ be an S-sorted signature and let A be a Σ algebra. Suppose $\Sigma_{w,s} = \emptyset$ for some sort s and all $w \in S^*$. Show, for any non-empty

subset $X \subseteq A_s$, that the algebra B obtained by replacing A_s with X in the algebra A is a Σ subalgebra of A.

6. The field of complex numbers $\mathbf{C}$ is the result of adjoining $i = \sqrt{-1}$ to the real numbers. Show that

$$\mathbf{Q}(i) = \{\ a + bi \mid a, b \in \mathbf{Q}\ \}$$

is the carrier set of a subfield of $\mathbf{C}$. Show that

$$\mathbf{Z}(i) = \{\ a + bi \mid a, b \in \mathbf{Z}\ \}$$

is the carrier set of a subring of $\mathbf{C}$. Is the set

$$\mathbf{N}(i) = \{\ a + bi \mid a, b \in \mathbf{N}\ \}$$

a subsemiring of $\mathbf{C}$?

7. Show that for any infinite subset $X \subseteq \mathbf{N}$, in the algebra $Pred(\mathbf{N})$

$$\langle\, X\, \rangle_{Pred(\mathbf{N})} = Pred(\mathbf{N}).$$

8. Let R be a commutative ring with 1 and of characteristic 0 and $a \in R$. Show that

$$\langle\, a\, \rangle_R = \{\ r_0 + r_1 a + r_2 a^2 + \ldots + r_n a^n : n \geq 0, r_0, \ldots, r_n \in \mathbf{Z}\ \}.$$

What role do the ring axioms play? Generalize this result to any Σ^R structure.

9. Generalize the result of Exercise (8) to $\langle\, X\, \rangle_R$ for X a finite subset of the ring R. What happens in the case that X is any subset of R?

10. Show that in the field of real numbers

$$\langle\, \sqrt{2}\, \rangle_{\mathbf{R}} = (\{\ a + b\sqrt{2} \mid a, b \in \mathbf{Q}\ \} : 0, 1, +, -, \cdot, {}^{-1}).$$

11. When is a term algebra $T(\Sigma)$ finitely generated?

12. Generalize the definition of $\langle\, X\, \rangle_A$ given in Definition 3.2.6 and systematically generalize the ancillary concepts and results to the many-sorted case. Pay special attention to the role of non-void signatures.

3.3 Congruences and quotient algebras

An important construction in any branch of mathematics is the *quotient construction* associated with an *equivalence relation* on a set, algebra or space. Recall the definition of an equivalence relation on a set A.

Definition 3.3.1. Let A be any non-empty set. An *equivalence relation* $\equiv$ on A is a binary, reflexive, symmetric and transitive relation on A. This means that $\equiv$ is a subset of A^2 which, writing $a \equiv b$ for $(a, b) \in \equiv$, satisfies for any a, b, $c \in A$ the following properties:

$$a \equiv a;$$

$$a \equiv b \Rightarrow b \equiv a;$$

and

$$a \equiv b \ \& \ b \equiv c \Rightarrow a \equiv c.$$

Given an equivalence relation $\equiv$ on a set A we can define for each element $a \in A$ the *equivalence class* $[a] \subseteq A$ of a with respect to $\equiv$ to be the subset

$$[a] = \{ \ b \in A \ | \ b \equiv a \ \}.$$

It is easily shown that the equivalence classes of A with respect to $\equiv$ form a *partition* of the set A, i.e. a collection of disjoint subsets whose union is the set A. We let $A/\equiv$ denote the set of all equivalence classes of members of A with respect to $\equiv$,

$$A/\equiv \ = \{ \ [a] \ | \ a \in A \ \}.$$

We call $A/\equiv$ the *quotient* or *factor* set of A with respect to the equivalence relation $\equiv$.

Given a quotient set $A/\equiv$ we can choose an element from each equivalence class in $A/\equiv$ as a representative of that class. A subset $T \subseteq A$ of elements such that every equivalence class $[a]$ has some representative $r \in T$ and no equivalence class has more than one representative $r \in T$, is known as a *transversal* or, in certain circumstances, a *set of canonical representatives* or *normal forms* for $A/\equiv$.

The special kind of equivalence relation on the carrier set of an algebra A which is 'compatible' with the operations of A is known as a *congruence*.

Definition 3.3.2. Let A be an S-sorted Σ algebra and $\equiv$ be an S-indexed family $\langle \equiv_s \ | \ s \in S \rangle$ of binary relations on A, such that for each $s \in S$ the relation $\equiv_s$ is an equivalence relation on the carrier set A_s. Then $\equiv$ is a Σ

congruence on A if, and only if, the relations $\equiv_s$ satisfy the following *sub-stitutivity condition*: for each non-empty word $w = s(1)\ldots s(n) \in S^+$, each sort $s \in S$, each function symbol $\sigma \in \Sigma_{w,s}$ and any $(a_1,\ldots,a_n)$, $(b_1,\ldots,b_n)$ $\in A^w$

$$a_i \equiv_{s(i)} b_i \text{ for } i = 1,\ldots,n \Rightarrow \sigma_A(a_1,\ldots,a_n) \equiv_s \sigma_A(b_1,\ldots,b_n).$$

We let $Con(A)$ denote the set of all Σ congruences on A. If $\equiv$ is a Σ congruence on A then for $a \in A_s$, $[a]$ denotes the *equivalence class* of a with respect to $\equiv_s$,

$$[a] = \{\ b \in A_s \ |\ b \equiv_s a\ \}.$$

Sometimes we index congruences $\equiv^\theta$, $\equiv^\psi$ etc., in which case we write $[a]_\theta$ for the equivalence class of a with respect to $\equiv^\theta$.

The substitutivity condition of Definition 3.3.2 is precisely what is required in order to perform the following construction.

Definition 3.3.3. Let $\equiv$ be a Σ congruence on an S-sorted Σ algebra A. The *quotient algebra* $A/\equiv$ of A by the congruence $\equiv$ is the Σ algebra with S-indexed family of carrier sets

$$A/\equiv \ = \langle\, (A/\equiv)_s \ |\ s \in S\,\rangle,$$

where for each sort $s \in S$,

$$(A/\equiv)_s = A_s/\equiv_s,$$

and the constants and algebraic operations of the quotient algebra are defined as follows.

For each sort $s \in S$ and each constant symbol $c \in \Sigma_{\lambda,s}$

$$c_{A/\equiv} = [c_A].$$

For each non-empty word $w = s(1)\ldots s(n) \in S^+$, each sort $s \in S$, each function symbol $\sigma \in \Sigma_{w,s}$ and any $(a_1,\ldots,a_n) \in A^w$

$$\sigma_{A/\equiv}([a_1],\ldots,[a_n]) = [\sigma_A(a_1,\ldots,a_n)].$$

As usual we denote the quotient algebra by its carrier set $A/\equiv$.

We check that what we have called the quotient algebra is indeed a Σ algebra.

Lemma 3.3.4. Let A be an S-sorted Σ algebra and $\equiv$ be a Σ congruence on A. Then the quotient algebra $A/\equiv$ is an S-sorted Σ algebra.

Proof. Since the carriers of A are non-empty and we have defined the constants, we need only check that, for each non-empty word $w = s(1) \ldots s(n) \in S^+$, each sort $s \in S$, and each function symbol $\sigma \in \Sigma_{w,s}$, the corresponding operation $\sigma_{A/\equiv}$ is well defined as a function

$$\sigma_{A/\equiv} : (A/\equiv)_{s(1)} \times \ldots \times (A/\equiv)_{s(n)} \to (A/\equiv)_s.$$

Consider any $a_i, b_i \in A_{s(i)}$ for $1 \le i \le n$ and suppose that $a_i \equiv_{s(i)} b_i$ for each $1 \le i \le n$. We must show

$$\sigma_{A/\equiv}([a_1], \ldots, [a_n]) = \sigma_{A/\equiv}([b_1], \ldots, [b_n]),$$

i.e. $\sigma_{A/\equiv}([a_1], \ldots, [a_n])$ does not depend upon the choice of representatives for the equivalence classes $[a_1], \ldots, [a_n]$. By assumption $\equiv$ is a Σ congruence. So by definition if $a_i \equiv_{s(i)} b_i$ for each $1 \le i \le n$ then

$$\sigma_A(a_1, \ldots, a_n) \equiv_s \sigma_A(b_1, \ldots, b_n),$$

i.e.

$$[\sigma_A(a_1, \ldots, a_n)] = [\sigma_A(b_1, \ldots, b_n)].$$

Then by definition of $\sigma_{A/\equiv}$

$$\sigma_{A/\equiv}([a_1], \ldots, [a_n]) = \sigma_{A/\equiv}([b_1], \ldots, [b_n]).$$

∎

Examples 3.3.5.

(i) For any S-sorted Σ algebra A the S-indexed family of relations $\langle =_s \mid s \in S \rangle$, where $=_s$ is the equality relation on A_s, is a Σ congruence on A known as the *equality, null* or *zero congruence* $=$. The S-indexed family of relations $A^2 = \langle A_s^2 \mid s \in S \rangle$, where $A_s^2 = A_s \times A_s$, is also a Σ congruence on A known as the *unit congruence* . As an exercise, we leave it to the reader to check that A/A^2 is a unit algebra. What is the relationship between A and $A/=$?

(ii) Consider the natural number algebra $PA(\mathbf{N})$ in 3.1.4.(iii). For any $n \in \mathbf{N}$ we define a relation

$$x \equiv^n y \Leftrightarrow x \bmod n = y \bmod n.$$

Equivalently, $x \equiv^n y$ means that

$$\text{if } x \ge y \text{ then } x - y = kn \text{ for some } k \in \mathbf{N},$$

$$\text{if } y \ge x \text{ then } y - x = kn \text{ for some } k \in \mathbf{N}.$$

It is easy to check that $\equiv^n$ is an equivalence relation and, indeed, a congruence on $PA(\mathbf{N})$. Consider the substitutivity condition for multiplication. Suppose $x_1 \equiv^n y_1$ and $x_2 \equiv^n y_2$. We want to show that

$$mult(x_1, x_2) \equiv^n mult(y_1, y_2)$$

or, equivalently, that

$$x_1 x_2 \equiv^n y_1 y_2.$$

Suppose that $x_1 > y_1$ and $x_1 - y_1 = k_1 n$; and that $x_2 > y_2$ and $x_2 - y_2 = k_2 n$. Thus

$$x_1 x_2 = (y_1 + k_1 n)(y_2 + k_2 n)$$

$$= y_1 y_2 + k_1 n y_2 + k_2 n y_1 + k_1 k_2 n^2,$$

and hence for $k = k_1 y_2 + k_2 y_1 + k_1 k_2 n$ we have

$$x_1 x_2 - y_1 y_2 = nk$$

and so

$$x_1 x_2 \equiv^n y_1 y_2.$$

The substitutivity condition for the other operations of successor and addition are proved similarly. Clearly the equivalence classes for $\equiv^n$ are

$$\{\ x \mid x \ mod \ n = r \ \}$$

for $r = 0, 1, \ldots, n - 1$ if $n \neq 0$.

(iii) Let Σ be a non-void signature and $T(\Sigma)$ the algebra of closed terms over Σ. Given any Σ algebra A we can evaluate or interpret any closed term $t \in T(\Sigma)$ as a sequence of operations of A which, applied to the constants of A, result in an element of A. For example, dropping the sort superscript *nat*, the terms over Σ^{PA}

$$add(mult(succ(0),\ succ(0)),\ succ(0)),$$

$$add(mult(succ(0),\ succ(add(0,\ 0))),\ succ(0))$$

denote the values in the interpretation $PA(\mathbf{N})$

$$((0 + 1) \cdot (0 + 1)) + (0 + 1) = 2,$$

$$((0 + 1) \cdot ((0 + 0) + 1)) + (0 + 1) = 2.$$

We formalize this evaluation as an S-indexed family of functions

$$V^A = \langle\, V_s^A : T(\Sigma)_s \to A_s \mid s \in S \,\rangle$$

defined by simultaneous induction on terms as follows. For each sort $s \in S$ and each constant symbol $c \in \Sigma_{\lambda, s}$

$$V_s^A(c^s) = c_A,$$

where $c_A \in A_s$. For each non-empty word $w = s(1) \ldots s(n) \in S^+$, each sort $s \in S$, each function symbol $\sigma \in \Sigma_{w,s}$ and any closed terms $t_i \in T(\Sigma)_{s(i)}$ for $1 \leq i \leq n$

$$V_s^A(\sigma^s(t_1, \ldots, t_n)) = \sigma_A(V_{s(1)}^A(t_1), \ldots, V_{s(n)}^A(t_n)),$$

where $\sigma_A : A^w \to A_s$.

Let us now define an S-indexed family of relations $\equiv^A = \langle \equiv_s^A \mid s \in S \rangle$ on $T(\Sigma)$ by

$$t \equiv_s^A t' \Leftrightarrow V_s^A(t) = V_s^A(t')$$

for each sort $s \in S$ and any terms $t, t' \in T(\Sigma)_s$. This family of relations $\equiv^A$ is called *A-equivalence*. It is easy to check that $\equiv^A$ is an S-indexed family of equivalence relations on A (note Exercise 3.3.25.(8)). To show that $\equiv^A$ is a Σ congruence suppose for some $w = s(1) \ldots s(n) \in S^+$, and terms $t_i, t_i' \in T(\Sigma)_{s(i)}$ for $1 \leq i \leq n$, that $t_i \equiv_{s(i)}^A t_i'$ for each $1 \leq i \leq n$. Now for any sort $s \in S$ and any function symbol $\sigma \in \Sigma_{w,s}$

$$V_s^A(\sigma^s(t_1, \ldots, t_n)) = \sigma_A(V_{s(1)}^A(t_1), \ldots, V_{s(n)}^A(t_n))$$

$$= \sigma_A(V_{s(1)}^A(t_1'), \ldots, V_{s(n)}^A(t_n'))$$

by assumption

$$= V_s^A(\sigma^s(t_1', \ldots, t_n')).$$

Hence $\sigma^s(t_1, \ldots, t_n) \equiv_s^A \sigma^s(t_1', \ldots, t_n')$. Since $t_1, t_1', \ldots, t_n, t_n'$ and σ were arbitrarily chosen then $\equiv^A$ satisfies the substitutivity property. Intuitively $t \equiv_s^A t'$ if, and only if, t and t' have the same value when evaluated in the algebra A. The quotient algebra $T(\Sigma)/\equiv^A$ is therefore an algebra that represents the A-equivalent terms.

(iv) Let Σ be a non-void signature. For simplicity we assume that Σ is single-sorted and omit sort superscripts on terms in $T(\Sigma)$. The *parse trees* for terms over Σ are given by a map

$$T : T(\Sigma) \to Tree(\Sigma)$$

which assigns to each term $t \in T(\Sigma)$ a tree $T(t)$, labelled by the constants and function symbols of Σ, such that

$$t = t' \Leftrightarrow T(t) = T(t').$$

The branching properties of $T(t)$ are determined by the arity of the function symbols appearing in t. The leaves of $T(t)$ are labelled by constants.

The parse trees suggest the following important notion of approximation for terms: for any $k = 0, 1, 2, \ldots$ we say that t and t' are *equal
to degree* or *depth* k if, and only if, $T(t)$ and $T(t')$ are the same tree
up to level k. Let us write $t \equiv^k t'$ in this case. For example

$$add(mult(succ(0), \, succ(0)), \, succ(0)),$$

$$add(mult(succ(0), \, succ(add(0, \, 0))), \, succ(0))$$

have the parse trees in Fig. 3.1.

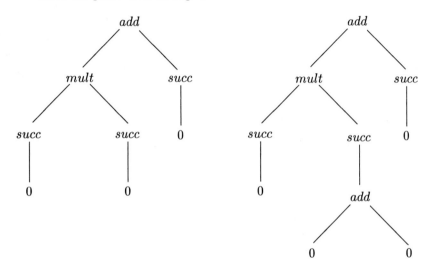

Fig. 3.1.

By inspection we see that for $k = 0, 1, 2$ these terms are equal to
depth k; but we see that they are not equal for $k = 3, 4, 5, \ldots$

Clearly, this notion of approximation has the property that $t \neq t'$ if,
and only if, for some $k \in \mathbf{N}$

$$t \equiv^n t' \text{ for } n < k \text{ and } t \not\equiv^n t' \text{ for } n \geq k.$$

Thus $t = t'$ if, and only if, $t \equiv^k t'$ for all $k \in \mathbf{N}$. We will define formally the family of approximation relations and prove they are a
family of *congruences* on the algebra $T(\Sigma)$.

We define a family of maps

$$\langle \, T_k : T(\Sigma) \to Tree(\Sigma) \mid k \in \mathbf{N} \, \rangle$$

by induction on the structure of terms. For $t \in T(\Sigma)$, the tree $T_k(t)$
is to be the tree obtained by stopping the construction of the parse
tree $T(t)$ at level k.

Let $t = c$ for some constant symbol $c \in \Sigma_0$. Then $T_k(t)$ is the single node labelled by c for all $k \in \mathbf{N}$:

$$\cdot c \ .$$

Let $t = \sigma(t_1, \ldots, t_n)$ for some n-ary function symbol $\sigma \in \Sigma_n$ and sequence of terms $t_1, \ldots, t_n \in T(\Sigma)^n$ and suppose that $T_k(t_1), \ldots, T_k(t_n)$ are defined for all $k \in \mathbf{N}$.

For $k = 0$, $T_k(t)$ is the single node labelled by σ.

For $k > 0$, $T_k(t)$ is a node labelled by σ with n edges to which the subtrees $T_{k-1}(t_1), \ldots, T_{k-1}(t_n)$ are attached by their roots.

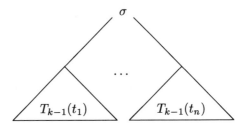

We now define for $t, t' \in T(\Sigma)$ and each $k \in \mathbf{N}$

$$t \equiv^k t' \Leftrightarrow T_k(t) = T_k(t').$$

Proposition 3.3.6. *For each $k \in \mathbf{N}$, the relation $\equiv^k$ is a Σ congruence on $T(\Sigma)$.*

Proof.   It is easy to see that for each $k \in \mathbf{N}$, the relation $\equiv^k$ is an equivalence relation on $T(\Sigma)$: note Exercise 3.3.25.(8). To prove that the substitutivity condition holds, consider an arbitrary n-ary function symbol $\sigma \in \Sigma_n$ and any terms $t_i, t_i' \in T(\Sigma)$ for $1 \leq i \leq n$. Suppose $t_i \equiv^k t_i'$ for $1 \leq i \leq n$. We must show that

$$\sigma(t_1, \ldots, t_n) \equiv^k \sigma(t_1', \ldots, t_n').$$

For $k = 0$, we note that the parse trees $T_k(\sigma(t_1, \ldots, t_n))$ and $T_k(\sigma(t_1', \ldots, t_n'))$ are equal to depth 0. For $k > 0$ we note that $T_k(\sigma(t_1, \ldots, t_n))$ and $T_k(\sigma(t_1', \ldots, t_n'))$ are the same provided that for $1 \leq i \leq n$

$$T_{k-1}(t_i) = T_{k-1}(t_i').$$

We know that $T_k(t_i) = T_k(t_i')$ by hypothesis; thus the proposition is proved using the following:

Lemma 3.3.7. *For any $k > 0$ and any t, $t' \in T(\Sigma)$, if $T_k(t) = T_k(t')$ then $T_{k-1}(t) = T_{k-1}(t')$.*

Proof. Exercise 3.3.25.(8). ∎

Definition 3.3.8. A class K of Σ algebras is said to be *closed under the formation of quotient algebras* if, and only if, whenever $A \in K$ and $B = A/\equiv$ then $B \in K$.

In Section 3.4 we return to this concept when we have established a connection between quotient algebras and homomorphisms.

In the remainder of Section 3.3, for simplicity, we consider only the single-sorted case. We leave it as an exercise for the reader to generalize the following definitions and results to the many-sorted case.

A Σ congruence on a Σ algebra A induces a Σ congruence on any subalgebra $B \leq A$ in the obvious way:

Definition 3.3.9. If $B \subseteq A$ is any subset of A then the restriction $\equiv |B$ of $\equiv$ to B is $\equiv \cap B^2$.

Lemma 3.3.10. *If $\equiv$ is a congruence on an algebra A and B is any subalgebra of A then $\equiv |B$ is a congruence on B.*

Proof. Exercise 3.3.25.(11). ∎

Definition 3.3.11. Given two congruences $\equiv^\phi$ and $\equiv^\theta$ on a Σ algebra A we say that $\equiv^\theta$ is a *finer congruence* than $\equiv^\phi$ if, and only if, $\equiv^\theta \subseteq \equiv^\phi$ (we also say $\equiv^\phi$ is *coarser congruence* than $\equiv^\theta$). Furthermore, we say that $\equiv^\theta$ is a *strictly finer congruence* than $\equiv^\phi$ if, and only if, $\equiv^\theta \subset \equiv^\phi$ (we also say $\equiv^\phi$ is *strictly coarser congruence* than $\equiv^\theta$). If $\equiv^\theta$ is a finer (strictly finer) congruence than $\equiv^\phi$ then each equivalence class $[a]_\theta$ is a subset (proper subset) of $[a]_\phi$.

Equality $=$ is the finest congruence on any Σ algebra A, i.e. equality is contained in *every* Σ congruence on A. The coarsest congruence on A is the unit congruence A^2, i.e. every Σ congruence on A is contained in A^2 and A/A^2 is a unit algebra.

An algebra A is *simple* if, and only if, the only congruences on A are equality $=$ and the unit congruence A^2.

Examples 3.3.12.

(i) In Example 3.3.5.(iv) above $\equiv^{k+1}$ is a finer congruence than $\equiv^k$ on $T(\Sigma)$.

(ii) For any signature Σ, any unit Σ algebra is simple.

(iii) The algebra $PA(\mathbf{N})$ is not simple since as we have seen in Example 3.3.5.(ii), the relations $\equiv^n$ are congruences on $PA(\mathbf{N})$ which are neither equality nor the unit congruence when $n \geq 2$.

(iv) Suppose that we augment the algebra $PA(\mathbf{N})$ with the predecessor function $Pred$ to give an algebra

$$PA_{pred}(\mathbf{N}) = (\mathbf{N}; \ 0; \ Succ, \ Pred, \ Add, \ Mult).$$

Then $PA_{pred}(\mathbf{N})$ is a simple algebra. For suppose that $\equiv$ is any congruence on $PA_{pred}(\mathbf{N})$ other than equality. Then for some $m, n \in \mathbf{N}$ such that $m \neq n$ it must be the case that $m \equiv n$ (otherwise $\equiv$ is equality). Suppose, without loss of generality, that $n \geq m$. Since $\equiv$ is a congruence then

$$Pred(m) \equiv Pred(n),$$

i.e. $m - 1 \equiv n - 1$ and by m repeated applications of $Pred$ we have

$$0 \equiv n - m.$$

Then

$$0 \equiv Pred(0) \equiv Pred^{m+1}(n),$$

i.e. $0 \equiv n - (m + 1)$, and by $(n - m) - 1$ applications of $Pred$ we have

$$0 \equiv 1.$$

From this it follows, by application of $Succ$, that for each $a \in \mathbf{N}$ we must have

$$0 \equiv a.$$

Therefore $\equiv$ is the unit congruence. By symmetry, the result also holds if $n < m$. Thus $PA_{pred}(\mathbf{N})$ has no congruences other than equality and the unit congruence, i.e. $PA_{pred}(\mathbf{N})$ is a simple algebra.

If $\equiv^\theta$ is a finer congruence than $\equiv^\phi$, then we can factor the congruence $\equiv^\phi$ by $\equiv^\theta$ to produce a congruence on the quotient algebra $A/\equiv^\theta$ as follows.

Definition 3.3.13. Let $\equiv^\phi$, $\equiv^\theta$ be any Σ congruences on a Σ algebra A such that $\equiv^\theta \subseteq \equiv^\phi$. We define the *factor congruence* $\equiv^{\phi/\theta}$ (also denoted by $\equiv^\phi / \equiv^\theta$) on $A/\equiv^\theta$ to be the binary relation given by

$$[a]_\theta \equiv^{\phi/\theta} [b]_\theta \Leftrightarrow a \equiv^\phi b.$$

We must now verify that what we have called a factor congruence is indeed a congruence.

Lemma 3.3.14. *If $\equiv^\phi$ and $\equiv^\theta$ are Σ congruences on a Σ algebra A such that $\equiv^\theta \subseteq \equiv^\phi$ then the factor congruence $\equiv^{\phi/\theta}$ is a Σ congruence on $A/\equiv^\theta$.*

Proof. It is easy to check that $\equiv^{\phi/\theta}$ is well defined as a relation, and is an equivalence relation on $A/\equiv^\theta$ since $\equiv^\phi$ is an equivalence relation on A. To prove that $\equiv^{\phi/\theta}$ satisfies the substitutivity condition consider any function symbol $\sigma \in \Sigma_n$ and $a_1, \ldots, a_n, b_1, \ldots, b_n \in A$. Suppose that for $i = 1, \ldots, n$

$$[a_i]_\theta \equiv^{\phi/\theta} [b_i]_\theta.$$

Then for $i = 1, \ldots, n$

$$a_i \equiv^\phi b_i,$$

and since $\equiv^\phi$ is a congruence then

$$\sigma_A(a_1, \ldots, a_n) \equiv^\phi \sigma_A(b_1, \ldots, b_n).$$

Thus

$$[\sigma_A(a_1, \ldots, a_n)]_\theta \equiv^{\phi/\theta} [\sigma_A(b_1, \ldots, b_n)]_\theta,$$

and hence

$$\sigma_{A/\equiv^\theta}([a_1]_\theta, \ldots, [a_n]_\theta) \equiv^{\phi/\theta} \sigma_{A/\equiv^\theta}([b_1]_\theta, \ldots, [b_n]_\theta).$$

So $\equiv^{\phi/\theta}$ satisfies the substitutivity condition and is therefore a Σ congruence on $A/\equiv^\theta$. ∎

Examples 3.3.15.

(i) Recall from Example 3.3.5.(iv) the family of congruences $\equiv^k$ on the term algebra $T(\Sigma)$. Clearly $\equiv^{k+1} \subseteq \equiv^k$, for given any $t, t' \in T(\Sigma)$, if $t \equiv^{k+1} t'$ then $t \equiv^k t'$. The quotient algebra $(T(\Sigma)/\equiv^{k+1})/\equiv^{k/k+1}$ is therefore well defined. We leave it to the reader, as an exercise, to prove that $[t] \equiv^{k/k+1} [t']$ if, and only if, $t \equiv^k t'$.

(ii) For any Σ congruence $\equiv^\theta$ on A the factor congruence $\equiv^{\theta/\theta}$ is just equality on $A/\equiv^\theta$.

Two basic facts on congruences which are frequently used are the following.

Lemma 3.3.16. *Let Σ be a non-void signature. Let $\equiv$ be any Σ congruence on a Σ algebra A. If A is generated by X then $A/\equiv$ is generated by $X/\equiv$.*

Proof. Recall from Theorem 3.2.17 that if Σ is non-void then

$$\langle X \rangle_A = E^\infty(X) = \bigcup_{n \in \mathbf{N}} E^n(X).$$

We show, by induction on n, that for each $n \in \mathbf{N}$

$$E^n(X/\equiv) = E^n(X)/\equiv.$$

Basis. Suppose that $n = 0$. By definition $E^0(X/\equiv) = X/\equiv = E^0(X)/\equiv$

Induction Step. Suppose that $n > 0$. By definition of E^n

$$E^n(X/\equiv) = E(E^{n-1}(X/\equiv))$$

$$= E^{n-1}(X/\equiv) \cup Cons^{A/\equiv}$$

$$\cup\{\ \sigma_{A/\equiv}([a_1],\dots,[a_n])\ |\ n \in \mathbf{N}^+, \sigma \in \Sigma_n$$

$$\text{and } [a_1],\dots,[a_n] \in E^{n-1}(X/\equiv)\ \};$$

by definition of E

$$= E^{n-1}(X)/\equiv \cup Cons^A/\equiv$$

$$\cup\{\ [\sigma_A(a_1,\dots,a_n)]\ |\ n \in \mathbf{N}^+, \sigma \in \Sigma_n \text{ and } a_1,\dots,a_n \in E^{n-1}(X)\ \};$$

by the induction hypothesis

$$= E^{n-1}(X)/\equiv \cup Cons^A/\equiv$$

$$\cup\{\ \sigma_A(a_1,\dots,a_n)\ |\ n \in \mathbf{N}^+, \sigma \in \Sigma_n \text{ and } a_1,\dots,a_n \in E^{n-1}(X)\ \}/\equiv$$

$$= E^n(X)/\equiv.$$

Thus, by Theorem 3.2.17.(ii),

$$\langle X/\equiv \rangle_{A/\equiv} = \bigcup_{n \in \mathbf{N}} E^n(X/\equiv)$$

$$= (\bigcup_{n \in \mathbf{N}} E^n(X))/\equiv$$

$$= \langle X \rangle_A/\equiv$$

$$= A/\equiv.$$

■

We are often interested in families of congruences on an algebra. For example these arise when considering various methods of approximating algebras, as we have seen in the case of $T(\Sigma)$ and the approximating congruences $\equiv^k$ in Example 3.3.5.(iv). Indeed the latter construction is typical of many forms of approximation in mathematics (compare Exercise 3.3.25.(14)).

Lemma 3.3.17. *Let A be a Σ algebra and $\{ \equiv^i \mid i \in I \}$ be any non-empty set of Σ congruences on A. Then $\cap_{i \in I} \equiv^i$ is a Σ congruence on A.*

Proof. Let $\equiv$ denote the relation $\cap_{i \in I} \equiv^i$. Since each congruence $\equiv^i$ is reflexive, symmetric and transitive then $\equiv$ is also reflexive, symmetric and transitive, i.e. an equivalence relation on A. To show that $\equiv$ satisfies the substitutivity property, consider any function symbol $\sigma \in \Sigma_n$ and any $a_1, \ldots, a_n, b_1, \ldots, b_n \in A$. Suppose that for $j = 1, \ldots, n$ we have

$$a_j \equiv b_j.$$

Then for $j = 1, \ldots, n$ and any $i \in I$ we have

$$a_j \equiv^i b_j.$$

Since $\equiv^i$ is a Σ congruence on A then

$$\sigma_A(a_1, \ldots, a_n) \equiv^i \sigma_A(b_1, \ldots, b_n).$$

Since $\equiv^i$ was arbitrarily chosen this holds for all $i \in I$. Hence

$$\sigma_A(a_1, \ldots, a_n) \equiv \sigma_A(b_1, \ldots, b_n).$$

So $\equiv$ is a Σ congruence on A. ∎

The result of this Lemma suggests the idea of *generating a congruence by a set*.

Definition 3.3.18. Let A be a Σ algebra and $X \subseteq A^2$ be any subset of ordered pairs from A. We define the Σ congruence $\equiv^X$ on A *generated* by the set X to be

$$\equiv^X = \cap\{ \equiv \mid X \subseteq \equiv \text{ and } \equiv \in Con(A) \}.$$

The set $\{ \equiv \mid X \subseteq \equiv \text{ and } \equiv \in Con(A) \}$ is non-empty since the unit Σ congruence A^2 exists. Thus $\equiv^X$ is a well defined Σ congruence by Lemma 3.3.17. Note that if $X = \emptyset$ then $\equiv^X$ is the intersection of *all* Σ congruences on A.

The congruence generated by a set X satisfies the following properties.

Lemma 3.3.19. *Let A be a Σ algebra and X, $Y \subseteq A^2$ be any subsets.*

(i) *If $X \subseteq Y$ then $\equiv^X \subseteq \equiv^Y$.*

(ii) *The congruences generated by the empty set and equality are the same.*

(iii) *Equality on A is the congruence $\cap Con(A)$.*

Proof. (i) Suppose for any a, $a' \in A$ that $a \equiv^X a'$. By definition of $\equiv^X$, for all congruences $\equiv \in Con(A)$ such that $X \subseteq \equiv$ we have $a \equiv a'$. For any congruence $\equiv \in Con(A)$ such that $Y \subseteq \equiv$ we have $X \subseteq \equiv$, since $X \subseteq Y$ and so $a \equiv a'$. Therefore $a \equiv^Y a'$. Since a and a' were arbitrarily chosen then $\equiv^X \subseteq \equiv^Y$.

Clearly (iii) is an immediate consequence of (ii). The proof of (ii) is left as an exercise (Exercise 3.3.25.(12)). ∎

We can give an inductive definition of the congruence $\equiv^X$ generated by a set X as follows.

Definition 3.3.20. Let A be any Σ algebra and let $P(A^2)$ denote the power set of $A \times A$. Define the map $I : P(A^2) \to P(A^2)$ by

$$I(X) = X \cup R(X) \cup S(X) \cup T(X) \cup Q(X),$$

where:

$$R(X) = \{ \ (a, a) \mid a \in A \ \}$$

$$S(X) = \{ \ (a, b) \mid (b, a) \in X \ \}$$

$$T(X) = \{ \ (a, c) \in A^2 \mid \quad \text{for some } b \in A, (a, b), (b, c) \in X \ \}$$

and

$$Q(X) = \{ \ (a, b) \in A^2 \mid \quad \text{for some } n \in \mathbf{N}^+, \sigma \in \Sigma_n \text{ and}$$

$$(a_1, b_1), \ldots, (a_n, b_n) \in X, (a, b) = (\sigma(a_1, \ldots, a_n), \sigma(b_1, \ldots, b_n)) \ \}.$$

We can iterate I to define a sequence of maps $I^n : P(A^2) \to P(A^2)$ for each $n \in \mathbf{N}$, where

$$I^0(X) = X$$

and

$$I^{n+1}(X) = I(I^n(X)).$$

Hence we can define

$$I^\infty(X) = \cup_{n \in \mathbf{N}} I^n(X).$$

Lemma 3.3.21. *Let A be any Σ algebra and $X \subseteq A^2$ be any subset. Then $I^\infty(X)$ is a Σ congruence on A.*

Proof. Exercise 3.3.25.(13). ∎

Theorem 3.3.22. *Let A be a Σ algebra and $X \subseteq A^2$ be any subset. Then*

$$\equiv^X = I^\infty(X).$$

Proof. Exercise 3.3.25.(13). ∎

The set $Con(A)$ of all Σ congruences on a Σ algebra A is partially ordered by the relation $\subseteq$ of set-theoretic inclusion. In fact $\langle Con(A), \subseteq \rangle$ forms a complete lattice. (See Exercise 3.3.25.(10).) The largest Σ congruence on A is the unit congruence, while the smallest Σ congruence on A is equality. Also of interest, when they exist, are the *maximal* and *minimal* congruences in this lattice, ignoring for obvious reasons the null and unit congruences.

Definition 3.3.23. A congruence $\equiv^\theta \in Con(A)$ is *maximal* if, and only if, there is no strictly coarser congruence $\equiv^\phi \in Con(A)$ such that

$$\equiv^\theta \subset \equiv^\phi \subset A^2.$$

Dually, a congruence $\equiv^\theta \in Con(A)$ is *minimal* if, and only if, there is no strictly finer congruence $\equiv^\phi \in Con(A)$ such that

$$= \subset \equiv^\phi \subset \equiv^\theta.$$

Theorem 3.3.24. *Let A be any Σ algebra and $\equiv$ be any Σ congruence on A. The quotient algebra $A/\equiv$ is a simple algebra if, and only if, $\equiv$ is a maximal congruence on A or $\equiv$ is the unit congruence A^2.*

Proof. $\Rightarrow$ Suppose that $A/\equiv$ is a simple algebra. Then the only congruences on $A/\equiv$ are equality $=$ and the unit congruence $(A/\equiv)^2$. If $(A/\equiv)^2$ is the equality relation on $A/\equiv$ then $A/\equiv$ must be a singleton set, so that $\equiv$ is A^2. If $(A/\equiv)^2$ is not the equality relation on $A/\equiv$ then there is no strictly coarser congruence $\equiv \subset \equiv^\phi \subset A^2$ on A such that

$$\equiv/\equiv \subset \equiv^\phi/\equiv \subset (A/\equiv)^2.$$

So there is no congruence $\equiv^\phi$ on A such that $\equiv \subset \equiv^\phi \subset A^2$. Therefore $\equiv$ is maximal.

$\Leftarrow$ Suppose that $\equiv$ is a maximal congruence on A or that $\equiv$ is the unit congruence A^2. If $\equiv$ is the unit congruence A^2 then $A/\equiv$ is a unit algebra and hence is simple. If $\equiv_C A^2$ is a maximal congruence on A then there is no congruence $\equiv^\phi$ on A such that

$$\equiv_C \equiv^\phi \subset A^2.$$

So there is no congruence $\equiv^\phi / \equiv$ on $A/\equiv$ such that

$$\equiv / \equiv_C \equiv^\phi / \equiv_C (A/\equiv)^2$$

and hence equality and the unit congruence are the only congruences on $A/\equiv$, i.e. $A/\equiv$ is simple. ∎

As we shall see in Section 5.1, there is a close relationship between the concepts of simple algebras, maximal congruences and *final algebras* which arise in the semantics of algebraic data type specifications.

Exercises 3.3.25.

1. Let A be a Σ algebra and A^2 be the unit congruence on A. Show that A/A^2 is a unit algebra.

2. Recall the congruence $\equiv^n$ on $PA(\mathbf{N})$ defined in Example 3.3.5.(ii). What is $|PA(\mathbf{N})/\equiv^n|$?

3. For which natural number algebras in Examples 3.1.4.(i) and 3.1.4.(iii) is the divisibility relation $\equiv^n$ on $\mathbf{N}$ a congruence?

4. Reconstruct the discussion of the congruence $\equiv^n$ on $PA(\mathbf{N})$ for the ring $\mathbf{Z}$ of integers.

5. Let A and B be two non-empty sets. Let $\phi : A \to B$ be any map. Define for $a, b \in A$

$$a \equiv^\phi b \Leftrightarrow \phi(a) = \phi(b).$$

Show that $\equiv^\phi$ is an equivalence relation. Suppose A is a Σ algebra. Formulate conditions on B and ϕ sufficient to ensure that $\equiv^\phi$ is a Σ congruence on A. Apply your results to $T_k(.)$ on $T(\Sigma)$.

6. Recall the definition of term evaluation (from Section 2.1.4 and Exercises 2.1.7) which formalize how a term t with variables contained in $\overline{x} = (x_1, \ldots, x_n)$ can be evaluated on $\overline{a} = (a_1, \ldots, a_n) \in A^n$ to yield $t(\overline{a}/\overline{x})$. Show that $\equiv$ is a Σ congruence on A if, and only if, for all

lists of variables $\bar{x} = (x_1, \ldots, x_n)$, any terms t in these variables, and all $\bar{a} = (a_1, \ldots, a_n)$, $\bar{b} = (b_1, \ldots, b_n)$ from A, we have

$$a_1 \equiv b_1, \ldots, a_n \equiv b_n \Rightarrow t(\bar{a}/\bar{x}) \equiv t(\bar{b}/\bar{x}).$$

7. Apply the results of (5) and (6) to the notions of compositionality in Section 2.3.4.

8. Show that $\equiv^k$ in Example 3.3.5.(iv) is an equivalence relation on terms. Prove the lemma in Example 3.3.5.(iv), i.e. show that

$$t \equiv^k t' \Rightarrow t \equiv^{k-1} t'$$

for all $k > 0$ to establish the proposition.

9. Define $PA(\mathbf{N})$ equivalence on $T(\Sigma^{PA})$.

10. Show that the set of all congruences on a Σ algebra A ordered by set inclusion forms a lattice, i.e. define join and meet operations on the congruences of $Con(A)$. Hence show that $Con(A)$ equipped with these operations forms a complete lattice.

11. Prove Lemma 3.3.10.

12. Prove Lemma 3.3.19.(ii).

13. Prove Lemma 3.3.21 and hence prove Theorem 3.3.22. Define $J : P(A^2) \to P(A^2)$ by $J(X) = X \cup T(X) \cup Q(X)$. Taking

$$J^0(X) = X \cup R(X) \cup S(X)$$

and

$$J^{n+1}(X) = J(J^n(X))$$

and

$$J^\infty(X) = \cup_{n \in \mathbf{N}} J^n(X)$$

show that $I^\infty(X) = J^\infty(X)$.

14. Let A be a non-empty set. Consider a set $E = \{ \equiv^k \mid k \in \mathbf{N} \}$ of equivalence relations on A. We call E an *approximating system* of equivalence relations on A if:

(i) for each $k = 0, 1, \ldots$ the relation $\equiv^{k+1}$ is finer than $\equiv^k$, i.e.

$$a \equiv^{k+1} b \Rightarrow a \equiv^k b;$$

(ii) if $a \neq b$ then $a \not\equiv^k b$ for some k.

Prove that condition (i) is equivalent with each of the following:

(iii) if $a \not\equiv^k b$ then $a \not\equiv^{k+1} b$;

(iv) if $a \equiv^k b$ then $a \equiv^n b$ for all $n \leq k$;

(v) if $a \not\equiv^k b$ then $a \not\equiv^n b$ for all $n \geq k$.

Prove that condition (ii) is equivalent with the condition that $\cap \{ \equiv^k \mid k \in \mathbf{N} \}$ is equality. Hence deduce that (i) and (ii) imply that if $a \neq b$ then $a \not\equiv^n b$ for some k and for all $n \geq k$.

15. Let $E = \{ \equiv^k \mid k \in \mathbf{N} \}$ be an approximating system of equivalence relations as defined above. Define $\mu : A \times A \to \mathbf{N}$ by

$$\mu(a, b) = \begin{cases} 0, & \text{if } a = b; \\ (\text{least } k)(a \not\equiv^k b), & \text{otherwise.} \end{cases}$$

Show that $d : A \times A \to \mathbf{R}$ defined by

$$d(a, b) = \begin{cases} 0, & \text{if } a = b; \\ 1/2^{\mu(a,\, b)}, & \text{otherwise} \end{cases}$$

is a metric on A.

3.4   Homomorphisms and isomorphisms

We will now formulate some general concepts for comparing a pair of Σ algebras A and B; in particular, we will define what it means for A and B to be equivalent or 'structurally identical' as Σ algebras. We consider the way that a mapping ϕ between carriers of A and B relates the operations of A and B. For generality we return to the many-sorted case.

Definition 3.4.1. Let A and B be any S-sorted Σ algebras. A Σ *homomorphism* $\phi : A \to B$ from A to B is an S-indexed family of mappings

$$\phi = \langle\, \phi_s : A_s \to B_s \mid s \in S \,\rangle,$$

such that: for each sort $s \in S$ and each constant symbol $c \in \Sigma_{\lambda,s}$

$$c_B = \phi_s(c_A);$$

and for each non-empty word $w = s(1) \ldots s(n) \in S^+$, each sort $s \in S$, each function symbol $\sigma \in \Sigma_{w,s}$ and any $(a_1, \ldots, a_n) \in A^w$

$$\phi_s(\sigma_A(a_1, \ldots, a_n)) = \sigma_B(\phi_{s(1)}(a_1), \ldots, \phi_{s(n)}(a_n)),$$

i.e. the following diagram commutes

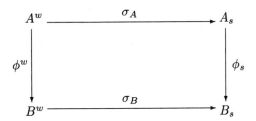

where $\phi^w : A^w \to B^w$ is defined by

$$\phi^w(a_1, \ldots, a_n) = (\phi_{s(1)}(a_1), \ldots, \phi_{s(n)}(a_n))$$

for any $(a_1, \ldots, a_n) \in A^w$. We sometimes omit reference to Σ and speak of a homomorphism.

If A and B are Σ algebras and ϕ is a Σ homomorphism from A to B, we let $im(\phi)$ denote the *image* of A under the homomorphism ϕ,

$$im(\phi) = \langle\, im(\phi)_s \mid s \in S \,\rangle,$$

where

$$im(\phi)_s = \{\, b \in B_s \mid b = \phi_s(a) \text{ for some } a \in A_s \,\}.$$

We may also use the notation $\phi(A) = \langle\, \phi_s(A_s) \mid s \in S \,\rangle$ for the image of A under ϕ. Shortly we will show that $\phi(A)$ is a Σ subalgebra of B.

For any Σ algebras A and B we let $Hom(A, B)$ denote the set of all Σ homomorphisms from A to B.

If each $\phi_s : A_s \to B_s$ is injective then A is said to be *embedded* in B and ϕ is termed a Σ *monomorphism* or a Σ *embedding*. If each $\phi_s : A_s \to B_s$ is surjective then B is said to be a *homomorphic image* of A and ϕ is termed a Σ *epimorphism*. If each $\phi_s : A_s \to B_s$ is bijective then ϕ is termed a Σ *isomorphism* and A and B are said to be *isomorphic*. If A and B are isomorphic then we write $A \cong B$. When $A = B$ then Σ homomorphisms and isomorphisms are termed Σ *endomorphisms* and Σ *automorphisms* of A respectively. For any Σ algebra A we let $End(A)$ denote the set of all Σ endomorphisms of A and $Aut(A)$ denote the set of all Σ automorphisms of A.

Examples 3.4.2.

(i) Clearly for any S-sorted Σ algebra A the S-indexed family of identity mappings $id^A = \langle\, id_s^A : A_s \to A_s \mid s \in S \,\rangle$ is a Σ isomorphism from

A to A, i.e. id is a Σ automorphism of A, and therefore A is isomorphic to itself. Furthermore for any Σ algebras A and B and any Σ homomorphism $\phi : A \to B$, we have

$$\phi \circ id^A = id^B \circ \phi = \phi.$$

For any Σ algebra B and Σ isomorphism $\phi : A \to B$, the S-indexed family $\phi^{-1} : B \to A = \langle\, \phi_s^{-1} : B_s \to A_s \mid s \in S \,\rangle$ of inverse mappings is a Σ isomorphism. (See Exercise 3.4.24.(1).) Thus we see that $\cong$ is reflexive and symmetric.

Let A be any S-sorted Σ algebra and $\mathbf{1}$ be a unit algebra containing a single element in each carrier set. Then the unique S-indexed family of maps $\phi = \langle\, \phi_s : A_s \to \mathbf{1}_s \mid s \in S \,\rangle$ is a Σ homomorphism.

(ii) Consider the algebra $PA(\mathbf{N})$ from 3.1.4.(iii). Any endomorphism

$$\phi : PA(\mathbf{N}) \to PA(\mathbf{N})$$

must satisfy

$$\phi(0) = 0,$$

$$\phi(Succ(x)) = Succ(\phi(x)).$$

Thus for example

$$\phi(1) = \phi(Succ(0)) = Succ(\phi(0)) = Succ(0) = 1$$

and

$$\phi(2) = \phi(Succ(1)) = Succ(\phi(1)) = Succ(1) = 2.$$

It is easily proved by induction that ϕ is the identity map. Clearly for each number algebra A in Example 3.1.4.(iii), equipped with 0 and $Succ$, we have $End(A) = \{\ id_A\ \}$.

(iii) Consider the additive group $\mathbf{Z}$ of integers and the map $\phi_z : \mathbf{Z} \to \mathbf{Z}$ defined by $\phi_z(x) = zx$ for $z \in \mathbf{Z}$. Then ϕ_z is a group endomorphism, but it is not a ring endomorphism. The group $im(\phi_z)$ is the additive subgroup of integers divisible by z.

(iv) Let Σ be a non-void S-sorted signature and A be a Σ algebra. The *evaluation mapping*

$$V^A = \langle\, V_s^A : T(\Sigma)_s \to A_s \mid s \in S \,\rangle,$$

that interprets closed terms over Σ as elements of A, is a Σ homomorphism between the Σ algebras $T(\Sigma)$ and A. This is evident from the definition of V^A wherein for each sort $s \in S$ and each constant symbol $c \in \Sigma_{\lambda,s}$, we have

$$V_s^A(c^s) = c_A,$$

and for each non-empty word $w = s(1)\ldots s(n) \in S^+$, each sort $s \in S$, each function symbol $\sigma \in \Sigma_{w,s}$ and any sequence of terms $(t_1,\ldots,t_n) \in T(\Sigma)^w$, we have

$$V_s^A(\sigma^s(t_1,\ldots,t_n)) = \sigma_A(V_{s(1)}^A(t_1),\ldots,V_{s(n)}^A(t_n)).$$

In particular, let us note that the idea of defining a function on terms by structural induction is intimately connected with the idea of a homomorphism. We will see that the minimal subalgebra of A has the carriers $im(V^A)$. Therefore A is minimal if, and only if, $A = im(V^A)$. Thus we see that an essential characteristic of a minimal algebra of non-void signature is the following:

Fact. Let Σ be a non-void signature. Then A is a minimal Σ algebra if, and only if, every element in the carrier of A can be named by a closed term over Σ.

(v) Let Σ be a non-void single-sorted signature. Consider the mapping

$$T : T(\Sigma) \to Tree(\Sigma)$$

that assigns a parse tree $T(t)$ to each term t. This was defined by induction on terms in 3.3.5.(iv). The definition can be improved by re-examining the method of constructing the tree. We will make $Tree(\Sigma)$ an algebra of trees with signature Σ such that the mapping T is a Σ homomorphism.

The constants of $Tree(\Sigma)$ are as follows: for each constant symbol $c \in \Sigma_0$ we take $c_{Tree(\Sigma)} \in Tree(\Sigma)$ to be the single node labelled by c:

$$\cdot c \ .$$

The operations of $Tree(\Sigma)$ are as follows: for each $n > 0$ and each function symbol $\sigma \in \Sigma_n$, we define $\sigma_{Tree(\Sigma)} : Tree(\Sigma)^n \to Tree(\Sigma)$ to be the operation that takes trees $T_1,\ldots,T_n$ and constructs a tree $\sigma_{Tree(\Sigma)}(T_1,\ldots,T_n)$ with root labelled σ and n edges connecting this root to the roots of the trees $T_1,\ldots,T_n$:

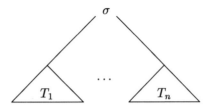

Symbolically the definition of T given in Example 3.3.5.(iv) can be rewritten as

$$T(c) = c_{Tree(\Sigma)}$$

and

$$T(\sigma(t_1,\ldots,t_n)) = \sigma_{Tree(\Sigma)}(T(t_1),\ldots,T(t_n)),$$

which state that T is a Σ homomorphism.

Lemma 3.4.3. *Let A and B be Σ algebras and $\phi : A \to B$ be a Σ homomorphism. Then $im(\phi)$ is a Σ subalgebra of B.*

Proof. Clearly $im(\phi)$ contains the constants of B since for each sort $s \in S$ and each constant symbol $c \in \Sigma_{\lambda,s}$,

$$c_B = \phi_s(c_A).$$

We must show that $im(\phi)$ is closed under the operations of B. Let $w = s(1)\ldots s(n) \in S^+$ be any non-empty word, $s \in S$ be any sort, $\sigma \in \Sigma_{w,s}$ be any function symbol and $(b_1,\ldots,b_n) \in im(\phi)^w$ be any elements of the image of A under ϕ . Then for some $(a_1,\ldots,a_n) \in A^w$ we have

$$\phi_{s(1)}(a_1) = b_1, \ \ldots, \ \phi_{s(n)}(a_n) = b_n.$$

Thus

$$\sigma_B(b_1,\ldots,b_n) = \sigma_B(\phi_{s(1)}(a_1),\ldots,\phi_{s(n)}(a_n))$$
$$= \phi_s(\sigma_A(a_1,\ldots,a_n))$$

since ϕ is a Σ homomorphism. Thus $\sigma_B(b_1,\ldots,b_n) \in im(\phi)_s$ because $\sigma_A(a_1,\ldots,a_n) \in A$. ∎

Definition 3.4.4. A class K of Σ algebras is said to be *closed under the formation of homomorphic images* if, and only if, for any Σ algebras A and B, whenever $A \in K$ and $\phi : A \to B$ is a Σ epimorphism from A to B then $B \in K$.

Examples 3.4.5.

(i) The class of all finite Σ algebras is closed under homomorphic images. However the class of all infinite Σ algebras is not closed under homomorphic images.

(ii) Later we will show that (by the First Homomorphism Theorem 3.4.18) if K is closed under isomorphism, then K is closed under the formation of homomorphic images if, and only if, K is closed under the formation of quotients.

(iii) The classes of groups, semigroups, rings and semirings are closed under the formation of homomorphic images. More generally, every class defined by equations is closed under homomorphic images.

(iv) Let K be any class of Σ algebras. Suppose the unit algebra $\mathbf{1} \notin K$. Then K is not homomorphism closed (cf. 3.4.2.(i)). Thus, if in fields the constants satisfy $0 \neq 1$ then the class of fields cannot be homomorphism closed.

A weaker condition on a class K of algebras than Definition 3.4.4 is the following.

Definition 3.4.6. A class K is said to be *closed under the formation of isomorphic images* (or simply *closed under isomorphism*) if, and only if, whenever $A \in K$ and $A \cong B$ then $B \in K$; such a class is called an *algebraic class*.

Obviously, if K is closed under homomorphic images then K is closed under isomorphism.

In algebra, if $A \cong B$ then A and B are intended to be indistinguishable, i.e. identical. Thus we define a property P of Σ algebras to be an *algebraic property* if, and only if, P determines a class $K = Alg(\Sigma, P)$ that is closed under isomorphism. Equivalently, if a property P of a specific algebra A is algebraic then P is true of A, and $A \cong B$ implies that P is true of B.

Many properties of computational interest are algebraic in this sense, as will be discussed later.

Examples 3.4.7.

(i) Finiteness is an algebraic property, since for any Σ algebra A, if A is finite and $B \cong A$ then B is finite. More generally, being an algebra of cardinality κ is an algebraic property.

(ii) Being a minimal Σ algebra is an algebraic property.

(iii) All first order definable properties are algebraic. Thus if Φ is any set of first order formulas over a signature Σ, and A is any Σ algebra such that $A \models \Phi$ and $B \cong A$ then $B \models \Phi$. We can rephrase this by simply saying that the class $Alg(\Sigma, \Phi)$ is closed under isomorphism.

(iv) Let S be a *while program* over a signature Σ. Let $\{\ p\ \}S\{\ q\ \}$ be a *correctness formula* over Σ and suppose that it is valid under *partial correctness*; in symbols

$$A \models \{\ p\ \}S\{\ q\ \}.$$

If $B \cong A$ then

$$B \models \{\ p\ \}S\{\ q\ \}.$$

The validity of correctness formulae under *total correctness*, *program termination*, and *program equivalence*, are three other important computational properties that are isomorphism invariants. For further information see Tucker and Zucker [1988].

(v) The various notions of effectively computable algebras, (e.g. computable, semicomputable and cosemicomputable) are algebraic properties. For further information see Meseguer and Goguen [1985], Bergstra and Tucker [1983], [1987].

In the remainder of Section 3.4, again for simplicity, we consider only the single-sorted case. We leave it as an exercise for the reader to generalize the following definitions and results to the many-sorted case.

The behaviour of a homomorphism is uniquely determined by its behaviour on the generators of an algebra.

Lemma 3.4.8. *Let A be any Σ algebra, $X \subseteq A$ be any subset and $\phi, \psi : A \to B$ be any homomorphisms. Suppose that either X is non-empty or that Σ is non-void. If A is generated by X then $\phi = \psi$ on A if, and only if, $\phi = \psi$ on X.*

Proof. Suppose that A is generated by a subset X. Clearly if $\phi = \psi$ on A then $\phi = \psi$ on X.

Conversely, suppose $\phi = \psi$ on X. Let $Y = \{\ a \in A \mid \phi(a) = \psi(a)\ \}$. We want to show that $Y = A$. Now $X \subseteq Y$ and $Cons^A \subseteq Y$. Thus by the hypothesis Y is non-empty. Furthermore Y is closed under the operations of A, since for any $n > 0$, $\sigma \in \Sigma_n$ and $a_1, \ldots, a_n \in Y$,

$$\phi(\sigma_A(a_1, \ldots, a_n)) = \sigma_B(\phi(a_1), \ldots, \phi(a_n))$$

$$= \sigma_B(\psi(a_1), \ldots, \psi(a_n))$$

since $a_1, \ldots, a_n \in Y$ and so

$$= \psi(\sigma_A(a_1, \dots, a_n)).$$

Thus Y is a carrier of a subalgebra of A. By Lemma 3.2.10.(i) we have that $\langle X \rangle_A \subseteq Y$. But $\langle X \rangle_A = A$ and so $A \subseteq Y$. ∎

Lemma 3.4.8 can be usefully reformulated as follows.

Corollary 3.4.9. *Let A, B be any Σ algebras and suppose that A is generated by a subset $X \subseteq A$. Let $\alpha : X \to B$ be any map. There is at most one Σ homomorphism $\phi : A \to B$ which agrees with α on X.*

If there is such a unique Σ homomorphism which extends α, we call this the *homomorphic extension* of α and denote this mapping by $\overline{\alpha} : A \to B$. Notice that it is not asserted that any extension exists, but that should it exist then it is unique. Later we will study the important property of an algebra A that an extension $\overline{\alpha}$ exists for *any* α on X; such an algebra is said to be *free* on its generating set. The property is immensely strong and immensely useful.

Lemma 3.4.10. *If A, B and C are Σ algebras and $\phi : A \to B$ and $\psi : B \to C$ are Σ homomorphisms then the composition $\psi \circ \phi : A \to C$ is a Σ homomorphism.*

Proof. For each constant symbol $c \in \Sigma_0$, since ϕ and ψ are Σ homomorphisms, $c_C = \psi(c_B)$ and $c_B = \phi(c_A)$ so

$$c_C = \psi(\phi(c_A)) = \psi \circ \phi(c_A).$$

For each $n > 1$, each function symbol $\sigma \in \Sigma_n$ and any $a_1, \dots, a_n \in A$,

$$\psi \circ \phi(\sigma_A(a_1, \dots, a_n)) = \psi(\phi(\sigma_A(a_1, \dots, a_n)))$$

$$= \psi(\sigma_B(\phi(a_1), \dots, \phi(a_n)))$$

$$= \sigma_C(\psi(\phi(a_1)), \dots, \psi(\phi(a_n)))$$

$$= \sigma_C(\psi \circ \phi(a_1), \dots, \psi \circ \phi(a_n)).$$

Therefore $\psi \circ \phi$ is a Σ homomorphism. ∎

Thus we see that $\cong$ is transitive. As we have already observed in Example 3.4.2.(i), the inverse of an isomorphism is again an isomorphism and the identity mapping on any Σ algebra A is an automorphism on A. These observations and Lemma 3.4.10 lead to the following result.

Proposition 3.4.11. *For any Σ algebra A, the set $End(A)$ of all endomorphisms on A forms a semigroup with 1 (or monoid) under composition;*

and the set $Aut(A)$ of all automorphisms on A forms a group under composition.

Proof. By Lemma 3.4.10, for any Σ algebra A the set $End(A)$ of all endomorphisms on A is closed under composition. Clearly function composition is associative. As we have already observed in Example 3.4.2.(i), the identity mapping id^A on A is a Σ endomorphism on A which acts as a neutral element under composition. Thus $End(A)$ forms a semigroup with 1 under composition.

The set $Aut(A)$ of all automorphisms on A is also closed under composition and contains id^A. By Exercise 3.4.24.(1), for every Σ automorphism ϕ on A there exists a Σ automorphism ϕ^{-1} on A such that $\phi^{-1} \circ \phi = \phi \circ \phi^{-1} = id^A$. Therefore $Aut(A)$ is a subsemigroup of $End(A)$ which forms a group. ∎

Thus we may examine any Σ algebra in terms of the structure of its automorphism group. However, for many algebras of interest to computer science this group structure is trivial.

Proposition 3.4.12. *Suppose Σ is a non-void signature. If A is any minimal Σ algebra then $End(A)$ and $Aut(A)$ are unit algebras.*

Proof. If A is minimal then, by Lemma 3.2.13, A is generated by the empty set. Every Σ endomorphism agrees with the identity map on the empty set and therefore, by Lemma 3.4.8, agrees with the identity map everywhere. So the only Σ endomorphism on A is the identity map id^A and hence $End(A)$ and $Aut(A)$ are unit algebras. ∎

We establish the important and useful fact that to prove $A \cong B$ for minimal Σ algebras A and B, it suffices to prove the existence of Σ homomorphisms $\phi : A \to B$ and $\psi : B \to A$.

Theorem 3.4.13. *Let Σ be a non-void signature. Let A and B be minimal Σ algebras. If there exist Σ homomorphisms $\phi : A \to B$ and $\psi : B \to A$ then $A \cong B$, ϕ and ψ are Σ isomorphisms, and $\phi = \psi^{-1}$, $\psi = \phi^{-1}$.*

Proof. By Lemma 3.4.10, $\psi \circ \phi : A \to A$ and $\phi \circ \psi : B \to B$ are endomorphisms on A and B. So by Proposition 3.4.12, $\phi \circ \psi = id^B$ and $\psi \circ \phi = id^A$. Since id^B is surjective then ϕ is surjective and since id^A is injective then ϕ is injective. Similarly, since id^A is surjective and id^B is injective, ψ is both injective and surjective. Hence both ϕ and ψ are isomorphisms and $A \cong B$. Since $\phi \circ \psi = id^B$ then

$$\psi = \phi^{-1} \circ \phi \circ \psi = \phi^{-1} \circ id^B = \phi^{-1}.$$

Similarly $\phi = \psi^{-1}$. ∎

There is a close relationship between Σ congruences and Σ epimorphisms: given any Σ homomorphism $\phi : A \to B$ we can construct a Σ congruence in a canonical way.

Definition 3.4.14. Let A and B be Σ algebras and $\phi : A \to B$ be a Σ homomorphism. The *kernel* of ϕ is the binary relation $\equiv^\phi$ on A defined by

$$a \equiv^\phi b \Leftrightarrow \phi(a) = \phi(b)$$

for all $a, b \in A$.

Lemma 3.4.15. *Let* $\phi : A \to B$ *be a* Σ *homomorphism. The kernel* $\equiv^\phi$ *of* ϕ *is a* Σ *congruence on* A.

Proof. Since equality on B is an equivalence relation on B it follows that $\equiv^\phi$ is an equivalence relation on A. To check that the substitutivity condition holds, consider any $n > 0$, any function symbol $\sigma \in \Sigma_n$ and $a_1, \ldots, a_n, b_1, \ldots, b_n \in A$. Suppose that for $i = 1, \ldots, n$ we have $a_i \equiv^\phi b_i$. Then for $i = 1, \ldots, n$ we have $\phi(a_i) = \phi(b_i)$. Since ϕ is a Σ homomorphism then

$$\phi(\sigma_A(a_1, \ldots, a_n)) = \sigma_B(\phi(a_1), \ldots, \phi(a_n))$$
$$= \sigma_B(\phi(b_1), \ldots, \phi(b_n))$$
$$= \phi(\sigma_A(b_1, \ldots, b_n)).$$

So by definition

$$\sigma_A(a_1, \ldots, a_n) \equiv^\phi \sigma_A(b_1, \ldots, b_n).$$

Therefore $\equiv^\phi$ satisfies the substitutivity condition. So $\equiv^\phi$ is a Σ congruence on A. ∎

Conversely, given any Σ congruence $\equiv$ on a Σ algebra A, we can construct a Σ homomorphism $\phi : A \to A/\equiv$ in a canonical way.

Definition 3.4.16. Let A be a Σ algebra and $\equiv$ be a Σ congruence on A, the *natural map* or *quotient map* of the congruence

$$nat : A \to A/\equiv$$

is defined by

$$nat(a) = [a].$$

For a superscripted congruence $\equiv^\phi$ we let nat^ϕ denote the corresponding natural mapping.

Lemma 3.4.17. Let A be a Σ algebra and $\equiv$ be any Σ congruence on A. The natural map of the congruence

$$nat : A \to A/\equiv$$

is a Σ epimorphism.

Proof. To check that nat is a Σ homomorphism, consider any constant symbol $c \in \Sigma_0$. Then

$$nat(c_A) = [c_A] = c_{A/\equiv}.$$

For any $n > 0$, any function symbol $\sigma \in \Sigma_n$ and any $a_1, \ldots, a_n \in A$ we have

$$nat(\sigma_A(a_1, \ldots, a_n)) = [\sigma_A(a_1, \ldots, a_n)]$$

$$= \sigma_{A/\equiv}([a_1], \ldots, [a_n])$$

$$= \sigma_{A/\equiv}(nat(a_1), \ldots, nat(a_n)).$$

So the natural mapping nat is a Σ homomorphism and clearly nat is surjective, i.e. a Σ epimorphism.   ∎

The fundamental results for homomorphisms are given by the First, Second and Third Homomorphism Theorems. The First Homomorphism Theorem asserts that for any Σ epimorphism $\phi : A \to B$ the homomorphic image $\phi(A)$ and the quotient algebra $A/\equiv^\phi$ are isomorphic, and hence, for the purposes of algebra, identical.

Theorem 3.4.18 (First Homomorphism Theorem). If $\phi : A \to B$ is a Σ epimorphism then there exists a Σ isomorphism $\psi : A/\equiv^\phi \to B$ such that the following diagram commutes

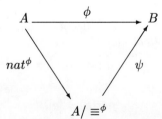

where $nat^\phi : A \to A/\equiv^\phi$ is the natural mapping associated with the kernel $\equiv^\phi$ of ϕ.

Proof. Define ψ by

$$\psi([a]_\phi) = \phi(a).$$

If $[a]_\phi = [b]_\phi$ then $\phi(a) = \phi(b)$. So $\psi([a]_\phi) = \psi([b]_\phi)$ and therefore $\psi([a]_\phi)$ is uniquely defined. To check that ψ is a Σ homomorphism, consider any constant symbol $c \in \Sigma_0$. Then

$$c_B = \phi(c_A) = \psi([c_A]_\phi) = \psi(c_{A/\equiv^\phi}).$$

Consider any $n > 0$, any function symbol $\sigma \in \Sigma_n$ and any $a_1, \ldots, a_n \in A$. Then

$$\sigma_B(\psi([a_1]_\phi), \ldots, \psi([a_n]_\phi)) = \sigma_B(\phi(a_1), \ldots, \phi(a_n))$$

$$= \phi(\sigma_A(a_1, \ldots, a_n))$$

$$= \psi([\sigma_A(a_1, \ldots, a_n)]_\phi)$$

$$= \psi(\sigma_{A/\equiv^\phi}([a_1]_\phi, \ldots, [a_n]_\phi)).$$

Therefore ψ is a Σ homomorphism. Since ϕ is surjective then ψ is surjective. For any $a, b \in A$, if $[a]_\phi \neq [b]_\phi$ then $\phi(a) \neq \phi(b)$ and so

$$\psi([a]_\phi) \neq \psi([b]_\phi).$$

Therefore ψ is injective and hence bijective. So ψ is a Σ isomorphism. ∎

We can apply this important theorem to obtain another result on minimal algebras. For minimal algebras any congruence, other than equality, is non-trivial in the following sense.

Proposition 3.4.19. *Suppose Σ is non-void. Let A be a minimal Σ algebra and $\equiv$ be any Σ congruence on A. Then $A \cong A/\equiv$ if, and only if, $\equiv$ is the equality relation on A.*

Proof. $\Leftarrow$ Suppose that $\equiv$ is the equality relation on A then clearly

$$A \cong A/\equiv .$$

$\Rightarrow$ Suppose that $A \cong A/\equiv$. Then there exists a Σ isomorphism $\phi : A \to A/\equiv$. Since A is minimal, by Lemma 3.2.13, A is generated by the constants named in Σ. Since ϕ must agree with the natural mapping $nat : A \to A/\equiv$ on the constants named in Σ, by Lemma 3.4.8 ϕ agrees with nat everywhere. So nat is injective. Now consider any $a, a' \in A$

and suppose that $a \equiv a'$. Then $nat(a) = nat(a')$ and since nat is injective, $a = a'$ and therefore

$$a \equiv a' \Leftrightarrow a = a'.$$

So $\equiv$ must be the equality relation on A. ∎

Recall from Definition 3.3.11 the idea of relaxing a congruence $\equiv^\theta$ to a congruence $\equiv^\phi$ on A, and the relevant construction of the factor congruence $\equiv^{\phi/\theta}$ on $A/\equiv^\theta$. We will answer the question: what is

$$(A/\equiv^\theta)/\equiv^{\phi/\theta}?$$

The following shows that it is $A/\equiv^\phi$ up to isomorphism.

Theorem 3.4.20 (Second Homomorphism Theorem). *Let $\equiv^\phi, \equiv^\theta$ be any Σ congruences on a Σ algebra A such that $\equiv^\theta \subseteq \equiv^\phi$. Then the map $\psi : (A/\equiv^\theta)/\equiv^{\phi/\theta} \to A/\equiv^\phi$ defined by*

$$\psi([[a]_\theta]_{\phi/\theta}) = [a]_\phi$$

is a Σ isomorphism.

Proof. By Definition 3.3.13, ψ is a well defined mapping. To show that ψ is a Σ homomorphism, consider any constant symbol $c \in \Sigma_0$. Then

$$\psi(c_{(A/\equiv^\theta)/\equiv^{\phi/\theta}}) = \psi([[c_A]_\theta]_{\phi/\theta})$$
$$= [c_A]_\phi$$
$$= c_{A/\equiv^\phi}.$$

Consider any $n > 0$, any function symbol $\sigma \in \Sigma_n$ and $[a_1]_\theta, \ldots, [a_n]_\theta \in A/\equiv^\theta$. Then

$$\psi(\sigma_{(A/\equiv^\theta)/\equiv^{\phi/\theta}}([[a_1]_\theta]_{\phi/\theta}, \ldots, [[a_n]_\theta]_{\phi/\theta})) = \psi([\sigma_{A/\equiv^\theta}([a_1]_\theta, \ldots, [a_n]_\theta)]_{\phi/\theta})$$
$$= \psi([[\sigma_A(a_1, \ldots, a_n)]_\theta]_{\phi/\theta})$$
$$= [\sigma_A(a_1, \ldots, a_n)]_\phi$$
$$= \sigma_{A/\equiv^\phi}([a_1]_\phi, \ldots, [a_n]_\phi)$$
$$= \sigma_{A/\equiv^\phi}(\psi([a_1]_\theta]_{\phi/\theta}), \ldots, \psi([[a_n]_\theta]_{\phi/\theta})).$$

So ψ is a Σ homomorphism. Clearly ψ is surjective. For any $[a]_\theta, [b]_\theta \in A/\equiv^\theta$, if

$$[[a]_\theta]_{\phi/\theta} \neq [[b]_\theta]_{\phi/\theta}$$

then $a \not\equiv^\phi b$. So $\psi([[a]_\theta]_{\phi/\theta}) \neq \psi([[b]_\theta]_{\phi/\theta})$. Therefore ψ is injective and hence bijective, i.e. a Σ isomorphism. ∎

The Third Homomorphism Theorem is considerably more technical than the preceding two. Intuitively, it asserts that factoring a subalgebra $B \leq A$

by the restriction ($\equiv |B$) to B of a congruence $\equiv$ on A is equivalent (up to isomorphism) to factoring an expansion $B(\equiv)$ of B by all elements $\equiv$ equivalent to elements of B using the corresponding restricted congruence ($\equiv |B(\equiv)$).

Definition 3.4.21. Let $B \subseteq A$ be any subset of a Σ algebra A and let $\equiv$ be any Σ congruence on A. Define the *closure* $B(\equiv)$ of B under $\equiv$ to be

$$B(\equiv) = \cup_{b \in B}[b].$$

The following shows that if $B \leq A$ is a subalgebra then the closure $B(\equiv)$ of B under $\equiv$ is also a carrier set for a subalgebra of A.

Lemma 3.4.22. Let B be any subalgebra of a Σ algebra A and let $\equiv$ be any Σ congruence on A. Then $B(\equiv) \leq A$.

Proof. It can easily be shown that $B(\equiv)$ contains the constants of A and is closed under the operations of A. ∎

Theorem 3.4.23 (Third Homomorphism Theorem). *Let* $B \leq A$ *be any Σ subalgebra of a Σ algebra A and let $\equiv \in Con(A)$ be any Σ congruence. Then*

$$B/(\equiv |B) \cong B(\equiv)/(\equiv |B(\equiv)).$$

Proof. Exercise 3.4.24.(9). ∎

Exercises 3.4.24.

1. Let $\phi : A \to B$ be a Σ isomorphism. Show that if $\phi^{-1} : B \to A$ is the inverse mapping of ϕ then ϕ^{-1} is a Σ isomorphism.

2. Let A and B be Σ algebras. Show that if $A \cong B$ then $End(A) \cong End(B)$ and $Aut(A) \cong Aut(B)$. Show that the converse is not the case.

3. Let R be a commutative ring with 1 and $n \in \mathbf{N}$. Then R is said to be of *characteristic* $n > 0$, if n is the least number such that the n-fold sum
$$1 + 1 + \ldots + 1$$
is 0. If no such n-fold sum is 0 we say R is of characteristic 0. Prove that if R has characteristic $n > 0$ then $\langle 1 \rangle_R \cong \mathbf{Z}_n$ and if R has characteristic 0 then $\langle 1 \rangle_R \cong \mathbf{Z}$.

4. Consider the rings $\mathbf{Z}$ of integers and $\mathbf{Z}_n$ of integers modulo n for $n \in \mathbf{N}$. Define $\phi_n : \mathbf{Z} \to \mathbf{Z}_n$ for any $n > 0$ by

$$\phi_n(x) = x \bmod n.$$

Show that ϕ_n is an epimorphism.

5. Let $\mathbf{Z}[x]$ be the set of all polynomials in indeterminate x

$$a_0 + a_1 x + a_2 x^2 + \ldots + a_n x^n$$

with $n \geq 0$, $a_0, \ldots, a_n \in \mathbf{Z}$. Equipped with the usual operations of addition and multiplication $\mathbf{Z}[x]$ is a ring. Prove that $\mathbf{Z}[x] \cong \mathbf{Z}[y]$ for y another indeterminate. Prove that $\mathbf{Z}[x]$ is generated by $\{x\}$ and that for any ring R and any map $\alpha : \{x\} \to R$ there is a unique homomorphic extension $\overline{\alpha} : \mathbf{Z}[x] \to R$. Prove that if $\alpha(x) = a$ then $im(\overline{\alpha}) = \langle a \rangle_R$. Deduce that

$$\mathbf{Z}[x]/ \equiv^{\overline{\alpha}} \cong \langle a \rangle_R.$$

6. Generalize the results in Exercise (5) to $\mathbf{Z}[X]$ where: (i) X is a finite set of indeterminates; (ii) X is any set of indeterminates.

7. Let A and B be any Σ algebras and $A' \leq A$ and $B' \leq B$ be any Σ subalgebras. If $\phi : A \to B$ is a Σ homomorphism show that $\phi(A')$ is a Σ subalgebra of B. Show that $\phi^{-1}(B')$ is a Σ subalgebra of A, where

$$\phi^{-1}(B') = \{ a \in A \mid \phi(a) \in B' \}.$$

8. Show that any finite algebra with n elements has at most $n!$ automorphisms and n^n endomorphisms.

9. Prove the Third Homomorphism Theorem 3.4.23.

10. Construct an infinite set of Σ^{PA} algebras of natural numbers that are isomorphic with $PA(\mathbf{N})$.

11. Recall the definitions of term evaluation (from Section 2.1.4) which formalize how a term with variables contained in $\overline{x} = (x_1, \ldots, x_n)$ can be evaluated on $\overline{a} = (a_1, \ldots, a_n) \in A^n$ to yield $t(\overline{a}/\overline{x})$. Show that $\phi : A \to B$ is a Σ homomorphism if, and only if, for all lists of variables $\overline{x} = (x_1, \ldots, x_n)$ and any terms in those variables and all $\overline{a} = (a_1, \ldots, a_n)$ from A, we have

$$\phi(t(\overline{a}/\overline{x})) = t(\phi(\overline{a})/\overline{x})$$

in B, where $\phi(\overline{a}) = (\phi(a_1), \ldots, \phi(a_n)) \in B^n$.

This important equation is often written without reference to variables, i.e. in the form

$$\phi(t(a_1, \ldots, a_n)) = t(\phi(a_1), \ldots \phi(a_n)).$$

Compare this exercise with 3.3.25.(6).

12. Let X be any set and $Sym(X)$ be the group of permutations of X, i.e. bijective maps $X \rightarrow X$. Let G be any group with multiplication operation. Show that for each $g \in G$, the map $\alpha_g : G \rightarrow G$ defined by $\alpha_g(a) = ag$ for all $a \in G$ is a permutation of the set G. Prove that the map $\phi : G \rightarrow Sym(G)$ defined by

$$\phi(g) = \alpha_g$$

is a monomorphism which embeds G in $Sym(G)$. Thus, every group is isomorphic to a group of permutations. This construction can be found in the work of A. Cayley.

3.5 Direct products

The direct product construction is a simple generalization to algebras of the Cartesian product of sets. Like the quotient construction, it can be found in many branches of mathematics.

Definition 3.5.1. Let A and B be S-sorted Σ algebras, the *direct product* of A and B is the Σ algebra with S-indexed family of carrier sets

$$A \times B = \langle (A \times B)_s \mid s \in S \rangle,$$

where $(A \times B)_s$ is the Cartesian product

$$(A \times B)_s = A_s \times B_s.$$

For each sort $s \in S$ and each constant symbol $c \in \Sigma_{\lambda, s}$

$$c_{A \times B} = (c_A, c_B).$$

For each non-empty word $w = s(1) \ldots s(n) \in S^+$, each sort $s \in S$, each function symbol $\sigma \in \Sigma_{w,s}$ and any $((a_1, b_1), \ldots, (a_n, b_n)) \in (A \times B)^w$

$$\sigma_{A \times B}((a_1, b_1), \ldots, (a_n, b_n)) = (\sigma_A(a_1, \ldots, a_n), \sigma_B(b_1, \ldots, b_n)).$$

As usual, we denote the direct product of algebras A and B by $A \times B$. We write A^2 for the direct product or *direct power* $A \times A$.

With A , B and the direct product $A \times B$ we associate S-indexed families of mappings

$$U^A = \langle U_s^A : (A \times B)_s \to A_s \mid s \in S \rangle,$$

$$U^B = \langle U_s^B : (A \times B)_s \to B_s \mid s \in S \rangle,$$

termed the (families of) *first* and *second projection functions* respectively. These are defined by

$$U_s^A(a,\ b) = a, \quad U_s^B(a,\ b) = b$$

for each sort $s \in S$ and each pair $(a,\ b) \in (A \times B)_s$. The algebras A and B are termed the *first* and *second factors* of $A \times B$.

Examples 3.5.2.

(i) Consider the ring of real numbers

$$R = (\mathbf{R},\ 0,\ 1,\ +,\ -,\ \cdot)$$

and its power R^2. The carrier of R^2 is the 2 dimensional plane $\mathbf{R} \times \mathbf{R}$. The constants of R^2 are

$$(0,\ 0) \text{ and } (1,\ 1).$$

The operations of R^2 are defined for $(x_1,\ y_1)$, (x_2, y_2) and $(x,\ y) \in \mathbf{R}^2$ by

$$(x_1,\ y_1) + (x_2,\ y_2) = (x_1 + x_2,\ y_1 + y_2)$$

$$-(x,\ y) = (-x,\ -y)$$

$$(x_1,\ y_1) \cdot (x_2,\ y_2) = (x_1 \cdot x_2,\ y_1 \cdot y_2).$$

Consider the geometric use of this algebra.

Clearly, vector addition and reflection are basic operations and vector subtraction can be defined easily by composition:

$$(x_1,\ y_1) - (x_2,\ y_2) = (x_1,\ y_1) + (-(x_2,\ y_2))$$

$$= (x_1 - x_2,\ y_1 - y_2).$$

However, to define further useful operations we need the projection functions

$$U_1(x,\ y) = x \text{ and } U_2(x,\ y) = y$$

which are *not* basic operations of the algebra R^2.

Given the basis points

$$(0,\ 1)\ \text{and}\ (1,\ 0),$$

which are *not* constants of the algebra R^2, then using addition we may define unit translations such as

$$x - step(x,\ y) = (x,\ y) + (1,\ 0) = (x+1,\ y)$$

$$y - step(x,\ y) = (x,\ y) + (0,\ 1) = (x, y+1).$$

Assume we have the diagonal embedding $i : \mathbf{R} \to \mathbf{R} \times \mathbf{R}$ defined by $i(x) = (x,\ x)$ which is not a basic operation of R^2. Then scalar multiplication can be defined for $\lambda \in R$ by

$$\lambda \cdot (x,\ y) = i(\lambda) \cdot (x,\ y)$$

$$= (\lambda,\ \lambda) \cdot (x,\ y)$$

$$= (\lambda \cdot x,\ \lambda \cdot y).$$

The point is that the use of R^2 for geometric calculations will require the addition of several constants and functions. This shows that the direct product is a rather primitive algebraic structure on cartesian products.

(ii) The algebraic constructions on the ring of real numbers in (i) can be made for any commutative ring with 1. Let R and S be commutative rings with 1, and consider the direct product $R \times S$. The algebra has constants

$$(0_R,\ 0_S)\ \text{and}\ (1_R,\ 1_S)$$

and operations defined by

$$(r_1,\ s_1) + (r_2,\ s_2) = (r_1 + r_2,\ s_1 + s_2)$$

$$-(r,\ s) = (-r,\ -s)$$

$$(r_1,\ s_1) \cdot (r_2,\ s_2) = (r_1 \cdot r_2,\ s_1 \cdot s_2)$$

for $r_1,\ r_2,\ r \in R$ and $s_1,\ s_2,\ s \in S$.

Most importantly, it can be shown that the direct product algebra $R \times S$ satisfies the ring axioms.

The additional operations corresponding with those in (i) are also useful. We must replace the diagonal embedding with

$$i_R : R \rightarrow R \times S \quad i_R(r) = (r,\ 0)$$

$$i_S : S \rightarrow R \times S \quad i_R(s) = (0,\ s)$$

which are monomorphisms embedding rings R and S into the ring $R \times S$.

(iii) The direct products of semirings, groups, semigroups and Boolean algebras are again semirings, groups, semigroups, and Boolean algebras, respectively. They also allow embeddings of the component algebras into the direct product. Thus, although the direct product algebra lacks many functions we desire (as we have seen in (i)), from given algebras it provides a construction of new algebras of the same kind, at least in many cases. We note that the direct product of two fields need not be a field.

Generalizing the observations above, we will see later that if a class K of algebras is axiomatically defined by equations then we have that $A, B \in K$ implies that $A \times B \in K$.

(iv) Consider the direct product B^2 of the algebra B of Booleans

$$B = (\{\ tt,\ ff\ \};\ tt,\ ff;\ not,\ and).$$

The carrier of B^2 is

$$\{\ (tt,\ tt),\ (tt,\ ff),\ (ff,\ tt),\ (ff,\ ff)\ \}.$$

The constants are
$$(tt,\ tt) \text{ and } (ff,\ ff)$$

and the operations are

$$not(a,\ b) = (not(a),\ not(b))$$

$$and((a_1,\ b_1),\ (a_2,\ b_2)) = (and(a_1,\ b_1),\ and(a_2,\ b_2)).$$

The algebra B is embedded in B^2 by $i : B \rightarrow B^2$ defined by

$$i(b) = (b,\ b).$$

Notice that the direct product of two algebras A_1 and A_2 containing the Booleans, i.e.

$$A_1|_\Sigma = B \text{ and } A_2|_\Sigma = B$$

for Σ the signature of B, does *not* contain the Booleans, i.e.

$$(A_1 \times A_2)|_\Sigma \neq B.$$

Notice however that the Σ subalgebra of $(A_1 \times A_2)|_\Sigma$

$$\langle\, (A_1 \times A_2)|_\Sigma \,\rangle_\Sigma$$

generated by the constants named in Σ, contains (an isomorphic copy) of B.

Lemma 3.5.3. *Let A and B be S-sorted Σ algebras. Then:*

(i) *the families of first and second projection mappings U^A and U^B are Σ homomorphisms;*

(ii) *for any Σ algebra C let $\phi^A : C \to A$ and $\phi^B : C \to B$ be any Σ homomorphisms. Then there is a unique Σ homomorphism $\phi : C \to A \times B$ such that $U_s^A \circ \phi_s = \phi_s^A$ and $U_s^B \circ \phi_s = \phi_s^B$, i.e. the following diagram commutes*

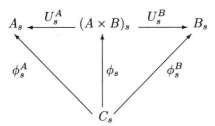

for each sort $s \in S$. Furthermore if each ϕ_s^A and ϕ_s^B is injective then each ϕ_s is injective;

(iii) *the S-indexed family of mappings*

$$\phi = \langle\, \phi_s : A_s \to (A \times A)_s \mid s \in S \,\rangle,$$

defined by $\phi_s(a) = (a, a)$ for each $s \in S$ and each $a \in A_s$, is a Σ embedding of A in $A \times A$.

Proof. (i) Consider any sort $s \in S$ and any constant symbol $c \in \Sigma_{\lambda,s}$. Then

$$U_s^A(c_{A \times B}) = U_s^A((c_A, \, c_B)) = c_A.$$

Consider any sort $s \in S$, any $w = s(1) \ldots s(n) \in S^+$, any n-ary function symbol $\sigma \in \Sigma_{w,s}$ and any $(\, (a_1, \, b_1), \ldots, (a_n, \, b_n)\,) \in (A \times B)^w$. Then

$$U_s^A(\,\sigma_{A \times B}((a_1,\,b_1),\ldots,(a_n,\,b_n))\,) =$$
$$U_s^A(\sigma_A(a_1,\ldots,a_n),\,\sigma_B(b_1,\ldots,b_n))$$
$$= \sigma_A(a_1,\ldots,a_n)$$
$$= \sigma_A(U_{s(1)}^A(a_1,\,b_1),\ldots,U_{s(n)}^A(a_n,\,b_n)).$$

Therefore U^A is a Σ homomorphism. The proof that U_B is a Σ homomorphism is similar.

(ii) Exercise 3.5.16.(8).

(iii) We show that ϕ is a Σ homomorphism as follows. Consider any sort $s \in S$ and any constant symbol $c \in \Sigma_{\lambda,s}$. Then

$$\phi_s(c_A) = (c_A,\,c_A) = c_{A \times A}.$$

For any sort $s \in S$, any $w = s(1)\ldots s(n) \in S^+$, any n-ary function symbol $\sigma \in \Sigma_{w,s}$ and any $(a_1,\ldots,a_n) \in A^w$

$$\phi_s(\sigma_A(a_1,\ldots,a_n)) = (\sigma_A(a_1,\ldots,a_n),\,\sigma_A(a_1,\ldots,a_n))$$
$$= \sigma_{A \times A}((a_1,\,a_1),\ldots,(a_n,\,a_n))$$
$$= \sigma_{A \times A}(\phi_{s(1)}(a_1),\ldots,\phi_{s(n)}(a_n)).$$

Thus ϕ is a Σ homomorphism. To show that each ϕ_s is injective, consider any sort $s \in S$ and any $a,\,a' \in A$. If $a \neq a'$ then $(a,\,a) \neq (a',\,a')$, i.e. $\phi_s(a) \neq \phi_s(a')$. Thus ϕ is a Σ embedding. ∎

Lemma 3.5.4.

(i) For any Σ algebras A, B and C

$$A \times B \cong B \times A \quad \text{and} \quad A \times (B \times C) \cong (A \times B) \times C.$$

(ii) If $\mathbf{1}$ is a unit Σ algebra then $A \times \mathbf{1} \cong \mathbf{1} \times A \cong A$.

(iii) For any Σ algebras A, A', B, B',

$$A \cong A' \text{ and } B \cong B' \text{ implies } A \times B \cong A' \times B'.$$

Proof. Exercise 3.5.16.(9). ∎

By part (i) of Lemma 3.5.4 we can omit the bracketing around binary direct products. Any method of restoring brackets does not affect the algebraic properties of the resulting product algebra. More generally, consider arbitrary finite direct products

$$A_1 \times (A_2 \times (\ldots \times A_n))$$

and direct powers $A^n = A \times (A \times (\ldots A))$. By induction and Lemma 3.5.4, we may omit brackets around any finite product.

The product construction provides one facility to build algebras of larger cardinality. However this facility is not always provided by the binary direct product defined above (unless the component algebras are finite and non-trivial). We usually require the following generalization.

Definition 3.5.5. Let $A = \langle\, A(i) \mid i \in I \,\rangle$ be an I-indexed family of S-sorted Σ algebras, for some (possibly empty) indexing set I. The *direct product* of A, is the S-sorted Σ algebra $\Pi A = \Pi_{i \in I} A(i)$ with S-indexed family of carrier sets

$$\Pi A = \langle\, (\Pi A)_s \mid s \in S \,\rangle,$$

where each carrier set $(\Pi A)_s$ is the Cartesian product of the sets $A(i)_s$ given by

$$(\Pi A)_s = \{\ f : I \to \cup_{i \in I} A(i)_s \mid f(i) \in A(i)_s \text{ for all } i \in I\ \}.$$

For each sort $s \in S$ and each constant symbol $c \in \Sigma_{\lambda,s}$,

$$c_{\Pi A}(i) = c_{A(i)}.$$

For each non-empty word $w = s(1)\dots s(n) \in S^+$, each sort $s \in S$ and each function symbol $\sigma \in \Sigma_{w,s}$ and any $(a_1,\dots,a_n) \in (\Pi A)^w$,

$$\sigma_{\Pi A}(a_1,\dots,a_n)(i) = \sigma_{A(i)}(a_1(i),\dots,a_n(i)).$$

The algebra $A(i)$ is termed the *ith factor* of ΠA. If $A(i) = B$ for each $i \in I$ then ΠA is termed the *direct power* of B, and may be denoted by B^I.

With each index $i \in I$ we associate an S-indexed family of mappings

$$U^i = \langle\, U^i_s : \Pi A_s \to A(i)_s \mid s \in S \,\rangle$$

termed the (family of) *ith projection functions*, defined by

$$U^i_s(a) = a(i)$$

for each sort $s \in S$ and each $a \in \Pi A_s$.

The fundamental properties of the general direct product construction are summarized in the following lemma.

Lemma 3.5.6. Let $A = \langle A(i) \mid i \in I \rangle$ be any I-indexed family of S-sorted Σ algebras:

(i) for each $i \in I$ the family of ith projection functions U^i is a Σ epimorphism;

(ii) for any Σ algebra B and each $i \in I$ let

$$\phi(i) = \langle \phi(i)_s : B_s \to A(i)_s \mid s \in S \rangle$$

be a Σ homomorphism, then there is a unique Σ homomorphism

$$\phi = \langle \phi_s : B_s \to \Pi A_s \mid s \in S \rangle$$

called the *product homomorphism* such that $U^i_s \circ \phi_s = \phi(i)_s$, i.e. the following diagram commutes

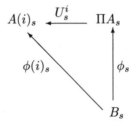

for each $s \in S$ and each $i \in I$. Furthermore, if each $\phi(i)_s$ is injective then each ϕ_s is injective;

(iii) if $A(i) = B$ and $id(i) : B \to A(i)$ is the identity mapping for each $i \in I$ then the unique product homomorphism is an embedding.

Proof. Exercise 3.5.16.(10).   ∎

Not only does the generalized direct product construction always allow us to construct algebras of larger cardinalities, it also allows us to combine infinite collections of algebras into a single algebra. This latter facility is used to define more elaborate constructions in Sections 4 and 5.

We can better understand the nature of this very general construction of the arbitrary direct product, if we look at the way that the direct power can be used to model algebras of functions.

Definition 3.5.7. Let X be any non-empty set and let A be any single-sorted Σ algebra. Let $F(X, A)$ be the set of all functions from X to the carrier of A. We define operations on $F(X, A)$ that make it a Σ algebra.

For each $c \in \Sigma_0$ define the function $c_{F(X, A)} : X \to A$ by

$$c_{F(X, A)}(x) = c_A$$

for all $x \in X$.

For each $n > 0$ and each $\sigma \in \Sigma_n$ define the functional

$$\sigma_{F(X, A)} : F(X, A)^n \to F(X, A)$$

by

$$\sigma_{F(X, A)}(f_1, \ldots, f_n)(x) = \sigma_A(f_1(x), \ldots, f_n(x))$$

for all $x \in X$. Then $F(X, A)$ is a Σ algebra.

We may advance this construction by assuming that X is an algebra and adding both X and $F(X, A)$ to A as follows: let $\overline{F(X, A)}$ be the three-sorted algebra consisting of carriers and operations of A, X and $F(X, A)$ with the further operation of $eval : F(X, A) \times X \to A$ defined by

$$eval(f, x) = f(x).$$

Examples 3.5.8. The algebra $F(X, A)$ is useful for organizing various types of function. If $X = \mathbf{N}$ then $F(X, A)$ contains the countable sequences of elements of A, and is equipped with the operations of A acting on sequences. For instance we met $\overline{F(\mathbf{N}, A)}$ in Section 2.4.

Consider the direct product of A with itself $|X|$ times. This is the direct power A^X obtained as the direct product of $\langle A_x \mid x \in X \rangle$, where $A_x = A$ for all $x \in X$.

Lemma 3.5.9. $A^X = F(X, A)$.

Proof. Consider precisely the definition of the direct product. The carrier of A^X is

$$\{ f : X \to \cup_{x \in X} A_x \mid f(x) \in A_x \text{ for all } x \in X \}.$$

But $A_x = A$ so this is the set of all functions $f : X \to A$, i.e. the carrier of $F(X, A)$. The operations correspond exactly. ∎

Let us note that binary direct products are just a special case of general direct products with the following relationship.

Lemma 3.5.10. Let $A = \langle A(i) \mid i = 1, \ldots, n \rangle$ be any finite family of Σ algebras then

$$\Pi A \cong A(1) \times (A(2) \times \ldots \times A(n)).$$

Proof. Exercise 3.5.16.(11). ∎

Definition 3.5.11. A class K of Σ algebras is said to be *closed under the formation of direct products* if, and only if, for any index set I, whenever

$A = \langle A(i) \in K \mid i \in I \rangle$ is any I-indexed family of Σ algebras in K then the direct product $\Pi A \in K$.

A class K of Σ algebras is said to be *closed under the formation of non-empty direct products* if, and only if, for any non-empty index set I, whenever $A = \langle A(i) \in K \mid i \in I \rangle$ is any I-indexed family of Σ algebras in K then the direct product $\Pi A \in K$.

Note that if the family A is empty, i.e. I is empty, then ΠA is a unit algebra. (See Exercise 3.5.16.(14).)

Examples 3.5.12.

(i) The classes of rings, semirings, groups, semigroups are closed under arbitrary direct products. Again, more generally, any equationally defined class of algebras is so closed.

(ii) In 3.5.2.(iv) we saw that classes of algebras with distinguished carriers (e.g. the Booleans) are not closed under direct products.

In Section 4 we will use direct products in making further constructions. Most of the useful results concerning direct products require some further construction, e.g. taking a subalgebra or a quotient of a direct product.

Direct products can be used as a canonical construction technique for certain kinds of algebras, and play a crucial role in classification theory.

Definition 3.5.13. A Σ algebra A is said to be *directly decomposable* if, and only if, A is isomorphic to a direct product of two non-unit Σ algebras, otherwise A is said to be *directly indecomposable*.

For example, every finite Σ algebra of prime cardinality is directly indecomposable. Finite directly indecomposable Σ algebras are the building blocks for all finite Σ algebras.

Lemma 3.5.14. *Every finite Σ algebra A is isomorphic to a finite direct product of directly indecomposable Σ algebras.*

Proof. By induction on the cardinality of A.
Basis. If $|A| = 1$ then A is directly indecomposable, since the product of two non-unit Σ algebras has cardinality > 1.
Induction Step. Suppose that $|A| > 1$. If A is directly indecomposable then the result follows. If A is not directly indecomposable then for some Σ algebras A_1 and A_2

$$A \cong A_1 \times A_2,$$

where $1 < |A_1|, |A_2| < |A|$. So by the induction hypothesis

$$A_1 \cong B_1 \times \ldots \times B_m,$$

$$A_2 \cong C_1 \times \ldots \times C_n,$$

where $B_1, \ldots, B_m, C_1, \ldots, C_n$ are directly indecomposable. Therefore by Lemma 3.5.4.(iii)

$$A \cong B_1 \times \ldots \times B_m \times C_1 \times \ldots \times C_n.$$

Hence A is isomorphic to a direct product of directly indecomposable Σ algebras. ∎

Consider the closure properties we have met in this section: they may be combined to form a very important concept:

Definition 3.5.15. A class K of Σ algebras is said to be a *variety* if, and only if, it is closed under the formation of:

(i) subalgebras;

(ii) homomorphic images; and

(iii) direct products.

Recall that if K is closed under isomorphisms then (ii) can be replaced by closure under quotient algebras.

Many of the best studied classes of algebras — rings, semirings, groups, semigroups, Boolean algebras, lattices, etc. — are varieties, though fields are not. We have stated, in examples, that classes axiomatically defined by equations are varieties. The proofs of this and, more remarkably, its converse are given in Section 5. Thus, the concept of a variety will turn out to be the semantics of equations, and find many applications in computer science, including most obviously, abstract data type theory, concurrent process theory and the logical theory of program verification.

Exercises 3.5.16.

1. Consider the power N^2 of the algebra

$$N = (\mathbf{N}; \ 0; \ Succ, \ Pred).$$

   The elements of N^2 can be used to label the integer points in the first quadrant of the 2 dimensional plane. What are the operations of N^2 in geometric terms? Add operations to N^2 to model simple geometric operations.

2. If A and B are minimal Σ algebras then is $A \times B$ a minimal Σ algebra?

3. If $A_1 < A_2$ and $B_1 < B_2$ then is $A_1 \times B_1 < A_2 \times B_2$?

4. Redefine the direct product $A \times B$ of A and B to be an algebra which includes A, B and the projection functions. What other constants

and functions can be added to enrich the algebra as a data type? Test your definitions using the integer grid and the Booleans. Generalize the definitions to the n-fold direct product.

5. Consider the general problem of embedding algebras A and B in $A \times B$. Try to formulate conditions on A and B so that this is possible.

6. Let $A = \langle A(i) \mid i \in I \rangle$ and $B = \langle B(i) \mid i \in I \rangle$ be families of Σ algebras. Let $\{ \phi_i : A(i) \rightarrow B(i) \mid i \in I \}$ be an indexed family of mappings. Define the *product map* $\phi : \Pi_{i \in I} A(i) \rightarrow \Pi_{i \in I} B(i)$ by

$$\phi(\bar{a})(i) = \phi_i(\bar{a}(i)).$$

Show that if each ϕ_i is a homomorphism then ϕ is a homomorphism. Show that if each ϕ_i is injective (respectively, surjective) then ϕ is injective (respectively, surjective).

7. Let Σ_0 be a subsignature of Σ. Let A_0 be a Σ_0 algebra and consider a class K of Σ algebras containing an isomorphic copy of A_0 as follows:

$$K = \{ A \mid A \text{ is a } \Sigma \text{ algebra and } A|_{\Sigma_0} \cong A_0 \}.$$

Starting from Example 3.5.2.(iv) discuss the direct product of K algebras. (Hint: consider minimality.)

8. Prove Lemma 3.5.3.(ii).

9. Prove Lemma 3.5.4 and generalize the results to infinite direct products.

10. Give proofs of the generalization of Lemma 3.5.3 to Lemma 3.5.6 for infinite direct products.

11. Prove Lemma 3.5.10.

12. Investigate Lemma 3.5.14 for infinite Σ algebras. What are the building blocks for infinite Σ algebras?

13. Consider the function $eval : F(X, A) \times X \rightarrow A$. Show that the maps $eval_x : F(X, A) \rightarrow A$ defined by

$$eval_x(f) = f(x)$$

are Σ homomorphisms. When is $eval_f : X \rightarrow A$ a Σ homomorphism? When is $eval : F(X, A) \times X \rightarrow A$ a Σ homomorphism?

14. Let A be the empty family of Σ algebras. Show that the empty product ΠA is a unit Σ algebra.

15. Exhibit a finite field F such that $F \times F$ is not a field.

3.6   Abstract data types

Consider the ideas that (i) a data type in a program should consist of both data *and* the operations and tests on data required in the program; (ii) a data type should be thought of abstractly, independently of the details of the representation of data, and hence of the algorithms implementing its operations; (iii) a data type may have many representations and a programmer does not know how the type is implemented.

The fundamental example is that of the integers. We think of the set of integers along with many of its operations (recall Sections 2.2 and 2.3) and use them, without information as to how these numbers are implemented. There are, of course, many representations of the integers and many algorithms for operations on the integers. A central theoretical and practical problem in computer science is to provide tools for designing, analysing and using all data types with the same degree of precision and abstraction as those for integers.

The simple algebraic ideas of this section provide a starting point for an algebraic theory of data types that is a solution to this problem.

Definition 3.6.1. A specific concrete representation or implementation of a data type is modelled or specified mathematically by an algebra A. A class of concrete implementations of a data type is modelled by a class K of algebras.

Two concrete implementations A and B of a data type are equivalent if A and B are isomorphic algebras.

An *abstract data type* is any class of algebras closed under isomorphism.

There are further conditions that one *may* require of a Σ algebra A for it to model a concrete implementation of a data type. A common property is that each datum in A can be constructed from the basic initial data by repeated applications of the operations. Equivalently, each element of A can be named by a closed term over the signature.

This property implies that the signature Σ is non-void and that A is a *minimal algebra*. It is possible to find more restrictive definitions of the notion of an abstract data type as:

(i) any class of minimal algebras closed under isomorphism;

(ii) the isomorphism type of a specific minimal algebra.

Other properties of algebras that reflect the semantics of data types are *finiteness*, *computability*, *semicomputability* and *cosemicomputability*.

In Section 5 we will discuss how abstract data types are specified by axioms.

4 Constructions

We will study the important constructions of *subdirect product, direct limit, inverse limit, reduced product* and *ultraproduct*. These constructions are used in connection with different methods for approximating an algebra A, by algebras selected from a class K of algebras. Each construction provides an algebra C in which A may be embedded, or with which A may be isomorphic, and we may ask is C in K? These constructions provide representation theorems for algebras characterized by approximation methods. Technically the constructions are based upon the direct product and its subalgebras and quotient algebras.

Approximation is an important theme in algebra which leads to several significant theoretical topics and applications. For example, we wish to approximate infinite algebras by finite and finitely generated algebras; noncomputable algebras by computable algebras; and uncountable algebras by countable and even finite algebras. We wish to build algebras from simpler algebras, perhaps to solve equations. These activities lead to different theories which apply to a wide range of problems.

In this section and the next we assume a greater mathematical maturity on the part of the reader. For simplicity, throughout the section we work with single-sorted algebras, except in Section 4.3 where we consider many-sorted reduced products and ultraproducts in order to study many-sorted Horn formulas in Section 5.3.

4.1 Subdirect products, residual and local properties

The subdirect product is a simple refinement of the direct product and provides a representation theorem (Theorem 4.1.9) for our first method of approximation, that of *separating congruences*.

Definition 4.1.1. An algebra B is said to be a *subdirect product* of an indexed family $A = \langle A(i) \mid i \in I \rangle$ of algebras if, and only if, $B \le \Pi A$ and each projection mapping $U^i : \Pi A \to A(i)$ restricted to B is surjective, i.e. $U^i(B) = A(i)$.

An embedding $\epsilon : B \to \Pi A$ is *subdirect* if, and only if, $\epsilon(B)$ is a subdirect product of A.

Definition 4.1.2. A non-empty family $\langle \equiv^i \mid i \in I \rangle$ of congruences on a Σ algebra A is *separating* if, and only if, $\cap_{i \in I} \equiv^i$ is the equality relation on A.

The way in which a separating family of congruences allows us to approximate A is given by the following reformulation of the definition:

Lemma 4.1.3. *A non-empty family $\langle \equiv^i \mid i \in I \rangle$ of Σ congruences on A is separating if, and only if, for any a, $a' \in A$, if $a \neq a'$ then $a \not\equiv^i a'$ for some $i \in I$.*

Proof. Simple exercise. ∎

Examples 4.1.4.

(i) Consider the divisibility congruence $\equiv^n$ on the commutative ring $\mathbf{Z}$ of integers defined by

$$x \equiv^n y \Leftrightarrow (x - y) \bmod n = 0$$

for each $n \in \mathbf{N}$ and all x, $y \in \mathbf{Z}$. The family $\langle \equiv^n \mid n \in \mathbf{N} \rangle$ is a separating family. For let x, $y \in \mathbf{Z}$ and $x \neq y$. Then $x - y$ is not divisible by every $n \in \mathbf{N}$, i.e. $x \not\equiv^n y$ for some $n \in \mathbf{N}$. By Proposition 4.1.3, we have a separating family. Notice that $\mathbf{Z}/\equiv^n$ is a finite algebra.

(ii) Subfamilies such as $\langle \equiv^n \mid n = 2^k$ for $k = 0, 1, \ldots \rangle$ and $\langle \equiv^n \mid n$ prime $\rangle$ of the family in (i) above are also separating families for the ring $\mathbf{Z}$. Indeed for any infinite set $S \subseteq \mathbf{N}$, $\langle \equiv^n \mid n \in S \rangle$ is a separating family.

(iii) Recall the approximating congruence $\equiv^k$ on terms defined on the term algebra $T(\Sigma)$ in Example 3.3.5.(iv); for t, $t' \in T(\Sigma)$, $t \equiv^k t'$ means that t and t' are identical up to height k. It is easy to see that $\langle \equiv^k \mid k \in \mathbf{N} \rangle$ is a separating family on $T(\Sigma)$; and that once again $T(\Sigma)/\equiv^k$ is a finite algebra.

(iv) By the First Homomorphism Theorem 3.4.18, every congruence is the kernel of a homomorphism. Thus the idea of a separating family of congruences, as defined in Definition 4.1.2 and Lemma 4.1.3, can be reformulated in terms of a separating family of homomorphisms.

Lemma 4.1.5. *Let $\langle \equiv^i \mid i \in I \rangle$ be any non-empty family of Σ congruences on a Σ algebra A and let $\equiv$ be*

$$\bigcap_{i \in I} \equiv^i .$$

(i) $A/\equiv$ *can be subdirectly embedded in* $\Pi_{i \in I}(A/\equiv^i)$.

(ii) If $\langle \equiv^i \mid i \in I \rangle$ *is separating then A can be subdirectly embedded in* $\Pi_{i \in I}(A/\equiv^i)$.

Proof.   (i) Recall Lemma 3.5.6: from a family of homomorphisms

$$\langle\, \phi(i) : B \to A(i) \mid i \in I \,\rangle$$

we can make the product homomorphism

$$\phi : B \to \Pi_{i \in I} A(i).$$

For each $i \in I$, let $A(i) = A/ \equiv^i$ and let $\phi(i) : A/ \equiv \to A/ \equiv^i$ be defined by

$$\phi(i)([a]_\equiv) = [a]_{\equiv^i}.$$

We know that $\phi(i)$ is a homomorphism by the Second Homomorphism Theorem 3.4.20, since $\equiv \subseteq \equiv^i$. Thus we construct the product homomorphism ϕ from the $\phi(i)$,

$$\phi : A/ \equiv \to \Pi_{i \in I}(A/ \equiv^i).$$

Then ϕ is injective since $\equiv$ is $\cap_{i \in I} \equiv^i$. To check that ϕ is a subdirect embedding, consider any $i \in I$. Then

$$U^i(\, \phi(A/ \equiv)\,) = U^i \circ \phi(A/ \equiv)$$

$$= \phi(i)(A/ \equiv),$$

by Lemma 3.5.6,

$$= A/ \equiv^i$$

by definition of $\phi(i)$.

(ii) Suppose $\langle\, \equiv^i \mid i \in I \,\rangle$ is separating. Then by definition $\equiv\, =\, \cap_{i \in I} \equiv^i$ is equality on A. Thus $A \cong A/ \equiv$ and the result follows from (i). ∎

We may refine the idea of separating congruences, in the style of Lemma 4.1.3, by relating it to a class K of algebras. For this we shall use homomorphisms rather than congruences (cf. Example 4.1.4.(iv)).

Definition 4.1.6. Let K be any class of Σ algebras. A Σ algebra A is said to be *residually K* if, and only if, for any $a, a' \in A$ there exists a K algebra B and a Σ epimorphism $\phi : A \to B$ such that

$$a \neq a' \Rightarrow \phi(a) \neq \phi(a'),$$

in this case we say that ϕ separates a and a'.

A class K of Σ algebras is said to be a *residual class* if, and only if, every algebra $A \in Alg(\Sigma)$ which is residually K is a K algebra.

Examples 4.1.7.

(i) If A is isomorphic with a K algebra B by $\phi : A \to B$ then A is residually K, since any pair of distinct elements of A can be separated by ϕ. The concept of an algebra being residually K is a generalization of the concept of it being isomorphic to a K algebra. The algebra A is approximated by K algebras by means of a separating family of homomorphisms; the kernels of these homomorphisms comprise a separating family of congruences on A.

(ii) A useful case of the concept is that of *residual finiteness*. An algebra is *residually finite* if for any a, $a' \in A$ such that $a \neq a'$ there exists a finite algebra B and an epimorphism $\phi : A \to B$ such that $\phi(a) \neq \phi(a')$. From our arguments in the Examples 4.1.4 we note that $\mathbf{Z}$ and $T(\Sigma)$ are residually finite. However the class of finite algebras is not a residual class since $\mathbf{Z}$ and $T(\Sigma)$ are infinite.

To relate the construction and the approximation method we must relativize the notion of congruence to a class of algebras.

Definition 4.1.8. Let Σ be a single-sorted signature, K be any class of Σ algebras and A be any Σ algebra. A congruence $\equiv$ on A is a K *congruence* if, and only if, $A/\equiv \in K$.

If K is closed under isomorphism then a congruence $\equiv$ on A is a K congruence if, and only if, $\equiv$ is the kernel of an epimorphism $\phi : A \to B$ for some $B \in K$.

The basic result linking the construction and the approximation method is this:

Theorem 4.1.9. *Let K be any class of Σ algebras closed under isomorphism and let A be any Σ algebra. Then the following are equivalent:*

(i) *A is residually K;*

(ii) *A has a separating family of K congruences;*

(iii) *A is isomorphic to a subdirect product of K algebras.*

Proof. (i) $\Rightarrow$ (ii) Suppose A is residually K. For each pair a, $a' \in A$ such that $a \neq a'$, let $\equiv^{(a, a')}$ be the kernel of an epimorphism $\phi : A \to B$ such that $\phi(a) \neq \phi(a')$. Then $a \not\equiv^{(a,a')} a'$. Since B is a K algebra, and K is closed under isomorphism, we have that $\equiv^{(a, a')}$ is a K congruence. Thus by Lemma 4.1.3, $\langle \equiv^{(a, a')} \mid a, a' \in A$ and $a \neq a' \rangle$ is a separating family of K congruences.

(ii) $\Rightarrow$ (iii) Suppose A has a separating family of K congruences $\langle \equiv^i \mid i \in I \rangle$. Then, by Lemma 4.1.5.(ii), $A/\equiv$ can be subdirectly embedded in $\Pi_{i \in I}(A/\equiv^i)$.

(iii) $\Rightarrow$ (i) Suppose A is isomorphic to a subdirect product of K algebras. Then for some family $A = \langle A(i) \mid i \in I \rangle$ of K algebras and subdirect product $S \leq \Pi A$ we have $A \cong S$. So there exists an isomorphism $\psi : A \to S$.

Furthermore by Lemma 3.5.6(i), for each $i \in I$, the projection mapping $U^i : S \to A(i)$ is a Σ homomorphism which is surjective since S is a subdirect product. So for each $i \in I$ the map $\phi_i = U^i \circ \psi : A \to A(i)$ is an epimorphism.

Now for any $a, a' \in A$ suppose $a \neq a'$. Since ψ is injective, $\psi(a) \neq \psi(a')$. So for some $i \in I$, $\psi(a)(i) \neq \psi(a')(i)$ and therefore for that i, $U^i \circ \psi(a) \neq U^i \circ \psi(a')$, i.e. $\phi_i(a) \neq \phi_i(a')$. So A is residually K. ∎

Example 4.1.10. According to Examples 4.1.4, $\mathbf{Z}$ and $T(\Sigma)$ have separating families of congruences that yield finite quotient algebras. Thus by Theorem 4.1.9, they are residually finite and are isomorphic to a subdirect product of finite algebras.

Corollary 4.1.11. *Let K be any class of Σ algebras. If K is a variety then K is a residual class.*

Proof. Suppose K is a variety, consider any $A \in K$ and suppose A is residually K. Then, by Theorem 4.1.9, A is isomorphic to a subdirect product of K algebras. Since K is closed under the formation of isomorphic images, subalgebras and direct products, $A \in K$. Therefore K is a residual class. ∎

We now turn our attention to three other kinds of approximation of algebras closely related to residuality, and for which the appropriate construction is the *direct limit* which will be discussed in the next section.

Definition 4.1.12. Let K be any class of Σ algebras. A Σ algebra A is said to be *weakly locally K* if, and only if, every finite subset of A is contained in a subalgebra of A which is a K algebra; equivalently, if every finitely generated subalgebra of A is contained in a subalgebra which is a K algebra.

We define a Σ algebra A to be *strongly locally K* if, and only if, every finitely generated subalgebra of A is a K algebra.

If A is weakly locally K then any number of distinct elements $a_1, \dots, a_n$ belonging to A can be separated by an inclusion map and a K algebra B contained in A. Calculations with these elements $a_1, \dots, a_n$ take place inside the subalgebra $\langle a_1, \dots, a_n \rangle_A$, which also lies in the K algebra B, but which need not be a K algebra itself. If A is strongly locally K then $\langle a_1, \ldots, a_n \rangle$ is itself a K algebra.

Lemma 4.1.13. *Let K be any class of Σ algebras and A be any Σ algebra.*

(i) *If A is a strongly locally K algebra then A is a weakly locally K.*

(ii) *If every finitely generated subalgebra of a K algebra is a K algebra then A is weakly locally K if, and only if, A is strongly locally K.*

Proof. (i) Suppose every finitely generated subalgebra of A is a K algebra. Then for any finite subset $X \subseteq A$ the subalgebra $\langle X \rangle_A$ is a K algebra and $X \subseteq \langle X \rangle_A$. So A is weakly locally K.

(ii) Suppose every finitely generated subalgebra of a K algebra is a K algebra

$\Rightarrow$ Suppose A is weakly locally K. Consider any finitely generated subalgebra $\langle X \rangle_A$ of A. Then the generators of X are contained in a subalgebra B of A which is a K algebra. Therefore $\langle X \rangle_A$ is a finitely generated subalgebra of a K algebra and so by hypothesis is a K algebra. Since $\langle X \rangle_A$ was arbitrarily chosen, every finitely generated subalgebra of A is a K algebra and A is strongly locally K.

$\Leftarrow$ Follows from (i). ∎

Often we approximate algebras by classes of algebras that are closed under the formation of subalgebras. Examples are the class of finite Σ algebras, and any class of Σ algebras defined by equations. Hence often the strong and weak notions coincide. Here is an important intermediate notion of approximation.

Definition 4.1.14. A set S of subalgebras of a Σ algebra A is a *local system* for A if, and only if:

(i) $A = \cup S$; and,

(ii) if $A, B \in S$ then for some $C \in S$ we have $A, B \subseteq C$.

Thus a local system for A is a non-empty set of subalgebras of A which is directed by inclusion and whose union is A.

An algebra A is said to be *locally K* if, and only if, it has a local system of K algebras. A class K of Σ algebras is said to be a *local class* if, and only if, every algebra $A \in Alg(\Sigma)$ which is locally K is a K algebra.

Example 4.1.15. Consider the set F of finitely generated subalgebras of an algebra A,

$$F = \{ \ \langle X \rangle_A \mid X \subseteq A \text{ and } X \text{ is finite } \}.$$

Clearly the properties (i) and (ii) are true of F. Thus F is a local system.

Lemma 4.1.16. *Let K be any class of Σ algebras and A be any algebra. Then*

(i) *if A is strongly locally K then A is locally K;*

(ii) *if A is locally K then A is weakly locally K.*

Furthermore, if every finitely generated subalgebra of a K algebra is a K algebra then these three concepts are equivalent.

Proof.   (i) Suppose A is strongly locally K so that every finitely generated subalgebra of A is a K algebra. Thus the set of all finitely generated subalgebras of A (as seen in Example 4.1.15) is a local system of K algebras and A is locally K.

(ii) Suppose A is locally K. Then A has a local system S of K algebras. Consider any finite subset $\{ \ a_1, \ldots, a_n \ \}$ of A. Since S is a local system, we can find $B \in S$ such that $a_1, \ldots, a_n \in B$ and B is a K algebra. Since the $a_1, \ldots, a_n$ were arbitrarily chosen, every finite subset of A is contained in a subalgebra B of A which is a K algebra. Thus A is weakly locally K.

Finally the last clause follows from Lemma 4.1.13.(ii).   ∎

Examples 4.1.17.

(i) We say that A is locally finite if, and only if, A is locally K for K the class of all finite Σ algebras. Thus, by Lemma 4.1.16, A is locally finite if, and only if, every finitely generated subalgebra of A is finite. For example, the infinite algebra $(\mathbf{N}, 0, pred)$ in Section 2.1.2 is locally finite.

(ii) Let X be any non-empty set and $Sym(X)$ the group of all permutations of X. A permutation $\phi : X \to X$ is called *finite* if the support

$$supp(\phi) = \{ \ x \in X \mid \phi(x) \neq x \ \}$$

of ϕ is finite. The set $Sym_f(X)$ of all finite permutations forms a subgroup of $Sym(X)$. The group $Sym_f(X)$ is (strongly) locally finite: every finitely generated subgroup of $Sym_f(X)$ is finite. This is proved as follows.

Consider $\{ \phi_1, \ldots, \phi_n \} \subseteq Sym_f(X)$. Every element ϕ of $\langle \phi_1, \ldots, \phi_n \rangle$ has the property that

$$supp(\phi) \subseteq \cup_{i=1}^{n} supp(\phi_i).$$

Define for any $Y \subseteq X$ the set

$$Sym_Y(X) = \{ \phi \in Sym(X) \mid supp(\phi) = Y \}$$

which forms a subgroup of $Sym(X)$ which is finite (what is its cardinality?). Clearly $\langle \phi_1, \ldots, \phi_n \rangle \leq Sym_Y(X)$ for $Y = \cup_{i=1}^{n} supp(\phi_i)$.

(iii) Suppose we have a class L of algebras that we would like to use to represent other algebras. For example we may wish to represent semigroups, groups and rings by means of semigroups, groups and rings of matrices.

By 'represent' let us mean 'embed': thus we consider algebras that can be embedded into an algebra in L. Let K be the class of all such algebras, i.e. algebras isomorphic to subalgebras of L algebras. Notice that K is subalgebra closed.

In view of Lemma 4.1.16, the notion that an algebra A is locally K therefore means that every finitely generated subalgebra of A can be embedded in some L algebra. We are interested in situations where K is a local class and therefore A is locally K implies that A is K; equivalently, this means that if every finitely generated subalgebra A can be embedded in an L algebra then A itself can be embedded in an L algebra. For example, this property is true in the case that L is the class of linear groups, namely those groups embeddable in matrix groups. We study this later in Section 4.4.

We conclude with some theorems.

Theorem 4.1.18. *Every residual class K which is closed under the formation of homomorphic images is a local class. Thus every variety is a local class.*

Proof. Suppose K is a residual class of Σ algebras which is closed under the formation of homomorphic images. Let A be any Σ algebra which is locally K. We must show that A is a K algebra. Since A is locally K, A

has a local system S of K algebras. Let $I = S$ and consider S as a family of algebras trivially indexed by I,

$$S = \langle\, S(i) = i \mid i \in I \,\rangle.$$

The indexing set I is partially ordered by the subalgebra relation $\leq$ on subalgebras of A; in fact I is a directed set under this ordering.

First we construct a subdirect product B of the $S(i)$ and an epimorphism $\phi : B \to A$.

Consider the direct product ΠS and the subset

$$B = \{\, b \in \Pi S \mid \text{ for some } k \in I \;\; b(i) = b(k) \text{ for } i \geq k \,\}.$$

The subset B is the set of all I-indexed sequences which are eventually constant. For $b \in B$ let $k(b) \in A$ be the constant value eventually reached by b. We show that B is a subdirect product of S.

For each constant symbol $c \in \Sigma_0$, the sequence $c_{\Pi S}$ is eventually constant, since $c_{S(i)} = c_{S(j)} = c_A$ for all i, $j \in I$. Therefore $c_{\Pi S} \in B$.

For any $n \in \mathbf{N}^+$, any function symbol $\sigma \in \Sigma_n$ and any $b_1, \ldots, b_n \in B$, where b_j eventually has the constant value $k(b_j)$ after some point $i_j \in I$. Since I is directed, after some point $i \geq i_1, \ldots, i_n$ the $b_1, \ldots, b_n$ all have constant values $k(b_1), \ldots, k(b_n)$ at each point $j \geq i$. Therefore, for all $j \geq i$

$$\sigma_{\Pi S}(b_1, \ldots, b_n)(j) = \sigma_{S(j)}(\, b_1(j), \ldots, b_n(j)\,)$$
$$= \sigma_A(\, k(b_1), \ldots, k(b_n)\,).$$

So $\sigma_{\Pi S}(b_1, \ldots, b_n)$ is eventually constant. Therefore $\sigma_{\Pi S}(b_1, \ldots, b_n) \in B$. It follows that B is a subalgebra of ΠS.

For any $i \in I$ and $a \in S(i)$, since S is a local system, $a \in S(j)$ for all $j \geq i$. So there is an element $b \in B$ such that $b(j) = a$ for all $i \leq j$, i.e. $U^i(b) = a$. Since i and a were arbitrarily chosen, each projection mapping $U^i : \Pi S \to S(i)$ is surjective on B. Therefore B is a subdirect product of S.

Now the mapping $\phi : B \to A$, defined by $\phi(b) = k(b)$, where $k(b)$ is the constant value eventually reached by b, is clearly a Σ homomorphism by the above argument. Since S is a local system, for any $a \in A$ we have $a \in S(i)$ for some $i \in I$ and hence $a \in S(j)$ for all $j \geq i$. Thus ϕ is surjective.

So ϕ is an epimorphism and A is a homomorphic image of B. Since B is a subdirect product of K algebras, by Theorem 4.1.9, B is residually K. Since K is a residual class, B is a K algebra. Since A is a homomorphic image of B, A is a K algebra by hypothesis.

By Corollary 4.1.11, if K is a variety then K is a residual class which is closed under the formation of homomorphic images and thus K is a local class. ∎

Theorem 4.1.19. *For any class K of Σ algebras closed under isomorphism, let $L(K)$ be the class of all locally K algebras and let $R(K)$ be the* class of all residually K algebras. Then $L(K)$ is a local class and $R(K)$ is a residual class.

Proof. For any Σ algebra A suppose A is locally $L(K)$. Then A has a local system S of $L(K)$ algebras. Since each $A \in S$ is an $L(K)$ algebra, A is locally K, i.e. A has a local system $S(A)$ of K algebras. Hence

$$S' = \bigcup_{A \in S} S(A)$$

is a local system of K algebras for A. So A is locally K, i.e. A is an $L(K)$ algebra. Since A was arbitrarily chosen, $L(K)$ is a local class.

For any Σ algebra A suppose A is residually $R(K)$. Then A has a separating family $\langle \equiv^i \mid i \in I \rangle$ of $R(K)$ congruences. For each $i \in I$, since $A/\equiv^i$ is residually K, there exists a separating family $\langle \equiv^{i,j} \mid j \in J(i) \rangle$ of K congruences, where $(A/\equiv^i)/\equiv^{i,j}$ is a K algebra for each $j \in J(i)$.

For each $j \in J(i)$ define the congruence $\equiv^{i(j)}$ on A by

$$a \equiv^{i(j)} b \Leftrightarrow [a]_i \equiv^{i,j} [b]_i.$$

Clearly $\equiv^{i(j)}$ is a congruence on A and by the Second Homomorphism Theorem 3.4.20,

$$A/\equiv^{i(j)} \cong (A/\equiv^i)/\equiv^{i,j} .$$

Since K is closed under isomorphisms $A/\equiv^{i(j)}$ is a K algebra. Consider any a, $a' \in A$ and suppose that $a \neq a'$. Then for some $i \in I$ we have $a \not\equiv^i a'$ so that $[a]_i \neq [a']_i$. Hence for some $j \in J(i)$ we have $[a]_i \not\equiv^{i,j} [a']_i$ so that $a \not\equiv^{i(j)} a'$.

Therefore

$$\bigcap_{i \in I} \bigcap_{j \in J(i)} \equiv^{i(j)}$$

is equality on A and the family $\langle \equiv^{i(j)} \mid i \in I$ and $j \in J(i) \rangle$ is a separating family of K congruences. So A is residually K, i.e. A is an $R(K)$ algebra. Since A was arbitrarily chosen, $R(K)$ is a residual class. ∎

Exercises 4.1.20.

1. Give a local system for the group $Sym_f(X)$ of finite permutations on a set X.

2. Prove that any finite group G can be embedded in the group $Sym_f(X)$ for X, any countably infinite set. (Recall Exercise 3.4.24.(12) on Cayley's theorem.)

3. A Σ algebra A is *subdirectly irreducible* if, and only if, A is not a unit algebra and for every subdirect embedding $\phi : A \to \Pi_{i \in I} A(i)$ in a family $\langle A(i) \mid i \in I \rangle$ of Σ algebras there exists $i \in I$ such that

$$U^i \circ \phi : A \to A(i)$$

is an isomorphism. Prove Birkhoff's Theorem, which states that every Σ algebra can be represented, up to isomorphism, as a subdirect product of subdirectly irreducible algebras $A(i)$ which are quotients of A (cf. Birkhoff [1944]).

4.2 Direct and inverse limits

The algebras that are direct or inverse limits of families of algebras include many familiar structures, and occur in most subjects of mathematics and computer science. Naturally occurring examples of direct limits include the algebras $T(\Sigma, X)$ of finite terms and algebraic closures of fields. Examples of inverse limits include the algebras $T_\infty(\Sigma, X)$ of finite and infinite terms and certain ultrametric algebras. The constructions are particularly useful and worth mastering.

The direct limit is a construction that provides a representation theorem for (a generalization of) the notion of approximation by local systems. Any algebra is isomorphic to a direct limit of any local system of subalgebras (Theorem 4.2.8, Corollary 4.2.9).

The inverse limit is a construction that complements the direct limit in a formal way, but has a different use. It is especially important because it defines algebras in which equations have solutions, providing they have approximate solutions (Theorem 4.2.22). The inverse limit is used in defining the semantics of computing systems specified by equations. We sketch its role in process algebra later on.

Definition 4.2.1. A partially ordered set $(I, \leq)$ is said to be *directed* if, and only if, $I \neq \emptyset$ and for all $i, j \in I$ there exists $k \in I$ such that $i, j \leq k$.

Definition 4.2.2. A *direct system* of Σ algebras consists of:

(i) a directed partially ordered set $(I, \leq)$;

(ii) an indexed family $A = \langle A(i) \mid i \in I \rangle$ of Σ algebras; and

(iii) an indexed family of Σ homomorphisms $\phi_{i,j} : A(i) \to A(j)$, for each $i \leq j$, such that
$$\phi_{j,k} \circ \phi_{i,j} = \phi_{i,k}$$
for all $i \leq j \leq k$ and $\phi_{i,i}$ is the identity map for each $i \in I$.

The *direct limit* $\overset{Lim}{\rightarrow} A$ of a direct system of Σ algebras $A = \langle A(i) \mid i \in I \rangle$ is the quotient algebra $C/\equiv$, where $C \leq \Pi A$ is the subalgebra with carrier set of *eventually consistent* elements

$$C = \{ \ a \in \Pi A \mid \quad \text{for some } k \in I \text{ and all } j \geq i \geq k, \ \ \phi_{i,j}(a(i)) = a(j) \ \},$$

and $\equiv$ is the congruence on C defined by

$$a \equiv a' \Leftrightarrow \text{ for some } k \in I \text{ and all } i \geq k, \ \ a(i) = a'(i).$$

It is left as an exercise for the reader to check that C is a Σ subalgebra of ΠA (remember: C must be non-empty) and that $\equiv$ is indeed a Σ congruence on C. For each $i \in I$ we can define a Σ homomorphism $\phi_{i,\infty} : A(i) \to \overset{Lim}{\rightarrow} A$ by

$$\phi_{i,\infty}(a) = [a'],$$

where $a' \in C$ is any element such that $a'(j) = \phi_{i,j}(a)$ for all $j \geq i$. Clearly such an element exists in C. Furthermore $\phi_{i,\infty}(a)$ does not depend upon the choice of a'. For suppose that $b \in C$ is any other element such that $b(i) = a$ and $b(j) = \phi_{i,j}(a)$ for all $i \leq j$. Then $b \equiv a'$. Again we leave it as an exercise for the reader to check that $\phi_{i,\infty}$ is a Σ homomorphism.

Examples 4.2.3. To explain the direct limit construction we begin with the special case of direct systems in which the index set is a chain and is countable.

(i) Consider a countable sequence $A = \langle A(n) \mid n \in \mathbf{N} \rangle$ of Σ algebras such that for all $n \in \mathbf{N}$,

$$A(0) \leq A(1) \leq \ldots \leq A(n) \leq A(n+1) \leq \ldots$$

Such a sequence is a chain of subalgebras. Notice we do *not* assume that all the $A(n)$ are subalgebras of a particular algebra B.

Such chains are direct systems (with inclusion homomorphisms) and occur in the process of building algebras by adding new elements. To illustrate, suppose any subset $S \subseteq X$ can be added by some means to a Σ algebra B to make a new Σ algebra $B(S)$. Then for $X = \{ \ x_1, x_2, \ldots \ \}$ we may imagine a chain defined by,

$$A(0) = B \text{ and } A(n) = B(x_1, \ldots, x_n).$$

A simple case is that of adding variables $X = \{ \ x_1, x_2, \ldots \ \}$ to term algebras,

$$A = T(\Sigma) \text{ and } A = T(\Sigma, \{ \ x_1, \ldots, x_n \ \}).$$

Often, extending algebras of interest can be complicated but important; the primary case is that of field extensions.

(ii) Let us examine the structure of the direct limit $\overset{Lim}{\to} A$ of a countably infinite chain

$$A = \langle\, A(n) \mid n \in \mathbf{N}\, \rangle$$

of subalgebras as described in (i). The associated mappings are inclusions: for $n \leq m$ we have

$$\ldots \leq A(n) \leq \ldots \leq A(m) \leq \ldots$$

and may define $\phi_{n,m} : A(n) \to A(m)$ by

$$\phi_{n,m}(a) = a.$$

The carrier of $\overset{Lim}{\to} A$ is defined to be

$$C = \{\, a \in \Pi A \mid \quad \text{for some } k \in \mathbf{N}$$

$$\text{and any } m \geq n \geq k,\ a(n) = a(m)\, \}$$

and, hence,

$$C = \{\, a \in \Pi A \mid \quad \text{for some } k \in \mathbf{N} \text{ and any } n \geq k,\ a(k) = a(n)\, \}.$$

That is, C is the set of all countably infinite sequences which have a constant value after some point; let this point and value be calculated by $l : C \to \mathbf{N}$,

$$l(a) = (\text{least } k)\,(a(k) = a(n) \text{ for all } n \geq k)$$

and $c : C \to \cup A$,

$$c(a) = a(l(a)).$$

The equivalence relation on C is defined to be

$$a \equiv a' \iff \text{for some } k \in \mathbf{N} \text{ and any } n \geq k,\ a(n) = a'(n).$$

That is, two sequences from C are identified if they have the same constant value,

$$c(a) = c(a').$$

Notice that we can associate this constant value $c(a)$ of a to the equivalence class $[a]$ by extending $c : C \to \cup A$ to $\hat{c} : C/\equiv\, \to \cup A$ by

$$\hat{c}([a]) = c(a).$$

Finally, the maps $\phi_{n,\infty} : A(n) \to \overset{Lim}{\to} A$ are defined by

$$\phi_{n,\infty}(a) = [a'],$$

where a' is any sequence such that $c(a') = a$. The maps $\phi_{n,\infty}$ are injections embedding each $A(n)$ in $\overset{Lim}{\to} A$.

(iii) In the previous example, from a chain A of subalgebras

$$A(0) \leq A(1) \leq \ldots \leq A(n) \leq A(n+1) \leq \ldots$$

we made $\overset{Lim}{\to} A$ into which they may be embedded. Consider the case that all the $A(n)$ are already subalgebras of a Σ subalgebra B. Does $\overset{Lim}{\to} A$ embed as a subalgebra of B?

Recall from Exercises 3.2.18(2) that the union of a chain A of subalgebras is a subalgebra $\cup A \leq B$. It may be shown that $\overset{Lim}{\to} A \cong \cup A$ under $\phi : \overset{Lim}{\to} A \to \cup A$ is defined by

$$\phi([a]) = \hat{c}([a]) = c(a).$$

Lemma 4.2.4. *Let $A = \langle A(i) \mid i \in I \rangle$ be a direct system of Σ algebras with direct limit $\overset{Lim}{\to} A$. If the $\phi_{i,j}$ are injective, for all $i \leq j$, then the $\phi_{i,\infty}$ are injective for all i and, in particular, each $A(i)$ embeds in $\overset{Lim}{\to} A$.*

Proof. Exercise 4.2.27.(4). ∎

Proposition 4.2.5. *Let $A = \langle A(i) \mid i \in I \rangle$ be a direct system of Σ algebras with direct limit $\overset{Lim}{\to} A$.*

(i) *For any i, $j \in I$, if $i \leq j$ then $\phi_{j,\infty} \circ \phi_{i,j} = \phi_{i,\infty}$.*

(ii) *Let $I' = I \cup \{ \infty \}$ be partially ordered by*

$$i \leq' j \Leftrightarrow j = \infty \text{ or } (i \neq \infty \text{ and } j \neq \infty \text{ and } i \leq j).$$

Let $A(\infty) = \overset{Lim}{\to} A$. Then the family of algebras $A' = \langle A(i) \mid i \in I' \rangle$ together with the homomorphisms $\phi_{i,j}$ for all $j \leq i$, the identity map $\phi_{\infty,\infty}$ on $\overset{Lim}{\to} A$ and the homomorphisms $\phi_{i,\infty}$ for all $i \in I$, forms a direct system of Σ algebras.

Proof. (i) By Definition 4.2.2, for any i, $j \in I$ such that $i \leq j$ and any $a \in A(i)$ we have

$$\phi_{j,\infty} \circ \phi_{i,j}(a) = [a'],$$

where $a' \in C$ is such that $a'(j) = \phi_{i,j}(a)$ and $a'(k) = \phi_{j,k}(\phi_{i,j}(a)) = \phi_{i,k}(a)$ for all $j \leq k$. Also $\phi_{i,\infty}(a) = [b']$, where $b' \in C$ is such that

$b'(i) = a$ and $b'(k) = \phi_{i,k}(a)$ for all $i \le k$. Since I is directed there exists $k \in I$ such that $i, j \le k$. Hence

$$a'(l) = \phi_{i,l}(a) = b'(l)$$

for all $l \ge k$, so that $a' \equiv b'$. Therefore $\phi_{j,\infty} \circ \phi_{i,j}(a) = \phi_{i,\infty}(a)$.

(ii) Follows immediately from (i) and Definition 4.2.2. ∎

The special characteristics of direct limits are its so called universal properties which are formulated as follows.

Definition 4.2.6. Let $A = \langle\, A(i) \mid i \in I \,\rangle$ be a direct system of Σ algebras with Σ homomorphisms $\phi_{i,j} : A(i) \to A(j)$. A Σ algebra B has the *direct limit property* with respect to A if, and only if, there exists a family of Σ homomorphisms $\langle\, \phi_i : A(i) \to B \mid i \in I \,\rangle$ such that the diagram of Fig. 4.1 commutes for any $i \le j$

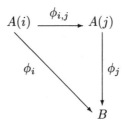

Fig. 4.1

and for every Σ algebra C and every family $\langle\, \psi_i : A(i) \to C \mid i \in I \,\rangle$ of homomorphisms satisfying

$$\psi_i = \psi_j \circ \phi_{i,j},$$

for each $i \le j$, there exists a unique Σ homomorphism $\psi : B \to C$ such that the diagram of Fig. 4.2 commutes for any $i \in I$

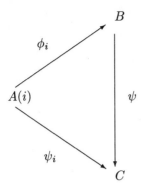

Fig. 4.2

The universal properties of direct limits are established in the following theorem, which also shows that being a direct limit of a direct system of Σ algebras is an algebraic property.

Theorem 4.2.7. *Let* $A = \langle\, A(i) \mid i \in I \,\rangle$ *be a direct system of* Σ *algebras with* Σ *homomorphisms* $\phi_{i,j} : A(i) \to A(j)$.

- (i) *The direct limit* $\overset{Lim}{\to} A$ *has the direct limit property with respect to* A.

- (ii) *If* B, C *are any* Σ *algebras which have the direct limit property with respect to* A *then* $B \cong C$.

Proof.   (i) Consider $\overset{Lim}{\to} A$ and the family $\langle\, \phi_i = \phi_{i,\infty} : A(i) \to \overset{Lim}{\to} A \mid i \in I \,\rangle$ of Σ homomorphisms. By Proposition 4.2.5.(i), the diagram of Fig. 4.1 of Definition 4.2.6 commutes.

Consider any Σ algebra C and any family $\langle\, \psi_i : A(i) \to C \mid i \in I \,\rangle$ of Σ homomorphisms satisfying

$$\psi_i = \psi_j \circ \phi_{i,j}$$

for any $i \leq j$. We must show that there exists a unique Σ homomorphism $\psi : \overset{Lim}{\to} A \to C$ which makes the diagram of Fig. 4.2 of Definition 4.2.6 commute.

Define ψ by

$$\psi([a]) = \psi_k(a(k)),$$

for any $a \in C$, where

$$C = \{ \ a \in \Pi A \mid \ \text{for some } k \in I$$

$$\text{and all } k \le i \le j \ \ \phi_{i,j}(a(i)) = a(j) \ \},$$

and $k \in I$ is any index such that $\phi_{i,j}(a(i)) = a(j)$ for all $k \le i \le j$.

To check that ψ is well defined, first consider any element $[a] \in \overset{Lim}{\rightarrow} A$ and any two choices of indices k, $k' \in I$ such that $\phi_{i,\,j}(a(i)) = a(j)$ for all $k \le i \le j$ and $\phi_{i,\,j}(a(i)) = a(j)$ for all $k' \le i \le j$. Since I is directed there exists $l \in I$ such that k, $k' \le l$ and hence

$$\psi_k(a(k)) = \psi_l(\phi_{k,\,l}(a(k))) = \psi_l(a(l)) = \psi_l(\phi_{k',\,l}(a(k'))) = \psi_{k'}(a(k')).$$

Secondly ψ does not depend upon the choice of representatives. For suppose for some a, $a' \in C$ that $a \equiv a'$. Since $a \in C$, for some $k \in I$ we have $\phi_{i,\,j}(a(i)) = a(j)$ for all $k \le i \le j$. Since $a \equiv a'$, for some $k' \in I$ we have $a(i) = a'(i)$ for all $i \ge k'$. Since I is directed there exists k, $k' \le l$ such that

$$\psi([a]) = \psi_l(a(l)) = \psi_l(a'(l)) = \psi([a']).$$

To check that ψ is a Σ homomorphism consider any constant symbol $c \in \Sigma_0$. Then

$$\psi(c_{\overset{Lim}{\rightarrow} A}) = \psi([c_{\Pi A}]) = \psi_k(c_{\Pi A}(k))$$

for any $k \in I$

$$= \psi_k(c_{A(k)}) = c_B.$$

For any $n \ge 1$, any function symbol $\sigma \in \Sigma_n$ and any $[a_1], \ldots, [a_n] \in \overset{Lim}{\rightarrow} A$,

$$\psi(\, \sigma_{\overset{Lim}{\rightarrow} A}([a_1], \ldots, [a_n])\,) = \psi(\, [\sigma_{\Pi A}(a_1, \ldots, a_n)]\,)$$

$$= \psi_k(\, \sigma_{\Pi A(i)}(a_1, \ldots, a_n)(k)\,)$$

for $k \in I$ such that $\phi_{i,j}(\, \sigma_{\Pi A}(a_1, \ldots, a_n)(i)\,) = \sigma_{\Pi A}(a_1, \ldots, a_n)(j)$ for all $k \le i \le j$

$$= \psi_k(\, \sigma_{A(k)}(a_1(k), \ldots, a_n(k))\,)$$

$$= \sigma_B(\, \psi_k(a_1(k)), \ldots, \psi_k(a_n(k))\,) \tag{1}.$$

For each $1 \leq l \leq n$ choose $k(l)$ such that

$$\phi_{i,j}(a_l(i)) = a_l(j)$$

for all $k(l) \leq i \leq j$ and choose $k' \in I$ such that $k, k(1), \ldots, k(n) \leq k'$. Then

$$\psi_k(a_i(k)) = \psi_{k'}(a_i(k')) = \psi([a_i]) \qquad (2).$$

So, by (1) and (2),

$$\psi(\sigma_{\underset{\to}{Lim} A}([a_1], \ldots, [a_n])) = \sigma_B(\, \psi([a_1]), \ldots, \psi([a_n]) \,).$$

Therefore ψ is a Σ homomorphism.

To show that ψ makes the diagram of Fig. 4.2 commute, i.e. for every $i \in I$, $\psi_i = \psi \circ \phi_i$, consider any $i \in I$ and $a \in A(i)$. Then

$$\psi \circ \phi_i(a) = \psi(\phi_{i,\infty}(a))$$

$$= \psi([a']),$$

where $a' \in C$ is any element such that $a'(j) = \phi_{i,j}(a)$ for all $j \geq i$,

$$= \psi_i(a'(i))$$

since for any $k \geq j \geq i$, $\phi_{j,k}(a'(j)) = \phi_{j,k}(\phi_{i,j}(a'(i))) = \phi_{i,k}(a'(i)) = a'(k)$,

$$= \psi_i(a).$$

To establish the uniqueness of ψ suppose $\theta : \underset{\to}{Lim} A \to B$ is any Σ homomorphism which makes the diagram of Fig. 4.2 of Definition 4.2.6 commute. Then for each $i \in I$ we have $\theta \circ \phi_{i,\infty} = \psi_i$. So for any $[a] \in \underset{\to}{Lim} A$ we have

$$\psi([a]) = \psi_k(a(k))$$

for $k \in I$ such that $\phi_{i,j}(a(i)) = a(j)$ for all $i, j \in I$ with $k \leq i \leq j$

$$= \theta(\phi_{k,\infty}(a(k))) = \theta([a]),$$

by definition of $\phi_{k,\infty}$. Hence $\psi = \theta$ and so ψ is unique.

(ii) Suppose that B and C have the direct limit property with respect to A. Then there exists a family $\langle \phi_i : A(i) \to B \mid i \in I \rangle$ such that for any $i, j \in I$ with $i \leq j$

$$\phi_i = \phi_j \circ \phi_{i,j} \qquad (1)$$

and for any Σ algebra D and family $\langle \theta_i : A(i) \to D \mid i \in I \rangle$ of homomorphisms satisfying

$$\theta_i = \theta_j \circ \phi_{i,j}$$

there exists a unique homomorphism $\theta : B \to D$ satisfying

$$\theta_i = \theta \circ \phi_i \qquad (2)$$

for each $i \in I$. Furthermore, since C has the direct limit property then there exists a family $\langle \psi_i : A(i) \to C \mid i \in I \rangle$ such that for any $i, j \in I$ with $i \le j$

$$\psi_i = \psi_j \circ \phi_{i,j} \qquad (3)$$

and for any Σ algebra D and family $\langle \theta_i : A(i) \to D \mid i \in I \rangle$ of homomorphisms satisfying

$$\theta_i = \theta_j \circ \phi_{i,j}$$

there exists a unique homomorphism $\theta : C \to D$ satisfying

$$\theta_i = \theta \circ \psi_i \qquad (4)$$

for each $i \in I$.

Now by (2) and (3) there is a unique homomorphism $\psi : B \to C$ satisfying

$$\psi_i = \psi \circ \phi_i$$

for each $i \in I$ and by (1) and (4) there is a unique homomorphism $\phi : C \to B$ satisfying

$$\phi_i = \phi \circ \psi_i$$

for each $i \in I$. For such ϕ and ψ we have

$$\phi_i = (\phi \circ \psi) \circ \phi_i$$

and

$$\psi_i = (\psi \circ \phi) \circ \psi_i$$

for each $i \in I$. But by (1) and (2) there is a unique homomorphism $\alpha : B \to B$ such that

$$\phi_i = \alpha \circ \phi_i$$

for all $i \in I$, namely the identity map $id_B : B \to B$. So $\phi \circ \psi = id_B$. Similarly by (3) and (4) $\psi \circ \phi = id_C$, where $id_C : C \to C$ is the identity map. Thus ψ and ϕ must be both injective and surjective, i.e isomorphisms and $B \cong C$. ∎

Here is the relationship between direct limits and local systems.

Theorem 4.2.8.

(i) *Let A be a local system for a Σ algebra B. Then A forms a direct system of Σ algebras and $\overset{Lim}{\to} A \cong B$.*

(ii) *If $\overset{Lim}{\to} A$ is the direct limit of a direct system $\langle A(i) \mid i \in I \rangle$ then*

$$\{ \ A'(i) = \phi_{i,\infty}(A(i)) \mid i \in I \ \}$$

is a local system for $\overset{Lim}{\to} A$.

Proof. (i) Choose the indexing set I of the proposed direct system to be the local system A; thus $i \in I$ is an algebra of the local system A. The set I can be partially ordered by the subalgebra relation $\leq$. We will consider the direct system A consisting of the family

$$A = \langle A(i) \mid i \in I \rangle,$$

where $A(i) = i$, the partial ordering $\leq$, where

$$i \leq j \ \Leftrightarrow \ A(i) \leq A(j),$$

and the homomorphisms

$$\phi_{i,j} : A(i) \to A(j)$$

for $i \leq j$, where $\phi_{i,j}$ is the inclusion map from $A(i)$ to $A(j)$.

Consider the direct limit $\overset{Lim}{\to} A = C/\equiv$, where $C \leq \Pi A$ is the subalgebra with carrier set

$$C = \{ \ a \in \Pi A \mid \phi_{i,j}(a(i)) = a(j) \ \text{for some} \ k \in I$$

$$\text{and all} \ j \geq i \geq k \ \}$$

and $a \equiv b \Leftrightarrow a(i) = b(i)$ for some $k \in I$ and all $i \geq k$.

Define the map $\psi : B \to \overset{Lim}{\to} A$ by $\psi(b) = [\bar{b}]$, where $\bar{b} \in C$ is any element such that $\bar{b}(i) = b$ for some $k \in I$ and all $i \geq k$. Since the algebras $A(i)$ form a local system, such an element $\bar{b}$ exists. Furthermore, for any two elements $\bar{b}, \bar{b}' \in C$ satisfying this condition we have $\bar{b} \equiv \bar{b}'$. Thus ψ is well defined.

To check that ψ is a Σ homomorphism, consider any $c \in \Sigma_0$. Then

$$\psi(c_B) = [\overline{c_B}] = [c_{\Pi A}] = c_{\underrightarrow{Lim} A}.$$

Consider any $n \geq 1$, any function symbol $\sigma \in \Sigma_n$ and any $b_1, \ldots b_n \in B$. Then

$$\psi(\sigma_B(b_1, \ldots, b_n)) = [\overline{\sigma_B(b_1, \ldots, b_n)}]$$
$$= [\sigma_C(\overline{b_1}, \ldots, \overline{b_n})]$$
$$= \sigma_{C/\equiv}([\overline{b_1}], \ldots, [\overline{b_n}])$$
$$= \sigma_{\underrightarrow{Lim} A}(\psi(b_1), \ldots, \psi(b_n)).$$

Thus ψ is a Σ homomorphism.

Consider any $b, b' \in B$ and suppose $\psi(b) = \psi(b')$. Then $\overline{b} \equiv \overline{b'}$, where $\overline{b}(i) = b$ for some $k_b \in I$, and all $k_b \leq i$, and $\overline{b'}(i) = b'$ for some $k_{b'} \in I$ and all $k_{b'} \leq i$. So for some $k \in I$, $\overline{b}(i) = \overline{b'}(i)$ for all $i \geq k$. Choosing $k_b, k_{b'}, k \leq l$ then

$$b = \overline{b}(l) = \overline{b'}(l) = b'.$$

Thus ψ is injective.

Consider any $a \in C$. Then for some $k \in I$, $a(i) = a(j)$ for all $j \geq i \geq k$. Let $b = a(k)$ for such a k. Then $\psi(b) = [\overline{a}]$ and hence ψ is surjective. Therefore ψ is a Σ isomorphism and $\underrightarrow{Lim} A \cong B$.

(ii) Suppose $\underrightarrow{Lim} A$ is the direct limit of a direct system of algebras $A = \langle A(i) \mid i \in I \rangle$. Consider any $i, j \in I$. Since I is directed, there exists $k \in I$ such that $i, j \leq k$. For any $a \in A(i)$ we have $\phi_{k,\infty}(\phi_{i,k}(a)) = \phi_{i,\infty}(a)$, by Proposition 4.2.5.(i). So $\phi_{i,\infty}(A(i)) \subseteq \phi_{k,\infty}(A(k))$. Similarly $\phi_{j,\infty}(A(i)) \subseteq \phi_{k,\infty}(A(k))$. Thus

$$\{ A'(i) = \phi_{i,\infty}(A(i)) \mid i \in I \}$$

is directed by inclusion. Consider any $a \in \Pi A$ and suppose $[a] \in \underrightarrow{Lim} A$. Then for some $k \in I$ we have

$$\phi_{i,j}(a(i)) = a(j)$$

for all $k \leq i \leq j$. Then $\phi_{k,\infty}(a(k)) = [a]$ and $a(k) \in A(k)$. Since a was arbitrarily chosen, $\underrightarrow{Lim} A = \cup_{i \in I} A'(i)$. Thus $\{ A'(i) = \phi_{i,\infty}(A(i)) \mid i \in I \}$ is a local system for $\underrightarrow{Lim} A$.

∎

Recalling one of the primary types of local system from 4.1.17, we have the following fact:

Corollary 4.2.9. *Any algebra is isomorphic to a direct limit of its finitely generated subalgebras.*

Definition 4.2.10. Let K be any class of Σ algebras. We say that K is *closed under the formation of direct limits* if, and only if, whenever A is a direct system of Σ algebras from K then $\overset{Lim}{\to} A \in K$.

Lemma 4.2.11. *Let K be any class of Σ algebras. If K is a variety then K is closed under the formation of direct limits.*

Proof. This is clear from an inspection of the construction in Definition 4.2.2. ∎

Two further consequences of Theorem 4.2.8 are the following:

Theorem 4.2.12. *Let K be any class of Σ algebras. If K is closed under the formation of isomorphic images and direct limits then K is a local class.*

Proof. Let B be any Σ algebra which is locally K. Then B has a local system A of K algebras. By Theorem 4.2.8.(i), A forms a direct system and
$$\overset{Lim}{\to} A \cong B.$$
By the closure properties on K, $B \in K$. Since B was arbitrarily chosen, K is a local class. ∎

Example 4.2.13. A group $G = (G;\ 0;\ +)$ is said to be *orderable* if, and only if, there exists a total ordering $\leq$ on the set G such that
$$x \leq x' \text{ and } y \leq y' \text{ imply } x + y \leq x' + y'.$$

The class of all orderable groups is a local class since a group G is orderable if, and only if, every finitely generated subgroup of G is orderable (see Fuchs [1963]).

Theorem 4.2.14. *Let K be a local class which is closed under the formation of homomorphic images. Then K is closed under the formation of direct limits.*

Proof. Let $A = \langle\, A(i) \mid i \in I\,\rangle$ be a direct system of K algebras with Σ homomorphisms $\phi_{i,j} : A(i) \to A(j)$ for all $i \leq j$. Let $\overset{Lim}{\to} A$ be the direct limit. Then $\overset{Lim}{\to} A = C/\equiv$, where $C \leq \Pi A$ is the subalgebra with carrier set
$$C = \{\, a \in \Pi A \mid \phi_{i,j}(a(i)) = a(j) \text{ for some } k \in I \text{ and all } j \geq i \geq k \,\},$$
and $a \equiv b \Leftrightarrow a(i) = b(i)$ for some $k \in I$ and all $i \geq k$.

For each $i \in I$ we have a Σ homomorphism $\phi_{i,\infty} : A(i) \to \overset{Lim}{\to} A$. By Theorem 4.2.8.(ii), $\{ \ B(i) = \phi_{i,\infty}(A(i)) \mid i \in I \ \}$ is a local system of K algebras for $\overset{Lim}{\to} A$. Since K is a local class, $\overset{Lim}{\to} A$ is a K algebra. Hence K is closed under the formation of direct limits. ∎

Corollary 4.2.15. *Let K be any class of Σ algebras which is closed under the formation of homomorphic images. Then K is a local class if, and only if, K is closed under the formation of direct limits.*

Proof. Immediate from Theorems 4.2.12 and 4.2.14. ∎

We will now turn our attention to inverse limits.

Definition 4.2.16. An *inverse system* of Σ algebras consists of:

  (i) a directed partially ordered set $(I, \leq)$;

 (ii) an indexed family $A = \langle \ A(i) \mid i \in I \ \rangle$ of Σ algebras; and

(iii) an indexed family of Σ homomorphisms $\phi_{i,j} : A(i) \to A(j)$, for each $i \geq j$, such that
$$\phi_{j,k} \circ \phi_{i,j} = \phi_{i,k}$$
for all $i \geq j \geq k$, and $\phi_{i,i}$ is the identity map for each $i \in I$.

The *inverse limit* $\overset{Lim}{\leftarrow} A$ of an inverse system of Σ algebras $A = \langle \ A(i) \mid i \in I \ \rangle$ is the subalgebra $\overset{Lim}{\leftarrow} A \leq \Pi A$ with carrier set

$$\overset{Lim}{\leftarrow} A = \{ \ a \in \Pi A \mid \quad \text{for all } i \text{ and } j \text{ with } i \geq j \ \ \phi_{i,j}(a(i)) = a(j) \ \},$$

if this set is non-empty (otherwise, undefined). It is left as an exercise for the reader to check that $\overset{Lim}{\leftarrow} A$ is a subalgebra of ΠA. Notice that if Σ has a constant symbol c then the interpretation $c_{\Pi A}$ of c in ΠA belongs to the carrier of $\overset{Lim}{\leftarrow} A$ and, in particular, $\overset{Lim}{\leftarrow} A$ exists. For each $i \in I$ we have a Σ homomorphism $\phi_{\infty,i} : \overset{Lim}{\leftarrow} A \to A(i)$, where $\phi_{\infty,i}$ is the ith projection function $U^i : \Pi A \to A(i)$ restricted to $\overset{Lim}{\leftarrow} A$, so that

$$\phi_{\infty,i}(a) = a(i).$$

Examples 4.2.17. To explain the inverse limit we focus on the special case of inverse systems in which the index sets are countable chains.

  (i) Let $A = \langle \ A(n) \mid n \in \mathbf{N} \ \rangle$ be an inverse system of Σ algebras with Σ homomorphisms $\phi_{n,m} : A(n) \to A(m)$ for $n, m \in \mathbf{N}$ and $n \geq m$; we may write

$$A(0) \xleftarrow{\phi_{1,0}} A(1) \longleftarrow \ldots \longleftarrow A(n) \xleftarrow{\phi_{n+1,n}} A(n+1) \longleftarrow \ldots$$

for inverse systems of this form. The inverse limit $\xleftarrow{Lim} A$ is the subalgebra of

$$\Pi A = A(0) \times A(1) \times \ldots \times A(n) \times \ldots$$

with carrier

$$\{ \ a \in \Pi A \mid \phi_{n,m}(a(n)) = a(m) \ \text{for all} \ n \geq m \ \}.$$

Thus, $\xleftarrow{Lim} A$ is an algebra of countable sequences of elements selected from the algebras of the family, such that later elements can be projected on to earlier elements in the sequences. Clearly, even if each of the algebras is finite then the inverse limit can be uncountable.

(ii) Now suppose that A is a particular Σ algebra and that $\langle \equiv_n \mid n \in \mathbf{N} \rangle$ is a countable family of Σ congruences on A satisfying this property: for any $n, m \in \mathbf{N}$, if $n > m$ and $a, a' \in A$ then

$$a \equiv_n a' \ \text{implies} \ a \equiv_m a'.$$

Thus, $\equiv_n$ is a finer congruence than $\equiv_m$ for $n > m$.

We think of this family as a system of congruences approximating elements of A in the sense that $a \equiv_n a'$ is interpreted as 'a and a' are approximately equal elements to degree n'. The above condition now means that if a and a' are approximately equal to degree n then they are approximately equal to any degree $m < n$.

Consider the inverse system consisting of the family

$$\langle A/\equiv_n \mid n \in \mathbf{N} \rangle$$

of factor algebras of A and the family

$$\langle \phi_{n,m} : A/\equiv_n \rightarrow A/\equiv_m \mid n, m \in \mathbf{N} \ \text{and} \ n > m \rangle$$

of homomorphisms where for any $a \in A$

$$\phi_{n,m}([a]_n) = [a]_m.$$

What is $\hat{A} = \xleftarrow{Lim} A/\equiv_n$? This algebra is the subalgebra of

$$\Pi \left(A/\equiv_n \right) \ = A/\equiv_0 \times A/\equiv_1 \times \ldots \times A/\equiv_n \times \ldots$$

with carrier set

$$C = \{\ \hat{a} \in \Pi(A/\equiv_n)\ |\ \phi_{n,\,m}(\hat{a}(n)) = \hat{a}(m)\ \}.$$

To further specify this set let us examine the nature of the elements of $\Pi(A/\equiv_n)$. For $\hat{a} \in \Pi(A/\equiv_n)$ we have

$$\hat{a}(n) = [a(n)]_n$$

for some choice of sequence $a : \mathbf{N} \to A$ of representatives for the equivalence classes of $\hat{a}$. The correspondence between sequences of representatives and sequences of equivalence classes is as follows.

Let $v_n : A \to A/\equiv_n$ be the natural homomorphism defined by $v_n(x) = [x]_n$ for $x \in A$. Then we may define the map

$$v : \Pi A \to \Pi(A/\equiv_n)$$

by

$$v(a)(n) = v_n(a(n)) = [a(n)]_n = \hat{a}(n)$$

which is a product map of the v_n. By Exercise 3.5.16.(6) v is an epimorphism.

We may now redefine the carrier C of $\overset{Lim}{\longleftarrow}\langle\, A/\equiv_n\ |\ n \in \mathbf{N}\,\rangle$ in $\Pi(A/\equiv_n)$:

$$C = \{\ \hat{a} \in \Pi(A/\equiv_n)\ |\ \phi_{n,\,m}([a(n)]_n) = [a(m)]_m\ \}$$

$$= \{\ \hat{a} \in \Pi(A/\equiv_n)\ |\ [a(n)]_m = [a(m)]_m\ \}.$$

Now the pre-image of C under v is a subalgebra (Exercise 3.4.24.(7)).

$$v^{-1}(C) = \{\ a \in \Pi A\ |\ v(a) \in C\ \}$$

$$= \{\ a \in \Pi A\ |\ a(n) \equiv_m a(m)\ \}.$$

Thus, since

$$(\Pi A)/\equiv_v\ \cong\ \Pi(A/\equiv_n)$$

we have

$$v^{-1}(C)/\equiv_v\ \cong\ \overset{Lim}{\longleftarrow}\langle\, A/\equiv_n\ |\ n \in \mathbf{N}\,\rangle.$$

Recall that by Lemma 3.5.6.(iii), $\phi : A \to \Pi A$ defined by

$$\phi(a)(n) = a$$

is an embedding. Further $im(\phi)$ is contained in $v^{-1}(C)$ and so we have a homomorphism $\psi = v \circ \phi$ mapping A into $\overset{Lim}{\longleftarrow}\langle\, A/\equiv_n\ |\ n \in \mathbf{N}\,\rangle$. If $\langle\equiv_n\ |\ n \in \mathbf{N}\,\rangle$ is a separating family then ψ is an embedding.

(iii) Consider the term algebra $T(\Sigma)$ and the family of congruences $\langle\equiv^k\ |\ k \in \mathbf{N}\,\rangle$ that approximate terms to each finite depth, as defined in

Example 3.3.5.(iv). We have observed in Example 4.1.4.(iii) that this is a separating family. The construction above in (ii) can be applied. The inverse system is

$$T(\Sigma)/ \equiv^0 \xleftarrow{\phi_{1,0}} T(\Sigma)/ \equiv^1 \xleftarrow{\phi_{2,1}} \ldots \longleftarrow T(\Sigma)/ \equiv^n \longleftarrow \ldots,$$

where $\phi_{n,m}([t]_n) = [t]_m$ for $t \in T(\Sigma)$. Notice that the algebras in the chain are finite algebras. The inverse limit $T_\infty(\Sigma) = \xleftarrow{Lim} T(\Sigma)/ \equiv^n$ can be constructed from infinite sequences of terms

$$v^{-1}(C) = \{ \underline{t} : \mathbf{N} \to T(\Sigma) \mid \underline{t}(n) \equiv_m \underline{t}(m) \}$$

factored by $\equiv^v$. Since the congruences are a separating family, $T(\Sigma)$ can be embedded in $\xleftarrow{Lim} T(\Sigma)/ \equiv^n$ by the map

$$\phi(t)(n) = t$$

for all $n \in \mathbf{N}$. This algebra is a precise formulation of the algebra of finite and infinite terms.

(iv) For an example where the carrier set of $\xleftarrow{Lim} A$ is empty, see Grätzer [1968] Chapter 3.

Lemma 4.2.18. *Let $A = \langle A(i) \mid i \in I \rangle$ be an inverse system of Σ algebras with inverse limit $\xleftarrow{Lim} A$. If the $\phi_{i,j}$ are injective, for all $i > j$, then the $\phi_{\infty,i}$ are injective.*

Proof. Easy. ∎

Proposition 4.2.19. *Let $A = \langle A(i) \mid i \in I \rangle$ be an inverse system of algebras with inverse limit $\xleftarrow{Lim} A$.*

(i) For all $i, j \in I$, if $i \geq j$ then $\phi_{i,j} \circ \phi_{\infty,i} = \phi_{\infty,j}$.

(ii) Let $I' = I \cup \{ \infty \}$ be partially ordered by

$$i \leq' j \Leftrightarrow j = \infty \text{ or } (i \neq \infty \text{ and } j \neq \infty \text{ and } i \leq j).$$

Let $A(\infty) = \xleftarrow{Lim} A$. Then the family of algebras $A' = \langle A(i) \mid i \in I' \rangle$ together with the homomorphisms $\phi_{i,j}$ for all $i \geq j$, the identity map

$\phi_{\infty,\infty}$ on $\overset{Lim}{\leftarrow} A$ and the homomorphisms $\phi_{\infty,i}$ for all $i \in I$, forms an inverse system of Σ algebras.

Proof.   (i) By Definition 4.2.16, for any i, $j \in I$ such that $i \geq j$ and any $a \in \overset{Lim}{\leftarrow} A$

$$\phi_{i,j} \circ \phi_{\infty,i}(a) = \phi_{i,j}(a(i)) = a(j) = \phi_{\infty,j}(a).$$

(ii) Follows immediately from (i) and Definition 4.2.16.   ■

The special characteristics of inverse limits are their so called *universal properties* which are formulated as follows.

Definition 4.2.20. Let $A = \langle A(i) \mid i \in I \rangle$ be an inverse system of Σ algebras with Σ homomorphisms $\phi_{i,j} : A(i) \to A(j)$. A Σ algebra B has the *inverse limit property* with respect to A if, and only if, there exists a family of Σ homomorphisms $\langle \phi_i : B \to A(i) \mid i \in I \rangle$ such that the diagram of Fig. 4.3 commutes for any $i \geq j$

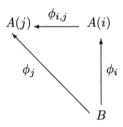

Fig. 4.3

and for every Σ algebra C and every family $\langle \psi_i : C \to A(i) \mid i \in I \rangle$ of homomorphisms satisfying

$$\psi_j = \phi_{i,j} \circ \psi_i,$$

for any $j \leq i$, there exists a unique Σ homomorphism $\psi : C \to B$ such that the diagram of Fig. 4.4 commutes for any $i \in I$.

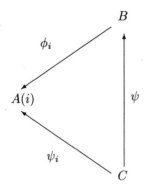

$$B$$
$$\phi_i$$
$$A(i)$$
$$\psi$$
$$\psi_i$$
$$C$$

Fig. 4.4

The universal properties of inverse limits are given in the following theorem, which also shows that being an inverse limit of an inverse system of Σ algebras is an algebraic property.

Theorem 4.2.21. *Let* $A = \langle\, A(i) \mid i \in I \,\rangle$ *be an inverse system of* Σ *algebras with inverse limit* $\overset{Lim}{\leftarrow} A$.

(i) *The inverse limit* $\overset{Lim}{\leftarrow} A$ *has the inverse limit property with respect to* A.

(ii) *If* B, C *are any* Σ *algebras which have the inverse limit property with respect to* A *then* $B \cong C$.

Proof. (i) Consider $\overset{Lim}{\leftarrow} A$ and the family $\langle\, \phi_i = \phi_{\infty,i} : \overset{Lim}{\leftarrow} A \to A(i) \mid i \in I \,\rangle$ of Σ homomorphisms. By Proposition 4.2.5, the diagram of Fig. 4.3 of Definition 4.2.20 commutes.

Consider any Σ algebra C and any indexed family $\langle\, \psi_i : C \to A(i) \mid i \in I \,\rangle$ of Σ homomorphisms satisfying for any $j \leq i$,

$$\psi_j = \phi_{i,j} \circ \psi_i.$$

We must show that there exists a unique Σ homomorphism $\psi : C \to \overset{Lim}{\leftarrow} A$ which makes the diagram of Fig. 4.4 of Definition 4.2.20 commute for any $i \geq j$. Let ψ be the product homomorphism of the ψ_i (cf. Lemma 3.5.6.(ii)), i.e.

$$\psi(c)(i) = \psi_i(c)$$

for any $c \in C$.

To show that ψ makes the diagram of Fig. 4.4 commute, i.e. for every $i \in I$, $\psi_i = \phi_i \circ \psi$, consider any $i \in I$ and any $c \in C$. Then

$$\phi_i \circ \psi(c) = \phi_{\infty,\,i}(\psi(c))$$

$$= \psi(c)(i)$$

$$= \psi_i(c).$$

To establish the uniqueness of ψ, suppose $\theta : C \to \overset{Lim}{\leftarrow} A$ is any Σ homomorphism which makes the diagram of Fig. 4.4 commute. Then for any $i \in I$ and $c \in C$ we have

$$(\,\theta(c)\,)(i) = \phi_{\infty,i} \circ \theta(c) = \psi_i(c) = \psi(c)(i).$$

Hence $\theta = \psi$ and so ψ is unique. Therefore $\overset{Lim}{\leftarrow} A$ has the inverse limit property with respect to A.

(ii) Let B be any Σ algebra which has the inverse limit property with respect to A. The proof that $B \cong \overset{Lim}{\leftarrow} A$ is similar to the proof of Theorem 4.2.7.(ii). ∎

We briefly consider equations in the inverse limit. An *equation* $e \equiv t = t'$ over signature Σ consists of two terms $t, t' \in T(\Sigma, X)$ for some set X of variables. Suppose e is

$$t(x_1, \ldots, x_k) = t'(x_1, \ldots, x_k)$$

for $x_1, \ldots, x_k \in X$. Let $Eqn(\Sigma, X)$ be the set of all equations over Σ and X. The terms can be evaluated at any $a_1, \ldots, a_k \in A$ for A a Σ algebra (Examples 2.1.4) and so the solutions to the equations in A are contained in the set

$$\{ (a_1, \ldots, a_k) \in A^k \mid t(a_1, \ldots, a_k) = t'(a_1, \ldots, a_k) \}.$$

Theorem 4.2.22. Let $A = \langle A(n) \mid n \in \omega \rangle$ be an inverse system of Σ algebras with inverse limit $\overset{Lim}{\leftarrow} A$ and let $e \in Eqn(\Sigma, X)$ be any equation.

(i) If e has a solution in $\overset{Lim}{\leftarrow} A$ then e has a solution in each $A(n)$.

(ii) *If e has at most finitely many (but not zero) solutions in each algebra $A(n)$ then e has a solution in $\overset{Lim}{\leftarrow}A$.*

Proof. Suppose e is

$$t(x_1,\ldots,x_k) = t'(x_1,\ldots x_k).$$

(i) Let $a_1,\ldots,a_k \in \overset{Lim}{\leftarrow}A$ be a solution of e. Then $t(a_1,\ldots,a_k) = t'(a_1, \ldots,a_k)$ in $\overset{Lim}{\leftarrow}A$. Apply $\phi_{\infty,n} : \overset{Lim}{\leftarrow}A \to A(n)$,

$$\phi_{\infty,n}(t(a_1,\ldots,a_k)) = \phi_{\infty,n}(t'(a_1,\ldots,a_n))$$

in $A(n)$. Since $\phi_{\infty,n}$ is a homomorphism

$$t(\phi_{\infty,n}(a_1),\ldots,\phi_{\infty,n}(a_k)) = t'(\phi_{\infty,n}(a_1),\ldots,\phi_{\infty,n}(a_k))$$

in $A(n)$ (using Exercise 3.4.24.(11)). Thus, if $a_1,\ldots,a_k$ is a solution of e in $\overset{Lim}{\leftarrow}A$ then $\phi_{\infty,n}(a_1),\ldots,\phi_{\infty,n}(a_n)$ is a solution of e in $A(n)$ for each $n \in \mathbf{N}$.

(ii) This is a tricky exercise involving König's Lemma.

∎

Definition 4.2.23. Let K be any class of Σ algebras. We say that K is *closed under the formation of inverse limits* if, and only if, whenever A is an inverse system of Σ algebras from K and $\overset{Lim}{\leftarrow}A$ exists then $\overset{Lim}{\leftarrow}A \in K$.

Lemma 4.2.24. *Any class closed under direct products and subalgebras is closed under the formation of inverse limits. In particular, any variety is closed under the formation of inverse limits.*

Proof. This is clear from the construction of $\overset{Lim}{\leftarrow}A$ in Definition 4.2.16. ∎

Theorem 4.2.25. *Let K be a residual class of Σ algebras which is closed under the formation of subalgebras. Then K is closed under the formation of inverse limits.*

Proof. Let $A = \langle A(i) \mid i \in I \rangle$ be an inverse system of K algebras with homomorphisms $\phi_{i,j} : A(i) \to A(j)$ for all $i \geq j$ and let $\overset{Lim}{\leftarrow}A$ be the inverse limit. Then $\overset{Lim}{\leftarrow}A \leq \Pi A$ is the subalgebra with carrier set

$$\overset{Lim}{\leftarrow}A = \{ a \in \Pi A \mid \phi_{i,j}(a(i)) = a(j) \text{ if } i \geq j \}.$$

For each $i \in I$, let $B(i) = \phi_{\infty,i}(\overset{Lim}{\leftarrow}A)$. Since K is closed under the formation of subalgebras, each $B(i)$ is a K algebra and $\phi_{\infty,i} : \overset{Lim}{\leftarrow}A \to B(i)$ is an epimorphism.

Consider any a, $a' \in \overset{Lim}{\leftarrow} A$ and suppose $a \neq a'$. Then $a(i) \neq a'(i)$ for some $i \in I$ and hence $\phi_{\infty,i}(a) \neq \phi_{\infty,i}(a')$. It follows that $\overset{Lim}{\leftarrow} A$ is residually K, and since K is a residual class then $\overset{Lim}{\leftarrow} A$ is a K algebra. So K is closed under the formation of inverse limits. ∎

Discussion of use of inverse limit 4.2.26
Inverse limits of the form

$$\hat{A} = \overset{Lim}{\leftarrow} A / \equiv_n,$$

for some separating family $\langle \equiv_n \mid n \in \mathbf{N} \rangle$ of congruences on A, are used in the study of models of computation and programming constructs. They allow us to solve equations which arise, for example, in giving a semantics to recursive definitions.

From the family of congruences we can make a metric space structure for A and for $\hat{A}$, such that $\hat{A}$ is the metric completion of A in which all Cauchy sequences converge and in which A is densely embedded. Thus, standard methods, such as Banach's Contraction Mapping Theorem for complete metric spaces, can be employed to solve algebraic equations in $\hat{A}$. This structure for $\hat{A}$ is developed in Exercises 4.2.27.(13), (14) and (15).

We have noted that the algebra $T_\infty(\Sigma)$ of finite and infinite terms or trees is an inverse limit of this kind. This algebra, together with its metric space structure, has been used in the semantical work of M. Nivat and his collaborators, on recursive program schemes: see Arnold and Nivat [1980a, 1980b]. Pure metric space methods have been much developed subsequently; for example, in de Bakker and Zucker [1984] to model concurrent processes.

The algebraic constructions discussed here have been employed in the algebraic theory of (concurrent and non-deterministic) processes. In process algebra, operations Σ for putting processes together are postulated and axiomatized by a set T of properties, and a process model semantics is taken to be any algebra A satisfying the axioms for T. An equation e over Σ is a specification for a set of processes. A semantics for a specification e is defined by finding an algebra A and a solution in A for the equation. The axioms governing processes are sometimes equations and the class of algebras is a variety. From this the algebra F of finite processes can be constructed (as an initial model) and can be used to generate a separating family $\langle \equiv_n \mid n \in \mathbf{N} \rangle$ of congruences that mean that concurrent processes in F can be approximately compared up to any finite complexity n. The algebra $\overset{Lim}{\leftarrow} F / \equiv_n$ is a huge algebra of infinite processes in which specifications, such as

$$X = pX,$$

which define infinite processes, such as $ppp\ldots$ can be expressed. This technique is explained in detail in Baeten and Weijland [1990].

In addition to its metric space structure, the inverse limit has a domain structure which allows us to solve equations by least fixed point methods. This has been studied in connection with so called *guarded equations* over inverse limits of the general form $\overset{Lim}{\leftarrow} A/\equiv_n$ in Stoltenberg-Hansen and Tucker [1991]. In addition, the general methods are illustrated for simple process algebras. Related work on inverse limits and domains can be found in Stoltenberg-Hansen and Tucker [1988].

Exercises 4.2.27.

1. Prove in detail that the set S and the relation $\equiv$ in Definition 4.2.2 are a subalgebra and congruence respectively.

2. Describe $\overset{Lim}{\rightarrow} A$ for $A = \langle A(i) \mid i \in I \rangle$ and $A(i) = B$ for all $i \in I$.

3. Consider the countable chain $A = \langle A(n) \mid n \in \mathbf{N} \rangle$ of subalgebras described in Examples 4.2.3. Rework the account of the direct limit of a chain in the following generalized cases:

   (a) $A = \langle A(i) \mid i \in I \rangle$ and I is a chain;

   (b) $A = \langle A(n) \mid n \in \mathbf{N} \rangle$ and the $\phi_{n,m}$ are derived from embeddings (not necessarily inclusions).

4. Prove Lemma 4.2.4 on the injectiveness of $\phi_{i,\infty}$.

5. If in a direct system the $\phi_{i,j}$ are surjective then are the $\phi_{i,\infty}$ also surjective?

6. Investigate the role of direct limits in the theory of field extensions. Show that the algebraic closure of a field is a direct limit construction.

7. Is every algebra the direct limit of a chain of finitely generated algebras?

8. Prove in detail that the set $\overset{Lim}{\leftarrow} A$ in Definition 4.2.16 is a subalgebra of ΠA.

9. What is the algebra $\overset{Lim}{\leftarrow} A$ for the inverse system $A = \langle A(i) \mid i \in I \rangle$ and $A(i) = B$, and $\phi_{i,\,j}$ the identity mapping for all $i \in I$?

10. Consider the polynomial ring $\mathbf{Z}[X]$ over the ring $\mathbf{Z}$ of integers. Define for $n \in \mathbf{N}$ and $p, q \in \mathbf{Z}[X]$,

$$p \equiv_n q \Leftrightarrow p - q \in (X^n),$$

where (X^n) is the ideal in $\mathbf{Z}[X]$ generated by X^n. In particular, $p \equiv_n q$ means that p and q are equal polynomials up to degree n. Show that $\langle \equiv_n \mid n \in \mathbf{N} \rangle$ is a separating family of congruences on $\mathbf{Z}[X]$, and prove that $\overset{Lim}{\longleftarrow}\mathbf{Z}[X]/\equiv_n$ is (isomorphic with) the ring $\mathbf{Z}[[X]]$ of all formal power series over $\mathbf{Z}$.

11. In what ways can Theorem 4.2.22 be generalized?

12. Analyse the validity of an equation in a general inverse limit in terms of its validity in the algebras of its inverse family. Perform a similar analysis for the general direct limit.

13. Consider an algebra A with a separating family $\langle \equiv_n \mid n \in \mathbf{N}\rangle$ of congruences, as used in Examples 4.2.17. Recall from Exercises 3.3.25.(14) and (15) the definition of a metric $d : A \times A \to \mathbf{R}$ on A. Show that d is an ultrametric, i.e. that the triangle rule for metric spaces can be replaced by an axiom

$$d(a,\, b) \leq max\{\ d(a,\, c),\, d(c,\, b)\ \}$$

for any a, b, $c \in A$. Show that the operations of A are non-expansive, i.e. that for each operation $\sigma_A : A^n \to A$

$$d(\,\sigma_A(a_1,\ldots,a_n),\, \sigma_A(b_1,\ldots,b_n)\,) \leq max\{\ d(a_i,\, b_i) \mid 1 \leq i \leq n\ \}$$

for any $a_1,\ldots,a_n,\, b_1,\ldots,b_n \in A$; deduce that the operations are continuous.

14. Let $\hat{A} = \overset{Lim}{\longleftarrow} A/\equiv_n$. Given the ultrametric $d : A \times A \to \mathbf{R}$ on A above, show how to extend d to $\hat{d} : \hat{A} \times \hat{A} \to \mathbf{R}$. Prove that:

   (a) $\hat{A}$ is a complete ultrametric space, i.e. each Cauchy sequence converges;

   (b) the embedding $\phi : A \to \hat{A}$ is an isometry; and,

   (c) A is dense in $\hat{A}$.

   Thus, the inverse limit $\hat{A}$ is the completion of the ultrametric algebra A.

15. Prove that for any ultrametric Σ algebra A with non-expansive operations, there exists a family $\langle \equiv_n \mid n \in \mathbf{N}\rangle$ of separating congruences on A such that the completion of A is $\overset{Lim}{\longleftarrow} A/\equiv_n$.

16. Investigate the complete metric space structure of the algebra $T_\infty(\Sigma)$ of finite and infinite terms over Σ and the power series ring $\mathbf{Z}[[X]]$.

4.3 Reduced products and ultraproducts

We have seen a number of algebraic constructions based upon the direct product ΠA of a family $A = \langle\, A(i) \mid i \in I \,\rangle$ of algebras. The constructions used subalgebras and factor algebras of ΠA defined by conditions involving I. The *reduced product* is a general type of construction using factor algebras of ΠA defined from sets of subsets of I called *filters*.

The basic idea is as follows. Let F be any set of subsets of I. We may define a relation $\equiv^F$ on ΠA by

$$a \equiv^F b \Leftrightarrow \{\, i \in I \mid a(i) = b(i) \,\} \in F$$

for $a, b \in \Pi A$. Thus, if $\equiv^F$ is an equivalence relation, the collection F may be used to measure the degree of approximation of elements of ΠA. For example, if F is a collection of 'large subsets' of F then $a \equiv^F b$ would mean that a and b are 'almost equal' (in a sense defined by F).

For $\equiv^F$ to be a congruence on ΠA it is sufficient (and almost necessary) for F to be a filter. Then $\Pi A/ \equiv^F$ is the direct product *reduced* by F.

We begin with an account of filters, before turning to the construction of reduced products.

Definition 4.3.1. Let I be any non-empty set. A *filter* F over I is a collection of subsets of I such that:

(i) $I \in F$ and $\emptyset \notin F$;

(ii) $X \in F$ and $Y \in F$ imply $X \cap Y \in F$;

(iii) $X \in F$ and $X \subseteq Y \subseteq I$ imply $Y \in F$.

Thus a filter over I is a non-empty collection of subsets of I which is closed under finite intersections, and supersets. Condition (iii) formalizes the idea that filters contain 'large sets'.

If Y is any non-empty subset of I then the set

$$\{\, X \subseteq I \mid Y \subseteq X \,\}$$

is a filter over I called the *principal filter generated by Y*. A filter F over I is *principal* if, and only if, it has the above form. It is easily shown that F is principal if, and only if, $\cap F \in F$ (Exercise 4.3.22.(1)).

Examples 4.3.2.

(i) The *trivial filter* on a non-empty set I is the set $\{\, I \,\}$ which is a principal filter.

(ii) Every finite filter $F = \{ X_1, \ldots, X_n \}$ over I is principal since F is closed under finite intersections so that $\cap_{i=1}^{n} X_i \in F$. Clearly, if I is finite then F is finite.

(iii) If I is infinite an example of a *non-principal filter* over I is the *cofinite* or *Fréchet filter*

$$C_I = \{ \ X \subseteq I \mid I - X \text{ is finite } \}.$$

This important filter is a simple example of a collection of 'large sets'.

Let S be a non-empty set of subsets of a non-empty set I. We say that S has the *finite intersection property* (FIP) if, and only if, the intersection of any finite set of members of S is non-empty. A useful test to check whether a collection of subsets may be extended to a filter is given by the following lemma.

Lemma 4.3.3. *Let S be any non-empty set of subsets of a non-empty set I. Then S extends to a filter over I if, and only if, S has the FIP.*

Proof. Suppose S extends to a filter F over I. Clearly F has the FIP since F is closed under finite intersections and $\emptyset \notin F$. Thus every subset of F, in particular S, has the FIP.

Conversely, suppose S has the FIP. Then the set

$$F = \{ \ Y \subseteq I \mid X_1 \cap \ldots \cap X_n \subseteq Y \text{ for some } X_1, \ldots, X_n \in S \ \}$$

is a filter containing S (indeed, F is the *smallest* filter containing S). ∎

The constructed filter F from S above is called the filter *generated* by S.

The following property is very useful in constructing algebras.

Definition 4.3.4. A filter F over a non-empty set I is an *ultrafilter* over I if, and only if, there is no filter G over I such that $F \subset G$, i.e. F is maximal under the partial ordering of the power set $P(P(I))$ by set-theoretic inclusion.

Theorem 4.3.5. *A filter F over a non-empty set I is an ultrafilter if, and only if, for every subset $X \subseteq I$ either $X \in F$ or $I - X \in F$.*

Proof. $\Rightarrow$ Suppose F is an ultrafilter over I and for some $X \subseteq I$ we have $X \notin F$. Then no subset $Y \subseteq X$ is a member of F. Thus $S = F \cup \{ \ I - X \ \}$ has the FIP. So by Lemma 4.3.3, S extends to a filter G over I. Now $F \subseteq G$ and since F is maximal, $F = G$. So $I - X \in F$.

$\Leftarrow$ Suppose for every subset $X \subseteq I$, either $X \in F$ or $(I - X) \in F$ (but clearly not both). Suppose for a contradiction F is not maximal. Then there exists a filter G over I such that $F \subset G$. Consider any set $X \in G - F$ then $(I - X) \in F$. So $X \cap (I - X) = \emptyset \in G$, which contradicts the fact that G is a filter. Therefore F is a maximal filter, i.e. an ultrafilter. ∎

Lemma 4.3.6. *Let F be a filter over a non-empty set I. Then there exists an ultrafilter U over I which contains F.*

Proof. We use Zorn's Lemma. The union of any non-empty chain in the set

$$\{\ G \mid G \text{ is a filter and } F \subset G\ \}$$

is again in the set. So by Zorn's Lemma the set of filters extending F has a maximal element which is thus an ultrafilter. ∎

We record some properties of ultrafilters reminiscent of logical connectives.

Lemma 4.3.7. *Let U be an ultrafilter over a non-empty set I. Then for any $X, Y \subseteq I$:*

(i) $X \cup Y \in U$ if, and only if, $X \in U$ or $Y \in U$;

(ii) $X \cap Y \in U$ if, and only if, $X \in U$ and $Y \in U$.

(iii) $I - X \in U$ if, and only if, $X \notin U$.

Proof.   (i) $\Leftarrow$ Suppose $X \in U$ or $Y \in U$. Since U is closed under supersets, $X \cup Y \in U$.

$\Rightarrow$ Suppose $X \cup Y \in U$. Since

$$(X \cup Y) \cap (I - (X \cup Y)) = \emptyset$$

and U is closed under finite intersections, $I - (X \cup Y) \notin U$. So by de Morgan's Law

$$(I - X) \cap (I - Y) \notin U$$

and then $(I - X) \notin U$ or $(I - Y) \notin U$ since U is closed under finite intersections. Since U is an ultrafilter, by Theorem 4.3.5, $X \in U$ or $Y \in U$.

(ii) True for all filters.

(iii) Immediate from Theorem 4.3.5.

In fact, (i) above is a *characterization* of ultrafilters; (Exercise 4.3.22.(4)).

Note that F is a principal ultrafilter if, and only if, F is the collection of all supersets of some singleton set $\{\ x\ \} \subseteq I$.

Lemma 4.3.8. *For any ultrafilter F over I, F is non-principal if, and only if, F contains the cofinite filter C_I.*

Proof. $\Rightarrow$ Suppose F is non-principal. Then for any non-empty finite

$$X = \{\ i_1, \ldots, i_n\ \} \subseteq I,$$

we have $X \notin F$. (Otherwise as $X = \{\ i_1\ \} \cup \ldots \cup \{\ i_n\ \}$, by continued application of 4.3.7.(i) there would be some $\{\ i_k\ \} \in F$, but then F would have to be the principal filter $\{\ Y \subseteq I \mid i_k \in Y\ \}$, which is a contradiction.) Since F is an ultrafilter, $I - X \in F$ so we have $C_I \subseteq F$.

$\Leftarrow$ Suppose $C_I \subseteq F$. Then for any $i \in I$, $\{\ i\ \} \notin F$, since $I - \{\ i\ \} \in F$. So F must be a non-principal ultrafilter. ∎

We now consider the reduced product construction. A filter F over a non-empty set I induces a Σ congruence on a direct product of S-sorted Σ algebras as follows.

Definition 4.3.9. Let I be a non-empty set and $A = \langle\ A(i) \mid i \in I\ \rangle$ be any I-indexed family of S-sorted Σ algebras. Let F be a filter over I. Define the S-indexed family of binary relations $\equiv^F = \langle \equiv_s^F \mid s \in S\ \rangle$ on the carriers of the direct product $\Pi_{i \in I} A(i)$ by

$$a \equiv_s^F b \Leftrightarrow \{\ i \in I \mid a(i) = b(i)\ \} \in F$$

for each $s \in S$ and $a, b \in \Pi_{i \in I} A(i)_s$. We will call $\equiv^F$ the Σ congruence on ΠA induced by F. This is justified by the following.

Lemma 4.3.10. *Let I be any non-empty set and $A = \langle\ A(i) \mid i \in I\ \rangle$ be any I-indexed family of S-sorted Σ algebras. Let F be a filter over I. Then $\equiv^F$ is a Σ congruence on ΠA.*

Proof. Since $I \in F$ then for any $s \in S$ and $a \in (\Pi A)_s$, $a \equiv_s^F a$. Therefore each $\equiv_s^F$ is reflexive.

Consider any $s \in S$ and $a, b \in (\Pi A)_s$ and suppose $a \equiv_s^F b$. Then

$$\{\ i \in I \mid a(i) = b(i)\ \} \in F$$

and hence $\{\ i \in I \mid b(i) = a(i)\ \} \in F$. So $b \equiv_s^F a$. Therefore each $\equiv_s^F$ is symmetric.

Consider any $s \in S$ and a, b, $c \in (\Pi A)_s$ and suppose $a \equiv_s^F b$ and $b \equiv_s^F c$. Then $\{\ i \in I \mid a(i) = b(i)\ \} \in F$ and $\{\ i \in I \mid b(i) = c(i)\ \} \in F$. Since F is closed under finite intersections,

$$\{\ i \in I \mid a(i) = b(i) \text{ and } b(i) = c(i)\ \} \in F$$

and since F is closed under supersets, $\{\ i \in I \mid a(i) = c(i)\ \} \in F$. So $a \equiv_s^F c$. Therefore each $\equiv_s^F$ is transitive.

Consider any $w = s(1) \ldots s(n) \in S^+$, any $s \in S$, any function symbol $\sigma \in \Sigma_{w,s}$ and any a_j, $b_j \in (\Pi A)_{s(j)}$ for $1 \leq j \leq n$. Suppose $a_j \equiv_{s(j)}^F b_j$ for $1 \leq j \leq n$. Then for each $1 \leq j \leq n$,

$$\{\ i \in I \mid a_j(i) = b_j(i)\ \} \in F.$$

Since F is closed under finite intersections,

$$\cap_{j=1}^n \{\ i \in I \mid a_j(i) = b_j(i)\ \} \in F.$$

Since F is closed under supersets,

$$\{\ i \in I \mid \sigma_{\Pi A}(a_1, \ldots, a_n)(i) = \sigma_{\Pi A}(b_1, \ldots, b_n)(i)\ \} \in F.$$

So $\sigma_{\Pi A}(a_1, \ldots, a_n) \equiv_s^F \sigma_{\Pi A}(b_1, \ldots, b_n)$. Hence $\equiv^F$ satisfies the substitutivity condition and is therefore a Σ congruence on ΠA. ∎

Definition 4.3.9 gives us a new construction on algebras.

Definition 4.3.11. Let I be any non-empty set, $A = \langle A(i) \mid i \in I \rangle$ be an I-indexed family of Σ algebras and let F be a filter over I. The *reduced product* of A by F is the quotient algebra

$$\Pi A / \equiv^F .$$

If $A(i) = B$ for all $i \in I$ then the reduced product $\Pi A / \equiv^F$ is termed a *reduced power*. If F is an ultrafilter over I then a reduced product (respectively, reduced power) is termed an *ultraproduct* (respectively, *ultrapower*).

Examples 4.3.12.

(i) Let $F = \{\ I\ \}$. Then

$$a \equiv^F b \Leftrightarrow \{\ i \in I \mid a(i) = b(i) \in I\ \}$$

$$\Leftrightarrow a(i) = b(i) \text{ for all } i.$$

Thus the reduced product $\Pi A / \equiv^F$ is trivially isomorphic to the direct product ΠA.

(ii) Let $R = \langle R_i \mid i \in I \rangle$ be a non-empty family of commutative rings with 1 and F any filter on I. Then the direct product ΠR and the reduced product $\Pi R / \equiv^F$ are commutative rings with 1.

(iii) If the rings of the family R in (ii) are fields then the direct product (and hence the reduced product) need not be a field because it may lack inverses.

Suppose, however, that U is an *ultrafilter* on I. We may now show the existence of inverses in $\Pi F / \equiv^U$ for $F = \langle F_i \mid i \in I \rangle$, a non-empty family of fields. Let 0_i and 1_i be the additive and multiplicative identities in F_i and let $0, 1 \in \Pi F$ be defined by $0(i) = 0_i$ and $1(i) = 1_i$ for all $i \in I$. Let $f \in \Pi F$ and suppose $f \not\equiv^U 0$ so that $[f]_F$ is a non-zero element of $\Pi F / \equiv^U$. Thus the set

$$X = \{\ i \in I \mid f(i) = 0_i\ \} \notin U$$

and, by Theorem 4.3.5, the complement

$$\overline{X} = \{\ i \in I \mid f(i) \neq 0_i\ \} \in U.$$

Define $g \in \Pi F$ by

$$g(i) = \begin{cases} f(i)^{-1}, & \text{if } i \in \overline{X}; \\ 0_i, & \text{if } i \in X\ . \end{cases}$$

Then $f(i)g(i) = 1_i$ for all $i \in \overline{X}$ and since $\overline{X} \in U$ we have that

$$f \cdot g \equiv^U 1.$$

That is, $[f]_F \cdot [g]_F = [1]_F$ in $\Pi F / \equiv^F$, i.e. $[g]_F$ is the inverse of $[f]_F$.

Thus the reduced product based on an ultrafilter is a field.

Definition 4.3.13. Let K be any class of Σ algebras. We say that K is *closed under the formation of reduced products* if, and only if, for any non-empty I-indexed family $A = \langle A(i) \mid i \in I \rangle$ of K algebras, and any filter F over I, the reduced product $\Pi A / \equiv^F \in K$.

We say that K is *closed under the formation of ultraproducts* if, and only if, for any non-empty I-indexed family $A = \langle A(i) \mid i \in I \rangle$ of K algebras, and any ultrafilter F over I, the ultraproduct $\Pi A / \equiv^F \in K$.

We have seen in 4.3.12.(i) that for any class K of Σ algebras, if K is closed under the formation of reduced products and isomorphic images then K is closed under the formation of non-empty direct products. Clearly, for any class K, if K is closed under the formation of reduced products then K is closed under the formation of ultraproducts. However the converse need not apply: a wide variety of counterexamples will be provided by a characterization result for reduced products that will be established in Section 5.3.

Lemma 4.3.14. *If K is a class closed under direct products and congruences then K is closed under reduced products; in particular, any variety is closed under reduced products.*

Proof. Immediate from the definition of reduced product. ∎

In fact, every first order axiomatizable class K of Σ algebras is closed under isomorphism and the formation of ultraproducts. However we have noted in the case of fields, that such a K is not necessarily closed under the formation of reduced products and direct products.

The importance of the distinction between principal and non-principal filters in algebra arises from the following lemma.

Lemma 4.3.15. *Let I be any non-empty set and $A = \langle A(i) \mid i \in I \rangle$ be an I-indexed family of S-sorted Σ algebras. Let F be a principal filter over I generated by $X = \cap F$. Then*

$$\Pi A/ \equiv^F \cong \Pi_{i \in X} A(i).$$

Proof. Define the S-indexed family

$$\phi : \Pi A/ \equiv^F \to \Pi_{i \in X} A(i)$$

of maps by

$$\big(\phi_s([a]_F) \big)(i) = a(i),$$

for each $s \in S$ and $a \in (\Pi A)_s$ and $i \in X$.

Consider any $s \in S$, to see that ϕ_s is well defined let $a, b \in \Pi A_s$ and suppose $a \equiv_s^F b$. Then $\{\ i \in I \mid a(i) = b(i)\ \} \in F$. But $X \subseteq \{\ i \in I \mid a(i) = b(i)\ \}$. So $a(i) = b(i)$ on each $i \in X$.

To check ϕ is a Σ homomorphism consider any $s \in S$ and any constant symbol $c \in \Sigma_{\lambda,s}$. Then for any $i \in I$,

$$\big(\phi_s(c_{\Pi A/\equiv^F}) \big)(i) = \big(\phi_s([c_{\Pi A}]_F) \big)(i) = c_{\Pi A}(i).$$

So $\phi_s(c_{\Pi A/\equiv^F}) = c_{\Pi A}$.

Consider any $w = s(1) \ldots s(n) \in S^+$, any function symbol $\sigma \in \Sigma_{w,s}$ and any $a_1, \ldots, a_n \in (\Pi A)^w$. Then

$$\left(\phi_s \left(\sigma_{\Pi A/\equiv^F}([a_1]_F, \ldots, [a_n]_F) \right) \right)(i) = \left(\phi_s \left([\sigma_{\Pi A}(a_1, \ldots, a_n)]_F \right) \right)(i)$$

$$= \sigma_{\Pi A}(a_1, \ldots, a_n)(i)$$

$$= \sigma_{A(i)}(a_1(i), \ldots, a_n(i))$$

$$= \sigma_{A(i)}\left((\phi_{s(1)}([a_1]_F))(i), \ldots, (\phi_{s(n)}([a_n]_F))(i) \right).$$

$$= \sigma_{\Pi A}(\phi_{s(1)}([a_1]_F), \ldots, \phi_{s(n)}([a_n]_F))(i).$$

So ϕ is a Σ homomorphism.

For any $s \in S$ and $a \in \Pi_{i \in X} A(i)_s$ there exists an element $b \in (\Pi A)_s$ such that $b(i) = a(i)$ for all $i \in X$. So $\phi_s([b]_F) = a$. Therefore ϕ is surjective. Consider any $s \in S$ and $a, b \in (\Pi A)_s$ and suppose $[a]_F \neq [b]_F$. Then $\{ i \in I \mid a(i) = b(i) \} \notin F$ and so $a \neq b$. So $a(i) \neq b(i)$ for some $i \in X$ since $X \in F$. Hence $\phi_s([a]_F) \neq \phi_s([b]_F)$. Therefore ϕ is injective. So ϕ is a Σ isomorphism and the result follows. ∎

We will relate the reduced product with some of the constructions in Sections 4.1 and 4.2. Since in Sections 4.1 and 4.2 we used single-sorted algebras, we will consider single-sorted reduced products, leaving the reader to verify the results for many-sorted reduced products.

Theorem 4.3.16. *Let* $A = \langle A(i) \mid i \in I \rangle$ *be a non-empty I-indexed family of Σ algebras and let F be a filter on I. Then the reduced product $\Pi A/ \equiv^F$ is isomorphic to the direct limit of a direct system formed from the $A(i)$.*

Proof. We form a direct system from the $A(i)$ as follows. Define the partial ordering $\leq$ on F by

$$X \leq Y \Leftrightarrow Y \subseteq X,$$

for any $X, Y \in F$. Then $(F, \leq)$ is a directed set, since for any $X, Y \in F$ we have $X \cap Y \in F$ and $X \cap Y \subseteq X$ and $X \cap Y \subseteq Y$, so that $X \leq X \cap Y$ and $Y \leq X \cap Y$.

For any $X, Y \in F$, such that $X \leq Y$, let $\phi_{X,Y} : \Pi_{i \in X} A(i) \to \amalg_{i \in Y} A(i)$ be the restriction map defined by

$$(\phi_{X,Y}(a))(j) = a(j),$$

for any $a \in \Pi_{i \in X} A(i)$ and any $j \in Y$. It is easily verified that $\phi_{X,Y}$ is a Σ homomorphism and that $\phi_{X,X}$ is the identity mapping on $\Pi_{i \in X} A(i)$. So the family

$$B = \langle A(X) = \Pi_{i \in X} A(i) \mid X \in F \rangle$$

of algebras together with the homomorphisms $\phi_{X,Y}$ for all $X \leq Y$, form a direct system with direct limit $\overset{Lim}{\to} B$. We show that $\Pi A / \equiv^F \cong \overset{Lim}{\to} B$. By definition $\overset{Lim}{\to} B = C / \equiv$, where $C \leq \Pi B$ is the subalgebra with carrier set

$$C = \{ \ a \in \Pi B \mid \phi_{X,Y}(a(X)) = a(Y) \ \text{ for some } \ Z \in F$$

$$\text{and all } Y \geq X \geq Z \ \}$$

and $a \equiv b$ if, and only if, for some $X \in F$ we have $a(Y) = b(Y)$ for all $Y \geq X$. Recall that for any $a, b \in \Pi A$ we have $a \equiv^F b$ if, and only if, $\{ \ i \in I \mid a(i) = b(i) \ \} \in F$.

Define the mapping $\theta : \Pi A \to C$ by

$$(\theta(a))(X)(i) = a(i),$$

for any $X \in F$ and $i \in X$. It is easily verified that θ is a Σ homomorphism. Define the mapping $\psi : \Pi A / \equiv^F \to \overset{Lim}{\to} B$ by

$$\psi([a]_F) = [\theta(a)].$$

This is valid since ψ does not depend upon the choice of representatives: if $a \equiv^F b$ then

$$X = \{ \ i \in I \mid a(i) = b(i) \ \} \in F.$$

So for all $Y \in F$ with $Y \subseteq X$ we have $\theta(a)(Y) = \theta(b)(Y)$ and hence $\theta(a) \equiv \theta(b)$. So $\psi([a]_F) = \psi([b]_F)$.

To check that ψ is a Σ homomorphism consider any constant symbol $c \in \Sigma_0$. Then

$$\psi(c_{\Pi A / \equiv^F}) = \psi([c_{\Pi A}]_F) = [\theta(c_{\Pi A})] = [c_{\Pi B}] = c_{\overset{Lim}{\to} B}.$$

Consider any $n \geq 1$, any function symbol $\sigma \in \Sigma_n$ and any $a_1, \ldots, a_n \in \Pi A$. Then

$$\psi(\sigma_{\Pi A / \equiv^F}([a_1]_F, \ldots, [a_n]_F)) = \psi([\sigma_{\Pi A}(a_1, \ldots, a_n)]_F)$$

$$= [\theta(\sigma_{\Pi A}(a_1, \ldots, a_n))]$$

$$= [\sigma_{\Pi B}(\theta(a_1), \ldots, \theta(a_n))]$$

$$= \sigma_{\Pi B}([\theta(a_1)], \ldots, [\theta(a_n)])$$

$$\sigma_{\overset{Lim}{\to} B}(\psi([a_1]_F), \ldots, \psi([a_n]_F)).$$

So ψ is a Σ homomorphism.

To check ψ is injective consider any a, $b \in \Pi A$ and let $X = \{\ i \in I \mid a(i) = b(i)\ \}$. Suppose $\theta(a) \equiv \theta(b)$. Then for some $Y \in F$ we have $(\theta(a))(Y) = (\theta(b))(Y)$. But $Y \subseteq X$ so that $X \in F$ and hence $a \equiv^F b$. To check ψ is surjective consider any $b \in C$. Then for some $Z \in F$ we have $\phi_{X,Y}(b(X)) = b(Y)$ for all $Z \leq X \leq Y$, i.e. for all $Y \subseteq X \subseteq Z$. Consider any $a \in \Pi A$ such that $a(i) = b(Z)(i)$ for all $i \in Z$. Then $\psi([a]_F) = [\theta(a)]$. So for any $X \subseteq Z$ we have $(\theta(a))(X) = b(X)$. Hence $\theta(a) \equiv b$ and so $\psi([a]_F) = [b]$. Therefore ψ is a Σ isomorphism and $\overset{Lim}{\rightarrow} B \cong \Pi A/ \equiv^F$. ∎

Definition 4.3.17. Let K be any class of Σ algebras. Then K is a quasivariety if, and only if, K is closed under the formation of isomorphic images, subalgebras and reduced products.

Examples 4.3.18.

(i) A semigroup $G = (G;\ 0,\ +)$ is said to be a *cancellation semigroup* if, and only if, G satisfies the laws

$$z + x = z + y \Rightarrow x = y, \quad x + z = y + z \Rightarrow x = y.$$

The class *CanGroup* of all cancellation semigroups is a quasivariety.

(ii) An abelian group $G = (G;\ 0,\ -,\ +)$ is said to be *torsion free* if, and only if, G satisfies the law

$$nx = 0 \Rightarrow x = 0$$

for each $n \geq 1$, where nx is the n-fold sum $x + x + \ldots + x$. The class *TFAbGroup* of all torsion free abelian groups is a quasivariety.

Quasivarieties have many characterizations, several of which we will study in depth in Section 5.3. For example, direct limits give an algebraic characterization of quasivarieties as follows.

Theorem 4.3.19. *Let K be any class of Σ algebras. Then K is a quasivariety if, and only if, K is closed under the formation of isomorphic images, subalgebras, non-empty direct products and direct limits.*

Proof. $\Leftarrow$ Suppose K is closed under the formation of isomorphic images, subalgebras, non-empty direct products and direct limits. Then the result follows by Theorem 4.3.16.

$\Rightarrow$ Suppose K is a quasivariety, i.e. K is closed under the formation of isomorphic images, subalgebras and reduced products. By Example 4.3.12.(i), K is closed under the formation of non-empty direct products.

Let $A = \langle A(i) \mid i \in I \rangle$ be a direct system of K algebras with Σ homo-morphisms $\phi_{i,j} : A(i) \to A(j)$ for all $i \leq j$. We show that $\overset{Lim}{\to} A$ can be embedded in a reduced product. Since K is closed under the formation of subalgebras, isomorphic images and reduced products, the result follows.

Recall that $\overset{Lim}{\to} A = C/ \equiv$, where $C \leq \Pi A$ is the subalgebra with carrier set

$$C = \{\ a \in \Pi A \mid \phi_{i,j}(a(i)) = a(j) \text{ for some } k \in I \text{ and all } j \geq i \geq k\ \},$$

and $a \equiv b \Leftrightarrow a(i) = b(i)$ for some $k \in I$ and all $i \geq k$.

For each $i \in I$, define $s(i) = \{\ j \in I \mid i \leq j\ \}$, and let $S = \{\ s(i) \mid i \in I\ \}$. Since I is a directed set, S has the finite intersection property. So by Lemma 4.3.3, S extends to a filter

$$F = \{\ X \subseteq I \mid s(i_1) \cap \ldots \cap s(i_n) \subseteq X \text{ for some } i_1, \ldots, i_n \in I\ \}.$$

It is routine to check that $\overset{Lim}{\to} A$ is a subalgebra of $\Pi A/ \equiv^F$. ∎

Theorem 4.3.20. *Every quasivariety is a local class.*

Proof. By definition, any quasivariety K is closed under isomorphisms and by Theorem 4.3.16, K is closed under direct limits. Thus by Theorem 4.2.12, K is a local class. ∎

Proposition 4.3.21. *Every quasivariety is closed under inverse limits.*

Proof. Follows from the definitions of quasivariety and inverse limit. ∎

Exercises 4.3.22.

1. Show that a filter F over a non-empty set I is principal if, and only if, $\cap F \in F$.

2. Give examples of finite filters over an infinite set I.

3. Prove Lemma 4.3.7 (ii).

4. Let F be a filter over a non-empty set I. Show that F is an ultrafilter if, and only if, for all $X, Y \subseteq I$, $X \cup Y \in F \Leftrightarrow X \in F$ or $Y \in F$.

5. Let $A = \langle A(i) \mid i \in I \rangle$ be a family of single-sorted algebras. Let F be any collection of subsets of I and define $\equiv^F$ on ΠA by

$$a \equiv^F b \Leftrightarrow \{\ i \mid a(i) = b(i)\ \} \in F.$$

Suppose that $|A(i)| > 2$ for all $i \in I$. Prove that if $\equiv^F$ is an equivalence relation then F is a filter on I. Is the assumption on the family

A necessary? Extend the result to the many-sorted case.

6. Show that if $F = \langle\, F_i \mid i \in I \,\rangle$ is a family of fields each of characteristic n then $\Pi F/ \equiv^U$ is a field of characteristic n for any ultrafilter U
on I.

7. Give examples of classes K of algebras where the reduced product
$\Pi A/ \equiv^F$ can be embedded in ΠA for each non-empty family $A = \langle\, A(i) \mid i \in I \,\rangle$ of K algebras and principal filter F over I.

4.4   Local and residual properties and approximation

In this last section we prove some theorems of the form: *if an algebra A can
be approximated by K algebras by some method then A can be embedded in a
K algebra*. The class K is assumed to be closed under ultraproducts so the
results apply to varieties and quasivarieties (and indeed any class of algebras axiomatized by a first order theory). The approximation methods for
an algebra A are: A is locally K (Corollary 4.4.3) and, a new concept that
generalizes residuality, A is uniformly K approximable (Corollary 4.4.6).
These pleasing theorems are easy consequences of results about ultraproducts.

Theorem 4.4.1. *For any Σ algebra A let $\{\, A(i) \mid i \in I \,\}$ be a local system for A. Let $B = \langle\, B(i) \mid i \in I \,\rangle$ be an I-indexed family of Σ algebras
such that for each $i \in I$ there exists an embedding (respectively homomor-
phism) $\phi_i : A(i) \to B(i)$. Then there exists an ultrafilter U on I and an
embedding (homomorphism) $\phi : A \to \Pi B/ \equiv^U$.*

Proof. For each $i \in I$ let $s(i) = \{\, j \in I \mid A(i) \subseteq A(j) \,\}$. Then $S = \{\, s(i) \mid i \in I \,\}$ has the finite intersection property since the algebras $A(i)$ form
a local system. So by Lemma 4.3.3, S can be extended to a filter and hence,
by Lemma 4.3.6, an ultrafilter on I. Define the map $\epsilon : A \to \Pi B$ by

$$\big(\epsilon(a)\big)(i) = \begin{cases} \phi_i(a), & \text{if } a \in A(i); \\ b_i, & \text{otherwise,} \end{cases}$$

where for each $i \in I$, the element b_i is an arbitrarily chosen element of $B(i)$.
Let $nat^U : \Pi B \to \Pi B/ \equiv^U$ be the natural mapping. Consider the map
$\phi : A \to \Pi B/ \equiv^U$, where

$$\phi = nat^U \circ \epsilon.$$

To show that ϕ is a Σ homomorphism consider any constant symbol $c \in \Sigma_0$. For each $i \in I$ we have $c_A \in A(i)$. So $\phi_i(c_A) = c_{B(i)}$. Therefore $\epsilon(c_A) = c_{\Pi B}$ and hence

$$\phi(c_A) = nat^U \circ \epsilon(c_A) = c_{\Pi B / \equiv^U}.$$

Consider any $n \geq 1$, any function symbol $\sigma \in \Sigma_n$ and any $a_1, \ldots, a_n \in A$. Then

$$\phi(\sigma_A(a_1, \ldots, a_n)) = [\ \epsilon(\ \sigma_A(a_1, \ldots, a_n)\ )\ ]_U$$

and

$$\sigma_{\Pi B / \equiv^U}(\ \phi(a_1), \ldots, \phi(a_n)\ ) = [\ \sigma_{\Pi B}(\epsilon(a_1), \ldots, \epsilon(a_n))\ ]_U.$$

We must show that

$$X = \{\ i \in I \ |\ \big(\epsilon(\sigma_A(a_1, \ldots, a_n))\big)(i) = \big(\sigma_{\Pi B}(\epsilon(a_1), \ldots, \epsilon(a_n))\big)(i)\ \} \in U.$$

Choose $k \in I$ such that $a_1, \ldots, a_n \in A(k)$. Then for each $i \in I$, such that $A(k) \subseteq A(i)$, we have

$$\big(\epsilon(\sigma_A(a_1, \ldots, a_n))\big)(i) = \phi_i(\sigma_A(a_1, \ldots, a_n))$$

$$= \sigma_{B(i)}(\ \phi_i(a_1), \ldots, \phi_i(a_n)\ )$$

$$= \big(\sigma_{\Pi B}(\epsilon(a_1), \ldots, \epsilon(a_n))\big)(i).$$

Hence $s(k) \subseteq X$ and so $X \in U$. Therefore

$$\phi(\ \sigma_A(a_1, \ldots, a_n)\ ) = \sigma_{\Pi B / \equiv^U}(\ \phi(a_1), \ldots, \phi(a_n)\ )$$

and ϕ is a Σ homomorphism.

Suppose all the maps ϕ_i are injective. Consider any $a, b \in A$ such that $a \neq b$. Choose $k \in I$ such that $a, b \in A(k)$. Then for each $i \in I$ such that $A(k) \subseteq A(i)$ we have

$$\big(\epsilon(a)\big)(i) = \phi_i(a) \neq \phi_i(b) = \big(\epsilon(b)\big)(i).$$

So $s(k) \subseteq \{\ i \in I \ |\ \big(\epsilon(a)\big)(i) \neq \big(\epsilon(b)\big)(i)\ \}$. Since $s(k) \in U$ then

$$\{\ i \in I \ |\ \epsilon(a)(i) \neq \epsilon(b)(i)\ \} \in U.$$

Since U is an ultrafilter, by Theorem 4.3.5, $\{\ i \in I \ |\ \big(\epsilon(a)\big)(i) = \big(\epsilon(b)\big)(i)\ \} \notin U$. Hence $\epsilon(a) \not\equiv^U \epsilon(b)$ and so $\phi(a) \neq \phi(b)$. Therefore ϕ is injective. ∎

Corollary 4.4.2. *If $\langle\, A(i) \ |\ i \in I \,\rangle$ is a local system for A then A embeds in the ultraproduct $\Pi_{i \in I} A(i) / \equiv^U$ for some ultrafilter U on I.*

Proof. Choose the $B(i)$ to be $A(i)$ in Theorem 4.4.1. ∎

Corollary 4.4.3. *Let K be a class of Σ algebras closed under the formation of ultraproducts. If A is locally K then A embeds in a K algebra.*

Proof. Suppose that K is ultraproduct closed and that A is locally K. Let $\langle A(i) \in K \mid i \in I \rangle$ be a local system of K algebras for A. By Corollary 4.4.2, A embeds in the ultraproduct $\Pi_{i\in I}A(i)/\equiv^U$ for some ultrafilter U on I and by hypothesis $\Pi_{i\in I}A(i)/\equiv^U$ is a K algebra. ∎

Definition 4.4.4. Let K be a class of Σ algebras. A Σ algebra A is said to be *n-K-approximable* if, and only if, for any n elements $a_1,\ldots,a_n \in A$ there exists an algebra $B \in K$ and an epimorphism $\phi : A \to B$ such that $a_i \neq a_j \Rightarrow \phi(a_i) \neq \phi(a_j)$ for all $1 \leq i,j \leq n$.

Clearly 1-K-approximability only asserts that A has a homomorphic image in K, while 2-K-approximability is equivalent to being residually K.

A Σ algebra A is said to be *uniformly K-approximable* if, and only if, A is n-K-approximable for all $n \in \mathbf{N}$.

A *uniform system S of K-approximations* for A is an indexed family of K algebras and Σ epimorphisms

$$S = \langle (B(i),\ \phi_i) \mid i \in I \rangle,$$

such that A is n-K-approximable for all $n \in \mathbf{N}$ by the algebras and morphisms of S.

Theorem 4.4.5. *If $S = \langle (B(i),\phi_i) \mid i \in I \rangle$ is a uniform system of K-approximations for a Σ algebra A then A embeds in the ultraproduct $\Pi_{i\in I}B(i)/\equiv^U$ for some ultrafilter U on I.*

Proof. Let $\epsilon : A \to \Pi_{i\in I}B(i)$ be the product homomorphism

$$\bigl(\epsilon(a)\bigr)(i) = \phi_i(a).$$

Clearly, ϵ is injective since for any $a,\ a' \in A$ if $a \neq a'$ then for some $i \in I$, $\phi_i(a) \neq \phi_i(a')$. So $\epsilon(a)(i) \neq \epsilon(a')(i)$ and hence $\epsilon(a) \neq \epsilon(a')$.

For each finite subset $A' \subseteq A$ let

$$s(A') = \{\ i \in I \mid \phi_i(a) \neq \phi_i(a')\ \text{for all}\ a,\ a' \in A'\ \text{such that}\ a \neq a'\ \}.$$

For each finite subset $A' \subseteq A$ the set $s(A')$ is non-empty since S is a uniform system of K-approximations. The set $S = \{\ s(A') \mid A' \subseteq A\ \text{is finite}\ \}$ has the finite intersection property since for any finite $A_i,\ A_j \subseteq A$,

$$\emptyset \neq s(A_i \cup A_j) \subseteq s(A_i) \cap s(A_j)$$

and therefore S extends to an ultrafilter U on I.

The mapping $\phi : A \to \Pi_{i \in I} B(i)/ \equiv^U$, where

$$\phi = nat^U \circ \epsilon$$

and $nat^U : \Pi_{i \in I} B(i) \to \Pi_{i \in I} B(i)/ \equiv^U$ is the natural mapping, is a Σ homomorphism. So we need only show ϕ is injective. Consider any a, $a' \in A$ and suppose $a \neq a'$. Since ϵ is injective, $\epsilon(a) \neq \epsilon(a')$. Consider the set

$$s(\{ a, a' \}) = \{ i \in I \mid \phi_i(a) \neq \phi_i(a') \}$$

$$= \{ i \in I \mid (\epsilon(a))(i) \neq (\epsilon(a'))(i) \}.$$

By definition, $s(\{ a, a' \}) \in U$ and since U is an ultrafilter,

$$\{ i \in I \mid (\epsilon(a))(i) = (\epsilon(a'))(i) \} \notin U.$$

So $\epsilon(a) \not\equiv^U \epsilon(a')$ and hence ϕ is injective, i.e. a Σ embedding. ∎

Corollary 4.4.6. *Let K be a class of Σ algebras closed under the formation of ultraproducts. If A has a uniform system of K-approximations then A embeds in a K algebra.*

Proof. Suppose that K is ultraproduct closed and that A has a uniform system of K-approximations $\langle (B(i), \phi_i) \mid i \in I \rangle$. By Theorem 4.4.5, A embeds in the ultraproduct $\Pi_{i \in I} B(i)/ \equiv^U$ for some ultrafilter U on I, and by hypothesis $\Pi_{i \in I} B(i)/ \equiv^U$ is a K algebra. ∎

Exercises 4.4.7.

1. Can Theorem 4.4.5 be deduced from Theorem 4.4.1?

2. Consider the representation of groups by groups of matrices over a field (first mentioned in 4.1.17.(iii)). Let $GL(n, F)$ be the group of all non-singular $n \times n$ matrices over the field F. A group G has a *faithful representation* over F of degree n if there exists a monomorphism

$$\rho : G \to GL(n, F).$$

Prove the following: let $n > 0$ be fixed and let K be the class of all groups having some faithful representation of degree n over some field F. Then K is closed under ultraproducts. (Hint: if the family $G = \langle G_i \mid i \in I \rangle$ is represented over fields $F = \langle F_i \mid i \in I \rangle$ then represent the ultraproduct $\Pi G/ \equiv_U$ over the ultraproduct $\Pi F_i/ \equiv_U$

for U any ultrafilter.)

3. Deduce from (2) the following: if each finitely generated subgroup of a group G has a faithful representation of degree n over some field then G has a faithful representation of degree n over some field.

4. (Hard) Show that the class K in Exercise (2) is not first order axiomatizable.

4.5   Remarks on references

Most of the constructions of this section were first used on particular types of algebras, especially groups. The study of special subclasses of the class of groups has shaped, and has been shaped by, the theory of classes of universal algebras that we record here and in the next section. We will give a small number of references to group theory to allow the reader a glimpse of these general ideas in operation in a fully mature mathematical theory. A good general introduction to group theory that is relevant to our concerns is Kargapolov and Merzljakov [1979].

The subdirect product of groups was studied by R. Remak in the early 1930s. The subdirect product of universal algebras and its basic properties (including Exercise 4.1.20.(3)) appear in Birkhoff [1944]. The general equivalences in Theorem 4.1.9 form a convenient signpost for many applications of the three important concepts of subdirect product, residual property and separating family of congruences. For example, we met the last in various places (in Sections 2, 3 and 4) to do with the algebra of infinite trees and of concurrency. The use of residuality is widespread in classifying algebras. The residually finite algebras are important in the theory of computable algebras (see the chapter on Effective Structures in this Handbook, volume 2).

Residual properties have been extensively studied in group theory: see Robinson [1972], and Lyndon and Schupp [1977]. Let us note that residuality can be generalized from its dependence on $=$ to other predicates. An interesting analysis of residual finiteness with respect to predicates in the case of groups is Malcev [1958].

The study of local properties of algebras was established by the work of A. I. Malcev. In Malcev [1941], the problem of proving local theorems for different classes of groups is tackled by considering the formal axiomatization of the classes. There, the first general statement of the Compactness Theorem for first order logic is made. The subject is taken further in Malcev [1959]. A further extension is in Cleave [1969].

The study of local properties of groups is a very well developed and integrated part of group theory: see Robinson [1972]. Our treatment has been influenced by the study of locally finite groups and locally linear groups, for which we have used Kegel and Wehrfritz [1973] and Wehrfritz [1973].

The direct and inverse limits arise in many situations. An account for groups appears in Eilenberg and Steenrod [1952]. We have used material in Grätzer [1979] and Cohn [1981], but have emphasized and added material relevant to our own interests (see Stoltenberg-Hansen and Tucker [1991]).

The ultraproduct construction is an important tool in the model theory of first order theories. Its basic property is this: let $A = \langle\, A(i) \mid i \in I \,\rangle$ be a non-empty family of Σ algebras and let U be an ultrafilter on I. Then for any sentence ϕ in the first order language of Σ,

$$\Pi A/ \equiv^{U} \models \phi \;\;\Leftrightarrow\;\; \{\, i \in I \mid A(i) \models \phi \,\} \in U.$$

Notice that this implies that any first order class is closed under ultraproducts. However, not every class closed under ultraproducts is first order.

A form of ultraproduct was used by T. Skolem in 1934, to construct a model of the complete first order theory of the structure $PA(\mathbf{N})$ of Peano arithmetic. The general tools of reduced products and ultraproducts were first shown in Los [1955]. The construction was further developed in Frayne *et al.* [1962]. An invaluable account of ultraproducts in model theory is Bell and Slomson [1969]. A useful survey of applications is Eklof [1977].

Our treatment of ultraproducts has ignored their role in first order model theory and concentrated on their use in algebraic contexts. This is seen in Section 4.4 and in the use in Section 5.3.

5 Classes of algebras

In this section we study classes of algebras axiomatized by equations, conditional equations and equational Horn formulas. The importance of these types of axioms is amply demonstrated by their widespread use in mathematics and computer science, as well as by specific results. In the simplest and most important case of equational axioms their practical use is enhanced by a simple, sound and complete calculus for reasoning with equations, which may be automated easily. The use of this calculus leads to the subject of *term rewriting*, which is about symbolic computation with terms and equations (see the chapter by J.W. Klop in this Handbook, volume 2). The most general semantics of a set of equations, like that of any other axiomatization, is the class of all algebras that satisfy the equations. Equationally axiomatized classes have an elegant theory which we will describe, together with some of its generalizations to the more complicated conditional equations and Horn formulas.

In Section 5.1 we begin by describing *free* and *initial algebras* for a class K and their connections with term algebras. In Section 5.2 we prove the basic results concerning equations, namely G. Birkhoff's Soundness and Completeness Theorems and his Variety Theorem. In Section 5.3 the algebraic properties of classes defined by conditional equations and Horn formulas are described. Finally, in Section 5.4 we discuss use of these algebraic ideas in the specification of abstract data types.

5.1   Free, initial and final algebras

In this section we introduce the important concept of a *free algebra*. Free algebras play a fundamental role in universal algebra as we shall see in this section. Initial algebras arise as a special case of free algebras and play a major role in the theory of abstract data types, where an initial algebra provides a standard model or semantics for certain kinds of axiomatic data type specification. Final algebras arise as the dual concept (in a category-theoretic sense) to initial algebras. They have also been studied as models of data type specifications.

Definition 5.1.1. Let Σ be an S-sorted signature. Let K be any class of Σ algebras. Let F be any Σ algebra (not necessarily in K) and let $X = \langle\, X_s \subseteq F_s \mid s \in S \,\rangle$ be any S-indexed family of subsets. We say that F is *free* for K on X if, and only if, for each algebra $A \in K$ and each S-indexed family of mappings

$$\alpha = \langle\, \alpha_s : X_s \to A_s \mid s \in S \,\rangle$$

there exists a unique Σ homomorphism $\overline{\alpha} = \langle\, \overline{\alpha_s} : F_s \to A_s \mid s \in S \,\rangle$ which agrees with α on X, i.e. the following diagram commutes for each sort $s \in S$

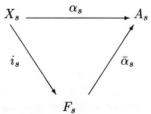

where $i_s : X_s \to F_s$ is the inclusion mapping.

If F is free for K on X and $F \in K$, we say that F is free *in* K on X. Clearly, if F is free for K and $K' \subseteq K$ then F is free for K'. So, for example, if F is free for $Alg(\Sigma)$ then F is free for any class K of Σ algebras.

We say that F is free for K (respectively in K) if F is free for K (respectively in K) on some family X of sets of subsets of F.

If F is free for a class K then we can construct *any* algebra in K, up to a certain cardinality, as a homomorphic image (or equivalently a quotient) of F. More precisely:

Theorem 5.1.2. *Let Σ be an S-sorted signature and K be any class of Σ algebras. Let $A \in K$ and $X = \langle X_s \subseteq F_s \mid s \in S \rangle$ be any S-indexed family of subsets. If F is free for K on X and $\mid X_s \mid \geq \mid A_s \mid$ for each sort $s \in S$, then A is a homomorphic image of F. Equivalently, A is isomorphic to a quotient algebra of F.*

Proof. Suppose $\mid X_s \mid \geq \mid A_s \mid$ for each sort $s \in S$. Then there exists an S-indexed family of surjections $\alpha = \langle \alpha_s : X_s \to A_s \mid s \in S \rangle$. By the freeness property, α has a unique homomorphic extension $\overline{\alpha} : F \to A$, which is a Σ epimorphism. Therefore $A = \overline{\alpha}(F)$. By the First Homomorphism Theorem, $A \cong F/\equiv^{\overline{\alpha}}$. ∎

Of course we have not yet addressed the question of whether free algebras for a class K exist. It is also important to ask whether they are unique. The latter question is more easily answered and will be addressed first. Free algebras in a class K on a family X of sets of particular cardinality when they exist, are unique up to isomorphism and therefore share exactly the same algebraic properties.

Theorem 5.1.3. *Let Σ be an S-sorted signature. Let K be any class of Σ algebras and let F and F' be free in K on the families X and X' respectively. If $\mid X_s \mid = \mid X'_s \mid$ for each $s \in S$ then $F \cong F'$.*

Proof. Let $\alpha = \langle \alpha_s : X_s \to X'_s \mid s \in S \rangle$ be a family of bijections with corresponding inverse family. Let $\alpha^{-1} = \langle \alpha_s^{-1} : X'_s \to X_s \mid s \in S \rangle$. Then the compositions of the respective homomorphic extensions $\overline{\alpha}$ and $\overline{\alpha^{-1}}$ given by

$$\overline{\alpha^{-1}} \circ \overline{\alpha} : F \to F, \quad \overline{\alpha} \circ \overline{\alpha^{-1}} : F' \to F',$$

are Σ endomorphisms on F and F' respectively, extending the families of identity maps $id = \langle id_s : X_s \to X_s \mid s \in S \rangle$ and $id' = \langle id'_s : X'_s \to X'_s \mid s \in S \rangle$. So by the freeness property

$$\overline{\alpha^{-1}} \circ \overline{\alpha} = \overline{id} \quad \text{and} \quad \overline{\alpha} \circ \overline{\alpha^{-1}} = \overline{id'}.$$

Now $\overline{id}$ and $\overline{id'}$ are the identity maps on F and F' respectively. Therefore $\overline{\alpha}$ and $\overline{\alpha^{-1}}$ are both surjective and injective, i.e. isomorphisms. Hence $F \cong F'$. ∎

Next we address the crucial question of the existence of free algebras. First let us again note that if F is free for K and $K' \subseteq K$, then F is free for K'.

Of course F need not be in K', nor indeed in K. An important consequence is that if F is free for the class $Alg(\Sigma)$ of *all* Σ algebras then F is free for *any* class K of Σ algebras (because $K \subseteq Alg(\Sigma)$). Thus it appears to be sufficient to consider the existence of free algebras for $Alg(\Sigma)$.

For $Alg(\Sigma)$, the simplest free algebras to construct are the *term algebras* $T(\Sigma, X)$ introduced in Example 3.1.4.(viii) and (ix). Because of their universality all our future constructions of free algebras begin with a term algebra. Let us recall the definition of $T(\Sigma, X)$.

Definition 5.1.4. Let Σ be an S-sorted signature and $X = \langle X_s \mid s \in S \rangle$ be an S-indexed family of sets X_s of *variable symbols* of sort s, satisfying the usual assumption that for each sort $s \in S$ the sets X_s and $\Sigma_{\lambda,s}$ are disjoint. Define by simultaneous induction the S-indexed family of sets

$$T(\Sigma, X) = \langle\, T(\Sigma, X)_s \mid s \in S \,\rangle$$

where $T(\Sigma, X)_s$ is the set of all *terms of sort s* over Σ and X, and for each term t the S-indexed family

$$Vars(t) = \langle\, Vars(t)_s \mid s \in S \,\rangle$$

of sets $Vars(t)_s$ of all *variables occurring in t* of sort s:

(i) for each sort $s \in S$ and each constant symbol $c \in \Sigma_{\lambda,\, s}$, $c^s \in T(\Sigma, X)_s$ and for each $s \in S$, $Vars(c^s)_s = \emptyset$;

(ii) for each sort $s \in S$ and each variable symbol $x \in X_s$, $x^s \in T(\Sigma, X)_s$, also $Vars(x^s)_s = \{\, x \,\}$ and $Vars(x^s)_{s'} = \emptyset$ for all $s' \in S$ such that $s' \neq s$;

(iii) for each sort $s \in S$, each non-empty word $w = s(1) \ldots s(n) \in S^+$, each function symbol $\sigma \in \Sigma_{w,\, s}$ and any terms $t_i \in T(\Sigma, X)_{s(i)}$, for $1 \leq i \leq n$,

$$\sigma^s(t_1, \ldots, t_n) \in T(\Sigma, X)_s$$

and $Vars(\, \sigma^s(t_1, \ldots, t_n)\,)_s = Vars(t_1)_s \cup \ldots \cup Vars(t_n)_s$.

The elements of $T(\Sigma)_s = T(\Sigma, \emptyset)_s$ are called *closed terms, ground terms* or *words* of sort s.

If for each sort $s \in S$ the set $T(\Sigma, X)_s$ is non-empty then we define the *term algebra* $T(\Sigma, X)$ to be the Σ algebra with S-indexed family of carrier sets $T(\Sigma, X)$, such that for each sort $s \in S$ and each constant symbol $c \in \Sigma_{\lambda,\, s}$,

$$c_{T(\Sigma,X)} = c^s.$$

Also for each sort $s \in S$, each non-empty word $w = s(1) \ldots s(n) \in S^+$, each function symbol $\sigma \in \Sigma_{w,\, s}$ and any terms $t_i \in T(\Sigma, X)_{s(i)}$, for $1 \leq i \leq n$,

$$\sigma_{T(\Sigma,X)}(t_1,\ldots,t_n) = \sigma^s(t_1,\ldots,t_n).$$

We will now define the evaluation of terms in an algebra, thus generalizing the material of Section 2.1.4. Let A be a Σ algebra and let $\alpha : X \to A = \langle \alpha_s : X_s \to A_s \mid s \in S \rangle$ be an S-indexed family of mappings that assign elements of A to the variables of X. For this α we define the *term evaluation mapping*,

$$V^\alpha = \langle V^\alpha : T(\Sigma,X)_s \to A_s \mid s \in S \rangle$$

that evaluates or interprets the terms of $T(\Sigma,X)$ as elements of A. The mappings V_s^α are defined by simultaneous induction as follows:
for each sort $s \in S$ and constant symbol $c \in \Sigma_{\lambda,s}$

$$V_s^\alpha(c^s) = c_A;$$

for each sort $s \in S$ and each variable symbol $x \in X_s$

$$V_s^\alpha(x^s) = \alpha_s(x);$$

for each non-empty word $w = s(1)\ldots s(n) \in S^+$, each sort $s \in S$, each function symbol $\sigma \in \Sigma_{w,s}$ and any terms $t_i \in T(\Sigma,X)_{s(i)}$ for $1 \le i \le n$,

$$V_s^\alpha(\sigma^s(t_1,\ldots,t_n)) = \sigma_A^s(V_{s(1)}^\alpha(t_1),\ldots,V_{s(n)}^\alpha(t_n)).$$

A common and useful notation for V^α is $\overline{\alpha} = \langle \overline{\alpha_s} : T(\Sigma,X)_s \to A \mid s \in S \rangle$.

Notice that if $X = \langle X_s \mid s \in S \rangle$ and each X_s is non-empty then each $T(\Sigma,X)_s$ is non-empty and so $T(\Sigma,X)$ forms an algebra. However, if for some $s \in S$, X_s is empty then $T(\Sigma,X)_s$ is non-empty if, and only if, there exists a closed term of sort s. In this case, recalling the discussion in Section 3.1, we say that Σ is *non-void in sort s*. It is common to work with an S-sorted signature Σ that is non-void in every sort $s \in S$.

Proposition 5.1.5. *For any S-sorted signature Σ with term algebra $T(\Sigma,X)$ over an S-indexed family X of sets of variable symbols, $T(\Sigma,X)$ is generated by X.*

Proof. Follows immediately from Definition 5.1.4 and Theorem 3.2.17. ∎

Theorem 5.1.6. *For any S-sorted signature Σ with term algebra $T(\Sigma,X)$ over an S-indexed family $X = \langle X_s \mid s \in S \rangle$ of sets of variable symbols, $T(\Sigma,X)$ is free in $Alg(\Sigma)$ on X.*

Proof. By Proposition 5.1.5, $T(\Sigma, X)$ is generated by X. For any Σ algebra A and S-indexed family of assignments $\alpha = \langle\, \alpha_s : X_s \to A_s \mid s \in S \,\rangle$, it is easily shown that the term evaluation mapping $\overline{\alpha} : T(\Sigma, X) \to A$, defined in Definition 5.1.4, is a homomorphism which extends α. Thus $\overline{\alpha}$ is the unique homomorphic extension of α by Lemma 3.4.8. Therefore $T(\Sigma, X)$ is free in $Alg(\Sigma)$ on X. ∎

Corollary 5.1.7. *For any S-sorted signature Σ with term algebra $T(\Sigma, X)$ over an S-indexed family X of sets of variables, and for any class K of Σ algebras, $T(\Sigma, X)$ is free for K on X. Thus for any $A \in K$ with $\mid A_s \mid \; \leq \; \mid X_s \mid$ for all $s \in S$, A is a quotient of $T(\Sigma, X)$.*

Proof. Exercise 5.1.19.(2). ∎

Although $T(\Sigma, X)$ is free for any class K of Σ algebras on X, it need *not* be the case that $T(\Sigma, X)$ is free *in* K on X. For example, for any class K of Σ algebras satisfying a non-trivial equation $t = t'$, i.e. t and t' are distinct terms, we have $T(\Sigma, X) \notin K$. Thus, by Theorem 5.1.6, $T(\Sigma, X)$ is free for K but not free in K. Let us consider the general problem of constructing a free algebra for a class K which is in K. We can construct free algebras, for any non-empty class K of algebras, by factoring a term algebra by a certain congruence constructed from K. This congruence can be defined in three different ways.

Definition 5.1.8.   Let Σ be an S-sorted signature with term algebra $T(\Sigma, X)$ over a family X of sets of variables. Let K be any non-empty class of Σ algebras. Define the following three S-indexed families of binary relations on the sets $T(\Sigma, X)_s$:

(i) $\equiv^K = \langle\, \equiv_s^K \mid s \in S \,\rangle$ where

$$\equiv_s^K = \bigcap \{\, \equiv_s^\phi \mid \phi : T(\Sigma, X) \to A \text{ is a } \Sigma \text{ homomorphism and}$$
$$A \in K \,\};$$

(ii) $\approx^K = \langle\, \approx_s^K \mid s \in S \,\rangle$ where

$$\approx_s^K = \bigcap \{\, \equiv_s \mid \equiv \text{ is a congruence on } T(\Sigma, X) \text{ and}$$
$$T(\Sigma, X)/\equiv \text{ is isomorphic to a subalgebra of some } A \in K \,\};$$

(iii) $\sim^K = \langle\, \sim_s^K \mid s \in S \,\rangle$, where

$$\sim_s^K = \{\, (t, t') \in T(\Sigma, X)_s^2 \mid \overline{\alpha}_s(t) = \overline{\alpha}_s(t') \text{ for all } A \in K \text{ and}$$
$$\alpha : X \to A \,\}.$$

Each of these three definitions simplifies the proof of a particular property of free algebras that will be required later on. First we verify that the three relations are identical Σ congruences on $T(\Sigma, X)$, and therefore give rise to three equivalent constructions of a free algebra.

Lemma 5.1.9. *For any non-empty class K of Σ algebras, the relations $\equiv^K$, $\approx^K$ and $\sim^K$ are identical Σ congruences on $T(\Sigma, X)$.*

Proof. By Lemma 3.3.17, both $\equiv^K$ and $\approx^K$ are well defined as Σ congruences on $T(\Sigma, X)$. Thus we need only prove the identity of $\equiv^K$, $\approx^K$ and $\sim^K$.

(i) First we show that $\equiv^K$ is equal to $\sim^K$. For any sort $s \in S$,

$$\sim^K_s = \{ \ (t, t') \in T(\Sigma, X)_s \ | \ t \equiv^{\overline{\alpha}} t' \text{ for all } A \in K \text{ and all}$$

$$\alpha : X \to A \ \} = \bigcap \{ \ \equiv^{\overline{\alpha}} \ | \ \alpha : X \to A \text{ for some } A \in K \ \}$$

$$= \bigcap \{ \ \equiv^{\phi} \ | \ \phi : T(\Sigma, X) \to A \text{ is a homomorphism for some}$$

$$A \in K \ \} = \equiv^K_s .$$

(ii) Now we show that $\equiv^K$ is equal to $\approx^K$. Suppose $\equiv^{\phi}$ is the kernel of a Σ homomorphism $\phi : T(\Sigma, X) \to A$ for some algebra $A \in K$. Then for the subalgebra $B \le A$ given by $B = \phi(T(\Sigma, X))$ the mapping

$$\phi : T(\Sigma, X) \to B$$

is a Σ epimorphism. By the First Homomorphism Theorem,

$$T(\Sigma, X)/ \equiv^{\phi} \ \cong \ B.$$

Hence $\equiv^{\phi} \in Con(T(\Sigma, X))$ and $T(\Sigma, X)/ \equiv^{\phi}$ is isomorphic to a subalgebra of some $B \in K$. Thus, $\approx^K$ is included in $\equiv^K$.

To prove the reverse inclusion, consider any congruence $\equiv$ on $T(\Sigma, X)$. Suppose that for some algebra $A \in K$ and subalgebra $B \le A$ we have an isomorphism $\psi : T(\Sigma, X)/ \equiv \ \to B$. Let

$$nat : T(\Sigma, X) \to T(\Sigma, X)/ \equiv$$

be the natural mapping associated with $\equiv$. Then

$$\psi \circ nat : T(\Sigma, X) \to B$$

is a Σ homomorphism by Lemma 3.4.10. Since ψ is a Σ isomorphism then

$$\equiv^{(\psi \circ nat)} \ = \ \equiv^{nat} \ = \ \equiv .$$

Hence $\equiv$ is the kernel of a Σ homomorphism $\phi : T(\Sigma, X) \to A$ for some $A \in K$. Thus, $\equiv^K$ is included in $\approx^K$, and so $\equiv^K$ and $\approx^K$ are identical. ∎

We can now construct free algebras for any non-empty class K of algebras as follows.

Definition 5.1.10. Let Σ be an S-sorted signature with term algebra $T(\Sigma, X)$ over an S-indexed family X of sets of variables. For any nonempty class K of Σ algebras define the Σ algebra $T_K(\Sigma, X)$ by

$$T_K(\Sigma, X) = T(\Sigma, X)/\equiv^K,$$

where $\equiv^K$ is the Σ congruence on $T(\Sigma, X)$ given by Definition 5.1.8. Let $\emptyset$ denote the S-indexed family of empty sets. We define $T_K(\Sigma) = T_K(\Sigma, \emptyset)$.

By Lemma 5.1.9, either one of the congruences, $\approx^K$ or $\sim^K$ gives rise to an identical algebra. This technical fact will be of use later on.

Theorem 5.1.11. *Let Σ be an S-sorted signature with term algebra $T(\Sigma, X)$ over the variables in $X = \langle X_s \mid s \in S \rangle$. Let K be any nonempty class of Σ algebras. Then $T_K(\Sigma, X)$ is free for K on $X/\equiv^K = \langle X_s/\equiv^K_s \mid s \in S \rangle$.*

Proof. By definition $T_K(\Sigma, X) = T(\Sigma, X)/\equiv^K$. Consider any algebra $A \in K$ and let $\alpha = \langle \alpha_s : X_s/\equiv^K_s \to A_s \mid s \in S \rangle$ be any S-indexed family of mappings. We show that there exists a unique homomorphic extension $\overline{\alpha} : T_K(\Sigma, X) \to A$ of α. Let

$$nat^K = \langle nat^K_s : X_s \to X_s/\equiv^K_s \mid s \in S \rangle$$

be the S-indexed family of natural mappings on X. Then we have an S-indexed assignment $\beta : X \to A$ given by $\beta = \alpha \circ nat^K$, with unique homomorphic extension $\overline{\beta} : T(\Sigma, X) \to A$. Define the map $\overline{\alpha} : T(\Sigma, X)/\equiv^K \to A$ by

$$\overline{\alpha}_s([t]_K) = \overline{\beta}_s(t).$$

To show that $\overline{\alpha}$ is well defined, consider any $s \in S$ and terms $t, t' \in T(\Sigma, X)_s$. If $t \equiv^K_s t'$ then since $\overline{\beta} : T(\Sigma, X) \to A$ is a homomorphic extension of β and $A \in K$, then

$$t \equiv^{\overline{\beta}}_s t'$$

by Definition 5.1.8, and so

$$\overline{\beta}_s(t) = \overline{\beta}_s(t').$$

Hence $\overline{\alpha}_s([t]_K) = \overline{\alpha}_s([t']_K)$ and so $\overline{\alpha}_s([t]_K)$ is uniquely defined.

Clearly $\overline{\alpha}_s$ agrees with α_s on each equivalence class of variables $[x]_K \in X_s/\equiv^K_s$, for each $s \in S$. Thus $\overline{\alpha}$ is an extension of α.

We show that $\overline{\alpha}$ is a Σ homomorphism as follows. For any sort $s \in S$ and constant symbol $c \in \Sigma_{\lambda, s}$,

$$\overline{\alpha}_s(c_{T_K(\Sigma, X)}) = \overline{\alpha}_s([c]_K) = \overline{\beta}_s(c) = c_A.$$

For any sort $s \in S$, any $w = s(1) \ldots s(n) \in S^+$, any function symbol $\sigma \in \Sigma_{w, s}$ and any $[t_i]_K \in T_K(\Sigma, X)_{s(i)}$, for $1 \le i \le n$, we have

$$\overline{\alpha}_s(\sigma_{T_K(\Sigma, X)}([t_1]_K, \ldots, [t_n]_K)) = \overline{\alpha}_s([\sigma(t_1, \ldots, t_n)]_K)$$

$$= \overline{\beta}_s(\sigma(t_1, \ldots, t_n)),$$

by definition of $\overline{\alpha}$,

$$= \sigma_A(\overline{\beta}_{s(1)}(t_1), \ldots, \overline{\beta}_{s(n)}(t_n))$$

since $\overline{\beta}$ is a homomorphism,

$$= \sigma_A(\overline{\alpha}_{s(1)}([t_1]_K), \ldots, \overline{\alpha}_{s(n)}([t_n]_K)),$$

by definition of $\overline{\alpha}$.

By Proposition 5.1.5, $T(\Sigma, X)$ is generated by X and so, by Lemma 3.3.16, $T(\Sigma, X)/ \equiv^K$ is generated by $X/ \equiv^K$. Now since the algebra $T(\Sigma, X)$ exists, we know that either Σ is non-void or X, and hence $X/ \equiv^K$, is non-empty. By Lemma 3.4.8, any homomorphism which agrees with $\overline{\alpha}$ on $X/ \equiv^K$ agrees with $\overline{\alpha}$. Hence $\overline{\alpha}$ is unique.

Since A and α were arbitrarily chosen, $T_K(\Sigma, X)$ is free for K on $X/ \equiv^K$. ∎

In the light of these facts we note that $T_K(\Sigma, X)$ is free for K on $X/ \equiv^K$, and that if $K' \subseteq K$ is any non-empty subclass of Σ algebras then $T_K(\Sigma, X)$ is free for K' on $X/ \equiv^K$. However, it need not be the case that $T_K(\Sigma, X) \cong T_{K'}(\Sigma, X)$, which is also free for K' on $X/ \equiv^{K'}$. This does not contradict the uniqueness of free algebras in K, up to isomorphism. We are simply forced to conclude that if $T_K(\Sigma, X) \not\cong T_{K'}(\Sigma, X)$ and $T_{K'}(\Sigma, X) \in K$ then $T_K(\Sigma, X)$ is not free *in* K'. For any class K, $T_K(\Sigma, X)$ is the free algebra for K which can be said to be 'closest' to lying in K. When is $T_K(\Sigma, X)$ actually a member of K? A general condition, which is sufficient but not necessary for $T_K(\Sigma, X)$ to be a member of K, can be given in terms of the closure conditions on classes of algebras introduced in Section 3.

Theorem 5.1.12 (G. Birkhoff 1935). *Let Σ be an S-sorted signature with term algebra $T(\Sigma, X)$ over an S-indexed family X of sets of variables.*

Let K be any non-empty class of Σ algebras. If K is closed under the formation of isomorphic images, subalgebras and non-empty direct products then

$$T_K(\Sigma,\, X) \in K.$$

Proof. By Lemma 5.1.9 and Definition 5.1.10,

$$T_K(\Sigma,\, X) = T(\Sigma, X)/ \equiv^K = T(\Sigma, X)/ \approx^K,$$

where

$$\approx^K = \bigcap C$$

and

$$C = \{\ \equiv \in Con(T(\Sigma, X))\ \mid$$

$T(\Sigma, X)/ \equiv$ is isomorphic to a subalgebra of some $A \in K$ $\}$.

Let I be any indexing set for the members of C so that $\approx^K = \cap_{i\in I} \equiv^i$. By Lemma 4.1.5, there exists a subdirect embedding

$$\epsilon : T(\Sigma, X)/ \approx^K \rightarrow \Pi_{i\in I}(T(\Sigma, X)/ \equiv^i).$$

Since K is closed under the formation of isomorphic images, subalgebras and non-empty direct products, we must have

$$\Pi_{i\in I}(T(\Sigma, X)/ \equiv^i) \in K.$$

Since $\epsilon\, (T(\Sigma, X)/ \approx^K) \leq \Pi_{i\in I}(T(\Sigma, X)/ \equiv^i)$ and K is closed under the formation of subalgebras, we have

$$\epsilon\, (T(\Sigma, X)/ \approx^K) \in K.$$

Since $T(\Sigma, X)/ \approx^K \cong \epsilon\, (T(\Sigma, X)/ \approx^K)$ and K is closed under isomorphism, it follows that

$$T_K(\Sigma,\, X) \in K.$$

∎

Thus if K satisfies the above conditions then K contains free algebras of all possible generator cardinalities.

Examples 5.1.13.

(i) Recall from Definition 3.5.15 that a class K of Σ algebras is said to be a variety if, and only if, K is closed under the formation of homomorphic images, subalgebras and direct products. Thus, by

Theorem 5.1.12, every variety contains free algebras of all possible generator cardinalities.

As we shall see in Section 5.2, varieties are precisely the classes of algebras which can be axiomatized by equations. Thus every class of algebras which can be equationally axiomatized contains free algebras of all possible generator cardinalities. In Section 5.3 we consider more general axiomatically defined classes of algebras containing free algebras.

By Theorem 5.1.3, we can speak of $T(\Sigma, X)$ as *the* free algebra for $Alg(\Sigma)$ of generator cardinality $\mid X_s \mid$ for each sort $s \in S$. The ground term algebra $T(\Sigma)$ occupies a unique position among all the term algebras. For any term algebra $T(\Sigma, X)$, given $\alpha : X \rightarrow A$, there is a unique homomorphic extension $\overline{\alpha} : T(\Sigma, X) \rightarrow A$. However, unless X is the S-indexed family of empty sets, there will in general be many different $\alpha : X \rightarrow A$, giving rise to many different homomorphisms $\overline{\alpha} : T(\Sigma, X) \rightarrow A$. For $T(\Sigma)$ there is one, and only one, homomorphism to any $A \in Alg(\Sigma)$. This motivates the following definition.

Definition 5.1.14. Let Σ be an S-sorted signature. Let K be any class of Σ algebras and A be any Σ algebra.

(i) We say that A is *initial* for K if, and only if, for every $B \in K$ there exists a unique Σ homomorphism from A to B.

(ii) We say that A is *final* for K if, and only if, for every $B \in K$ there exists a unique Σ homomorphism from B to A.

If A is initial (respectively final) for K and $A \in K$ then we say that A is initial (respectively final) *in K*.

We leave it to the reader, as an exercise, to check that initial and final algebras in any class K are unique up to isomorphism (Exercise 5.1.19.(4)).

Examples 5.1.15.

(i) For any S-sorted signature Σ with ground term algebra $T(\Sigma)$, $T(\Sigma)$ is initial in $Alg(\Sigma)$. Hence if K is any class of Σ algebras then $T(\Sigma)$ is initial for K.

(ii) Recall the algebra $Succ(\mathbf{N}) = (\mathbf{N}, 0, succ)$ with signature Σ^{Succ} of Example 3.1.4.(iii). Consider $T(\Sigma^{Succ})$. This algebra has as carrier set

$$\{ \ succ^n(0) \mid n = 0, 1, 2, \dots \ \}$$

with constant 0 and operation

$$succ_{T(\Sigma^{Succ})}(\ succ^n(0)\ ) = succ^{n+1}(0).$$

It is easy to see $T(\Sigma^{Succ}) \cong Succ(\mathbf{N})$ under the map $\phi : Succ(\mathbf{N}) \to T(\Sigma^{Succ})$ given by

$$\phi(n) = succ^n(0).$$

Thus $Succ(\mathbf{N})$ is also initial in $Alg(\Sigma^{Succ})$.

(iii) Recall the algebra $PA(\mathbf{N}) = (\mathbf{N},\ 0,\ Succ,\ Add,\ Mult)$ with signature Σ^{PA} of Example 3.1.4.(iii). Consider the class $K = Alg(\Sigma^{PA},\ E)$ of all Σ^{PA} algebras satisfying the set E of equations consisting of

$$add(0,\ x) = x$$

$$add(x,\ succ(y)) = succ(add(x,\ y))$$

$$mult(x,\ 0) = 0$$

$$mult(x,\ succ(y)) = add(mult(x,\ y),\ x).$$

Clearly Σ^{PA} is non-void and the algebra $PA(\mathbf{N})$ satisfies these axioms. We will show that $T_K(\Sigma^{PA}) \cong PA(\mathbf{N})$.

Since both algebras are minimal, by Theorem 3.4.13, it is enough to show the existence of homomorphisms $\phi : T_K(\Sigma^{PA}) \to PA(\mathbf{N})$ and $\psi : PA(\mathbf{N}) \to T_K(\Sigma^{PA})$. We know that by construction $T_K(\Sigma^{PA})$ is free for K and since $PA(\mathbf{N})$ is in K there exists a homomorphism $\phi : T_K(\Sigma^{PA}) \to PA(\mathbf{N})$. Now define $\psi : PA(\mathbf{N}) \to T_K(\Sigma)$ by

$$\psi(n) = [succ^n(0)]_K.$$

It is left to the reader, as an exercise, to show that ψ is a homomorphism.

(iv) Notice $T(\Sigma^{PA})$ is initial in $Alg(\Sigma^{PA})$ and therefore is initial for $K = Alg(\Sigma^{PR},\ E)$ in (iii). However $T(\Sigma^{PA}) \not\cong PA(\mathbf{N})$.

(v) For any S-sorted signature Σ, each unit Σ algebra is final in $Alg(\Sigma)$.

(vi) Recall the algebra $Count(\mathbf{N}) = (\mathbf{N},\ 0,\ Succ,\ Pred)$ with signature Σ^{Count} of Example 3.1.4.(iii). This is both initial and final for $K = Alg(\Sigma^{Count},\ E)$ where E consists of the single equation

$$pred(succ(x)) = x.$$

The proof is left to the reader as an exercise.

The fact that $T(\Sigma)$ is initial for $Alg(\Sigma)$ can be strengthened as follows.

Theorem 5.1.16. *Let* Σ *be any non-void signature and* K *be any non-empty class of* Σ *algebras. If* K *is closed under the formation of isomorphic images, subalgebras and non-empty direct products then* $T_K(\Sigma)$ *is initial in* K.

Proof. Immediate from Theorems 5.1.11 and 5.1.12. ∎

Recall the concept of a *simple* algebra, introduced in Definition 3.3.11. This concept can be relativized to any non-empty class K of algebras, to provide a characterization of final algebras using the concept of a K congruence introduced in Definition 4.1.8.

Definition 5.1.17. *Let* Σ *be an* S-*sorted signature,* K *be any class of* Σ *algebras and* A *be any* Σ *algebra. Then* A *is said to be* K-*simple if, and only if, there is no* K *congruence on* A *other than possibly equality or the unit congruence.*

Theorem 5.1.18. *Let* Σ *be a non-void* S-*sorted signature. Let* K *be any class of* Σ *algebras, closed under isomorphism, and let* F *be a minimal algebra which is final in* K. *Then for any algebra* $A \in K$, *if* $A \cong F$ *then* A *is* K-*simple. Furthermore if* F *is not a unit algebra then the converse holds.*

Proof. Suppose $A \cong F$ and that $\psi : F \to A$ is an isomorphism. Suppose for a contradiction that A is not K-simple. Then there exists a K congruence $\equiv$ on A, which is neither equality nor the unit congruence.

Since F is final in K and $A/\equiv \in K$, there exists a Σ homomorphism $\phi : A/\equiv \to F$ and hence a homomorphism $\psi \circ \phi : A/\equiv \to A$. Furthermore the natural mapping $nat : A \to A/\equiv$ is a homomorphism. Since F is minimal so is A, hence so is $A/\equiv$. So $A \cong A/\equiv$ by Lemma 3.4.13. Therefore, by Proposition 3.4.19, $\equiv$ is the equality relation on A which contradicts the assumption that $\equiv$ is not equality. So A is K-simple.

For the converse, suppose F is not a unit algebra and that A is K-simple. Since F is final in K, there exists a Σ homomorphism $\phi : A \to F$. Since F is minimal, ϕ is surjective. Consider the kernel $\equiv$ of ϕ. Since $F \in K$ and $F \cong A/\equiv$ and K is closed under isomorphism, $A/\equiv \in K$. Thus $\equiv$ is a K-congruence. Since $F \cong A/\equiv$ is not a unit algebra but A is K-simple, $\equiv$ must be equality on A. So $A \cong A/\equiv \cong F$. ∎

Note that in the previous theorem, if F is a unit algebra and A is K-simple then it need not be the case that $A \cong F$, since A could have arbitrary cardinality depending upon K.

Exercises 5.1.19.

1. Let Σ be an S-sorted signature and let K be any class of Σ algebras. If F is free for K on $X \subseteq F$, is F generated by X?

2. Let A be a Σ algebra. Construct a term algebra $T(\Sigma, X)$ and an epimorphism $\phi : T(\Sigma, X) \to A$ so that $A \cong T(\Sigma, X)/\equiv^\phi$. To what extent is ϕ unique?

3. Let Σ be an S-sorted signature which is void in some sort $s \in S$. Let X be an S-indexed family of sets of variables and let K be any non-empty class of Σ algebras. Is $T_K(\Sigma, X)$ free for K on $X/\equiv^K$?

4. Let Σ be an S-sorted signature. Let K be any class of Σ algebras. Let I and I' be initial in K and let F and F' be final in K. Show that $I \cong I'$ and $F \cong F'$. What happens if the algebras do not belong to K?

5. For any class K of Σ algebras and any Σ algebra A, if A is free for K on the S-indexed family $\emptyset$ of empty sets then A is initial for K. Does the converse hold?

6. Complete the proof that $T_K(\Sigma^{PA}) \cong PA(\mathbf{N})$ for Example 5.1.15.(iii).

7. Recalling Example 5.1.15.(iii), add each of the following functions to the algebra $PA(\mathbf{N})$ to make a new algebra A, and give equations over a suitable signature such that A is isomorphic to the initial algebra of the class of all algebras satisfying the equations:

   (a) $Square(x) = x^2$;

   (b) $Exp(x, y) = x^y$;

   (c) $Fact(x) = x \cdot (x - 1) \cdot \ldots \cdot 1$.

   (Hint: in cases (ii) and (iii) what are the recursive definitions of these functions?)

8. Can one define a system of equations such that the algebra

$$(\mathbf{N}; 0; Succ, Square)$$

   is an initial algebra in the class of all algebras satisfying the equations? Can this system be finite?

9. Show that $Count(\mathbf{N})$ of Example 5.1.15.(vi) is both initial and final in $K = Alg(\Sigma^{Count}, E)$. (Hint: use Theorem 5.1.18.)

10. Consider the signature Σ^R for rings with 1 of Example 3.1.2.(ii), and let E be the set of ring axioms and $K = Alg(\Sigma^R, E)$ be the class

of all rings with 1. Let $X = \{\, x_1, \ldots, x_n \,\}$ be a set of n variables. Show that $T(\Sigma^R, X)/ \equiv^K$ is isomorphic with the ring $\mathbf{Z}[x_1, \ldots, x_n]$ of polynomials in n indeterminates.

5.2 Equational logic

In this section, and in Section 5.3, we consider three important fragments of first order logic which can be studied using the algebraic methods and results developed so far in this chapter of the Handbook. The fragment of first order logic we shall study in this section is known as *equational logic*. It is the simplest fragment of first order logic with equality.

We begin by defining the concept of an equation over a many-sorted signature Σ.

Definition 5.2.1. Let Σ be an S-sorted signature and $X = \langle\, X_s \mid s \in S \,\rangle$ be an S-indexed family of sets of variable symbols satisfying the usual assumption that for each sort $s \in S$ the sets X_s and $\Sigma_{\lambda,s}$ are disjoint. By an *equation* or *identity* of sort $s \in S$ over Σ and X we mean an expression of the form

$$t = t',$$

where $t, t' \in T(\Sigma, X)_s$ are terms of the same sort $s \in S$. We let $Eqn(\Sigma, X)_s$ denote the set of all equations over Σ and X of sort $s \in S$, and we let $Eqn(\Sigma, X) = \cup_{s \in S} Eqn(\Sigma, X)_s$. An equation $t = t' \in Eqn(\Sigma, X)_s$ is said to be a *ground* or *closed equation* if, and only if, t and t' are ground terms, i.e. $t, t' \in T(\Sigma)_s$ are variable free terms. We let $Eqn(\Sigma)_s = Eqn(\Sigma, \emptyset)_s$ denote the set of all ground equations over Σ of sort s. By an *equational theory* over Σ we mean a set of equations $E \subseteq Eqn(\Sigma, X)$; an element $e \in E$ is sometimes termed an *equational axiom*.

Much of the significance of equational logic comes from the large number of interesting equational theories which arise in mathematics and computer science.

Examples 5.2.2.

(i) Recall the single-sorted signatures Σ^{SG}, Σ^G and Σ^R of Example 3.1.2.(ii), for semigroups, groups and rings with unity, respectively.

The equational theory *Semigroup* of semigroups consists of the single equation

$$x + (y + z) = (x + y) + z.$$

The equational theory *Group* of groups consists of *Semigroup* together with the equations

$$x + 0 = x \quad 0 + x = x$$
$$x + (-x) = 0 \quad (-x) + x = 0.$$

The equational theory *AbGroup* of Abelian groups consists of *Group* augmented by the equation

$$x + y = y + x.$$

The equational theory *Ring* for rings with unity consists of *AbGroup* together with the equations

$$x \cdot (y \cdot z) = (x \cdot y) \cdot z,$$

$$x \cdot (y + z) = (x \cdot y) + (x \cdot z), \quad (x + y) \cdot z = (x \cdot z) + (y \cdot z),$$

$$x \cdot 1 = x, \quad 1 \cdot x = x.$$

The equational theory *ComRing* for commutative rings with unity consists of *Ring* together with the equation

$$x \cdot y = y \cdot x.$$

(ii) Recall from Example 3.1.2.(iii) the single-sorted signatures Σ^L and Σ^{Bool} for lattice and Boolean algebras. The equational theory *Latt* of lattices consists of the equations

$$x \vee y = y \vee x, \quad x \wedge y = y \wedge x,$$

$$x \vee (y \vee z) = (x \vee y) \vee z, \quad x \wedge (y \wedge z) = (x \wedge y) \wedge z,$$

$$x \vee x = x, \quad x \wedge x = x,$$

$$x \vee (x \wedge y) = x, \quad x \wedge (x \vee y) = x.$$

The equational theory *DistLatt* of distributive lattices consists of *Latt* together with the equations

$$x \wedge (y \vee z) = (x \wedge y) \vee (x \wedge z), \quad x \vee (y \wedge z) = (x \vee y) \wedge (x \vee z).$$

The equational theory *Bool* of Boolean algebras consists of *DistLatt* together with the equations

$$x \wedge 0 = 0, \quad x \vee 1 = 1,$$

$$x \wedge (\neg x) = 0, \quad x \vee (\neg x) = 1.$$

Next we define the notions of *truth* and *validity* for equations with respect to an algebra A, and the satisfaction relation $\models$ between algebras or classes of algebras, and equations or sets of equations.

Definition 5.2.3. Let Σ be an S-sorted signature with term algebra $T(\Sigma, X)$ over an S-indexed family X of sets of variables. Let A be a Σ algebra. An *assignment* from X to A is an S-indexed family of mappings

$$\alpha : X \to A = \langle\, \alpha_s : X_s \to A_s \mid s \in S \,\rangle.$$

For any assignment $\alpha : X \to A$, recall from Definition 5.1.4 the *term evaluation map* $\overline{\alpha} : T(\Sigma, X) \to A$. By Theorem 5.1.6, $\overline{\alpha}$ is the unique homomorphic extension of α.

An equation $t = t' \in Eqn(\Sigma, X)_s$ is said to be *true in A* under an assignment $\alpha : X \to A$ if, and only if,

$$\overline{\alpha}_s(t) = \overline{\alpha}_s(t').$$

We write $(A, \alpha) \models t = t'$ if $t = t'$ is true in A under the assignment α. We say that $t = t'$ is *valid* in A and write $A \models t = t'$ if, and only if, $t = t'$ is true in A under every assignment $\alpha : X \to A$. If $t = t'$ is not valid in A then we write $A \not\models t = t'$. If E is any set of equations then we write $A \models E$ if $A \models e$ for each equation $e \in E$. If $A \models E$ then we say that A is a *model* of E or a (Σ, E) *algebra*. We let

$$Alg(\Sigma, E) = \{\, A \in Alg(\Sigma) \mid A \models E \,\}$$

denote the class of all models of E.

If K is any class of Σ algebras then we write $K \models t = t'$ if $A \models t = t'$ for each algebra $A \in K$. We write $K \models E$ if $K \models e$ for each equation $e \in E$. We let

$$Eqn_K(\Sigma, X) = \{\, e \in Eqn(\Sigma, X) \mid K \models e \,\}$$

denote the *equational theory of K*.

The satisfaction relation $\models$ for equations leads to new and important classes of algebras, namely those classes that are axiomatizable by equations.

Definition 5.2.4. Let Σ be an S-sorted signature. A class K of Σ algebras is an *equational class* if, and only if, K is the class of all models of an equational theory $E \subseteq Eqn(\Sigma, X)$, i.e.

$$K = Alg(\Sigma, E).$$

In these circumstances E is said to be an *equational axiomatization* over Σ of the class K.

Examples 5.2.5.

(i) For any S-sorted signature Σ, the class $Alg(\Sigma)$ of *all* Σ algebras is an equational class, since the empty theory $\emptyset$ is an equational axiomatization of $Alg(\Sigma)$.

(ii) The class of all semigroups is the equational class $Alg(\Sigma^{SG}, Semigroup)$. Similarly, the classes of all groups, Abelian groups, rings with unity and commutative rings with unity are the equational classes $Alg(\Sigma^G, Group)$, $Alg(\Sigma^G, AbGroup)$, $Alg(\Sigma^R, Ring)$ and $Alg(\Sigma^R, ComRing)$ respectively.

(iii) The classes of all lattices, distributive lattices and Boolean algebras are the equational classes $Alg(\Sigma^L, Latt)$, $Alg(\Sigma^L, DistLatt)$ and $Alg(\Sigma^{Bool}, Bool)$ respectively.

(iv) The class of all integral domains and the class of fields are not equational classes. The class of all torsion-free Abelian groups and the class of all cancellation semigroups are not equational classes. (See Exercise 5.3.28.(1).)

In this section we shall study two basic topics in equational logic. The first of these concerns a standard question in mathematical logic: *does there exist a logical calculus for equations, i.e. a deductive system of inference rules, which allows us to generate from any equational theory E precisely those equations which are valid in all models of E?* Such a calculus, if it exists, is said to be both *sound* (meaning no invalid equations are derived) and *complete* (meaning all valid equations are derived).

The many-sorted first order predicate calculus with equality provides one such example of a sound and complete calculus, by the Completeness Theorem of K. Gödel and L. Henkin. However, from the point of view of algebra, this solution is unsatisfactory. The inference rules of the predicate calculus with equality refer to first order formulas which are not equations, while the proof of the Completeness Theorem uses methods that are not purely algebraic.

We shall study a much simpler inference system known as the *equational calculus*, which uses only the language of equational formulas and which can be proved sound and complete by purely algebraic means. This calculus is based on the principles that equality is a reflexive, symmetric and transitive relation and that we may substitute equals for equals.

To provide a notation for substitutions we make the following definition.

Definition 5.2.6.   Let Σ be an S-sorted signature with term algebra $T(\Sigma, X)$ over an S-indexed family $X = \langle\, X_s \mid s \in S \,\rangle$ of sets X_s of variable symbols of sort $s \in S$. For any sort $s' \in S$, any variable symbol $x \in X_{s'}$ and any term $t_0 \in T(\Sigma, X)_{s'}$ of sort s', there is a unique assignment $\alpha : X \to T(\Sigma, X)$ satisfying, for each $s \in S$ and $y \in X_s$,

$$\alpha_s(y) = \begin{cases} t_0, & \text{if } s = s' \text{ and } y = x; \\ y, & \text{otherwise.} \end{cases}$$

This assignment α has a unique homomorphic extension $\overline{\alpha} : T(\Sigma, X) \to T(\Sigma, X)$. Let $s \in S$ be any sort and $t \in T(\Sigma, X)_s$ be any term of sort $s \in S$. Then $\overline{\alpha}_s(t)$ is the term obtained by *substituting* the term t_0 for every occurrence of the variable symbol x in the term t. It is conventional to denote the term $\overline{\alpha}_s(t)$ by $t[x/t_0]$ and we speak of *substituting t_0 for x in t*.

We are now ready to give the ordinary deduction rules of equational logic. At this point we will assume that all signatures are non-void. Most examples of signatures and equational theories satisfy this hypothesis. Without this assumption the ordinary deduction rules can still be applied but may not be sound in the case of two or more sorts: see Exercises 5.2.20.(2), (3), (4) and (5) for an analysis of this situation.

Definition 5.2.7. The *ordinary deduction rules of equational logic* are the following.

(i) For any equation $t = t \in Eqn(\Sigma, X)$,

$$\overline{\emptyset \vdash t = t}$$

is a *reflexivity rule*.

(ii) For any $E \subseteq Eqn(\Sigma, X)$, any sort $s \in S$ and any terms $t_0, t_1 \in T(\Sigma, X)_s$,

$$\frac{E \vdash t_0 = t_1}{E \vdash t_1 = t_0}$$

is a *symmetry rule*.

(iii) For any $E_0, E_1 \subseteq Eqn(\Sigma, X)$, any sort $s \in S$ and any terms $t_0, t_1, t_2 \in T(\Sigma, X)_s$,

$$\frac{E_0 \vdash t_0 = t_1, \quad E_1 \vdash t_1 = t_2}{E_0 \cup E_1 \vdash t_0 = t_2}$$

is a *transitivity rule*.

(iv) For any $E_0, E_1 \subseteq Eqn(\Sigma, X)$, each sort $s \in S$, any terms $t, t' \in T(\Sigma, X)_s$, any sort $s' \in S$, any variable $x \in X_{s'}$ and any terms $t_0, t_1 \in T(\Sigma, X)_{s'}$,

$$\frac{E_0 \vdash t = t', \quad E_1 \vdash t_0 = t_1}{E_0 \cup E_1 \vdash t[x/t_0] = t'[x/t_1]}$$

is a *substitution rule*.

Note that unlike the deduction rules of the predicate calculus, the deduction rules of the equational calculus refer only to equations in their assumptions and conclusions.

Let us now make precise what we mean by a *proof* using the rules of equational deduction.

Definition 5.2.8. Let Σ be an S-sorted signature and $E \subseteq Eqn(\Sigma, X)$ be an equational theory. By a *proof tree* (or simply *proof*) using E and the ordinary deduction rules of equational logic, we mean a non-empty finite tree P with each node n labelled by a pair (E_n, e_n) consisting of a finite set E_n of equations and an equation e_n. For each node n in P, either: (i) n has no antecedent nodes in P and $E_n = \{ e_n \}$ and $e_n \in E$ is an axiom; or, (ii) n has exactly k antecedent nodes $n_1, \dots, n_k$ in P labelled $(E_{n_1}, e_{n_1}), \dots, (E_{n_k}, e_{n_k})$ respectively, and

$$\frac{E_{n_1} \vdash e_{n_1}, \dots, E_{n_k} \vdash e_{n_k}}{E_n \vdash e_n}$$

is an ordinary deduction rule of equational logic. If the proof tree P has root r labelled by (E_r, e_r), then we say that P is a *proof of e_r using E and the ordinary deduction rules of equational logic*.

Notice that if P is a proof with root (E_r, e_r) then P is a proof of e_r using the finite subset $E_r \subseteq E$. Examples of formal proofs using the deduction rules of equational logic occur in Exercises 5.2.20.

Definition 5.2.9. The rules of equational logic induce an *inference relation* $\vdash$ between equational theories $E \subseteq Eqn(\Sigma, X)$ and equations $e \in Eqn(\Sigma, X)$, defined by $E \vdash e$ if, and only if, there exists a proof of e using E and the ordinary deduction rules of equational logic.

As noted above, if $E \vdash e$ then for some finite subset $E' \subseteq E$ we have $E' \vdash e$ since only finitely many of the axioms of E are used in a proof of e.

The first main result of this section states that the equational calculus is sound and complete. We shall give a simple algebraic proof due to Birkhoff [1935]. To prove the completeness of the equational calculus, we will show that the inference relation $\vdash$ induces a congruence called *provable equivalence* on the term algebra $T(\Sigma, X)$.

Definition 5.2.10. Let Σ be an S-sorted signature and X be an S-indexed family of sets of variables. Let E be any equational theory over Σ and

X. Define the S-indexed family $\equiv^E = \langle \equiv^E_s \mid s \in S \rangle$ of binary relations $\equiv^E_s \subseteq T(\Sigma, X)^2_s$ by

$$t \equiv^E_s t' \Leftrightarrow E \vdash t = t',$$

for each sort $s \in S$ and terms $t, t' \in T(\Sigma, X)_s$. We term $\equiv^E$ the relation of *provable equivalence* on $T(\Sigma, X)$ induced by E.

Lemma 5.2.11. *Let Σ be an S-sorted signature with term algebra $T(\Sigma, X)$ over an S-indexed family X of sets of variables. For any equational theory E over Σ and X, the relation $\equiv^E$ of provable equivalence is a Σ congruence on $T(\Sigma, X)$.*

Proof. The deduction rules of reflexivity, symmetry and transitivity ensure that $\equiv^E$ is an equivalence relation. The substitution rule ensures that $\equiv^E$ satisfies the substitutivity condition and is therefore a Σ congruence. ∎

Theorem 5.2.12 (Soundness and Completeness Theorem (G. Birkhoff 1935)). *Let Σ be a non-void S-sorted signature and $X = \langle X_s \mid s \in S \rangle$ be an S-indexed family of sets of variables. Let E be an equational theory over Σ and X. For any equation $e \in Eqn(\Sigma, X)$,*

$$E \vdash e \Leftrightarrow Alg(\Sigma, E) \models e.$$

Proof. $\Rightarrow$ Exercise 5.2.20.(1).

$\Leftarrow$ Consider the quotient algebra $T(\Sigma, X)/ \equiv^E$ any sort $s \in S$ and any equation $t = t' \in Eqn(\Sigma, X)$ of sort s. We first establish that

$$E \vdash t = t' \Leftrightarrow T(\Sigma, X)/ \equiv^E \models t = t' \tag{1}.$$

First suppose $E \vdash t = t'$. Any assignment $\alpha : X \to T(\Sigma, X)/ \equiv^E$ can be factored (not necessarily uniquely) into a composition

$$nat^E \circ \alpha' = \alpha,$$

where $nat^E : T(\Sigma, X) \to T(\Sigma, X)/ \equiv^E$ is the natural mapping and $\alpha' : X \to T(\Sigma, X)$ is an assignment. This composition extends uniquely to a Σ homomorphism

$$\overline{nat^E \circ \alpha'} : T(\Sigma, X) \to T(\Sigma, X)/ \equiv^E$$

identical to the term evaluation map $\overline{\alpha} : T(\Sigma, X) \to T(\Sigma, X)/ \equiv^E$ derived from α. Furthermore,

$$\overline{nat^E \circ \alpha'} = nat^E \circ \overline{\alpha'}$$

since both homomorphisms agree on the family X of generators for $T(\Sigma, X)$. So

$$\overline{\alpha}_s(t) = \overline{nat^E \circ \alpha'}_s(t) = nat_s^E \circ \overline{\alpha'}_s(t) = nat_s^E(\overline{\alpha'}_s(t))$$

and similarly

$$\overline{\alpha}_s(t') = nat_s^E(\overline{\alpha'}_s(t')).$$

Now by assumption, $E \vdash t = t'$ and for all $x \in X_s$, $E \vdash \overline{\alpha'}_s(x) = \overline{\alpha'}_s(x)$ by reflexivity. So by the substitution rule $E \vdash \overline{\alpha'}_s(t) = \overline{\alpha'}_s(t')$ and hence $nat_s^E(\overline{\alpha'}_s(t)) = nat_s^E(\overline{\alpha'}_s(t'))$. Therefore $\overline{\alpha}_s(t) = \overline{\alpha}_s(t')$. Since α was arbitrarily chosen,

$$T(\Sigma, X)/\equiv^E \models t = t'.$$

Next suppose $E \not\vdash t = t'$. Then $t \not\equiv_s^E t'$. Consider the family $id = \langle\, id_s : X_s \to X_s \mid s \in S \,\rangle$ of identity mappings. This induces an assignment

$$nat^E \circ id : X \to T(\Sigma, X)/\equiv^E$$

with unique homomorphic extension

$$\overline{nat^E \circ id} : T(\Sigma, X) \to T(\Sigma, X)/\equiv^E.$$

Now $\overline{nat^E \circ id} = nat^E \circ \overline{id}$ by Lemma 3.4.8, since Σ is non-void and both homomorphisms agree on the family X of generators. So

$$\overline{nat^E \circ id}_s(t) = nat_s^E \circ \overline{id}_s(t) = nat_s^E(\overline{id}_s(t)) = nat_s^E(t)$$

and similarly

$$\overline{nat^E \circ id}_s(t') = nat_s^E(t').$$

Since $t \not\equiv_s^E t'$, it follows that $\overline{nat^E \circ id}_s(t) \neq \overline{nat^E \circ id}_s(t')$. Therefore

$$T(\Sigma, X)/\equiv^E \not\models t = t'.$$

So (1) above holds.

Since $E \vdash e$ for each $e \in E$, by (1) we have $T(\Sigma, X)/\equiv^E \models E$ and thus $T(\Sigma, X)/\equiv^E \in Alg(\Sigma, E)$. Furthermore, if $E \not\vdash e$ then by the equivalence (1) above $T(\Sigma, X)/\equiv^E \not\models e$, and hence $Alg(\Sigma, E) \not\models e$. ∎

The Soundness and Completeness Theorem for the equational calculus reveals the interesting and useful fact that the more powerful inference rules of the predicate calculus with equality generate no equations from an equational theory which cannot already be derived by equational reasoning alone. This fact makes the much simpler equational calculus an interesting and attractive calculus to try to automate by mechanical theorem proving techniques. Such attempts have been fairly successful; some of the techniques used are described elsewhere in this Handbook, for example in the chapter on Term Rewriting, volume 2.

The second topic in equational logic that we wish to consider concerns *equational axiomatizability*. In particular we shall address the question: *which classes of algebras can be axiomatized by equational means alone?* We shall present a characterization result for equational classes known as the Variety Theorem, which is again from Birkhoff [1935]. The Variety Theorem 5.2.16 characterizes equational classes of algebras in terms of the closure operations on classes of algebras introduced in Section 3. Recall the concept of a variety.

Definition 5.2.13. Let Σ be an S-sorted signature. A class K of Σ algebras is said to be a *variety* if, and only if, K is closed under the formation of subalgebras, homomorphic images and direct products.

The two key technical ideas in the proof of the Variety Theorem are that an equational class K contains algebras free for K, and that K is closed under the formation of homomorphic images. In fact K satisfies rather more than the sufficient conditions of Theorem 5.1.12.

Theorem 5.2.14. *Let Σ be an S-sorted signature. Every equational class $K \subseteq Alg(\Sigma)$ is closed under the formation of subalgebras, homomorphic images and direct products. Thus every equational class is a variety.*

Proof. By assumption, $K = Alg(\Sigma, E)$ for some equational theory $E \subseteq Eqn(\Sigma, X)$ over variable family X. Without loss of generality we may assume that the algebra $T(\Sigma, X)$ exists, i.e. for each sort $s \in S$, either X_s is non-empty or Σ is non-void in sort s.

(i) Consider any algebra $A \in K$ and any Σ subalgebra $B \leq A$. For any assignment $\alpha : X \to B$, we have $\alpha : X \to A$. So for any equation $t = t' \in E$ of sort $s \in S$, since $A \models E$,

$$\overline{\alpha}_s(t) = \overline{\alpha}_s(t').$$

Hence $B \models t = t'$. Since $t = t'$ was arbitrarily chosen, $B \models E$. Since A and B were arbitrarily chosen, K is closed under the formation of subalgebras.

(ii) Consider any $A \in K$, any $B \in Alg(\Sigma)$ and any Σ epimorphism $\phi : A \to B$. For any assignment $\beta : X \to B$ consider an assignment $\alpha : X \to A$ such that
$$\phi \circ \alpha = \beta.$$

Such an assignment exists since ϕ is surjective. Since $T(\Sigma, X)$ is generated by X, then by Lemma 3.4.8,

$$\overline{\beta} = \overline{\overline{\phi} \circ \alpha} = \phi \circ \overline{\alpha}.$$

So for any equation $t = t' \in E$ of sort $s \in S$,

$$\overline{\alpha}_s(t) = \overline{\alpha}_s(t').$$

Hence

$$\phi_s \circ \overline{\alpha}_s(t) = \phi_s \circ \overline{\alpha}_s(t').$$

Thus

$$\overline{\phi \circ \alpha}_s(t) = \overline{\phi \circ \alpha}_s(t'),$$

so that

$$\overline{\beta}_s(t) = \overline{\beta}_s(t').$$

Since β was arbitrarily chosen, $B \models t = t'$. Since $t = t'$ was arbitrarily chosen, $B \models E$. Since A, B and ϕ were arbitrarily chosen, K is closed under the formation of homomorphic images.

(iii) Let I be any indexing set and

$$A = \langle\, A(i) \in K \mid i \in I \,\rangle,$$

be any I-indexed family of K algebras. Suppose $A(i) \models t = t'$ for each $i \in I$. For any assignment $\alpha : X \to \Pi A$ define the I-indexed family of assignments $\langle\, \alpha^i : X \to A(i) \mid i \in I \,\rangle$ by

$$\alpha_s^i(x) = \alpha_s(x)(i),$$

for each sort $s \in S$ and each $x \in X_s$. For any equation $t = t' \in E$ of sort $s \in S$, and for each $i \in I$,

$$\overline{\alpha}_s(t)(i) = U_s^i \circ \overline{\alpha_s}(t) = \overline{\alpha^i}_s(t), \quad \overline{\alpha}_s(t')(i) = U_s^i \circ \overline{\alpha_s}(t') = \overline{\alpha^i}_s(t')$$

by Lemma 3.4.8, since for each sort $s \in S$, either X_s is non-empty or Σ is non-void in sort s, and $\overline{\alpha}_s$ and $U_s^i \circ \overline{\alpha_s}$ agree on X_s. Since $A(i) \models t = t'$ then

$$\overline{\alpha^i}_s(t) = \overline{\alpha^i}_s(t')$$

for each $i \in I$. So

$$\overline{\alpha}_s(t)(i) = \overline{\alpha}_s(t')(i)$$

for all $i \in I$. Hence

$$\overline{\alpha}_s(t) = \overline{\alpha}_s(t').$$

Since α was arbitrarily chosen

$$\Pi A \models t = t',$$

and since $t = t'$ was arbitrarily chosen $\Pi A \models E$. Since I and A were arbitrarily chosen, K is closed under the formation of direct products.

■

By Theorem 5.1.12, we have:

Corollary 5.2.15. *Let Σ be an S-sorted signature and let E be an equational theory over Σ. Then:*

(i) *The class $K = Alg(\Sigma, E)$ contains the free algebras $T_K(\Sigma, X)$ for any S-indexed family $X = \langle X_s \mid s \in S \rangle$ of sets of variables with $\mid X_s \mid \geq 1$ for all $s \in S$.*

(ii) *If Σ is non-void then the class $K = Alg(\Sigma, E)$ contains an initial algebra $T_K(\Sigma)$.*

If Σ is non-void then the algebra $T_K(\Sigma)$ in an equational class $K = Alg(\Sigma, E)$ will be denoted by $T(\Sigma, E)$.

The Variety Theorem establishes that the converse of Theorem 5.2.14 holds.

Theorem 5.2.16 (Variety Theorem. (G. Birkhoff 1935)). *Let Σ be an S-sorted signature and let K be any class of Σ algebras. Then K is a variety if, and only if, K is an equational class.*

Proof. $\Leftarrow$ Immediate from Theorem 5.2.14.

$\Rightarrow$ Suppose K is a variety. Then K is non-empty since K contains the empty product $\Pi\emptyset$. Consider any S-indexed family $X = \langle X_s \mid s \in S \rangle$ of *infinite* sets X_s of variables and the class

$$K^* = Alg(\Sigma, Eqn_K(\Sigma, X)).$$

Clearly for every algebra $A \in K$, $A \models Eqn_K(\Sigma, X)$ and so

$$K \subseteq K^* \tag{1}.$$

Thus it suffices to show that $K^* \subseteq K$.

Consider any S-indexed family $Y = \langle Y_s \mid s \in S \rangle$ of sets Y_s of variables. We first show

$$Eqn_K(\Sigma, Y) = Eqn_{K^*}(\Sigma, Y) \tag{2}.$$

Consider any $s \in S$ and any equation $t = t' \in Eqn_{K^*}(\Sigma, Y)$ of sort $s \in S$. Then $K^* \models t = t'$. So by (1), $K \models t = t'$ and hence $t = t' \in Eqn_K(\Sigma, Y)$. Therefore $Eqn_{K^*}(\Sigma, Y) \subseteq Eqn_K(\Sigma, Y)$.

Conversely consider any equation $t = t' \in Eqn_K(\Sigma, Y)$ of sort $s \in S$.

Then $K \models t = t'$. Let $\phi : Vars(t) \cup Vars(t') \to X$ be any S-indexed family of injective mappings. Such a ϕ exists since each X_s is infinite. Since each ϕ_s is an injection

$$K \models \overline{\phi}_s(t) = \overline{\phi}_s(t'),$$

but $\overline{\phi}_s(t) = \overline{\phi}_s(t') \in Eqn_K(\Sigma, X)$ and since $K^* = Alg(\Sigma, Eqn_K(\Sigma, X))$,

$$K^* \models \overline{\phi}_s(t) = \overline{\phi}_s(t')$$

again by injectivity of each ϕ_s. Hence $K^* \models t = t'$, i.e. $t = t' \in Eqn_{K^*}(\Sigma, Y)$. Therefore $Eqn_K(\Sigma, Y) \subseteq Eqn_{K^*}(\Sigma, Y)$ and so (2) above holds.

Now we will show that $K^* \subseteq K$. Consider any $A \in K^*$ and an S-indexed family $Y = \langle Y_s \mid s \in S \rangle$ of sets of variables such that $|Y_s| \geq |A_s|$ for each sort $s \in S$. Let $\alpha : Y \to A$ be any surjection. Then by Theorems 5.1.2 and 5.1.6, there exists a unique homomorphic extension $\overline{\alpha} : T(\Sigma, Y) \to A$ and $A = \overline{\alpha}(T(\Sigma, Y))$. Let $\equiv^{\overline{\alpha}}$ be the kernel of $\overline{\alpha}$. Recalling the relation $\sim^{K^*}$ of Definition 5.1.8 then $\sim^{K^*} \subseteq \equiv^{\overline{\alpha}}$ by Lemma 5.1.9. Thus letting $\equiv^{\overline{\alpha}/K^*}$ denote the factor congruence $\equiv^{\overline{\alpha}} / \sim^{K^*}$, and $\equiv^{\overline{\alpha}/K}$ denote the factor congruence $\equiv^{\overline{\alpha}} / \sim^K$ we have

$$A = \overline{\alpha}(T(\Sigma, Y)) \cong T(\Sigma, Y)/\equiv^{\overline{\alpha}}$$

$$\cong (T(\Sigma, Y)/\sim^{K^*})/\equiv^{\overline{\alpha}/K^*},$$

by the Second Homomorphism Theorem 3.4.20,

$$\cong (T(\Sigma, Y)/\sim^K)/\equiv^{\overline{\alpha}/K},$$

by (2) above, and

$$\cong nat^{\overline{\alpha}/K}(T_K(\Sigma, Y))$$

by Lemma 5.1.9.

Since K is a variety, by Theorem 5.1.12, $T_K(\Sigma, Y) \in K$. Since K is closed under the formation of homomorphic images, $nat^{\overline{\alpha}/K}(T_K(\Sigma, Y)) \in K$ and hence $A \in K$. Since A was arbitrarily chosen, $K^* \subseteq K$ as required. ∎

The Variety Theorem is one of many results of the same type, known as *characterization theorems*, which characterize syntactic subclasses of first order formulas in terms of closure of their classes of models under certain algebraic constructions. We shall encounter two more characterization theorems in Section 5.3.

We see from Theorem 5.2.16 and its proof that for any class K, K is a variety if, and only if, the equational theory $Eqn_K(\Sigma, X)$ of K is an equational axiomatization of K over Σ, i.e.

$$K = Alg(\Sigma, Eqn_K(\Sigma, X))$$

where each variable set of X is infinite. Notice that this equational axiomatization of K is not finite. The following is an interesting and important question: *if K is equationally axiomatizable, does K possess a finite equational axiomatization?* This question is relevant for data type theory (see Section 5.4).

To conclude this section we return to the subject of equational calculus. We shall consider two further results which have applications to the theory of data type specification.

By Corollary 5.2.15.(ii) every equational class $Alg(\Sigma, E)$ admits an initial model $I(\Sigma, E)$. This fact is a starting point for the theory of algebraic data type specification. An *algebraic data type specification* consists of a pair (Σ, E), where Σ is an S-sorted signature and E is an equational theory over Σ. The initial model $I(\Sigma, E)$ may then be viewed as the *intended* or *standard model* of the specification (Σ, E). In the process of evaluating data type specifications and verifying programs on the basis of data type axioms, it is natural to try to deduce from E equations which are valid in the initial model $I(\Sigma, E)$. Unfortunately, on theoretical grounds, establishing the validity of all but the simplest kinds of equations in the initial model turns out to be intractable. First let us establish one positive result.

Theorem 5.2.17. *Let Σ be a non-void S-sorted signature and E be a set of equations over Σ. For any ground equation $e \in Eqn(\Sigma)$,*

$$E \vdash e \Leftrightarrow I(\Sigma, E) \models e.$$

Proof. Exercise 5.2.20.(10). ∎

Definition 5.2.18. Let Σ be a non-void S-sorted signature and $E \subseteq Eqn(\Sigma, X)$ be any set of equations. If every equation over Σ, which is valid in the initial model $I(\Sigma, E)$, is provable from E by the rules of equational deduction, then E is said to be an *ω-complete theory;* otherwise E is said to be *ω-incomplete*.

The notion of ω-completeness is related to notions of induction when viewed proof theoretically (compare Exercise 5.2.20.(13)).

Trivial examples of ω-incomplete theories are easily found. Clearly if an equation e is valid in the initial model $I(\Sigma, E)$ of a theory E, but not valid in all models of E, then by the Soundness Theorem for the equational calculus, $E \not\vdash e$. For example, consider the single-sorted signature Σ with only three constant symbols c_1, c_2, c_3 and the equational theory E consisting of two equations $c_1 = c_2$, $c_1 = c_3$. In the initial algebra $I(\Sigma, E)$ we have $I(\Sigma, E) \models c_1 = x$. However this equation is not valid in every non-minimal Σ algebra. Therefore by the soundness of the equational calculus it cannot be proved from E using the rules of equational deduction. More interesting examples of ω-complete and ω-incomplete theories are the following. (Recall the algebras of numbers in Section 2.1.2.)

Lemma 5.2.19.

(i) *The equational theory*

$$add(x, 0) = x, \quad mult(x, 0) = 0,$$

$$add(x,\ succ(y)) = succ(add(x,\ y)),$$
$$mult(x,\ succ(y)) = add(x,\ mult(x,\ y)),$$
$$add(x,\ y) = add(y,\ x), \quad mult(x,\ y) = mult(y,\ x),$$
$$add(x,\ add(y,\ z)) = add(add(x,\ y),\ z),$$
$$mult(x,\ mult(y,\ z)) = mult(mult(x,\ y),\ z),$$
$$mult(x,\ add(y,\ z)) = add(mult(x,\ y),\ mult(x,\ z)),$$

over the signature $(nat; 0;\ succ, add, mult)$ is ω-complete for its initial model

$$(\mathbf{N};\ 0;\ Succ,\ Add,\ Mult).$$

(ii) The algebra

$$(\mathbf{N};\ 0;\ Succ,\ Add,\ Mult,\ Sub),$$

has no recursive ω-complete equational axiomatization over the signature

$$(nat;\ 0;\ succ,\ add,\ mult,\ sub).$$

Proof. For (i) see Heering [1986]. For (ii) see Davis *et al.* [1976]. ∎

Further examples and general results about ω-completeness can be found in Heering [1986], Bergstra and Heering [1989] and in Groote [1990].

To what extent is the existence of ω-incomplete theories a limitation? We have seen that ω-incompleteness is a property of an equational axiomatization and its initial algebra. If the set of equations valid in $I(\Sigma, E)$ is not recursively enumerable then no finitary proof system using E will suffice to derive all valid equations. Nevertheless, adding new axioms and rules, such as induction rules, may yield further useful equations that are valid in $I(\Sigma, E)$. Often, only the introduction of an *infinitary deduction rule* (i.e. a deduction rule with infinitely many premises) yields a complete calculus for the initial algebra. One such rule which allows us to derive all equations valid in $PA(\mathbf{N})$ is the ω-rule

$$\frac{t(succ^{k_1}(0),\dots,succ^{k_n}(0)) = t'(succ^{k_1}(0),\dots,succ^{k_n}(0)) \mid k_1,\dots,k_n \in \mathbf{N}}{t(x_1,\dots,x_k) = t'(x_1,\dots,x_k).}$$

Exercises 5.2.20.

1. Complete the proof of the Soundness and Completeness Theorem 5.2.12 for equational logic by showing that for any equation $e \in Eqn(\Sigma, X)$, if $E \vdash e$ then $Alg(\Sigma, E) \models e$.

2. We will show that the hypothesis that Σ is non-void is necessary for the soundness of the ordinary rules of equational logic. Recall that

the role of this hypothesis is to ensure that the carrier sets of the term algebra $T(\Sigma)$ are non-empty. If we allowed the carrier sets of algebras to be empty then this hypothesis, which has been heavily used in Section 3 and in this section, would not seem necessary. However empty carriers give rise to many problems of which the soundness of equational logic is one of the most surprising and interesting.

Consider the following two-sorted signature: let $S = \{\ r,\ s\ \}$ be a set of sorts. Let Σ be the S-sorted signature consisting of:

$$\Sigma_{\lambda,\,r} = \emptyset, \quad \Sigma_{\lambda,\,s} = \{\ p,\ q\ \},$$

$$\Sigma_{r,\,s} = \{\ f,\ g\ \}, \quad \Sigma_{w,\,s} = \Sigma_{w,\,r} = \emptyset \ \text{for all}\ w \in S^{+},\ \ w \neq r.$$

Let E be the set of equations

$$f(x) = p, \quad g(x) = q, \quad f(x) = g(x).$$

(a) Show that, using the ordinary rules of equational deduction,

$$E \vdash p = q.$$

(b) What is the initial algebra $T(\Sigma,\ E)$ in $Alg(\Sigma,\ E)$?

(c) Show that
$$T(\Sigma,\ E) \not\models p = q.$$

Hence conclude that the ordinary rules are not sound when applied to $(\Sigma,\ E)$.

3. We can solve the problem of soundness in the general many-sorted case by redefining the ideas of an equation and its validity. Let Σ be any S-sorted signature and let $X = \langle\, X_s \mid s \in S\,\rangle$ be any S-indexed family of sets of variable symbols.

By an *equation of sort s with a variable declaration* over Σ and X, or simply a *declared equation of sort s*, we mean a pair $(Y, t = t')$, where Y is an S-indexed family of finite sets $Y_s \subseteq X_s$ of variables and $t, t' \in T(\Sigma, Y)_s$. Let $Deqn(\Sigma, X)_s$ denote the set of all declared equations over Σ and X of sort $s \in S$ and $Deqn(\Sigma, X) = \bigcup_{s \in S} Deqn(\Sigma, X)_s$.

Let $(Y, t = t')$ be a declared equation over Σ and X and let A be a Σ algebra. We say $(Y, t = t')$ is *valid* in A and write $A \models (Y, t = t')$

if, and only if, for every assignment $\alpha = \langle \alpha_s : Y_s \to A_s \mid s \in S \rangle$ we have

$$\overline{\alpha}_s(t) = \overline{\alpha}_s(t').$$

Show that:

(a) $I(\Sigma, E) \models (\{\ Y_r,\ Y_s\ \}, p = q)$, where $Y_r = \{\ x\ \}$ and $Y_s = \emptyset$; and that,

(b) $I(\Sigma, E) \not\models (\{\ Y_r,\ Y_s\ \}, p = q)$, where $Y_r = \emptyset$ and $Y_s = \emptyset$.

4. Consider the following natural extension of the ordinary rules of equational logic to deal with declared equations over any signature Σ and S-indexed collection X of sets of variables.

(a) For any sort $s \in S$ and term $t \in T(\Sigma, X)_s$,

$$\frac{}{\emptyset \vdash (\ Y, t = t\ ),}$$

where $Y = Vars(t)$, is a *reflexivity* rule.

(b) For any set $E \subseteq Deqn(\Sigma, X)$ of declared equations and any declared equation $(\ Y, t_0 = t_1\ ) \in Deqn(\Sigma, X)$,

$$\frac{E \vdash (\ Y, t_0 = t_1\ )}{E \vdash (\ Y, t_1 = t_0\ )}$$

is a *symmetry* rule.

(c) For any sets $E_0, E_1 \subseteq Deqn(\Sigma, X)$ of declared equations, any sort $s \in S$ and any declared equations $(\ Y, t_0 = t_1\ )$, $(\ Y, t_1 = t_2\ ) \in Deqn(\Sigma, X)_s$,

$$\frac{E_0 \vdash (\ Y, t_0 = t_1\ ), \quad E_1 \vdash (\ Y, t_1 = t_2\ )}{E_0 \cup E_1 \vdash (\ Y, t_0 = t_2\ )}$$

is a *transitivity* rule.

(d) For any sets $E_0, E_1 \subseteq Deqn(\Sigma, X)$ of declared equations, any declared equation $(\ Y, t = t'\ ) \in Deqn(\Sigma, X)_s$, any sort $s' \in S$, any variable $x \in Y_{s'}$ and any declared equation $(\ Z,\ t_0 = t_1\ ) \in Deqn(\Sigma,\ X)_{s'}$,

$$\frac{E_0 \vdash (\ Y,\ t = t'\ ), \quad E_1 \vdash (\ Z,\ t_0 = t_1\ )}{E_0 \cup E_1 \vdash (\ W,\ t[x/t_0] = t'[x/t_1]\ ),}$$

where for each $s \in S$,

$$W_s = \begin{cases} Y_s - \{\ x\ \} \cup Z_s, & \text{if } s = s'; \\ Y_s \cup Z_s, & \text{otherwise,} \end{cases}$$

is a *substitution* rule.

Using the obvious extension of the notion of proof based on these rules, define an inference relation $\vdash$ between sets of declared equations. Show that this calculus is sound in the sense that

$$E \vdash e \quad \Rightarrow \quad Alg(\Sigma,\ E) \models e$$

for E a declared equational theory and e any declared equation over Σ.

5. Show that the calculus of declared equations described in (4) is incomplete. Add new rules which overcome this incompleteness and prove the completeness of the resulting calculus. (Hint: consider the addition and deletion of declared variables from declared equations.)

6. Let Σ be an S-sorted signature and X be an S-indexed family of infinite sets of variables. An equational theory E over Σ is said to be *consistent* if, and only if, there exists an equation $e \in Eqn(\Sigma,\ X)$ which is not provable from E. Show that E is consistent if, and only if, for each sort $s \in S$ and any distinct variables $x,\ y \in X_s$ the equation $x = y$ is not provable from E. Hence show that E is consistent if, and only if E has a non-unit model.

7. The notion of consistency for equational logic is similar to, but not the same as, the notion of consistency for first order logic. For example we have the

   Compactness Theorem for Equational Logic. *For any S-sorted signature Σ and any equational theory $E \subseteq Eqn(\Sigma,\ X)$, E is consistent if, and only if, every finite subset of E is consistent.*

   Prove this result.

8. Consider the equational theory *Latt* of lattices and the two additional distribution laws of distributive lattices (Example 5.2.2.(ii)). Give

formal proofs using the ordinary deduction rules of equational logic to show that

$$Latt \cup \{ \ x \wedge (y \vee z) = (x \wedge y) \vee (x \wedge z) \ \} \vdash x \vee (y \wedge z) = (x \vee y) \wedge (x \vee z)$$

and

$$Latt \cup \{ \ x \vee (y \wedge z) = (x \vee y) \wedge (x \vee z) \ \} \vdash x \wedge (y \vee z) = (x \wedge y) \vee (x \wedge z).$$

9. Consider the equational theory *Group* of groups. Give a formal proof using the ordinary deduction rules of equational logic to show that

$$Group \vdash -(x + y) = (-y) + (-x).$$

10. Let Σ be a signature with term algebra $T(\Sigma, X)$ over an S-indexed family X of sets of variables. Let $E \subseteq Eqn(\Sigma, X)$ be any equational theory and let $K = Alg(\Sigma, E)$. Using the Soundness and Completeness Theorem 5.2.12, show that for any equation $e \in Eqn(\Sigma, X)$,

$$E \vdash e \ \Leftrightarrow \ T_K(\Sigma, X) \models e$$

and thus that

$$Alg(\Sigma, E) \models e \ \Leftrightarrow \ T_K(\Sigma, X) \models e.$$

Hence prove Theorem 5.2.17.

11. Prove that an equational theory E over Σ is ω-complete if, and only if, for all equations $t(x) = t'(x) \in Eqn(\Sigma, X)$, $E \vdash t(x) = t'(n)$ if, and only if, for all substitutions *sub* of ground terms over Σ for the variables X, $E \vdash sub(t(x)) = sub(t'(x))$.

12. Let E' be the equational theory in 5.2.19.(i) with the associativity, commutativity and distribution laws removed. Show that E' is ω-incomplete.

13. Generalize the ω-rule for $PA(\mathbf{N})$ to an infinitary deduction rule yielding a complete infinitary equational calculus for the initial model $I(\Sigma, E)$, for an arbitrary equational theory E.

5.3 Equational Horn logic

So far the axiomatizations of classes of algebras that we have investigated have been quite basic. We have studied equations $t = t'$ and axiomatizations made of finite or infinite sets of equations,

$$\{ \ t_1 = t'_1, \ \ldots, \ t_n = t'_n \ \}$$

$$\{ \ t_i = t'_i \ | \ i \in I \ \},$$

these being equivalent to finite and infinite conjunctions of equations respectively. What other types of axiomatizations are useful in the study of universal algebra?

At the heart of universal algebra is the notion of closure of a class of algebras under a family of constructions. For example, in addition to the closure properties based on constructions of Section 4, we are especially interested in classes that admit initial and final algebras. Recall from Theorem 5.1.11 that $T_K(\Sigma, X)$ is free for K, but we have observed that it need not be in K. We will extend the range of axiomatizations by generalizing the sets of equations to sets of the following:

  (i) *conditional equations* ; and,

  (ii) *equational Horn formulas.*

We will also treat the use of negated equations in axiomatizations. In each case, the classes of algebras we obtain using these axiomatizations enjoy many of the agreeable properties of equationally defined classes.

Definition 5.3.1.   Let Σ be an S-sorted signature with term algebra $T(\Sigma, X)$ over an S-indexed family $X = \langle \ X_s \ | \ s \in S \ \rangle$ of sets of variables. By a *conditional equation* over Σ and X we mean an expression of the form

$$e_1 \wedge \ldots \wedge e_n \Rightarrow e,$$

where $e_1, \ldots, e_n, e \in Eqn(\Sigma, X)$ are any equations. Let $Cond(\Sigma, X)$ denote the set of all conditional equations over Σ and X. A *conditional equational theory* is a set of conditional equations.

Let A be a Σ algebra and let $\alpha : X \to A = \langle \ \alpha_s : X_s \to A_s \ | \ s \in S \ \rangle$ be any assignment with homomorphic extension given by the term evaluation mapping $\overline{\alpha} = \langle \ \overline{\alpha_s} : T(\Sigma, X)_s \to A_s \ | \ s \in S \rangle$. A conditional equation $\phi \in Cond(\Sigma, X)$ of the form

$$t_1 = t'_1 \wedge \ldots \wedge t_n = t'_n \Rightarrow t_{n+1} = t'_{n+1},$$

where $t_i, t'_i \in T(\Sigma, X)_{s(i)}$ for $1 \leq i \leq n + 1$, is *true in A* under the assignment $\alpha : X \to A$ if, and only if,

$$\overline{\alpha}_{s(1)}(t_1) = \overline{\alpha}_{s(1)}(t_1') \text{ and } \dots \text{ and } \overline{\alpha}_{s(n)}(t_n) = \overline{\alpha}_{s(n)}(t_n')$$

imply

$$\overline{\alpha}_{s(n+1)}(t_{n+1}) = \overline{\alpha}_{s(n+1)}(t_{n+1}').$$

We write $(A, \alpha) \models \phi$ if ϕ is true in A under α. We say that ϕ is *valid* in A and write $A \models \phi$ if, and only if, ϕ is true in A under every assignment $\alpha : X \to A$. If ϕ is not valid in A then we write $A \not\models \phi$. If Φ is any set of conditional equations then we write $A \models \Phi$ if $A \models \phi$ for each $\phi \in \Phi$. If $A \models \Phi$ we may say that A is a *model* of Φ or a (Σ, Φ) algebra. We let

$$Alg(\Sigma, \Phi) = \{ A \in Alg(\Sigma) \mid A \models \Phi \}$$

denote the class of all models of Φ. If K is any class of Σ algebras then we write $K \models \phi$ if $A \models \phi$ for each algebra $A \in K$. We write $K \models \Phi$ if $K \models \phi$ for each conditional equation $\phi \in \Phi$. We let

$$Cond_K(\Sigma, X) = \{ \phi \in Cond(\Sigma, X) \mid K \models \phi \}$$

denote the *conditional equational theory* of K.

The satisfaction relation $\models$ for conditional equations leads to further classes of algebras, namely thoses classes that can be axiomatized by conditional equations.

Definition 5.3.2. Let Σ be an S-sorted signature. A class K of Σ algebras is a *conditional equational class* if, and only if, K is the class of all models of a conditional equational theory $\Phi \subseteq Cond(\Sigma, X)$, i.e.

$$K = Alg(\Sigma, \Phi).$$

Under these conditions Φ is said to be a *conditional equational axiomatization* over Σ of the class K.

Example 5.3.3.

(i) The class K of all *cancellation semigroups* is a conditional equational class since $K = Alg(\Sigma^{SG}, CanSemigroup)$ where

$$CanSemigroup = Semigroup \cup$$

$$\{ z + x = z + y \Rightarrow x = y, \quad x + z = y + z \Rightarrow x = y \}.$$

However K is not a variety (see Exercise 5.3.28.(1)).

(ii) The class K of all *torsion-free Abelian groups* is a conditional equational class since $K = Alg(\Sigma^G, TFAbGroup)$ where

$$TFAbGroup = AbGroup \cup \{ \; nx = 0 \Rightarrow x = 0 \mid n \geq 1 \; \},$$

where nx is the n-fold sum $x + x + \ldots + x$. Again K is not a variety (see Exercise 5.3.28.(1)).

Conditional equations represent a true generalization of equations since every equation $t = t' \in Eqn(\Sigma, X)$ is logically equivalent to a conditional equation of the form

$$x = x \Rightarrow t = t',$$

but not conversely (because of the examples above).

Let us consider which results of Section 5.2 for equations generalize to conditional equations. Our first lemma generalizes Theorem 5.2.14 on closure properties of equational classes to closure properties of conditional equational classes.

Lemma 5.3.4. *Every conditional equational class $K \subseteq Alg(\Sigma)$, is closed under the formation of subalgebras, isomorphic images and direct products (including the empty product).*

Proof. Suppose that $K = Alg(\Sigma, \Phi)$, for some $\Phi \subseteq Cond(\Sigma, X)$. Without loss of generality we may assume that the algebra $T(\Sigma, X)$ exists.

We leave it to the reader as an exercise to show that K is closed under the formation of: (i) subalgebras and (ii) isomorphic images.

(iii) To show that K is closed under the formation of direct products consider any (possibly empty) family A of K algebras

$$A = \langle \; A(i) \mid i \in I \text{ and } A(i) \in K \; \rangle$$

and the direct product ΠA. We must show that $\Pi A \models \Phi$.

Consider an arbitrary conditional equation $\phi \in \Phi$, where ϕ is

$$t_1 = t_1' \wedge \ldots \wedge t_n = t_n' \Rightarrow t_{n+1} = t_{n+1}'$$

and $t_i, t_i' \in T(\Sigma, X)_{s(i)}$ for $1 \leq i \leq n+1$. Consider any assignment $\alpha : X \to \Pi A$. Suppose that for each $1 \leq j \leq n$

$$\overline{\alpha}_{s(j)}(t_j) = \overline{\alpha}_{s(j)}(t_j').$$

Then for each $1 \leq j \leq n$ and each $i \in I$

$$\overline{\alpha}_{s(j)}(t_j)(i) = \overline{\alpha}_{s(j)}(t_j')(i).$$

Since $A(i) \models \phi$,

$$\overline{\alpha}_{s(n+1)}(t_{n+1})(i) = \overline{\alpha}_{s(n+1)}(t'_{n+1})(i).$$

Therefore $\overline{\alpha}_{s(n+1)}(t_{n+1}) = \overline{\alpha}_{s(n+1)}(t'_{n+1})$. Since α and ϕ were arbitrarily chosen, $\Pi A \models \Phi$. Since the algebras $A(i)$ were arbitrarily chosen then K is closed under the formation of direct products.

∎

An important corollary of Lemma 5.3.4 is the following result.

Corollary 5.3.5. *Let* Σ *be an S-sorted signature and let* Φ *be a conditional equational theory over* Σ.

(i) *The class* $K = Alg(\Sigma, \Phi)$ *contains free algebras* $T_K(\Sigma, X)$ *for any S-indexed family* $X = \langle X_s \mid s \in S \rangle$ *of sets of variables with* $\mid X_s \mid \geq 1$ *for all* $s \in S$.

(ii) *If* Σ *is non-void then the class* $K = Alg(\Sigma, \Phi)$ *contains an initial algebra* $T_K(\Sigma)$.

Proof. Clearly K is non-empty since it contains all unit algebras. Then (i) and (ii) follow immediately from Lemma 5.3.4 and Theorem 5.1.12 on the existence of free algebras. ∎

If Σ is non-void then the initial algebra $T_K(\Sigma)$ which is unique up to isomorphism in a conditional equational class $K = Alg(\Sigma, \Phi)$ will be denoted by $I(\Sigma, \Phi)$.

Another class of formulas of interest are *negated equations*.

Definition 5.3.6. Let Σ be an S-sorted signature with term algebra $T(\Sigma, X)$ over an S-indexed family $X = \langle X_s \mid s \in S \rangle$ of sets of variables. By a *negated equation* over Σ and X we mean an expression of the form

$$\neg(t = t')$$

where $t = t' \in Eqn(\Sigma, X)$ is any equation. We let $Neqn(\Sigma, X)$ denote the set of all negated equations over Σ and X, and $NE(\Sigma, X)$ denote the set of all equations *and* negated equations over Σ and X. By a *negated equational theory* we mean a set of equations and negated equations. We may write $t \neq t'$ or $\neg t = t'$ to denote the negated equation $\neg(t = t')$.

Let A be a Σ algebra and let $\alpha : X \to A = \langle \alpha_s : X_s \to A_s \mid s \in S \rangle$ be any assignment with homomorphic extension given by the term evaluation mapping $\overline{\alpha} = \langle \overline{\alpha_s} : T(\Sigma, X)_s \rightarrow A_s \mid s \in S \rangle$. A negated equation $\neg e \in$ $Neqn(\Sigma, X)$ of the form

$$\neg t = t',$$

where $t, t' \in T(\Sigma, X)_s$, is *true in* A under the assignment $\alpha : X \rightarrow A$ if, and only if,

$$\overline{\alpha}_s(t) \neq \overline{\alpha}_s(t').$$

We write $(A, \alpha) \models \neg e$ if $\neg e$ is true in A under α. We say that $\neg e$ is *valid* in A and write $A \models \neg e$ if, and only if, $\neg e$ is true in A under every assignment $\alpha : X \rightarrow A$. If $\neg e$ is not valid in A then we write $A \not\models \neg e$. If Φ is any set of equations and negated equations then we write $A \models \Phi$ if $A \models \phi$ for each $\phi \in \Phi$. If $A \models \Phi$ we say that A is a *model* of Φ or a (Σ, Φ) algebra. We let

$$Alg(\Sigma, \Phi) = \{ A \in Alg(\Sigma) \mid A \models \Phi \}$$

denote the class of all *models* of Φ. If K is any class of Σ algebras then we write $K \models \phi$ if $A \models \phi$ for each algebra $A \in K$. We write $K \models \Phi$ if $K \models \phi$ for each $\phi \in \Phi$. We let

$$Neqn_K(\Sigma, X) = \{ \phi \in Neqn(\Sigma, X) \mid K \models \phi \}$$

denote the *negated equational theory* of K.

The satisfaction relation $\models$ for both equations and negated equations leads to further classes of algebras, namely thoses classes that can be axiomatized by sets of equations and negated equations.

Definition 5.3.7. Let Σ be an S-sorted signature. A class K of Σ algebras is an *NE class* if, and only if, K is the class of all models of a set of equations and negated equations $\Phi \subseteq NE(\Sigma, X)$, i.e.

$$K = Alg(\Sigma, \Phi).$$

Under these conditions Φ is said to be an *NE axiomatization* over Σ of the class K.

Examples 5.3.8.

(i) The class K of all non-unit Boolean algebras is an NE class since

$$K = Alg(\Sigma^{Bool}, \Phi),$$

where $\Phi = Bool \cup \{ 0 \neq 1 \}$.

(ii) The class of all non-unit rings is an NE class since

$$K = Alg(\Sigma^R, \Psi),$$

where $\Psi = Ring \cup \{ 0 \neq 1 \}$.

Notice that negated equations are sufficiently expressive to allow inconsistent theories, i.e. theories which have *no* models such as $\{ x \neq x \}$ or, combined with equations, $\{\ t = t',\ \neg t = t'\ \}$. For this reason an NE class is not necessarily closed under the empty direct product $\Pi\emptyset$ which always guarantees a unit model.

Now we can study sets (conjunctions) of negated equations and equations. The possibility also arises to generalize conditional equations $e_1 \wedge \ldots \wedge e_n \Rightarrow e_{n+1}$ by allowing $e_1, \ldots, e_n, e_{n+1}$ to vary over equations and negated equations. For example, the class of all fields is axiomatizable using the formula

$$x \neq 0 \Rightarrow xx^{-1} = 1.$$

However we shall not pursue such a generalization. Instead we concentrate on a different generalization of equations and conditional equations obtained by allowing disjunctions of negated equations. This gives the class of so called *equational Horn formulas*.

Definition 5.3.9. Let Σ be an S-sorted signature with term algebra $T(\Sigma, X)$ over an S-indexed family $X = \langle\, X_s \mid s \in S \,\rangle$ of sets of variables. By a (*universal*) *equational Horn formula* over Σ and X we mean an expression of the form

$$\neg e_1 \vee \neg e_2 \vee \ldots \vee \neg e_n, \tag{i}$$

or of the form

$$\neg e_1 \vee \neg e_2 \vee \ldots \vee \neg e_n \vee e_{n+1} \tag{ii}.$$

Let $Horn(\Sigma, X)$ denote the set of all equational Horn formulas over Σ and X. An *equational Horn theory* is a set of equational Horn formulas.

Equational Horn formulas of the form (ii) are often known as *positive* or *strict equational Horn formulas*, since they contain an equation.

Lemma 5.3.10. *Let Σ be an S-sorted signature and $\Phi \subseteq NE(\Sigma, X)$ be a set of equations and negated equations over Σ. Then Φ is an equational Horn theory.*

Proof. Observe that $n = 0$ is permissible in (ii) of Definition 5.3.9, i.e. every equation is an equational Horn formula. Furthermore $n = 1$ is permissible in (i) of Definition 5.3.9, i.e. every negated equation is an equational Horn formula. ∎

Let A be a Σ algebra and let $\alpha : X \to A = \langle\, \alpha_s : X_s \to A_s \mid s \in S \,\rangle$ be any assignment with homomorphic extension given by the term evaluation mapping $\overline{\alpha} = \langle\, \overline{\alpha_s} : T(\Sigma, X)_s \to A_s \mid s \in S \,\rangle$. An equational Horn formula $\phi \in Horn(\Sigma, X)$ of the form (i)

$$\neg t_1 = t_1' \vee \ldots \vee \neg t_n = t_n',$$

with $t_i, t_i' \in T(\Sigma, X)_{s(i)}$ for $1 \leq i \leq n$, is *true in A* under the assignment $\alpha : X \rightarrow A$ if, and only if,

$$\overline{\alpha}_{s(1)}(t_1) \neq \overline{\alpha}_{s(1)}(t_1') \text{ or } \ldots \text{ or} \overline{\alpha}_{s(n)}(t_n) \neq \overline{\alpha}_{s(n)}(t_n').$$

An equational Horn formula $\phi \in Horn(\Sigma, X)$ of the form (ii)

$$\neg t_1 = t_1' \vee \ldots \vee \neg t_n = t_n' \vee t_{n+1} = t_{n+1}',$$

with $t_i, t_i' \in T(\Sigma, X)_{s(i)}$ for $1 \leq i \leq n+1$, is *true in A* under the assignment $\alpha : X \rightarrow A$ if, and only if,

$$\overline{\alpha}_{s(1)}(t_1) \neq \overline{\alpha}_{s(1)}(t_1') \text{ or } \ldots \text{ or } \overline{\alpha}_{s(n)}(t_n) \neq \overline{\alpha}_{s(n)}(t_n')$$
$$\text{or } \overline{\alpha}_{s(n+1)}(t_{n+1}) = \overline{\alpha}_{s(n+1)}(t_{n+1}').$$

We write $(A, \alpha) \models \phi$ if ϕ is true in A under α. We say that ϕ is *valid in A* and write $A \models \phi$ if, and only if, ϕ is true in A under every assignment. If ϕ is not valid in A then we write $A \not\models \phi$. If Φ is any set of equational Horn formulas then we write $A \models \Phi$ if $A \models \phi$ for each $\phi \in \Phi$. If $A \models \Phi$ we say that A is a *model* of Φ or a (Σ, Φ) algebra. We let

$$Alg(\Sigma, \Phi) = \{ A \in Alg(\Sigma) \mid A \models \Phi \}$$

denote the class of all models of Φ. If K is any class of Σ algebras then we write $K \models \phi$ if $A \models \phi$ for each algebra $A \in K$. We write $K \models \Phi$ if $K \models \phi$ for each formula $\phi \in \Phi$. We let

$$Horn_K(\Sigma, X) = \{ \phi \in Horn(\Sigma, X) \mid K \models \phi \}$$

denote the *equational Horn theory* of K.

The satisfaction relation $\models$ for equational Horn formulas leads to further classes of algebras, namely thoses classes that can be axiomatized by equational Horn formulas.

Definition 5.3.11. Let Σ be an S-sorted signature. A class K of Σ algebras is an *equational Horn class* if, and only if, K is the class of all models of an equational Horn theory $\Phi \subseteq Horn(\Sigma, X)$, i.e.

$$K = Alg(\Sigma, \Phi).$$

Under these conditions Φ is said to be an *equational Horn axiomatization* over Σ of the class K.

Lemma 5.3.12. *Let Σ be an S-sorted signature with term algebra $T(\Sigma, X)$ over an S-indexed family X of sets of variables. Let $\Phi \subseteq Cond(\Sigma, X)$ be* a conditional equational theory over Σ and X. Then there exists an equational Horn theory $\Psi \subseteq Horn(\Sigma, X)$ such that

$$Alg(\Sigma, \Phi) = Alg(\Sigma, \Psi).$$

Proof. Clearly a conditional equation ϕ of the form

$$t_1 = t_1' \wedge \ldots \wedge t_n = t_n' \Rightarrow t_{n+1} = t_{n+1}',$$

is logically equivalent to a positive equational Horn formula ψ of the form (ii)

$$\neg t_1 = t_1' \vee \ldots \vee \neg t_n = t_n' \vee t_{n+1} = t_{n+1}',$$

i.e. for any $A \in Alg(\Sigma)$ we have $A \models \phi$ if, and only if, $A \models \psi$. ∎

Therefore equational Horn formulas are a generalization of conditional equations and every conditional equational class K is an equational Horn class. Under what conditions does the converse hold?

Lemma 5.3.13. For any equational Horn class $K \subseteq Alg(\Sigma)$, K is a conditional equational class if, and only if, K contains a unit algebra.

Proof. $\Rightarrow$ Trivial.

$\Leftarrow$ Suppose that K contains a unit algebra. Since K is an equational Horn class then $K = Alg(\Sigma, \Phi)$ for some equational Horn theory $\Phi \subseteq Horn(\Sigma, X)$.

Suppose for a contradiction that some formula $\phi \in \Phi$ is of the form

$$\neg t_1 = t_1' \vee \ldots \vee \neg t_n = t_n',$$

with $t_i, t_i' \in T(\Sigma, X)_{s(i)}$ for $1 \leq i \leq n$. For any unit algebra $A \in K$, with carrier sets $A_s = \{ a_s \}$ for each $s \in S$, we have $A \models \phi$. So for any assignment $\alpha : X \to A$ and some $1 \leq i \leq n$ we have

$$\overline{\alpha}_{s(i)}(t_i) \neq \overline{\alpha}_{s(i)}(t_i')$$

which contradicts the fact that

$$\overline{\alpha}_{s(i)}(t_i) = \overline{\alpha}_{s(i)}(t_i') = a_{s(i)}.$$

So Φ must consist entirely of equational Horn formulas of the form $\neg e_1 \vee \ldots \vee \neg e_n \vee e_{n+1}$, which are logically equivalent to conditional equations of the form $e_1 \wedge \ldots \wedge e_n \Rightarrow e$. Therefore K is a conditional equational class. ∎

Example 5.3.14. By Lemma 5.3.10, the classes of non-unit Boolean algebras and non-unit rings of Examples 5.3.8 are equational Horn classes.

However by Lemma 5.3.13 these classes are not conditional equational classes.

Once again we may ask what results for equations and conditional equations generalise to equational Horn formulas. A generalization of Lemmas 5.2.14 and 5.3.4 about the basic closure operations and free algebra constructions is the following.

Lemma 5.3.15. *Every equational Horn class* $K \subseteq Alg(\Sigma)$ *is closed under the formation of subalgebras, isomorphic images, non-empty direct products and reduced products.*

Proof. Suppose that $K = Alg(\Sigma, \Phi)$, for some equational Horn theory Φ. Without loss of generality we may assume that the algebra $T(\Sigma, X)$ exists.

We leave it to the reader, as an exercise, to show that K is closed under the formation of: (i) subalgebras and (ii) isomorphic images.

(iii) To show that K is closed under the formation of reduced products, consider any non-empty family A of K algebras

$$A = \langle\, A(i) \mid i \in I \text{ and } A(i) \in K \,\rangle,$$

any filter F over I and the reduced product $B = \Pi A/ \equiv^F$. We must show that $B \models \Phi$.

Consider any equational Horn formula $\phi \in \Phi$ and any assignment $\beta : X \rightarrow B$. Let $\alpha : X \rightarrow \Pi A$ be an assignment such that for each $s \in S$

$$(nat^F \circ \alpha)_s = \beta_s,$$

where nat^F is the natural mapping associated with the congruence $\equiv^F$. Such an assignment exists since nat^F is surjective.

If ϕ is of the form (i)

$$\neg t_1 = t_1' \vee \ldots \vee \neg t_n = t_n',$$

where $t_i, t_i' \in T(\Sigma, X)_{s(i)}$ for $1 \leq i \leq n$, then for each $i \in I$, $A(i) \models \phi$. So for each $i \in I$,

$$\overline{\alpha}_{s(1)}(t_1)(i) \neq \overline{\alpha}_{s(1)}(t_1')(i) \text{ or } \ldots \text{ or } \overline{\alpha}_{s(n)}(t_n)(i) \neq \overline{\alpha}_{s(n)}(t_n')(i).$$

Therefore

$$\bigcap_{j=1}^{n} \{\, i \in I \mid \overline{\alpha}_{s(j)}(t_j)(i) = \overline{\alpha}_{s(j)}(t_j')(i) \,\} = \emptyset.$$

Since F is closed under finite intersections and $\emptyset \notin F$ then for some $1 \leq j \leq n$,

$$\{\ i \in I \mid \overline{\alpha}_{s(j)}(t_j)(i) = \overline{\alpha}_{s(j)}(t'_j)(i)\ \} \notin F.$$

Hence $\overline{\alpha}_{s(j)}(t_j) \neq^F_{s(j)} \overline{\alpha}_{s(j)}(t'_j)$ and therefore $\overline{\beta}_{s(j)}(t_j) \neq \overline{\beta}_{s(j)}(t'_j)$. Thus

$$\overline{\beta}_{s(1)}(t_1) \neq \overline{\beta}_{s(1)}(t'_1) \ \text{ or } \ \ldots \ \text{ or } \ \overline{\beta}_{s(n)}(t_n) \neq \overline{\beta}_{s(n)}(t'_n).$$

Since β was arbitrarily chosen, $B \models \phi$.

If ϕ is of the form (ii)

$$\neg t_1 = t'_1 \vee \ldots \vee \neg t_n = t'_n \vee t_{n+1} = t'_{n+1}$$

with $t_i, t'_i \in T(\Sigma, X)_{s(i)}$ for $1 \leq i \leq n+1$, suppose that for each $1 \leq j \leq n$, $(B, \beta) \not\models \neg t_j = t'_j$, i.e.

$$\overline{\beta}_{s(j)}(t_j) = \overline{\beta}_{s(j)}(t'_j).$$

Then we must show $(B, \beta) \models t_{n+1} = t'_{n+1}$. By assumption, for each $1 \leq j \leq n$,

$$(\overline{nat^F \circ \alpha})_{s(j)}(t_j) = (\overline{nat^F \circ \alpha})_{s(j)}(t'_j)$$

and hence

$$nat^F_{s(j)} \circ \overline{\alpha}_{s(j)}(t_j) = nat^F_{s(j)} \circ \overline{\alpha}_{s(j)}(t'_j).$$

Therefore

$$\overline{\alpha}_{s(j)}(t_j) \equiv^F_{s(j)} \overline{\alpha}_{s(j)}(t'_j).$$

So for each $1 \leq j \leq n$,

$$\{\ i \in I \mid \overline{\alpha}_{s(j)}(t_j)(i) = \overline{\alpha}_{s(j)}(t'_j)(i)\ \} \in F.$$

By the finite intersection property of F,

$$S = \bigcap_{j=1}^{n} \{\ i \in I \mid \overline{\alpha}_{s(j)}(t_j)(i) = \overline{\alpha}_{s(j)}(t'_j)(i)\ \} \in F.$$

For each $i \in S$,

$$\overline{\alpha}_{s(1)}(t_1)(i) = \overline{\alpha}_{s(1)}(t'_1)(i) \ \text{ and } \ \ldots \ \text{ and } \ \overline{\alpha}_{s(n)}(t_n)(i) = \overline{\alpha}_{s(n)}(t'_n)(i),$$

and since $A(i) \models \phi$,

$$\overline{\alpha}_{s(n+1)}(t_{n+1})(i) = \overline{\alpha}_{s(n+1)}(t'_{n+1})(i).$$

So $S \subseteq \{ i \in I \mid \overline{\alpha}_{s(n+1)}(t_{n+1})(i) = \overline{\alpha}_{s(n+1)}(t'_{n+1})(i) \}$ and since F is closed under supersets,

$$\{ i \in I \mid \overline{\alpha}_{s(n+1)}(t_{n+1})(i) = \overline{\alpha}_{s(n+1)}(t'_{n+1})(i) \} \in F.$$

Therefore

$$\overline{\alpha}_{s(n+1)}(t_{n+1}) \equiv^F_{s(n+1)} \overline{\alpha}_{s(n+1)}(t'_{n+1}).$$

So $\overline{\beta}_{s(n+1)}(t_{n+1}) = \overline{\beta}_{s(n+1)}(t'_{n+1})$. Since β was arbitrarily chosen, $B \models \phi$.

Since ϕ was arbitrarily chosen, $B \models \Phi$. Since the algebras $A(i)$ and filter F were arbitrarily chosen, K is closed under the formation of reduced products.

(iv) Since K is closed under reduced products it follows, by Example 4.3.12.(i), that K is closed under non-empty direct products.

∎

An important corollary of Lemma 5.3.15 is the following result:

Corollary 5.3.16. Let Σ be an S-sorted signature and let Φ be an equational Horn theory over Σ. Suppose $K = Alg(\Sigma, \Phi)$ is non-empty. Then:

(i) K contains free algebras $T_K(\Sigma, X)$ for any S-indexed family $X = \langle X_s \mid s \in S \rangle$ of sets of variables with $\mid X_s \mid \geq 1$ for all $s \in S$.

(ii) If Σ is non-void then K contains an initial algebra $T_K(\Sigma)$.

Proof. Follows immediately from Lemma 5.3.15 and Theorem 5.1.12 on the existence of free algebras. ∎

If Σ is non-void then the initial algebra $T_K(\Sigma)$ which is unique up to isomorphism in an equational Horn class $K = Alg(\Sigma, \Phi)$ will be denoted by $I(\Sigma, \Phi)$.

In the remainder of this section we shall develop a characterization theorem for equational Horn classes which is known as the *Quasivariety Theorem*. It says: *the equational Horn axiomatizable classes of algebras are precisely the quasivarieties of algebras*. This theorem, due to A.I. Malcev, is an analogue of the Variety Theorem of Section 5.2 for equational Horn formulas.

To prove the theorem we employ the following definition and two results which introduce a useful technique known as the method of *equational diagrams*. The following Extension Lemma is a special case for algebra of a result from first order model theory, known as the Diagram Lemma. (See Chang and Keisler [1990] for the general result.)

Definition 5.3.17. Let Σ be an S-sorted signature and $X = \langle\, X_s \mid s \in S \,\rangle$ be any S-indexed family of sets. Then $\Sigma[X]$ denotes the S-sorted signature which is the disjoint sum of Σ and X obtained by adding a distinct new constant symbol $\hat{x}$ to $\Sigma_{\lambda,\,s}$ for each sort $s \in S$ and each $x \in X_s$.

Having extended Σ to $\Sigma[X]$, we wish to interpret the new constant symbols given by X. If A is any Σ algebra and $\alpha : X \to A$ is any S-indexed family of mappings then $A[\alpha]$ denotes the unique $\Sigma[X]$ algebra such that $A[\alpha]|_\Sigma = A$ and $\hat{x}_{A[\alpha]} = \alpha_s(x)$ for each sort $s \in S$ and $x \in X_s$.

If K is any class of Σ algebras then $K[X]$ denotes the class

$$K[X] = \{\, A[\alpha] \in Alg(\Sigma[X]) \mid A \in K \text{ and } \alpha : X \to A \text{ is any mapping } \}.$$

If A is a Σ algebra we may choose X to be the S-indexed family A of carrier sets of A.

Proposition 5.3.18. *Let K be any class of Σ algebras and let $A \in K$ be any algebra.*

(i) *If K is closed under the formation of non-empty direct products then $K[A]$ is closed under the formation of non-empty direct products.*

(ii) *If K is closed under the formation of ultraproducts then $K[A]$ is closed under the formation of ultraproducts.*

Proof. Exercise 5.3.28.(14). ∎

Recall, for any S-sorted signature Σ and any Σ algebra A, that $Eqn_A(\Sigma)$ (respectively, $Neqn_A(\Sigma)$) denotes the set of all ground equations (respectively, ground negated equations) over Σ which are true in A. If X is an S indexed family of subsets of A that generates A then the set

$$Eqn_{A[id_X]}(\Sigma[X]) \cup Neqn_{A[id_X]}(\Sigma[X])$$

may be termed the *equational diagram* of A. The set

$$Eqn_{A[id_X]}(\Sigma[X])$$

may be termed the *positive equational diagram* of A.

Given S-sorted Σ algebras A and B, an S-indexed family X of subsets of A and an S-indexed family $\beta : X \to B$ of mappings we may ask: *when does β uniquely extend to a homomorphism or an embedding $\overline{\beta} : A \to B$?*

Lemma 5.3.19 (Extension Lemma). *Let A and B be any Σ algebras and let X be an S-indexed family of subsets which generates A. Let $\beta : X \to B$ be any map and let $\alpha : X \to X$ be the identity map.*

(i) *If $B[\beta] \models Eqn_{A[\alpha]}(\Sigma[X])$ then β extends uniquely to a $\Sigma[X]$ homomorphism*

$$\overline{\beta} : A[\alpha] \to B[\beta].$$

(ii) *If $B[\beta] \models Eqn_{A[\alpha]}(\Sigma[X]) \cup Neqn_{A[\alpha]}(\Sigma[X])$ then β extends uniquely to a $\Sigma[X]$ embedding*

$$\overline{\beta} : A[\alpha] \to B[\beta].$$

Proof. (i) Suppose that $B[\beta] \models Eqn_{A[\alpha]}(\Sigma[X])$. Define the map $\overline{\beta} : A \to B$ by

$$\overline{\beta}_s(V_s^{A[\alpha]}(t)) = V_s^{B[\beta]}(t)$$

for each closed term $t \in T(\Sigma[X])_s$ and each $s \in S$, where $V^{A[\alpha]} : T(\Sigma[X]) \to A[\alpha]$ and $V^{B[\beta]} : T(\Sigma[X]) \to B[\beta]$ are the valuation maps for $T(\Sigma[X])$ with respect to $A[\alpha]$ and $B[\beta]$.

It is easily checked that $\overline{\beta}$ is well defined. For any $s \in S$ and closed terms t, $t' \in T(\Sigma[X])_s$, if

$$V_s^{A[\alpha]}(t) = V_s^{A[\alpha]}(t')$$

then $t = t' \in Eqn_{A[\alpha]}(\Sigma[X])_s$ and since $B[\beta] \models Eqn_{A[\alpha]}(\Sigma[X])$, it follows that

$$V_s^{B[\beta]}(t) = V_s^{B[\beta]}(t').$$

Furthermore, since X generates A then $V^{A[\alpha]}$ is surjective so that each $\overline{\beta}_s$ is defined everywhere.

To show that $\overline{\beta}$ is a $\Sigma[X]$ homomorphism consider any $s \in S$ and constant symbol $c \in \Sigma[X]_{\lambda, s}$. Then

$$\overline{\beta}_s(c_{A[\alpha]}) = \overline{\beta}_s(V_s^{A[\alpha]}(c)) = V_s^{B[\beta]}(c) = c_{B[\beta]}.$$

Consider any $w = s(1) \dots s(n) \in S^+$, $s \in S$ any function symbol $\sigma \in \Sigma[X]_{w,s} = \Sigma_{w,s}$ and any $t_i \in T(\Sigma[X])_{s(i)}$ for $1 \le i \le n$. Then

$$\overline{\beta}_s(\, \sigma_{A[\alpha]}(V_{s(1)}^{A[\alpha]}(t_1), \dots, V_{s(n)}^{A[\alpha]}(t_n))\,)$$

$$= \overline{\beta}_s(\, V_s^{A[\alpha]}(\sigma(t_1, \dots, t_n))\,)$$

$$= V_s^{B[\beta]}(\sigma(t_1, \dots, t_n))$$

$$= \sigma_{B[\beta]}(V^{B[\beta]}_{s(1)}(t_1), \ldots, V^{B[\beta]}_{s(n)}(t_n))$$

$$= \sigma_{B[\beta]}(\overline{\beta}_{s(1)}(V^{A[\alpha]}_{s(1)}(t_1)), \ldots, \overline{\beta}_{s(n)}(V^{A[\alpha]}_{s(n)}(t_n))).$$

Therefore $\overline{\beta}$ is a $\Sigma[X]$ homomorphism and clearly $\overline{\beta}$ extends ϕ. Since X generates A, $\overline{\beta}$ is unique.

(ii) Suppose $B[\beta] \models Eqn_{A[\alpha]}(\Sigma[X]) \cup Neqn_{A[\alpha]}(\Sigma[X])$. Consider the map $\overline{\beta}$ as defined in (i) above. For any $s \in S$ and $t, t' \in T(\Sigma[X])_s$, suppose that $V^{A[\alpha]}_s(t) \neq V^{A[\alpha]}_s(t')$. Then $t \neq t' \in Neqn_{A[\alpha]}(\Sigma[X])$. Since $B[\beta] \models Neqn_{A[\alpha]}(\Sigma[X])$ it follows that $V^{B[\beta]}_s(t) \neq V^{B[\beta]}_s(t')$. Therefore $\overline{\beta}_s(V^{A[\alpha]}_s(t)) \neq \overline{\beta}_s(V^{A[\alpha]}_s(t'))$. So $\overline{\beta}$ is injective, i.e. a $\Sigma[X]$ embedding.

∎

Our final technical tool is a special case of a fundamental theorem on ultraproducts, from Łos [1955], that relates the truth of an equational Horn formula in an ultraproduct to its truth in the component algebras. This result will be used to prove the Compactness Theorem for equational Horn formulas.

Theorem 5.3.20 (Ultraproduct Theorem (J. Łos 1955)). *Let $A = \langle A(i) \mid i \in I \rangle$ be any non-empty family of Σ algebras and U be an ultrafilter on I. Let $w = s(1) \ldots s(n) \in S^+$, $x_1, \ldots, x_n \in X^w$, and $a_1, \ldots, a_n \in \Pi A^w$. Then for any equational Horn formula*

$$\phi(x_1, \ldots, x_n) \in Horn(\Sigma, X),$$

$$\Pi A/ \equiv^U \models \phi([a_1]_U, \ldots, [a_n]_U) \Leftrightarrow$$
$$\{ i \in I \mid A(i) \models \phi(a_1(i), \ldots, a_n(i)) \} \in U.$$

Proof. Since ϕ is a disjunction of equations and negated equations we prove the result by induction on the length k of the disjunction.

Basis. Suppose $k = 1$. Let $\alpha : X \to \Pi A$ be any assignment with $a_i = \alpha_{s(i)}(x_i)$, for $1 \leq i \leq n$.

(i) Suppose ϕ is a single equation $t = t' \in Eqn(\Sigma, X)$ of sort $s \in S$. Then

$$\Pi A/ \equiv^U \models \phi([a_1]_U, \ldots, [a_n]_U) \Leftrightarrow \overline{nat^U \circ \alpha_s}(t) = \overline{nat^U \circ \alpha_s}(t')$$

$$\Leftrightarrow nat^U_s \circ \overline{\alpha}_s(t) = nat^U_s \circ \overline{\alpha}_s(t')$$

$$\Leftrightarrow \overline{\alpha}_s(t) \equiv^U_s \overline{\alpha}_s(t')$$

$$\Leftrightarrow \{ \ i \in I \ | \ A(i) \models \overline{\alpha}_s(t)(i) = \overline{\alpha}_s(t')(i) \ \} \in U$$

$$\Leftrightarrow \{ \ i \in I \ | \ A(i) \models \phi(a_1(i), \ldots, a_n(i)) \ \} \in U.$$

(ii) Suppose ϕ is a single negated equation $\neg t = t'$ with $t, t' \in T(\Sigma, X)_s$. Then

$$\Pi A/ \equiv^U \models \phi([a_1]_U, \ldots, [a_n]_U) \Leftrightarrow \overline{nat^U \circ \alpha_s}(t) \neq \overline{nat^U \circ \alpha_s}(t')$$

$$\Leftrightarrow nat_s^U \circ \overline{\alpha}_s(t) \neq nat_s^U \circ \overline{\alpha}_s(t')$$

$$\Leftrightarrow \overline{\alpha}_s(t) \neq_s^U \overline{\alpha}_s(t')$$

$$\Leftrightarrow \{ \ i \in I \ | \ A(i) \models \overline{\alpha}_s(t)(i) = \overline{\alpha}_s(t')(i) \ \} \notin U$$

$$\Leftrightarrow \{ \ i \in I \ | \ A(i) \models \overline{\alpha}_s(t)(i) \neq \overline{\alpha}_s(t')(i) \ \} \in U,$$

by Theorem 4.3.5,

$$\Leftrightarrow \{ \ i \in I \ | \ A(i) \models \phi(a_1(i), \ldots, a_n(i)) \ \} \in U.$$

Induction Step. Suppose that ϕ is a disjunction of equations and negated equations for $k > 1$ of the form $\neg e_1 \vee \ldots \vee \neg e_k \vee \phi$, where ϕ is either an equation or negated equation. Let ψ be the formula $\neg e_2 \vee \ldots \vee \neg e_k \vee \phi$, then ψ is also an equational Horn formula. Now

$$\Pi A/ \equiv^U \models \phi([a_1]_U, \ldots, [a_n]_U) \Leftrightarrow \Pi A/ \equiv^U \models \psi([a_1]_U, \ldots, [a_n]_U)$$

$$\text{or } \Pi A/ \equiv^U \models \neg e_1([a_1]_U, \ldots, [a_n]_U)$$

$$\Leftrightarrow \{ \ i \in I \ | \ A(i) \models \psi(a_1(i), \ldots, a_n(i)) \ \} \in U$$

$$\text{or } \{ \ i \in I \ | \ A(i) \models \neg e_1(a_1(i), \ldots, a_n(i)) \ \} \in U,$$

by the induction hypothesis,

$$\Leftrightarrow \{ \ i \in I \ | \ A(i) \models \phi(a_1(i), \ldots, a_n(i)) \ \} \in U,$$

by Lemma 4.3.7. ∎

Theorem 5.3.21 (Compactness Theorem for Horn Logic). *Let K be any non-empty class of Σ algebras which is closed under the formation of ultraproducts, and let $\Phi \subseteq Horn(\Sigma, X)$ be any equational Horn theory. Then Φ has a model in K if, and only if, every finite subset of Φ has a model in K.*

Proof. $\Rightarrow$ Trivially, if Φ has a model in K then every finite subset of Φ has a model in K.

$\Leftarrow$ Let I be the collection of all finite subsets of Φ and suppose that for each $i = \{ \phi_1, \ldots, \phi_n \} \in I$, there exists a model $A(i) \models i$ in K.

Let $\uparrow i = \{ j \in I \mid i \subseteq j \}$ for each $i \in I$ and let

$$B = \{ \uparrow i \mid i \in I \}.$$

Clearly B has the finite intersection property since for any $i, j \in I$,

$$\uparrow i \cap \uparrow j = \uparrow (i \cup j).$$

So by Lemmas 4.3.3 and 4.3.6, B extends to a filter and hence an ultrafilter U on I.

Let $A = \langle A(i) \mid i \in I \rangle$ and consider the ultraproduct $\Pi A / \equiv^U$ and any equational Horn formula $\phi(x_1, \ldots, x_n) \in \Phi$. Clearly $i_\phi = \{ \phi \} \in I$, so that $\uparrow i_\phi \in U$. Now

$$A(j) \models \phi(x_1, \ldots, x_n)$$

for each $j \in \uparrow i_\phi$, since $i_\phi \subseteq j$.

So for any $a_1, \ldots, a_n \in \Pi A^w$ and the set

$$S = \{ i \in I \mid A(i) \models \phi(a_1(i), \ldots, a_n(i)) \}$$

we have $\uparrow i_\phi \subseteq S$ and since U is closed under supersets, $S \in U$.

By Theorem 5.3.20,

$$\Pi A / \equiv^U \models \phi([a_1]_U, \ldots, [a_n]_U).$$

Since $a_1, \ldots, a_n$ and ϕ were arbitrarily chosen, $\Pi A / \equiv^U \models \Phi$. Since K is closed under the formation of ultraproducts, Φ has a model in K. ∎

The general version of the Ultraproduct Theorem 5.3.20, to be proved in Exercise 5.3.28.(15), gives a proof of the Compactness Theorem 5.3.21 for arbitrary first order theories.

The easy part of the Quasivariety Theorem has already been established in Lemma 5.3.15. The converse of Lemma 5.3.15 is established in the following lemma, using the method of equational diagrams and the Compactness Theorem for Horn formulas. Recall that for any class K of Σ algebras, $Horn_K(\Sigma, X)$ denotes the equational Horn theory of K over Σ and X.

Lemma 5.3.22. *Let Σ be an S-sorted signature and let K be any class of Σ algebras. If K is closed under the formation of subalgebras, isomorphic images, non-empty direct products and ultraproducts then K is an equational Horn class, in fact*

$$K = Alg(\Sigma, Horn_K(\Sigma, X))$$

for X an S-indexed family of infinite sets of variables.

Proof. Suppose K is closed under the formation of subalgebras, isomorphic images, non-empty direct products and ultraproducts. If K is empty then, trivially, K is an equational Horn class since

$$K = Alg(\Sigma, \{ x \neq x \}).$$

Thus $K = Alg(\Sigma, Horn_K(\Sigma, X))$. So suppose K is non-empty. Let $X = \langle X_s \mid s \in S \rangle$ be an S-indexed family of infinite sets of variables, and let

$$K^* = Alg(\Sigma, Horn_K(\Sigma, X)).$$

Clearly $K \models Horn_K(\Sigma, X)$ and so

$$K \subseteq' K^*.$$

It suffices to prove the converse, namely

$$K^* \subseteq K.$$

Consider any $A \in K^*$ then $A \models Horn_K(\Sigma, X)$. Recall that $\Sigma[A]$ denotes the S-sorted signature obtained by augmenting each set $\Sigma_{\lambda, s}$ with a distinct new constant symbol $\hat{a}$ for each $s \in S$ and element $a \in A_s$. Let $\alpha : A \to A$ be the identity mapping. Then $A[\alpha]$ is a $\Sigma[A]$ algebra. Let

$$\Delta(A) = Eqn_{A[\alpha]}(\Sigma[A]) \cup Neqn_{A[\alpha]}(\Sigma[A])$$

denote the equational diagram of A. We will show that there exists a Σ algebra $B \in K$ and assignment $\beta : A \to B$ such that $B[\beta] \models \Delta(A)$. Since K is closed under the formation of ultraproducts it will suffice to show that for every finite subset $\Delta \subseteq \Delta(A)$, there exists a Σ algebra $B \in K$ and assignment $\beta : A \to B$ such that $B[\beta] \models \Delta$.

Let $w = s(1) \ldots s(k) \in S^+$ and $\bar{x} = x_1, \ldots, x_k \in X^w$ be any sequence of variables. Consider any finite set of equations and negated equations

$$\{ e_1(\bar{x}), \ldots, e_m(\bar{x}), \neg e_1'(\bar{x}), \ldots, \neg e_n'(\bar{x}) \}.$$

For any sequence of constant symbols $\bar{a} = \hat{a}_1, \ldots, \hat{a}_k$, with $a_i \in A_{s(i)}$ for $1 \leq i \leq k$, suppose

$$\{ e_1(\bar{a}), \ldots, e_m(\bar{a}), \neg e_1'(\bar{a}), \ldots, \neg e_n'(\bar{a}) \} \subseteq \Delta(A).$$

Then

$$A \models \{ e_1(\bar{a}), \ldots, e_m(\bar{a}), \neg e_1'(\bar{a}), \ldots, \neg e_n'(\bar{a}) \} \tag{1}$$

Suppose that $n = 0$. Then the formula

$$\neg e_1(\overline{x}) \vee \ldots \vee \neg e_m(\overline{x})$$

is an equational Horn formula which is not valid in A by (1). Since

$$A \models Horn_K(\Sigma, X),$$

it follows that

$$K \not\models \neg e_1(\overline{x}) \vee \ldots \vee \neg e_m(\overline{x}).$$

So for some algebra $B \in K$,

$$B \models (\exists x_1), \ldots, (\exists x_k)(e_1(\overline{x}) \wedge \ldots \wedge e_m(\overline{x})),$$

and family $\beta : A \to B$ of mappings we have

$$B[\beta] \models \{ \ e_1(\overline{a}), \ \ldots, \ e_m(\overline{a}) \ \}.$$

Suppose that $n \geq 1$. Then for each $1 \leq i \leq n$ the formula

$$\neg e_1(\overline{x}) \vee \ldots \vee \neg e_m(\overline{x}) \vee e_i'(\overline{x})$$

is an equational Horn formula which is not valid in A by (1). So by the same reasoning as above for some algebra $B(i) \in K$, and family $\beta(i) : A \to B(i)$ of mappings we have

$$B(i)[\beta(i)] \models \{ \ e_1(\overline{a}), \ \ldots, \ e_m(\overline{a}), \ \neg e_i'(\overline{a}) \ \}.$$

So for $B = \Pi_{i=1}^n B(i)$ there exists $\beta : A \to B$ such that

$$B[\beta] \models \{ \ e_1(\overline{a}), \ \ldots, \ e_m(\overline{a}), \ \neg e_1'(\overline{a}), \ \ldots, \ \neg e_n'(\overline{a}) \ \}.$$

Since K is closed under the formation of non-empty direct products then, by Proposition 5.3.18.(i), $K[A]$ is closed under the formation of non-empty direct products. Hence $B[\beta] \in K[A]$. Since K is closed under the formation of ultraproducts then by Proposition 5.3.18.(ii), $K[A]$ is closed under the formation of ultraproducts. Since each finite subset of $\Delta(A)$ has a model in $K[A]$, by Theorem 5.3.21, $\Delta(A)$ has a model $B[\beta]$ in $K[A]$ where $B \in K$. Since A is generated by A, by the Extension Lemma 5.3.19, there exists a $\Sigma[A]$ embedding $\phi : A[\alpha] \to B[\beta]$, and hence a Σ embedding $\psi : A \to B$.

Now $\psi(A)$ is a Σ subalgebra of B and since K is closed under the formation of Σ subalgebras, $\psi(A) \in K$. Since $A \cong \psi(A)$ and K is closed under the formation of isomorphic images, $A \in K$.

Since A was arbitrarily chosen, $K^* \subseteq K$. So $K^* = K$. Hence K is an equational Horn class

$$K = Alg(\ \Sigma,\ Horn_K(\Sigma,\ X)\ ).$$

∎

Once again, let us note that in general the axiomatization $Horn_K(\Sigma,\ X)$ for K, delivered by Lemma 5.3.22, is an infinite set of Horn formulas.

Recall from Section 4.3 the following idea.

Definition 5.3.23. Let K be any class of Σ algebras. Then K is a *quasivariety* if, and only if, K is closed under the formation of subalgebras, isomorphic images and reduced products.

Theorem 5.3.24 (Quasivariety Theorem (A.I. Malcev 1966)). *Let K be any class of Σ algebras. The following are equivalent:*

 (i) *K is an equational Horn class;*

 (ii) *K is quasivariety;*

(iii) *K is closed under the formation of subalgebras, isomorphic images, non-empty direct products and ultraproducts.*

Proof.   (i) $\Rightarrow$ (ii) By Lemma 5.3.15, every equational Horn class is a quasivariety.

 (ii) $\Rightarrow$ (iii) If K is closed under the formation of subalgebras, isomorphic images and reduced products then, by Example 4.3.12.(i), K is closed under the formation of subalgebras, isomorphic images, non-empty direct products and ultraproducts.

(iii) $\Rightarrow$ (i) By Lemma 5.3.22, if K is closed under the formation of subalgebras, isomorphic images, non-empty direct products and ultraproducts then K is the class of all models of an equational Horn theory.

∎

A further characterization involving direct limits is available in Exercise 5.3.28.(6).

Corollary 5.3.25. *For any class K of Σ algebras, K is a conditional equational class if, and only if, K is a quasivariety which contains a unit algebra.*

Proof. ⇐ If K is a quasivariety then by the Quasivariety Theorem 5.3.24, K is an equational Horn class. Since K admits unit algebras then by Lemma 5.3.13, K is a conditional equational class.

⇒ If K is a conditional equational class, since conditional equations are logically equivalent to Horn formulas of the form $\neg e_1 \vee \ldots \vee \neg e_n \vee e_{n+1}$, then K is an equational Horn class. So by the Quasivariety Theorem 5.3.24, K is a quasivariety and by Lemma 5.3.13, K admits unit algebras. ∎

Examples 5.3.26.

(i) By Theorem 5.3.24, the classes of all non-unit Boolean algebras and non-unit rings are quasivarieties.

(ii) By Corollary 5.3.25, the classes of all cancellation semigroups and torsion-free Abelian groups are quasivarieties which admit unit algebras.

Definition 5.3.27 (Discussion of First Order Languages). The equations, negated equations, conditional equations and equational Horn formulas encountered in this section, and in Section 5.2, are all simple fragments of the more expressive language of *many-sorted first order logic*. However, in this chapter we have assumed little familiarity with first order logic and most of our definitions and proofs have been made starting from first principles.

To compare these sublanguages of Sections 5.2 and 5.3 with full first order logic, we must consider the general definition of a *many-sorted first order language* over a many-sorted signature Σ.

Let Σ be an S-sorted signature and $X = \langle X_s \mid s \in S \rangle$ be an S-indexed family of sets of variable symbols of sort s. The set $\mathcal{L}(\Sigma, X)$ of all *many-sorted first order formulas with equality* over Σ and X is defined inductively as follows.

(i) For each sort $s \in S$ and any terms $t_1, t_2 \in T(\Sigma, X)_s$, the expression

$$t_1 = t_2 \in \mathcal{L}(\Sigma, X),$$

also known as an *atomic formula, identity* or *equation*.

(ii) If $\phi, \psi \in \mathcal{L}(\Sigma, X)$ are first order formulas then

$$(\neg\phi),\ (\phi \wedge \psi),\ (\phi \vee \psi),\ (\phi \Rightarrow \psi),\ (\phi \Leftrightarrow \psi) \in \mathcal{L}(\Sigma, X).$$

The symbols ¬ (negation) ∧ (conjunction), ∨ (disjunction), ⇒ (implication) and ⇔ (bi-implication) are known as *propositional connectives*.

(iii) If $\phi \in \mathcal{L}(\Sigma, X)$ is a first order formula and $x \in X_s$ is any variable symbol then

$$(\forall x)(\phi), \ (\exists x)(\phi) \in \mathcal{L}(\Sigma, X).$$

The symbols $\forall$ and $\exists$ are known as the *universal* and *existential quantifier* respectively.

First order theories considerably increase the range of classes of algebras that we may axiomatize. These first order or elementary classes are closed under several important constructions such as isomorphisms and ultraproducts (recall remarks in 4.5). However, characterization theorems for general first order axiomatisations are much less algebraic, involving logical notions in essential ways. (See for example Chang and Keisler [1990].)

It is also interesting to reconsider the results of this chapter in the context of first order classes. For example: *If a first order theory* $T \subseteq \mathcal{L}(\Sigma, X)$ *and each of its consistent extensions* $T \subseteq T' \subseteq \mathcal{L}(\Sigma, X)$ *admit an initial model, then* T *is equivalent to a universal Horn theory.*

Variations of this theorem are considered in Malcev [1971] and Mahr and Makowsky [1983]. At this point we have arrived at the border between universal algebra and the branch of mathematical logic known as *model theory*. Model theory concerns itself with first order languages and their semantics. A proper treatment of many-sorted model theory is beyond the scope of this chapter and we can only encourage the reader to pursue this subject further. We have occasionally made remarks or set exercises that concern first order languages, for instance in our discussion of ultraproducts in Section 4.3 and this section, but our aim has been to remain within the discourse of algebra.

Elementary introductions to many-sorted first order logic may be found in Enderton [1972] and Manzano [1992]. A survey of many-sorted logic and its applications in computing science is Meinke and Tucker [1992]. Elementary introductions and advanced works on the subject of model theory are discussed in the section on Further Reading.

Exercises 5.3.28.

1. Show that the classes of all cancellation semigroups and all torsion-free Abelian groups of Example 5.3.3 are not varieties. Is the class of all integral domains a quasivariety?

2. Let Σ be a many-sorted signature and let $\Phi \subseteq NE(\Sigma, X)$ be any set of equations and negated equations. Show that $Alg(\Sigma, \Phi)$ is closed under homomorphic images if, and only if, there exists an equational theory $E \subseteq Eqn(\Sigma, X)$ such that

$$Alg(\Sigma, \Phi) = Alg(\Sigma, E).$$

Give a set of equations and negated equations which is not closed under homomorphic images. Thus conclude that sets of equations and negated equations are strictly more expressive than sets of equations.

3. Let Σ be a many-sorted signature and consider a set of formulas $\Phi = E \cup N$, where $E \subseteq Eqn(\Sigma, X)$ is a set of equations and $N \subseteq Neqn(\Sigma, X)$ is a set of negated equations. Suppose $Alg(\Sigma, \Phi)$ is non-empty. By Corollary 5.3.16, $Alg(\Sigma, \Phi)$ contains an initial algebra $I(\Sigma, \Phi)$ and by Corollary 5.2.15, $Alg(\Sigma, E)$ contains an initial algebra $I(\Sigma, E)$. Is

$$I(\Sigma, \Phi) \cong I(\Sigma, E)?$$

4. Reconsider Exercise (3) in the case that $\Phi = C \cup N$ where $C \subseteq Cond(\Sigma, X)$ is a set of conditional equations and $N \subseteq Horn(\Sigma, X)$ is a set of disjunctions of negated equations. If $Alg(\Sigma, \Phi)$ is non-empty, is

$$I(\Sigma, \Phi) \cong I(\Sigma, C)?$$

5. Let R be a commutative ring. We say that R is *partially orderable* if, and only if, there exists a partial ordering $\leq$ on R such that for all $x, x', y, y' \in R$ and

$$x \leq x' \text{ and } y \leq y' \text{ implies } x + y \leq x' + y'$$

and

$$x \leq x' \text{ and } y \geq 0 \text{ implies } x \cdot y \leq x' \cdot y.$$

Show that the class K of all partially orderable rings forms a quasi-variety. Can you give an equational Horn axiomatization of K?

6. Generalize Theorem 4.3.16 to the case of a many-sorted signature Σ. Hence show that for any many-sorted signature Σ and any class K of Σ algebras, K is an equational Horn class if, and only if, K is closed under the formation of subalgebras, isomorphic images, direct products and direct limits.

7. Let $\Phi \subseteq \mathcal{L}(\Sigma, X)$ be any set of first order formulas. Show that $Alg(\Sigma, \Phi)$ is closed under the formation of isomorphic images and ultraproducts. (Hence if Φ is a set of equational Horn formulas then $Alg(\Sigma, \Phi)$ is closed under the formation of isomorphisms and ultraproducts.)

8. Let $\Phi \subseteq \mathcal{L}(\Sigma, X)$ be any set of universal first order formulas, each of the form

$$(\forall x_1), \ldots, (\forall x_n)\phi(x_1, \ldots, x_n),$$

where $\phi(x_1, \ldots, x_n)$ is a quantifier free formula. Show that $Alg(\Sigma, \Phi)$ is closed under the formation of subalgebras. (Hence if Φ is a set of equational Horn formulas then $Alg(\Sigma, \Phi)$ is closed under the formation of subalgebras.)

9. For each $n \geq 0$ we define the sets of all $\Sigma^n - Horn(\Sigma, X)$ and $\Pi^n - Horn(\Sigma, X)$ of all Σ^n and Π^n equational Horn formulas, by induction as follows:

$$\Sigma^0 - Horn(\Sigma, X) = \Pi^0 - Horn(\Sigma, X) = Horn(\Sigma, X)$$

and for $n \geq 1$,

$$\Sigma^n - Horn(\Sigma, X) = \{ \exists x_1 \ldots \exists x_n \, \phi \mid \phi \in \Pi^{n-1} - Horn(\Sigma, X) \}$$

and

$$\Pi^n - Horn(\Sigma, X) = \{ \forall x_1 \ldots \forall x_n \, \phi \mid \phi \in \Sigma^{n-1} - Horn(\Sigma, X) \}.$$

Show that if Φ is any set of Σ^n or Π^n equational Horn formulas then $Alg(\Sigma, \Phi)$ is closed under the formation of reduced products.

10. By a (*universal*) *infinitary conditional equation* over an S-sorted signature Σ and S-indexed family $X = \langle X_s \mid s \in S \rangle$ of sets X_s of variables, we mean an expression of the form

$$\bigwedge_{i \in I} t_i = t'_i \Rightarrow t = t',$$

where I is any set and for each $i \in I$, $t_i, t'_i \in T(\Sigma, X)_{s(i)}$ and $t, t' \in T(\Sigma, X)_s$ for some sorts $s, s(i) \in S$. We can extend the notions of truth and validity to infinitary conditional equations in the obvious way. Show that if Φ is any set of infinitary conditional equations then $Alg(\Sigma, \Phi)$ is closed under the formation of subalgebras, isomorphic images and direct products. Deduce that:

(a) every infinitary conditional equational class K admits free algebras of all generator cardinalities; and,

(b) every infinitary conditional equational class K admits an initial algebra.

What conditions on Σ and X are needed for (a) and (b)?

11. Establish the converse to (10) by showing that if K is any class of Σ algebras which is closed under subalgebras, isomorphisms and direct

products then $K = Alg(\Sigma, \Phi)$ for some set Φ of infinitary conditional equations.

12. Develop logical calculi for many-sorted conditional equations and equational Horn formulas which are sound and complete. (Hint: consider Selman [1972]).

13. Let K be a quasivariety of Σ algebras. Let Σ' be a subsignature of Σ and let K' be the class of all Σ' algebras that can be Σ' embedded in a Σ algebra of K. Show that K' is a quasivariety.

14. Prove Proposition 5.3.18.

15. Generalize the Ultraproduct Theorem 5.3.20 to arbitrary first order formulas $\phi \in \mathcal{L}(\Sigma, X)$. Hence prove the Compactness Theorem 5.3.21 for an arbitrary first order theory $\Phi \subseteq \mathcal{L}(\Sigma, X)$ which states that: *for any set $\Phi \subseteq \mathcal{L}(\Sigma, X)$, Φ has a model if, and only if, every finite subset of Φ has a model.*

16. Using the Compactness Theorem for first order logic, prove the Upward Löwenheim Skolem Theorem which states: *let $\Phi \subseteq \mathcal{L}(\Sigma, X)$ be a first order theory. If Φ has a countably infinite model then for every infinite cardinal κ, Φ has a model of cardinality κ.*

5.4   Specification of abstract data types

After introducing the basic definitions of algebra we were able, in Section 3.6, to formulate the basic ideas of the theory of abstract data types. In particular, we explained that a many-sorted algebra models a specific representation of a data type, a class of algebras models a class of implementations, and isomorphisms between algebras are used to discuss data in a way that is independent of specific representations.

With the basic definitions and results about equational theories and their initial models of Section 5.2, we are now able to formulate further basic ideas concerning the axiomatic specification of abstract data types.

To specify a data type to a potential user or to a language implementor, we may first list the names of the sets of data, distinguished elements, and operations belonging to the type, in a many-sorted signature Σ. Then we may describe their behaviour by means of a set T of axioms that the elements and operations must obey. The pair (Σ, T) we will call an *axiomatic specification* of a data type.

The semantics of this specification (Σ, T) we may take to be the class $Alg(\Sigma, T)$ of all Σ algebras satisfying the axioms of T. To reflect our conception of an abstract data type, we must assume that the axioms of T allow $Alg(\Sigma, T)$ to be closed under isomorphism. The type of axioms that will be of most use to us are equations (and conditional equations and equational Horn formulas), for which this assumption is true.

Consider this axiomatic approach in the simple case of the data type of natural numbers defined by equations. Recalling Example 5.1.15.(iii), we redefine the data type specification (Σ^{PA}, E) as follows

data type	PA;
sorts	nat;
constants	$0 : \rightarrow nat$;
operations	$succ : nat \rightarrow nat$;
	$add : nat \times nat \rightarrow nat$;
	$mult : nat \times nat \rightarrow nat$;
variables	$x, y : nat$;
axioms	
	$add(0, x) = x$;
	$add(x, succ(y)) = succ(add(x, y))$;
	$mult(x, 0) = 0$;
	$mult(x, succ(y)) = add(mult(x, y), x)$;
end	

Now, the semantics of (Σ^{PA}, E) is the class $Alg(\Sigma^{PA}, E)$ of *all* algebras satisfying the equations. This class is a variety so it is closed under subalgebras, homomorphisms and direct products (as well as a host of constructions like direct and inverse limits, of course). In particular, in addition to the standard algebra

$$PA(\mathbf{N}) = (\mathbf{N}; \ 0; \ Succ, \ Add, \ Mult)$$

where $\mathbf{N} = \{\ 0, \ 1, \ 2, \ \ldots \ \}$ is the set of natural numbers (in decimal notation), $Alg(\Sigma^{PA}, E)$ contains many other algebras which are not isomorphic with $PA(\mathbf{N})$. In this case the specification may be considered to have rather loose semantics. The standard algebra of natural numbers can be recovered from the semantics of (Σ^{PA}, E) by noting that the initial model of (Σ^{PA}, E) is isomorphic with $PA(\mathbf{N})$.

We may summarize the general ideas about data type specification as follows. The desired semantics of an axiomatic data type specification (Σ, T) is some class $K \subseteq Alg(\Sigma, T)$ closed under isomorphism. The algebras of K may need to satisfy certain properties, they are likely to be minimal and/or computable, for example. Often, as in the case of the natural numbers, the class K will be the isomorphism class of a specific algebra.

In this latter case, a method M is needed that assigns to an appropriate specification (Σ, T) an algebra $M(\Sigma, T)$ that is unique up to isomorphism. According to the results in this section, a convenient method is to use an initial algebra $I(\Sigma, T)$ for the semantics of (Σ, T) in the cases when T is a set of:

   (i) equations (Theorem 5.2.15);

  (ii) conditional equations (Corollary 5.3.5);

 (iii) equational Horn formulas (Corollary 5.3.16).

In these cases, and with good reasons, we may think of the initial algebra $I(\Sigma, T)$ as a standard algebra satisfying the axioms of (Σ, T). It is also possible to use a final algebra as a semantics, however this is less convenient because existence theorems for final algebras are less general. At the heart of algebraic specification theory for data types are the following definitions.

Definition 5.4.1. Let A be a many-sorted minimal algebra with signature Σ. Then

   (i) A has an *equational specification under initial algebra semantics* if there is a set E of equations over Σ such that

$$I(\Sigma, E) \cong A.$$

  (ii) A has an *equational specification with hidden functions and hidden sorts under initial algebra semantics* if there is a signature Σ' and a set E' of equations over Σ' such that Σ is a subsignature of Σ' and

$$I(\Sigma', E')|_{\Sigma} \cong A.$$

Similar definitions can be made in the cases where E and E' contain conditional equations and equational Horn formulas. Furthermore, final algebras may be substituted for initial algebras to obtain further methods for specifying data types.

To illustrate the use of case (i) of the above definition we need only recall that

$$PA(\mathbf{N}) \cong I(\Sigma^{PA}, E)$$

and so (Σ^{PA}, E) is an equational specification under initial algebra semantics of $PA(\mathbf{N})$.

To illustrate the use of case (ii) of the above definition, consider the algebra

$$SQ = (\mathbf{N}; \; 0; \; Succ, \; Square)$$

where $Square(x) = x^2$; let SQ have signature Σ^{SQ}. This algebra can be defined by extending $PA(\mathbf{N})$ and (Σ^{PA}, E) with $Square$ and an axiom defining $Square$ in terms of $Mult$. Thus we may choose (Σ', E') where

$$\Sigma' = \Sigma^{PA} \cup \{ \; square \; \}$$

and

$$E' = E \cup \{ \; square(x) = mult(x, x) \; \}$$

and rewrite

data type	SQ;
sorts	nat;
constants	$0 : \; \to nat$;
operations	$succ : nat \to nat$;
	$square : nat \to nat$;
hidden operations	$add : nat \times nat \to nat$;
	$mult : nat \times nat \to nat$;
variables	$x, \, y : nat$;
axioms	
	$add(0, \, x) = x$;
	$add(x, \, succ(y)) = succ(add(x, \, y))$;
	$mult(x, \, 0) = 0$;
	$mult(x, \, succ(y)) =$
	$\quad add(mult(x, \, y), \, x)$;
	$square(x) = mult(x, \, x)$;
end	

The point is that

$$A \cong I(\Sigma', E')|_{\Sigma^{SQ}}.$$

However it is impossible to give a finite equational specification (Σ^{SQ}, E) such that

$$I(\Sigma^{SQ}, E) \cong A$$

(see Bergstra and Tucker [1987]). Thus hidden functions are both natural and necessary in making algebraic specifications.

For convenience, we give a few suggestions for further reading on the algebraic theory of data types.

General references are Ehrig and Mahr [1985, 1989] and an invaluable bibliography is Kutzler and Lichtenberger [1983].

Some of the early papers remain excellent introductions to the basic ideas and results of the subject. We recommend Goguen *et al.* [1978]. The extension of axiomatic specifications from equations to include conditional equations, hidden functions, and even hidden sorts, arises because many algebras have very simple specifications using these methods or cannot be specified without them.

An important topic is the scope and limits of the specification methods: do algebraic specifications under initial algebra semantics define all and only the algebras one desires? In order to make precise and answer questions of this form, in addition to constructing examples, we must turn to the theory of effectively computable algebras. Detailed surveys of the topic are Bergstra and Tucker [1987] also Meseguer and Goguen [1985].

Case studies in software design and related developments can be seen in Bergstra *et al.* [1989] and Wirsing and Bergstra [1989]. Interesting work on the formal verification of compilers can be seen in Thatcher *et al.* [1980]. Substantial work on algebraic methods for hardware has been discussed in Section 2.4.

The algebraic methods described have led to extensive developments in the design and construction of specification languages, programming languages and theorem proving systems. An early representative example is the system OBJ which may be charted in Goguen and Tardo [1979], Goguen and Winkler [1988], Goguen [1989a] and Goguen [1990].

These practical concerns have led to new algebra. For example, higher order universal algebra in Meinke [1992a, 1992b], and the universal algebra of modules in Bergstra *et al.* [1990].

The mathematical theory, software tools and practical applications of these algebraic methods for abstract data types are too substantial to sketch adequately in this Handbook chapter on Universal Algebra. A chapter on Algebraic Specification is planned for a later volume of this Handbook, and significant topics will arise in other chapters, for example, that of *computable data types* and the existence of *complete term rewriting systems* in the Handbook chapters on Effective Structures in a later volume, and Term Rewriting in volume 2.

5.5   Remarks on references

Most of the material of Sections 5.1 and 5.2 (in the single-sorted case) is due to G. Birkhoff and first appeared in Birkhoff [1935]; this includes free algebras, equational logic, soundness and completeness theorems, and the Variety Theorem. This important work has been treated in several subsequent books (e.g. Grätzer [1979] and Burris and Sankappanavar [1981]). The many-sorted case was considered in P.J. Higgins [1963], motivated by the need to apply a many-sorted theory to generators and relations for categories, and also in Birkhoff and Lipson [1970].

In the many-sorted case, the problem of empty sorts in equational logic was pointed out recently, in Goguen and Meseguer [1982]. The solution involving declared equations is based on Goguen and Meseguer [1982]. The technique of declared variables appears in the work of A. Mostowski on a version of single-sorted first order logic called *free logic*. The delightful counterexample in 5.2.20.(2) is due to H. Simmons. Like the closure constructions on classes in Section 4, the theory of equational classes enjoys a special historical relationship with groups. Some references that explore this are: Magnus *et al.* [1976] and Lyndon and Schupp [1977].

The role of initial algebras in semantics was developed by J.A. Goguen, J.W. Thatcher, E. Wagner and J.B. Wright in Goguen [1975], Goguen and Thatcher [1974], Goguen *et al.* [1977].

The study of simple generalization of equations, such as conditional equations and Horn formulas, begins in McKinsey [1943]. The main results in Section 5.3 are observations of A.I. Malcev who thoroughly studied the algebra and logic of conditional equations and other formulas in a series of papers starting 1956: see Chapters 4, 10 and 31 of Malcev [1971] and Malcev [1973]. Applications of the characterization theorems in Sections 5.2 and 5.3 to prove Craig interpolation theorems for equational and conditional equational logic appear in Rodenburg [1991, 1992]. Interesting characterization theorems for infinitary conditional equations are given in Wechler [1991]. Some of these have been set as exercises in 5.3.28.

6 Further reading

Starting from this chapter, there are four paths that can be usefully explored, in the directions of the following subjects:

1. universal algebra;

2. model theory;

3. specific algebraic theories in mathematics;

4. specific algebraic theories in computer science.

In Section 2, we have already suggested some further reading for (3), on semigroups, groups, rings, fields, Boolean algebras, lattices and so on. Each section contains some further information for (4); for example, on abstract data types (in 3.6 and 5.4) and process algebra (in Discussion 4.2.26). We will conclude with some comments on general literature for (1) and (2). These notes are not intended to be comprehensive or historically sensitive.

6.1 Universal algebra

This chapter of the Handbook is an introduction to many-sorted universal algebra that is based on topics that are relevant to computer science. Thus, there are many subjects that have been neglected. The reader interested in a thorough knowledge of universal algebra is encouraged to study the following (most of which are concerned with single-sorted algebras).

The primary references for the subject are the books of Cohn [1981] and Grätzer [1979], first published in 1965 and 1968 respectively, and both essential reading. Useful bibliographies can be found in these works including the survey Taylor [1979] which also appears in Grätzer [1979]. Another essential work is the book Malcev [1973], and the collected papers Malcev [1971]. Special subjects are treated in Freese and McKenzie [1987], Hobby and McKenzie [1988], Lausch and Nöbauer [1973], McKenzie and Valeriote [1989] and Plotkin [1972].

Two mathematical textbooks are Burris and Sankappanavar [1981] and McKenzie *et al.* [1987]. A recent monograph on universal algebra for computer science is Wechler [1992].

6.2 Model theory

A number of the subjects we have treated are best understood from the point of view of the theory of algebras or models, satisfying axioms written in first order many-sorted theories. The subjects include ultraproducts (in Section 4.3) and the characterization theorems for conditional and Horn axiomatizations (in Section 5.3). Some of the reasons have been indicated in various places.

The primary references for the model theory of first order (single-sorted) theories are Chang and Keisler [1990] and Barwise and Feferman [1985]; some survey articles in Barwise [1977] are useful. An important bibliography is Ebbinghaus [1987].

For students new to model theory, the following textbooks are suitable for serious study: Barnes and Mack [1975]; Bell and Slomson [1969]; Bridge [1977]; Hodges [1985]; Kreisel and Krivine [1967]. It is also important to examine early books and papers on the subject. For example, the monograph Robinson [1963]; and the collected papers Robinson [1979] and Malcev [1971]; and the conference Addison *et al.* [1965].

References

[Addison *et al.*, 1965] J. Addison, L. Henkin, and A. Tarski, editors. *The*

Theory of Models. North-Holland, Amsterdam, 1965.

[Arnold and Nivat, 1980a] A. Arnold and M. Nivat. Metric interpretations of infinite trees and semantics of nondeterministic recursive programs. *Theoretical Computer Science*, 11:181–205, 1980.

[Arnold and Nivat, 1980b] A. Arnold and M. Nivat. The metric space of infinite trees. algebraic and topological properties. *Fundamentae Informaticae*, 4:445–476, 1980.

[Asveld and Tucker, 1982] P. Asveld and J. V. Tucker. Complexity theory and the operational structure of algebraic programming systems. *Acta Informatica*, 17:451–476, 1982.

[Back, 1981] R. J. R. Back. On correct refinement of programs. *J. Computer and System Sciences*, 23:49–68, 1981.

[Baeten and Weijland, 1990] J. C. M. Baeten and P. Weijland. *Process Algebra*. Cambridge University Press, 1990.

[Baeten, 1990] J. C. M. Baeten. *Applications of Process Algebra*. Cambridge University Press, 1990.

[Barnes and Mack, 1975] D. W. Barnes and J. M. Mack. *An Algebraic Introduction to Mathematical Logic*. Springer, Berlin, 1975.

[Barwise and Feferman, 1985] J. Barwise and S. Feferman, editors. *Model-Theoretic Logics*. Springer Verlag, Berlin, 1985.

[Barwise, 1977] J. Barwise, editor. *Handbook of Mathematical Logic*. North-Holland, Amsterdam, 1977.

[Bell and Slomson, 1969] J. L. Bell and A. B. Slomson. *Models and Ultraproducts: an Introduction*. North-Holland, Amsterdam, 1969.

[Bergstra and Heering, 1989] J. A. Bergstra and J. Heering. Which data types have ω-complete initial algebra specifications? Technical Report CS-R8958, Dept of Software Technology, Centrum voor Wiskunde en Informatica, Amsterdam, 1989.

[Bergstra and Tucker, 1980a] J. A. Bergstra and J. V. Tucker. A characterisation of computable data types by means of a finite equational specification method. In J. W. de Bakker and J. van Leeuwen (Editors), editors, *Automata, Languages and Programming, Seventh Colloquium*, pages 76–90. Lecture Notes in Computer Science 81. Springer, Berlin, 1980.

[Bergstra and Tucker, 1980b] J. A. Bergstra and J. V. Tucker. A natural data type with a finite equational final semantics specification but no effective equational initial semantics specification. *Bulletin of the European Association for Theoretical Computer Science*, 11:23–33, 1980.

[Bergstra and Tucker, 1983] J. A. Bergstra and J. V. Tucker. Initial and final algebra semantics for data type specifications: two characterisation theorems. *SIAM Journal on Computing*, 12:366–387, 1983.

[Bergstra and Tucker, 1985] J. A. Bergstra and J. V. Tucker. Top-down design and the algebra of communicating processes. *Science of Computer Programming*, 5:171–199, 1985.

[Bergstra and Tucker, 1987] J. A. Bergstra and J. V. Tucker. Algebraic specifications of computable and semicomputable data types. *Theoretical Computer Science*, 50:137–181, 1987.

[Bergstra and Tucker, 1988] J. A. Bergstra and J. V. Tucker. The inescapable stack: an exercise in algebraic specification with total functions. Technical Report Report 13. 88, Centre for Theoretical Computer Science, University of Leeds, 1988.

[Bergstra *et al.*, 1982] J. A. Bergstra, J. Tiuryn, and J. V. Tucker. Floyd's principle, correctness theories and program equivalence. *Theoretical Computer Science*, 17:113–149, 1982.

[Bergstra *et al.*, 1989] J. A. Bergstra, J. Heering, and P. Klint, editors. *Algebraic Specification*. Addison-Wesley Publishing Company, England, 1989.

[Bergstra *et al.*, 1990] J. A. Bergstra, J. Heering, and P. Klint. Module algebra. *Journal of the Association for Computing Machinery*, 37:335–372, 1990.

[Birkhoff and Lipson, 1970] G. Birkhoff and J. D. Lipson. Heterogeneous algebras. *Journal of Combinatorial Theory*, 8:115–133, 1970.

[Birkhoff and MacLane, 1965] G. Birkhoff and S. MacLane. *A Survey of Modern Algebra*. Macmillan, third edition, 1965.

[Birkhoff, 1933] G. Birkhoff. On the combination of subalgebras. *Proceedings of the Cambridge Philosophical Society*, 29:441–464, 1933.

[Birkhoff, 1935] G. Birkhoff. On the structure of abstract algebras. *Proceedings of the Cambridge Philosophical Society*, 31:433–454, 1935.

[Birkhoff, 1944] G. Birkhoff. Subdirect unions in universal algebra. *Bulletin of the American Mathematical Society*, 50:764–768, 1944.

[Birkhoff, 1967] G. Birkhoff. *Lattice Theory*, volume 25 of *Colloquium Publications*. American Mathematical Society, Providence, third edition, 1967.

[Bridge, 1977] J. Bridge. *Beginning Model Theory: The Completeness Theorem and Some Consequences.* Oxford: Clarendon Press, 1977.

[Broy and Wirsing, 1982] M. Broy and M. Wirsing. Partial abstract data types. *Acta Informatica,* 80:47–64, 1982.

[Burmeister, 1986] P. Burmeister. *A Model Theoretic Oriented Approach to Partial Algebras,* volume 31 of *Mathematical Research.* Akademie-Verlag, Berlin, 1986.

[Burris and Sankappanavar, 1981] S. Burris and H. P. Sankappanavar. *A Course in Universal Algebra.* Springer, Berlin, 1981.

[Burstall and Landin, 1969] R. Burstall and P. Landin. Programs and their proofs: an algebraic approach. In B. Meltzer and D. Michie, editors, *Machine Intelligence 4,* pages 17–43. Edinburgh University Press, 1969.

[Chang and Keisler, 1990] C. C. Chang and H. J. Keisler. *Model Theory.* North-Holland, Amsterdam, third edition, 1990.

[Cleave, 1969] J. P. Cleave. Local properties of systems. *J. London Mathematical Society,* 1(44):121–130, 1969. Addendum J. LMS (2) 1 (1969), 384.

[Cleave, 1992] J. P. Cleave. *A Study of Logic.* Oxford University Press, 1992.

[Codd, 1968] E. F. Codd. *Cellular Automata.* Academic Press, 1968.

[Cohn, 1981] P. M. Cohn. *Universal Algebra.* D. Reidel, Dordrecht, second edition, 1981.

[Cohn, 1982] P. M. Cohn. *Algebra, Volume 1.* Wiley, second edition, 1982.

[Davis et al., 1976] M. Davis, Y. Matijasevic, and J. Robinson. Hilbert's tenth problem; positive aspects of a negative solution. In F. E. Browder, editor, *Mathematical Developments Arising from Hilbert Problems,* pages 323–378. American Mathematical Society, Providence, RI, 1976.

[de Bakker and Zucker, 1984] J. W. de Bakker and J. I. Zucker. Denotational semantics of concurrency. *Information and Control,* 69:109–137, 1984.

[Dubbey, 1977] J. M. Dubbey. Babbage, peacock and modern algebra. *Historia Mathematica,* 4:295–302, 1977.

[Dubbey, 1978] J. M. Dubbey. *The mathematical work of Charles Babbage.* Cambridge University Press, Cambridge, 1978.

[Ebbinghaus, 1987] H. D. Ebbinghaus. Ω-*Bibliography of Mathematical Logic: Model Theory.* Springer, Berlin, 1987.

[Ehrig and Mahr, 1985] H. Ehrig and B. Mahr. *Fundamentals of Algebraic Specification 1: Equations and Initial Semantics.* EATCS Monographs on Theoretical Computer Science 6. Springer Verlag, Berlin, 1985.

[Ehrig and Mahr, 1989] H. Ehrig and B. Mahr. *Fundamentals of Algebraic Specification 2: Module Specifications and Constraints.* EATCS Monographs on Theoretical Computer Science 21. Springer Verlag, Berlin, 1989.

[Eilenberg and Steenrod, 1952] S. Eilenberg and N. E. Steenrod. *Foundations of Algebraic Topology.* Princeton University Press, Princeton, 1952.

[Eilenberg, 1974] S. Eilenberg. *Automata, Languages and Machines, Volume A.* Academic Press, New York, 1974.

[Eilenberg, 1976] S. Eilenberg. *Automata, Languages and Machines, Volume B.* Academic Press, New York, 1976.

[Eker and Tucker, 1989] S. M. Eker and J. V. Tucker. Specification and verification of synchronous concurrent algorithms: a case study of the pixel planes architecture. In P. M. Dew, R. A. Earnshaw, and T. R. Heywood, editors, *Parallel Processing for Computer Vision and Display*, pages 16–49. Addison-Wesley, 1989.

[Eker, 1990] S. M. Eker. *Foundations for the Design of Rasterisation Algorithms and Architectures.* PhD thesis, School of Computer Studies, University of Leeds, 1990.

[Eklof, 1977] P. C. Eklof. *Ultraproducts for algebraists*, pages 105–135. in: J. Barwise, 1977.

[Enderton, 1972] H. B. Enderton. *A Mathematical Introduction to Logic.* Academic Press, New York, 1972.

[Engeler, 1988] E. Engeler. Representation of varieties in combinatory algebras. *Algebra Universalis*, 25:85–95, 1988.

[Fisher, to appear] M. Fisher. Abstractness and compositionality. *Formal Aspects of Computing*, to appear.

[Frayne *et al.*, 1962] T. E. Frayne, A. C. Morel, and D. S. Scott. Reduced direct products. *Fundamentae Mathematica*, 51:195–228, 1962.

[Freese and McKenzie, 1987] R. Freese and R. McKenzie. Commutator theory for congruence modular varieties. *London Mathematical Society Lecture Note Series*, 125, 1987.

[Fuchs, 1963] L. Fuchs. *Partially ordered algebraic systems.* Pergamon Press, Oxford, 1963.

[Goguen and J.Thatcher, 1974] J. A. Goguen and J.Thatcher. Initial algebra semantics. In *Proc. Fifteenth Symp. on Switching and Automata Theory*, pages 63–77. IEEE, 1974.

[Goguen and Meseguer, 1982] J. A. Goguen and J. Meseguer. Completeness of many–sorted equational logic. *Association for Computing Machinery SIGPLAN Notices*, 17:9–17, 1982.

[Goguen and Tardo, 1979] J. A. Goguen and J. Tardo. An introduction to obj: a language for writing and testing software specifications. In M. Zelkowitz, editor, *Specification of Reliable Software*, pages 18–29. IEEE Press, 1979.

[Goguen and Winkler, 1988] A. Goguen and T. Winkler. Introducing OBJ3. Technical Report SRI–CSL–88–9, Computer Science Laboratory, SRI International, Menlo Park, 1988.

[Goguen *et al.*, 1975] J. A. Goguen, J. Thatcher, E. Wagner, and J. B. Wright. Abstract data types as initial algebras and the correctness of data representations. In A. Klinger, editor, *Computer Graphics, Pattern Recognition and Data Structure*, pages 89–93. IEEE, 1975.

[Goguen *et al.*, 1977] J. A. Goguen, J. Thatcher, E. Wagner, and J. B. Wright. Initial algebra semantics and continuous algebras. *J. Association for Computing Machinery*, 24:68–95, 1977.

[Goguen *et al.*, 1978] J. A. Goguen, J. Thatcher, and E. Wagner. An initial algebra approach to the specification, correctness and implementation of abstract data types. In R. T. Yeh, editor, *Current Trends in Programming Methodology IV*, pages 80–149. Prentice Hall, 1978.

[Goguen, 1975] J. A. Goguen. Semantics of computation. In E. G. Manes, editor, *Proc. First. Int. Symp. on Category Theory Applied to Computation and Control*, pages 151–163, Berlin, 1975. Lecture Notes in Computer Science, 25. Springer.

[Goguen, 1989a] J. A. Goguen. Obj as a theorem prover with applications to hardware verification. In V. P. Subrahmanyam and G. Birtwhistle, editors, *Current Trends in Hardware Verification and Automated Theorem Proving*, pages 218–267. Springer, Berlin, 1989a.

[Goguen, 1989b] J. A. Goguen. Memories of ADJ. *Bulletin of the European Association for Theoretical Computer Science*, 39:97–102, 1989b.

[Goguen, 1990] J. A. Goguen. Proving and rewriting. In *Proceedings of the Second International Conference on Algebraic and Logic Programming*, 1990.

[Grätzer, 1979] G. Grätzer. *Universal Algebra*. Springer, Berlin, second edition, 1979.

[Gries, 1978] D. Gries, editor. *Programming Methodology*. Springer, Berlin, 1978.

[Groote, 1990] J. F. Groote. A new strategy for proving ω–completeness applied to process algebra. In J. C. M. Baeten and J. W. Klop, editors, *CONCUR' 90 Theories of Concurrency: Unification and Extension*, pages 314–331. Lecture Notes in Computer Science. Springer, Berlin, 1990.

[Guessarian, 1981] I. Guessarian. *Algebraic Semantics*. Lecture Notes in Computer Science 99. Springer, Berlin, 1981.

[Guttag and Horning, 1978] J. V. Guttag and J. J. Horning. The algebraic specification of abstract data types. *Acta Informatica*, 10:27–52, 1978.

[Guttag, 1975] J. V. Guttag. *The specification and application to programming of abstract data types*. PhD thesis, University of Toronto, 1975.

[Hajnal and Nemeti, 1979] A. Hajnal and I. Nemeti. Applications of universal algebra, model theory and categories in computer science. *Computational Linguistics and Computer Languages*, 13:251–282, 1979.

[Halmos, 1963] P. Halmos. *Lectures on Boolean Algebras*. Van Nostrand, Princeton, 1963.

[Harman, 1989] N. A. Harman. *Formal Specifications for Digital Systems*. PhD thesis, School of Computer Studies, University of Leeds, 1989.

[Heering, 1986] J. Heering. Partial evaluation and ω-completeness of algebraic specifications. *Theoretical Computer Science*, 43:149–167, 1986.

[Henkin, 1977] L. Henkin. The logic of equality. *American Mathematical Monthly*, 83:597–612, 1977.

[Hennessy, 1988] M. Hennessy. *Algebraic Theory of Processes*. MIT Press, 1988.

[Herstein, 1964] I. N. Herstein. *Topics in Algebra*. Wiley, New York, first edition, 1964.

[Higgins, 1963] P. J. Higgins. Algebras with a scheme of operators. *Mathematische Nachrichten*, 27:115–132, 1963.

[Hoare, 1969] C. A. R. Hoare. An axiomatic basis for computer programming. *Communications of the Association for Computing Machinery*, 12:576–580, 1969.

[Hobby and McKenzie, 1988] D. Hobby and R. McKenzie. The structure of finite algebras. *Contemporary Mathematics*, 76, 1988. American Mathematical Society.

[Hobley et al., 1988] K. Hobley, B. C. Thompson, and J. V. Tucker. Specification and verification of synchronous concurrent algorithms: a case study of a convolution algorithm. In G. Milne, editor, *Proceedings of IFIP Working Group 10. 2 Working Conference on The Fusion of Hardware Design and Verification*, pages 375–374. North-Holland, Amsterdam, 1988.

[Hodges, 1985] W. Hodges. *Building Models by Games*. London Mathematical Society Student Texts. Cambridge University Press, Cambridge, 1985.

[Horn, 1951] A. Horn. On sentences which are true of direct unions of algebras. *Journal of Symbolic Logic*, 16:14–21, 1951.

[Huet and Oppen, 1980] G. Huet and D. C. Oppen. Equations and rewrite rules: a survey. In R. Book, editor, *Formal Languages, Perspectives and Open Problems*. Academic Press, New York, 1980.

[Jacobson, 1951 1953 1964] N. Jacobson. *Lectures in Abstract Algebra*, volume I, II, III. van Nostrand, 1951, 1953, 1964.

[Jones, 1980] N. D. Jones, editor. *Semantics-directed compiler generation*. Lecture Notes in Computer Science 94. Springer, Berlin, 1980.

[Kargapolov and Merzljakov, 1979] M. I. Kargapolov and J. I. Merzljakov. *Fundamentals of the Theory of Groups*. Graduate Texts in Mathematics. Springer, Berlin, 1979.

[Kegel and Wehrfritz, 1973] O. H. Kegel and B. A. F. Wehrfritz. *Locally Finite Groups*. North-Holland, Amsterdam, 1973.

[Keisler, 1979] H. J. Keisler. *Model Theory for Infinitary Logic*. North-Holland, Amsterdam, second edition, 1979.

[Klop, 1992] J. W. Klop. Term rewriting systems. In *Handbook of Logic in Computer Science, Volume 2*, pages 1–111. Oxford University Press, 1992.

[Kneale and Kneale, 1962] W. Kneale and M. Kneale. *The Development of Logic*. Oxford University Press, Oxford, 1962.

[Knuth and Bendix, 1970] D. E. Knuth and P. B. Bendix. Simple word problems in universal algebras. In J. Leech, editor, *Computational problems in abstract algebra*. Pergamon, Oxford, 1970.

[Kreisel and Krivine, 1967] G. Kreisel and J. L. Krivine. *Elements of Mathematical Logic: Model Theory*. North-Holland, Amsterdam, 1967.

[Krohn and Rhodes, 1965] K. B. Krohn and J. L. Rhodes. Algebraic theory of machines I: the main decomposition theorem. *Trans. American Mathematical Society*, 116:450–464, 1965.

[Kung, 1982] H. T. Kung. Why systolic architectures? *Computer*, 15:37–46, 1982.

[Kutzler and Lichtenberger, 1983] B. Kutzler and F. Lichtenberger. Bibliography on abstract data types. *Informatik Fachberichte*, 68, 1983. Springer Verlag, Berlin.

[Lallement, 1979] G. Lallement. *Semigroups and Combinatorial Applications*. J. Wiley, Chichester, 1979.

[Lang, 1965] S. Lang. *Algebra*. Addison Wesley, 1965.

[Lausch and Nöbauer, 1973] H. Lausch and W. Nöbauer. *Algebra of Polynomials*. North-Holland, Amsterdam, 1973.

[Liskov and Zilles, 1975] B. H. Liskov and S. N. Zilles. Specification techniques for data abstractions. *IEEE Transactions on Software Engineering*, SE-1:7–19, 1975.

[Lyndon and Schupp, 1977] R. C. Lyndon and P. E. Schupp. *Combinatorial Group Theory*. Springer, Berlin, 1977.

[Lyndon, 1959] R. C. Lyndon. Properties preserved in subdirect products. *Pacific Journal of Mathematics*, 9:155–164, 1959.

[Magnus et al., 1976] W. Magnus, A. Karass, and D. Solitar. *Combinatorial group theory*. Dover, second edition, 1976.

[Mahr and Makowsky, 1983] B. Mahr and J. A. Makowsky. Characterising specification languages which admit initial semantics. In *Proceedings of 8th CAAP*, pages 300–316. Lecture Notes in Computer Science 159. Springer, Berlin, 1983.

[Malcev, 1941] A. I. Malcev. A general method for obtaining local theorems in group theory, 1941. Chapter 2 in Malcev [1971].

[Malcev, 1958] A. I. Malcev. On homomorphisms onto finite groups. *Uč. zap. Ivanovskogo ped. in-ta*, 18:49–60, 1958.

[Malcev, 1959] A. I. Malcev. Model correspondences, 1959. Chapter 11 in Malcev [1971].

[Malcev, 1971] A. I. Malcev. *The Metamathematics of Algebraic Systems: Collected papers 1936–1967.* Translated and edited by B. F. Wells III. North-Holland, Amsterdam, 1971.

[Malcev, 1973] A. I. Malcev. *Algebraic Systems.* Grundlehren der Mathematischen Wissenschaften, Volume 192. Springer, Berlin, 1973.

[Manzano, 1992] M. Manzano. Introduction to many-sorted logic. In K. Meinke and J. V. Tucker, editors, *Many-sorted Logic and its Applications.* John Wiley, 1992. to appear.

[Martin, 1989] A. Martin. *Specification and Simulation of Synchronous Concurrent Algorithms.* PhD thesis, School of Computer Studies, University of Leeds, 1989.

[McConnell and Tucker, 1992] B. McConnell and J. V. Tucker. Infinite synchronous concurrent algorithms: the specification and verification of a hardware stack. In H. Schwichtenberg, editor, *Logic and Algebra for Specification.* Springer, Berlin, 1992.

[McCulloch and Pitts, 1943] W. S. McCulloch and W. Pitts. A logical calculus of ideas imminent in nervous activity. *Bulletin of Mathematical Biophysics*, 2(5):115–133, 1943.

[McEvoy and Tucker, 1990] K. McEvoy and J. V. Tucker. *Theoretical Foundations of VLSI design.* Cambridge Tracts in Theoretical Computer Science 10. Cambridge University Press, Cambridge, 1990.

[McKenzie and Valeriote, 1989] R. McKenzie and M. Valeriote. *The Structure of Decidable Locally Finite Varieties.* Birkhäuser, Basel, 1989.

[McKenzie et al., 1987] R. N. McKenzie, G. F. McNulty, and W. F. Taylor. *Algebras, Lattices, Varieties*, volume 1. Wadsworth and Brookes Cole, Monterey, 1987.

[McKinsey, 1943] J. C. C. McKinsey. The decision problem for some classes of sentences without quantifiers. *Journal of Symbolic Logic*, 8:61–76, 1943.

[McNulty, 1989] G. F. McNulty. An equational logic sampler. In N. Dershowitz, editor, *Rewriting Techniques and Applications*, pages 234–262. Lecture Notes in Computer Science 255. Springer, Berlin, 1989.

[Meinke and Tucker, 1988] K. Meinke and J. V. Tucker. The scope and limits of synchronous concurrent computation. In F. H. Vogt, editor,

Concurrency '88, pages 163–180. Lecture Notes in Computer Science 335. Springer, Berlin, 1988.

[Meinke and Tucker, 1992] K. Meinke and J. V. Tucker, editors. *Many-sorted Logic and its Applications*. John Wiley, 1992. to appear.

[Meinke, 1988] K. Meinke. *A Graph-theoretic Model of Synchronous Concurrent Algorithms*. PhD thesis, School of Computer Studies, University of Leeds, 1988.

[Meinke, 1992a] K. Meinke. Universal algebra in higher types. *Theoretical Computer Science*, 99, 1992a. to appear.

[Meinke, 1992b] K. Meinke. Subdirect representation of higher type algebras, 1992b. To appear in: Meinke and Tukcer, [1992].

[Meseguer and Goguen, 1985] J. Meseguer and J. A. Goguen. Initiality, induction and computation. In M. Nivat and J. Reynolds, editors, *Algebraic Methods in Semantics*, pages 459–541. Cambridge University Press, Cambridge, 1985.

[Minsky, 1967] M. Minsky. *Computation: Finite and Infinite Machines*. McGraw-Hill, 1967.

[Neumann, 1967] H. Neumann. *Varieties of Groups*. Springer, Berlin, 1967.

[os, 1955] J. Łos. Quelques remarques, théorèms et problèms sur les classes définissables d'algébres. In *Mathematical Interpretation of Formal Systems*, pages 98–113. North-Holland, Amsterdam, 1955.

[Peacock, 1830] G. Peacock. *A Treatise on Algebra*. Cambridge, 1830.

[Pin, 1986] J. Pin. *Varieties of Formal Languages*. North Oxford/Plenum, London/New York, 1986.

[Plotkin, 1972] B. I. Plotkin. *Groups of Automorphisms of Algebraic Systems*. Walters-Noordhoff, Groningen, 1972.

[Rasiowa and H., 1970] R. Rasiowa and Sikorski H. *The Mathematics of Metamathematics*. PWN, Warsaw, third edition, 1970.

[Reichel, 1987] H. Reichel. *Initial Computability, Algebraic Specifications, and Partial Algebras*. Clarendon Press, Oxford, 1987.

[Robinson, 1963] A. Robinson. *Introduction to Model Theory and to the Metamathematics of Algebra*. North-Holland, Amsterdam, 1963.

[Robinson, 1972] D. J. S. Robinson. *Finiteness Conditions and Generalised Soluble Groups*. Part 1. Springer, Berlin, 1972.

[Robinson, 1979] A. Robinson. *Model Theory and Algebra: Selected Papers*, volume 1. Yale University Press, New Haven, 1979.

[Rodenburg and van Glabbeek, 1988] P. H. Rodenburg and R. J. van Glabbeek. An interpolation theorem in equational logic. Technical Report CWI Report CS-R8838, CWI, Amsterdam, 1988.

[Rodenburg, 1991] P. H. Rodenburg. A simple algebraic proof of the equational interpolation theorem. *Algebra Universalis*, 28:48–51, 1991.

[Rodenburg, 1992] P. H. Rodenburg. Interpolation in conditional equational logic. *Fundamenta Informaticae*, 1992. to appear.

[Rotman, 1973] J. J. Rotman. *The Theory of Groups: An Introduction*. Allyn and Bacon, Inc, Boston, second edition, 1973.

[Salomaa and Soittola, 1978] A. Salomaa and M. Soittola. *Automata-theoretic Aspects of Formal Power Series*. Springer, Berlin, 1978.

[Selman, 1972] A. Selman. Completeness of calculi for axiomatically defined classes of algebras. *Algebra Universalis*, 2:20–32, 1972.

[Smith, 1982] G. C. Smith. *The Boole-de Morgan Correspondence*. Oxford University Press, Oxford, 1982.

[Soulie *et al.*, 1987] F. Fogelman Soulie, Y. Robert, and M. Tchuente. *Automata Networks in Computer Science: Theory and Applications*. Manchester University Press, 1987.

[Stoltenberg-Hansen and Tucker, 1988] V. Stoltenberg-Hansen and J. V. Tucker. Complete local rings as domains. *Journal of Symbolic Logic*, 53:603–624, 1988.

[Stoltenberg-Hansen and Tucker, 1991] V. Stoltenberg-Hansen and J. V. Tucker. Algebraic equations and fixed-point equations in inverse limits. *Theoretical Computer Science*, 87:1–24, 1991.

[Tarlecki, 1986] A. Tarlecki. Quasivarieties in abstract algebraic institutions. *J. Computer and System Sciences*, 33:333–360, 1986.

[Tarski, 1968] A. Tarski. Equational logic and equational theories of algebras. In H. A. Schmidt, K. Schütte, and H. J. Thiele, editors, *Logic Colloquium '66*, pages 275–289. North-Holland, Amsterdam, 1968.

[Taylor, 1979] W. Taylor. Equational logic. *Houston Journal of Mathematics*, pages 1–83, 1979. also in: G. Grätzer, *Universal Algebra*, Second Edition, Springer, Berlin, 1979

[Thatcher *et al.*, 1980] J. W. Thatcher, E. G. Wagner, and J. B. Wright. More advice on structuring compilers and proving them correct. In N. D.

410   *References*

Jones, editor, *Semantics-directed compiler generation*, Lecture Notes in Computer Science 94, pages 163–188. Springer, Berlin, 1980.

[Thompson and Tucker, 1985] B. C. Thompson and J. V. Tucker. Theoretical considerations in algorithm design. In R. A. Earnshaw, editor, *Fundamental Algorithms for Computer Graphics*. Springer, Berlin, 1985.

[Thompson and Tucker, 1988] B. C. Thompson and J. V. Tucker. A parallel deterministic language and its application to synchronous concurrent algorithms. In *Proceedings 1988 UK IT Conference, Institute of Electrical Engineers*, pages 500–503, 1988. Extended Abstract.

[Thompson and Tucker, 1991] B. C. Thompson and J. V. Tucker. Equational specification of synchronous concurrent algorithms and architectures. Technical Report Report CSR 9–91, Department of Mathematics and Computer Science, University College of Swansea, 1991.

[Thompson, 1987] B. C. Thompson. *Mathematical Theory of Synchronous Concurrent Algorithms*. PhD thesis, School of Computer Studies, University of Leeds, 1987.

[Tucker and Zucker, 1988] J. V. Tucker and J. I. Zucker. *Program Correctness over Abstract Data Types with Error-state Semantics*. North-Holland, Amsterdam, 1988.

[Tucker and Zucker, 1992] J. V. Tucker and J. I. Zucker. Algebraic specifications for computable functions and selection functions over abstract data types. Technical Report Report CSR 1–92, Department of Mathematics and Computer Science, University College of Swansea, 1992.

[Tucker, 1980] J. V. Tucker. Computing in algebraic systems. In S. S. Wainer F. R. Drake, editor, *Recursion Theory, its Generalisations and Applications*, pages 215–236. London Mathematical Society Lecture Note 45. Cambridge University Press, Cambridge, 1980.

[Tucker, 1991] J. V. Tucker. Theory of computation and specification on abstract data types and its applications. In F. L. Bauer, editor, *Logic, Algebra and Computation*, pages 1–37. Springer, Berlin, 1991.

[van der Waerden, 1970] B. L. van der Waerden. *Algebra*, volume I and II. Ungar, New York, 1970.

[van Wijngaarden, 1966] A. van Wijngaarden. Numerical analysis as an independent science. *BIT*, 6:68–81, 1966.

[Wechler, 1992] W. Wechler. *Universal Algebra for Computer Scientists*. EATCS Monographs on Theoretical Computer Science. Springer, Berlin, 1992.

[Wehrfritz, 1973] B. A. F. Wehrfritz. *Infinite Linear Groups*. Springer, Berlin, 1973.

[Wirsing and Bergstra, 1989] M. Wirsing and J. A. Bergstra, editors. *Algebraic Methods: Theory, Tools and Applications*. Lecture Notes in Computer Science. Springer, Berlin, 1989.

[Wirsing and Broy, 1980] M. Wirsing and M. Broy. Abstract data types as lattices of finitely generated models. In P. Dembinski, editor, *Proc. Int. Symp. on Mathematical Foundations of Computer Science*, pages 673–685. Lecture Notes in Computer Science 88. Springer, Berlin, 1980.

[Wirsing, 1991] M. Wirsing. Algebraic specification. In J. van Leeuwen, editor, *Handbook of Theoretical Computer Science: Vol B Formal Methods and Semantics*, pages 675–788. North-Holland, Amsterdam, 1991.

[Wolfram, 1990] S. Wolfram. *Theory and Applications of Cellular Automata*. World Scientific, Singapore, 1990.

Basic Category Theory

Axel Poigné

Contents

Preface

If asked for a single reason for the attention that category theory, at least
as a language, enjoys in some areas of computer science, I would guess that
its attraction stems from being a foundational theory of functions which
provides a sound basis for (functional) programming and programming
logic. If asked for more reasons I would recollect the familiar arguments,
namely that category theory

- formalizes otherwise vague concepts,

- provides a language that brings to the surface common basic concepts
  in ostensibly unrelated areas,

- allows us to translate problems from one area to another where a
  solution may be more easily achieved,

or more specifically with regard to computer science, category theory

- allows easier access to various areas of mathematics in that it provides
  a core of properties to be looked for,

- offers a rich language in which to axiomatize, differentiate and compare structures in computer science and mathematics.

In the presence of such masterpieces as MacLane's book only our particular perspective justifies this introduction to categories, namely to consider category theory in the first instance as a discipline of functional programming. A similar effort took place in the tutorial part of [Pitt *et al.*, 1985] where David Pitt, David Rydeheard and myself joined forces. Then Samson Abramsky gave advice strongly arguing for the approach taken here, an approach for which ultimately Dana Scott should be credited.

Of course, functional programming touches only one aspect. We try to draw on experience from other areas of theoretical computer science such as denotational semantics, programming logic, automata theory, data type specification. However, at the end, category is a formal game of combining the structure of monoids with that of partial orders, which is a remark the reader may appreciate having read this exposition.

In general, we emphasize motivation of concepts rather than elaborating proofs to the last detail. So if not happy about a specific twist in a proof, the reader may find help in the standard textbooks such as MacLane's book from which we take most of our notation.

The exposition is divided in three main parts:

- The first part introduces *categories, functors* and *natural transformations* (Section 1) and some categorical infrastructure such as *products, coproducts* and *Cartesian closure* (Section 2). Moreover, typical styles of categorical reasoning are discussed. In these two sections the analogy to functional programming is used for motivation.

- The second part focuses on core concepts of category theory such as adjunctions and limits which are indispensable prerequisites for a deeper understanding of categorical reasoning and modelling. We again try to exploit intuitions and paradigms of computer science for motivation. Section 3 introduces the idea of definitions in terms of *universal properties* ending with the definition of *adjunctions*. Section 5 considers a special class of such universal definitions, *limits* and *colimits*, under the name of 'data structures' which is what the section is about. Section 6, which is quite advanced, translates well-known mathematical constructions like term generation into the language of categories introducing concepts such as *Kan extensions* which play an important role in *enriched category theory*. The link to enriched category theory is provided in Section 4 where we discuss properties of *representable functors*.

- Finally, we have sections concerned with more specific issues. Section 7 studies category theory as a tool for *axiomatization of programming structures.* Section 8 develops parts of *universal algebra*, while Section 9 introduces *categorical logic*. Familiarity with Sections 1 and 2 will enable the reader to read substantial parts of these sections, and the reader is invited to do so before embarking on Sections 3–6.

Quite clearly, the presentation and the examples reflect my own experiences and tastes, but even within these limits, by no means all applications of category theory in computer science are touched, due to restrictions on space and time. I apologize to everybody whose work related to the subject is not mentioned or, which I do not hope, misinterpreted. I have very much appreciated the comments of Eric Wagner which have helped to improve the text in every respect, though I have probably failed to meet his standards.

For the format of presentation, each section is divided into several subsections which consist of numbered paragraphs. Paragraphs are the smallest logical units, typically with a definition or a proposition at their centre; they sometimes have headings to stress a certain aspect.

1   Categories, functors and natural transformations

1.1   Types, composition and identities

1.1.1 *Application* is the basic operation of functional programming. Let us use $f \cdot x$ to express that the program f is applied to an argument x. Programs are built from *primitive* functions using *functional forms,*, e.g. $f \circ g$, $\langle f, g \rangle$. The (operational) meaning is expressed by *applicative laws,* e.g.

$$id \cdot x = x$$
$$g \circ f \cdot x = g \cdot (f \cdot x)$$
$$first \cdot \langle x, y \rangle = x$$
$$\langle f, g \rangle \cdot x = \langle f \cdot x, g \cdot x \rangle.$$

We share the enthusiasm of most programmers for a type discipline and attach *types* to arguments and elementary functions

$$id_A : A \to A$$
$$first_{A,B} : A \times B \to A$$

and restrict application and functional forms by type information:

$$first_{A,B} \cdot \langle a, b \rangle \qquad \text{if } a : A, b : B$$
$$g \circ f : A \to C \qquad \text{if } f : A \to B, g : B \to C$$
$$\langle f, g \rangle : A \to B \times C \quad \text{if } f : A \to B, g : A \to C.$$

where $a : A$ stands for 'a is of type A'.

One can visualize the type information of functions by graphs of the form

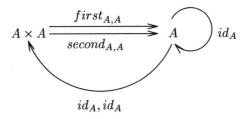

The picture could be filled in by adding all the functional forms obtained by composition, however, it is more convenient, and less confusing, to assume the existence of all composites and of all identities as a standard feature.

1.1.2 The applicative laws induce *functional laws*, that is equalities between functional forms

$$id_B \circ f = f$$
$$f \circ id_A = f$$
$$h \circ (g \circ f) = (h \circ g) \circ f$$
$$first_{A,B} \circ \langle f, g \rangle = f$$

in that functional forms can yield the same result if applied to the same argument

$$
\begin{aligned}
f \circ id_A \cdot x &= f \cdot (id_A \cdot x) = f \cdot x \\
h \circ (g \circ f) \cdot x &= h \cdot (g \circ f \cdot x) = h \cdot (g \cdot (f \cdot x)) = (h \circ g)(f \cdot x) = \\
&\quad (h \circ g) \circ f \cdot x \\
first_{A,B} \circ \langle f, g \rangle \cdot x &= first_{A,B} \cdot (\langle f, g \rangle \cdot x) = \\
&\quad first_{A,B} \cdot (\langle f \cdot x, g \cdot x \rangle) = f \cdot x
\end{aligned}
$$

In fact, if we focus on composition and identities, the applicative laws and the functional laws are equivalent in that every equality of functional forms can be derived from the applicative laws, if and only if, the equality follows from the functional laws (consult Section 1.3 for a formal discussion).

This observation suggests a new perspective, namely that we distinguish between

- functional programs and the functional laws they satisfy, and

- the meaning or semantics of programs as defined by application.

We will argue that *categories* provide an axiomatization of (very simple) functional programs while the notion of *functors* (or homomorphisms between categories) provides the corresponding axiomatization of meaning.

1.2   Categories

1.2.1 Functional programming at least requires composition of functional programs and existence of identities as codified by the definition of a category.

Definition 1.2.1. A *category* **C** is given by the following data

- $Obj(\mathbf{C})$, a class (called the *objects* of **C**).

- $Mor(\mathbf{C})$, a class (called the *morphism* or *arrows* of **C**).

- $dom, cod : Mor(\mathbf{C}) \to Obj(\mathbf{C})$, functions (for $f \in Mor(\mathbf{C}), dom(f)$ is called the *domain*, or *source*, of f, and $cod(f)$ is called the *codomain* or *target*, of f).

- $1_\_ : Obj(\mathbf{C}) \to Mor(\mathbf{C})$, a function (for $A \in Obj(\mathbf{C}), 1_A$ is called the *identity morphism* for A).

- $\circ : Mor(\mathbf{C}) \times Mor(\mathbf{C}) \to Mor(\mathbf{C})$, a partial function (called *composition*, for $f, g \in Mor(\mathbf{C})$, we write $g \circ f$ for the *composite* of f and g).

These data are subject to the conditions

- $dom(1_A) = A = cod(1_A)$.

- $g \circ f$ is defined if, and only if, $cod(g) = dom(f)$.

- If $g \circ f$ is defined, then $dom(g \circ f) = dom(f)$ and $cod(g \circ f) = cod(g)$.

- If dom(f) $= A$ and $cod(f) = B$ then $1_B \circ f = f, f \circ 1_A = f$.

- If $h \circ g$ and $g \circ f$ are defined then $h \circ (g \circ f) = (h \circ g) \circ f$.

It is convenient, and conventional, to write $f : A \to B$ for '$f \in Mor(\mathbf{C})$ with $dom(f) = A$ and $cod(f) = B$'. We generally write $\mathbf{C}[A, B]$ for the class of morphisms $f : A \to B$. If, for all $A, B \in Obj(\mathbf{C}), \mathbf{C}[A, B]$ is a set, we refer to these sets as *Hom*(omorphism) *sets*. If $Obj(\mathbf{C})$ is a set, and if, for all $A, B \in Obj(\mathbf{C}), \mathbf{C}[A, B]$ is a set then we say **C** is a *small category*, otherwise we we generally say **C** is a *large category*.

An alternative, but equivalent definition of categories is based on Hom sets:

A *category* **C** consists of

Obj(**C**), a class (called the *objects* of **C**).

For each pair $A, B \in Obj(\mathbf{C})$ a class $\mathbf{C}[A, B]$ (called the class of *morphisms of* **C** *with domain A and codomain B*).

For each triple $A, B, C \in Obj(\mathbf{C})$, a function $\circ_{A,B,C} : \mathbf{C}[B, C] \times \mathbf{C}[A, B] \rightarrow \mathbf{C}[A, C]$ (called the *composition* operator at A, B, C. For $f \in \mathbf{C}[A, B]$ and $g \in \mathbf{C}[B, C]$ we write their *composite* as $g \circ f$).

For each $A \in Obj(\mathbf{C})$ we have a distinguished element $1_A \in \mathbf{C}[A, A]$ (called the *identity* at A).

These data are subject to the following conditions

$\mathbf{C}[A, B]$ and $\mathbf{C}[X, Y]$ are disjoint unless $A = X$ and $B = Y$.

If $f \in \mathbf{C}[A, B]$, $g \in \mathbf{C}[B, C]$ and $h \in \mathbf{C}[C, D]$ then $h \circ_{A,C,D} (g \circ_{A,B,C} f) = (h \circ_{B,C,D} g) \circ_{A,B,D} f$.

If $f \in \mathbf{C}[A, B]$ then $f \circ_{A,A,B} 1_A = f$ and $1_B \circ_{B,B,A} f = f$.

Though equivalent, the spirit of this definition is rather different: composition is partial in our original definition, here partiality is resolved by introduction of a suitable type structure. We will elaborate the differences further below.

We are deliberately vague about foundational issues such as what set theory we refer to, since naïve set theory will do in most cases. However the distinction between sets and classes plays a significant role in our development, and the reader who is not familiar with this distinction should consult [MacLane, 1971], I.6, or some set theory reference.

1.2.2 Category theory as functional programming only provides a certain, though important aspect. The subsequent examples should broaden the perspective, and demonstrate that categories should be thought of in general as structures which consist of arrows (in a rather pictorial way) which are closed under composition. However, functional programming appears a good vehicle to provide *at least one* interpretation of categorical concepts when addressing the computer science community.

- **Set** is the category of sets and functions. Here the objects are sets, and a morphism f between sets A and B is, of course, a function from A to B, *the identity morphism for a set A* is the identity function on A, and composition is the usual composition of functions. However, to satisfy either definition of category we have to be careful about what we mean by a function. It is not good enough to give the common set theoretic definition of a function $f : A \rightarrow B$ as a set $f \subseteq A \times B$ which is *single-valued* (if $(a, b), (a, b') \in f$ then $b = b'$) and *totally defined* ($A = \{a \mid a \in A, \exists b \in B.(a, b) \in f\}$), for these conditions are not sufficient to uniquely determine the codomain of f. Thus a precise definition of **Set** must define a function f from A to B as something like a triple $(A, \text{graph}(f), B)$ where $\text{graph}(f)$ is the *graph* of f, i.e. the appropriate single-valued, totally defined subset of $A \times B$.

- If we drop the requirement of total definedness we get a category **Pfn** of sets and partial functions.

- Dropping the requirement of single-valuedness and total-definedness gives us the category **Rel** of sets and relations.

- The category **Set**$_\perp$ of *pointed sets* is closely related to the category of partial function. Here objects are pairs $(X, \perp_X)$ where X is a set and $\perp_X \in X$. Morphisms are triples $((X, \perp_X), f, (Y, \perp_Y))$ where $f : X \rightarrow Y$ is a function such that $f(\perp_X) = \perp_Y$. The composition is that of functions.

- Define a (large) *directed graph* as being specified by two sets (classes) E, and N (E for *edges*, N for *nodes*) and two functions $d, c : E \rightarrow N$ (edge $e \in E$ goes from $d(e)$ to $c(e)$). Then we get a category **Graph** in which the objects are graphs and a morphism from $G = (N, E, c, d)$ to $G' = (N', E', c', d')$ is determined by a pair of functions $h_E : E \rightarrow E'$ and $h_N : N \rightarrow N'$ such that $h_N \circ d = d' \circ h_E$ and $h_N \circ c = c' \circ h_E$. Here again, a precise specification of a morphism from G to G' must be something like a 4-tuple (g, h_E, h_N, G') since, while we can determine E, E', N and N' from h_E and h_N we cannot determine d, d', c, and c'.

- Define a *monoid* to be a structure specified by a set M, an associative operation $* : M \times M \rightarrow M$, and a designated element, *unit*, $e \in M$, with the property that $a * e = a = e * a$ for all $a \in M$. Then we get a category **Mon** of monoids, in which a morphism from $M = (M, *, e)$ to $M' = (M', *', e')$ is given by a function $f : M \rightarrow M'$ such that $f(a * b) = f(a) *' f(b)$ and $f(e) = e'$. Again, a precise definition of morphisms must also contain the domain and codomain.

A very useful monoid is the *monoid of words* X^* over some set X, with concatenation vw as operation, and the empty ϵ as unit.

Those readers familiar with modern algebra will recognize the morphisms between monoids as the familiar monoid homomorphisms. Indeed, any class of algebras and their associated homomorphisms (e.g. groups and group homomorphisms, or, ring and ring homomorphisms) will give us a corresponding category. In general, any notion of *sets with structure* equipped with mappings preserving that structure, will define a category, not least small categories as we will learn below. These categories are all large categories.

As a variation:

- The category **SLat** of *semilattices* consists of semilattices $(X, +)$ where X is a set and where $+ : X \times X \rightarrow X$ is an associative, commutative and idempotent operation, i.e.

$$
\begin{aligned}
(x + y) + z &= x + (y + z) \\
x + y &= y + x \\
x + x &= x
\end{aligned}
$$

Semilattice homomorphisms $h : X \rightarrow X'$ are mappings $h : X \rightarrow X'$ such that $h(x + y) = h(x) +' h(y)$.

- Here are simple 'categories of automata':

Let $M = (M, *, e)$ be a monoid. Then define an *M-automaton* or *M-module* (or *M-dynamics*) to consists of a set S of *states*, and a *transition function* $\delta : M \times S \rightarrow S$ such that $\delta(x * y, s) = \delta(x, \delta(y, s))$ and $\delta(e, s) = s$ for all $x, y \in M$ and all $s \in S$.

Often, M is chosen as a word monoid Act^* of *actions*, and the transition function is presented by a *next-state function* $\delta : Act \times S \rightarrow S$ (taking *inputs* or *actions* and states to a *resulting state*). The next-state function extends to a transition function $\delta^* : Act^* \times S \rightarrow S$ by $\delta^*(wa, s) = \delta^*(w, \delta(a, s))$ and $\delta^*(\varepsilon, s) = s$. An *M*-automaton may come along with an *output function* $\lambda : S \rightarrow O$ (taking states to *outputs*).

A morphism of *M*-automata from (S, δ, λ) to (S', δ', λ') is specified by giving a mapping $f : S \rightarrow S'$ such that $f(\delta(x, s)) = \delta'(x, f(s))$ and $\lambda(s) = \lambda'(f(s))$. All the usual, more complicated notions of automata and morphisms between them, also yield categories.

- Define a (large) *preorder X* to consist of a set (class) X with a reflexive and transitive relation $\sqsubseteq \subseteq X \times X$, i.e. $x \sqsubseteq x$ and $\sqsubseteq z$ whenever $x \sqsubseteq y$ and $y \sqsubseteq z$. If the relation is also antisymmetric, i.e. $x \sqsubseteq y$ and $y \sqsubseteq x$ implies $x = y$, we speak of a *partially ordered set*, or *poset* for short. A mapping $f : X \rightarrow X'$ between preorders is called *monotone* if $f(x) \sqsubseteq' f(y)$ whenever $x \sqsubseteq y$ for $x, y \in X$.

Then we get a category **Pre** (resp. **Pos**) with preorders (posets) as objects, and morphisms being triples $((X, \sqsubseteq), f, (X', \sqsubseteq'))$ where $f : X \to Y$ is a monotone mapping.

Let $(X, \sqsubseteq)$ be a poset. An element $z \in X$ is said to be an *upper bound* of a subset $Y \subseteq X$ if $y \sqsubseteq z$ for all $y \in Y$. It is a *least upper bound* of Y if $z \sqsubseteq z'$ for all upper bounds z' of Y. We use the notation $\bigsqcup Y$ for least upper bounds. A subset $S \subseteq X$ is *directed* if every finite set $Y \subseteq X$ has an upper bound in S. Note that directed sets are non-empty.

- A category of substantial interest for denotational semantics resp. domain theory, has objects being posets $(X, \sqsubseteq)$ such that all directed subsets have a least upper bound and that X has a least element $\bot$.

  Homomorphisms $h : (X, \sqsubseteq) \to (X', \sqsubseteq')$ are triples $((X, \sqsubseteq), h, (X', \sqsubseteq'))$ where $h : X \to Y$ is a *continuous* mapping, i.e. a monotone mapping such that $h(\bigsqcup S) = \bigsqcup \{h(s) \mid s \in S\}$ for all directed sets S.

  Such homomorphisms do not preserve the least element which is a source of some anomalies. We use **Dcpo** to denote this category.

- The category ω**Pos** of ω-*complete posets* is rather similar. A poset $(X, \sqsubseteq)$ is ω-*complete* if every (non-empty) ω-chain $x_1 \sqsubseteq x_2 \sqsubseteq \ldots \sqsubseteq x_n \sqsubseteq \ldots$ has a least upper bound $\bigsqcup_n x_n$. Morphisms are ω-*continuous* functions $f : (X, \sqsubseteq) \to (Y, \sqsubseteq)$, i.e. monotone functions such that
  $$f(\bigsqcup_n x_n) = \bigsqcup_n f(x_n).$$

  ω denotes the first limit ordinal, resp. the poset $\omega = 0 \leq 1 \leq 2 \leq \cdots \leq n \leq \cdots$.

Now for small categories.

- The smallest category is **0** with no objects and no morphisms. Not far behind is the category **1**, the *trivial category*, with only one object and the identity morphism on that object.

- Given any set S we get a category called the discrete category for S whose objects are the elements of S, and in which the only morphisms are the identity morphisms, one for each $s \in S$.

- Any monoid $(M, *, e)$ corresponds to a category with just one object, with morphism set M, composition $*$, and identity morphism e.

- Given a set S we get a monoid/category whose morphisms are the set S^* of strings on S, whose composition operation is concatenation of strings, and whose identity morphism is the empty string.

- Let $G = (N, E, d, c)$ be a graph. We construct a category G^*, called the *path category* over G, with objects being the nodes in N and morphisms $f : A \rightarrow B$ being either strings $e_n \ldots e_2 e_1$ on E such that $d(e_1) = A$ and $c(e_n) = B$, and $d(e_{i+1}) = c(e_i)$ for $i = 1, \ldots, n-1$, or of the form 1_A with domain and codomain A. Compostion $g \circ f$ is concatenation of strings, except for the case that $f = 1_A$ or $g = 1_B$ where $g \circ f = g$, or if $g \circ f = f$ respectively.

- Every preorder $X = (X, \leq)$ defines a category with objects being the elements of X, exactly one morphism $f : x \rightarrow y$ between objects x, y iff $x \leq y$. Such a category is called an *order category*.

- Sometimes order categories may be thought of as comprising formulas of some logic with a preorder stating 'entailment' $\varphi \vdash \psi$.

  This interpretation carries over to arbitrary categories providing us with a different view of categories (taken in [Lambek and Scott, 1986]): consider objects as propositions and a morphism $f : A \rightarrow B$ as a proof of a proposition B from the 'assumption' A. Then there is an 'identity proof' $1_A : A \rightarrow A$, and proofs compose. The equalities state that certain proofs are considered as equivalent. In fact, these equivalences correspond to equivalences of proof trees as for instance discussed in [Prawitz, 1965]. We will elaborate this theme later, and we will see that category theory also embodies a substantial part of **proof theory**.

We can construct new categories from given ones:

- The direction of morphisms can be reversed without affecting the axioms:

  Given a category $\mathbf{C}$ we define the category $\mathbf{C}^{\mathrm{op}}$ to be the category with the same objects and morphisms as $\mathbf{C}$, but with $dom^{\mathrm{op}}(f) = cod(f)$ and $cod^{\mathrm{op}}(f) = dom(f)$, and with $f \circ_{\mathrm{op}} g = g \circ f$. We say that $\mathbf{C}^{\mathrm{op}}$ is the **dual category** to $\mathbf{C}$.

- Let $\mathbf{C}$ and $\mathbf{D}$ be categories. The *product category* $\mathbf{C} \times \mathbf{D}$ has objects (A, B), A being an object of $\mathbf{C}$, and B being an object of $\mathbf{D}$, and morphism $(f, g) : (A, B) \rightarrow (C, D)$ where $f : A \rightarrow C$ is in $\mathbf{C}$ and $g : B \rightarrow D$ is in $\mathbf{B}$. $(1_A, 1_B) : (A, B) \rightarrow (A, B)$ is the identity for the object (A, B), and composition is defined in the components, $(f', g') \circ (f, g) = (f' \circ f, g' \circ g)$.

- *Comma categories* are another example of this kind.

Let **C** be a category and A be an object of **C**. We construct a category *of morphisms out of A*, denoted by $(A, \mathbf{C})$ with

objects (f, X) where $X \in Obj(\mathbf{C})$, and $f : A \to X \in \mathbf{C}$, and

morphisms $h : (f, X) \to (g, Y)$ are triples $((f, X), h, (g, Y))$ where $h : X \to Y$ is a morphism of **C** such that $h \circ f = g$,

identities and composition are that of **C**.

Similarly, comma categories $(\mathbf{C}, B)$ of *morphisms into B* can be constructed with

objects (X, f) where $X \in Obj(\mathbf{C})$, and $f : X \to B \in \mathbf{C}$, and

morphisms $h : (X, f) \to (Y, g)$ are triples $((X, f), h, (Y, g))$ where $h : X \to Y$ is a morphism of **C** such that $g \circ h = f$.

Though somewhat unusual, comma categories are quite important technically. You should notice that this is not the most general definition of a comma category (cf. [MacLane, 1971]).

1.3 Relating functional calculus and category theory

1.3.1 The motivation of category theory in terms of functional programming may appear as rather *ad hoc*. However, the claim of 'category theory as functional programming' has found substantial witnesses in recent work on foundations of programming where, for instance, categorical axiomatizations of diverse λ-calculi have been given. As a sort of appetizer, and to prepare the grounds for this subject (we will look at it more closely in Section 7) we introduce a rather rudimentary language of 'procedures', and establish various links between this more familiar idiom and category theory. In particular, we show that the language itself defines a category and that there exists a systematic way to interpret the language in an arbitrary category, the latter being the very rudimentary form of what is called a *denotational semantics*.

1.3.2 Our language consists of *functional terms* ('procedures') of the form $f = \langle x^A \rangle M : A \to B$ where x^A is a *formal parameter* of type A, and where x^A is the only free variable in the 'body' M (The restriction to only one parameter reflects that category theory is a theory of unary functions). The basic operation is *parameter passing* for which we use the notation $f \cdot N$. Parameter passing is defined by syntactical substitution $(\langle x^A \rangle M) \cdot N = M[x^A/N]$, meaning that all free occurences of the formal parameter x^A in M are replaced by the actual parameter N.

More precisely, we assume that a set T of *types* and a set P of *elementary operations* of the form $f : A \to B$ are given a priori. *Functional terms* $f : A \to B$ are introduced by the following rules

- every elementary operation $f : A \to B$ is a functional term,

- $\langle x^A \rangle x^A : A \to A$ is a functional term if A is a type,

- $\langle x^A \rangle (f \cdot N) : A \to C$ is a functional term if $\langle x^A \rangle N : A \to B$ and $f : B \to C$ are so.

  (One should note that the parameter x^A can only occur free in N by construction.)

These data are subject to the equalities

- $\langle x^A \rangle (\langle y^B \rangle M \cdot N) = \langle x^A \rangle M[y^B/N] : A \to C$

- $\langle x^A \rangle (f \cdot x^A) = f : A \to B$.

We consider functions as equal if one is obtained from the other by renaming of the parameter (of course only if the binding structure is preserved), e.g. $\langle x^A \rangle (\langle x^A \rangle f \cdot x^A) \cdot x^A = \langle y^A \rangle (\langle z^A \rangle f \cdot z^A) \cdot y^A$. Formally, the rules of α-conversion as in [Barendregt, 1984] apply.

Moreover, equality is reflexive, symmetric and transitive, and compatible with the structure, e.g.

$$\langle x^A \rangle (f \cdot M) = \langle x^A \rangle (g \cdot N) : A \to B$$

if

$$\langle x^A \rangle M = \langle x^A \rangle N : A \to B$$

and

$$f = g : B \to C.$$

1.3.3 A natural notion of composition is defined in functional terms by

$$g \circ f := \langle x^A \rangle (g \cdot (f \cdot x^A)) : A \to C$$

where $f : A \to B$, and $1_A := \langle x^A \rangle x^A$ is clearly a candidate for an identity. This composition is associative

$$
\begin{aligned}
h \circ (g \circ f) &= \langle x^A \rangle h \cdot ((g \circ f) \cdot x^A) \\
&= \langle x^A \rangle h \cdot (((\langle y^A \rangle g \cdot (f \cdot y^A)) \cdot x^A) \\
&= \langle x^A \rangle h \cdot (g \cdot (f \cdot x^A)) && \text{by definition of} \\
& && \text{composition} \\
&= \langle x^A \rangle h \cdot (g \cdot (f \cdot x^A)) && \text{by application of the} \\
& && \text{first of the equalities} \\
&= \langle x^A \rangle (h \circ g) \cdot (f \cdot x^A) && \text{ditto but in the other} \\
& && \text{direction} \\
&= (h \circ g) \circ f && \text{ditto}
\end{aligned}
$$

and, similarly, the identities satisfy the respective properties

$$f \circ 1_A = \langle x^A \rangle f \cdot (1_A \cdot x^A) = \langle x^A \rangle f \cdot (((\langle y^A \rangle y^A) \cdot x^A) = \langle x^A \rangle f \cdot x^A = f$$

$$1_B \circ f = \langle x^A \rangle 1_B \cdot (f \cdot x^A) = \langle x^A \rangle (\langle y^A \rangle y^A) \cdot (f \cdot x^A) = \langle x^A \rangle f \cdot x^A = f.$$

Hence we expect that a category might surface; define a category $\mathbf{P}^*$ by the following data:

- $Obj(\mathbf{P}^*) = T$

- $Mor(\mathbf{P}^*) = \{[f] \mid f : A \to B \text{ functional term}\}$ where $[f] = \{g : A \to B \mid f = g : A \to B \text{ is derivable}\}$.

- $dom([f]) = A$ and $cod([f]) = B$ if $f : A \to B$

- $[g] \circ [f] = [\langle x^A \rangle (g \cdot (f \cdot x))]$, and

- $1_A = [\langle x^A \rangle x^A]$.

This is well defined because of the rules for compatibility. The axioms for categories hold by the computations above.

1.3.4   A more interesting phenomenon, though, is that a (denotational) semantics of our language can be defined in every category $\mathbf{C}$. Types A are interpreted as objects $[\![A]\!]$ of $\mathbf{C}$, and programs $f : A \to B$ as morphisms $[\![f]\!] : [\![A]\!] \to [\![B]\!]$ by

$$[\![\langle x^A \rangle x]\!] = 1_{[\![A]\!]}$$
$$[\![\langle x^A \rangle (f \cdot N)]\!] = [\![f]\!] \circ [\![\langle x^A \rangle N]\!].$$

What should be surprising is the claim that parameter passing or, more mathematically, substitution on free variables corresponds to composition in a category. However:

Proposition 1.3.1. *The semantics is sound, i.e.* $[\![f]\!] = [\![g]\!]$ *holds whenever* $f = g$ *is derivable.*

Proof. [Sketch] By induction on the length of deductions.
E.g. If $[\![A]\!]$ is well defined, so is $1_{[\![A]\!]}$, or

if $[\![\langle x^A \rangle N]\!] : [\![A]\!] \to [\![B]\!]$ and $[\![f]\!] : [\![B]\!] \to [\![C]\!]$ are well defined,
so is $[\![\langle x^A \rangle (f \cdot N)]\!] : [\![A]\!] \to [\![C]\!]$.

The only difficulties are given by the equality judgements. For instance to prove

$$[\![\langle x^A\rangle(((\langle y^B\rangle)M)\cdot N)]\!] = [\![\langle y^B\rangle)M]\!]\circ[\![\langle x^A\rangle N]\!]? =?[\![\langle x^A\rangle M[y^B/N]]\!]$$

one needs to use induction on the structure of M.

- $[\![\langle x^A\rangle(((\langle y^B\rangle y^B)\cdot N]\!]$
  $= [\![\langle y^B\rangle y]\!]\circ[\![\langle x^A\rangle N]\!]$
  $= 1_B\circ[\![\langle x^A\rangle N]\!] = [\![\langle x^A\rangle N]\!]$
  $= [\![\langle x^A\rangle y^B[y^B/N]]\!]$
- $[\![\langle x^A\rangle(((\langle y^B\rangle(f\cdot M))\cdot N)]\!]$
  $= [\![\langle y^B\rangle(f\cdot M)]\!]\circ[\![\langle x^A\rangle N]\!]$
  $= [\![f]\!]\circ[\![\langle y^B\rangle M]\!]\circ[\![\langle x^A\rangle N]\!]$
  $= [\![f]\!]\circ[\![\langle x^A\rangle M[y^B/N]]\!]$ by inductive assumption
  $= [\![\langle x^A\rangle(f\cdot M[y^B/N])]\!]$
  $= [\![\langle x^A\rangle(f\cdot M)[y^B/N]]\!].$

The latter uses the fact that y^B does not occur free in f which is an assumption which needs to be established by the same proof technique. ∎

This rather simple proof technique is more or less varied when denotational semantics are given to λ-calculi, type theories or programming languages which should make the exercise worthwhile. In fact, one can prove a much stronger fact, and we will do so in Section 7, namely that the equational logic of the procedural language and the equational logic of categories are equivalent. So the old conflict of pure functional programming versus procedural programming is resolved; the paradigms are equivalent, at least in as much as one is concerned with theories.

1.4 Compositionality is functorial

1.4.1 Application tells us how to evaluate a functional program given some (input) data. Since we favour a rigid type discipline, application should only be defined if data have the correct format:

$f\cdot a$ is defined iff $f:A\to B$ and if a is a data of type A,

e.g.

$$first_{A,B}\cdot\langle a,b\rangle = a \qquad apply_{A,A}\cdot\langle\lambda\langle x^A\rangle x,a\rangle = (\lambda\langle x^A\rangle x)\cdot a.$$

More precisely, if *Prog* is a typed functional language with programs of the form $f:A\to B$, let *Data* be a set of arguments and let *type : Data →Types* be a typing function. Then application is a partial function

$$\_\cdot\_: Prog\times Data\to Data$$

such that $f\cdot a$ is defined if, and only if, for some type B, $f:type(a)\to B$.

Additional requirements are imposed by the applicative rules such as

$$id_A \cdot a = a \qquad g \circ f \cdot a = g \cdot (f \cdot a)$$

$$first_{A,B} \cdot \langle a, b \rangle = a \qquad \langle f, g \rangle \cdot a = \langle f \cdot a, g \cdot a \rangle$$

but otherwise data and operations on data are not restricted. Every suitable choice of data and application provides the programming language with a *semantics* or a *model*.

1.4.2 A somewhat different perspective results from observing the effect of a single program with regard to data of suitable type. Application induces a function

$$Sem(f) : Sem(A) \to Sem(B), a \mapsto f \cdot a$$

for each program $f : A \to B$ where $Sem(A) = \{a \in Data \mid type(a) = A\}$. Every model is fully determined by a choice of sets $Sem(A)$ and functions $Sem(f) : Sem(A) \to Sem(B)$.

The style of this semantic is often referred to as *denotational* or *compositional* by programmers; a meaning or *denotation* is given to every syntactic entity, here types and functional programs, and the meaning of composed programs is defined in terms of the denotations of the components. For example, the denotation $Sem(g \circ f)$ is expressed in terms of the denotations $Sem(f)$ and $Sem(g)$ of the components if defined as composition $Sem(g) \circ Sem(f) : A \to C$, assuming that the denotations of the components are functions $Sem(f) : A \to B$ and $Sem(g) : B \to C$. Typically, the syntactic operation on programs finds a semantical counterpart as above for the composition, or, for the identity, is given by $Sem(id_A) = 1_{Sem(A)}$. Mathematicians would be more likely to say, *homomorphic semantics*, instead of compositional semantics.

A denotational semantics which preserves composition and identities is a special case of a *functor* or homomorphism of categories.

1.4.3 Definition 1.4.1. A *(covariant) functor* $F : \mathbf{C} \to \mathbf{D}$ consists of an object mapping $F_{Obj} : Obj(\mathbf{C}) \to Obj(\mathbf{D})$ and a morphism mapping $F : Mor(\mathbf{C}) \to Mor(\mathbf{D})$ such that

- $dom(F(f)) = F_{Obj}(dom(f))$ and $cod(F(f)) = F_{Obj}(cod(f))$ for all $f \in Mor(\mathbf{C})$

- $F(1_A) = 1_{F(A)}$ for all $A \in Obj(\mathbf{C})$

- $F(g \circ f) = F(g) \circ F(f)$ for all $f, g \in Mor(\mathbf{C})$ such that $g \circ f$ is defined.

We typically drop the subscript $_{-Obj}$.

As with categories we have an alternative definition of functors orientated at Hom sets: a *functor* $F : \mathbf{C} \to \mathbf{D}$ comprises a function $F : Obj(\mathbf{C}) \to Obj(\mathbf{D})$ and a family of functions $F[A, B] : \mathbf{C}[A, B] \to \mathbf{D}[F(A), F(B)]$ indexed by pairs (A, B) of objects of $\mathbf{C}$ such that

$$F[A, A](1_A) = 1_{F(A)}.$$
$$F[A, C](g \circ_{A,B,C} f) = F[B, C](g) \circ_{F(A),F(B),F(C)} F[A, B](f).$$

Functors compose in the obvious way : $G \circ F(A) := G(F(A))$ and $G \circ F(f) := G(F(f))$. Identity functors $\mathrm{Id}_{\mathbf{C}}(f) = f$ are units with regard to composition. These data define a category $\mathbf{CAT}$. We write $\mathbf{Cat}$ to denote the full subcategory (see Section 1.4.4) of $\mathbf{CAT}$ with objects being small categories.

Strictly speaking $\mathbf{CAT}$ is not a category since the objects do not form a class. However, $\mathbf{CAT}$ is rather well behaved, so we again do not worry about foundational problems.

1.4.4 Functors as semantics focuses on a very particular aspect. The nature of functors, of course, varies as much as that of categories:

- There exist *inclusion* functors $I : \mathbf{Set} \to \mathbf{Pfn}$ and $I : \mathbf{Pfn} \to \mathbf{Rel}$ which are identities on objects, and where $\mathbf{Set}(X, Y) \subseteq \mathbf{Pfn}(X, Y) \subseteq \mathbf{Rel}(X, Y)$.

- A functor $\_ \perp : \mathbf{Pfn} \to \mathbf{Set}_\perp$ maps a set X to the disjoint set $X_\perp = X + \{\perp\}$, and a partial function $f : X \to Y$ to the function $f_\perp : X_\perp \to Y_\perp$ such that $f_\perp(x) = f(x)$ for $x \in X$ and $f_\perp(\perp) = \perp$. In the other direction, a functor $D : \mathbf{Set}_\perp \to \mathbf{Pfn}$ is given by $D(X) = X \backslash \{\perp\}$ and $D(f)(x) = f(x)$ if $f(x) \neq \perp$, and $D(f)(x)$ undefined otherwise.

- The functor $V : \mathbf{Cat} \to \mathbf{Graph}$ forgets about compositon and identity; $V(\mathbf{C}) = (Obj(\mathbf{C}), Mor(\mathbf{C}), dom, cod)$. Every functor $F : \mathbf{C} \to \mathbf{D}$ is a graph morphism by definition.

- $\_^* : \mathbf{Graph} \to \mathbf{Cat}$ maps a graph G to the path category G^*. Every graph homomorphism $h : G \to G'$ induces a functor $h^* : G^* \to G'^*$ which is the identity on objects, and which maps a sequence $f_n \ldots f_1$ of morphisms to the sequence $h^*(f_n) \ldots h^*(f_1)$.

- Let $\mathbf{Mon}$ be the category of monoids. We can map every monoid $(M, *, e)$ to its underlying set M, and every monoid homomorphism to its underlying set mapping. This defines a functor $U : \mathbf{Mon} \to \mathbf{Set}$. We say that U is a *forgetful functor* since it forgets about the monoid structure. The same idea works for all categories of sets with structure.

- Since every Abelian monoid (i.e. a monoid satisfying $x * y = y * x$) is a monoid there is an *inclusion functor* $I : \mathbf{AbMon} \to \mathbf{Mon}$ of the category of Abelian monoids to the category of monoids.

- For each set X we can construct the word monoid X^* with concatenation and empty word λ as unit. Every mapping $f : X \to Y$ extends to a monoid homomorphism $f^* : X^* \to Y^*$ by $f^*(\varepsilon) = \varepsilon$ and $f^*(vw) = f(v)f^*(w)$ where $v, w \in X^*$. These data define a functor $\_^* : \mathbf{Set} \to \mathbf{Mon}$.

- $U : \mathbf{Aut}(M) \to \mathbf{Set}$ maps an M-automaton (S, δ) to S. A homomorphism $f : (S, \delta) \to (S', \delta')$ is defined by a function $f : S \to S'$ anyway.

- There are two functors in the other direction. $\_^* : \mathbf{Set} \to \mathbf{Aut}(M)$ maps X to the automaton X^* with states $M \times X$ and transition function $\delta^* : M \times (M \times X) \to M \times X, \delta^*(m, (m', x)) = (m * m', x)$. Homomorphism $f^* : X^* \to Y^*$ is defined by $f^*(m, x) = (m, f(x))$. (Remark on the notation: we uniformly use the notation $\_^*$ for a certain kind of functors which we will classify as 'free functors' below. So the specific interpretation of for instance X^* depends on the context; here we refer to the 'free automaton' $X^* = (M \times X, \delta^*)$, in the context of monoids we refer to the word monoid $X^* = (X^*, conc, \varepsilon)$. This should cause not too much confusion since the context will always be clearly stated, but has the merit of providing a uniform notation for a fundamental notion in mathematics.)

  The second functor $\__* : \mathbf{Set} \to \mathbf{Aut}(M)$ maps X to the automaton X_* with states X^M, and transition function $\delta^* : M \times X^M \to X^M, \delta_*(m, g) = M \to X$ is defined by $\delta^*(m, g)(m') = g(m * m')$. The homomorphism $f_* : X_* \to Y_*$ is defined by $f_*(g)(m) = f(g(m))$.

- Every M-automaton $\mathcal{D}$ with transition function $\delta : M \times S \to S$ defines a functor $\mathcal{D} : M \to \mathbf{Set}$ which maps the only object to S, and $\mathcal{D}(x) : S \to S$ is defined by $\mathcal{D}(x)(s) = \delta(x, s)$.

- Given a preorder $(X, \sqsubseteq)$ define a partially ordered set X^{po} with elements $[x] = \{y \in X \mid x \sqsubseteq y \text{ and } y \sqsubseteq x\}$ for $x \in X$, and order relation $[x] \sqsubseteq [y]$ if $x \sqsubseteq y$. This defines the object part of a functor $\_^{po} : \mathbf{Pre} \to \mathbf{Pos}$ and $f^{po}([x]) = [f(x)]$ defines the monotone function.

- There exists only one functor $I : \mathbf{0} \to \mathbf{C}$ and one functor $T : \mathbf{C} \to \mathbf{1}$ whatever category $\mathbf{C}$ is.

- Every functor between discrete categories is basically a function.

- Monoid homomorphism $f : M \to M'$ exactly correspond to functors between the monoids as categories.

- Let $(X, \sqsubseteq)$ and $(Y, \sqsubseteq)$ be partially ordered sets considered as categories. A functor $f : (X, \sqsubseteq) \to (Y, \sqsubseteq)$ is a function $f : X \to Y$ such that $f(x) \sqsubseteq f(x')$ if $x \sqsubseteq x'$. In other words, functors on partially ordered sets are *monotone functions*.

- The functor *List* : **Set** $\to$ **Set** maps a set X to the list X^*. The function $List(f)(x_1 \ldots x_n) = f(x_1) \ldots f(x_n)$ is known as 'maplist' in programming.

- Cartesian products of sets define a functor $\_ \times \_ : \textbf{Set} \times \textbf{Set} \to \textbf{Set}$. A pair (X, Y) of sets is mapped to its Cartesian product, and a pair $(f : X \to Y, g : X' \to Y')$ of functions to the function $f \times g : X \times X' \to Y \times Y'$ such that $f \times g(x, x') = (f(x), g(x'))$.

- There are two functors related to power sets:

  The *covariant power set functor* P : **Set** $\to$ **Set** maps a set X to its power set $P(X) = \{U \mid U \subseteq X\}$, and a mapping $f : X \to Y$ to the *direct image function* $P(f) : P(X) \to P(Y), U \mapsto \{f(x) \mid x \in U\}$.

  The *contravariant power set functor* $P^{\bullet}$: **Set**^{op} $\to$ **Set** maps a set X to its power set $P(X)$, and a mapping $f : X \to Y$ to the *inverse image function* $P^*(f) : P(Y) \to P(X), U \mapsto \{x \mid f(x) \in U\}$.

- The function space functor $\_ \Rightarrow \_ : \textbf{Set}^{\mathrm{op}} \times \textbf{Set} \to \textbf{Set}$ maps a pair (X, Y) to the function space $X \Rightarrow Y = \{f \mid f \text{ is a function from } X \text{ to } Y\}$. A pair $(f : Y \to X, g' : X' \to Y')$ of functions is mapped to the function $f \Rightarrow g : X \Rightarrow X' \to Y \Rightarrow Y'$ such that $f \Rightarrow g(h) = g \circ h \circ f$.

- The Hom-sets of a category give rise to a variety of functors.

$$\mathbf{C}[A, \_] : \mathbf{C} \to \textbf{Set}$$
$$\mathbf{C}[A, g] : \mathbf{C}[A, B] \to \mathbf{C}[A, C], f : A \to B \mapsto g \circ f : A \to C$$

defines a functor, called *covariant Hom functor*, for every object A where $g : B \to C$. Composition on the right hand side defines functors, the so-called *contravariant Hom functors*:

$$\mathbf{C}[\_, C] : \mathbf{C}^{\mathrm{op}} \to \textbf{Set}$$
$$\mathbf{C}[f, C] : \mathbf{C}[B, C] \to \mathbf{C}[A, C], g : B \to C \mapsto f \circ g : A \to C.$$

Combination yields a (bi-)functor $\mathbf{C}[\_, \_] : \mathbf{C}^{\mathrm{op}} \times \mathbf{C} \to \textbf{Set}$.

- The functors $First : \mathbf{C} \times \mathbf{D} \to \mathbf{C}$ and $Second : \mathbf{C} \times \mathbf{D} \to \mathbf{D}$ map the the first or second component of the product $\mathbf{C} \times \mathbf{D}$ of categories.

- Given a comma category $(X, \mathbf{C})$, the functor $Q : (X, \mathbf{C}) \to \mathbf{C}$ maps (f, X) to X and $h : (f, X) \to (g, Y)$ to $h : X \to Y$.

Here are some definitions related to functors:

Definition 1.4.2. A functor $F : \mathbf{C}^{\mathrm{op}} \to \mathbf{D}$ is called *contravariant* because the direction of morphisms is reversed in that $f : A \to B$ is mapped to $F(f) : F(B) \to F(A)$. For contrast, functors $F : \mathbf{C} \to \mathbf{D}$ are called *covariant*.

$\mathbf{C}$ is a *subcategory* of $\mathbf{D}$ if there exists an *inclusion* functor $I : \mathbf{C} \to \mathbf{D}$ such that $I(f : A \to B) = f : A \to B$. A functor $F : \mathbf{C} \to \mathbf{D}$ is *faithful* if $F(f : A \to B) = F(g : A \to B)$ implies $f = g : A \to B$, and it is *full* if for all $g : F(A) \to F(B)$ there exists a morphism $f : A \to B$ such that $F(f) = g$.

Categories $\mathbf{C}$ and $\mathbf{D}$ are *isomorphic* if there exist functors $F : \mathbf{C} \to \mathbf{D}$ and $G : \mathbf{D} \to \mathbf{C}$ such that $G \circ F = 1_{\mathbf{C}}$ and $F \circ G = 1_{\mathbf{D}}$.

The functors $I : \mathbf{Set} \to \mathbf{Pfn}$, $I : \mathbf{Pfn} \to \mathbf{Rel}$, $I : \mathbf{AbMon} \to \mathbf{Mon}$ are inclusion functors. $U : \mathbf{Mon} \to \mathbf{Set}$, $V : \mathbf{Cat} \to \mathbf{Graph}$ are faithful, and the functor $\_^{\mathrm{po}} : \mathbf{Pre} \to \mathbf{Pos}$ is full.

1.4.5 The semantics of the language $\mathbf{P}^*$ in Section 1.3 defines a functor $[\![\_]\!] : \mathbf{P}^* \to \mathbf{C}$. The equalities

$$[\![g \circ f]\!] = [\![g]\!] \circ [\![f]\!] \text{ and } [\![1_A]\!] = 1_{[\![A]\!]}$$

hold as checked by computation

$$
\begin{aligned}
[\![g \circ f]\!] &= [\![\langle x^A \rangle g \cdot (f \cdot x))]\!] = [\![g]\!] \circ [\![\langle x^A \rangle f \cdot x]\!] \\
&= [\![g]\!] \circ [\![f]\!] \circ [\![\langle x^A \rangle x]\!] = [\![g]\!] \circ [\![f]\!] \circ 1_{[\![A]\!]} \\
&= [\![g]\!] \circ [\![f]\!], \\
[\![1_A]\!] &= [\![\langle x^A \rangle x]\!] = 1_{[\![A]\!]}.
\end{aligned}
$$

But the more interesting observation is that this functor is totally determined if the denotations of the types and the elementary functions are given. This is a prominent property of the category $\mathbf{P}^*$.

Proposition 1.4.3. *Let an object $Sem(A)$ and morphism*

$$Sem(f) : Sem(A) \to Sem(B)$$

in a category $\mathbf{C}$ be given for every type A and every program $f : A \to B$ in P. Then there exists only one functor $[\![\_]\!]_{Sem} : \mathbf{P}^* \to \mathbf{C}$ such that $[\![A]\!]_{Sem} = Sem(A)$ and $[\![f : A \to B]\!]_{Sem} = Sem(f)$.

Proof. By induction on length of terms. For types and elementary programs $[\![\_]\!]_{Sem}$ is already defined. For the other programs use the definition of Section 1.3.4. This is a functor as demonstrated above.

Assume that there is a second functor $F : \mathbf{P}^* \to \mathbf{C}$ such that $F(A) = Sem(A)$ and $F(f) = Sem(f)$. F must agree with $[\![\_]\!]_{Sem}$ on types and elementary programs by assumption. Moreover

$$F(\langle x^A \rangle x) = F(1_A) = 1_{Sem(A)} = [\![\langle x^A \rangle x]\!]_{Sem}$$
$$F(\langle x^A \rangle f \cdot M) = F(f \circ \langle x^A \rangle M)$$

since $\qquad f \circ \langle x^A \rangle M = \langle x^A \rangle f \cdot ((\langle x^A \rangle M) \cdot x) = \langle x^A \rangle f \cdot M$

$$= F(f) \circ F(\langle x^A \rangle M)$$
$$= [\![f]\!]_{Sem} \circ [\![\langle x^A \rangle M]\!]_{Sem} \text{ by inductive assumption}$$
$$= [\![\langle x^A \rangle f \cdot M]\!]_{Sem}.$$

∎

It might look like cheating to claim compositionality of this functorial semantics since the inherent operation of the programming language is substitution which is mapped to composition. However, these are really two sides of the same coin as will be argued in Section 7.

1.5 Natural transformations

1.5.1 While even a casual reader gets used to categories and functors, natural transformations are probably a first stumbling block, simply because the examples tend to raise the level of mathematical sophistication. But natural transformation arises naturally even in the world of programming as demonstrated by an example.

A *data type* comprises, by common consent, a type and some characteristic operations. A data type may be parameterized, that is, defined as a function of types, in which case all the operations are parameterized. A typical example is that of $list(A)$, of *lists of type A*, together with operations

$$insert_A : A \to list(A) \qquad\qquad append_A : A \times list(A) \to list(A).$$

These operations are *polymorphic*, or uniform, in the sense that whatever A is chosen, application of a function f, to data of type A is compatible with the operations *insert* and *append* in that the equalities

$$list(f) \circ insert_A = insert_B \circ f$$

and

$$list(f) \circ append_A = append_B \circ (f \times list(f))$$

hold, or, equivalently, the diagrams

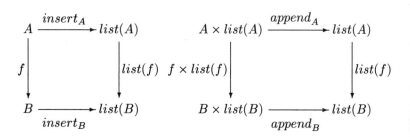

commute. Similarly, other functions on lists behave uniformly if they do not depend on the specific nature of the parameter, e.g. a function which reverses a list (definition left to the reader)

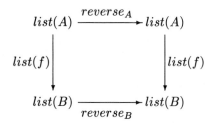

Remark 1.5.1. Reasoning in terms of diagrams is suggestive, and typical for category theory. If we say that a diagram, e.g.

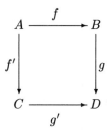

commutes, we mean that whatever sequence of arrows we pass through from a specific node to another, the composition of every such sequence of arrows is equal. So $g \circ f = g' \circ f'$ since both morphisms have the same source and target. So commutative diagrams are just a convenient way to visualize equalities of morphisms.

This implies that commutative diagrams are closed under composition in that a diagram commutes if all the subdiagrams commute. E.g. if the inner squares commute in the diagram

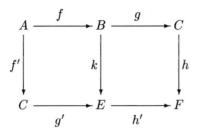

then the outer squares commute as well; $h \circ g = h' \circ k$ and $k \circ f = g' \circ f'$ implies $h \circ g \circ f = h' \circ k \circ f = h' \circ g' \circ f'$.

Some authors prefer the notation $f; g$ instead of $g \circ f$ for composition, since this notation is compatible with reading the names of the arrows in a path from the diagrams in the right order, i.e. from the source to the target (sink) of the path, e.g.

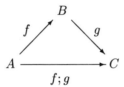

We stick to the 'traditional' notation $g \circ f$ for the sake of compatibility with most textbooks.

There are other familiar functions which are naturally polymorphic:

- the projections $first : X \times Y \to X$ and $second : X \times Y \to Y$: for every pair $f : X \to Y, g : X' \to Y'$ the equalities

$$first_{Y,Y'} \circ (f \times g) = f \circ first_{X,X'} \quad second_{Y,Y'} \circ (f \times g) = g \circ second_{X,X'}$$

hold.

- the embedding $\eta_X : X \to X^*, x \mapsto x$ of a set X into the monoid of words X^*: for every function $f : X \to Y$ the equality

$$f^* \circ \eta_X = \eta_Y \circ f$$

holds.

1.5.2 The notion of natural transformations abstracts the kind of polymorphism of these examples.

Definition 1.5.2. Let $F : \mathbf{C} \to \mathbf{D}$ and $G : \mathbf{C} \to \mathbf{D}$ be functors. A collection $\sigma = \langle \sigma_A : F(A) \to G(A) \mid A \in Obj(\mathbf{C}) \rangle$ of morphisms, $\sigma_A : F(A) \to$

$G(A)$ of **D**, indexed by objects A of **C** is called a *natural transformation* if
the diagram

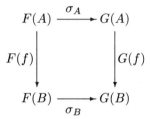

commutes for all $f : A \to B$ in **C**.

We shall say that σ is a natural transformation from $F : \mathbf{C} \to \mathbf{D}$ to
$G : \mathbf{C} \to \mathbf{D}$, and use the notation $\sigma : F \to G : \mathbf{C} \to \mathbf{D}$ to indicate the
domains and codomains involved.

Passing through the list of functors more well-known concepts appear
as natural transformations:

- Every homomorphism $h : \mathcal{D} \to \mathcal{D}'$ of M-automata defines a nat-
  ural transformation $h : \mathcal{D} \to \mathcal{D}' : \mathbf{M} \to \mathbf{Set}$ since $h(\mathcal{D}(x)(s)) =$
  $h(\delta(x, s)) = \delta'(x, h(s))$.

- Let $f, g : (X, \sqsubseteq) \to (Y, \sqsubseteq)$ be monotone functions considered as functors. Then there exists exactly one natural transformation $\sigma : f \to g$
  iff $f \sqsubseteq g$, i.e. $f(x) \sqsubseteq (x)$ for all $x \in X$.

We will learn about many more examples of natural transformation.
In most cases the idea of a polymorphic function is quite appropriate.
The other interpretation which fits this bill is that of a homomorphism of
functors as indicated by the example of automata. The latter is a theme
we dwell on for some time in Section 8.

1.5.3 Functors and natural transformations interact in various ways.

- Functors and natural transformations of the same kind form a category:

  Let $\sigma : F \to G$ and $\tau : G \to H$ be natural transformations where
  $F, G, H : \mathbf{C} \to \mathbf{D}$. Then $(\tau \circ \sigma)_A := \tau_A \circ \sigma_A$ is a natural transformation since $H(f) \circ \tau_A \circ \sigma_A = \tau_B \circ G(f) \circ \sigma_A = \tau_B \circ \sigma_B \circ F(f)$.
  $(1_F)_A = 1_A$ defines an identity $1_F : F \to F : \mathbf{C} \to \mathbf{D}$.

  These data define a *functor category*, often denoted by $\mathbf{C} \to \mathbf{D}$ or by
  $\mathbf{D}^\mathbf{C}$, with functors $F : \mathbf{C} \to \mathbf{D}$ as objects and natural transformations
  $\sigma : F \to G$ as morphisms. Composition is defined by $\tau \circ \sigma := \langle \tau_A \circ \sigma_A :$

$F(A) \to H(A) \mid A \in Obj(\mathbf{C})\rangle$ if $\sigma : F \to G$ and $\tau : G \to H$ where $F, G, H : \mathbf{C} \to \mathbf{D}$.

- Suppose there are functors $H : \mathbf{B} \to \mathbf{C}$, $F, G : \mathbf{C} \to \mathbf{D}$ and $K : \mathbf{C} \to \mathbf{D}$, and a natural transformation $\sigma : F \to G$ as in the diagram

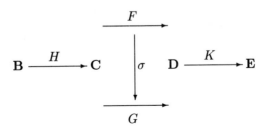

(which suggests a diagrammatic presentation for natural transformations). Then σ induces two natural transformations $\sigma * H : F \circ H \to G \circ H : \mathbf{B} \to \mathbf{D}$ and $K * \sigma : K \circ F \to K \circ G : \mathbf{C} \to \mathbf{E}$ defined by $(\sigma * H)_A = \sigma_{H(A)} : F(H(A)) \to F(G(A))$ and $(K * \sigma)_B = K(\sigma_B) : K(F(B)) \to K(G(B))$.

Composition of natural transformations and pre- and post-composition with a functor, intertwine to give another composition $\tau * \sigma : H \circ F \to K \circ G : \mathbf{C} \to \mathbf{D}$ of natural transformations

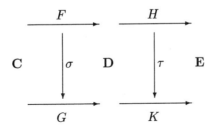

defined by $(\tau * \sigma)_A = \tau_{G(A)} * H(\sigma_A) = K(\sigma_A) * \tau_{F(A)}$.

We will also use the notation $\tau * \sigma : H * F \to K * G$ for composition of functors if used in this context. For motivation see Section 4.3.6.

2 On universal definitions: products, disjoint sums and higher types

2.1 Product types

2.1.1 The notion of a category on its own is obviously rather meagre. Thought of as a functional language, only single-argument programs are considered, though the type of the argument and of the result may vary. Functions with a number of arguments of possibly different types add some sophistication. Such an extension is achieved by introduction of *product types* (or *types of pairs*) as specified by the following rules

$$A \times B \text{ is a type} \quad \text{if } A, B \text{ are types}$$
$$\langle a, b \rangle : A \times B \quad \text{if } a : A \text{ and } b : B.$$

Typically, product types come along with two primitive functions

$$first_{A,B} : A \times B \to A \qquad second_{A,B} : A \times B \to B$$

the behaviour of which is defined by

$$first_{A,B} \cdot \langle a, b \rangle = a \qquad second_{A,B} \cdot \langle a, b \rangle = b.$$

Adding the equality

$$\langle first_{A,B} \cdot x, second_{A,B} \cdot x \rangle = x \qquad \text{for } x : A \times B$$

guarantees that pairs $\langle a, b \rangle$ are the only objects of type $A \times B$.

Pairing is extended to functional programs by

$$\langle f, g \rangle : A \to A \times B \qquad \text{if } f : A \to B, g : A \to C$$
$$\langle f, g \rangle \cdot x = \langle f \cdot x, g \cdot x \rangle.$$

We step back to observe a few consequences of these equations (axioms) (dropping subscripts for convenience):

$$first \circ \langle f, g \rangle \cdot x = first \cdot (\langle f, g \rangle \cdot x) = first \cdot \langle f \cdot x, g \cdot x \rangle = f \cdot x$$
$$second \circ \langle f, g \rangle = second \cdot (\langle f, g \rangle \cdot x) = g \cdot x$$
$$\langle first \circ h, second \circ h \rangle \cdot x = \langle first \circ h \cdot x, second \circ h \cdot x \rangle =$$
$$= \langle first \cdot (h \cdot x), second \cdot (h \cdot x) \rangle = h \cdot x$$
$$\langle f, g \rangle \circ h \cdot x = \langle f, g \rangle \cdot (h \cdot x) = \langle f \cdot (h \cdot x), g \cdot (h \cdot x) \rangle$$
$$= \langle f \circ h \cdot x, g \circ h \cdot x \rangle = \langle f \circ h, g \circ h \rangle \cdot x.$$

In the terminology of functional programming, the applicative laws induce functional laws

$$first \circ \langle f, g \rangle = f$$
$$second \circ \langle f, g \rangle = g$$
$$\langle first \circ h, second \circ h \rangle = h$$

which axiomatize 'pairing' since all the other functional laws are induced, e.g.

$$\langle f, g \rangle \circ h = \langle first \circ (\langle f, g \rangle \circ h), second \circ (\langle f, g \rangle \circ h) \rangle$$
$$= \langle (first \circ \langle f, g \rangle) \circ h, (second \circ \langle f, g \rangle) \circ h \rangle = \langle f \circ h, g \circ h \rangle.$$

2.1.2 In category theory, pairing is captured by the concept of products.

Definition 2.1.1. Let A and B be objects of a category **C**. A (binary) *product* of A and B consists of a *product object* $A \Pi B$ and two *projection* morphisms $p_{A,B} : A \Pi B \to A$ and $q_{A,B} : A \Pi B \to B$. These data satisfy the property that for all objects C and all morphisms $f : C \to A, g : C \to B$ there exists a unique morphism $\langle f, g \rangle : C \to A \Pi B$ such that the equations

$$p_{A,B} \circ \langle f, g \rangle = f$$
$$q_{A,B} \circ \langle f, g \rangle = g$$

hold. This is often expressed by saying that the diagram

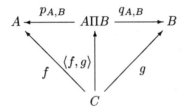

commutes. (For convenience we use $p_{A,B}$ and $q_{A,B}$ instead of $first_{A,B}$ and $second_{A,B}$ in the categorical context. Note that the equation $\langle first_{A,B} \circ h, second_{A,B} \circ h \rangle = h$ guarantees that h is uniquely determined.)

Terminal objects allow the introduction of 'constant functions' to our functional calculus.

Definition 2.1.2.

- An object 1 of a category **C** is called *terminal object* if for all objects A there exists a unique morphism $\langle \rangle_A : A \to 1$.

- A category **C** has *finite products* if it has a terminal object and a binary product for each ordered pair of objects.

The terminology reflects the notion of 'finite tuples'. However, one should be warned not to associate terminal objects too closely with 'singleton-set-like-objects' or to associate binary products with 'Cartesian products' of sets. The example shows that a variety of concepts subsume to the definition which do not necessarily agree with a naïve idea of tupling.

2.1.3 Again functional programming only offers one perspective. Other examples demonstrate a variety of familiar concepts to be products.

- *Cartesian products* in the category **Set** of sets and total functions, of course, are products. The projections $p_{X,Y} : X \times Y \to X$ and

$q_{X,Y} : X \times Y \to Y$ map to the components, $p_{X,Y}(x,y) = x$ and $q_{X,Y}(x,y) = y$. An induced function $\langle f,g \rangle : Z \to X \times Y$ is defined by $\langle f,g \rangle(z) = (f(z), g(z))$. Every one-point set is a terminal object.

- Products in the category of sets and partial functions are slightly more elaborate. Given sets X and Y a product object is given by $X \Pi Y := X \cup X \times Y \cup Y$. Projections are defined by

$$p(z) = \begin{cases} x & \text{if } z = \langle x,y \rangle \in X \times Y \text{ or } z = x \in X \\ \text{undefined} & \text{else} \end{cases}$$

$$q(z) = \begin{cases} y & \text{if } z = \langle x,y \rangle \in X \times Y \text{ or } z = y \in Y \\ \text{undefined} & \text{else} \end{cases}$$

The empty set is the only terminal object.

The Cartesian product does not satisfy the requirements of a categorical product since no proper definition of $\langle f,g \rangle : Z \to X \times Y$ can be given if f or g is not totally defined.

- Binary products in the category $\mathbf{Set}_\perp$ of pointed sets are defined by $X \Pi Y := X \times Y$ with projections $p(x,y) = x$ and $q(x,y) = y$. The set $\{\perp\}$ is the only terminal object. (Remark: relate this definition of products to that for partial functions.)

- A product of monoids $\mathcal{M}_1 = (M_1, *_1, e_1)$ and $\mathcal{M}_2 = (M_2, *_2, e_2)$ is defined by $M_1 \times M_2$ with multiplication $(x_1, x_2) * (y_1, y_2) = (x_1 *_1 y_1, x_2 * y_2)$. The projections to the components are monoid homomorphisms. Products for other sets with structures, such as groups etc., are defined in the same way. The reader might try to define products of automata and of graphs.

- Let $X = (X, \sqsubseteq)$ be a poset, and S be a subset of X. Then $x \in X$ is called a *lower bound* of S if $x \sqsubseteq s$ for all $s \in S$. $\sqcap S \in X$ is a *greatest lower bound* of S if $y \sqsubseteq \sqcap S$ for all lower bounds y of S.

If X is considered as an order category, the greatest lower bound $\sqcap\{x,y\}$, or the *meet* $x \sqcap y$, is a binary product. $\sqcap \varnothing$ is a terminal object, which is the greatest element of X.

Specifically, if $\mathcal{X}$ is a set of subsets of a set X ordered by inclusion, $U \cap V$ is the product object of the subsets U and V (if $U \cap V \in \mathcal{X}$). The projections are the inclusion $U \cap V \subseteq U, U \cap V \subseteq V$. $W \subseteq U$ and $W \subseteq V$ implies that $W \subseteq U \cap V$. Finally X itself is a terminal object if element of $\mathcal{X}$.

- Given an order category $(\Gamma, \vdash)$ with objects being propositions, product objects define conjunctions with elimination rules as projections

$$\varphi \wedge \psi \vdash \varphi \qquad\qquad \varphi \wedge \psi \vdash \psi$$

and pairing of morphisms yields a sequent version of the introduction rule

$$\frac{\gamma \vdash \varphi \qquad \gamma \vdash \psi}{\gamma \vdash \varphi \wedge \psi}$$

The terminal object defines 'truth' in that $\varphi \vdash \mathsf{tt}$.

- The product category $\mathbf{C} \times \mathbf{D}$ is a product in the category **Cat**.

- Let A be a set, then a product in the comma category $(\mathbf{Set}, A)$ is of the following form: given objects (X, f) and (Y, g) define a $\mathrm{set}(X, f)\Pi(Y, g) = \{(x, y) \in X \times Y \mid f(x) = g(x)\}$ and a function $h : (X, f)\Pi(Y, g) \to A, (x, y) \mapsto f(x)$. We claim that $((X, f)\Pi(Y, g), h)$ determines a product in $(\mathbf{Set}, A)$ with projections defined by $p : (X, f)\Pi(Y, g) \to X, (x, y) \mapsto x$ and $q : (X, f)\Pi(Y, g) \to Y, (x, y) \mapsto y$.

2.1.4 Here is a rather syntactical category of *procedures*. A language of procedures extending that of Section 1.3.1 by lists of parameters is specified by

Types
- 1 is a type
- $A \times B$ is a type if A and B are types

Parameter lists
- $\langle x^A \rangle : A$ where x is a variable,
- $\langle \langle p \rangle, \langle q \rangle \rangle : A \times B$ if $\langle p \rangle : A$ and $\langle q \rangle : B$ such that the variables in $\langle p \rangle$ and $\langle q \rangle$ are disjoint

Functional terms
- $\langle x^A \rangle \langle \rangle : A \to 1$
- $\langle p \rangle x^B : A \to B$ if x^B occurs in $\langle p \rangle : A$
- $\langle p \rangle \langle M, N \rangle : A \to B \times C$ if $\langle p \rangle M : A \to B$ and $\langle p \rangle N : A \to C$
- $\langle p \rangle (f \cdot M) : A \to C$ if $\langle p \rangle M : A \to B$ and $f : B \to C$

Axioms

$\langle p \rangle (((\langle q \rangle M) \cdot N) = \langle p \rangle M[\langle q \rangle / N]$
$\langle p \rangle (f \cdot \langle p \rangle) = f$
$\langle p \rangle \langle first \cdot M, second \cdot M \rangle = \langle p \rangle M$ where $first = \langle x^A, y^B \rangle \cdot x^A$
$\qquad\qquad\qquad\qquad\qquad\qquad\qquad$ and $second = \langle x^A, y^B \rangle \cdot y^B$
$f = \langle p \rangle \langle \rangle$ where $f : A \to 1$
$f = f$
$f = g$ if $f = g$
$f = h$ if $f = g$ and $g = h$
$\langle p \rangle f \cdot M = \langle p \rangle g \cdot N$ if $f = g$ and $\langle p \rangle M = \langle p \rangle N$

$$\langle p\rangle\langle M_1, N_1\rangle = \langle p\rangle\langle M_2, N_2\rangle \quad \text{if } \langle p\rangle M_i = \langle p\rangle N_i \text{ for } i = 1, 2.$$

We assume that parameters can be renamed for free provided that the binding structure is preserved (technically: α-conversion holds [Barendregt, 1984]. $M[\langle p\rangle/N]$ denotes simultaneous substitution of all variables). A motivation for the specific choice of parameter lists is given in Section 7.1.

A *signature* Σ consists of a collection B of *base types*, and a collection of elementary operations $f : A \to B$ where A, B are types.

As an example, consider the following signature Σ_{nat} of natural numbers as given by the following data: Σ_{nat}

Base types	$\underline{nat}, \underline{bool}$
Elementary operations	$0 :\to \underline{nat}$
	$suc, pred : \underline{nat} \to \underline{nat}$
	$add : \underline{nat} \times \underline{nat} \to \underline{nat}$
	$true, false : 1 \to \underline{bool}$
	$and : \underline{bool} \times \underline{bool} \to \underline{bool}$
	$not : \underline{bool} \to \underline{bool}$
	$iszero : \underline{nat} \to \underline{bool}$

We can then form functional terms such as

$$\langle x^{\underline{nat}}\rangle iszero \cdot \langle add \cdot \langle x^{\underline{nat}}, 0 \cdot \langle\rangle\rangle\rangle : \underline{nat} \to \underline{bool}$$
$$\langle x^{\underline{nat}}, y^{\underline{nat}}\rangle add \cdot \langle pred \cdot x^{\underline{nat}}, suc \cdot x^{\underline{nat}}\rangle : \underline{nat} \times \underline{nat} \to \underline{nat}$$
$$\langle x^1\rangle iszero \cdot (0 \cdot x^1) : 1 \to \underline{bool}.$$

Given a signature Σ, we define a category $\mathbf{T}_\Sigma$ as follows:

Objects	A
Morphisms	$[f] : A \to B$ where
	$[f] = \{g : A \to B \mid f = g : A \to B \text{ is derivable}\}$
Composition	$[g] \circ [f] = [g \circ f],$
Identities by	$1_A = [\langle x^A\rangle x].$

Proposition 2.1.3. $\mathbf{T}_\Sigma$ *is a category with finite products.*

Proof. [Exercise] Compute all the properties along the lines of Section 1.3.4. ∎

We obtain a sound semantics by extending the semantics of Section 1.4.5:

$$\llbracket \langle x^A \rangle \rrbracket = \llbracket A \rrbracket$$
$$\llbracket \langle \langle p \rangle, \langle q \rangle \rangle \rrbracket = \llbracket \langle p \rangle \rrbracket \times \llbracket \langle q \rangle \rrbracket$$
$$\llbracket \langle p \rangle \langle \rangle \rrbracket = \langle \rangle_{\llbracket \langle p \rangle \rrbracket}$$
$$\llbracket \langle x^A \rangle x \rrbracket = 1_{\llbracket A \rrbracket}$$
$$\llbracket \langle \langle p \rangle, \langle q \rangle \rangle x \rrbracket = \begin{cases} \llbracket \langle p \rangle x \rrbracket \circ p_{\llbracket \langle p \rangle \rrbracket, \llbracket \langle q \rangle \rrbracket} & \text{if } x \text{ occurs in } \langle p \rangle \\ \llbracket \langle q \rangle x \rrbracket \circ q_{\llbracket \langle p \rangle \rrbracket, \llbracket \langle q \rangle \rrbracket} & \text{if } x \text{ occurs in } \langle p \rangle \end{cases}$$
$$\llbracket \langle p \rangle \langle M, N \rangle \rrbracket = \langle \llbracket \langle p \rangle M \rrbracket, \llbracket \langle p \rangle N \rrbracket \rangle.$$

Proposition 2.1.4. *Let an object $Sem(A)$ and morphism $Sem(f)$: $\llbracket A \rrbracket_{Sem}$ $\to \llbracket B \rrbracket_{Sem}$ in a category $\mathbf{C}$ be given for every type A and every program $f : A \to B$ in Σ. Then there exists only one functor $\llbracket \_ \rrbracket_{Sem} : \mathbf{T}_\Sigma \to \mathbf{C}$ which preserves finite products and is such that $\llbracket A \rrbracket_{Sem} = Sem(A)$ and $\llbracket f : A \to B \rrbracket_{Sem} = Sem(f)$, where:*

A functor $F : \mathbf{C} \to \mathbf{D}$ preserves finite products if $F(1)$ is a terminal object in $\mathbf{D}$, and if $F(A \Pi B)$ with projections $F(p_{A,B})$ and $F(q_{A,B})$ is a binary product in $\mathbf{D}$.

Proof. [Exercise] Along the lines of Section 1.4.5. ∎

A semantics for a signature Σ is usually called a Σ-*algebra*. Hence the result states that every algebra can be extended to a functor. This observation leads to a categorical analysis of algebra under the name of *functorial semantics*, which will be discussed more thoroughly in Section 8.

2.2 Coproducts

2.2.1 Disjoint sums are another example for a universal definition of a data type. In set theory, a disjoint sum $X + Y$ is typically defined as the set $X + Y = \{1\} \times X \cup \{2\} \times Y$. The prominent feature, however, is that every pair of functions $f : X \to Z, g : Y \to Z$ can be extended to a function $[f, g] : X + Y \to Z$ by case distinction $[f, g](1, x) = f(x)$ and $[f, g](2, y) = g(y)$. Obviously, this is the only function $h : X + Y \to Z$ such that $h(1, x) = f(x)$ and $h(2, y) = g(y)$.

Category theory transfers the pattern.

Definition 2.2.1. Let A and B be objects of a category $\mathbf{C}$. A (binary) *coproduct* of A and B consists of a *coproduct object* $A \Pi B$ and two *injection morphisms* $inl_{A,B} : A \to A \Pi B$, $inr_{A,B} : B \to A \Pi B$. These data satisfy the property that for all objects C and for all morphisms $f : A \to C, g : B \to C$ there exists a unique morphism $[f, g] : A \Pi B \to C$ such that the diagram

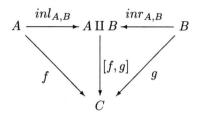

commutes.

An object 0 of a category $\mathbf{C}$ is called *initial object* if for all objects A of $\mathbf{C}$ there exists a unique morphism $0_A : 0 \to A$.

Disjoint sums are closely related to pattern matching as found in programming languages where a notation might be

$$h = \langle z^{A+B} \rangle case \; z, inl \cdot x \to f, inr \cdot y \mapsto g \; end$$

where either f or g is used according to the pattern of the argument

$$h \cdot (inl \cdot a) = f \cdot a \qquad \text{and} \qquad h \cdot (inr \cdot b) = g \cdot b$$

(Pattern matching will be looked at more closely below).

2.2.2   Examples

- Coproducts in the category **Set** of sets and total functions are, of course, disjoint sums. The empty set is the only initial object.

- Coproducts in categories of algebras are more sophisticated. We consider the example of monoids:

  Let $\mathcal{M}_1 = (M_1, *_1, e_1), \mathcal{M}_2 = (M_2, *_2, e_2)$ be monoids. We define a monoid $\mathcal{M}_1 \amalg \mathcal{M}_2$ with a carrier $M_1 \amalg M_2 \subseteq (M_1 + M_2)^*$ such that $w \in M_1 \amalg M_2$ iff $i \neq j$ whenever $w = u(i,x)(j,y)v$.

  Multiplication is defined by

  $w * \varepsilon = \varepsilon * w = w,$

  $v(i,x) * (j,y)w = v * (i, x *_i y) * w$ if $i = j$

  $v(i,x) * (j,y)w = v * (i,x)(j,y) * w$ if $i \neq j$

  Initial objects in categories of algebras will be discussed in Section 3.6.

- The least element and binary joins are the finite coproducts in a partial order as a category.

- If we interpret objects as propositions and morphisms as entailment, finite coproducts correspond to 'falseness' and 'disjunction':

$$\mathrm{ff} \vdash \varphi$$

$$\varphi \vdash \varphi \vee \psi \qquad \psi \vdash \varphi \vee \psi \qquad \dfrac{\varphi \vdash \gamma \qquad \psi \vdash \gamma}{\varphi \vee \psi \vdash \gamma}$$

- A non-trivial case of non-existence of coproducts:

The category **Dpos** of directed complete posets does not have finite coproducts. Assume that there exists a coproduct $X \amalg Y$ of directed complete posets X and Y. Define the *separated sum* $X +_{\perp} Y$ of X and Y to consist of the set $X \cup Y \cup \{\perp\}$ with partial order $z \leq z'$ if $z \leq z'$ in X or Y or if $z = \perp$ (we assume without restriction of generality that X and Y are disjoint). The inclusions $u_X : X \subseteq X +_{\perp} Y$ and $u_Y : Y \subseteq X +_{\perp} Y$ are continuous.

A unique continuous mapping $[u_X, u_Y] : X \amalg Y \to X +_{\perp} Y$ is induced such that $[u_X, u_Y] \circ inl = u_X$ and $[u_X, u_Y] \circ inr = u_Y$. We claim that $[u_X, u_Y](\perp) = \perp$; the only other possibilities are $[u_X, u_Y](\perp) = u_X(\perp)$ or $[u_X, u_Y](\perp) = u_Y(\perp)$ which cannot be monotone (take $\perp \leq u_Y(\perp)$ for instance).

On the other hand, inl and inr induce a mapping $case(inl, inr) : X +_{\perp} Y \to X \amalg Y$ such that $case(inl, inr)(x) = inl(x)$ for $x \in X$, $case(inl, inr)(y) = inr(y)$ for $y \in Y$, and $case(inl, inr)(\perp) = \perp$. Then $case(inl, inr) \circ [u_X, u_Y] = 1_{X \amalg Y}$ by coproduct properties since

$$case(inl, inr) \circ [u_X, u_Y] \circ inl = case(inl, inr) \circ u_X = inl$$

and similarly for inr.

The separated sum has the property that for every two continuous mappings $f : X \to Z, g : Y \to Z$ there exists a unique continuous mapping $[f, g] : X +_{\perp} Y \to Z$ such that $[f, g] \circ u_X = f$, $[f, g] \circ u_Y = g$, and $[f, g](\perp) = \perp$. Use this to prove $[u_X, u_Y] \circ case(inl, inr) = 1_{X +_{\perp} Y}$. Hence $X +_{\perp} Y$ and $X \amalg Y$ are isomorphic and $(X +_{\perp} Y, u_X, u_Y)$ should be a coproduct (see Section 2.5.4).

But this cannot be true since the condition $[f, g](\perp) = \perp$ is necessary for uniqueness of the induced morphism. As a counterexample consider $1 +_{\perp} 1$ where two continuous mappings into $(1 +_{\perp} 1)_{\perp}$ exist. $X_{\perp}$ is *lifted* meaning that a new least element is added.

- We extend the procedural language of Section 2.1.4 by pattern matching.

Types
$$A + B$$

Terms

$\langle p \rangle inl \cdot M : A \to B + C \qquad$ if $\langle p \rangle M : A \to B$

$\langle p \rangle inr \cdot N : A \to B + C \qquad$ if $\langle p \rangle N : A \to C$

$\langle p \rangle case\ x^A in\ inl \cdot M \mapsto f \cdot M, inr \cdot N \mapsto g \cdot N\ end$
$$\text{if } f : A \to B \text{ and } g : A \to C$$

Axioms

$\langle q \rangle (\langle p \rangle case\ x^A in\ inl \cdot y^B \mapsto f \cdot y^B,$
$$inr \cdot z^C \mapsto g \cdot z^C end) \cdot (inl \cdot M) = \langle q \rangle f \cdot M$$

$\langle q \rangle (\langle p \rangle case\ x^A in\ inl \cdot y^B \mapsto f \cdot y^B,$
$$inr \cdot z^C \mapsto g \cdot z^C end) \cdot (inr \cdot N) = \langle q \rangle g \cdot N$$

$\langle p \rangle case\ x^A in\ inl \cdot y^B \mapsto h \cdot (inl \cdot y^B),$
$$inr \cdot z^C \mapsto h \cdot (inr \cdot z^C))\ end\ = h$$

plus axioms for compatibility with the structure.

Then $\mathbf{T}_\Sigma$ constructed as in Section 2.1.4 defines a category with products and coproducts (but products and coproducts do not commute).

2.2.3 The definition of coproducts *dualizes* the definition of products in that the direction of all morphisms is reversed. This dualization is familiar for partially ordered sets where joins become meets, and vice versa, if we reverse the order. However, the full generality of the duality principle was probably first appreciated in category theory where all definitions dualize in that they apply to the dual category. The dual definition makes sense in almost every case which only reveals the predominance of symmetry in mathematical thinking.

Duality has the obvious advantage that it cuts down the work invested in proofs by half, since the direction of arrows may be reversed in every categorical proof.

Certain definitions are *self-dual* in that the dual definition does not define a new concept. Clearly, the definition of isomorphism is self-dual as well as the definition of categories itself. One should distinguish the self-duality of the *definition* of categories from the concept of the dual category $\mathbf{C}^{op}$.

2.3 Higher types

2.3.1 Higher types, sometimes called function types, introduce 'functions' as 'first class citizens' in that functions may have arguments and results which are themselves functions. If we introduce higher types by

$$A \Rightarrow B \qquad \text{is a type if } A, B \text{ are types,}$$

here are some typical examples of (polymorphic) functional programs employing higher types

$$apply : (A \Rightarrow B) \times A \to B$$
$$apply \cdot \langle f, a \rangle = f \cdot a$$
$$Y : (A \Rightarrow A) \to A$$
$$Y \cdot f = f \cdot (Y \cdot f)$$
$$maplist : [A \Rightarrow B] \times list(A) \to list(B)$$
$$maplist \cdot \langle f, nil \rangle = nil$$
$$maplist \cdot \langle f, append \cdot \langle insert \cdot a, x \rangle \rangle = append \cdot \langle insert \cdot (f \cdot a),$$
$$maplist \cdot \langle f, x \rangle \rangle.$$

The applicative definitions confuse the level of arguments and programs in that the same symbol f is used for the *function* $f : A \to B$ and the *element* $f \in A \Rightarrow B$ of the function space. This usage reflects set theory, where the function space $A \Rightarrow B$ can be identified with the *set of functions* $\mathbf{Set}(A, B) = \{f \mid f : A \to B\}$. However, there is a clear distinction in terms of the operators which can be applied; functions $f : A \to B$ and $g : B \to C$ can be composed to yield a function $g \circ f : A \to C$, while the function $apply : (A \Rightarrow B)\Pi A \to B, (f, a) \mapsto f \cdot a$ maps a function and an argument to the respective result. The difference may appear as rather subtle but is made precise by the isomorphism $A \times B \to C \cong A \to (B \Rightarrow C)$, well known from set theory and functional programming, which allows one to turn every binary function $f : A \times B \to C$ into a unary function $\lambda(f) : A \to (B \Rightarrow C)$ into a function space. This bijection can be axiomatized as follows:

For each choice of A, B and C let there be operators

$$\lambda_{A,B,C} : (A \times B \to C) \to (A \to (B \Rightarrow C)) \quad f \mapsto \lambda(f)$$
$$\lambda^{-1}_{A,B,C} : (A \to (B \Rightarrow C)) \to (A \times B \to C) \quad g \mapsto \lambda^{-1}(g)$$

satisfying the applicative laws

$$(\lambda(f) \cdot a) \cdot b = f \cdot \langle a, b \rangle \qquad \lambda^{-1}(g) \cdot \langle a, b \rangle = (g \cdot a) \cdot b$$

and thus the functional laws

$$\lambda^{-1}(\lambda(f)) = f \qquad \lambda(\lambda^{-1}(g)) = g. \tag{$*$}$$

These functional laws provide exactly what is needed to axiomatize higher types within the language of categories.

As an alternative, we may define

$$apply \cdot \langle f, a \rangle := f \cdot a$$

and use

$$apply \circ (g \times 1_B) \cdot \langle a, b \rangle = apply \cdot \langle g \cdot a, b \rangle = (g \cdot a) \cdot b = \lambda(g) \cdot \langle a, b \rangle \qquad (**)$$

to restate the equations $(*)$ as

$$apply \circ (\lambda(f) \times 1_B) = f \qquad\qquad \lambda(apply \circ (g \times 1_B)) = g$$

which is an equivalent axiomatization (exercise). The above ideas are neatly packaged in the concept of a Cartesian closed category.

Definition 2.3.1. We say a category **C** is a *Cartesian closed category* if for all ordered pairs A, B of objects

- there is a designated product $A\Pi B$, (with designated projections $p_{A,B}$ and $q_{A,B}$).

- there is a designated object $B \Rightarrow C$, and

- a designated morphism $apply_{B,C} : (B \Rightarrow C)\Pi B \to C$

where these data satisfy the requirement that for all objects C and each morphism $f : A\Pi B \to C$ there exists a unique morphism $\lambda(f) : A \to (B \Rightarrow C)$ such that $apply_{B,C} \circ (\lambda(f)\Pi 1_B) = f$, or equivalently, the diagram

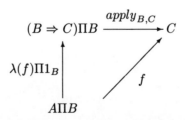

commutes (where, for morphisms $f : A \to B$ and $g : C \to D$ the morphism $f\Pi g : A\Pi C \to B\Pi D$ is defined by $f\Pi g = \langle f \circ p_{A,C}, g \circ p_{A,C} \rangle$. The second of the equations $(**)$ guarantees uniqueness of $\lambda(f)$.).

2.3.2 Examples

- The category of sets and total functions is Cartesian closed where $(X \Rightarrow Y) = \{f \mid f : X \to Y \text{ total function}\}$. The equation $apply_{X,Y}(f,x) = f(x)$ defines the function $apply_{X,Y} : (X \Rightarrow Y) \times X \to Y$.

- The category **Pfn** of sets and partial functions is not Cartesian closed: assume that there exists a 'function space set' $[Y \Rightarrow Z]$ such that $\mathbf{Pfn}(\varnothing, (Y \Rightarrow Z)) \cong \mathbf{Pfn}(\varnothing \Pi Y, Z)$ (where $\mathbf{Pfn}(X, X') = \{f \mid f : X \to X' \text{ partial function}\}$. As $\varnothing \Pi Y \equiv (\varnothing \times Y) + \varnothing + Y \cong Y$ (where $\times$ denotes the Cartesian product), $\mathbf{Pfn}(\varnothing, (Y \Rightarrow Z)) \cong \{\varnothing\} \cong \mathbf{Pfn}(Y, Z)$ which yields a contradiction for $Y = Z = \{\varnothing\}$.

(Remark: the Cartesian product $X \times Y$ is, however, well behaved with regard to the set $Y \Rightarrow Z := \{f \mid f : X \to Y \text{ partial function}\}$ in that the following property holds: for every *partial* function $f : X \times Y \to Z$ there exists a unique *total* function $\lambda(f) : X \to (Y \Rightarrow Z)$ such that $apply_{Y,Z} \circ (\lambda(f) \times 1_Y) = f$ where $apply_{X,Y}$ is defined as above.

Suitable abstractions of this example have recently found some attention under the name of *partially Cartesian closed categories* [Rosolini and Robinson, 1988], and have been used to model higher types with call-by-name parameter passing mechanism.)

- The category of monoids is not Cartesian closed (consider $f : 1 \times M \to M'$).

(Remark: the subcategory of commutative monoids have a rather similar property. Those readers familiar with modern algebra will know of the notion of a *tensor product* $M_1 \otimes M_2$ of (commutative) monoids which has the property that every bilinear mapping $f : M_1 \times M_2 \to M$ uniquely extends to a homomorphism $f^{\#} : M_1 \otimes M_2 \to M$ ($f : M_1 \times M_2 \to M$ is *bilinear* if $f(m_1 *_1 m'_1, m_2) = f(m_1, m_2) * f(m'_1, m_2)$ and $f(m_1, m_2 *_2 m'_2) = f(m_1, m_2) * f(m_1, m'_2)$). Let $(M_2 \Rightarrow M_3)$ be the Abelian monoid with elements $\{f \mid f : M_2 \to M_3 \text{ homomorphism }\}$ and multiplication $f * g(m_2) := f(m_2) *_3 g(m_2)$. Since the mapping $apply : (M_2 \Rightarrow M_3) \times M_2 \to M_3$, $(f, m) \mapsto f(m)$ is bilinear, a mapping $apply^{\#} : (M_2 \Rightarrow M_3) \otimes M_2 \to M_3$ is induced. Then for every homomorphism $f : M_1 \otimes M_2 \to M_3$ there exists a unique homomorphism $\lambda(f) : M_1 \to (M_2 \Rightarrow M_3)$ such that $apply^{\#} \circ (\lambda(f) \otimes 1_{M_2}) = f$. Abstraction of this set-up yields the notion of a *monoidal closed category* [MacLane, 1971].

- Function spaces $X \Rightarrow Y$ in the category **Pos** of posets have elements being monotone functions $f : X \to Y$ ordered pointwise, i.e. $f \leq g$ if $f(x) \leq g(x)$ for all $x \in X$. Application is functional application.

The category **Dcpo** has function spaces $X \Rightarrow Y$ which consist of continuous functions which are ordered pointwise. Least upper bounds of directed sets $F \subseteq (X \Rightarrow Y)$ are defined by $(\bigsqcup F)(x) = \bigsqcup\{f(x) \mid f \in F\}$.

- In a meet semilattice P, the *pseudo complement* of x relative to y is, if it exists, the greatest element of the set $\{z \mid x \sqcap z \leq y\}$. P is Cartesian closed if all pseudo complements exist. P is a *Heyting algebra* if additionally all finite joins exist.

  In the category $\mathbf{P}(X)$ of subsets of a set X, pseudo complements are of the form $(U \Rightarrow V) = (X \backslash U) \cup V$. Clearly, $(U \Rightarrow V)\sqcap U = ((X\backslash U) \cup V) \cap U = V \cap U \subseteq V$.

- As a category of propositions and entailment, a Heyting algebra formalizes intuitionistic propositional logic. The pseudo complements define implication

$$\frac{\varphi \wedge \psi \vdash \gamma}{\varphi \vdash (\psi \Rightarrow \gamma)} \qquad \frac{}{(\varphi \Rightarrow \psi) \wedge \varphi \vdash \psi}$$

  (Intuitionistic) negation is defined by $\neg\varphi := (\varphi \Rightarrow \text{ff})$ where ff is the initial object or least element.

- The comma category $(\mathbf{Set}, A)$ is Cartesian closed. The function space $(X, f) \Rightarrow (Y, g)$ of objects (X, f) and (Y, g) is defined by $(\{(a, s) \mid a \in A, s : f^{-1}(a) \to Y$ is a function$\}, p)$ where $p(\langle a, s\rangle) = a$. Application is defined by $apply_{(X,f),(Y,g)}((a, s), x) = s(x)$ which is well defined as $f(x) = p((a, s)) = a$. Let $k : (Z, h)\Pi(X, f) \to (Y, g)$ be a morphism of $(\mathbf{Set}, A)$. Then abstraction is given by $\lambda(k)(z) = \langle h(z), k_z\rangle$ where $k_z : f^{-1}(h(z)) \to Y, x \mapsto k(z, x)$.

  (Remark: A category $\mathbf{C}$ such that a terminal object exists and that all comma categories $(\mathbf{C}, A)$ are Cartesian closed is called *locally Cartesian closed*. Such categories play an important role in the categorical interpretation of *dependent types* [Seely, 1984]. (See Section 9.4.1))

- The language of procedures is extended by higher types as follows:

Types	$A \Rightarrow B$	
Terms	$\langle p\rangle(\lambda\langle q\rangle M) : A \to (B \Rightarrow C)$	if $\langle\langle p\rangle, \langle q\rangle\rangle M : A \times B \to C$
	$\langle p\rangle(MN) : A \to C$	if $\langle p\rangle M : A \to (B \Rightarrow C)$
		and $\langle p\rangle N : A \to B$
Axioms	$\langle p\rangle((\lambda\langle q\rangle M)N) = \langle p\rangle M[\langle q\rangle/N]$	
	$\langle p\rangle(\lambda\langle q\rangle(M\langle q\rangle)) = \langle p\rangle M$	if no variable of the parameter list $\langle q\rangle$ occurs free in M

plus axioms for compatibility with the structure.

Let a *higher type signature* Σ consist of a set B of base types, and a set of operators $f : A \to B$ where A and B are types (generated from base types). Define a category Λ_Σ which consists of types A and of [equivalence classes] of functional terms $[\langle p \rangle M] : A \to B$ which are generated from Σ in analogy to Section 2.1.4.

Proposition 2.3.2 (Exercise). Λ_Σ *is a Cartesian closed category.*

This is a variant of typed λ-calculus, the equivalence of which, with Cartesian closure, will be shown in Section 7.

2.3.3   Exercises

(A) Check that $\lambda(f) \circ g = \lambda(f \circ (g \Pi id_A))$ where $f : C\Pi A \to B$, $g : D \to C$.

(B) In the category of sets there is a function '$comp : (A \Rightarrow B)\Pi(B \Rightarrow C) \to (A \to C)$' which composes functions as elements of function spaces. Define this morphism in an arbitrary Cartesian closed category.

(C) $\ulcorner f \urcorner := \lambda(f \circ q_{1,A}) : 1 \to (A \Rightarrow B)$ for $f : A \Rightarrow B$.

Prove the equality $comp \circ (\ulcorner id_A \urcorner \Pi id_{[A \Rightarrow B]}) = q_{1,[A \Rightarrow B]}$ where '$comp$' is defined as under (B). One may establish 'associativity' of '$comp$'.

(D) Given the fixpoint operator $Y : [B \Rightarrow B] \to B$ define a 'fixpoint operator with parameter' $Y : (A\Pi B \Rightarrow B) \to (A \Rightarrow B)$.

(E) Define the morphism $(A \Rightarrow f) : (A \Rightarrow B) \to (A \Rightarrow C)$ for a given $f : B \to C$.

2.4   Reasoning by universal arguments

2.4.1   The definitions of products, coproducts and Cartesian closure follow the scheme

'X is ...' if '$\forall Y$ similar to X' $\exists ! ... : ...$

where $\exists !$ stands for 'there exists a unique'. As we will see, many other important concepts have definitions of this form which, being so common in category theory, are referred to as **universal definitions** or **universal solutions**. We shall be exploring these concepts more fully in Section 3. But we want to give some indication here as to how they are typically exploited to

• define special morphisms

• prove equality of morphisms.

2.4.2 Defining morphisms

Products resp. Cartesian closure induce morphisms such as

- $reverse_{A,B} := \langle q_{A,B}, p_{A,B} \rangle : A\Pi B \to B\Pi A$

- $f\Pi g := \langle f \circ p_{A,B}, g \circ q_{A,B} \rangle : A\Pi B \to C\Pi D$

- $A \Rightarrow g := \lambda(g \circ apply_{A,B}) : (A \Rightarrow B) \to (A \Rightarrow C)$

  where $g : B \to C$

- $f \Rightarrow C := \lambda(apply_{A,C} \circ (1_{B \Rightarrow C}\Pi f)) : (B \Rightarrow C) \to (A \Rightarrow C)$

  where $f : A \to B$

- $\circ := \lambda(apply_{B,C} \circ \langle p_{A \Rightarrow B, B \Rightarrow C} \circ p..., A, apply_{A,B} \circ \langle p_{A \Rightarrow B, B \Rightarrow C} \circ p..., A, q..., A \rangle \rangle : (A \Rightarrow B)\Pi(B \Rightarrow C) \to (A \Rightarrow C)$

where the latter three morphisms state 'the function $f \in A \Rightarrow B$ is mapped to $g \circ f \in A \Rightarrow C$', '$g \in B \Rightarrow C$ is mapped to $f \circ g \in A \Rightarrow C$' and '$f \in A \Rightarrow B$ and $g \in B \Rightarrow C$ are mapped to $g \circ f \in A \Rightarrow C$'.

Diagrams very often help to illustrate such definitions. For example, the morphism $reverse_{A,B}$ given above is the unique morphism making the diagram

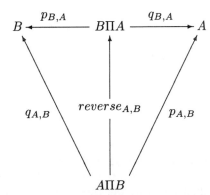

commute. (Note that $(B\Pi A, p_{B,A}, q_{B,A})$ is a product of B and A, in contrast to $(A\Pi B, p_{A,B}, q_{A,B})$ which is a product of A and B. The definition motivates the subscripts for projections.)

From the definition of Cartesian closure we see that the morphism $A \Rightarrow g$ is the unique morphism making the diagram

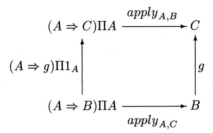

commute.

2.4.3 Proving equalities of morphisms

To prove the equality $reverse_{B,A} \circ reverse_{A,B} = 1_{A\Pi B}$ we need only to
note that, by the (universal) definition of products there is exactly one
morphism $h : A\Pi B \rightarrow A\Pi B$ such that $p_{A,B} \circ h = p_{A,B}$ and $q_{A,B} \circ h = q_{A,B}$, and that the commutativity of the following diagrams show that both
$reverse_{B,A} \circ reverse_{A,B}$ and $1_{A\Pi B}$ satisfy this property.

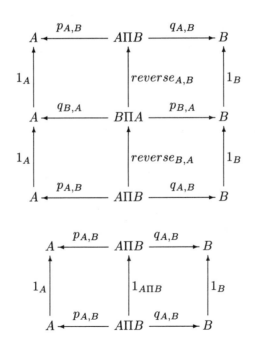

Similarly, to prove the equality $\langle g, h \rangle \circ f = \langle g \circ f, h \circ f \rangle$ we note that there
exists only one morphism $k : A \rightarrow C\Pi D$ such that $p_{C,D} \circ k = g \circ f$ and
$q_{C,D} \circ k = h \circ f$. But the following commuting diagrams claim the existence
of two such morphisms, namely $\langle g, h \rangle \circ f$ and $\langle g \circ f, h \circ f \rangle$ which thus must
be equal.

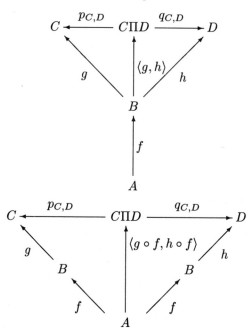

One often uses the term *diagram chase* to refer to this style of argument on diagrams.

Even though we have only discussed the specific concept of finite products, we hope to have demonstrated that reasoning about functional programming is similar in style to the kind of reasoning one does in certain kinds of category theoretic arguments (such as those found in diagram chasing). This may help to diminish some of the prejudices about category theory and categorical reasoning as being 'general abstract nonsense'. In fact, the categorical style of reasoning even appears as more intuitive and versatile since, once one is familiar with categorical arguments, proofs develop in a structured way with a natural choice of subgoals.

2.4.4 All structures introduced by universal definitions are only determined up to isomorphism. This will be demonstrated for binary products:

Definition 2.4.1. A morphism $f : A \to B$ is an *isomorphism* if there exist morphisms $g : B \to A$ such that $g \circ f = 1_A$ and $f \circ g = 1_B$. It is easy to see that if f is an isomorphism then g is uniquely determined, and so we call it the *inverse* of f. It is often convenient to write f^{-1} for the inverse of f. Objects A and B are *isomorphic* if there exists an isomorphism $f : A \to B$.

Proposition 2.4.2.

- Let $(C, p : C \to A, q : C \to B)$ and $(D, p' : D \to A, q' : D \to B)$ be products of A and B. Then there exists an isomorphism $f : C \to D$ such that the diagram

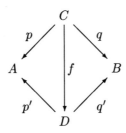

commutes.

- Let $(C, p : C \to A, q : C \to B)$ be a product of A and B, and let $f : C \to D$ be an isomorphism. Then $(D, p \circ f^{-1} : D \to A, q \circ f^{-1} : D \to B)$ is a product of A and B.

Proof. • The equality $f^{-1} \circ f = 1_C$ follows from the universal property of products in the commutative diagrams

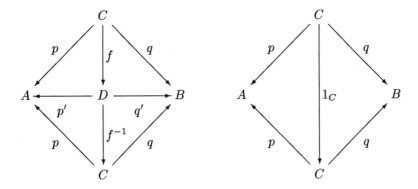

- Left to the reader.

∎

We note the difference from functional programming where products are **specified** since a specific object (and projections) is chosen, or where, in jargon, products are defined 'on the nose'.

2.5   Another 'universal definition': primitive recursion

2.5.1   We touch on the topic of primitive recursion which provides very good examples of definitions in terms of universal properties, which are of interest to computer science, and which do not fit in elsewhere since they are not in the main line of reasoning of this exposition. We assume everybody to be familiar with primitive recursion.

The property of natural numbers, we want to abstract, is the following. Let N denote the natural numbers. Then for any sets X and Y and any functions $g : X \to Y$ and $h : X \times Y \to Y$ there exists a unique function $f : N \times X \to Y$ defined by the scheme

$$f(0, x) = g(x)$$
$$f(n + 1, x) = h(x, f(n, x)).$$

Definition 2.5.1. Let **C** be a category with finite products. An object N together with two morphism $0 : 1 \to N, suc : N \to N$ is called *primitive recursion object* if for each pair of morphisms $g : X \to Y, h : X \Pi Y \to Y$ there exists a unique morphism $f : N \Pi X \to Y$ such that the diagrams

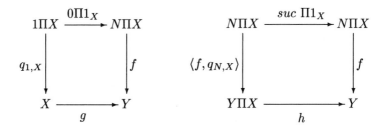

commute.

Existence of such an object allows one to define functions in terms of primitive recursion whatever the category is. As an example one might consider the category **Set** $\times$ **Set** where the pair (N, N) of natural numbers gives a primitive recursion object. A sort of simultaneous primitive recursion is defined.

We do not bother to give specific examples for the definition of functions since translation is straightforward.

2.5.2   If the category is Cartesian closed we have only to look for the following property of natural numbers.

Definition 2.5.2. A *natural number object* (NNO) in a category **C** consists of a pair of morphisms $0 : 1 \to N, suc : N \to N$ which satisfy the property that for every pair of morphisms $a : 1 \to A, h : X \to X$ there exists a unique $f : N \to X$ such that

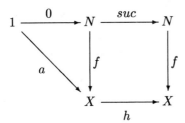

commutes.

Proposition 2.5.3. *Let* $\mathbf{C}$ *be a Cartesian closed category. Then* $(N, 0, suc)$ *is a natural number object if, and only if, it is a primitive recursion object.*

Proof. A bit of programming is needed. We have morphisms $\langle h, q_{X,Y} \rangle :$ $Y \Pi X \to Y \Pi X$ and $1_X \Rightarrow \langle h, q_{X,Y} \rangle : (X \Rightarrow (Y \Pi X)) \to X \Rightarrow (Y \Pi X))$, $\langle g, 1_X \rangle \circ q_{1,X} : X \to Y \Pi X$ and $\lambda(\langle g, 1_X \rangle \circ q_{1,X}) : 1 \to X \Rightarrow (Y \Pi X)$. The morphisms $1_X \Rightarrow \langle h, q_{X,Y} \rangle$ and $\lambda(\langle g, 1_X \rangle \circ q_{1,X})$ induce a morphism $f : N \to X \Rightarrow (Y \Pi X)$ by the property of NNO's. We claim that

$$apply_{X,X\Pi Y} \circ \langle f, 1_X \rangle : N \Pi X \to Y \Pi X$$

fits the bill, and leave it to the reader to check the properties. For the other direction use $X = 1$. $\blacksquare$

2.6 The categorical abstract machine

Functional programmers may appreciate the rigidity and maybe the elegance of the categorical approach to functional programming but keep at least one reserve: category theory does not run on a computer. Quite wrong! Category theory runs on a computer, the abstract machine even matches the elegance of the categorical language and most surprisingly the implementation runs quite efficiently. The machine is introduced in the paper [Cosineau *et al.*, 1987].

The programming language of the machine is the categorical language but with the minor difference that composition is concatenation in diagrammatic order, e.g. $f_n \circ \ldots \circ f_1$ becomes $f_1 \ldots f_n$. The language is then specified by an abstract syntax:

$$C ::= \quad t \mid f \mid 1 \mid first \mid second \mid apply \mid inl \mid inr \mid C1\ C2 \mid \langle C1, C2 \rangle \mid$$
$$\lambda(C) \mid [C1, C2] \mid Y(C)$$

(f ranges over some base functions), Y is a fixpoint operator which only applies to morphisms of type $f : X \to [A \Rightarrow A]$ to yield a morphism $Y(f) : X \to A$ such that $apply \circ \langle f, Y(f) \rangle = Y(f)$.

The machine is a stack machine which, given a program, a value and a stack, processes the first symbol (from left) of the program. In each step the program code, the value and the stack is changed according to the following table:

program	value	stack	→	program	value	stack
t C	v	s		C	0	s
f C	v	s		C	$f(v)$	s
1 C	v	s		C	v	s
⟨C	v	s		C	v	vs
, C	v	$v'\,s$		C	v'	vs
)C	v'	$v\,s$		C	$\langle v,v'\rangle$	s
first C	$\langle v,v'\rangle$	s		C	v	s
second C	$\langle v,v'\rangle$	s		C	v'	s
λ(C1) C2	v	s		C2	C1:v	s
apply C2	(C1:v,v')	s		C1 C2	$\langle v,v'\rangle$	s
inl C	v	s		C	$inl(v)$	s
inr C	v	s		C	$inr(v)$	s
[C1,C2] C	$inl(v)$	s		C1 C	v	s
[C1,C2] C	$inr(v)$	s		C2 C	v	s
Y(C1)C2	v	s		runc(C1)C2	(x=runc(C1):x,v)	s

where $C, C1, C2$ range over the code, and where

$$runc(C) = \langle second\ C, first\ C\rangle apply.$$

We notice the following properties:

- The machine is untyped since the basic operations of application and projection are natural (or parametric in computer terminology). If wanted, a type checking of the categorical program takes place at compile time.

- Abstraction is modelled by closure; $C : v$ states that 'v is the first argument for C the second argument of which is still to be computed'.

- The notation $x = C : x$ states that x can be expanded to $C : (x = C : x)$. This expansion takes care of recursive calls. The trick here is that the closure for λ-abstraction is used to deal with recursion at the same time; due to the type restriction, recursive calls are only necessary if an '$apply$' is to be executed.

We compute the factorial as an example. We assume existence of a type equation $\varphi : Nat \cong 1 + Nat$ which induces morphism $0 : 1 \rightarrow Nat, suc :$

Nat → *Nat*. We leave it as an exercise to define an operator *times* : *Nat* × *Nat* → *Nat*. The categorical code for the factorial is given by

$$apply \circ \langle Y(fak) \circ \langle \rangle, 1 \rangle$$

where

$fak = \lambda(\lambda(apply \circ \langle [\lambda(if), \lambda(else)] \circ \varphi \circ second, first \rangle) \circ second)$
$if = suc \circ 0 \circ first : 1 \times [Nat \Rightarrow Nat] \to Nat$
$else = times \ \circ \langle suc \circ first, apply \circ reverse \rangle : Nat \times [Nat \Rightarrow Nat] \to Nat$

(a simpler code is obtained if a branching operator is used, but we want to demonstrate that we can do without it.)

This translates to machine code

$$\langle tY(\lambda(second \ \lambda(\langle second \ \varphi[\lambda(first \ 0 \ suc), \lambda(\langle \langle second, first \rangle apply,$$
$$first \ suc \rangle times)], first \rangle apply)), 1 \rangle apply.$$

Assume that the input is 1 which corresponds to the code $\varphi^{-1} \circ inr \circ \varphi^{-1} \circ inl \circ 0$ and that the stack is empty. Then we have the following computation

$\langle t \ Y(\ldots), 1 \rangle apply$	1	
		λ
$t \ Y(\ldots), 1 \rangle apply$	1	
		1
$Y(\ldots), 1 \rangle apply$	()	
		1
$\langle second \ \lambda(\ldots), first \rangle apply, 1 \rangle apply$	$(x = runc(\ldots) : x, ())$	
		1
$second \ \lambda(\ldots), first \rangle apply, 1 \rangle apply(x = runc(\ldots) : x, ())$		
		$(x = runc(\ldots) : x, ())1$
$\lambda(\ldots), first \rangle apply, 1 \rangle apply$	()	
		$(x = runc(\ldots) : x, ())1$
$\ldots$		
$apply, 1 \rangle apply$	$((\ldots) : (), x = runc(\ldots) : x)$	
		1
$second \ \lambda(\ldots), 1 \rangle apply$	$((), x = runc(\ldots) : x)$	
		1
$\lambda(\ldots), 1 \rangle apply$	$x = runc(\ldots) : x$	
		1
$\ldots$		
$apply$	$((\ldots) : (x = runc(\ldots) : x), 1)$	
		λ
$\langle second \ \varphi[\ldots], first \rangle apply$	$(x = runc(\ldots) : x), 1)$	
		λ
$second \ \varphi[\ldots], first \rangle apply$	1	
		$(x = runc(\ldots) : x, 1)$
$[\ldots], second \rangle apply, 1 \rangle apply$	$inr \ \circ \varphi^{-1} \circ inl \ \circ 0$	
		$(x = runc(\ldots) : x, 1)$

$\lambda(\langle\langle\ldots\rangle times), second\rangle apply, 1\rangle apply$

$$\varphi^{-1} \circ inl \circ 0$$
$$(x = runc(\ldots) : x, 1)$$

$, second\rangle apply, 1\rangle apply$ $(\langle\ldots\rangle times) : \varphi^{-1} \circ inl \circ 0$
$$(x = runc(\ldots) : x, 1$$

$second\rangle apply, 1\rangle apply$ $(x = runc(\ldots) : x, 1)$
$$(\langle\ldots\rangle times) : \varphi^{-1} \circ inl \circ 0$$

$\langle apply, 1\rangle apply$ 1
$$(\langle\ldots\rangle times) : \varphi^{-1} \circ inl \circ 0$$

$apply, 1\rangle apply$ $(\langle\ldots\rangle times) : \varphi^{-1} \circ inl \circ 0, 1)$
$$\lambda$$

$\langle\langle second, first\rangle apply, first\ suc\rangle times$

$$(\varphi^{-1} \circ inl \circ 0, x = (\ldots : x))$$
$$\lambda$$

$\ldots$

$apply, first\ suc\rangle times$ $(x = (\ldots : x), \varphi^{-1} \circ inl \circ 0)$
$$(\varphi^{-1} \circ inl \circ 0, x = (\ldots : x))$$

'recursive call'
$\langle second\ \lambda(\ldots), first\rangle apply, first\ suc\rangle times$

$$(x = (\ldots : x), \varphi^{-1} \circ inl \circ 0)$$
$$(\varphi^{-1} \circ inl \circ 0, x = (\ldots : x))$$

$\ldots$

$, first\rangle apply, first\ suc\rangle times$ $(\ldots) : (x = (\ldots : x))$
$$(x = (\ldots : x), \varphi^{-1} \circ inl \circ 0)(\varphi^{-1} \ldots)$$

$first\rangle apply, first\ suc\rangle times$ $(\varphi^{-1} \circ inl \circ 0, x = (\ldots : x))$
$$(\ldots) : (x = (\ldots : x))(\varphi^{-1} \ldots)$$

$\rangle apply, first\ suc\rangle times$ $\varphi^{-1} \circ inl \circ 0$
$$(\ldots) : (x = (\ldots : x))(\varphi^{-1} \ldots)$$

$apply, first\ suc\rangle times$ $((\ldots) : (x = (\ldots : x) \circ \varphi^{-1} \circ inl \circ 0)$
$$(\varphi^{-1} \circ inl \circ 0, x = (\ldots : x))$$

$\langle second\ \varphi[\ldots], first\rangle apply, first\ suc\rangle times$

$$(x = (\ldots : x), \varphi^{-1} \circ inl \circ 0)$$
$$(\varphi^{-1} \circ inl \circ 0, x = (\ldots : x))$$

$\ldots$

Evaluation is inefficient, partly because of lack of optimization, partly because the code is extreme, not using a branching conditional. A much better machine is used in the original paper.

3   Universal problems and universal solutions

If some single concept of category theory should be declared as most important, then adjunctions come close to qualifying. Adjunctions crystallize the idea of a universal solution, meaning roughly that among all the possible entities satisfying some specified properties there exists a maximal or minimal solution. This idea is used extensively in all areas of mathematics including the mathematics of computer science, but category theory defines the precise framework to formalize a concept otherwise conveyed solely by examples.

The abstraction involved makes it difficult to approach the topic. Usually, the approach is descriptive, looking at a variety of examples typically of mathematical nature, not in every case easily explained against a background in computer science. We attempt a more prescriptive approach exploiting certain paradigms, which should be familiar to a computer scientist. The notions of observability and data abstraction and of data generation are used to suggest a rationale of adjunctions, though in the very end, we rely on examples as does everybody else. Even more likely, the text may fail to convey any substantial intuition at a first reading. However, I sincerely hope that, after consulting other textbooks, the reader may at least appreciate my effort.

We start with a conceptual analysis of abstraction and generation which leads to the definition of adjunction. The development is illustrated by only a few examples for the sake of stringency. Other examples are given in Section 3.6, which may be looked at at any point for further substantiation of the arguments.

3.1 On observation and abstraction

3.1.1 *Abstraction* and *generation* (or *construction*) are very basic principles upon which any kind of formal activity is based. Roughly, abstraction identifies structure up to inequalities perceived by observation, while generation means to construct some entity from given components according to given rules. By analogy, the perspectives are those of a physicist and an engineer; the physicist analyses what is given, while the engineer constructs according to requirements. The position of a physicist is maximalist abstracting from all unnecessary details, while the engineer is minimalist striving for the 'cheapest' solution.

Example 3.1.1. Let an automaton be given with transition function δ : $Act^* \times S \to S$, initial states $In \subseteq S$ and output function $\lambda : S \to Out$.

- States $s, s' \in S$ have different capabilities or are 'observable unequal' if they yield different outputs, for some input sequence $w \in Act^*$, or dually, states s and s' are *observably equivalent* if $\lambda(\delta(w,s)) = \lambda(\delta(w,s'))$ for all $w \in Act^*$.

  Abstraction identifies states which cannot be distinguished or *separated* by such observations. An automaton is separated or *fully abstract* if all states can be distinguished by observation.

- Starting in initial states one may pass on to other states by providing inputs $w \in Act^*$. This procedure (re-)constructs or generates a 'reachable' part of the automaton which is minimal in that it contains the initial states and is closed under transition (i.e. a set $Gen(X) = \{\delta(w, s) \in X \mid s \in X \text{ and } w \in Act^*\}$).

Identification of structures which behave equally under all observations (or in all contexts) is a typical form of abstraction, just as closure under given operations is a typical form of generation. We will encounter more complex interpretations of abstraction and generation though the constructions for automata are archetypal (as seen in Section 8).

3.1.2 The following analysis leads to a categorical interpretation of the closely linked notions of observation and *abstraction*. For the informal idea, consider a class of objects, the structure of which is unknown a priori, but which is to be determined by means of observation. For instance, imagine an astronomer observing stars or a biologist using a microscope. Though objects such as stars or viruses have an identity as such, the knowledge about their very nature is restricted by what can be observed. This is vaguely the sort of intuition we would like to appeal to. Mathematical objects are, of course, of a more precise nature.

For the beginning, let us consider sets A with structure (such as automata), and let us say that observations Φ separate elements of such sets, that is, permit us to observe that $a \neq_\Phi a'$, for selected $a, a' \in A$. In fact, we may go a step further and identify 'observation' with its extension, i.e. the inequalities observed.

Examples 3.1.2.

(A) Observation of a set A via an (output) map $\lambda : A \to B$ distinguishes elements $a \neq_\lambda a'$ if $\lambda(a) \neq \lambda(a')$.

(B) Given an automaton with states S, inputs Act, a transition function $\delta : Act^* \times S \to S$ and an output function $\lambda : S \to O$, different outputs may be obtained only after execution of a sequence $w \in Act^*$ of inputs. Hence, observation should be stated in a more complicated way by $s \neq_\lambda s'$ if $\exists w \in Act^*.\lambda(\delta(w, s)) \neq \lambda(\delta(w, s'))$.

(C) Let there be two output functions $\lambda_1 : A \to B, \lambda_2 : A \to C$ to observe a set A, and define $a \neq_\lambda a'$ if $\lambda_1(a) \neq \lambda_1(a')$ or $\lambda_2(a) \neq \lambda_2(a')$.

(D) A modification of (A) – (C) restricts the notion of observation functions by means of certain predicates, e.g.

(a) $f(\lambda(a)) = g(\lambda(a))$ where $f, g : B \to C$, or

(b) $f(\lambda_1(a)) = g(\lambda_2(a))$ where $f : B \to D$ and $g : C \to D$.

These 'running' examples anticipate a 'functional' point of view in that 'observation' is specified in terms of 'output functions'.

In analogy to the *Gedankenspiel* of the astronomer/biologist, we may argue that observation turns 'real' objects into entities of a new kind, i.e. *objects under observation* (A, Φ), which reflect 'reality' as well as the specific means of observation. It is quite obvious that the same object may exhibit different properties if observed in different ways and that different objects may appear to have the same structure under observation. Going to the extreme, given an Act^*-automaton (S, δ) the identity function $id_S : S \to S$ (as observation function) distinguishes all states while under 'constant' observation $\lambda : S \to 1$ all automata have the same behaviour.

We use this perspective, for instance, when minimizing automata: given an Act^*-automaton (S, δ) and an output function $\lambda : S \to O$ we can identify states s and s' which cannot be distinguished under observation, defining a congruence $s \sim s'$ iff $\lambda(\delta(w, s)) = \lambda(\delta(w, s'))$ for all inputs $w \in Act^*$. Typically, new states $[s]$ are defined as congruence classes $[s] = \{s' \mid s \sim s'\}$, and new transition and output functions are defined by $\delta_{\min}(w, [s]) = [\delta(w, s)]$ resp. $\lambda_{\min}([s]) = \lambda(s)$ (one might check that this is well defined. This construction is often referred to as Nerode construction). The newly constructed, minimal automaton behaves as the original did in terms of observing the outputs; for every state in one automaton we have a state in the other automaton such that whatever the input might be the output will be the same, formally $\lambda(\delta(w, s')) = \lambda_{\min}(\delta_{\min}(w, [s]))$ for every $s' \in [s]$. Hence, the minimal automaton is equivalent to the original one if we restrict attention to the input/output behaviour.

3.1.3 In category theory, objects should come along with an appropriate notion of morphism. Given sets A and B with structure, we may say that a homomorphism $h : A \to B$ *preserves observation* if $h(a) \neq_\Psi h(a')$ whenever $a \neq_\Phi a'$ for $a, a' \in A$, where Φ and Ψ are the observation on A and B respectively.

Minimization as above provides a natural example. The map $\varepsilon : S \to S_{\min}, s \mapsto [s]$ is a homomorphism of automata which preserves observation since $\lambda_{\min}(\delta_{\min}(w, [s])) \neq \lambda_{\min}(\delta_{\min}(w, [s']))$ whenever $\lambda(\delta(w, s)) \neq \lambda(\delta(w, s'))$ for $w \in Act^*$.

One might tentatively speak of such functions as *abstractions* since they identify what cannot be distinguished by the chosen notion of observation anyway. Note that, in every specific case, a new category of 'objects under observation' and 'abstractions' is generated which, so to speak, embodies a particular concept of observation. For instance, such a category is defined if we consider Act^*-automata $D = (S, \delta)$ with observation functions $\lambda : S \to O$ from a set of states to a fixed set of outputs and abstractions $h : (D, \lambda) \to (D', \lambda')$ being homomorphisms $h : D \to D'$ such that $\lambda'(h(s)) = \lambda(s)$ for

$s \in S$ (this is equivalent to $h(s) \neq_{\lambda'} h(s')$ if $s \neq_\lambda s'$).

It is this, maybe still vague idea of observation, that we want to express in categorical terms below.

3.1.4 Let us call those abstractions *maximal* (or *full*) where no further abstraction is possible, meaning that observable inequality coincides with the inherent inequality, i.e. $a \neq_\Phi a'$ iff $a \neq a'$ for $a, a' \in A$. In such a case we say that A is *separated*[1] *by* Φ.

Examples 3.1.3. (We continue the running examples started above.)

(A) A is separated by $\lambda : A \to O$ if $\lambda(a) \neq \lambda(a')$ whenever $a \neq a'$. Hence, separation coincides with injectivity of the observation function.

(B) Here separation means that any two states $s, s' \in S$ can be distinguished by some observation, i.e. there exists some $w \in Act^*$ such that $\lambda(\delta(w, s)) \neq \lambda(\delta(w, s'))$. Separated automata are usually called minimal.

(C) Here separated objects are basically subsets of the Cartesian product $B \times C$, the observation maps being projections to the components. One might say that the observation functions are 'jointly injective'.

(D) The sets $\{a \in A \mid f(a) = g(a)\}$ are separated by the inclusion to A, resp. $\{(b, c) \in B \times C \mid f(b) = g(c)\}$ by the projections to the components.

At least the examples of injective functions and minimal automata suggest separation as a rather natural concept. With separation as a yardstick we can precisely specify the problem of minimization: given an object A and a mode of observation Φ, one looks for a maximal abstraction which is separated by definition. The hopefully familiar *minimization problem* will be a first stepping stone on our way to catching the idea of universal problems.

3.1.5 One should note the distinction between the statement of a problem and the actual construction which solves the problem. The distinction is by analogy the same as that between *specification* and *program/implementation*, the *what* and the *how* of computer science. We want to emphasize that, in this section, we are only concerned with the specification aspect, namely to identify a class of problems of rather universal nature. In Section 6 we will ask for strategies to find solutions.

---

[1] The terminology is somewhat unconventional. But it seems to coincide with the use in mathematics, and it seems to be more neutral than to speak of *full abstraction* as used in computer science for particular instances.

Nevertheless, constructions should help to *illustrate* the concepts as in the running examples.

Examples 3.1.4.

(A) Let $a \approx a'$ if $\lambda(a) = \lambda(a')$. Define $A_{\text{sep}} := \{[a] \mid a \in A\}$ where $[a] := \{a' \in A \mid a \approx a'\}$. Then $\lambda_{\text{sep}} : A_{\text{sep}} \to B$ with $\lambda_{\text{sep}}([a]) := \lambda(a)$ is a maximal abstraction of λ.

Alternatively, the inclusion $\lambda(A) \subseteq B$ is a maximal abstraction of $\lambda : A \to B$ where $\lambda(A) := \{\lambda(a) \mid a \in A\}$ is the image of A under λ.

(B) The Nerode construction above defines a separated automaton. The image construction does not work in this case.

(C) The constructions are minor modifications of (A): either one uses the equivalence $a \approx a'$ if $\lambda_1(a) = \lambda_1(a')$ and $\lambda_2(a) = \lambda_2(a')$, or the image $\lambda_1(A) \times \lambda_2(A) \subseteq B \times C$

(D) Either we modify the image construction to give (a) $\{\lambda(a) \mid f(\lambda(a)) = g(\lambda(a))\}$ and (b) $\{(\lambda_1(a), \lambda_2(a)) \mid f(\lambda_1(a)) = g(\lambda_2(a))\}$ as maximal abstractions, or one uses a Nerode-like congruence by factorization as in (A) and (C).

3.2 A more categorical point of view

3.2.1 Our programme is to rephrase the previous section in categorical language. We propose a definition somewhat by brute force, the merits of which are discussed a posteriori.

Not by chance the running examples have a specific structure in that observation is expressed in terms of functions. We notice further that 'preservation of observation' can be expressed diagrammatically (as may be checked by the reader)

(A)

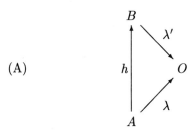

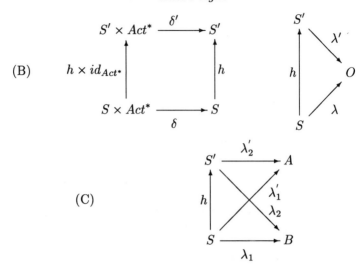

(B)

(C)

The uniform underlying pattern may still not be obvious, however, the examples are instantiations of the following definition, which expresses observational inequality in the language of functions (as opposed to a definition in terms of elements).

Definition 3.2.1.

- An *observation* is a (generalized) morphism $\lambda : F(A) \to B$ where $F : \mathbf{C} \to \mathbf{D}$ is a functor relating the structures of the observed object A and the observer B.

- An *abstraction* is a morphism $h : A \to A'$ such that the diagram

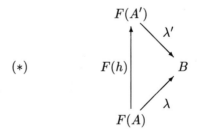

(∗)

commutes.

- A is *separated by* $\lambda : F(A) \to B$ if there is at most one morphism $h : A \to A'$ such that (∗) commutes.

Remark 3.2.2.

- The definition of separation does not seem to exist in the categorical vernacular except for the special case of limits and colimits (see below) where the terms 'mono source' (and 'epi sink' for the dual) are used in [Herrlich and Strecker, 1973; Herrlich and Strecker, 1991]. However, there is the notion of a *separator*, i.e. an object C such that for all morphisms $f \neq g : A \to B$ there exists a morphism $h : B \to C$ such that $h \circ f \neq h \circ g$.

- Observations may be thought of as presentations of separated objects and abstractions as homomorphisms between such presentations.

Examples 3.2.3. (We continue the running examples.)

(A) Observer and the observed live in the same category **Set**, the functor is the identity $1_{\mathbf{Set}}$.

(B) The output functions for automata are defined on the underlying sets which are obtained by the 'forgetful' functor $U : \mathbf{Aut}(M) \to \mathbf{Set}$ which maps automata (S, δ) to the set S.

(C) The observation functions are packed together as morphisms in **Set** × **Set**. The functor is the diagonal functor $\Delta : \mathbf{Set} \to \mathbf{Set} \times \mathbf{Set}$ which maps a set A to the pair (A, A).

(D) The functor is not so obvious here.

(a) Define a category $\mathbf{Set}^{\mathrm{eq}}$ with objects B being pairs of functions $(f_B : B_1 \to B_2, g_B : B_1 \to B_2)$ and morphisms being pairs $(h_1 : B_1 \to C_1, h_2 : B_2 \to C_2)$ such that the diagrams

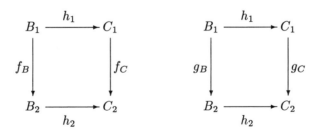

commute. Define a functor $\Delta : \mathbf{Set} \to \mathbf{Set}^{\mathrm{eq}}$ which maps sets A to $(1_A : A \to A, 1_A : A \to A)$ and functions $f : A \to B$ to the pair $(f : A \to B, f : A \to B)$. Then an observation morphism $\lambda : \Delta(A) \to B$ corresponds to the diagrams

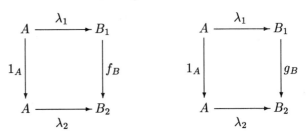

which state that $f_B \circ \lambda_1 = g_b \circ \lambda_1$ which, forgetting about the subscripts, is exactly the requirement stated in Section 3.1.2.

(b) Here one has to use a category $\mathbf{Set}^{\mathrm{pb}}$, the objects of which are pairs of functions $(f : B_1 \to B_3, g : B_2 \to B_3)$ and where morphisms are triples of functions such that the diagram

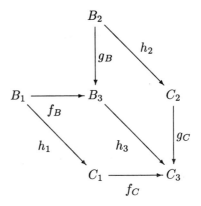

commutes. Again one uses the 'diagonal functor' $\Delta : \mathbf{Set} \to \mathbf{Set}^{\mathrm{pb}}$ which maps a set A to the pair of identity functions to capture the observation.

These examples and generalizations thereof will be discussed at length in Section 5 under the title of equalizers (a) and pullbacks (b).

Specification of observation in terms of functions is certainly a restriction at the first glance. However, we cannot do better if 'functions' are the underlying paradigm, and, as a claim, observation is in most cases naturally expressed in terms of functions.

3.2.2 In order to appreciate the definition of separation we notice that an alternative definition can be given, which very much reflects 'observational inequality' as discussed in the previous section: define $a \neq_\lambda a'$ iff $\lambda \circ F(a) \neq \lambda \circ F(a')$ for $a, a' : X \to A$ where $\lambda : F(A) \to B$. Then A is *separated by* Φ if $a \neq a'$ whenever $a \neq_\lambda a'$ for all $a, a' : X \to A$.

In fact, no difference can be noted with regard to the running examples.

(A) The new definition of separation trivially coincides with the one given in Section 3.1.4 since a function $\lambda : A \to B$ is injective iff, for all functions $a, a' : X \to A$, it holds that $\lambda \circ a \neq \lambda \circ a'$ whenever $a \neq a'$.

(B) The argument is not as simple in the case of automata. Let $a, a' : X \to D$ be homomorphisms of automata, and define $a \neq_\lambda a'$ if $\lambda \circ a \neq \lambda \circ a'$. Then there exists some $x \in S_X$ such that $\lambda(\delta(\varepsilon, a(x))) = \lambda(a(x)) \neq \lambda(a'(x)) = \lambda(\delta(\varepsilon, a'(x)))$. Hence $a(x) \neq a''(x) \in S$ by assumption. Vice versa, let $\lambda(\delta(w, s)) \neq \lambda(\delta(w, s'))$. We construct an automaton D^* with states Act^* and transition function being concatenation $\delta^*(a, w) = wa$, and define homomorphisms $a, a' : D^* \to D$ by $a(v) := \delta(v, s)$ and $a'(v) := \delta(v, s')$. Then $a \neq_\lambda a'$ as $\lambda(a(w)) \neq \lambda(a'(w))$, implies $a \neq a'$ by assumption. Hence there exists a $v \in Act^*$ such that $\delta(v, s) = a(v) \neq a'(v) = \delta(v, s')$ which implies $s \neq s'$.

$)$–(D) are left to the reader.

Now one might argue that we can do very well with the set-theoretic definition in terms of elements and that the functional view complicates the argument unnecessarily. Another example might tell the contrary:

Let the 'observation map' $\lambda : A \to O$ be a partial function, and define $a \neq_\lambda a'$ if $\lambda(a)$ and $\lambda(a')$ exist, and if $\lambda(a) \neq \lambda(a')$. The property '$a \neq_\lambda a'$ iff $a \neq a'$' does here not necessarily imply maximal abstraction; for instance, if λ is the totally undefined map, $a \neq_\lambda a'$ holds for all a, a' in A, but the mapping $\varnothing : A \to \varnothing$ is a maximal abstraction (why?). We happily agree to any objection of the kind 'your definition is just stupid', but we want to score the point that the new definition in terms of functions is 'correct' even in this case (as the reader may check).

Certainly, one example is not conclusive, but experience proves that definitions in terms of elements often obscure the picture: for instance, a naïve interpretation of 'injectivity' in **Pfn** '$h(x) \neq h(y)$ if $x \neq y$ (where $h(x)$ and $h(y)$ are defined)' does not satisfy the expected cancellation property, namely that $h \circ f = h \circ g$ implies $f = g$ (which is, of course, only a reformulation of the previous paragraph).

The categorical or functional definition of 'injectivity', *monomorphism*,[2] encodes the cancellation property a priori. A characterization in terms of elements may then be given a posteriori if wanted and possible (the reader may find out what the proper 'injective' partial functions are).

An aside: one may expect as another property of injective functions $h : X \to Y$ that there exists an inverse surjective function $h^{-1} : Y \to X$

---

[2] A morphism $m : B \to C$ is called *monomorphism* if, for all morphisms $f, g : A \to B$, $f = g$ whenever $h \circ f = h \circ g$. Dually, a morphism $e : A \to B$ is called *epimorphism* if, for all morphisms $f, g : B \to C$, $f = g$ whenever $f \circ e = e \circ g$.

meaning that $h^{-1} \circ h = 1_X$ (such a morphism h is called a *coretract*). Unfortunately, the cancellation property '$f = g$ if $h \circ f = h \circ g$' does not imply the coretraction property as demonstrated by an example:

A homomorphism $h : D \to D'$ of automata is a monomorphism if $h(s) \neq h(s')$ implies $s \neq s'$ but does not need to be a coretraction: consider the automata presented by

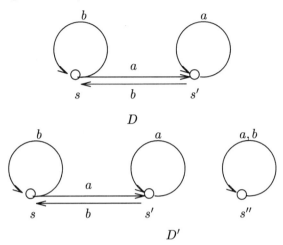

$$D$$

$$D'$$

where the one step transition function $\delta(a, s) = s'$ is specified by arrows $s \xrightarrow{a} s'$. The embedding of D into D' is injective, but there exists no homomorphism $h : D' \to D$.

3.3 Universal morphisms

3.3.1 Let us for a moment ramble about on another line of thought. Moving between languages one would appreciate transferring knowledge from one language to another. If language stands for categories, functors are a means to relate categories. But the mere existence of a functor is not very illuminating with regard to transfer of structure. This is why functors are characterized in terms of additional properties such as preservation of products.

Structure is not only transported along a functor but also in the opposite direction. Consider the forgetful functor $U : \mathbf{Mon} \to \mathbf{Set}$ from monoids to the underlying sets. Products of monoids are constructed as Cartesian products of the underlying sets with the monoid structure being defined componentwise. Hence we know that $first : \mathcal{M} \to \mathcal{M}_1, second : \mathcal{M} \to \mathcal{M}_2$ is a product of monoids if $U(\mathcal{M}) = U(\mathcal{M}_1) \times U(\mathcal{M}_2)$, and if $U(first)$ and $U(second)$ are the projection of the Cartesian product. Similarly, factorization by a congruence relation on the underlying set defines the quotient of a monoid.

This fragmentary discussion may put forward the idea that functors, as a means to compare categories, should be classified according to what infrastructure is transported to and from. One might for instance ask, whether, given an object B in a category $\mathbf{D}$ which satisfies some specified property, there exists an object A in a category $\mathbf{C}$ which satisfies the same property if seen through a functor $F : \mathbf{C} \to \mathbf{D}$. Obviously, categories are most closely related by a functor which is an isomorphism (neglecting identity functors). Then for every object in one category there exists an object in the other category with exactly the same properties. *Equivalence*[3] of categories is a slightly weaker relation in that more isomorphic copies of objects may live in the other category. We will learn below that adjunctions define a considerably weaker relation between categories.

3.3.2 Morphisms of the form $\lambda : F(A) \to B$ characterize B's capability to observe or to separate structure in $\mathbf{C}$ as viewed through the functor $F : \mathbf{C} \to \mathbf{D}$. This capability is transported along the functor if there exists an object B_* in $\mathbf{C}$ which observes the same inequalities in $\mathbf{C}$ as does B. B_* should be separated by some morphism $\varepsilon_B : F(B_*) \to B$ to ensure that every inequality observed in B_* can be observed in B as well. Moreover, every observation (morphism) $\lambda : F(A) \to B$ should induce an abstraction (morphism) $\langle \lambda \rangle : A \to B_*$ which, according to our argumentation, preserves the inequalities observable by $\lambda : F(A) \to B$.

Definition 3.3.1. Let $F : \mathbf{C} \to \mathbf{D}$ be a functor, and let B be an object of $\mathbf{D}$. A pair $(B_*, \varepsilon_B : F(B_*) \to B)$ is called *universal morphism from F to B* if for all morphisms $(A, f : F(A) \to B)$ from F to A there exists a unique morphism $\langle f \rangle : A \to B_*$ such that the diagram

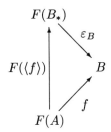

commutes. (The definition has a dual which will be looked at more closely in the next section.)

The terminology is that of [MacLane, 1971]. In line with the development we may say that B_* is *fully separated (by $\varepsilon_B : G(B_*) \to B$)*.

---

[3] Categories $\mathbf{C}$ and $\mathbf{D}$ are equivalent if there exist functors $F : \mathbf{C} \to \mathbf{D}$ and $G : \mathbf{D} \to \mathbf{C}$ such that $G \circ F \cong 1_{\mathbf{C}}$ and $F \circ G \cong 1_{\mathbf{D}}$ naturally.

Examples 3.3.2. (The running examples for the last time.)

(A) is boring since an identity functor does not change the character of objects. $1_A : 1_{\mathbf{C}}(A) \to A$ is universal.

(B) Existence of a fully separated M-automaton Out_* over the set Out might be familiar: Out_* is given by $S_* := (Act^* \Rightarrow Out), \delta_* : S_* \times Act \to S_*$ with $\delta_*(f, x)(w) = f(wx)$ and with $\lambda_*(f) := f(\varepsilon)$ as universal morphism.

(C) The Cartesian product $O_1 \times O_2$ is fully separated by the projections $first : O_1 \times O_2 \to O_1$ and $second : O_1 \times O_2 \to O_2$.

We have already encountered other witnesses for full separation which are more abstract:

- Let $\Delta : \mathbf{C} \to \mathbf{C} \times \mathbf{C}$ be the diagonal functor with $\Delta(A) = (A, A)$. We leave it as an exercise for the reader to check that the product $A\Pi B$ is separated by the projections $(p_{A,B}, q_{A,B}) : (A\Pi B, A\Pi B) \to (A, B)$.

- A terminal object 1 is fully separated with regard to the functor $\varnothing : \mathbf{0} \to \mathbf{C}$ where $\mathbf{0}$ is the empty category.

- Let $\mathbf{C}$ be a category with finite products. Define a functor $\_ \Pi A : \mathbf{C} \to \mathbf{C}$ by $B \mapsto B\Pi A$ and $f : B \to C \mapsto f\Pi 1_A : B\Pi A \to C\Pi A$. Then $A \Rightarrow B$ is fully separated by $apply : (A \Rightarrow B) \times A \to B$.

We once more emphasize that the definition of 'universal morphism' specifies a property, or a 'problem'. Hence the example of products and function spaces provides an alternative specification of these concepts. Stating a problem does not imply existence of a solution, which, however, might exist for specific cases, as for instance in the case of the examples (A)–(C). A universal morphism, for example, does not exist from the forgetful functor $U : \mathbf{Mon} \to \mathbf{Set}$ to the empty set since existence of a function $U(\varnothing_*) \to \varnothing$ implies $U(\varnothing_*) \cong \varnothing$. But every monoid has at least one element, the unit. We stress the point so much because one sometimes finds that specific constructions are considered as specifications, which can confuse the issue.

3.3.3 Inequality is preserved by isomorphisms. Hence separation should only be determined up to isomorphism. The following is a generalization of the proof that products are defined only up to isomorphisms.

Proposition 3.3.3 (Exercise). *Let* $(U_i, u_i : F(U_i) \to A, i = 1, 2$, *be universal morphisms from F to A. Then*

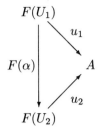

commutes where $\alpha : U_1 \to U_2$ is an isomorphism in **C**. *Moreover, given such an isomorphism $(U_1, u_2 \circ F(\alpha) : F(U_1) \to A)$ is universal from F to A if $(u_2, u_2 : F(U_2) \to A)$ is so.*

Definition up to isomorphism only is inherent in a functional setting, where we cannot talk about the internal structure of an object, but only about its interaction with the environment which, per definition, should be invariant under isomorphisms.

3.4 Adjunction

3.4.1 Full separation means to transport (observable) inequality along a functor $F : \mathbf{C} \to \mathbf{D}$ though in the opposite direction. This is a characteristic property of functors in the light of the discussion above — and one would expect, a fundamental one, since categorical reasoning is basically equational — which merits a name. If for every object in **D** there exists a fully separated object in **C** we may pick a specific such object $(F_*(A), \varepsilon_A)$ and define a functor $F_* : \mathbf{D} \to \mathbf{C}$ by $A \mapsto F_*(A)$ and $f : A \to B \mapsto F_*(f) = \langle f \circ \varepsilon_A \rangle$. Note that this defines a natural transformation $\varepsilon : F \circ F_* \to 1_\mathbf{D} : \mathbf{D} \to \mathbf{D}$ since the diagram

$$
\begin{array}{ccc}
F(F_*(f)) & \xrightarrow{\ F(F_*(f))\ } & F(F_*(f)) \\
{\scriptstyle \varepsilon_A}\big\downarrow & & \big\downarrow{\scriptstyle \varepsilon_B} \\
A & \xrightarrow{\ \ f\ \ } & B
\end{array}
$$

commutes.

Definition 3.4.1. A functor $F : \mathbf{C} \to \mathbf{D}$ has a *right adjoint* $F_* : \mathbf{D} \to \mathbf{C}$ if for all objects B of **D** there exists a natural transformation ε_B :

$F(F_*(B)) \to B$ in **D** such that for all objects A of **C** and all morphisms $f : F(A) \to B$ of **D** there exists a unique morphism $\langle f \rangle : A \to F_*(B)$ such that the diagram

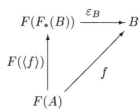

commutes.

Note that all the functors in the 'running examples' have right adjoints. More examples will be discussed in Section 3.6.

3.4.2 Right adjoints have useful properties, the most prominent one, according to MacLane, being that right adjoints preserve definitions stated in terms of separation.

Proposition 3.4.2. *Let* $(B_*, \varepsilon_B : F(B_*) \to B$, *be a universal morphism from* $F : \mathbf{C} \to \mathbf{D}$ *to* B, *and let* B *be separated by* $\lambda : H(B) \to C$ *where* $H : \mathbf{D} \to \mathbf{E}$. *Then* $F(B_*)$ *is separated by* $\lambda \circ H(\varepsilon_B) : H(F(B_*)) \to C$. *Moreover* $(B_*, \lambda \circ H(\varepsilon_B))$ *is universal from* $H \circ F$ *to* C *if* (B, λ) *is universal from* H *to* C.

Proof. Let $f, g : X \to B_*$ such that $\lambda \circ H(\varepsilon_B) \circ H(F(f)) \neq \lambda \circ H(\varepsilon_B) \circ H(F(g))$. Then $\varepsilon_B \circ F(f) \neq \varepsilon_B \circ F(g)$ since $\lambda \circ H(\varepsilon_B) \circ H(F(f)) = \lambda \circ H(\varepsilon_B \circ (F(f)))$ and similarly for $f \neq g$.

Universality holds since observations $f : H(F(X)) \to C$ induce abstractions $\langle f \rangle : F(X) \to B$ and $\langle\langle f \rangle\rangle : X \to B_*$). (Note that abstractions satisfy certain diagrams per definitionem.) ∎

The argument is trivial though applications are not so. One might for instance try to prove the following corollary by a straightforward computation.

Corollary 3.4.3.

- Let $F_* : \mathbf{D} \to \mathbf{C}$ be a right adjoint to $F : \mathbf{C} \to \mathbf{D}$, and let $A\Pi B$ be a product of A and B in **D** with projections $p_{A,B} : A\Pi B \to A, q_{A,B} : A\Pi B \to B$. Then $F_*(A\Pi B)$ is a product of $F_*(A)$ and $F_*(B)$ in **C** with projections $F_*(p_{A,B}) : F_*(A\Pi B) \to F_*(A), F_*(q_{A,B}) : F_*(A\Pi B) \to F_*(B)$.

- $A \Rightarrow (B\Pi C)$ is naturally isomorphic to $(A \Rightarrow B)\Pi(A \Rightarrow C)$ in a Cartesian closed category (for instance in **Set**).

- $F_*(f)$ *is a monomorphism if* $f : A \to B$ *is so.*

Proof. • Products are universal as indicated above. $A \Pi B$ with projections is universal from $\Delta_{\mathbf{D}}$ to (A, B). Then $(p_{A,B}, q_{A,B}) \circ \Delta_{\mathbf{D}}(\varepsilon_{A\Pi B})$ is universal from $\Delta_{\mathbf{D}} \circ F$ to (A, B). Now we use that $(F \times F) \circ \Delta_{\mathbf{C}} = \Delta_{\mathbf{D}} \circ F$ to obtain that $(p_{A,B}, q_{A,B}) \circ \Delta_{\mathbf{D}}(\varepsilon_{A\Pi B}) = (F_*(p_{A,B}) \circ F_*(q_{A,B}))$ is universal from $(F \times F) \circ \Delta_{\mathbf{C}}$ to (A, B) where the equality follows because of naturality of ε.

- Essentially the same.

- Immediate consequence.

∎

3.4.3 To summarize the section: a variety of concepts such as products, 'injective functions' and function spaces turn out to be defined in terms of separation by observation (which might substantiate the fundamental role of separation, we claim, in mathematics or, more generally, formal reasoning). Universal morphisms transfer an object's ability to separate from one category to another with the consequence that properties which are expressed in terms of separation are transferred as well. Existence of a right adjoint $F_* : \mathbf{D} \to \mathbf{C}$ states that this transfer can be achieved for all objects in $\mathbf{C}$.

3.5 On generation

3.5.1 We shift perspective in that functions $f : A \to B$ are considered as a means to express *reference*; elements a in A refer to elements $f(a)$ in B, or: the elements $f(a)$ are *generated* by f. The codomain may bear additional structure, for instance being the set of states of an automaton or the carrier set of a monoid. In such a case the inherent operations provide additional generative capability in that, given some input $w \in Act^*$, the element $\delta(w, f(a))$ can be referred to resp. is generated as well, or in case of monoids, the element $f(a) * f(a')$ is generated. The idea is found in familiar definitions.

Examples 3.5.1.

- Let $\mathcal{D} = (S, \delta)$ be an automaton, and let the inclusion mapping $i :$ $X \subseteq S$ define a set of initial states. States $\delta(w, x)$ such that $x \in X$ and $w \in Act$ are usually called *reachable*. The reachable states define a (sub-)automaton $\langle X \rangle_{\mathcal{D}}$ with states $S_{\mathrm{gen}} := \{\delta(w, x) \in S \mid x \in$

$X, w \in Act^*$} with the transition function restricted to the reachable states.

- Similarly, a subset $i : X \subseteq M$ of a monoid $\mathcal{M} = (M, *, e)$ *generates* a submonoid $\langle X \rangle_{\mathcal{M}}$ with elements being the least subset containing X which is closed under the monoid operations, i.e. $x \in \langle X \rangle_{\mathcal{M}}$ if $x \in X$, $e \in \langle X \rangle_{\mathcal{M}}$, and $x * y \in \langle X \rangle_{\mathcal{M}}$ whenever $x, y \in \langle X \rangle_{\mathcal{M}}$.

The terminology captures in all cases the same phenomenon: some part $f(A)$ of a structure B is referred to by a function $f : A \rightarrow G(B)$. Then $f(A)$ has to be closed under the inherent operations to define the *generated structure*.

3.5.2 A functional characterization of generation is less overt than in the case of separation. The construction above involves the internal structure of objects which is not visible a priori in a purely functional setting.

Generation comprises two components: the definition of a set of generators in terms of a function $f : A \rightarrow G(B)$ and closure under the internal structure of B. Whatever this structure is, in a reasonably defined category, we expect morphisms to preserve the structure, or rather under functional bias: there is no other internal structure as preserved by morphisms. Hence the behaviour of a morphism on the generated part should be totally determined by its behaviour on the generators. In other words, if an object is generated, outgoing morphisms must be equal if they agree on the generators.

Definition 3.5.2. Let **C**, **D** be categories, and $G : \mathbf{D} \rightarrow \mathbf{C}$ be a functor. A and B are objects in **C** and **D** respectively.

- A morphism $f : A \rightarrow G(B)$ is called a *generator*.

- B is *generated* by $f : A \rightarrow G(B)$ if $G(g) \circ f = G(h) \circ f$ implies $g = h$ for all $g, h : B \rightarrow B'$ in **D**, or equivalently, for every $f' : A \rightarrow F(B')$ there exists at most one morphism $h : B \rightarrow B'$ such that

$(*)$

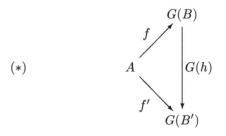

commutes.

A morphism $h : B \to B'$ such that the diagram $(*)$ commutes may be thought of as a homomorphism of generators.

Here are some examples:

- For $1_{\mathbf{Set}} : \mathbf{Set} \to \mathbf{Set}$ generation coincides with surjectivity.

- An automaton is generated if all states are reachable in that $s = \delta(w, \iota(x))$ for some $w \in Act^*, x \in In$ where $\iota : In \to S$.

- The union $f(A) \cup g(A')$ is generated by the functions

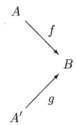

or in the categorical way of speaking: $\Delta : \mathbf{Set} \to \mathbf{Set} \times \mathbf{Set}$ is the diagonal functor and $f(A) \cup g(A')$ is the $\mathbf{Set} \times \mathbf{Set}$ structure generated by $(f, g) : (A, A') \to \Delta(B)$.

- $\langle X \rangle_{\mathcal{M}}$ is a monoid generated by $\iota : X \subseteq U(\langle X \rangle_{\mathcal{M}})$ where $U : \mathbf{Mon} \to \mathbf{Set}$.

- Let $U : \mathbf{AbMon} \to \mathbf{Mon}$ map commutative monoids to monoids forgetting about commutativity. $M \sqsubseteq U(M_{ab})$ ('is a submonoid') generates the least commutative submonoid of M_{ab} which includes M.

3.5.3 The previous paragraph implicitly claims that morphisms $f : A \to G(B)$ characterize the 'generative capability' of A with regard to $\mathbf{D}$ as viewed through G. This capability is transported along the functor if there exists an object A^* in $\mathbf{D}$ which generates the same structures in $\mathbf{D}$ as does A meaning that morphisms $f : A \to G(B)$ are in bijective correspondence to morphisms $g : A^* \to B$ to ensure that A^* can generate whatever A can.

Definition 3.5.3. $(A^*, \eta : A \to G(A^*))$ is called a *universal morphism* from A to G if for all morphisms $(B, g : A \to F(B))$ from A to F there exists a unique morphism $[f] : A^* \to B$ such that the diagram

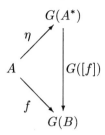

commutes.

The pair (A^*, η) is often called *free*(ly generated).
We reconsider the examples:

- The identity $1_A : A \to A$ is universal with regard to $1_\mathbf{C} : \mathbf{C} \to \mathbf{C}$.

- An automaton with states $X \times Act^*$ and transition function $\delta^*(w, \langle x, v \rangle) = \langle x, wv \rangle$ is freely generated by a set X. The universal morphism is defined by $\eta(x) = \langle x, \varepsilon \rangle$.

- The disjoint sum $A + A'$ is freely generated with inclusions as universal morphisms.

Two more abstract examples for free generation may be checked by the reader:

- The coproduct $A \amalg B$ is freely generated w.r.t. $\Delta : \mathbf{C} \to \mathbf{C} \times \mathbf{C}$ with inclusion morphisms $inl_{A,B} : A \to A \amalg B, inr_{A,B} : B \to A \amalg B$.

- An initial object is freely generated with regard to the functor $\varnothing : \mathbf{0} \to \mathbf{C}$.

- $B \amalg A$ is freely generated with regard to the functor $A \Rightarrow \_ : \mathbf{C} \to \mathbf{C}$ given by $B \mapsto A \Rightarrow B$ and $f \mapsto \lambda(f \circ apply_{A,B})$. The universal morphism is $\lambda(1_{B \amalg A})$.

- The most familiar example: the word monoid $A^* = (A^*, conc, \varepsilon)$ with concatenation is freely generated by the set A. The function $\eta : A \to A^*$ maps an element a to the word a. Every function $f : A \to M$ extends to a monoid homomorphism $[f] : A^* \to \mathcal{M}$ where $\mathcal{M} = (M, *, e)$.

Note that our notation is as lax as usually to be found, namely without explicit reference to the forgetful functor $U : \mathbf{Mon} \to \mathbf{Set}$.

3.5.4 Existence of universal morphisms from A to G for all objects A of $\mathbf{C}$ again induces a functor $G^* : \mathbf{C} \to \mathbf{D}$ the properties of which are coded in the following definition.

Definition 3.5.4. A functor $G : \mathbf{D} \to \mathbf{C}$ has a *left adjoint* to $G^* : \mathbf{C} \to \mathbf{D}$ if for all objects A of $\mathbf{C}$ there exists a natural transformation (the *unit*) $\eta_A : A \to G(G^*(A))$ in $\mathbf{C}$, such that for all objects X and all morphisms $f : A \to G(X)$ of $\mathbf{D}$ there exists a unique morphism $[f] : G^*(A) \to X$ such that the diagram

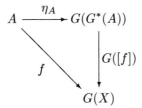

commutes.

3.5.5 The reader may be well aware of the duality of separation and generation. Still I believe it to be somewhat surprising that the analysis of two a priori different concepts leads to the same definition.

Proposition 3.5.5. $F : \mathbf{C} \to \mathbf{D}$ *is left adjoint to* $G : \mathbf{D} \to \mathbf{C}$ *iff* G *is right adjoint to* F.

Proof. $\varepsilon_A = [1_{G(A)}] : F(A) \to A$ is the universal morphism from F to A if F is left adjoint. The reader may check the other direction. ∎

Definition 3.5.6. We speak of such a pair of functors as an *adjunction*, and use the notation $F \dashv G$. The universal natural transformation $\eta : 1_{\mathbf{C}} \to G \circ F$ is called *unit* of the adjunction, and the universal natural transformation $\varepsilon : F \circ G \to 1_{\mathbf{D}}$ is called *counit*.

This duality is not so apparent in most examples. Typically, either an analysis in terms of separation or in terms of generation appears as more natural: a disjoint union is clearly generated by its components while a Cartesian product is separated. To observe that $A \amalg A$ is separated by $[1_A, 1_A] : A \amalg A \to A$ and that $A \amalg A$ is generated by $\langle 1_A, 1_A \rangle : A \to A \amalg A$ is certainly less convincing.

3.5.6 The above proposition gives rise to an equational characterization of adjunctions in terms of the universal generator and observation morphisms.

Proposition 3.5.7. *The diagrams*

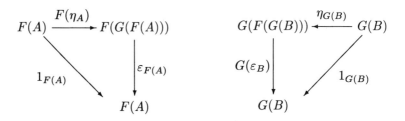

commute for the unit $\eta_A : A \to G(F(A))$ *and the counit* $\varepsilon_B : F(G(B)) \to B$
of an adjunction $F \dashv G : \mathbf{C} \to \mathbf{D}$. *Equivalently, the triangle equalities*

$$(F * \eta) \circ (\varepsilon * F) = F \qquad (\eta * G) \circ (G * \varepsilon) = G$$

hold (for the notation see Section 1.5.3).

Vice versa, every such equality of natural transformations defines an adjunction.

Proposition 3.5.8. *Let* $\eta : 1_{\mathbf{C}} \to G * F$ *and* $\varepsilon : F * G \to 1_{\mathbf{D}}$ *be natural transformations such that the triangle equalities hold. Then* $F \dashv G : \mathbf{C} \to \mathbf{D}$ *with unit* η *and counit* ε.

Proof. Let $f : A \to G(B)$ be a morphism. The extension $[f] : F(A) \to B$ is defined by $[f] = \varepsilon_B \circ F(f)$. The diagrams

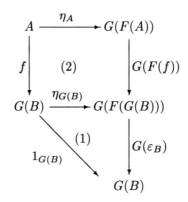

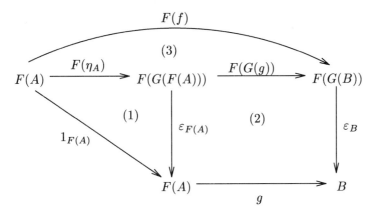

commute because of the triangle equations (1), the naturality of (co-)units (2) and the assumption $G(g) \circ \eta_A = f$ for $g : F(A) \to B$. ∎

Hence, adjunctions or universality reduces to an equational definition.

3.5.7 For the sake of completeness we state the dual of Proposition 3.3.3.

Proposition 3.5.9. *Let $(A^*, \eta_A : A \to G(A^*))$ be a universal morphism from A to $G : \mathbf{D} \to \mathbf{C}$, and let A be separated by $\iota : C \to H(A)$ where $H : \mathbf{C} \to \mathbf{E}$. Then A^* is separated by $H(\eta_A) \circ \iota : C \to H(G(A^*))$. $(A^*, (H(\eta_A) \circ \iota))$ is universal from C to $H \circ G$ if (A, ι) is universal from C to H.*

Thus an adjunction transfers generative capability in one direction and separative capability in the other direction.

3.6 More examples for separation and generation

The examples below are relevant for computer science, but there are many more. Every textbook on categories (and every textbook on mathematics implicitly) provides a variety of other examples.

3.6.1 (Freely) Generated algebras
Universal algebra provides the probably most familiar examples of (free) generation. For the sake of self-containedness we start with

A short primer on universal algebra
(For more details have a look at the chapter on 'Universal Algebra'.)

A *signature* $\Sigma = (S, \Sigma)$ is specified by a set (class) B of (*base*) *types* or *sorts*, and a family of sets (classes) $\Sigma_{w,b}$ (of *operators*) indexed by $w \in B^*, b \in B$. We use the notation $f : w \to b$ for $f \in \Sigma_{w,b}$.

A Σ-*algebra* $\mathcal{A}$ is an assignment of a designated one-element set $\underline{1} = \varepsilon^{\mathcal{A}}$ to ε, a set $b^{\mathcal{A}}$ to each base type $b \in B$, and a function $f^{\mathcal{A}}$ to each operator f in Σ, where $f^{\mathcal{A}} : \underline{1} \to b^{\mathcal{A}}$ for $f : \varepsilon \to b$, and $f^{\mathcal{A}} : b_1^{\mathcal{A}} \times \ldots \times b_n^{\mathcal{A}} \to b_{n+1}^{\mathcal{A}}$ for $f : b_1 \times \ldots \times b_n \to b_{n+1}$. A Σ-*homomorphism* $h : \mathcal{A} \to \mathcal{B}$ is a B-indexed family of function $h_b : b^{\mathcal{A}} \to b^{\mathcal{B}}$ such that the diagram

$$
\begin{array}{ccc}
b_1^{\mathcal{A}} \times \ldots \times b_n^{\mathcal{A}} & \xrightarrow{\;f^{\mathcal{A}}\;} & b^{\mathcal{A}} \\[2mm]
\Big\downarrow{\scriptstyle h_{b_1} \times \ldots \times h_{b_n}} & & \Big\downarrow{\scriptstyle h_b} \\[2mm]
b_1^{\mathcal{B}} \times \ldots \times b_n^{\mathcal{B}} & \xrightarrow[\;f^{\mathcal{B}}\;]{} & b^{\mathcal{B}}
\end{array}
$$

commutes for all $f : b_1 \times \ldots \times b_n \to b$ of Σ. Σ-algebras and Σ-homomorphisms form a category we refer to by $\mathbf{Alg}(\Sigma)$.

Example 3.6.1.

  Σ_{int} for 'integers'.

Base types	$\underline{int}, \underline{bool}$
Operators	$0 :\to \underline{int}$
	$suc, pred : \underline{int} \to \underline{int}$
	$add : \underline{int}\ \underline{int} \to \underline{int}$
	$true, false :\to \underline{bool}$
	$iszero : \underline{int} \to \underline{bool}$
Some Σ_{int}-terms	
	$iszero(add(x, 0))$
	$add(pred(x), suc(x)) : \underline{int}$
	$iszero(0) : \underline{bool}$

Σ_{int}-algebras are given by

  $\mathbb{Z}$ $\underline{int}^{\mathbb{Z}} = \mathbb{Z}$ $\underline{bool}^{\mathbb{Z}} = \mathbb{B}$ with the obvious functions.
  $\mathbb{N}$ $\underline{int}^{\mathbb{N}} = \mathbb{N}$ $\underline{bool}^{\mathbb{N}} = \mathbb{B}$ again with the obvious functions, except for $pred^{\mathbb{N}}$ where $pred^{\mathbb{N}}(0) = 0$

Σ_{int}-homomorphism

  $abstr : \mathbb{N} \to \mathbb{Z}$ $n \mapsto n$
  There exists no homomorphism $h : \mathbb{Z} \to \mathbb{N}$ (why?).

Let X be a sorted subset of $\mathcal{A}$, i.e. $X_b \subseteq \mathcal{A}_b$ for all $b \in B$. The subalgebra $\langle X \rangle_{\mathcal{A}}$ of $\mathcal{A}$ is *generated by* X if carriers are inductively defined by

(i) $x \in \langle X \rangle_{A,b}$ if $x \in X_b$

(ii) $f^A(a_1, \ldots, a_n) \in \langle X \rangle_{A,b}$ if $f : b_1 \times \ldots \times b_n \to b \in \Sigma$ and $a_i \in \langle X \rangle_{A,b_i}$ for $i = 1, \ldots, n$

Operations of A are restricted to $\langle X \rangle_A$ in that $f^{\langle X \rangle} A(a_1, \ldots, a_n) = f^A(a_1, \ldots, a_n)$.

Proposition 3.6.2. $\langle X \rangle_A$ *is generated with the generator being the inclusion map* $\iota : X \subseteq \langle X \rangle_A$.

Σ-*terms* $T_\Sigma(X)_b$ of sort b are inductively defined by:

- $x \in T_\Sigma(X)_b$ if $x \in X_b$,

- $f \in T_\Sigma(X)_b$ if $f : \varepsilon \to b$,

- $f(t_1, \ldots, t_n) \in T_\Sigma(X)_b$ if $f : b_1 \times \ldots \times b_n \to b \in \Sigma$ and $t_i \in T_\Sigma(X)_{b_i}$ for $i = 1, \ldots, n$

where $X = (X_b \mid b \in B)$ is a typed set of *generators* or *variables*. Terms carry a natural algebraic structure: the *term algebra* $T_\Sigma(X)$ has carriers $T_\Sigma(X)_b$ for $b \in B$ and operations $f^{\Sigma T_\Sigma(X)}(t_1, \ldots, t_n) = f(t_1, \ldots, t_n)$.

Term construction has the fundamental property that every *variable assignment* $\Theta = (\Theta_b : X_b \to b^A \mid b \in B)$ with values in a Σ-algebra A extends to all terms by

- $[\Theta](x) = \Theta_b(x)$ for $x \in X_b$

- $[\Theta](f(t_1, \ldots, t_n)) = f^A([\Theta]t_1, \ldots, [\Theta]t_n)$

defining a family $[\Theta]_b : T_\Sigma(X)_b \to A_b$ of functions. This extension is the unique one which preserves the algebraic structure:

Free Algebra Theorem 1 (standard format). *For all variable assignments* $\Theta : X \to A$ *there exists a unique homomorphism* $[\Theta] : T_\Sigma(X) \to A$ *such that* $[\Theta]_b(x) = [\Theta](x)$ *for all* $x \in X_b$.

Proof. Let $h : T_\Sigma(X) \to A$ be another homomorphism such that $h(x) = \Theta(x)$. We use induction on terms:

- $h_b(x) = \Theta_b(x) = [\Theta](x)$

- $h_{b_{n+1}}(f(t_1, \ldots, t_n)) = f^A(h_{b_1}(t_1), \ldots, h_{b_n}(t_n)) = f^A([\Theta]t_1, \ldots, [\Theta]t_n)$ (by inductive assumption)
  $= [\Theta]f(t_1, \ldots, t_n)$. ∎

$T_\Sigma(X)$ defines the object part of a left adjoint to the forgetful functor $U : \mathbf{Alg}(\Sigma) \to \mathbf{Set}^B$ which maps a Σ-algebra A to the B-sorted family $(A_b \mid b \in B)$ of carriers.

Free Algebra Theorem 1 (categorical format). $\mathcal{T}_\Sigma(X)$ *with unit* $\eta_X : X \to U(\mathcal{T}_\Sigma(X))$ *which maps the element x to the term x is a universal morphism from X to U.*

A *congruence relation* on an algebra $\mathcal{A}$ consists of a B-sorted family $\sim_b \subseteq A_b \times A_b$ of equivalence relations such that

$$f^{\mathcal{A}}(a_1, \ldots, a_n) \sim_b f^{\mathcal{A}}(a'_1, \ldots, a'_n) \quad \text{if} \quad a_i \sim_{b_i} a'_i \quad \text{for all} \quad i = 1, \ldots, n$$

where $f : b_1 \times \ldots \times b_n \to b \in \Sigma$.

The quotient $\mathcal{A}_{/\sim}$ of $\mathcal{A}$ under the congruence relation $\sim$ has elements $[a] \in \mathcal{A}_{/\sim_b}$ which are congruence classes $[a] = \{a' \in \mathcal{A}_b \mid a \sim a'\}$ and operations $f^{\mathcal{A}/\sim}([a_1], \ldots, [a_n]) = [f^{\mathcal{A}}(a_1, \ldots, a_n)]$.

Proposition 3.6.3. *Let $h : \mathcal{A} \to \mathcal{A}'$ be a homomorphism such that $h(a) = h(a')$ if $a \sim a'$. Then there exists a unique homomorphism $[h] : \mathcal{A}_{/\sim} \to \mathcal{A}'$ such that $[h]([a]) = h(a)$. In other words, $\mathcal{A}_{/\sim}$ is generated by $[\_] : \mathcal{A} \to \mathcal{A}_{/\sim}, a \mapsto [a]$.*

Proof. $[h]$ is a homomorphism as

$$
\begin{aligned}
[h](f^{\mathcal{A}/\sim})([a_1], \ldots, [a_n]) &= [h]([f^{\mathcal{A}}(a_1, \ldots, a_n)]) \\
&= h(f^{\mathcal{A}}(a_1, \ldots, a_n)) \\
&= f^{\mathcal{A}'}(h(a_1), \ldots, h(a_n)) \\
&= f^{\mathcal{A}'}([h]([a]_1), \ldots, [h]([a]_n)).
\end{aligned}
$$

Use induction for uniqueness.   ∎

A Σ-*equation* $t_1 =_{X,b} t_2$ consists of two terms $t_1, t_2 \in \mathcal{T}_\Sigma(X)_b$ where X is a finite B-sorted set of variables, and where $b \in B$. A *specification* $Sp = (\Sigma, E)$ consists of a signature Σ and a set E of equations.

An algebra $\mathcal{A}$ *satisfies* an equation $t_1 =_{X,b} t_2$ if $[\Theta](t_1) = [\Theta](t_2)$ for all variable assignments $\Theta : X \to \mathcal{A}$. A Σ-algebra $\mathcal{A}$ is called a *Sp-algebra*, if it satisfies all the equations in E. The category **Alg**(Sp) consists of Sp-algebras as objects, and Σ-homomorphisms as morphisms.

Let a congruence relation be defined by : $t_1 \sim t_2$ iff $\Theta(t_1) = \Theta(t_2)$ for all Sp-algebras $\mathcal{A}$ and for all variable assignments $\Theta : X \to \mathcal{A}$.

Free Algebra Theorem 2. $\mathcal{T}_\Sigma(X)_{/\sim}$ *with embedding* $\eta : X \to \mathcal{T}_\Sigma(X)_{/\sim}$, $x \mapsto [x]$ *is a universal morphism from $\mathcal{A}$ to* $U : \textbf{Alg}(Sp) \to \textbf{Set}^B$.

Proof. Every variable assignment $\Theta : X \to \mathcal{A}$ extends to a unique homomorphism $[\Theta] : \mathcal{T}_\Sigma(X) \to \mathcal{A}$ such that $[\Theta](x) = \Theta(x)$. $[\Theta](t_1) = [\Theta](t_2)$

implies $t_1 \sim t_2$ by definition of the congruence. Hence we can apply the 'congruence proposition'. ∎

Remark 3.6.4. $\mathcal{T}_\Sigma(\varnothing)_{/\sim}$ is an *initial Sp-algebra*.

A *signature morphism* $H : \Sigma \to \Sigma'$ maps sorts to sorts, and operators to operators. It consists of a mapping $H : B \to B'$ and a family of mappings $H_{w,b} : \Sigma_{w,b} \to \Sigma'_{H(w),H(b)}$ indexed by $w \in B^*, b \in B$. Signatures and signature morphisms form a category **Sign** with composition being defined for the components.

A *specification morphism* $H : Sp \to Sp'$ is a signature morphism $H : \Sigma \to \Sigma'$ such that $H(t_1) =_{H(X),H(b)} H(t_2)$ for all equations $t_1 =_{X,b} t_2$ in E where $H(x) = x$ and $H(f(t_1, \ldots, t_n)) = H(f)(H(t_1), \ldots, H(t_n))$.

For every Σ-algebra $\mathcal{A}$, a Σ'-algebra $\mathcal{T}_H(\mathcal{A}) = \mathcal{T}_{\Sigma'}(\mathcal{A})_{/\sim}$ is constructed as follows: a Σ'-term algebra $\mathcal{T}_{\Sigma'}(\mathcal{A})$ is generated by

- $a \in \mathcal{T}_{\Sigma'}(\mathcal{A})_{H(b)}$ if $a \in \mathcal{A}_b$ where $b \in B$

- $f \in \mathcal{T}_{\Sigma'}(\mathcal{A})_b$ if $f : \varepsilon \to b \in \Sigma'$

- $f(t_1, \ldots, t_n) \in \mathcal{T}_{\Sigma'}(\mathcal{A})_b$ if $f : b_1 \times \ldots \times b_n \to b \in \Sigma'$ and $t_i \in \mathcal{T}_{\Sigma'}(\mathcal{A})_{b_i}$ for $i = 1, \ldots, n$.

Let $t_1 \sim t_2$ be the least congruence relation on $\mathcal{T}_{\Sigma'}(\mathcal{A})$ such that $f(a_1, \ldots, a_n) \sim f^{\mathcal{A}}(a_1, \ldots, a_n)$ where the a_i's are elements of $\mathcal{A}$.

Free Algebra Theorem 3. $\mathcal{T}_H(\mathcal{A})$ with embedding $\eta : \mathcal{A} \to (\mathcal{T}_H(\mathcal{A}))_H$ given by $\eta(a) = [a]$ is a universal morphism from $\mathcal{A}$ to $_{-H} : \mathbf{Alg}(\Sigma) \to \mathbf{Alg}(\Sigma)$ which maps $\mathcal{B}$ to $(H(b)^{\mathcal{B}}, (H_{w,b}(\sigma)^{\mathcal{B}} \mid \sigma \in \Sigma_{w,b}))$. ($\mathcal{B}_H$ is called the reduct of $\mathcal{B}$ under H.)

Proof. Let $h : \mathcal{A} \to \mathcal{B}_H$ be a Σ-homomorphism where $\mathcal{B}$ is a Σ'-algebra: h extends to a unique homomorphism $[h] : \mathcal{T}_{\Sigma'}(\mathcal{A}) \to \mathcal{B}$ such that $[h](a) = h(a)$ because of the Free Algebra Theorem 1, $[h]$ preserves the congruence since

$$
\begin{aligned}
[h](f(a_1, \ldots, a_n)) &= f^{\mathcal{B}}([h](a_1), \ldots, [h](a_1)) \\
&= f^{\mathcal{B}}(h(a_1), \ldots, h(a_1)) \\
&= h(f^{\mathcal{B}}(a_1, \ldots, a_n)),
\end{aligned}
$$

hence uniquely extends to the quotient. ∎

Let $H : Sp \to Sp'$ be a specification morphism. Combination of (C) and (D) yields the result, of which all the others are specializations.

Free Algebra Theorem 4 (Exercise).   *The reduct functor* $_{-H}$: **Alg**(Sp')
$\rightarrow$ **Alg**(Sp) *has a left adjoint.*

Here are some more specific examples for *algebraic adjoints.*

Let the functor *Graph* : **Cat** $\rightarrow$ **Graph** map a category **C** $= (Obj, Mor,$
$dom, cod, \circ, 1)$ to its underlying graph *Graph*(**C**) $= (Obj, Mor, dom, cod)$.

Proposition 3.6.5 (Exercise). *The 'path category' G^* (see Section 1.2.2)
with $\eta : G \rightarrow Graph(G^*)$, $f \mapsto f$ is universal from G to Graph (η maps
edges to the corresponding sequence of length one). (Hint: consider path
categories as 'generalized monoids'.)*

Let **Th** be the category of all categories with finite products and of
functors which preserve finite products (cf. Section 2.1.4). Define a functor
Sign : **Th** $\rightarrow$ **Sign**, the object part of which is given by

$$Sign(\mathbf{T}) = (Obj(\mathbf{T}), (Sign(\mathbf{T})_{W,B} \mid W \in Obj(\mathbf{T})^*, B \in Obj(\mathbf{T})\}$$

where $Sign(\mathbf{T})_{\varepsilon,B} = \mathbf{T}[1, B]$ and $Sign(\mathbf{T})_{W,B} = \mathbf{T}[A_1\Pi \ldots \Pi A_n, B]$ if $W =$
$A_1 \ldots A_n$. Then the proposition of Section 2.1.4 can be rephrased to:

Proposition 3.6.6. *The functor Sign : **Th** $\rightarrow$ **Sign** has a left adjoint
which maps each signature Σ to the theory $\mathbf{T}_\Sigma$ (cf. 2.1.4) where $\eta : \Sigma \rightarrow$
$Sign(\mathbf{T}_\Sigma)$ maps sorts b to the object b, and the operator $f : b_1 \ldots b_n \rightarrow b$
to the morphism $[f] : b_1 \times \ldots \times b_n \rightarrow b$.*

Proposition 3.6.7. *The forgetful functor U : **Slat** $\rightarrow$ **Set**, which maps a
semilattice to the underlying set, has a left adjoint P_f : **Set** $\rightarrow$ **Slat** which
maps a set X to the set of finite, non-empty subsets of X. Union is the
semilattice operation.*

Proposition 3.6.8. *The forgetful functor U : **Clat** $\rightarrow$ **Set** of complete
lattices to the underlying sets has a left adjoint, the object part of which
is defined by the power set $P(X)$ with union as operation.*

Proposition 3.6.9. *The contravariant power set functor $P^\bullet$: **Set** $\rightarrow$
Set^{op} (see Section 1.4.4) is left adjoint to the covariant power set functor
P : **Set** $\rightarrow$ **Set**.*

3.6.2   Adjunctions on partially ordered sets

Partially ordered sets (as categories) provide an excellent testbed for categorical definitions, not only because definitions specialize to familiar concepts, but also because certain universal constructions for arbitrary categories are a generalization of constructions in partially ordered sets.

A pair of monotone functions $f : A \rightarrow B, g : B \rightarrow A$ between partially
ordered sets A, B is an adjunction if $a \le g(b) \Leftrightarrow f(a) \le b$ for all $a \in A, b \in$

B (by specialization of the definition) where f is left adjoint to g. This is called a *Galois connection* in the context of partially ordered sets.

A left adjoint preserves least upper bounds as an easy computation proves: if $f(a) \leq c$ for all $a \in X \subseteq A$, then $g(f(a)) \leq g(c)$, hence $a \leq g(f(a)) \leq g(c)$ by adjointness. Then $\bigsqcup X \leq g(c)$, and $f(\bigsqcup X) \leq f(g(c))$ by monotonicity, and finally $f(\bigsqcup X) \leq f(g(c)) \leq c$ by adjointness. Now choose $c = \bigsqcup\{f(a) \mid a \in X\}$. Dually, a right adjoint preserves greatest lower bounds.

The interesting observation here is that Galois connections are characterized in terms of preservation properties:

Proposition 3.6.10. *Let B be complete (i.e. all greater lower bounds exist), and let $g : B \rightarrow A$ be a monotone function which preserves greatest lower bounds. Then g is a right adjoint.*

Proof. Define the left adjoint by $f(a) := \sqcap\{b \in B \mid a \leq g(b)\}$. Then $a \leq g(b)$ implies $f(a) \leq b$ by definition of greatest lower bounds, and $f(a) \leq b$ implies $g(\sqcap\{b \in B \mid a \leq g(b)\}) = \sqcap\{g(b) \in B \mid a \leq g(b)\} \leq g(b)$ because of monotonicity and assumption, hence $a \leq g(b)$. ∎

Suitable translation of this proof to arbitrary categories provides existence criteria for adjunctions which come under the name of the 'Adjoint Functor Theorem' (cf. Section 6.1).

If partially ordered sets are good for giving guidelines for the proof of general results about categories, they are even better for providing counter examples.

One might cherish the idea that products distribute over coproducts, meaning that there exists a natural isomorphism $A\Pi(B \amalg C) \cong (A\Pi B) \amalg (A\Pi C)$. The partially ordered set

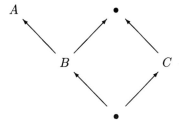

is a counterexample.

Galois connections as such may appear to have little bearing on computer science. Some examples may prove the contrary.

(A) Let $D = (D, \delta : Act \times S \rightarrow S)$ be an automaton and $X \subseteq S$. The problem is to construct a *subautomaton* $\langle X \rangle_D$ of D generated by X where

- D' is a *subautomaton* of D (notation $D' \sqsubseteq D$) if $S' \subseteq S$ and $\delta'(a, s) = \delta(a, s)$ for all $s \in S$.

- $D' \sqsubseteq D$ is *generated by* X if

  (i) $X \subseteq S'$, and

  (ii) whenever $D'' \sqsubseteq D$ and $X \subseteq S''$ then $D' \sqsubseteq D''$.

The definition is nothing but a Galois connection:

The subautomata of an automaton $D = (S, \delta)$ ordered by $D' \sqsubseteq D''$ form a partially ordered set, denoted by $\mathbf{Sub}(D)$. In fact, the subautomata form a complete lattice with the greatest lower bound given by the intersection of the carrier sets (*i.e.* $\bigcap \{D_i \mid i \in I\}$ is the set of states). The least upper bound is, of course, given by $\bigsqcup \{D \mid \forall i \in I.D_i \sqsubseteq D\}$. Similarly, the subsets of S form a complete lattice $\mathbf{Sub}(S)$ if ordered under inclusion.

Mapping a subautomaton $D' = (S', \delta')$ to its set of states S' then determines a monotone function $g : \mathbf{Sub}(D) \to \mathbf{Sub}(S)$. Then a subdymanics $\langle X \rangle_D$ is generated by X if $\langle X \rangle_D \sqsubseteq D'$ for all $D' \in \mathbf{Sub}(D)$ such that $X \subseteq g(D')$. As g preserves greatest lower bounds, $\langle X \rangle_D = \bigsqcap \{D' \in \mathbf{Sub}(D) \mid X \subseteq g(D')\}$ by the proposition above. This is the familiar construction of generated algebras by means of intersection.

Automata have been chosen because little formal effort is necessary to follow the argument. Monoids work as well as every other algebraic structure (cf. chapter on 'Universal Algebra'). Note, however, that in general the least upper bound is *not* given as union of the underlying sets, hence preservation of greatest lower bounds does not imply preservation of least upper bounds. It may be an illuminating exercise to check the details for monoids.

(B) Let $\mathcal{A}$ be a Σ-algebra. Define a preorder with elements being pairs $(\mathcal{B}, h)$ with $\mathcal{B}$ being a Σ-algebra and $h : \mathcal{A} \to \mathcal{B}$ being a surjective homomorphism. Let $(\mathcal{B}, h) \sqsubseteq (\mathcal{B}', h')$ if there exists a homomorphism $k : \mathcal{B} \to \mathcal{B}'$ such that $k \circ h = h'$. An order $\mathbf{Epi}(\mathcal{A})$ is induced if we factorize by antisymmetry; the elements are isomorphism classes $[(\mathcal{B}, h)] = \{(\mathcal{B}', h') \mid (\mathcal{B}, h) \sqsubseteq (\mathcal{B}', h') \text{ and } (\mathcal{B}', h') \sqsubseteq (\mathcal{B}, h)\}$, and the order is $[(\mathcal{B}, h)] \sqsubseteq [(\mathcal{B}', h')]$ if $(\mathcal{B}, h) \sqsubseteq (\mathcal{B}', h')$.

$\mathbf{Epi}(\mathcal{A})$ has least upper bounds: let $([(\mathcal{B}_i, h_i)], i \in I)$ be a family of elements, and let $a \sim a'$ be the transitive closure of $\{(b, b') \mid \exists i \in I.h_i(b) = h_i(b')\}$. Then $[(\mathcal{A}_{/\sim}, h_\sim)]$ is a least upper bound where $h_\sim : \mathcal{A} \to \mathcal{A}_{/\sim}$ is the quotient mapping.

Now let $H : \Sigma \to \Sigma'$ be a signature morphism. Given a surjective mapping $h : T_H(\mathcal{A}) \to \mathcal{B}$ define a Σ-algebra $Im_H(\mathcal{B})$ with carriers

$Im_H(\mathcal{B})_b = \{h_H(\eta(a)) \mid a \in \mathcal{A}_b\}$ where $h_H : T_H(\mathcal{A})_H \to \mathcal{B}_H$ is the reduct of h. Operations are induced by $\mathcal{A}$.

Use this to define a monotone function

$$_{-H} : \mathbf{Epi}(T_H(\mathcal{A})) \to \mathbf{Epi}(\mathcal{A}), [(\mathcal{B}, h)] \mapsto [(Im_H(\mathcal{B}), \pi)]$$

where $\pi(a) = h_H(\eta(a))$. We claim that the function preserves least upper bounds, and hence has a right adjoint.

In fact, this right adjoint has been used in computing, for instance under the name of *final realization* [Goguen and Meseguer, 1982]. The right adjoint can also be obtained by 'Nerode factorization':

A Σ'-term $t \in T'_\Sigma(X)_{H(b)}$ is called a *context* where X is a finite list of variables. Given a Σ-algebra $\mathcal{B}$, define a congruence relation on $T_H(\mathcal{B})$ by $b \sim b'$ if $[\Theta]t = [\Theta']t$ for all contexts t and for all assignments $\Theta, \Theta' : X \to T_H(\mathcal{B})$ such that $\Theta(x) = b$ and $\Theta'(x) = b'$ for some $x \in X$ and $\Theta(y) = \Theta'(y)$ for all $y \in X\backslash\{x\}$. This defines a functor $FIN : \mathbf{Alg}(\Sigma) \to \mathbf{Alg}(\Sigma'), \mathcal{B} \mapsto T_H(\mathcal{B})_{/\sim}$. Reduction of this functor to $\mathbf{Epi}(\mathcal{A}) \to \mathbf{Epi}(T_H(\mathcal{A}))$ is the right adjoint to $_{-H} : \mathbf{Epi}(T_H(\mathcal{A})) \to \mathbf{Epi}(\mathcal{A})$.

(C) A *non-deterministic automaton* $ND = (S, \delta)$ consists of a set S of states and a non-deterministic transition function $\delta : Act \times S \to 2^S$. Homorphisms $h : ND \to ND'$ are functions $h : S \to S'$ such that $\delta'(a, h(s)) = \{h(s') \mid s' \in \delta(a, s)\}$.

Define partial orders $\mathbf{Epi}(ND)$ and $\mathbf{Epi}(S)$ in analogy to (B). A monotone function $f : \mathbf{Epi}(ND) \to \mathbf{Epi}(S)$ maps (isomorphism classes of) non-deterministic automata to the underlying (isomorphism classes of) sets of states. The function has a right adjoint which can be defined as follows:

Let $ND = (S, \delta)$ be a non-deterministic automaton, and $f : S \to Y$ be a surjective mapping. A *bisimulation relation* $p \approx_f q$ on S is defined as the greatest relation which satisfies

$p \approx_f q$ iff $f(p) = f(q)$ and if $\forall a \in Act.\forall p'.(p' \in \delta(a, p) \Rightarrow \exists q'.q' \in \delta(a, q) \wedge p' \approx_f q')$, and if $\forall a \in Act.\forall q'.(q' \in \delta(a, q) \Rightarrow \exists p'.p' \in \delta(a, p) \wedge p' \approx_f q)$.

The right adjoint $g : \mathbf{Epi}(S) \to \mathbf{Epi}(ND)$ is specified by $g([Y, \pi]) = [(ND_{/\approx_f}, \pi)]$ where $\pi : ND \to ND_{/\approx_f}$ is the quotient mapping. We obtain the strong bisimulation of [Milner, 1980] in the special case where $f : S \to 1$.

The least upper bound of sets of surjective mappings (epimorphisms) is often called a *cointersection*. Hence the we can reinterpret the Galois connection as 'bisimulation is the cointersection of epis'. The automaton obtained by cointersection of all surjective homomorphisms is called *reduced* in the book of [Ehrig *et al.*, 1974]. (See also [Poigné, 1989].)

(D) Every mapping $f : X \to Y$ induces a monotone inverse $f^{-1} : \mathbf{Sub}(Y) \to \mathbf{Sub}(X)$ on the poset of subsets. This inverse has a left adjoint $\exists_f : \mathbf{Sub}(X) \to \mathbf{Sub}(X)$ and a right adjoint $\forall_f : \mathbf{Sub}(X) \to \mathbf{Sub}(Y)$ defined by $\exists_f(\varphi) = \{y \in Y \mid \exists x \in X.f(x) = y \wedge x \in \varphi\}$ and $\forall_f(\varphi) = \{y \in Y \mid \forall x \in X.f(x) = y \Rightarrow x \in \varphi\}$.

The notation stems from the special case of projections $p : X \times Y \to X$ where $\exists_f(\varphi) = \{x \in X \mid \exists y \in Y.(x,y) \in \varphi\}$ and $\forall_f(\varphi) = \{x \in X \mid \forall y \in Y.(x,y) \in \varphi\}$.

It is a fundamental insight by Lawvere that quantification can be expressed by adjunctions [Lawvere, 1969a]. We inspect these adjunctions more closely in Section 9.

3.6.3   Completion

Define a poset $(X, \sqsubseteq)$ to be *complete*, if every subset $Y \subseteq X$ has a least upper bound $\bigsqcup Y$. A *continuous* mapping is a monotone mapping between complete posets such that

$$f(\bigsqcup Y) = \bigsqcup\{f(y) \mid y \in Y\}.$$

Complete posets and continuous maps define a category **Cpos**.

The inclusion functor $I : \mathbf{Cpos} \to \mathbf{Pos}$ of complete posets to posets has a left adjoint $Compl : \mathbf{Pos} \to \mathbf{Cpos}$ defined as follows.

Let $(X, \sqsubseteq)$ be a poset. Define a preorder on subsets U, V of X by

$$U \sqsubseteq V \text{ if } \forall u \in U \exists v \in V.u \sqsubseteq v.$$

Define $C(X, \sqsubseteq)$ to consist of elements $[U] := \{V \subseteq X \mid U \sqsubseteq V \text{ and } V \sqsubseteq U\}$ where $U \subseteq X$, and define a partial order $[U] \sqsubseteq [V]$ if $U \sqsubseteq V$.

Proposition 3.6.11. $C(X, \sqsubseteq)$ *is freely generated with unit* $\eta_X(x) = [\{x\}]$.

Proof. We leave it to the reader to check that

$$\bigsqcup W = [\bigcup_{[U] \in W} U]$$

is the least upper bound of $W \subseteq C(X, \sqsubseteq)$, and that $[f]([U]) := \bigsqcup\{f(u) \mid u \in U\}$ defines the uniquely induced continuous function $[f] : C(X, \sqsubseteq) \to$

$(Y, \sqsubseteq)$ which extends a monotone map $f : (X, \sqsubseteq) \to (Y, \sqsubseteq)$ into a complete poset. ∎

This construction has a number of variations according to the type of least upper bound. For example, let $DC(X, \sqsubseteq)$ consist of elements $[U]$ such that $U \subseteq X$ is directed. Then

Proposition 3.6.12. $DC(X, \sqsubseteq)$ *is freely generated with unit* $\eta_X(x) = [\{x\}]$ *where* $I : \mathbf{Dcpo} \to \mathbf{Pos}$ *is the inclusion functor.*

Proof. [Hint] $\bigcup_{[U] \in W} U$ is directed; if $x \in U_1$ and $y \in U_2$ then there exists a U_3 such that $U_1, U_2 \sqsubseteq U_3$ since W is directed, hence elements $x', y' \in U_3$ such that $x \sqsubseteq x'$ and $y \sqsubseteq y'$, and, as U_3 is directed, an element $z \in U_3$ such that $x' \sqsubseteq z$ and $y' \sqsubseteq z$. ∎

Or: let $\omega(X, \sqsubseteq)$ consists of elements $[(x_n)_{n \in \omega}]$ where $(x_n)_{n \in \omega}$ is an ω-chain (i.e. $x_1 \sqsubseteq \ldots \sqsubseteq x_n \sqsubseteq \ldots$).

Proposition 3.6.13. $\omega(X, \sqsubseteq)$ *is freely generated with unit* $\eta_X(x) = [(x)_{n \in \omega}]$ *where* $I : \omega\mathbf{Pos} \to \mathbf{Pos}$ *is the inclusion functor.*

Proof. [Hint] Let $[c_1] \sqsubseteq \ldots \sqsubseteq [c_n] \sqsubseteq \ldots$ be an ω-chain $\omega(X, \sqsubseteq)$. Define a chain d with elements d_n being the least element of the chain c_n such that $c_{n,n} \sqsubseteq d_n$ and that $d_i \sqsubseteq d_n$ for all $i \sqsubseteq n$. Then $[d]$ is the least upper bound of the given chain. ∎

In the proofs of these propositions we have used yet another adjunction.

Proposition 3.6.14 (Exercise). *The inclusion functor* $I : \mathbf{Pos} \to \mathbf{Pre}$ *has a left adjoint which turns a preorder* $(X, \leq)$ *into the partial order with elements* $[x] := \{y \in X \mid x \sqsubseteq y, y \sqsubseteq x\}$ *and order relation* $[x] \sqsubseteq [y]$ *if* $x \sqsubseteq y$.

All these completion functors preserve finite products which allows one to construct completions of algebras which are enriched by order.

Proposition 3.6.15. *The completion functors* $C : \mathbf{Pos} \to \mathbf{Cpos}, DC : \mathbf{Pos} \to \mathbf{Dcpo}, \omega : \mathbf{Pos} \to \omega\mathbf{Pos}$ *preserve finite limits.*

Proof. [Hint] Check that a continuous isomorphism (in $\mathbf{Cpos}$, etc.)

$$C(X, \sqsubseteq) \times C(Y, \sqsubseteq) \cong C(X \times Y, \sqsubseteq)$$

is given by $([U], [V]) \mapsto [U \times V]$. ∎

Define an *ordered (complete, directed complete, ω-complete)* Σ-**algebra** $\mathcal{A}$ to consist of a (complete, directed complete, ω-complete) poset $b^{\mathcal{A}}$ for every sort $b \in B$, and a monotone (continuous of various degree) function $f^{\mathcal{A}} : b_1^{\mathcal{A}} \times \ldots \times b_n^{\mathcal{A}} \to b^{\mathcal{A}}$ for every operator $f : b_1 \times \ldots \times b_n \to b$. With homomorphisms being monotone (continuous ...) mappings this defines categories $\mathbf{PosAlg}(\Sigma)$, $\mathbf{CposAlg}(\Sigma)$, $\mathbf{DcpoAlg}(\Sigma)$ and $\omega\mathbf{PosAlg}(\Sigma)$. We then have the following corollary of the previous proposition.

Proposition 3.6.16. *The forgetful functor* $U : \mathbf{CposAlg}(\Sigma) \to \mathbf{PosAlg}(\Sigma)$ *which maps a complete Σ-algebra to the underlying ordered Σ-algebra has a left adjoint* $C : \mathbf{PosAlg}(\Sigma) \to \mathbf{CposAlg}(\Sigma)$ *where* $b^{C(\mathcal{A})} = C(b^{\mathcal{A}})$ *and where*

$$ f^{C(\mathcal{A})} = b_1^{C(\mathcal{A})} \times \ldots \times b_n^{C(\mathcal{A})} \cong C(b_1^{\mathcal{A}} \times \ldots \times b_n^{\mathcal{A}}) \xrightarrow{C(f^{\mathcal{A}})} C(b^{\mathcal{A}}). $$

Similarly in the other cases.

4 Elements and beyond

If anything, this exposition, so far, has been a *tour de force* of how to replace elements by functions. But in spite of all the claims, substantial use is still made of set theory. All the basic definitions reflect set theoretic properties, and are, in fact, stated in terms of set theory by means of the Hom-sets. Proper insight into this dependency proves fruitful from two points of view: the understanding grows of how to transfer set theoretic concepts into the language of categories, and, somewhat as a consequence, an idea is conceived of how to develop category theory independent of set theory with 'universality' being a sort of driving force.

4.1 Variable elements, variable subsets and representabl functors

4.1.1 A function is expected to be determined by its behaviour on elements. A background in set theory suggests considering morphisms $a : 1 \to A$ as elements, but most categories do not have enough such elements if they have a terminal object at all (for instance there is only one partial function $\varnothing : \varnothing \to A$ in the category $\mathbf{Pfn}$ of partial functions where $\varnothing$ is terminal). The standard intuition of functions mapping elements to elements is retained if every morphism $a : X \to A$ is considered as an element of A; then every morphism $f : A \to B$ induces a family of mappings, in fact a natural transformation, $f_X := C[X, f] : C[X, A] \to C[X, B]$ indexed by $Obj(\mathbf{C})$ (for the definition of Hom functors $\mathbf{C}[\_, A] : \mathbf{C}^{\mathrm{op}} \to \mathbf{Set}$ see Section 1.4.4).

Proposition 4.1.1. *There is a bijection between morphisms $f : A \to B$ and natural transformations*

$$\mathbf{C}[\_, f] : \mathbf{C}[\_, A] \to \mathbf{C}[\_, B] : \mathbf{C}^{\mathrm{op}} \to \mathbf{Set}.$$

We take advantage of the bijection and identify f and $\mathbf{C}[\_, f]$ notationally.

Proof. We claim that for a morphism $f : A \to B$ in $\mathbf{C}$ the corresponding natural transformation is $\mathbf{C}[\_, f]$, while for a natural transformation $f : \mathbf{C}[\_, A] \to \mathbf{C}[\_, B]$ the corresponding natural transformation is $f_A(1_A) : A \to B$. To show that this yields a bijection we proceed as follows:

1. Let $f : A \to B$ be a morphism in $\mathbf{C}$, then

$$\mathbf{C}[\_, f](1_A) = f \circ 1_A = f.$$

2. Let $f : \mathbf{C}[\_, A] \to \mathbf{C}[\_, B]$ be a natural transformation, and let $g : X \to A$, then

$$
\begin{aligned}
\mathbf{C}[\_, f_A(1_A)](g) &= f_A(1_A) \circ g \\
&= (\mathbf{C}[g, B] \circ f_A)(1_A) \\
&= (f_X \circ \mathbf{C}[g, A])(1_A) \\
&= f_A(1_A \circ g) = f_A(g).
\end{aligned}
$$

(Remark: despite its brevity, the proof is fundamental and should be looked at very carefully.) ∎

The perception of morphisms as elements is not so much of a *tour de force* if we interpret $a : X \to A$ as a *variable element* (or *parameterized element*) of *type* A with X as *domain of variation*.

A few examples may justify the terminology:

- Let a circle be given by x- and y-coordinates and a radius. Each circle is a (constant) element $c : 1 \to \underline{circle} = \mathbb{R} \times \mathbb{R} \times \mathbb{R}$. If we fix the radius, say r, the map $c : \mathbb{R} \times \mathbb{R} \to \underline{circle}$, $(x, y) \mapsto (x, y, r)$ may be considered as a 'circle of radius r which varies over the coordinates, or with the coordinates as parameters'.

- Consider a list $l : \{1, \ldots, n\} \to A$ as a variable element $l(x)$ of type A parameterized by $x \in \{1, \ldots, n\}$. This thought is probably not so unfamiliar since operations on lists are typically generated by operations on A, e.g. $(l + l')(x) = l(x) + l'(x)$. Certainly, a programmer may have his doubts about the 'variable element of even numbers'

$\_^{*}2 : \mathbb{N} \to \mathbb{N}$, or the 'variable element of factorials' $\_! : \mathbb{N} \to \mathbb{N}$. However, one might argue that $x!$ denotes an element of $\mathbb{N}_0$, the value of which depends on the actual argument.

The proposition suggests the notation $a \in A$ for $a : X \to A$, and $f \cdot a$ for $\mathbf{C}[\_, f](a)$ which is often quite convenient in that some categorical concepts involving separation obtain a 'set theoretic' touch in terms of variable elements:

Proposition 4.1.2.

- $f : A \to B$ is an isomorphism iff $f \cdot a = f \cdot b \Leftrightarrow a = b$ (i.e. $\mathbf{C}[X, f]$ is bijective for all X).

- $f : A \to B$ is a monomorphism iff $f \cdot a = f \cdot b$ implies $a = b$ for all $a, b \in A$ (i.e. $f : \mathbf{C}[\_, A] \to \mathbf{C}[\_, B]$ is (naturally) injective).

- $A \amalg B$ is a binary product of A and B if for all $a \in A$ and $b \in B$ there exists a unique element $c \in A \amalg B$ such that $p_{A,B} \cdot c = a$ and $q_{A,B} \cdot c = b$.

  (which states existence of a natural (in X) isomorphism $\langle \_, \_ \rangle_X :$ $\mathbf{C}[X, A] \times \mathbf{C}[A, B] \cong \mathbf{C}[X, A \amalg B]$).

The notation $f(a)$ is probably more appealing to mathematicians. We here prefer application $f \cdot a$ of functional programming for notational uniformity.

4.1.2 One should not get carried away by referring to elements; coproducts, for instance, cannot be defined in terms of variable elements since $\mathbf{C}[X, A \amalg B]$ is in general not isomorphic to the disjoint union $\mathbf{C}[X, A] + \mathbf{C}[X, B]$ (not even in the category of sets, take $X = \varnothing$).

But coproducts induce a natural isomorphism $\mathbf{C}[A \amalg B, X] \cong \mathbf{C}[A, X] \times \mathbf{C}[B, X]$ which agrees with set theoretic intuitions if we interpret morphisms $f : A \to X$ as *partitions* (or — to match the terminology of the previous section — *variable subsets*).

A family $P = (P_x \mid x \in \Xi)$ of sets is called a *partition* of a set A if for all $a \in A$ there exists a unique $x \in X$ such that $a \in P_x$. A partition is *proper* if $P_x \neq \varnothing$ for all $x \in X$.

Every function $f : A \to B$ maps a partition P of B to a partition $[f]P := (P_{f(x)} \mid x \in X)$, or, if partitions are thought of as functions $P : B \to X, [f]P = P \circ f$. Every function is determined by its behaviour on partitions:

Proposition 4.1.3. *There exists a bijection between morphisms $f : A \to B$ and natural transformations $[f] : \mathbf{C}[B, \_] \to \mathbf{C}[A, \_] : \mathbf{C} \to \mathbf{Set}$.*

Let us use $P :: B$ to state that P is a partition of B, and abbreviate $\mathbf{C}[f, \_](P)$ by $[f]P$, the similarity to the notation for predicate transformation being not by chance. Then

Proposition 4.1.4. *$A \amalg B$ is a binary coproduct of A and B if for all $P :: A$ and $Q :: B$ there exists a unique $R :: A \amalg B$ such that $[inl_{A,B}]R = P$ and $[inr_{A,B}]R = Q$.*

states that partitions of a 'disjoint union' are pairs of partitions of the components which is a — maybe unusual — characterization of disjoint unions in the category of sets.

As a final example: a function $f : A \to B$ is surjective iff the induced function on partitions is injective:

Proposition 4.1.5. *A morphism $f : A \to B$ is an epimorphism if $[f] : \mathbf{C}[B, \_] \to \mathbf{C}[A, \_]$ is injective.*

4.1.3 All these observations crucially depend on the fact that the Hom functors $\mathbf{C}[A, \_]$ and $\mathbf{C}[\_, A]$ are fully determined by the identity 1_A.

Definition 4.1.6. Let $F : \mathbf{C} \to \mathbf{Set}$ be a functor. An element $u \in F(U)$ is called *universal* if for all elements $a \in F(A)$ there exists a unique morphism $f : U \to A$ such that $F(f)(u) = a$.

In other words, all elements in the 'model' $F : \mathbf{C} \to \mathbf{Set}$ are generated from U by application of the operations $f : A \to B$. Up to isomorphisms, Hom functors are the only functors which have universal elements.

Proposition 4.1.7. *A functor $F : \mathbf{C} \to \mathbf{Set}$ has a universal element $u \in F(U)$ iff $F \cong \mathbf{C}[U, \_]$ (dually, $F \cong \mathbf{C}[\_, U]$ for $F : \mathbf{C}^{op} \to \mathbf{Set}$).*

Proof. The isomorphism is determined by $u \leftrightarrow 1_U$. ∎

Definition 4.1.8. A functor with a universal element is also called a *representable functor*.

The elements of a representable functor $F : \mathbf{C}^{op} \to \mathbf{Set}$ may be thought of as *cogenerated* in that every element is obtained by inverse application $f \mapsto v$ where $v \in F(V)$ is the universal element, and where $f : A \to V$ in $\mathbf{C}$.

Here are some examples:

- The functors $\mathbf{C}[\_, A] \times \mathbf{C}[\_, B] : \mathbf{C}^{op} \to \mathbf{Set}$ is representable iff $\mathbf{C}$ has finite products.

- The forgetful functor $\mathbf{Mon} \to \mathbf{Set}$ is represented by $\mathbf{Mon}[1^*, \_]$. Similarly, for all other algebraic structures.

The following examples are here restricted to sets for systematic reasons. The next section will consider the examples in their full glory.

- Let **Pfn** be the category of partial functions. The functor **Pfn**$[E(\_), A]$: **Set** $\rightarrow$ **Set**, where E : **Set** $\rightarrow$ **Pfn** is the canonical embedding, is representable: **Pfn**$[E(\_), A] \cong$ **Set**$[\_, A_\perp]$ with $A_\perp := A + \{\perp\}$.

- Let **Rel** be the category of relations with $R \in$ **Rel**$[A, B]$ if $R \subseteq A \times B$ and with composition $S \circ R := \{(a, c) \mid \exists b \in B.(a, b) \in R, (b, c) \in S\}$. The functor **Rel**$[\_, A]$: **Set** $\rightarrow$ **Set** is representable : **Rel**$[\_, A] \cong$ **Set**$[\_, P(A)]$ where $P(A)$ is the power set of A.

- Let Sub : **Set**$^{op} \rightarrow$ **Set** be defined by $Sub(A) = \{X \mid X \subseteq A\}$ and $Sub(f)(X) = f^{-1}(X)$. Sub is representable : $Sub \cong$ **Set**$[\_, 2]$ or, every subset is defined by a characteristic function, and vice versa.

4.1.4 The concept of representable functors is an equivalent formalization of 'universality'.

Proposition 4.1.9.

- Let $G : \mathbf{D} \rightarrow \mathbf{C}$ be a functor. G has a *left adjoint* iff $\mathbf{C}[A, G(\_)]$: $\mathbf{D} \rightarrow$ **Set** *is representable for all objects A.*

- $u \in G(U)$ *is universal iff* $u : 1 \rightarrow G(U)$ *is universal from 1 to G.*

However, representable functors offer a slightly different perspective if considered as *specification* of types in a category. Types are typically introduced by means of *introduction* and *elimination* rules, e.g.

$$(intro) \quad \frac{x : A \qquad y : B}{\langle x, y \rangle : A \times B} \qquad (elim) \quad \frac{z : A \times B}{first(z) : A} \qquad \frac{z : A \times B}{second(z) : B}$$

plus (equational) axioms which state the properties of the operators. Replacing elements by variable elements does not essentially change the setup:

$$(intro) \quad \frac{x : X \rightarrow A \qquad y : X \rightarrow B}{\langle x, y \rangle : X \rightarrow A \times B}$$

$$(elim) \quad \frac{z : X \rightarrow A \times B}{first(z) : X \rightarrow A} \qquad \frac{z : X \rightarrow A \times B}{second(z) : X \rightarrow B}$$

The introduction rule may be split into two steps, the construction of a new term $\langle x, y \rangle$, and the typing of this terms $\langle x, y \rangle : A \times B$. Construction is captured by the functor $\mathsf{Pair}(A, B) : \mathbf{C}^{op} \rightarrow$ **Set** with $\mathsf{Pair}(A, B)(X) =$

$\{\langle x, y \rangle \mid x : X \to A, y : X \to B\}$ (note that the structural rule of substitution for the type systems corresponds to composition), or, up to isomorphism, $\mathbf{C}[\_, A] \times \mathbf{C}[\_, B] : \mathbf{C}^{\mathrm{op}} \to \mathbf{Set}$. The typing is achieved by representabilty of the functor. Elimination rules and equalities are implicitly given; for example, the natural transformation $first : \mathbf{C}[\_, A] \times \mathbf{C}[\_, B] \to \mathbf{C}[\_, A]$ defines a morphism $first : A\Pi B \to A$, etc.

More generally, a representable functor $T : \mathbf{C}^{\mathrm{op}} \to \mathbf{Set}$ specifies a type T the A-elements of which are $T(A)$. The specification is thus given in terms of set theoretic constructions, the specified types reflecting properties of the respective construction; the product $A\Pi B$ is presented by the Cartesian product $\mathbf{C}[\_, A] \times \mathbf{C}[\_, B]$ of variable elements, or similarly the function space $B \Rightarrow C$ is defined by the functor $\mathbf{C}[\_\Pi B, C]$.

As another example, an equality type $\mathsf{Eq}(f, g)$ is specified by a functor $\mathsf{Eq}(f, g) : \mathbf{C}^{\mathrm{op}} \to \mathbf{Set}, \mathsf{Eq}(f, g)(X) := \{x : X \to A \mid f \cdot x = g \cdot x\}$.

Proposition 4.1.10. *The natural isomorphism* $\mathbf{C}[\_, \mathsf{Eq}(f, g)] \cong \mathsf{Eq}(f, g)$ *defines an equalizer (see Section 5.1.3).*

(Check that $\mathbf{C}[\_, \mathsf{Eq}(f, g)]$ is the equalizer of $f, g : \mathbf{C}[\_, A] \to \mathbf{C}[\_, B]$.)

The reflection of set theoretic constructions in a category has been used extensively in the first part of the exposition where categorical definitions have been motivated by the corresponding set theoretical constructions (though a motivation in terms of type theory probably might have been more appropriate).

It might be said that the preceding realizes Martin-Löf's claim that types are introduced by introduction rules and that the elimination rules are of a secondary nature because they are induced.

4.1.5 Dually, representable functors $F : \mathbf{C} \to \mathbf{Set}$ specify types in terms of partitions, e.g. coproducts are specified by a construction on partitions $\mathbf{C}[A + B, \_] \cong \mathbf{C}[A, \_] \times \mathbf{C}[B, \_]$, or:

The *quotient type* $A_{/R}$ induced by a relation $R \subseteq A \times A$ is defined as 'minimal' partition $\varepsilon : A \to A_{/R}$ of A such that '$R \leq \varepsilon$'. More precisely, for every partition $P : A \to Y$ there exists a unique morphism $P_* : A_{/R} \to Y$ such that $P_* \circ \varepsilon = P$. In other words: $\mathbf{C}[A_{/R}, \_] \cong \mathsf{Coeq} : \mathbf{C} \to \mathbf{Set}$ where $\mathsf{Coeq}(B) = \{P : A \to Y \mid [f]P = [g]P\} = \{P : A \to Y \mid P \circ f = P \circ g\}$.

Specification in terms of partitions is probably disturbing for every reader biased by set theory who would like to see the categorical definitions rather as characteristic properties. The type theoretic definition of disjoint sums, for instance,

$$(intro) \quad \frac{x : A}{inl(x) : A + B} \qquad \frac{y : B}{inr(y) : A + B}$$

$$\text{(elim)} \quad \frac{z : A + B \quad M : C \quad N : C}{case \; z, M, N \; esac : C}$$

$$(x : A) \quad (y : B)$$

reflects the set theoretic predominance of elements; the supposedly more important introduction rule specifies the universal element from which the representing functor is not trivially induced, if at all (one might doubt here Martin-Löf's dictum that the elimination rule is induced). In contrast, the categorical definition specifies the representing functor $\mathbf{C}[A, \_] \times \mathbf{C}[B, \_]$, which naturally induces the universal element. Hence we believe that 'generated types' are best specified in terms of partitions, at least in a functional language.

4.2   Yoneda's heritage

4.2.1   Apart from providing a sort of 'extensional' view of the internal structure of categories, representable functors behave like *term models* if a category $\mathbf{C}$ is interpreted as a (programming) language and functors $F : \mathbf{C} \to \mathbf{Set}$ as models:

Terms are defined to be generated by application of operators. For instance $(3 * x) + (4 * y)$ is generated by the operators $3, 4, \_ + \_, \_ * \_$ from the variables x, y. Terms have the fundamental property that their value can be computed, once values are assigned to variables, and the semantics of the operators are known, the value being uniquely determined provided that the computation preserves the structure. This is the message of the

Free Algebra Theorem (cf. Section 3.6.1).   *Let* $\Sigma = (B, \Sigma)$ *be a signature and* A *be a* Σ-*algebra. Then for every variable assignment* $\Theta = (\Theta_b : X_b \to b^A \mid b \in B)$ *there exists a unique homomorphism* $[\Theta] : T_\Sigma(X) \to A$ *such that* $[\Theta]_b(x) = \Theta_b(x)$ *for* $x \in X_b$.

4.2.2   Terms are constructed by formal application of operators to arguments of suitable type where 'formal application' means that the arguments are prefixed by the operator without further evaluation. As the functional language does not distinguish between operators and derived operators (every 'function' $f : A \to B$ is a derived operator with parameter of type A), the only means to construct terms is by formal application of functions to arguments.

Let us introduce the notation

$$f \otimes x$$

for formal application, where $f : A \to B$ and $x \in X(A)$, $X(A)$ being a set (of variables) for each type A. We define a functor

$$\mathbf{C}(X) : \mathbf{C} \to \mathbf{Set} \qquad \text{with } \mathbf{C}(X)(B) = \{f \otimes x \mid f : A \to B, x \in X(A)\}$$

$$\text{and } \mathbf{C}(X)(g)(f \otimes x) = (g \circ f) \otimes x$$

which corresponds to the term algebra, $\mathbf{C}(X)(B)$ being the set of terms of type B with variables X.

Proposition 4.2.1. *Let* $\mathbf{C}$ *be a category and* $F : \mathbf{C} \to \mathbf{Set}$ *be a functor. Then for every variable assignment* $\Theta = (\Theta_A : X(A) \to F(A) \mid A \in | \mathbf{C} |)$ *there exists a unique natural transformation* $[\Theta] : \mathbf{C}(X) \to F$ *such that* $[\Theta]_A(1_A \otimes x) = \Theta_A(x)$ *for* $x \in X(A)$.

Proof. This is recommended as an exercise. $[\Theta]$ is defined by $[\Theta](f \otimes x) = F(f)(\Theta(x))$. ∎

There is an alternative, more categorical formulation of the result.

Proposition 4.2.2. *Let* $Obj(\mathbf{C})$ *be the discrete category of objects of* $\mathbf{C}$. *The inclusion functor* $\mathsf{Inc} : Obj(\mathbf{C}) \to \mathbf{C}$ *induces a functor* $\mathbf{Set}^{\mathsf{Inc}} : \mathbf{Set}^{\mathbf{C}} \to \mathbf{Set}^{Obj(\mathbf{C})}, F \to F \circ \mathsf{Inc}$ *on functor categories.* $\mathbf{Set}^{\mathsf{Inc}}$ *has a left adjoint* $\mathbf{C}(\_) : \mathbf{Set}^{Obj(\mathbf{C})} \to \mathbf{Set}^{\mathbf{C}}$ *which maps a functor (of variables)* $X : Obj(\mathbf{C}) \to \mathbf{Set}$ *to* $\mathbf{C}(X) : \mathbf{C} \to \mathbf{Set}$.

The result corresponds to the Free Algebra Theorem, though in a restricted form (i.e. the Free Algebra Theorem 2 of Section 3.6.1). We will learn below (6.2–6.3) that formal application of derived operators to arguments, or *tensoring*, is a fundamental construction of mathematics for generation of structure, including generation of free algebras.

4.2.3 Hom functors $\mathbf{C}[A, \_]$ are (up to isomorphism) a particular case of functors $\mathbf{C}(X)$, namely where $X(B) = \varnothing$ for $B \neq A$ and $X(A) = 1$. Then we obtain as a corollary the

Yoneda Lemma. *For every* $a \in F(A)$ *there exists a unique natural transformation* $\alpha : \mathbf{C}[A, \_] \to F$ *such that* $\alpha(1_A) = a$, *or: there exists a natural isomorphism (in* A) $\mathbf{Set}^{\mathbf{C}}[\mathbf{C}[A, \_], F] \cong F(A)$.

Further specialization yields $\mathbf{Set}^{\mathbf{C}}[\mathbf{C}[A, \_], \mathbf{C}[B, \_]] \cong \mathbf{C}[B, A]$ which is the isomorphism of Section 4.1.1, and which is often referred to as *Yoneda embedding*.

The following proposition is a particular application of the Yoneda Lemma. Assume that $\mathbf{C}$ has finite products, and that $F : \mathbf{C} \to \mathbf{Set}$ preserves the finite products. Covariant Hom functors $\mathbf{C}[A, \_]$ preserve finite products by definition: $\mathbf{C}[A, B\Pi C] \cong \mathbf{C}[A, B] \times \mathbf{C}[A, C]$ and $\mathbf{C}[A, 1] \cong 1$.

Initial Model Theorem.   *The Hom functor* $\mathbf{C}[1, \_] : \mathbf{C} \to \mathbf{Set}$ *is initial in the category* $Fun_\Pi(\mathbf{C}, \mathbf{Set})$ *of functors* $F : \mathbf{C} \to \mathbf{Set}$ *which preserve finite products, and of natural transformations between such functors.*

Proof. There exists a unique natural transformation $\sigma : \mathbf{C}[1, \_] \to F$ by the Yoneda Lemma since $F(1) \cong 1$.   ∎

The name is paradigmatic in that the proposition covers the construction of initial algebras as demonstrated in Section 6.2, and — as a thesis: whatever notion of models is considered (in **Set**), an initial such model must be representable.

4.2.4   The fundamental nature of representable functors and of the Yoneda Lemma is difficult to appreciate though reflected by basically every definition and every proof in category (since 'true' categorical arguments are universal). We have ventured two interpretations of representable functors, as 'specification' of types and as 'term models', which hopefully cover some ground.

4.3   Towards an enriched category theory

We warn the reader that a totally new chapter of category theory is opened which s/he should definitely skip at a first reading, if not familiar with the subjects discussed so far. However, there is some rationale to position this subsection here, since subsequent definitions and observations extend the main theme of this section on the fundamental nature of the Hom functors. We just give an introduction for the reader to apprehend the basic ideas, and hint at several applications already to be found in the computer science literature.

4.3.1   If a substantial part of category theory reflects set theory via Hom sets, this essential link must be cut if we are to develop a category theory without assuming the existence of set theory a priori. The starting point of such an undertaking is the 'alternative' definition of categories in terms of Hom sets as introduced in Section 1:

A *category* $\mathbf{C}$ consists of

$Obj(\mathbf{C})$, a class (called the *objects* of $\mathbf{C}$),

for each pair $A, B \in Obj(\mathbf{C})$ a class $\mathbf{C}[A, B]$,

for each triple $A, B, C \in Obj(\mathbf{C})$, a function $\circ_{A,B,C} : \mathbf{C}[B, C] \times \mathbf{C}[A, B] \to \mathbf{C}[A, C]$, and

for each $A \in Obj(\mathbf{C})$ we have a distinguished element $1_A \in \mathbf{C}[A, A]$

where the data are subject to the following conditions

$\mathbf{C}[A, B]$ and $\mathbf{C}[X, Y]$ are disjoint unless $A = X$ and $B = Y$,

if $f \in \mathbf{C}[A, B], g \in \mathbf{C}[B, C]$, and $h \in \mathbf{C}[C, D]$ then $h \circ_{A,C,D} (g \circ_{A,B,C} f) = (h \circ_{B,C,D} g) \circ_{A,B,D} f$,

if $f \in \mathbf{C}[A, B]$ then $f \circ_{A,A,B} 1_A = f$ and $1_B \circ_{B,B,A} f = f$.

What does it mean to replace **Set** by some category **V**? *Hom sets* $\mathbf{C}[A, B]$ become *Hom objects* $\mathbf{C}[A, B] \in Obj(\mathbf{V})$, composition and identities are given by morphisms $* : \mathbf{C}[B, C]\Pi\mathbf{C}[A, B] \to \mathbf{C}[A, C]$ resp. $e_A : 1 \to \mathbf{C}[A, A]$ in **V**, and the axioms are rephrased in the language of diagrams.

Definition 4.3.1 (Preliminary Definition). Let **V** be a category with finite products. A **V**-*category* **C** is defined by the following data

$Obj(\mathbf{C})$, a class,

for each pair $A, B \in Obj(\mathbf{C})$, an object $\mathbf{C}[A, B]$ of **V**,

for each triple $A, B, C \in Obj(\mathbf{C})$, a morphism $*_{A,B,C} : \mathbf{C}[B, C] \times \mathbf{C}[A, B] \to \mathbf{C}[A, C]$ of **V**,

for each $A \in Obj(\mathbf{C})$, a morphism $e_A : 1 \to \mathbf{C}[A, A]$ of **V**.

These data are subject to the condition that the following diagrams commute (subscripts omitted)

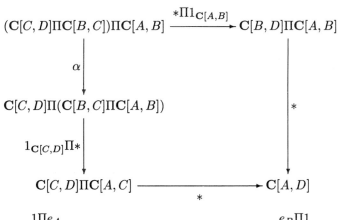

where $\alpha_{A,B,C} := \langle \langle p_{A,B\Pi C}, p_{B,C} \circ q_{A,B\Pi C}\rangle, q_{B,C} \circ q_{A,B\Pi C}\rangle$ and $\rho_A = p_{A,1}$, $\lambda_A = q_{1,A}$.

The following are examples to be found in the computer science literature:

- The reader may check that ordinary categories are **Set**-enriched.

- Posets are **2**-enriched where $\mathbf{2} = (0 \to 1)$. Let $x \sqsubseteq y$ in $\mathbf{C}$ if $\mathbf{C}[x, y] = 1$.

- **Pos**-enriched categories have Hom objects being posets where composition is monotone . Note that **Pos** is **Pos**-enriched if the Hom-sets are ordered pointwise, i.e. $f \sqsubseteq g$ if $f(x) \sqsubseteq g(x)$ for all $x \in X$.

- **Cat** is **Cat**-enriched if we consider functor categories as Hom objects.

4.3.2 In fact, the definition of categories does need much less infrastructure. The only property of products used is existence of the natural isomorphisms $\alpha_{A,B,C} : A\Pi(B\Pi C) \to (A\Pi B)\Pi C$, $\lambda_A : 1\Pi A \to A$ and $\rho_A : A\Pi 1 \to A$, which satisfy certain *coherence axioms*, necessary for equational reasoning with regard to the categorical axioms.

Definition 4.3.2. A *monoidal category* **V** consists of the following data:

- a category $\mathbf{V}_o$,

- a functor $\otimes : \mathbf{V} \times \mathbf{V} \to \mathbf{V}$,

- a distinguished object I of **V**,

- and natural isomorphisms $\alpha_{A,B,C} : (A \otimes B) \otimes C \to A \otimes (B \otimes C)$, $\lambda_A : I \otimes A \to A$ and $\rho_A : A \otimes I \to A$.

These data satisfy the *coherence axioms*, meaning that the diagrams

$$((A \otimes B) \otimes C) \otimes D \xrightarrow{\ \alpha\ } (A \otimes B) \otimes (C \otimes D) \xrightarrow{\ \alpha\ } A \otimes (B \otimes (C \otimes D))$$

$$\alpha \otimes 1 \downarrow \qquad\qquad\qquad\qquad\qquad\qquad\qquad\qquad \uparrow 1 \otimes \alpha$$

$$(A \otimes (B \otimes C)) \otimes D \xrightarrow[\qquad\qquad \alpha \qquad\qquad]{} A \otimes ((B \otimes C) \otimes D)$$

$$(A \otimes I) \otimes C \xrightarrow{\ \alpha\ } A \otimes (I \otimes C)$$

$$\rho \otimes 1 \searrow \qquad\qquad \swarrow 1 \otimes \lambda$$

$$A \otimes B$$

commute.

A monoidal category is *symmetric* if additionally a natural transformation $\gamma_{A,B} : A \otimes B \to B \otimes A$ exists such that the diagrams

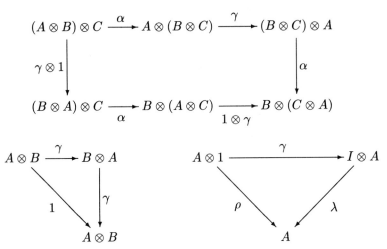

commute.

(For convenience we use α, λ, ρ and γ for both directions of the isomorphisms.)

The *coherence theorem* ([MacLane, 1971], VII.2) states that every diagram involving only the natural transformations above and their inverses commutes. Informally, one thinks of the transformations as induced by products (where $\gamma_{A,B} = \langle q_{A,B}, p_{A,B}\rangle$); then the diagrams capture exactly those equalities between composite transformations which are induced by product properties.

Definition 4.3.3. Let $\mathbf{V}$ be a monoidal category. A $\mathbf{V}$-*(enriched) category* is defined as in Section 4.3.1 but replace the products $A \Pi B$ by $A \otimes B$, and 1 by I.

Examples 4.3.4.

- Examples for the more general concept typically are of mathematical nature, e.g. Abelian groups, rings, modules where the usual tensor product defines the monoidal structure. We consider an example with a more computer science touch in Section 8.6 under the name of *non-deterministic algebras*.

- Here are other example of interest: let $\mathbf{X} = (X, \sqsubseteq)$ be a poset with a least upper bound $x \sqcup y$ and a least element $\bot$. $\mathbf{X}$-categories $\mathbf{C}$ consist of a set of partial orders $A \sqsubseteq_x B$ indexed by $x \in X$ which are compatible in that $A \sqsubseteq_y B$ if $A \sqsubseteq_x B$ and $x \sqsubseteq y$, and $A \sqsubseteq_{x \sqcup y} C$ if $A \sqsubseteq_x B \sqsubseteq_y C$ or $A \sqsubseteq_y B \sqsubseteq_x C$.

There are various other monoidal structures available on a poset. For instance in the case of $\mathbf{n} = (0 \sqsubseteq 1 \sqsubseteq \ldots \sqsubseteq n-1)$ there exist 2^{n-1} symmetric monoidal structures which are idempotent, i.e. $x \otimes x = x$, half of which are closed, namely those satisfying $x \otimes 0 = 0$ (see [Casley *et al.*, 1989]). The various monoidal structures on **3** have been used to model concurrency [Gaifmann, 1989; Gaifmann and Pratt, 1987].

- The non-negative reals including ∞ form a monoidal category with $\otimes$ to be *max* and I to be 0. The corresponding enriched categories are basically *ultrametric spaces*. A second monoidal structure on the reals takes $\otimes$ to be addition. Then the enriched categories correspond to metric spaces [Lawvere, 1973].

4.3.3   Let us 'enrich' the other basic notions of category theory.

Definition 4.3.5.

- Let **C** and **D** be **V**-categories. A **V**-*functor* $F : \mathbf{C} \to \mathbf{D}$ consists of a function $F : Obj(\mathbf{C}) \to Obj(\mathbf{D})$ and a family of morphisms $F_{A,B} : \mathbf{C}[A, B] \to \mathbf{D}[F(A), F(B)]$ indexed by objects A, B of **C** such that the diagrams

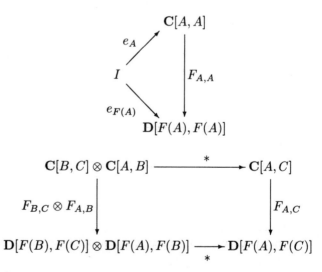

commute.

- Let $F : \mathbf{C} \to \mathbf{D}$ and $G : \mathbf{C} \to \mathbf{D}$ be **V**-functors. A **V**-*natural transformation* $\sigma : F \to G$ is defined by morphism $\sigma_A : I \to \mathbf{D}[F(A), G(A)]$ indexed by objects A of **C** such that the diagram

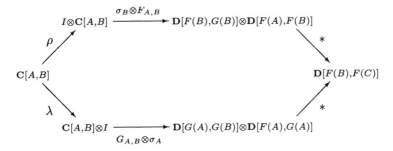

commutes.

The reader may wish to check the following examples:

- Functors are **Set**-functors and natural transformations are **Set**-natural transformations.

- **2**-Functors are monotone mappings of posets, a natural transformation $\sigma : f \to g : X \to Y$ exists iff $f \sqsubseteq g$ pointwise.

- **Pos**- functors are monotone on Hom sets, natural transformations satisfy essentially the same diagrams as ordinary natural transformations except that they are monotone mappings. Similarly for **Dcpo** etc.

Every morphism $f : 1 \to \mathbf{C}[A, B]$ induces morphisms $\mathbf{C}[X, f] : \mathbf{C}[X, A] \to \mathbf{C}[X, B]$ and $\mathbf{C}[f, Y] : \mathbf{C}[B, Y] \to \mathbf{C}[A, Y]$ where $\mathbf{C}[X, f] = {*}{\circ}(f{\otimes}1_{\mathbf{C}[X,A]}){\circ} \lambda^{-1}$ and $\mathbf{C}[f, Y] = * \circ (1_{\mathbf{C}[B,Y]} \otimes f) \circ \rho^{-1}$.

Thus it makes sense to speak of morphisms $f : A \to B$ of $\mathbf{C}$ though, strictly speaking, there is nothing like a morphism of $\mathbf{C}$.

Then the diagram for natural transformations turns into

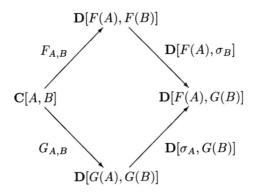

which may be more readable.

Following [Lawvere, 1973], [Casley *et al.*, 1989] advocate the use **V**-enriched categories as generalized metric spaces in the theory of concurrency. The idea is to consider objects of a **V**-enriched category as *events* and the Hom objects $C[A, B]$ as a 'distance' between the events A and B. Distance may be a causal or temporal order if **V** is an order category, or a real-time delay if **V** is a monoidal category over the reals.

4.3.4 This newly enriched world would make little sense if we could not transpose and generalize concepts we so dearly love in the world of, if not sets, **Set**-enriched categories. This exposition does not have space to do so, or to ponder about the new fundamental insights which can be learned by doing so. The only remark, which may be helpful though, is that much of the theory is based on the notion of representable functors, which are the means to define **V**-enriched structure in terms of structure available in **V**, just as above in the case of sets. Consider the example of **V**-enriched products.

Definition 4.3.6. Let **C** be a **V**-enriched category, and let A, B be objects of **C**. A **V**-*product* of A and B consists of an object $A\Pi B$ of **C**, and morphisms $p_{A,B} : I \to C[A\Pi B, A], q_{A,B} : I \to C[A\Pi B, B]$ in **V** (!) such that $C[X, A\Pi B]$ with projections $C[X, p_{A,B}] : C[X, A\Pi B] \to C[X, A]$ and $C[X, q_{A,B}] : C[X, A\Pi B] \to C[X, B]$ is a product in **V**. Hence $C[X, A\Pi B] \cong C[X, A]\Pi C[X, B]$ 'naturally'.

We should note that the approach is rather naïve. Enriched limits and colimits should be indexed as defined in the book [Kelly, 1982].

4.3.5 As another example, an adjunction of **V**-functors $F, G : C \to D$ should be defined by a 'natural isomorphism' $\varphi : D[F(A), B] \cong C[A, G(B)]$. However, naturality should be **V**-enriched, and, consequently, the Hom functor should be **V**-enriched. We need more structure.

Definition 4.3.7. A (symmetric) monoidal category **V** is called (*symmetric*) *monoidal closed* if for all objects A of **V**, the functor $\_ \otimes A : V \to V$ has a right adjoint $[A \Rightarrow \_] : V \to V$.

For motivation, it is probably best to consider Cartesian closed categories.

Definition 4.3.8. Let **V** be a monoidal closed category. Define a **V**-category, called **V**, whose objects are those of the category **V**, and whose Hom objects $V[A, B]$ are $[A \Rightarrow B]$. The composition $* : [B \Rightarrow C] \otimes [A \Rightarrow B] \to [A \Rightarrow C]$ is uniquely induced by monoidal closure in the diagram

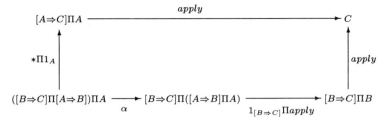

and $e_A = \lambda(\rho_A) : I \to [A \Rightarrow A]$.

(One might want to check that $\mathbf{V}$ is a $\mathbf{V}$-category. Hint: it might be helpful to do the diagrammatic computations alongside a proof that $\mathbf{Set}$ is $\mathbf{Set}$-enriched.)

Definition 4.3.9.

- The dual $\mathbf{V}$-category $\mathbf{C}^{op}$ has the same objects as $\mathbf{C}$ and Hom sets $\mathbf{C}^{op}[A, B] = \mathbf{C}[B, A]$. Composition is given by $*^{op} = * \circ \gamma : \mathbf{C}[C, B] \otimes \mathbf{C}[B, A] \to \mathbf{C}[C, A]$.

- Given $\mathbf{V}$-categories $\mathbf{C}_1$ and $\mathbf{C}_2$, define a $\mathbf{V}$-category $\mathbf{C}_1 \otimes \mathbf{C}_2$ with objects being pairs (A_1, A_2) of objects A_1 of $\mathbf{C}_1$ and A_2 of $\mathbf{C}_2$, and Hom objects $\mathbf{C}_1 \otimes \mathbf{C}_2[(A_1, A_2), (B_1, B_2)] = \mathbf{C}_1[A_1, B_1] \otimes \mathbf{C}_2[A_2, B_2]$. (We leave it as an exercise in programming without variables to define the composition which is componentwise.)

- A $\mathbf{V}$-functor $Hom_{\mathbf{C}} : \mathbf{C}^{op} \otimes \mathbf{C} \to \mathbf{V}$ is defined by $Hom_{\mathbf{C}}(A, B) = \mathbf{C}[A, B]$ and by $Hom_{\mathbf{C}}(B, C), (A, D) : \mathbf{C}^{op}[B, A] \otimes \mathbf{C}[C, D] \to (\mathbf{C}[B, C] \Rightarrow \mathbf{C}[A, D])$ which is the morphism obtained by adjunction from the obvious morphism $(\mathbf{C}[A, B] \otimes \mathbf{C}[C, D]) \otimes \mathbf{C}[B, C] \to \mathbf{C}[A, D]$.

Now we are in the position to define an adjunction of $\mathbf{V}$-functors in terms of 'representable functors'.

Definition 4.3.10. Let $F : \mathbf{C} \to \mathbf{D}$ and $G : \mathbf{D} \to \mathbf{C}$ be $\mathbf{V}$-functors. F is said to be left adjoint to G if there exists a $\mathbf{V}$-natural isomorphism $\varphi_{A,B} : Hom_{\mathbf{D}}[F(A), B] \cong Hom_{\mathbf{C}}[A, G(B)]$.

It may suffice here to substantiate the claim that the structure of a 'base category' $\mathbf{V}$ is transported via 'Hom functors'. We refer to the book [Kelly, 1982] on *Basic Concepts of Enriched Category Theory* for further reading, and we give examples below of bits and pieces of enriched category theory which have surfaced in the mathematics of computer science (Section 8.2).

4.3.6 *2-Categories* are an example of enriched categories which deserve special attention.

Axel Poigné

Definition 4.3.11. *2-Categories* are **Cat**-enriched categories where the monoidal structure on **Cat** is the product structure. Explicitly, a 2-category **C** has Hom objects $\mathbf{C}[A, B]$ as categories, composition being functors $* : \mathbf{C}[B, C] \times \mathbf{C}[A, B] \to \mathbf{C}[A, C]$ and a unit functor $I : \mathbf{1} \to \mathbf{C}[A, A]$.

The objects of a 2-category **C** are called *0-cells*, the objects of any Hom category $\mathbf{C}[A, B]$ are called *1-cells*, and the morphisms of $\mathbf{C}[A, B]$ are called *2-cells*.

The archetypal 2-category is **cat** the category of small categories where $\mathbf{cat}[A, B]$ is the functor category, so 0-cells are categories, 1-cells are functors, and 2-cells are natural transformations. Of course, **Cat** has the same structure, but care is needed because **Cat** is not a large category.

A special case is obtained if all categories $C[A, B]$ are partial orders. This subclass of 2-categories corresponds exactly to the **Pos**-enriched categories.

2-Categories have been used in computer science for the analysis of rewriting [Rydeheard and Stell, 1987; Power, 1989]. As an example consider the language of 'procedures with higher types' (cf. Section 2.3)

Types
  - 1 is a type
  - $A \times B$ is a type if A and B are types
  - $A \Rightarrow B$ is a type if A and B are types

Functional terms
  - $\langle x^A \rangle \langle \rangle : A \to 1$
  - $\langle p \rangle x^B : A \to B$ if x^B occurs in $\langle p \rangle : A$
  - $\langle p \rangle \langle M, N \rangle : A \to B \times C$ if $\langle p \rangle M : A \to B$ and $\langle p \rangle N : A \to C$
  - $\langle p \rangle (f \cdot M) : A \to C$ if $\langle p \rangle M : A \to B$ and $f : B \to C$
  - $\langle p \rangle (\lambda \langle q \rangle M) : A \to (B \Rightarrow C)$ if $\langle \langle p \rangle, \langle q \rangle \rangle M : A \times B \to C$
  - $\langle p \rangle (MN) : A \to C$ if $\langle p \rangle M : A \to (B \Rightarrow C)$ and $\langle p \rangle N : A \to B$

where only the following axioms hold:

$(\beta \cdot)$ $\qquad \langle p \rangle ((\langle q \rangle M) \cdot N) = \langle p \rangle M[\langle q \rangle / N]$
$(\eta \cdot)$ $\qquad \langle p \rangle (f \cdot \langle p \rangle) = f$

plus axioms for compatibility with the structure, e.g.

$$\frac{\langle p \rangle M = \langle p \rangle N \qquad f = g}{\langle p \rangle (f \cdot M) = \langle p \rangle (g \cdot N))}$$

Define *reduction* or *rewriting* by

(π) $\qquad \langle p \rangle \langle first \cdot M, second \cdot M \rangle \leq \langle p \rangle M$

where $first = \langle x^A, y^B \rangle \cdot x^A$ and $second = \langle x^A, y^B \rangle \cdot y^B$

(τ) $f \leq \langle p \rangle \langle \rangle$ where $f : A \to 1$

(β) $\langle p \rangle ((\lambda \langle q \rangle M) N) \leq \langle p \rangle M[\langle q \rangle / N]$

(η) $\langle p \rangle (\lambda \langle q \rangle (M \langle q \rangle)) \leq \langle p \rangle M$

if no variable of the parameter list $\langle q \rangle$ occurs free in M.

plus axioms for reflexivity, antisymmetry and transitivity of reduction, and for compatibility with the structure, e.g.

$$\frac{\langle p \rangle M \ \leq \ \langle p \rangle N \qquad f \leq g}{\langle p \rangle (f \cdot M) \ \leq \ \langle p \rangle (g \cdot N))}$$

These data define a 2-category with 0-cells being the types, 1-cells being $[f] : A \to B$ where $[f] = \{g \mid f = g\}$, and 2-cells being determined by the partial order $f \leq g$. The rules for compatibility ensure that the composition $*$ is well defined.

By this reformalization we gain powerful tools such as results about 'pasting', which allow us to prove 'confluence' or 'Church–Rosser' results [Power, 1989], which we cannot present here due to lack of space (in fact, this has already been used to prove certain coherence results in higher type λ-calculi with inheritance [Curien and Ghelli, 1990], and there are prospects to implement completion algorithms based on this framework).

Similarly, algebraic term rewriting can be given a 2-categorical interpretation where the language is stripped off the higher type part (cf. [Rydeheard and Stell, 1987]).

For convenience of presentation, reduction is specified as a partial order. The reduction rules could be named, e.g.

$$\beta(\langle p \rangle, \lambda \langle q \rangle M, N) : \langle p \rangle ((\lambda \langle q \rangle M) N) \to \langle p \rangle M[\langle q \rangle / N],$$

and closed under composition to give proper 2-cells. This gives an additional handle for termination conditions in that 'reduction orders' are defined with regard to such 2-cells or *reduction proofs* and not with regard to terms. This technique has already been used in [Bachmayr *et al.*, 1986]; not of course in the 2-categorical framework.

4.3.7 Category theoreticians have naturally not been interested in 2-categories because of applications with regard to rewriting. 2-Categories offer once again a new, rather syntactic perspective of category theory. Let us consider an example:

According to Section 3.5.6 adjunctions can be specified by natural transformation $\eta : 1_{\mathbf{C}} \to G * F$ and $\varepsilon : F * G \to 1_{\mathbf{D}}$, such that where $F : \mathbf{C} \to \mathbf{D}$ and $G : \mathbf{D} \to \mathbf{C}$ are functors. The definition works in every 2-category.

Definition 4.3.12. Let **C** be a 2-category. Define an *adjunction of 1-cells* $f : C \to D$ and $g : D \to C$ to be specified by 2-cells $\eta : 1_C \to g * f$ and $\varepsilon : f * g \to 1_D$ such that $(f * \eta) \circ (\varepsilon * f) = f$ and $(\eta * g) \circ (g * \varepsilon) = g$.

Adjunctions then give rise to pictures of the form

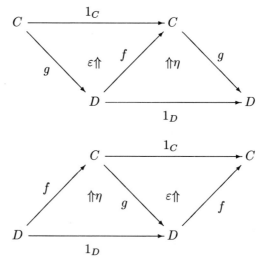

which may be pasted together to give 'directed surfaces' (which is an intuition exploited by Vaughan Pratt in new work on 'Modelling concurrency with geometry' [Pratt, 1991]).

Starting with this definition one can rephrase all the basic concepts of category theory in a rather abstract way, and, of course, we may consider enriched 2-categories, and so on. I have to admit that I am not aware of applications of such 'formal category theory' in computer science (except maybe in Curien's work [Curien, 1990]).

4.3.8 The perspective of 2-categories gives additional insight with regard to ordinary (**Set**-enriched) category theory. Consider for instance the category **Dcpo** of directed complete posets. We have seen that **Dcpo** does not have coproducts (Section 2.2.2). However, separated sums have the following property:

Observation 4.3.13. Let $X +_\perp Y$ be the separated sum of the dcpo's X and Y with injections $inl : X \to X +_\perp Y$ and $inr : Y \to X +_\perp Y$. For all continuous functions $f : X \to Z$ and $g : Y \to Z$ there exists a unique (!) continuous function $[f, g] : X +_\perp Y \to Z$ such that

- $f \sqsubseteq [f, g] \circ inl$ and $g \sqsubseteq [f, g] \circ inr$, and

- $[f, g] \sqsubseteq h$ for all $h : X +_\perp Y \to Z$ with $f \sqsubseteq h \circ inl$ and $g \sqsubseteq h \circ inr$.

This is a special instance of the following notion.

Definition 4.3.14. A *lax coproduct* of 0-cells A, B (in a 2-category) is defined to consist of a 0-cell $A \oplus B$ and two 1-cells $inl : A \to A \oplus B, inr : B \to A \oplus B$ with the universal property that for all 1-cells $f : A \to C$, $g : B \to C$ there exists a 1-cell $[f, g] : A \oplus B \to C$ and 2-cells $inl_{f,g} : [f, g] * inl \to f$, $inr_{f,g} : [f, g] * inr \to g$ such that given any other 1-cell $h : A \oplus B \to C$ and 2-cells $\sigma : h * inl \to f$, $\tau : h * inr \to g$ there exists a unique 2-cell $[\sigma, \tau] : [f, g] \to h$ such that $[\sigma, \tau] \circ inl_{f,g} = \sigma$ and $[\sigma, \tau] \circ inr_{f,g} = \tau$.

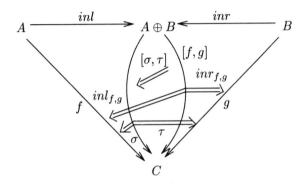

If the direction of the 2-cells is reversed we speak of an *oplax coproduct*.

The example may indicate how to turn all basic concepts of category theory into (op)lax concepts, where separated sums give some motivation as to why these concepts may be useful, though a thorough investigation in this direction still seems to be missing.

5 Data structures

Products and disjoint sums are basic data structures provided by any reasonable programming language. Other principles of data structuring such as restriction to subsets and factorization are as fundamental for formal reasoning, though in general not supported as a programming construct because of difficulties with implementation. The difficulty, however, is not an inherent one, since the data structures can be accommodated in a functional environment.

5.1 Subtypes

5.1.1 Subsets are typically introduced by comprehension;

$$\{a \mid a \in A, \varphi(a)\}$$

denotes the 'subset of elements of A which satisfy the predicate φ'.

This subset is equivalently characterized by the property that functions $f : X \to A$ such that $f \models \varphi$ are in one-to-one correspondence to functions $f : X \to \{a \mid a \in A, \varphi(a)\}$. The notation $f \models \varphi$ states that the function $f : X \to A$ *satisfies* a predicate φ *of type* A, meaning that $\varphi(f(x))$ for all $x \in X$. Note that these functions are closed under composition in that $g \circ f \models \varphi$ if $g \models \varphi$, in other words, every predicate φ defines a contravariant functor $\varphi : \mathbf{Set}^{\mathrm{op}} \to \mathbf{Set}$ which maps objects A to sets $\{f : X \to A \mid f \models \varphi\}$ of functions, and morphisms $g : X \to Y$ to functions $g \circ \_ : \{f : Y \to A \mid f \models \varphi\} \to \{f : X \to A \mid f \models \varphi\}, f \mapsto g \circ f$.

5.1.2 By abstraction, a contravariant functor $\varphi : \mathbf{C}^{\mathrm{op}} \to \mathbf{Set}$ may be taken as a specification of a subtype in a category $\mathbf{C}$, the specified subtype being an object $(A \mid \varphi)$ of $\mathbf{C}$ which represents the functor (if it exists), i.e. $\varphi \cong \mathbf{C}[\_, (A \mid \varphi)]$ naturally, or equivalently;

Definition 5.1.1. Let $\varphi : \mathbf{C}^{\mathrm{op}} \to \mathbf{Set}$ be a functor.

$$(\langle A \mid \varphi \rangle, \iota_\varphi : \langle A \mid \varphi \rangle \to A)$$

is called the *subtype* of A *determined* by φ if

- $\iota_\varphi \models \varphi$, and
- for all $f : X \to A$ such that $f \models \varphi$ there exists a unique

$$h : X \to \langle A \mid \varphi \rangle$$

such that

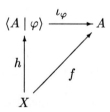

commutes.

Note that $\iota_\varphi : \langle A \mid \varphi \rangle \to A$ is a monomorphism, since $f \circ \iota_\varphi$ and $g \circ \iota_\varphi$ satisfy φ.

5.1.3 Equality types

The equality type $Eq(g, h)$ of two functions $g, h : A \to B$ consists of the subset of elements $a \in A$ such that $f(a) = g(a)$, or, equivalently, is specified by the natural isomorphism $\mathbf{Set}[X, Eq(g, h)] \cong \{x : X \to A \mid g \circ x = h \circ x\}$. Textbooks on category theory give an explicit definition of equality types for an arbitrary category $\mathbf{C}$.

Definition 5.1.2. An object $Eq(g, h)$ of $\mathbf{C}$ together with a morphism $\varepsilon : Eq(g, h) \to A$ is called *equalizer* of the morphisms $g, h : A \to B$ if

- $g \circ \varepsilon = h \circ \varepsilon$, and

- if for all objects X and all morphisms $f : X \to A$ there exists a unique morphism $\langle f \rangle : X \to Eq(g, h)$ such that the diagram

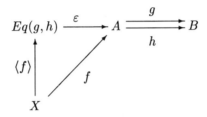

commutes.

We propose the notation $(A \mid f = g)$ for equalizer objects which is consistent with our notation for subobjects.

Equalizers increase expressivity considerably as seen by the following examples. (Exercise: compute the subsequent definitions in $\mathbf{Set}$.)

- The *graph of a function* $f : A \to B$ is defined by

$$graph(f) = (A \Pi B \mid f \circ first = second).$$

- The inverse image $f^{-1}(m)$ of a subtype (subobject) $m : X \subseteq B$ is defined by $f^{-1}(m) = (A \Pi X \mid f \circ first = m \circ second)$.

- If relations $R : A \to B$ are represented by a pair $dom_R : R \to A$ and $cod_R : R \to B$, composition of relations $R : A \to B$, $S : B \to C$ is defined by $S \circ R = (R \Pi S \mid cod_R \circ first = dom_S \circ second)$ with $dom_{S \circ R} = dom_R \circ first$ and $cod_{S \circ R} = cod_S \circ second$.

  This defines a category $\mathbf{Rel}$ with Hom sets $\mathbf{Rel}[A, B]$ of relations $R : A \to B$.

- *Partial functions* $f : A \to B$ can be introduced as total functions on a subtype of A, i.e. a partial function $f : A \to B$ is determined by a monomorphism $m : dom(f) \to A$ and a morphism $f : dom(f) \to B$. This being a relation, composition is that of relations. Note that the domain of the composed relation is a monomorphism (exercise), hence composition of partial functions is well-defined.

- Let O and M be sets (of objects and morphisms) with functions $dom, cod : M \to O$. The type of 'composable morphisms' is defined by $(M \times M \mid cod \circ first = dom \circ second)$. 'Composition' becomes a (total) function

$$\circ : (M \times M \mid cod \circ first = dom \circ second) \to M$$

which provides a natural interpretation for the *conditional declaration*

$$f, g : M, dom(f) = cod(g) \Rightarrow g \circ f : M.$$

Here are further examples:

- Equalizers in $\mathbf{Alg}(\Sigma)$ are basically defined as in $\mathbf{Set}$. Given homomorphisms $g, h : A \to B$ the equalizer has a carrier $b^{Eq} = Eq(g_b, h_b)$ of sort b where $g_b, h_b : b^A \to b^B \in \mathbf{Set}$ are the b-components of the homomorphisms. The operations are induced by the diagram

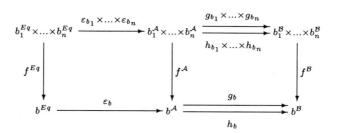

Similarly, equalizers in 'algebraic categories' such as $\mathbf{Mon}$, $\mathbf{Graph}$, $\mathbf{Cat}$ etc. can be computed in the underlying categories. In fact, this is a component of 'algebraicity' (see Section 8.3.3).

- Equalizers $f, g : (X, \sqsubseteq) \to (Y, \sqsubseteq)$ in the category of posets are obtained as equalizers $\{x \in X \mid f(x) = g(x)\}$ of underlying set functions, with order being inherited from $(X, \sqsubseteq)$.

- The category $\mathbf{Dcpo}$ fails to have equalizers since the continuous functions are not strict on the least element, e.g. the two functions $\bot, \top : \{\bot\} \to \{\bot, \top\}$ do not have an equalizer.

5.1.4 An exercise on representable functors: toposes

The category of sets has the prominent property that all subsets are equationally defined: subsets $X \subseteq A$ are in bijective correspondence to characteristic functions $\chi : A \to 2$, the subsets being determined by a characteristic function $\chi^{-1}(1) = (A \mid \chi = \mathit{true} \circ \langle \rangle_A)$ where $\mathit{true} : 1 \to 2$.

It is one of the deep insights of Lawvere that this property is characteristic for 'set-like' categories.

Definition 5.1.3.

- An object Ω together with a morphism $\mathit{true} : 1 \to \Omega$ is called *subobject classifier* if for every subobject $m : X \to A$ there exists a *characteristic morphism* $\chi_m : A \to \Omega$ such that $(X, m : X \to A)$ is an equalizer of χ_m and $\mathit{true} \circ \langle \rangle_A$.

- A Cartesian closed category with equalizers and a subobject classifier is called a *topos*.

The subobject classifier might as well be defined in terms of a representable functor.

Proposition 5.1.4. *Let* $\mathbf{C}$ *be a category with finite limits. Define a functor* $Sub : \mathbf{C}^{\mathrm{op}} \to \mathbf{Set}$ *by*
$Sub(A) = \{m : X \to A \mid m \ mono\}$ *and by* $Sub(f)(m) = f^{-1}(m)$.
Sub is representable iff $\mathbf{C}$ *has a subobject classifier.*

An alternative characterization of a topos states that relations can be presented as functions into the power type.

Proposition 5.1.5. *Let* $\mathbf{C}$ *be a category with finite limits.* $\mathbf{C}$ *is a topos iff* $\mathbf{Rel}[\_, A]$ *is presentable for every* A. *The presented object* $\mathbf{C}[\_, P(A)] \cong \mathbf{Rel}[\_, A]$ *is called* **power object**.

Proof. [Exercise] The subobject classifier is given by $P(1)$ and the power object by $A \Rightarrow \Omega$ (see [Goldblatt, 1979], [Johnstone, 1977]). ∎

It may be difficult to fully appreciate this definition, and here is not the right place to attempt a serious discussion which will be found elsewhere in these volumes. However, the definition of toposes by Lawvere and Tierney should be considered as one of the most fundamental insights of category theory.

5.2   Limits

5.2.1   Pullbacks

Limits may be thought of as equationally defined subtypes of products though of a slightly more general nature: instead of

$$\{x \mid x \in A \times B, f \circ first(x) = g \circ second(x)\} \qquad (*)$$

one may use

$$\{(a,b) \mid a \in A, b \in B, f(a) = g(b)\} \qquad (**)$$

which appears only as a shift of notation in set theory. However, existence of products is assumed a priori in $(*)$ but not in $(**)$. We like to refer to the latter definition as *conditional pairing* since, conceptually, the pairing operation is restricted by an equational condition. Conditional pairing is called a *pullback* in category theory.

Definition 5.2.1. An object $Pb(f,g)$ together with two morphisms $p : Pb(f,g) \to A$, $q : Pb(f,g) \to B$ is called *pullback* of the morphisms $f : A \to C$, $g : B \to C$ if $f \circ p = g \circ q$, and if for all objects D and for all morphisms $p' : D \to A$ and $q' : D \to B$ such that $f \circ p' = g \circ q'$, there exists a unique morphism $\langle p', q' \rangle : D \to Pb(f,g)$ such that the diagram

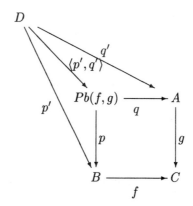

commutes.

The overloading of notation is deliberate since binary products are a special case where the equational condition is trivial.

Proposition 5.2.2. *The pullback*

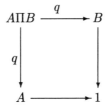

determines a binary product (Exercise).

Thus the subtle difference between the definition $(*)$ and $(**)$ disappears in the presence of terminal objects.

Proposition 5.2.3.

- *A category has pullbacks if it has binary products and equalizers.*

- *A category has finite products and equalizers if it has pullbacks and terminal objects.*

Proof.   • The equalizer of $f \circ first, g \circ second : A \times B \to C$ defines a pullback with 'projections' $first \circ eq$ and $second \circ eq$ (this is $(*)$ in the categorical language).

- In order to obtain an equalizer, take the pullback of $\langle 1_A, f \rangle : A \to A \times B$ and $\langle 1_A, g \rangle : A \to A \times B$ (in terms of set theory $\{(x, y) \mid x \in A, y \in A, x = y, f(x) = g(y)\}$).   ■

Definition 5.2.4. A category which has finite products and equalizers, or, equivalently, pullbacks and a terminal object is said to have *finite limits* or is called a *Cartesian category.*

5.2.2   Limits in general

Limits are just a generalization of conditional pairing to *conditional tupling.* However, a specific format is used for equational restriction. In set theoretic terms, conditions are restricted to equations of the form $f(x) = y$ without losing expressivity, e.g.

$$\{(a, b, c) \mid a \in A, b \in B, c \in C, f(a) = c, g(b) = c\}$$

is isomorphic to

$$\{(a, b) \mid a \in A, b \in B, f(a) = g(b)\}.$$

Similarly

$$\{(a,b) \mid a \in A, b \in B, f(a) = b, g(a) = b\}$$

is isomorphic to the equalizer

$$\{a \mid a \in A, f(a) = g(a)\}.$$

The formatting of conditions supports a uniform formalization of conditional tupling in the categorical language. If a tuple $\langle x_i \mid i \in I \rangle$ is restricted by the equality $x_i = f(y_i)$ this may be expressed by a commuting diagram

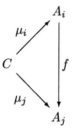

where C is supposed to be the type of tuples and $\mu_i : C \to A_i$ and $\mu_j : C \to A_j$ are projections to the components. Then every such conditional restriction is fully specified by the morphism $f : A_i \to A_j$, and a limit type is specified by a set of objects A_i and a set of morphisms $f : A_i \to A_j$ between these objects. In fact, we can assume without restriction of generality that these data are given by a functor $D : \mathbf{J} \to \mathbf{C}$.

The data type which is specified by such a functor $D : \mathbf{J} \to \mathbf{C}$ should then have the following properties:

(i) it consists of an object *lim D* and (projection) morphisms $\varepsilon_i : lim\ D \to A_i$ for all objects i of $\mathbf{J}$,

(ii) these data *satisfy* the specification $D : \mathbf{J} \to \mathbf{C}$, or are called a *cone*, if the diagram

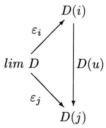

commutes for all morphisms $u : i \to j$ in $\mathbf{J}$.

(iii) for every other object X with morphisms $\mu_i : X \to A_i$ such that

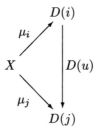

(i.e. which satisfies the specification) there exists a unique (abstraction) morphism $\langle \mu \rangle : X \to lim\ D$ such that

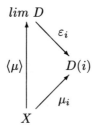

commutes.

Definition 5.2.5.

- The object $lim\ D$ together with morphisms $\varepsilon_i : lim\ D \to A_i$ which satisfy the conditions (i) – (iii) is called a limit of the *diagram D* : **J** → **C**.

- A category *has (small, finite) limits* if a limit exists for all (small, finite) diagrams, where a diagram $D : \mathbf{J} \to \mathbf{C}$ is small if **J** is a small resp. finite category.

The conditions (i) – (iii) naturally cover the bijective correspondence of, in set theoretical language, functions $f : X \to \{\langle x_i \rangle \mid \varphi_D(\langle x_i \rangle)\}$ with families $\mu = (\mu_i : X \to D(i) \mid i \in I)$ of functions such that $\mu \models \varphi_D$ where the predicate is given by $\varphi_D(\langle x_i \rangle)$ iff $x_i \in D(i)$ and $D(u)(x_i) = x_j$ for all objects i, j and morphisms $u : i \to j$ of **J**. This correspondence characterizes the set $\{\langle x_i \rangle \mid \varphi_D(\langle x_i \rangle)\}$ up to isomorphism.

We rephrase the specific instances encountered so far:

- An equalizer is specified by a pair $f, g : A \to B$ of morphisms of **C**. If we define the category **J** to consist of two objects 0 and 1 and two morphisms $u, v : 0 \to 1$ (except for identities), then every such pair of morphisms defines a functor $D : \mathbf{J} \to \mathbf{C}$ by $D(0) = A, D(1) = B$, and $D(u) = f, D(v) = g$, and vice versa. Condition (ii) then implies commutativity of the diagram

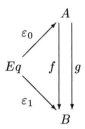

which is equivalent to $f \circ \varepsilon_0 = g \circ \varepsilon_0$.

- For pullbacks **J** is given by the diagram

- Products are specified by diagrams $D : \mathbf{J} \to \mathbf{C}$ where **J** is a discrete category (with identities being the only morphisms). Note that $D : \varnothing \to \mathbf{C}$ specifies the terminal object.

Here is another example:

- The type A^ω of infinite sequences of 'elements' of a given type A is specified by the diagram

$$1 \xleftarrow{\;\langle\rangle_A\;} A \xleftarrow{\;first\;} (A \times A) \times A \longleftarrow \cdots \longleftarrow A^n \xleftarrow{\;first\;} A^n \times A \longleftarrow \cdots$$

5.2.3   Objects X together with the morphisms $\mu_i : X \to D(i)$ define a notion of observation, in the terminology of Section 3. A limit is, of course, fully separated presuming that all definitions are put into proper categorical terms:

Let the *diagonal functor* $\Delta : \mathbf{C} \to \mathbf{C}^{\mathbf{J}}$ map objects A to functors $\Delta(A) : \mathbf{J} \to \mathbf{C}$ such that $\Delta(A)(i) = A$, and $\Delta(A)(u) = 1_A$. Then every cone $(X, \mu_i : X \to D(i))$ defines a natural transformation $\mu : \Delta(X) \to D$.

Proposition 5.2.6.

- A limit defines a universal morphism $(lim\ D, \varepsilon : \Delta(lim\ D) \to D)$ from Δ to $lim\ D$.

- A category **C** has limits of type **J** if and only if the diagonal functor $\Delta : \mathbf{C} \to \mathbf{C}^{\mathbf{J}}$ has a right adjoint.

5.2.4 All limits can be expressed in terms of products and equalizers provided these exist. The intuitive idea is to put all the conditions $D(u)(x_i) =
x_j$ in parallel : $\langle D(u)(x_{u^-})\rangle_u = \langle x_{u^+}\rangle_u$ where $u : u^- \to u^+$ ranges over all
the morphisms in $\mathbf{J}$. The equality is represented by the diagram

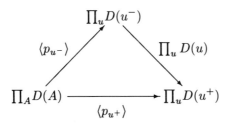

where A and u range over the objects and morphisms of $\mathbf{J}$. The morphisms
$\langle p_{u^-}\rangle$ resp. $\langle p_{u^+}\rangle$ are induced by product property from the projections
$p_{u^-} : \prod_A D(A) \to D(u^-)$ resp. $p_{u^+} : \prod_A D(A) \to D(u^+)$.

Proposition 5.2.7. *Let* $D : \mathbf{J} \to \mathbf{C}$ *be a diagram. Then the equalizer*
(Eq, ε) *of* $\prod_u D(u) \circ \langle p_{u^-}\rangle$ *and* $\langle p_{u^+}\rangle$ *determines the limit* $(Eq, (p_A \circ \varepsilon \mid
A \in \mathbf{D}))$ *where the morphisms* $p_A : \prod_A D(A) \to D(A)$ *are the projection.*

Proof. Exercise (see [MacLane, 1971], V.2 Th.1). ∎

The proposition substantiates the claim above that limits are basically
equationally defined subtypes of products.

5.2.5 An exercise in pulling back — internal categories
We define *internal categories* in an arbitrary category $\mathbf{C}$ with finite limits as
a starting point for another generalization of category theory. The exercise
demonstrates the gain in expressivity obtained by limits.

The basic observation is that the composition $\circ : Mor(\mathbf{C}) \times Mor(\mathbf{C}) \to
Mor(\mathbf{C})$ of a category can be defined as a total function on the set $\{(f, g) \in
Mor(\mathbf{C}) \times Mor(\mathbf{C}) \mid cod(f) = dom(g)\}$, which is obtained as the pull-
back $Pb(cod, dom)$ of the mappings $dom : Mor(\mathbf{C}) \to Ob(\mathbf{C})$ and $cod :
Mor(\mathbf{C}) \to Ob(\mathbf{C})$. The following definitions are straightforward general-
izations.

Definition 5.2.8.

- Let $\mathbf{E}$ be a category with finite limits. Define an *internal graph* G to
  consist of two morphisms $dom, cod : E \to N$. A morphism $h : G \to G'$
  of such graphs is given by a pair $h_E : E \to E'$ and $h_N : N \to N'$ such

that $h_N \circ dom = dom' \circ h_E$ and $h_N \circ cod = cod' \circ h_E$. This defines a category **Graph(E)** of internal graphs in **E**.

- Let G_1 and G_1 be graphs, both with nodes N. Define the internal graph $G_2 \otimes G_1 = (E_2 \otimes E_1, N, dom_1 \circ p, cod_2 \circ q)$ by the pullback

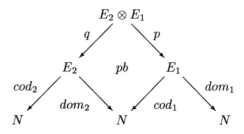

Definition 5.2.9. An *internal category* C in **E** consists of the following data

- an internal graph (M, O, dom, cod)
- a morphism $* : M \otimes M \to M$, and
- a morphism $e : O \to M$

such that the diagrams

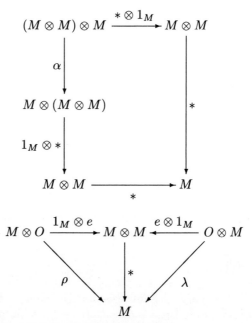

commute (α, λ, ρ are the obvious isomorphisms).

(The dual internal category $\mathbf{C}^{\mathrm{op}}$ is obtained if domain and codomain mappings are exchanged. We leave it as an exercise to define the product $C \times D$ of internal categories.)

- An *internal functor* $F : C_1 \to C_2$ is a homomorphism of the underlying graphs such that the diagrams

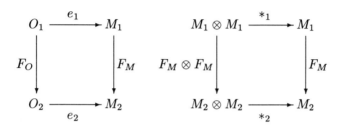

  commute.

- An *internal natural transformation* $\sigma : F \to G : C_1 \to C_2$ is given by a morphism $\sigma : O_1 \to M_2$ in E such that the diagrams

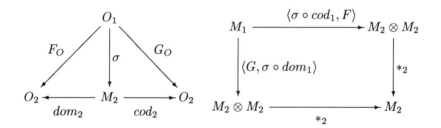

commute (the arrows into the pullbacks are induced).

(The reader should analyse the definition for the case $\mathbf{E} = \mathbf{Set}$.)

As with enriched category theory, Hom functors will be the handle to internalize the other concepts of category theory. But how to define 'functors' $F : \mathbf{C} \to \mathbf{E}$ where $\mathbf{C}$ is an internal category ? Let us return to the very first discussion of functors where application $f \cdot a$ of 'functions' to 'arguments' were used to define functors $\mathbf{C} \to \mathbf{Set}$ (Section 1.4.1), and let us rephrase the argument in the language of categories.

Definition 5.2.10. Let $\mathbf{C} = (M, O, dom, cod)$ be an internal category in $\mathbf{E}$. A 'functor', often called *internal presheaf*, $D : \mathbf{C} \to \mathbf{E}$ consists of an object S and a morphism $type : S \to O$ and a morphism $\delta : M \otimes S \to S$ such that the diagrams

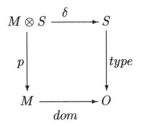

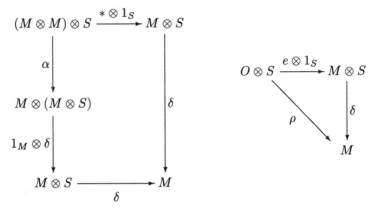

commute where the obvious pullbacks are used to match domain and type information, e.g. $M \otimes F$ is the pullback of $dom : M \to O$ and $type : F \to O$, $p : M \otimes F \to M$ is the projection to the first component.

Much of the theory should appear familiar if one notes that internal categories are basically monoids, while the definitions of internal functors and internal presheaves are those of monoid homomorphisms and (monoid) automata. Hence we leave it as an exercise to define the notion of a 'morphism of presheaves' along the lines of homomorphisms of automata.

Definition 5.2.11. The *internal Hom functor* is defined as a presheaf $Hom_{\mathbf{C}} : \mathbf{C}^{\mathrm{op}} \times \mathbf{C} \to \mathbf{E}$ with $\langle cod, dom \rangle : M \to O\Pi O$ and $\delta : (M\Pi M) \otimes M \to M$ as defined in

$$
\begin{array}{ccc}
(M\Pi M) \otimes M & \overset{\delta}{\dashrightarrow} & M \\
{\scriptstyle \langle p' \circ p,\, q' \otimes 1_M \rangle} \downarrow & & \uparrow {\scriptstyle *} \\
M \otimes (M \otimes M) & \underset{1_M \otimes *}{\longrightarrow} & M \otimes M
\end{array}
$$

where

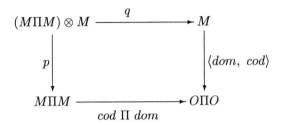

is a pullback, p', q' being the projections out of the product $M\Pi M$. Informally, $\delta = \lambda fgh.h \circ g \circ f$.

Adjunctions are characterized by the natural isomorphism $\mathbf{D}[F(A), B] \cong \mathbf{C}[A, G(B)]$ which should be written more precisely $\mathbf{D}[\_, \_] \circ (F^{\mathrm{op}} \times 1_{\mathbf{D}}) \cong \mathbf{C}[\_, \_] \circ (1_{\mathbf{C}^{\mathrm{op}}} \times G)$. Thus an internal version requires composition of an internal functor with a presheaf.

Definition 5.2.12. Let $D_2 : \mathbf{C}_2 \to \mathbf{E}$ be a presheaf, and let $H : \mathbf{C}_1 \to \mathbf{C}_2$ be an internal functor. Define a presheaf $D_1 : \mathbf{C}_1 \to \mathbf{E}$ by (the defined morphisms are dotted)

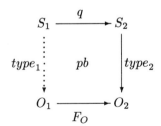

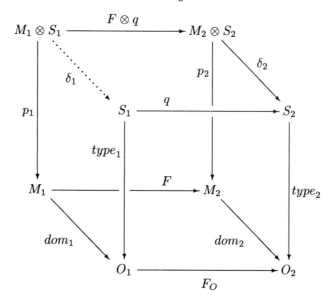

We use the notation $F * D = D_1 : \mathbf{C}_1 \to \mathbf{E}$.

All the definitions so far, come together in the notion of an internal adjunction.

Definition 5.2.13. An *internal adjunction* consists of internal functors $F : \mathbf{C} \to \mathbf{D}$ and $G : \mathbf{D} \to \mathbf{C}$ and an isomorphism $\varphi : (F^{\mathrm{op}} \times 1_{\mathbf{D}}) * Hom_{\mathbf{D}} \cong (1_{\mathbf{C}^{\mathrm{op}}} \times G) * Hom_{\mathbf{C}}$ of presheaves.

With the basic notion of internal adjunction one can start to redevelop the whole of category theory in an internal version. This programme is an alternative to enriched category theory in that a category theory is developed which does not depend on set theory. However esoteric internal category theory may appear to a computer scientist, the theory has found substantial applications in computer science providing a sound semantics for *second* (and higher) *order λ-calculus* which also comes under the name of *polymorphic λ-calculus* [Girard, 1972; Reynolds, 1974; Pitts, 1987; Asperti and Martini, 1989], cf. also Section 9.4.3).

5.3   Colimits

5.3.1   Colimits is the dual concept to limits, which is not a very helpful remark, though it provides the general definition and the dual 'equivalence results', where the duals of pullbacks are called *pushouts*, and the duals of equalizers are called *coequalizers*.

Definition 5.3.1. Let $D : \mathbf{J} \to \mathbf{C}$ be a diagram. A *colimit* of D consists of an object *colim D* of **C** and a morphism $\eta_i : D(i) \to colim\ D$ in **C** for each object i of **J** which satisfy the following conditions:

(i) The diagram

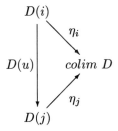

commutes for every morphism $u : i \to j$ in **J**.

(ii) For every *cocone* (Y, τ) (i.e. an object Y of **C** with morphisms $\tau_i : D(i) \to Y$ such that

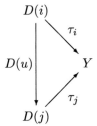

commutes) there exists a unique morphism $[\tau] : colim\ D \to Y$ such that

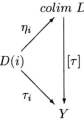

commutes.

Remark 5.3.2. A *colimit* is a universal morphism $(colim\ D, \eta : D \to \Delta(colim\ D))$ from D to the diagonal $\Delta : \mathbf{C} \to \mathbf{C}^{\mathbf{J}}$.

5.3.2 Quotients

As in the case of limits there are some basic types of colimits which are of specific interest: coproducts (which have already been discussed), coequalizers, and pushouts and a new concept, filtered colimits.

Axel Poigné

Coequalizers are roughly the equivalent to factorization by an equivalence (or rather congruence) relation $E \subseteq A \times A$. An equivalence relation states that certain elements are considered as equal, which is naturally expressed by the axioms of reflexivity, symmetry and transitivity. Factorization identifies the elements considered as equal; each set $[a] := \{a' \in A \mid (a, a') \in E\}$ of equal elements in A has one representative in the quotient $A_{/E}$, usually the *equivalence class* $[a]$. If A is carrier of an algebra, let's say a monoid, the equivalence relation must be compatible with the operators: $(a * b, a' * b') \in E$ if $(a, a'), (b, b') \in E$. Then the operator is well defined on the equivalence classes in that $[a] * [b] := [a * b]$ is independent of the choice of a and b. Equivalence relations are typically *presented* by a relation $R \subseteq A \times A$, the equivalence relation presented being the least equivalence relation $R^* \subseteq A \times A$ such that $R \subseteq R^*$.

Every function $h : A \to B$ determines an equivalence relation $ker(h) := \{(a, a') \in A \times A \mid h(a) = h(a')\}$, the *kernel* of h. A function $h : A \to B$ *satisfies* an (equivalence) *relation* $R \subseteq A \times A$ if $ker(h) \subseteq R$. The quotient $A_{/R^*}$ is *freely generated* in that for every function $h : A \to B$ which satisfies R there exists a unique function $[h] : A_{/R^*} \to B$ such that

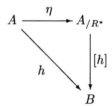

commutes where $\eta(a) = [a]$.

Let us justify the terminology of 'free generation': if the relation R is presented by a pair of functions $dom_R, cod_R : R \to A$, then $f : A \to B$ satisfies R if $f \circ dom_R = f \circ cod_R$, and the quotient is obtained as a colimit, namely as a dual to equalizers.

Definition 5.3.3. An object $Coeq(f, g)$ together with a morphism $\eta : B \to Coeq(g, h)$ is called *coequalizer* of the morphisms $f, g : A \to B$ if

- $\eta \circ f = \eta \circ g$, and

- if for all objects Y and all morphisms $h : B \to Y$ there exists a unique morphism $[h] : Coeq(f, g) \to Y$ such that the diagram

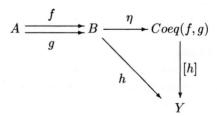

commutes.

Fact 5.3.4. $Coeq(f, g)$ is freely generated by $(\eta, \eta \circ f) : (f, g) \to \Delta(Coeq(f, g))$ where $\Delta : \mathbf{C} \to \mathbf{C}^{\Rightarrow}$.

We discuss some (hopefully) familiar examples:

- Coequalizers in categories of algebras are defined similarly to those in sets. For instance, every monoid homomorphism $h : \mathcal{M}_1 \to \mathcal{M}_2$ has a kernel $ker(h) = \{(x, y) \mid h(x) = h(y)\} \subseteq M \times M$ which is a congruence relation, i.e. $(x * x', y * y') \in ker(h)$ if $(x, y) \in ker(h)$ and $(x', y') \in ker(h)$, and $(e, e') \in ker(h)$. Congruence relations are closed under intersection, hence for every relation $R \subseteq M \times M$ there exists a least congruence R^* such that $R \subseteq R^*$. Now monoid homomorphisms $g, h : \mathcal{M}_1 \to \mathcal{M}_2$ define a relation $R_{g,h} = \{(g(z), h(z)) \mid z \in M_1\}$. The quotient of $(\mathcal{M}_{2/R^*_{g,h}})$ with quotient map $\pi : \mathcal{M}_2 \to \mathcal{M}_{2/R^*_{g,h}}, x \mapsto [x]$ defines a coequalizer.

  The construction is similar for most 'algebraic' categories. (For a more abstract account of the construction cf. Section 8.3.4.)

- A rather different construction is used in the category of partial orders and monotone functions. Let $f : (X, \sqsubseteq) \to (Y, \sqsubseteq)$ be a monotone mapping. Define on $Im(f) = \{f(x) \mid x \in X\}$ a partial order $f(x) \sqsubseteq_{Im(f)} f(y)$ as transitive closure of $\{(f(x), f(y)) \mid x, y \in X\}$. The induced mapping $f^* : (X, \sqsubseteq) \to (Im(f), \sqsubseteq_{Im(f)})$ is monotone.

  Now let $g, h : (X, \sqsubseteq) \to (Y, \sqsubseteq)$ be monotone mappings, and let $CE = \{f : (Y, \sqsubseteq) \to (Z, \sqsubseteq) \in \mathbf{Pos} \mid f \circ g = f \circ h\}$. As the cardinality of $\{Im(f) \mid f : (Y, \sqsubseteq) \to (Z, \sqsubseteq)\}$ is smaller than $\bigcup_{U \subseteq Y} 2^{U \times Y}$, the product $\prod_{f \in CE} Im(f)$ is well defined, and a monotone mapping $\langle f^* \mid f \in CE \rangle : (Y, \sqsubseteq) \to \prod_{f \in CE} Im(f)$ is induced. $\langle f^* \mid f \in CE \rangle^* : (Y, \sqsubseteq) \to Im(\langle f^* \mid f \in CE \rangle)$ is a coequalizer. (For a more abstract account of the construction cf. Section 8.3.4.)

- Let Σ be a signature, and let $t_1, t_2 \in T_\Sigma(X)$ be Σ-terms. A variable assignment $\Theta : X \to T_\Sigma(Y)$ is said to be a *unifier* of (t_1, t_2) if $[\Theta](t_1) = [\Theta](t_2)$. Θ is the *most general unifier* if for all other unifiers $\partial : X \to T_\Sigma(Z)$ there exists a homomorphism $h : T_\Sigma(Y) \to T_\Sigma(Z)$ such that $h \circ \Theta = \partial$.

  Define a category $\mathbf{T}_\Sigma^{op}$ with objects being finite sets X of variables, and with morphisms $\Theta : X \to Y$ being variable assignments $\Theta : X \to T_\Sigma(Y)$ and with composition $\partial \circ \Theta := [\partial] \circ \Theta$ where $[\partial] : T_\Sigma(Y) \to T_\Sigma(Z)$ is the uniquely induced homomorphism. This is a category with finite coproducts given by the (sorted) disjoint sum $X + Y$.

Coequalizers in $\mathbf{T}_\Sigma^{op}$, if they exist, are most general unifiers (of a set of pairs of terms) (see [Rydeheard and Stell, 1987]). However, most general unifiers are not necessarily coequalizers, e.g.

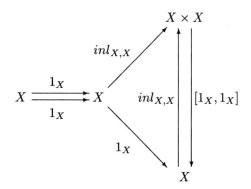

commutes.

(By the way, $\mathbf{T}_\Sigma^{op}$ is closely related to the category $\mathbf{T}_\Sigma$ defined in Section 2.1.4. It is almost dual, except that all isomorphic products are collapsed. This is the original definition of *algebraic theories* to be found in Lawvere's thesis [Lawvere, 1963] for the one-sorted case. Cf. also Section 8.)

5.3.3 Amalgamated sums

Amalgamated sums combine features of unions and disjoint unions. Let B and C be sets with a common subset $A \subseteq B, A \subseteq C$. Define a set $B +_A C := A \cup (B \backslash A + C \backslash A)$ which is the disjoint union except for the specified subset A of which only one copy is taken. Union and disjoint union are obtained as specific instances if A is chosen to be $B \cap C$ resp. $A = \varnothing$.

$B +_A C$ has the characteristic property that two functions $h : B \to Y$ and $k : C \to Y$ which agree on the common subset A can be extended to a function $[h, k] : B +_A C \to Y$ by means of case distinction,

$$[h, k](z) = \begin{cases} h(b) & \text{if } z = (1, b) \\ k(c) & \text{if } z = (2, c) \\ h(a) = k(a) & \text{if } z = a \in A \end{cases}$$

(where $B \backslash A + C \backslash A = (\{1\} \times B \backslash A) \cup (\{2\} \times C \backslash A)$).

The definition of a pushout is a mild generalization in that common parts are specified by functions $f : A \to B$ and $g : A \to C$.

Definition 5.3.5. An object $Po(f, g)$ together with two morphisms $inl : A \to Po(f, g)$, $inr : A \to Po(f, g)$ is called *pushout* of the morphisms

$f : A \to B$, $g : A \to C$ if $inl \circ f = inr \circ g$, and if for all objects D and for all morphisms $h : B \to D$, $k : C \to D$ such that $h \circ f = k \circ g$, there exists a unique $[h, k] : Po(f, g) \to D$ such that the diagram

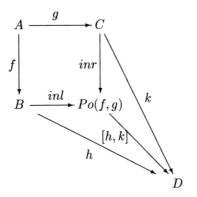

commutes.

While the simple example above conveys the nature of pushouts in the category of sets, the general case does need more attention. A pair of functions $f : A \to B$ and $g : A \to C$ may be seen as specifying an equivalence relation R^* on the disjoint set $B + C$ which is the least equivalence relation containing the relation $R = \{((1, f(a)), (2, g(a))) \mid a \in A\}$.

Fact 5.3.6. The quotient $(A + B)_{/R^*}$ is a pushout with injections $inl(b) = [(1, b)]$ and $inr(b) = [(2, b)]$.

Proof. This follows from the dual of the second proposition in 5.2.1 but an explicit computation may be helpful. The equation $inl \circ f = inr \circ g$ holds as $((1, f(a)), (2, g(a))) \in R \subseteq R^*$. Let $h : B \to Y$ and $k : C \to Y$ be functions. Then $R \subseteq ker([h, k])$ where $[h, k] : B + C \to Y$ is the only function such that $[h, k](1, b) = h(b)$ and $[h, k](2, c) = k(c)$. By Section 5.3.2 there exists a unique function $[h, k]^* : (A + B)_{/R^*} \to Y$ such that $[h, k]^*([z]) = [h, k](z)$. This is quite clearly the only function such that $[h, k]^* \circ inl = h$ and $[h, k]^* \circ inr = k$. ∎

The construction in the category of sets gives the right kind of insight for two major applications of pushouts in computer science.

Example 5.3.7 ([Ehrig and Mahr, 1985]). A *parameterized specification* consists of a pair of specifications (Sp_1, Sp_2) such that Sp_1 is a subspecification of Sp_1, i.e. $B_1 \subseteq B_2$, $\Sigma_1 \subseteq \Sigma_2$, and $E_1 \subseteq E_2$. The specification Sp_1 is called the *formal parameter*. *Parameter passing* updates the formal parameter by means of a *view* to an *actual parameter*, i.e. a specification morphism $H : Sp_1 \to Sp_3$ to give a result specification Sp_4

where the formal parameter is replaced by the actual parameter. This is modelled by a pushout

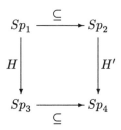

in the category of specifications and specification morphisms. Explicitly, Sp_4 is given by $B_4 = B_3 + B_2 \backslash B_1$, $\Sigma_4 = \Sigma_3 + \Sigma_2 \backslash \Sigma_1$, $E_4 = E_3 + \{H'(t_1) = H'(t_2) \mid t_1 = t_2 \in E_2\}$ where $H'(b) = $ *if* $b \in B_2 \backslash B_1$ *then* b *else* $H(b)$, $H'(f) = $ *if* $f \in \Sigma_2 \backslash \Sigma_1$ *then* f *else* $H(f)$, and where $H'(t)$ extends this translation to terms, i.e. $H'(x^b) = x^{H'(b)}$, $H'(f(t_1, \ldots, t_n)) = H'(f)(H'(t_1), \ldots, H'(t_n))$.

(Exercise: check the existence of pushouts in the category of specifications. Hint: construct the pushout of all the components of specification morphisms in **Set**.)

Example 5.3.8 ([Ehrig, 1978]). A *graph grammar production* consists of two graph morphisms

$$G_L \xleftarrow{\;P_L\;} G_I \xrightarrow{\;P_L\;} G_R$$

Given such a production and a graph morphism $H : G_I \to D$, (a *context*), a *direct derivation* $G_1 \Rightarrow G_2$ is defined by two pushouts

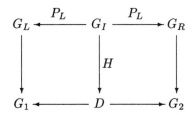

The idea is that the pattern of the left hand side of the production is found as a substructure in G_1 and then replaced by the right hand side where the pushouts define the gluing conditions, i.e. how to 'glue' dandling edges to the new subgraph.

   Pushouts of graphs are a mild generalization of pushouts in the category of sets; just construct the pushouts of the functions on nodes and on edges, domain and codomain functions are then induced by the dotted arrows due to the universal properties of the outer pushout square

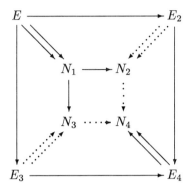

5.3.4 Filtered colimits

Filtered colimits are generalizations of least upper bounds of directed sets.

Definition 5.3.9.

- A category $\mathbf{C}$ is *filtered* if

  - it is non-empty,
  - for all pairs A, B of objects, there exist morphisms

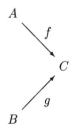

  - and if for every pair $f : A \to B$, $g : A \to B$ of morphisms there exists a morphism $h : B \to C$ such that

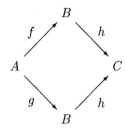

commutes.

- A colimit *colim D* is called *filtered* if the category **J** in the diagram
  $D : \mathbf{J} \to \mathbf{C}$ is filtered.

The construction of filtered colimits in the category of sets gives probably the best idea what they are about. Let $D : \mathbf{J} \to \mathbf{Set}$ be a diagram such that **J** is filtered and small. Define an equivalence relation on the disjoint sum

$$\bigcup_i \{i\} \times D(i) \text{ by } (i, x) \sim (j, y)$$

if $D(u)(x) = D(v)(y)$ for some morphisms $u : i \to k, v : j \to k$.

Proposition 5.3.10. *colim* $D = ([(i, x)]_\sim \mid i \in \mathbf{J}, x \in D(i)\}$ *is a colimit with injections* $\eta_i : D(i) \to$ *colim* $D, x \mapsto [(i, x)]$.

Proof. The important bit is that $\sim$ is transitive. Transitivity may be induced by patterns

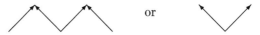   or

in **J**. The conditions for **J** being filtered ensure the patterns

   resp.

which are just what is needed. Otherwise the proof is straightforward. Define $[\sigma]([i, x]) := \sigma_i(x)$ for another natural transformation $\sigma_i : D(i) \to X$. ∎

The proof nicely demonstrates that filtered colimits really generalize least upper bounds of directed sets in that no inconsistent or 'non-deterministic' information is obtained by branching (of course, directed sets are filtered in a poset).

A particular example of filtered colimits, widely used in *domain theory* (cf. the corresponding chapter, Vol. 3) is that of ω-colimits of particular chains of the form

$$X_1 \xrightarrow{f_1} X_2 \longrightarrow \cdots \longrightarrow X_i \xrightarrow{f_n} X_{i+1} \longrightarrow \cdots$$

in the category **Dcpo** of directly complete partially ordered sets.

As an intermediate step we re-examine the construction of filtered colimits in the category of sets, but we will only be interested in the construction of ω-chains of injective functions. Here the elements $x_i \in X_i$ and $x_j \in X_j$ are identified in the colimit if $f_{j,i}(x_i) = x_j$ where $f_{j,i} = f_{j-1} \circ \ldots \circ f_i$ for $i \leq j$. So, elements of the colimit may be thought of as representing 'threads' $(x, f_i(x), f_{i+1}(f_i(x)), \ldots)$ which are 'maximal', meaning that there is no element $y \in X_{i-1}$ such that $f_{i-1}(y) = x$. Maximal threads can be represented by infinite tuples $(x_i, x_{i+1}, \ldots, x_j, \ldots)$ such that $x_j \in X_j$, $x_j = f_j^{-1}(x_{j+1})$ for all j, and $f_{i-1}^{-1}(x_i)$ is not defined where f^{-1} denotes the partial inverse of f. The colimit cone is then given by $\eta_i(x) = (x, f_i(x), f_{i+1}(f_i(x)), \ldots)$ where the colimit object consists of all maximal threads.

Restriction to injective functions reflects the application we have in mind, namely to construct solutions for type equations (see Section 7.3.2). In line with the argument for continuous functions, consider the X_i's as better and better approximations of the intended least solution. To make this intuition work, the functions $f_i : X_i \to X_{i+1}$ should be injections as a natural notion of approximation in the category of sets.

The following definition is widely accepted to capture approximation of dcpo's.

Definition 5.3.11. A pair $f : (X, \sqsubseteq) \to (Y, \sqsubseteq)$, $g : (Y, \sqsubseteq) \to (X, \sqsubseteq)$ of continuous mappings of dcpo's is called an *approximation* (sometimes called a *projection pair* otherwise) if $g \circ f = 1_{(X,\sqsubseteq)}$ and if $f \circ g \sqsubseteq 1_{(Y,\sqsubseteq)}$.

The first condition ensures injectivity as a set function, the second states that x is the 'best approximation' of $f(x)$ in $(X, \sqsubseteq)$, at least in that f is a left adjoint of g.

Proposition 5.3.12. *Given an ω-chain of approximations*

$$(X_1, \sqsubseteq) \underset{g_1}{\overset{f_1}{\rightleftarrows}} (X_2, \sqsubseteq) \rightleftarrows \cdots \rightleftarrows (X_i, \sqsubseteq) \underset{g_n}{\overset{f_n}{\rightleftarrows}} (X_{i+1}, \sqsubseteq) \rightleftarrows \cdots$$

the set $\{(x_0, x_1, \ldots, x_i, \ldots) \mid x_i = f_i^{-1}(x_{i+1})\}$ *ordered componentwise defines a colimit dcpo X^ω with colimit cone given by*

$$\eta_i(x) = (x, f_i(x), f_{i+1}(f_i(x)), \ldots).$$

6 Universal constructions

Category theory provides a variety of constructions of universal morphisms in terms of other available infrastructure, for instance limits and colimits.

We outline some of these constructions which reflect and generalize those well-known from set theory.

We are particularly concerned with constructions by *closure properties* and by *inductive definitions* for generation, and with *Nerode-like congruences* for abstraction, which will lead to the *adjoint functor theorem*, and with constructions for *left Kan extensions* and *right Kan extensions*. *Kan extensions* are a particular sort of adjoints which, in a sense, subsume all the other universal concepts.

6.1   The adjoint functor theorem

6.1.1   Freyd's adjoint functor theorem generalizes existence criteria for Galois connections (cf. Section 3.6.2), translating greatest lower bounds to limits. Apart from the purely formal argument that Galois connections are adjunctions restricted to partially ordered sets, some other evidence for the feasibility of such an undertaking can be found. For instance, generation of subalgebras is achieved by a limit construction, being the greatest lower bound of all subalgebras which contain the generating set, explicitly: $\langle X \rangle_{\mathcal{A}} = \sqcap \{ \mathcal{A}' \in \mathbf{Sub}(\mathcal{A}) \mid X \subseteq \mathcal{A}' \}$. Free Σ-algebras X^* behave like greatest lower bounds by formal analogy if we replace order by morphisms; X^* is the 'least' algebra among those which are 'greater' than the sorted set X, meaning that some $[f] : X^* \to \mathcal{A}'$ exists for all $\mathcal{A}$ such that $f : X \to G(\mathcal{A}')$. The uniqueness conditions, of course, pose a problem not so easily resolved.

Hypothetically, the development of the adjoint functor theorem may have started with the following conjecture.

Conjecture 6.1.1.   Let $\mathbf{D}$ be a category with limits, and let $G : \mathbf{D} \to \mathbf{C}$ be a functor which preserves the limits. Then G has a left adjoint $F : \mathbf{C} \to \mathbf{D}$ defined by $F(X) = lim\ Q$ where the diagram $Q : (X, G) \to \mathbf{D}$ is defined as follows:

Let (X, G) denote the comma category with

> objects (A, f) where $X \in \mathbf{C}, A \in \mathbf{D}$, and $f : X \to G(A) \in \mathbf{C}$, and
>
> morphisms $h : (A, f) \to (B, g)$ being morphisms $h : A \to B \in \mathbf{D}$ such that $G(h) \circ f = g$.

Then $Q : (X, G) \to \mathbf{D}$ maps $h : (A, f) \to (B, g)$ to the underlying morphism $h : A \to B$.

(Remark: we use a slightly more general version of *comma categories*. The previously given definition of Section 1.2.2 is obtained if $G = 1_{\mathbf{C}} : \mathbf{C} \to \mathbf{C}$.)

Proof.   The proof works deceptively well, though, I am afraid it is technically involved because of heavy use of machinery.

Let $(lim\ Q, \mu)$ be the limit of the diagram Q. G preserves limits. Hence $G(lim\ Q)$ with natural transformation $G(\varepsilon_{(A,f)}) : G(lim\ Q) \to G(A)$ is a limit of the diagram $G \circ Q : (X, G) \to \mathbf{C}$. Moreover $\tau_{(A,f)} := f : X \to G(A)$ defines a natural transformation $\tau : \Delta(X) \to G \circ Q : (X, G) \to \mathbf{C}$, where $\Delta : \mathbf{C} \to \mathbf{C}^{(X,G)}$ is the diagonal functor. The natural transformation τ induces a unique morphism $\langle \tau \rangle : X \to G(lim\ Q)$ such that $G(\varepsilon_{(A,f)}) \circ \langle \tau \rangle = \tau_{(A,f)}$.

We claim that $(lim\ Q, \langle \tau \rangle : X \to G(lim\ Q))$ is a universal morphism from X to G: if $f : X \to G(A)$ then $\varepsilon_{(A,f)} : lim\ Q \to A$ and $G(\varepsilon_{(A,f)}) \circ \langle \tau \rangle = \tau_{(A,f)} = f$. Let $k : lim\ Q \to A$ be another morphism such that $G(k) \circ \langle \tau \rangle = f$. The following diagrams commute

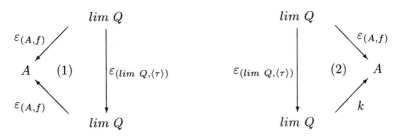

since $\varepsilon : \Delta(lim\ Q) \to Q : (X, G) \to \mathbf{D}$ is a natural transformation and $\varepsilon_{(A,f)} : (lim\ Q, \langle \tau \rangle) \to (A, f)$ (for (1)), and $k : (lim\ Q, \langle \tau \rangle) \to (A, f)$ (for (2)) are morphisms of (X, G). ∎

6.1.2 However, there is a shortcoming: the diagrams $Q : (X, G) \to \mathbf{D}$ may be large, meaning that (X, G) is a large category, i.e. the objects do not form a set but a proper class. 'So what?', one might ask, 'just modify the definition of limits so as to allow large diagrams, and we are well off'. However:

Proposition 6.1.2. *Let $\mathbf{C}$ be a category such that all large limits exist. Then every Hom set $\mathbf{C}[A, B]$ has at most one element.*

Proof. Let $Q : (X, \mathbf{C}) \to \mathbf{C}$ map a morphism $h : (A, f) \to (B, g)$ to the underlying morphism $h : A \to B$. We claim that X is a limit with cone $\varepsilon_{(A,f)} = f$ being the components of the natural transformation: if $\tau : \Delta(Y) \to Q$ is another cone, then $\tau_{(X, 1_X)} : Y \to X$ satisfies the equality $\varepsilon_{(A,f)} \circ \tau_{(X, 1_X)} = f \circ \tau_{(X, 1_X)} = \tau_{(A,f)}$ as τ is a natural transformation, and it is the unique such morphism since $\varepsilon \circ \Delta(k) = \varepsilon$ implies $k = 1_{(X, 1_X)} \circ k = \tau_{(X, 1_X)}$. ∎

So the conjecture falls short of our expectations, just providing sufficient criteria for the existence of a Galois connection, though for large partially ordered sets (i.e. a proper partially ordered class).

6.1.3   Partial orders, however, can contribute to the solution. In order to find a greatest lower bound of some subset X in a partial order $(Y, \sqsubseteq)$, it is sufficient to find a set $Z \subseteq X$ which is 'below' X in that for all elements $x \in X$ there exists an element $z \in Z$ such that $z \sqsubseteq x$. Then $\sqcap Z = \sqcap X$.

We use this observation to relate the existence of a Galois connection between large partially ordered sets to the existence and preservation of small limits.

Proposition 6.1.3.   *Let $(X, \sqsubseteq)$ and $(Y, \sqsubseteq)$ be large partially ordered sets and $g : (Y, \sqsubseteq) \to (X, \sqsubseteq)$ be a monotone mapping where $(Y, \sqsubseteq)$ has small least upper bounds.*

Then there exists a monotone mapping $f : (X, \sqsubseteq) \to (Y, \sqsubseteq)$ such that $x \sqsubseteq g(y)$ iff $f(X) \sqsubseteq y$, i.e. a Galois connection, if and only if g preserves small limits and if there exists a subset Z of Y such that the following condition holds

- *$x \sqsubseteq g(z)$ for all $z \in Z$, and*

- *for all $y \in Y$ such that $x \sqsubseteq g(y)$ there exists a $z \in Z$ with $z \sqsubseteq y$.*

Proof.   Same as that for Galois connections between small partially ordered sets using that $\sqcap Z = \sqcap\{y \mid x \sqsubseteq g(y)\}$. ∎

6.1.4   The adjoint functor theorem transfers the idea to arbitrary categories.

Theorem 6.1.4.   *Let $\mathbf{D}$ be a category with all small limits. Then a functor $G : \mathbf{D} \to \mathbf{C}$ has a left adjoint if and only if all small limits are preserved and if the following solution set condition holds:*

> *For each object A of $\mathbf{C}$ there exists a set I and an I-indexed family of morphisms $f_i : X \to G(A_i)$ such that for every morphism $h : X \to G(A)$ there exists an index i and some morphism $t : A_i \to A$ such that $h = G(t) \circ f_i$.*

Proof.   Right adjoints preserve limits (Section 3.5.2), and the universal morphisms $(X, \eta : X \to G(F(X)))$ provide the solution set.

The other direction is a modification of Section 6.1.2. Let $(X, G)_I$ be the full subcategory of (X, G) with objects $f_i : X \to G(A_i)$, the inclusion being denoted by $E : (X, G)_I \subseteq (X, G)$. Let $(lim \ Q \circ E, \varepsilon)$ be the limit of the functor $Q \circ E : (X, G)_I \to \mathbf{D}$. Hence $G(lim \ Q \circ E)$ with natural transformation $G(\varepsilon_i) : G(lim \ Q \circ E) \to G(A_i)$ is a limit of the diagram $G \circ Q \circ E : (X, G) \to \mathbf{C}$ where we use ε_i instead of $\varepsilon_{(A_i, f_i)}$ for convenience. Moreover $\tau_i := f_i : X \to G(A_i)$ defines a natural transformation

$\tau : \Delta(X) \to G \circ Q \circ E : (X, G)_I \to \mathbf{C}$ which induces a unique morphism $\langle \tau \rangle : X \to G(lim\ Q \circ E)$ such that $G(\varepsilon_i) \circ \langle \tau \rangle = \tau_i$.

We claim that $(lim\ Q, E), \langle \tau \rangle)$ is a universal morphism from X to G:

Let $h : X \to G(A)$. Then there exists an index i and some morphism $t : A_i \to A$ such that $h = G(t) \circ f_i$ by the solution set condition. We compose $t \circ \varepsilon_i : G(lim\ Q, E) \to G(A)$ to obtain a morphism $[h]$ which satisfies $G([h]) \circ \langle \tau \rangle = h$ since $G(t \circ \varepsilon_i) \circ \langle \tau \rangle = G(t) \circ G(\varepsilon_i) \circ \langle \tau \rangle = G(t) \circ \tau_i = G(t) \circ f_i = h$ by assumption.

The proof of uniqueness of the induced morphisms is even more involved: there exists an index j and some morphism $t' : A_j \to lim\ Q \circ E$ such that $\langle \tau \rangle = G(t') \circ f_i$ by the solution set condition. We construct the a pullback

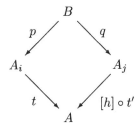

in $\mathbf{D}$ which is preserved by $\mathbf{G}$, and which induces a morphism $k : X \to G(B)$ by universal property in

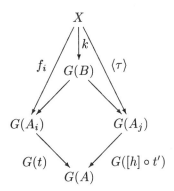

The outer square commutes since $G([h]) \circ \langle \tau \rangle = h = G(t) \circ f_i$. Again the solution set condition is used to generate an index m and a morphism $t'' : A_m \to B$ such that $k = G(t'') \circ f_m$. Then the diagram

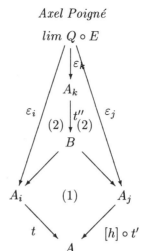

commutes: (1) is a pullback square, and (2) commutes as $(X, G)_I$ is a full subcategory of (X, G) and as ε is a natural transformation.

Finally, $t' \circ \varepsilon_j = 1_{lim\ E \circ Q}$ by universal property of the limit as $\varepsilon_i \circ t' \circ \varepsilon_j = \varepsilon_i$ for all i. The latter holds as $t' \circ \varepsilon_j$ is a morphism in $(X, G)_I$ which follows from the commutativity of the diagram

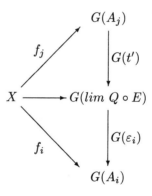

Then $t \circ \varepsilon_i = [h] \circ t' \circ \varepsilon_j = [h]$.   ∎

6.1.5   The following is a non-trivial application of the adjoint functor theorem. Let **SLωPos** be the category of ω-complete semilattices where objects are semilattices $(X, +)$ such that X is an ω-complete poset, and $+ : X \times X \to X$ is ω-continuous. Morphisms are semilattice homomorphisms which are ω-continuous. The forgetful functor $U : \textbf{SL}\omega\textbf{Pos} \to \omega\textbf{Pos}$ maps ω-complete semilattices $(X, +)$ to the underlying ω-complete poset X. This functor has a left adjoint $P : \omega\textbf{Pos} \to \textbf{SL}\omega\textbf{Pos}$ which maps every ω-complete poset to the *power domain* $P(X)$. This left adjoint has been constructed explicitly by [Plotkin, 1976], [Smyth, 1978] (for an alternative account see [Poigné, 1985]), and plays an important role in the semantics of non-determinism (see Chapter on Domain theory, Vol. 3). Existence of this left adjoint can be shown using the adjoint functor theorem.

Proposition 6.1.5. $SL\omega Pos$ *has all small limits and* $U : SL\omega Pos \to \omega Pos$ *preserves them.*

Proof. [Hint] Compute the limits as in **Set**. Order is induced. ∎

For the solution set condition we factor ω-complete functions $f : X \to Y$ as follows:

Define a poset $Im_{\text{mon}}(f) = \{f(x) \mid x \in X\}$ with order $f(x) \sqsubseteq_f f(x')$ being the transitive closure of $f(x) \sqsubseteq f(x')$ if $x \sqsubseteq x'$ (this is an epimorphism in **Pos**).

Let $Im(f) = \{\bigsqcup y_n \mid (y_n)_{n \in \omega}$ is an ω-chain in $Im_{\text{mon}}(f)\}$ where the least upper bound exists in Y. Define a partial order by $\bigsqcup y_m \sqsubseteq \bigsqcup y'_n$ if $\forall m \exists n. y_m \sqsubseteq_f y'_n$.

Lemma 6.1.6.

- $Im(f)$ *is* ω-*complete.*

- $f = f_2 \circ f_1$ *where* $f_1 : X \to Im(f), x \mapsto f(x)$ *and* $f_2 : Im(f) \to Y, y \mapsto y$.

- $card(Im(f)) \leq card(X^\omega)$.

Proof. [Hint] ω-Completeness follows by the same technique as used for ω-completion (Section 3.6.3). ∎

Proposition 6.1.7. $U : SL\omega Pos \to \omega Pos$ *has a left adjoint.*

Proof. By the adjoint functor theorem, the solution set condition being satisfied by the lemma above. ∎

This proof was first given in the paper [Hennessy and Plotkin, 1979], in fact as a **Pos**-enriched adjunction.

6.2 Generation as partial evaluation

6.2.1 By a *partial evaluation* we mean that a (functional) program applied to some data fails to execute all the code. This may for instance occur if application is only defined for a sublanguage, say **C**, of a language **D**.

Application of a program $f : A \to B$ of $\mathbf{D}$ to data a of type A may then have the following potential results

- if f is a program of $\mathbf{C}$, application of f to a is defined and yields the result $f \cdot a$.

- if f is not a program of $\mathbf{C}$, then either

  — some code can be executed: there exists some f_1 in $\mathbf{C}$ such that $f = f_2 \circ f_1$. Then application of f_1 to a yields $f_1 \cdot a$ to which f_2 is still to be applied, or

  — no code can be executed.

Let us distinguish between $f \otimes a$ as a syntactic expression for application, and $f \cdot a$ as the result of the application (cf. Section 4.2). Then all the cases above are captured by the conditional equation

$$(f_2 \circ f_1) \otimes a = f_2 \otimes (f_1 \cdot a) \quad \text{if } f_1 \text{ is in } \mathbf{C}.$$

Obviously, the second case is subsumed; f_2 is still to be applied to the result of $f_1 \cdot a$ since $f_1 \otimes a$ can be executed. Otherwise the result will be $f \otimes a$ which states that f is to be applied to a but the code cannot be executed. To cover the first case, the language needs to have identities; one computes $f \otimes a = (1_B \circ f) \otimes a = 1_B \otimes (f \cdot a)$ which is just a different presentation of $f \cdot a$.

Languages are categories in our context, so $\mathbf{C}$ is a subcategory of $\mathbf{D}$ with inclusion $E : \mathbf{C} \subseteq \mathbf{D}$, and restriction of application to $\mathbf{C}$ is determined by functor $M : \mathbf{C} \to \mathbf{Set}$. Partial evaluation then defines a functor $M \otimes E : \mathbf{D} \to \mathbf{Set}$. In fact, we immediately consider a more general case where the one 'language' is translated into the other by means of a functor $T : \mathbf{C} \to \mathbf{D}$. Translation will retain the meaning in that application of the translated function $T(f) : T(A) \to T(B)$ is defined in terms of the original function, as expressed by the conditional equation

$$(f_2 \circ T(f_1)) \otimes a = f_2 \otimes (f_1 \cdot a) \quad \text{if } f_1 \text{ is in } \mathbf{C}.$$

Definition 6.2.1. Let $T : \mathbf{C} \to \mathbf{D}$ and $M : \mathbf{C} \to \mathbf{Set}$ be functors.

- A term of the form $f \otimes a$ where $f : T(A) \to B \in \mathbf{D}$ and $a \in M(A)$ is called a *partial evaluation*.

Let $[f \otimes a]$ denote the congruence class of $f \otimes a$ with regard to the least congruence generated by the conditional equation

$$g \circ T(f) \otimes a = g \otimes (f \cdot a) \quad \text{if } f : A \to B \text{ is in } \mathbf{C}$$

where $f \cdot a$ is a shorthand notation for $M(f)(a)$.

- The functor $M \otimes T : \mathbf{D} \to \mathbf{Set}$ is defined by

  — $M \otimes T(B) = \{[f \otimes a] \mid f : T(A) \to B \in \mathbf{D}, a \in M(A)\}$

  — application $g \cdot [f \otimes a] = [(g \circ f) \otimes a]$.
  (More precisely, $M \otimes T(g)([f \otimes a]) = [(g \circ f) \otimes a]$.)

A few remarks may be appropriate:

- The tensor $\otimes$ is used for traditional reasons to distinguish partial application from application. The programmer may overload notation replacing $f \otimes a$ by $f \cdot a$ as partial application is subject to the same axioms, except for $1_A \cdot a = a$.

- The tradition stems from tensoring bimodules with left modules (cf. for instance [Chevalley, 1956]). A reader with a mathematical background might have noticed the connection between rings and categories, and left modules and functors.

- $T \otimes M$ defines a functor since $1_A \cdot [f \otimes a] = [(1_A \circ f) \otimes a] = [f \otimes a]$ and $h \circ g \cdot [f \otimes a] = [h \circ g \circ f \otimes a] = h \cdot [g \circ f \otimes a] = h \cdot (g \cdot [f \otimes a])$.

- It is also a tradition not to distinguish terms and congruence classes, meaning that $f \otimes a$ denotes also the term as the congruence class. We will adhere to this practice for convenience.

6.2.2 An example may help to enhance understanding. Though a forward pointer, probably the most useful observation is that the tensoring corresponds to the construction of free algebras (it may be a good idea to have a look at Sections 8.1 and 8.2 first).

We will show in Section 8.1.2 that Σ-algebras $\mathcal{A}$ can be represented by a product preserving functor $\mathcal{A} : \mathbf{T}_\Sigma \to \mathbf{Set}$ where $\mathbf{T}_\Sigma$ is the free algebraic theory over the signature Σ. Roughly, base types are mapped to $b^{\mathcal{A}}$, and base operations $\sigma : b_1 \times \ldots \times b_n \to b$ to functions $\sigma^{\mathcal{A}} : b_1^{\mathcal{A}} \times \ldots \times b_n^{\mathcal{A}} \to b^{\mathcal{A}}$. This interpretation extends uniquely to $\mathbf{T}_\Sigma$ (cf. Section 2.1.4). Moreover every signature morphism $H : \Sigma_1 \to \Sigma_2$ induces a product preserving functor $H : \mathbf{T}_{\Sigma_1} \to \mathbf{T}_{\Sigma_2}$. Tensoring $\mathcal{A}$ with H yields (up to equivalence of categories) a free Σ_2-algebra $T_H(\mathcal{A})$ generated by $\mathcal{A}$.

Proposition 6.2.2. $\mathcal{A} \otimes H : \mathbf{T}_{\Sigma_2} \to \mathbf{Set}$ *is naturally isomorphic to* $T_H(\mathcal{A}) : \mathbf{T}_{\Sigma_2} \to \mathbf{Set}$.

Proof. Functors $F : \mathbf{T}_{\Sigma_2} \to \mathbf{Set}$ are uniquely determined by their behaviour on types $b \in B$ and morphisms $f : b_1\Pi \ldots \Pi b_n \to b \in \Sigma$. By definition $T_H(\mathcal{A})_b = \{[t] \mid t \in T_\Sigma(\mathcal{A})_b\}$ (cf. Section 5.6.1). Each term $t \in T_\Sigma(\mathcal{A})_b$ corresponds to a formal application defined by induction

$$a \mapsto 1_b \otimes a$$
$$f(t_1,\ldots,t_n) \mapsto f \circ (f_1\Pi \ldots \Pi f_n) \otimes \langle a_1,\ldots,a_n\rangle$$
$$\text{if } t_i \mapsto f_i \otimes a_i \text{ for } i = 1,\ldots,n.$$

On the other hand every formal application $f \otimes a$ corresponds to a term in $T_\Sigma(A)$ defined by substitution

$$\langle p\rangle M \otimes \langle \overline{a}\rangle \mapsto M[\langle\langle p\rangle\rangle/\langle\overline{a}\rangle].$$

This defines a bijection. ∎

6.2.3 The example suggests that tensoring is of a universal nature with reduction as a right adjoint, where *reduction* here just means precomposition with a functor.

Definition 6.2.3. Let $T : \mathbf{C} \to \mathbf{D}$ and $M : \mathbf{D} \to \mathbf{Set}$ be functors. $M \circ T : \mathbf{C} \to \mathbf{D}$ is called the *reduct* of M along T.

The terminology stems from the specific case that $E : \mathbf{C} \subseteq \mathbf{D}$. Then $M \circ T$ is the model M *reduced* to the sublanguage $\mathbf{C}$.

Proposition 6.2.4. *Let* $T : \mathbf{C} \to \mathbf{D}$ *be a functor. Then natural transformations* $\alpha : M_1 \to M_2 \circ T$ *are in (natural) bijective correspondence with natural transformations* $\beta : M_1 \otimes T \to M_2$ *where* $M_1 : \mathbf{C} \to \mathbf{Set}$, $M_2 : \mathbf{D} \to \mathbf{Set}$.

Proof. We define the isomorphism by $\varphi(\alpha)(f \otimes a) := f \cdot \alpha_A(a)$ and $\varphi^{-1}(\beta)(a) := \beta_A(1_A \otimes a)$. Bijectivity follows from $\varphi^{-1}(\varphi(\alpha))(a) = \varphi(\alpha)(1_A \otimes a)_A = 1_A \cdot \alpha(a) = \alpha(a)$ and $\varphi(\varphi^{-1}(\beta))(f \otimes a) = f \cdot \varphi^{-1}(\beta)_A(a) = f \cdot \beta(1_A \otimes a) = \beta(f \cdot 1_A \otimes a) = \beta(f \otimes a)$. Naturality of $\varphi(\alpha)$ means that $T(g)\circ\varphi(\alpha)(f\otimes a) = \varphi(\alpha)\circ M_2\circ T(g)(f\otimes a)$: we compute $T(g)\circ\varphi(\alpha)(f\otimes a) = g \cdot \varphi(\alpha)(f \otimes a) = g \cdot (f \cdot \alpha(a)) = (g \circ f) \cdot \alpha(a) = \varphi(\alpha)((g \circ f) \otimes a)) = \varphi(\alpha)(g \cdot (f \otimes a)) = \varphi(\alpha) \circ M_2 \circ T(g)(f \otimes a)$. Naturality of $\varphi^{-1}(\beta)$ is checked just as easily, as is the naturality of the bijection (Exercise). ∎

Corollary 6.2.5. *The functor* $\_ \circ T : \mathbf{Set}^\mathbf{D} \to \mathbf{Set}^\mathbf{C}, M \mapsto M \circ F$ *has a left adjoint* $\_ \otimes T : \mathbf{Set}^\mathbf{C} \to \mathbf{Set}^\mathbf{D}$.

6.2.4 We obtain a categorical proof of the free algebra theorem using the following proposition.

Proposition 6.2.6. *Let* **C** *and* **D** *be categories with finite products, and let the functors* $T : \mathbf{C} \to \mathbf{D}$ *and* $M : \mathbf{C} \to \mathbf{D}$ *preserve finite products. Then* $M \otimes T : \mathbf{D} \to \mathbf{Set}$ *preserves finite products.*

Proof. Let $f \otimes a \in M \otimes T(A)$ and $g \otimes b \in M \otimes T(B)$. Then $f\Pi g \otimes (a, b) \in M \otimes T(A \Pi B)$. Vice versa, if $h \otimes c \in M \otimes T(A \Pi B)$ then $p \cdot (h \otimes c) = (p \circ h) \otimes c \in M \otimes T(A)$ and $q \cdot (h \otimes c) = (q \circ h) \otimes c \in M \otimes T(B)$. This determines a natural isomorphism $M \otimes T(A)\Pi M \otimes T(B) \cong M \otimes T(A \Pi B)$, since $(p \circ h)\Pi(q \circ h) \otimes (c, c) = (p \circ h)\Pi(q \circ h) \otimes (\Delta \cdot c) = (p \circ h)\Pi(q \circ h) \circ \Delta \otimes c = (p\Pi q) \circ (h\Pi h) \circ \Delta \otimes c = (p\Pi q) \circ \langle h, h \rangle \otimes c = \langle p \circ h, q \circ h \rangle \otimes c = h \otimes c$.
$M \otimes T(1)$ has only one element $1_1 \otimes a$ since $\langle \rangle_A \otimes a = 1_1 \otimes (\langle \rangle_A \cdot a) = 1_1 \otimes \langle \rangle$. ∎

This proposition holds if finite products are replaced by arbitrary limits (Exercise).

Both the propositions put together prove the existence of free algebras with regard to an arbitrary specification morphism $H : Sp_1 \to Sp_2$ (again with a forward reference to Section 8)

Corollary 6.2.7. *The forgetful functor* $_{-H} : \mathbf{Alg}(Sp_2) \to \mathbf{Alg}(Sp_1)$ *has a left adjoint.*

Proof. By the following chain of natural isomorphisms (cf. Section 8.2)

$$\begin{aligned} \mathbf{Alg}(Sp_1)[A, B_H] &\cong \mathbf{T}_{Sp_1}{}^* \Rightarrow_\Pi \mathbf{Set}[A, B \circ H^*] \\ &\cong [\mathbf{T}_{Sp_2}{}^* \Rightarrow_\Pi \mathbf{Set}][A \otimes H, B] \\ &\cong \mathbf{Alg}(Sp_2)[[A \otimes H, B] \end{aligned}$$

$\mathbf{C} \Rightarrow_\Pi \mathbf{Set}$ denotes the full subcategory of $\mathbf{Set}^{\mathbf{C}}$ with objects being product preserving functors. We overload notation identifying algebras and corresponding functors. ∎

6.3 Left Kan extensions, tensors and coends

6.3.1 The adjunction and the tensoring construction give rise to some new fundamental concepts, if the category of sets is replaced by a category **E** with suitable infrastructure. In fact, the development so far was only meant to prepare the grounds for a more categorical treatment.

The reduction functor $_- \circ T : \mathbf{E}^{\mathbf{D}} \to \mathbf{E}^{\mathbf{C}}$ is defined for an arbitrary functor $T : \mathbf{C} \to \mathbf{D}$. If this functor has a left adjoint, it is called a *left Kan extension*. Typically a more explicit definition is given.

Definition 6.3.1. Let $T : \mathbf{C} \to \mathbf{D}$ and $M : \mathbf{C} \to \mathbf{E}$ be functors. Then a functor $Lan_T M : \mathbf{D} \to \mathbf{E}$ together with a natural transformation $\eta : M \to Lan_T M * T$ is called a *left Kan extension* of M along T if for every functor $G : \mathbf{D} \to \mathbf{E}$ and every natural transformations $\sigma : M \to G * T$, there exists a unique natural transformation $[\sigma] : Lan_T M \to G$ such that $([\sigma] * T) \circ \eta = \sigma$, diagrammatically:

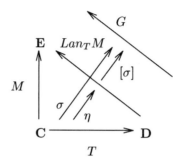

(We use the notation $G * F$ of Section 1.5.3 for composition $G \circ F$ of functors because of the interaction of functors and natural transformations.)

The notion of a right Kan extension $Ran_{T-} : \mathbf{E}^{\mathbf{C}} \to \mathbf{E}^{\mathbf{D}}$, meaning a right adjoint to $\_ \circ T : \mathbf{E}^{\mathbf{D}} \to \mathbf{E}^{\mathbf{C}}$, will be discussed in the next section.

6.3.2   As MacLane points out, the notion of Kan extensions subsume all other fundamental notions of category theory.

Proposition 6.3.2. *A functor $D : \mathbf{C} \to \mathbf{E}$ has a colimit if the left Kan extension along the functor $T : \mathbf{C} \to \mathbf{1}$ exists. The colimit object colim D is given by $Lan_T D(0)$.*

Proof. Let $\sigma_A : C \to D(A)$ define a natural transformation $\sigma : C \to D : \mathbf{C} \to \mathbf{E}$ (we identify the object C and the constant functor $C : \mathbf{1} \to \mathbf{D}$ with image C). Then there exists a unique natural transformation $\langle \sigma \rangle : C \to Lan_T D : \mathbf{1} \to \mathbf{D}$, basically a morphism $\langle \sigma \rangle : C \to Lan_T D$, such that $\langle \sigma \rangle \circ \eta = \sigma$. The latter means that $\langle \sigma \rangle \circ \eta_A = \sigma_A$, which establishes that $(Lan_T D \circ \eta)$ is a colimit.   ∎

It might be interesting to consider the case where $\mathbf{E} = \mathbf{Set}$; the colimit object then consists of elements $[1_1 \otimes a]$ where $a \in D(A)$ and $[1_1 \otimes a]$ is the congruence class with regard to the least congruence containing the equality $1_1 \otimes (f \cdot a) = 1_1 \otimes a$.

For the proof of the following proposition relating adjunctions and Kan extensions we direct the reader to the book of MacLane ([MacLane, 1971], X.7).

Proposition 6.3.3. *A functor* $T : \mathbf{C} \to \mathbf{D}$ *has a right adjoint if and only if the left Kan extension* $Lan_T 1_{\mathbf{C}}$ *exists and if* T *preserves the Kan extension. Then the functor* $Lan_T 1_{\mathbf{C}} : \mathbf{D} \to \mathbf{C}$ *is the right adjoint, and* $\eta : 1_{\mathbf{C}} \to (Lan_T 1_{\mathbf{C}}) * T$ *is the unit of the adjunction.*

(A functor $F : \mathbf{E} \to \mathbf{E}'$ *preserves a left Kan extension* $(Lan_T M, \eta)$ if $F * Lan_T M$ with natural transformation $F * \eta_A : F * T(A) \to F * Lan_T M(A)$, where $A \in \mathbf{C}$, is the left Kan extension of $M * F : \mathbf{C} \to \mathbf{E}'$ along $T : \mathbf{C} \to \mathbf{D}$.)

6.3.3 The tensor construction in a suitable abstract form provides existence criteria for left Kan extensions as will be demonstrated subsequently. The set-based construction consists of two steps,

- the formation of formal applications $f \otimes a$, and

- a factorization specified by the equation $(g \circ f) \otimes a = g \otimes (f \cdot a)$.

The formation of terms of type B can be represented by the formation of *copowers* $\mathbf{C}[A, B] \cdot M(A)$ meaning a coproduct with injections $in_f :$ $M(A) \to \mathbf{C}[A, B] \cdot M(A)$ indexed by morphisms $f : A \to B$. A slightly more tailor-made definition is the following.

Definition 6.3.4. Let $\mathbf{C}$ be a category, A be an object of $\mathbf{C}$, and X be a set. An object $X \otimes A$ together with a natural transformation $\varphi_A :$ $\mathbf{C}[X \otimes A, \_] \cong \mathbf{Set}[X, \mathbf{C}[A, \_]]$ is called the *tensor of* X *with* A.

Note that $X \otimes A = \coprod_{x \in X} A$ if $\mathbf{C}$ has coproducts.

As an immediate consequence we obtain the existence of two functors.

Proposition 6.3.5 (Exercise). *If tensors exists for all sets* X *and all objects* A *then*

- *the functor* $\mathbf{C}[A, \_] : \mathbf{C} \to \mathbf{Set}$ *has a left adjoint* $\_ \otimes A : \mathbf{Set} \to \mathbf{C}$, *and*

- *a functor* $X \otimes \_ : \mathbf{C} \to \mathbf{C}$ *is defined as follows :*

  let $\eta : X \to \mathbf{C}[B, X \otimes B]$ *be defined by* $\varphi_B(1_{X \otimes B})$. *Then* $h : X \to$ $\mathbf{C}[A, X \otimes B]$, $x \mapsto \eta(x) \circ f$ *induces the morphism* $X \otimes f := \varphi_A^{-1}(h) :$ $X \otimes A \to X \otimes B$.

6.3.4 All the terms $f \otimes a$ are to be collected, but only modulo the equality $(g \circ f) \otimes a = g \otimes (f \cdot a)$. This is achieved by a colimit construction with regard to a diagram which consists of all the subdiagrams of the form

$$\mathbf{C}[A, C] \otimes M(A)$$

$$\mathbf{C}[f, C] \otimes M(1_A) \nearrow$$

$$\mathbf{C}[B, C] \otimes M(A)$$

$$\mathbf{C}[1_B, C] \otimes M(f) \searrow$$

$$\mathbf{C}[B, C] \otimes M(B)$$

where $f : A \to B$ is in $\mathbf{C}$. Here the lower morphism corresponds to $g \otimes (f \cdot a)$, and the upper morphism to $(g \circ f) \cdot a$.

Obviously, this constructions becomes somewhat messy when written in full detail. Yet another concept will pave the way, and even add considerable elegance.

6.3.5 Above we want something like a colimit of a functor $F : \mathbf{C}^{\mathrm{op}} \times \mathbf{C} \to \mathbf{D}$, but with the requirement that the diagrams of the form

$$F(A, A)$$

$$F(f, 1_A) \nearrow \qquad \searrow \eta_{A,A}$$

$$F(B, A) \qquad \qquad \text{`colim'}$$

$$F(1_B, f) \searrow \qquad \nearrow \eta_{B,B}$$

$$F(B, B)$$

commute, where $\eta_{A,A} : F(A, A) \to \text{`colim'}$ are components of the supposed colimit cone.

Definition 6.3.6. Let $F : \mathbf{C}^{\mathrm{op}} \times \mathbf{C} \to \mathbf{D}$ be a functor.

- An *extranatural transformation* or *wedge* $\beta : F \xrightarrow{..} W$ consists of morphisms $\beta_A : F(A, A) \to W$ indexed by objects A of $\mathbf{C}$ such that the diagram

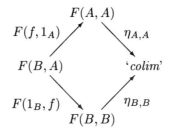

commutes for all $f : A \to B$ in **C**.

- An object CE together with a wedge $\eta : F \dot\to CE$ is called a *coend* if for every wedge $\omega : F \dot\to W$ there exists a unique morphism $[\omega] : CE \to W$ such that $[\omega] \circ \eta_A = \omega_A$ for all objects A of **C**. The notation for the coend (object) is $CE = \int^X F(X,X)$.

(A family of functions $\omega_A : W \to F(A,A)$ such that $F(1_A, f) \circ \omega_A = F(f, 1_B) \circ \omega_B$ will also be called a wedge. Notation is $\omega : W \dot\to F$.)

Proposition 6.3.7. *Coends exist if* **C** *is cocomplete.*

Proof. See [MacLane, 1971], IX.5. The remarks motivating coends should give an idea. ∎

6.3.6 Finally, the 'tensoring' construction is captured by means of tensors and coends.

Proposition 6.3.8. *Let $T : \mathbf{C} \to \mathbf{D}$ and $M : \mathbf{C} \to \mathbf{E}$ be functors such that all the tensors $\mathbf{D}[T(X), A] \otimes M(X)$ exist. Then there exists a left Kan extension $Lan_T M$ of M along T if for all objects B of **D** the following coend exists*

$$Lan_T M(B) = \int^X \mathbf{D}[T(X), B] \otimes M(X)$$

which defines the object mapping of $Lan_T M : \mathbf{D} \to \mathbf{Set}$.

Proof. [Sketch] Our proof will be rather explicit, lacking the elegance of the proof in MacLane's book (Chapter X.4) which should be appreciated. However, the explicit proof illustrates the prescriptive power of categorical definitions as claimed above.

We have to find a natural transformation $\eta_A : M(A) \to Lan_T M(T(A))$, and for every functor $G : \mathbf{D} \to \mathbf{E}$ and every natural transformation $\sigma_A : M(A) \to G(T(A))$ a natural transformation $[\sigma]_B : Lan_T M(B) \to G(B)$.

Since $Lan_T M(T(A)) = \int^X \mathbf{D}[T(X), T(A)] \otimes M(X)$, the morphism η'_A of the coend wedge

$$M(A) \xrightarrow{\eta_A} \int^X \mathbf{D}[T(X), T(A)] \otimes M(X)$$

$$\eta'_A \uparrow$$

$$? \searrow \qquad \mathbf{D}[T(A), T(A)] \otimes M(A)$$

is a suitable candidate to find an η_A. Looking for the missing morphism, the definition of the tensor should be used; the isomorphism $\varphi : \mathbf{E}[\mathbf{D}[T(A),$

$T(A)]\otimes M(A), \mathbf{D}[T(A), T(A)]\otimes M(A)]\cong\mathbf{Set}[\mathbf{D}[T(A), T(A)]\mathbf{E}[M(A), \mathbf{D}[T(A),$
$T(A)] \otimes M(A)]]$ yields

$$\varphi_A(1...)(1_{T(A)}) : M(A) \to \mathbf{D}[T(A), T(A)] \otimes M(A).$$

To find a morphism $[\sigma]_B : \int^X \mathbf{D}[T(X), B]\otimes M(X) \to G(B)$, the obvious idea is to define another wedge $\omega_X : \mathbf{D}[T(X), B] \otimes M(X) \to G(B)$ which should fall out of the isomorphism

$$\psi : \mathbf{E}[\mathbf{D}[T(X), B] \otimes M(X), G(B)] \cong \mathbf{Set}[\mathbf{D}[T(X), B], \mathbf{E}[M(X), G(B)].$$

So for every $g : T(X) \to B$ a morphism $h : M(X) \to G(B)$ has to be found. We have $\sigma_X : M(X) \to G(T(X))$ which gives $h = G(g) \circ \sigma_X : M(X) \to G(B)$.

We have to prove the equality

$$\omega_X \circ \mathbf{D}[T(f), B] \otimes M(1_X) = \omega_Y \circ \mathbf{D}[T(1_Y), B] \otimes M(f)$$

for every $f : X \to Y$ to establish ω as a wedge. Naturality of ψ implies that this follows from the equality $G(g \circ T(f)) \circ \sigma_X = G(g) \circ G(T(f)) \circ \sigma_X = G(g) \circ \sigma_Y \circ M(f)$ where $g : T(Y) \to B$. The second equality holds as σ is a natural transformation.

We leave it to the reader to check that the natural transformations satisfy the required properties. ∎

6.3.7 The result of Section 6.2.4 extends to categories $\mathbf{E}$ with finite products such that all functors $\_\Pi B, A\Pi\_ : \mathbf{E} \to \mathbf{E}$ preserve tensors and coends, i.e. $A\Pi(X \otimes B) \cong X \otimes (A\Pi B)$ and $A\Pi \int^X F(X, X) \cong \int^X A\Pi F(X, X)$ and symmetric.

Proposition 6.3.9. *Let* $\mathbf{E}$ *be Cartesian closed. Then*

$$Lan_T M = \int^X \mathbf{D}[T(X), B] \otimes M(X) : \mathbf{D} \to \mathbf{E}$$

preserves (finite) products if $T : \mathbf{C} \to \mathbf{D}$ *and* $M : \mathbf{C} \to \mathbf{E}$ *do so.*

Proof. [Sketch] $\_\Pi A : \mathbf{E} \to \mathbf{E}$, being a left adjoint, preserves tensors and coends since they are universal concepts (along the lines of Section 3.4.2).

Define natural transformations $\varphi : (R[X, A]\otimes M(X))\Pi(R[Y, B]\otimes M(Y)) \to R[X\Pi Y, A\Pi B] \otimes M(X\Pi Y)$ and $\psi : R[X\Pi Y, A\Pi B] \otimes M(X\Pi Y) \to (R[X, A]\otimes M(X))\Pi(R[Y, B]\otimes M(Y))$. For this use $(R[X, A]\otimes M(X))\Pi(R[Y, B] \otimes M(Y)) \cong (R[X, A]\Pi R[Y, B]) \otimes (M(X)\Pi M(Y))$. Composition of these transformations with the wedges of $\int^X \int^Y (R[X, A] \otimes M(X))\Pi(R[Y, B] \otimes$

$M(Y)) \cong \int^X (R[X,A] \otimes M(X))\Pi \int^Y (R[Y,B] \otimes M(Y))$ and $\int^{X,Y} R[X\Pi Y,$ $A\Pi B]\otimes M(X\Pi Y)$ induce an isomorphism $\int^X (R[X,A]\otimes M(X))\Pi \int^Y (R[Y,B]$ $\otimes M(Y)) \cong \int^{X,Y} R[X\Pi Y, A\Pi B]\otimes M(X\Pi Y)$. Finally prove that $\int^X \mathbf{D}[T(X),$ $A\Pi B] \otimes M(X) \cong \int^{X,Y} R[X\Pi Y, A\Pi B] \otimes M(X\Pi Y)$ which is obtained by universal argument using that the respective wedges can be embedded into each other. ∎

The construction indicates how to transfer the ideas of this section into an enriched world.

6.4 Separation by testing

6.4.1 It is not a particularly helpful remark to say that reversing all the arrows in the previous section gives a construction of right Kan extension in terms of cotensors and ends. So let us try to develop a scenario which might appeal to computer scientists.

Let $\delta : Act^* \times S \to S$ be an Act^*-automaton with output function $\lambda : S \to O$. Any input sequence $w \in Act^*$ is considered as a *test*. If the automaton can be reset so that all tests can be applied to the same state, then a function $\lambda_s : Act^* \to O$ is defined by $\lambda_s(w) = \lambda(\delta(w,s))$ for every state $s \in S$, called the *behaviour* at s. States s, s' are considered as equivalent if they show the same behaviour under testing, meaning that $\lambda_s = \lambda_{s'}$. Of course, this is just the idea of Nerode congruence where $s \sim s'$ iff $\lambda(\delta(w,s)) = \lambda(\delta(w,w))$ for all $w \in Act^*$. Thus behaviours define a separated automaton $Beh(D,\lambda) = \{\lambda_s \mid s \in S\}$ with states $Beh(D,\lambda)$ and transition function $\delta_{Beh}(w,f)(v) = f(vw)$. The observation function is $\lambda_{Beh}(f) = f(\varepsilon)$.

A fully separated automaton over O should comprise all 'possible behaviours', which it does since O_* has states O^{Act^*} and a transition function $\delta_* : Act^* \times O^{Act^*} \to O^{Act^*}$ defined by $\delta_*(w,f)(v) = f(vw)$. Function $\lambda_* : O^{Act^*} \to O, f \mapsto f(\varepsilon)$ is the observation function.

Every automaton defines a functor $D : Act^* \to \mathbf{Set}$ as $E : 1 \to Act^*$, $\varnothing \mapsto \varepsilon$ and $O : 1 \to \mathbf{Set}, \varnothing \mapsto O$ do (Act^* is here considered as a category). Then the fully separated automaton $O_* : Act^* \to \mathbf{Set}$ has the property that there exists a natural transformation (mapping) $\lambda_* : O_* \to O$ such that, for every other automaton $D \circ E$ and every natural transformation (mapping) $\lambda : D \circ E \to O$, there exists a unique natural transformation (homomorphism) $\langle\lambda\rangle : D \to O_*$ such that $\lambda_* \circ (\langle\lambda\rangle \circ \varepsilon) = \lambda$. This is an example for a right Kan extension.

6.4.2 Definition 6.4.1. A functor $Ran_T M : \mathbf{D} \to \mathbf{E}$ together with a natural transformation $\varepsilon : Ran_T M *T \to M$ is called a *right Kan extension* of $M : \mathbf{C} \to \mathbf{E}$ along $T : \mathbf{C} \to \mathbf{D}$ if for every functor $G : \mathbf{D} \to \mathbf{E}$

and every natural transformations $\sigma : G * T \to M$ there exists a unique natural transformation $\langle\sigma\rangle : G \to Ran_T M$ such that $\varepsilon \circ (\langle\sigma\rangle * T) = \sigma$, diagrammatically:

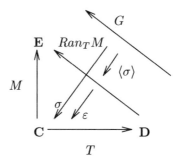

(Remark: the correspondences are $T \cong E$, $M \cong O$, $G \cong D$, $Ran_T M \cong O_*$ in the example above.)

6.4.3 The example of automata provides guidelines for the construction of a right Kan extension in case of $\mathbf{E} = \mathbf{Set}$.

Let $M : \mathbf{C} \to \mathbf{Set}$, $G : \mathbf{D} \to \mathbf{Set}$ be functors, and let $\lambda : G{\circ}T \to M$ be a natural transformation. Using morphisms $f : A \to T(B)$ as tests or inputs, elements $a, a' \in G(A)$ can be distinguished if $\lambda_B(G(f)(a)) \neq \lambda_B G(f)(a)$, or: let the *behaviour* $\lambda_{B,a} : \mathbf{C}[A, T(B)] \to M(B)$ of a at B be defined by $\lambda_{B,a}(f) = \lambda_B(G(f)(a))$. Then a and a' are *behaviourally equivalent* if $\lambda_{B,a} = \lambda_{B,a'}$ for all B. Hence the product $\prod_X M(X)^{\mathbf{D}[A,T(X)]}$ appears as a candidate for the 'set of behaviours' of type A.

But the components of the product are not independent because

$$
\begin{aligned}
\lambda_{C,a}(T(g) \circ f) &= \lambda_C(G(T(g)) \circ G(f)(a)) \\
&= \lambda_C \circ G(T(g))(G(f)(a)) \\
&= M(g) \circ \lambda_B(G(f)(a)) \\
&= M(g) \circ \lambda_{C,a}(G(f))
\end{aligned}
$$

the postcomposition with $T(g)$ does not contribute to separation; if

$$\lambda_{C,a}(T(g) \circ f) \neq \lambda_{C,a'}(T(g) \circ f)$$

then already $\lambda_B(G(f)(a)) \neq \lambda_B G(f)(a)$. Hence, the smaller, 'more abstract' set

$$
\begin{aligned}
M_*(A) = \quad &\{\langle h_X\rangle \in \prod_X M(X)^{\mathbf{D}[A,T(X)]} \mid h_Y(T(g) \circ f) = M(g) \circ h_X(G(f)) \\
&\text{where } f : A \to T(X), g : X \to Y\}
\end{aligned}
$$

specifies all possible behaviours.

$M_*(A)$ extends to a functor $M_* : \mathbf{D} \to \mathbf{Set}$ by

$$M_*(f)(\langle h_X \rangle) = \langle h_X \circ \mathbf{D}[A, f] \rangle,$$

and a natural transformation $\varepsilon_B : \prod_X M(X)^{\mathbf{C}[T(B), T(X)]} \to M(B)$ is defined by $\varepsilon_B(\langle h_X \rangle) = h_B(1_{T(B)})$.

Proposition 6.4.2 (Exercise). (M_*, ε) *is a right Kan extension of* $M :$ $\mathbf{C} \to \mathbf{Set}$ *along* $T : \mathbf{C} \to \mathbf{D}$.

6.4.4 In abstract terms, the construction above splits into two components, the construction of *powers* $M(B)^{\mathbf{D}[A,T(B)]}$ and the restriction of the product $\prod_X M(X)^{\mathbf{D}[A,T(X)]}$. Powers will be introduced as *cotensors*, the restrictions is obtained by means of *ends*, dualizing the proceeding for left Kan extensions.

Definition 6.4.3. Let $\mathbf{C}$ be a category, A be an object of $\mathbf{C}$, and X be a set. An object B^X together with a natural transformation $\varphi_B : \mathbf{C}[\_, B^X] \cong$ $\mathbf{Set}[X, \mathbf{C}[\_, B]]$ is called the *cotensor of* X *with* B.

Definition 6.4.4. Let $F : \mathbf{C}^{\mathrm{op}} \times \mathbf{C} \to \mathbf{D}$ be a functor. An object E together with a wedge $\varepsilon : E \xrightarrow{\cdot} F$ is called an *end* if for every wedge $\omega : W \xrightarrow{\cdot} F$ there exists a unique morphism $\langle \omega \rangle : W \to E$ such that $\varepsilon_A \circ \langle \omega \rangle = \omega_A$ for all objects A of $\mathbf{C}$.
The notation for the coend (object) is $E = \int_X F(X, X)$.

Proposition 6.4.5. $\mathrm{Ran}_T M(A) = \int_X M(X)^{\mathbf{D}[A,T(X)]}$ *provided the necessary cotensors and ends exist.*

6.5 On bimodules and density

6.5.1 A *bimodule* relates categories $\mathbf{C}_1$ and $\mathbf{C}_2$ by means of generalized functions $g : B \to C$ where $A \in Obj(\mathbf{C}_1)$ and $B \in Obj(\mathbf{C}_2)$. $\mathbf{C}_1$ may be thought of as 'import' and $\mathbf{C}_2$ as 'output' languages, which are related by the bimodule in that the generalized functions 'generate output data from input data'. This is roughly the idea of 'modules' in the theory of algebraic data type specification [Ehrig and Mahr, 1990], which is an example for bimodules in computer science.

Definition 6.5.1. Let $\mathbf{C}$ and $\mathbf{D}$ be categories. A *bimodule* $R : \mathbf{C} \nrightarrow \mathbf{D}$ is presented by a functor $R : \mathbf{C}^{\mathrm{op}} \times \mathbf{D} \to \mathbf{Set}$, or equivalently, by functors $R : \mathbf{C} \to (\mathbf{Set}^{\mathbf{D}})^{\mathrm{op}}$ or $R : \mathbf{D} \to \mathbf{Set}^{\mathbf{C}^{\mathrm{op}}}$. (Bimodules are sometimes called *profunctors* in category theory.)

The terminology stems from algebra (for instance [Chevalley, 1956]), the bimodules of which are an instance of enriched (over the category of Abelian groups with tensor) bimodules in our sense.

The following notation is convenient in proofs:

$$g * f := \mathbf{C}[f, B](g) : A \to C \in R \quad \text{if } f : A \to B \in \mathbf{C}, \text{ and}$$
$$h * g := \mathbf{D}[C, h](g) : B \to D \in R \quad \text{if } h : C \to D \in \mathbf{D}.$$

Then we have the equalities

$$h * (g * f) = (h * g) * f$$
$$1_C * g = g = g * 1_B$$

which correspond to the axioms of categories.

Observation 6.5.2.

- Hom functors $\mathbf{C}[\_, \_] : \mathbf{C}^{\mathrm{op}} \times \mathbf{C} \to \mathbf{C}$ are bimodules.

- Every functor $F : \mathbf{C} \to \mathbf{D}$ defines bimodules

$$F_* : \mathbf{D} \nrightarrow \mathbf{C} \quad \text{with } F_*[C, B] = \mathbf{D}[B, F(C)]$$
$$F^* : \mathbf{C} \nrightarrow \mathbf{D} \quad \text{with } F^*[B, C] = \mathbf{D}[F(B), C]$$

- $R[A, \_] : \mathbf{D} \to \mathbf{Set}, R[\_, B] : \mathbf{C}^{\mathrm{op}} \to \mathbf{Set}$ are functors.

The observations suggest considering bimodules as a sort of 'generalized Hom sets' which explains much of the subsequent development.

Proposition 6.5.3.

- Let $R : \mathbf{C} \nrightarrow \mathbf{D}$ and $S : \mathbf{D} \nrightarrow \mathbf{E}$ be bimodules. Define a composition $S * R : \mathbf{C} \nrightarrow \mathbf{E}$ by $S * R[A, C] = \{g * f \mid f \in R[A, B], g \in S[B, C]\}$ where $(g * h) * f = g * (h * f)$ for $h : B \to B'$ (more precisely, $S[h, C](g) * f = g * R[B, h](f)$). This composition is associative and Hom functors are the units.

- Bimodules define a (meta) 2-category with 0-cells being categories, 1-cells being bimodules, and 2-cells being natural transformations $\sigma : R \to S : \mathbf{C}^{\mathrm{op}} \times \mathbf{C} \to \mathbf{Set}$.

(We overload the notation in that $g * f$ is used for the congruence class modulo $(g * h) * f = g * (h * g)$ (cf. Section 6.2.1).)

We might define composition less explicitly

Proposition 6.5.4. $S * R[A, B] = \int^X S[X, B] \otimes R[A, X]$.

Proof. Compute the right hand side along the lines of Section 6.2.3. ∎

Observation 6.5.5. The composition of bimodules $F_* : \mathbf{C} \nrightarrow \mathbf{D}, G_* : \mathbf{D} \nrightarrow \mathbf{E}$ corresponds to the composition of functors in that $(G * F)_* \cong G_* \circ F_*$. Similarly, $(G * F)^* \cong G^* \circ F^*$.

Proof. Let $f : A \to G(B) \in G_*$ and $g : B \to F(C) \in F_*$. Then $G(g) \circ f : A \to G(F(C)) \in (G * F)_*$. If $h : A \to G(F(C)) \in (G * F)_*$ then $h : A \to G(F(C)) \in G_*$, and we have $F(1_C) * h \in G_* \circ F_*$. This defines an isomorphism. ∎

6.5.2 *Partial evaluation* as defined above extends to bimodules.

Definition 6.5.6. Let $R : \mathbf{C} \nrightarrow \mathbf{D}$ be a module, and let $M : \mathbf{C} \to \mathbf{Set}$ be a functor. Define functors $Lan_T M, Ran_R M : \mathbf{D} \to \mathbf{Set}$ by

$$Lan_R M(B) = \int^X R[X, B] \otimes M(X)$$
$$Ran_R M(A) = \int_X M(X)^{R[A,X]}$$

(An explicit definition can be given along the lines of Sections 6.2.3 and 6.4.3.)

These functors are left resp. right adjoint to the following 'reduct functor'.

Definition 6.5.7. Let $R : \mathbf{C} \nrightarrow \mathbf{D}$ be a module, and let $M : \mathbf{D} \to \mathbf{Set}$ be a functor. Define a functor $M_R : \mathbf{C} \to \mathbf{Set}$ by $M_R = \mathbf{Set}^{\mathbf{D}}[M, \_] \circ R$ where $R : \mathbf{C} \to (\mathbf{Set}^{\mathbf{D}})^{\mathrm{op}}$. Explicitly, $M_R(A) = \{\sigma \mid \sigma : R[A, \_] \to M : \mathbf{C} \to \mathbf{Set}$ natural transformation$\}$ and by $M_R(f)(\sigma) = \sigma \circ R[f, \_]$.

In the case that R is a Hom functor we have the isomorphism $M(A) \cong M_R(A)$ which is the source of the construction. A natural transformation $\sigma : R[A, \_] \to M$ may be thought of as 'generalized assignment'.

Proposition 6.5.8. $\__R : \mathbf{Set}^{\mathbf{D}} \to \mathbf{Set}^{\mathbf{C}}$ has a left and a right adjoint, the object parts of which are given by Lan_R and Ran_R.

Proof. Along the lines of Sections 6.2.3 and 6.4.3. ∎

6.5.3 As a prominent property, bimodules characterize all adjunctions between functor categories. This result needs an auxiliary definition of some importance.

Definition 6.5.9. A functor $H : \mathbf{C} \to \mathbf{D}$ is called *dense* if the canonical morphism $colim\ (H \circ P^C) \to C$ is an isomorphism. Dually, H is *codense* if the canonical morphism $C \to lim\ (H \circ Q^C)$ is an isomorphism.

(Remark: $P^C : (H, C) \to \mathbf{C}$ maps $f : F(A) \to C$ to A, and $Q^C : (C, H) \to \mathbf{C}$ maps $g : C \to H(B)$ to B. The objects $f : H(A) \to C$ in (H, C), resp. $g : C \to H(B)$ in (C, H) are cones with regard to $H \circ P^C$ resp. $H \circ Q^C$, which induce the morphisms $colim(H \circ P^C) \to C, C \to lim\ (H \circ Q^C)$.)

Here are some familar examples:

- $1 : \mathbf{1} \to \mathbf{Set}$ which maps to a one-point set is dense. The category of free Σ-algebras $\mathbf{FreeAlg}(\Sigma)$ is dense in $\mathbf{Alg}(\Sigma)$ via the inclusion functor since every algebra can be presented as a coequalizer of free algebras (see for instance [Manes, 1974]) or even as a filtered colimit over free algebras [Johnstone, 1982].

- The completion of a poset to a complete poset is dense.

Proposition 6.5.10. *Let $H : \mathbf{C} \to \mathbf{D}$ be codense, and $G : \mathbf{C} \to \mathbf{E}$ be a functor where $\mathbf{E}$ is complete. Then there exists a unique (up to isomorphism) functor $[G] : \mathbf{D} \to \mathbf{E}$ which preserves limits such that $[G] \circ H \cong G$.*

Proof. Define $[G](C) = lim\ G \circ Q^C$. Let $F : \mathbf{D} \to \mathbf{E}$ be another functor which satisfies the conditions. Then $F(C) \cong F(lim\ (H \circ Q^C)) \cong lim\ (F \circ H \circ Q^C) \cong lim\ (G \circ Q^C)$. ∎

The Yoneda Lemma turns out to be a result about (co-)density.

Proposition 6.5.11. *The functor $Y_{\mathbf{C}} : \mathbf{C} \to (\mathbf{Set}^{\mathbf{C}})^{\mathrm{op}}, A \mapsto \mathbf{C}[A, \_]$ is codense.*

Proof. We show that $F : \mathbf{C} \to \mathbf{Set}$ is a limit of the diagram $Y_{\mathbf{C}} \circ Q^{\mathbf{C}} : (F, Y_{\mathbf{C}}) \to (\mathbf{Set}^{\mathbf{C}})^{\mathrm{op}}$ where the limit cone is $\varepsilon_{(A,\sigma)} = \sigma : Y_{\mathbf{C}}(Q^{\mathbf{C}}(A, \sigma)) = \mathbf{C}[A, \_] \to F$. If $\mu : Y_{\mathbf{C}} \circ Q^{\mathbf{C}} \to G$ is another cone, define a natural transformation by $[\mu]_A : F(A) \to G(A)$ by $[\mu]_A(x) = \mu_{(A,\sigma_x)}(1_A)$ where $\sigma_x : \mathbf{C}[A, \_] \to F$ is the natural transformation which corresponds to $x \in F(A)$ by the Yoneda lemma. (The proof is very explicit. A more elegant proof can be given using the mechanism in [MacLane, 1971], X.6–7.) ∎

Similarly, one can prove that the functor $\mathbf{C} \to \mathbf{Set}^{\mathbf{C}^{\mathrm{op}}}, A \mapsto \mathbf{C}[\_, A]$ is dense (Use $(Y_{\mathbf{C}^{\mathrm{op}}})^{\mathrm{op}}$).

Proposition 6.5.12. *A functor $F : \mathbf{Set}^{\mathbf{C}} \to \mathbf{Set}^{\mathbf{D}}$ has a right adjoint iff $F \cong Lan_R$ for some bimodule $R : \mathbf{C} \nrightarrow \mathbf{D}$.*

Proof. Let $R : \mathbf{C}^{\mathrm{op}} \times \mathbf{D} \to \mathbf{Set}$ which corresponds to the functor $F^{\mathrm{op}} \circ Y_{\mathbf{C}} : \mathbf{C} \to (\mathbf{Set}^{\mathbf{D}})^{\mathrm{op}}$. The functor F^{op} preserves limits since F has a right adjoint. Hence $F = Lan_R$ by the results above. ∎

Corollary 6.5.13. *$Lan_T M(C) \cong colim(T \circ P^C)$ defines a left Kan extension if the right hand side exists.*

This result is fundamental from a technical point of view, being prescriptive in that other results are obtained by variation, for instance the following characterization of all adjoints between categories of algebras (again as a forward reference to Section 8 where we prove that algebras correspond to functors $F : \mathbf{C} \to \mathbf{Set}$ which preserve the finite products of $\mathbf{C}$).

Definition 6.5.14. Let $\mathbf{C}$ and $\mathbf{D}$ be categories with finite products. Call a bimodule $R : \mathbf{C} \nrightarrow \mathbf{D}$ *algebraic* if the functors $R[\_, B] : \mathbf{C}^{\mathrm{op}} \to \mathbf{Set}$ preserve finite coproducts, and if the functors $R[A, \_] : \mathbf{D} \to \mathbf{Set}$ preserve finite products.

Proposition 6.5.15.

- $M_R : \mathbf{Set}^{\mathbf{D}} \to \mathbf{Set}^{\mathbf{C}}$ and $Lan_R : \mathbf{Set}^{\mathbf{C}} \to \mathbf{Set}^{\mathbf{D}}$ restrict to product preserving functors if $R : \mathbf{C} \nrightarrow \mathbf{D}$ is algebraic.

- *[Wraith, 1970]* A functor $F : [\mathbf{C} \Rightarrow_\Pi \mathbf{Set}] \to [\mathbf{D} \Rightarrow_\Pi \mathbf{Set}]$ has a right adjoint iff $F \cong Lan_R$ for some algebraic bimodule $R : \mathbf{C} \nrightarrow \mathbf{D}$.

Proof. • $\mathbf{Set}^{\mathbf{D}}[M, \_] \circ R$ preserves finite products since $R : \mathbf{C} \to (\mathbf{Set}^{\mathbf{D}})^{\mathrm{op}}$ does so.

The real crunch is to prove that $Lan_R M(B) = \int^X R[X, B] \otimes M(X)$ preserves finite products. One follows the lines of Section 6.2.4 for an explicit computation.

- Since $\mathbf{D}[A \Pi B, \_]$ is a coproduct, and since $\mathbf{D}[1, \_]$ is initial in $\mathbf{Set}^{\mathbf{D}}$, $Y_{\mathbf{C}}$ preserves finite products, and hence restricts to a functor $Y_{\mathbf{C}} : \mathbf{C} \to [\mathbf{D} \Rightarrow_\Pi \mathbf{Set}]^{\mathrm{op}}$ which is codense. $F^{\mathrm{op}} \circ Y_{\mathbf{C}} : \mathbf{C} \to [\mathbf{D} \Rightarrow_\Pi \mathbf{Set}]^{\mathrm{op}}$ is algebraic. We here use that every algebraic bimodule $R : \mathbf{C} \nrightarrow \mathbf{D}$ induces a functor $R : \mathbf{C} \to [\mathbf{D} \Rightarrow_\Pi \mathbf{Set}]^{\mathrm{op}}$, $A \mapsto R(A, \_)$ which preserves finite products. ∎

6.5.4 Bimodules may appear to be of mathematical interest only, but we claim that, if not the results, at least the implicit methodology has been used in computer science. Let us start with a sort of example of minor importance.

Bimodules can be presented by a pair of functors $\mathsf{Imp} : \mathbf{In} \to \mathbf{Body}$ and $\mathsf{Exp} : \mathbf{Out} \to \mathbf{Body}$, the bimodule presented being $\mathsf{Exp}_* * \mathsf{Imp}^*$. Vice versa, bimodules define a module presentation as follows:

Definition 6.5.16. Let $R : \mathbf{C} \nrightarrow \mathbf{D}$ be a bimodule. Define a category $\mathbf{R}$ as follows: objects are are those of $\mathbf{C}$ and $\mathbf{D}$ and morphisms $f : A \to B$

are the elements of $\mathbf{C}[A, B]$, $\mathbf{D}[A, B]$ or $R[A, B]$ (we assume disjointness of $Obj(\mathbf{C})$ and $Obj(\mathbf{D})$ for convenience). Composition is either $g \circ f$ or $g * f$.

Proposition 6.5.17. *Let* $R : \mathbf{C} \nrightarrow \mathbf{D}$ *be a bimodule. Then* $R \cong \mathsf{Exp}_* * \mathsf{Imp}^*$ *where* $\mathsf{Imp} : \mathbf{C} \to \mathbf{R}$, *and* $\mathsf{Exp} : \mathbf{D} \to \mathbf{R}$ *are the inclusion functors.*

Proof. Let $h : A \to C \in R$. Then $h : A \to C \in \mathbf{R}$ resp. $h : \mathsf{Imp}(A) \to \mathsf{Exp}(C) \in \mathbf{R}$, hence $1_C * h * 1_A \in R$. Vice versa, if $f : \mathsf{Imp}(A) \to B$ and $g : B \to \mathsf{Exp}(C)$ then $g * f : \mathsf{Imp}(A) \to \mathsf{Exp}(C) \in R$. ∎

Proposition 6.5.18. $Lan_{\mathsf{Exp}_* * \mathsf{Imp}^*} M \cong Lan_{\mathsf{Imp}} M \circ \mathsf{Exp}$ *for* $\mathsf{Imp} : \mathbf{In} \to \mathbf{B}$ *and* $\mathsf{Exp} : \mathbf{Out} \to \mathbf{Body}$.

Proof. $\_ \circ \mathsf{Exp}$ preserves generating universal morphism by Section 3.5.7 since it has a left adjoint. ∎

Specifically, let $\mathbf{C} = \mathbf{T}_{Import}$ and $\mathbf{D} = \mathbf{T}_{Export}$ be algebraic theories generated by specifications $Import$ and $Export$ (cf. Sections 2.14, 8.1). A 'module' is given by specification morphisms $\mathsf{Imp} : Import \to Body$ and $\mathsf{Exp} : Export \to Body$ which induce functors $\mathsf{Imp} : \mathbf{T}_{Import} \to \mathbf{T}_{Body}$ and $\mathsf{Exp} : \mathbf{T}_{Export} \to \mathbf{T}_{Body}$ (cf. Section 8.3.2). Since all the results of this section are valid for algebraic bimodules, the isomorphism $Lan_{\mathsf{Exp}_* * \mathsf{Imp}^*} M \cong Lan_{\mathsf{Imp}} M \circ \mathsf{Exp}$ states that the Kan construction $Lan_{\mathsf{Exp}_* * \mathsf{Imp}^*} M$ corresponds to the semantics of a module in the sense of [Ehrig and Mahr, 1990] which is defined by $Lan_{\mathsf{Imp}} M \circ \mathsf{Exp}$.

6.5.5 We suspect the existence of more substantial witnesses when bimodules and Kan extensions are enriched. We refrain here from introducing all the enriched machinery, and compute the example directly. The reader should consult Kelly's book [Kelly, 1982] or Lawvere's paper [Lawvere, 1973] where I learned about bimodules, and which probably is the most readable introduction to enriched category theory.

If we consider posets $\mathbf{C}$ as 2-enriched categories ($\mathbf{2} = 0 \sqsubseteq 1$), 2-enriched bimodules $R : \mathbf{C}^{op} \otimes \mathbf{D} \to \mathbf{2}$ correspond to *approximable mappings* [Scott, 1982] between partially ordered sets: $R : \mathbf{C} \nrightarrow \mathbf{D}$, i.e. relation $R \subseteq \mathbf{C} \times \mathbf{D}$ such that $(x, x') \in R$ if $x \sqsubseteq y$, $(y, y') \in R$, and $y' \sqsubseteq x'$. 'Models' $\varphi : \mathbf{C} \to \mathbf{2}$ are just monotone maps, or equivalently, *upwards closed subsets*, i.e. $\varphi \subseteq \mathbf{C}$ such that $y \in \varphi$ if $x \in \varphi$ and $x \sqsubseteq y$.

Proposition 6.5.19. *There exists a bijective correspondence between approximable mappings* $R : \mathbf{C} \nrightarrow \mathbf{D}$ *and functions* $f : \mathbf{2}^{\mathbf{C}} \to \mathbf{2}^{\mathbf{D}}$ *which preserve all least upper bounds.*

Proof. Given $f : \mathbf{2}^{\mathbf{C}} \to \mathbf{2}^{\mathbf{D}}$ define a bimodule $(x, y) \in R_f \Leftrightarrow y \in f(x\uparrow)$ where $x\uparrow = \{z \mid x \sqsubseteq z\}$, and, given a bimodule $R : \mathbf{C} \nrightarrow \mathbf{D}$ define $y \in$

$f_R(\varphi) \Leftrightarrow \exists z \in \varphi.(z,y) \in R$. Then $(x,y) \in R_{f_R} \Leftrightarrow y \in f_R(x\uparrow) \Leftrightarrow \exists z \in$ $x\uparrow.zRy \Leftrightarrow (x,y) \in R$ since $x \sqsubseteq z$ and R is approximable. Vice versa $y \in f_{R_f}(\varphi) \Leftrightarrow \exists z \in \varphi.(z,y) \in R_f \Leftrightarrow \exists z \in \varphi.y \in f(z\uparrow) \Leftrightarrow y \in \bigcup_{z \in \varphi} f(z\uparrow) \Leftrightarrow$ $y \in f(\bigcup_{z \in \varphi} z\uparrow) \Leftrightarrow y \in f(z)$. (Once again, the proof is not properly stated in the **2**-enriched world, but the **2**-enriched construction yields $Lan_R\varphi(y) = \top$ if $\exists z \in \varphi.(z,y) \in R$.)   ∎

We claim that the underlying tensor construction is prototypical for constructions used in a variety of 'duality results' as in the work of Scott and of Abramsky where the functors (monotone functions) are restricted to continuous functions or prime proper filters, and where bimodules are restricted accordingly [Scott, 1982; Abramsky, 1987]. Other examples are to be found in Johnstone's book on Stone spaces [Johnstone, 1982]. Similarly, bimodules lurk through in an exposition of Winskel when analysing the relationship between Scott's thesis (that computable functions are continuous) and the axiom of finite causes in event structures [Winskel, 1987]. I wonder if systematic study might give here more insight into the nature of duality results, and leave the study of these examples to the reader.

7 Axiomatizing programming languages

7.1 Relating theories of λ-calculus

7.1.1 Dana Scott suggests in his paper with this title [Scott, 1980] that 'a theory in typed λ-calculus is just the same as a Cartesian closed category'. If we accept typed λ-calculus as a style of functional programming, Cartesian closure appears as an alternative axiomatization of functional programming with higher types (to be precise, functional programming with call-by-name parameter passing which is the programming terminology for the mathematical or rather logical notion of substitution). We elaborate Scott's thesis giving an account of the categorical interpretation of various constructs used in programming languages.

7.1.2 We consider a typed λ-calculus which has the following scheme for generation of types and terms:

Types

      1 is a type

      $A \times B$ is a type if A and B are types

      $A \Rightarrow B$ is a type if A and B are types

Terms

$$\langle\rangle : 1$$

$x^A : A$	where x^A is a variable of type A
$\langle M, N \rangle : A \times B$	if $M : A$ and $N : B$
$\lambda x^A.M : A \Rightarrow B$	if $M : B$
$M N : B$	if $M : A \Rightarrow B$ and $N : A$.

The product type or 'surjective pairing' corresponds to the product structure of Cartesian closed categories.

7.1.3 Such terms induce functions if interpreted in a Cartesian closed category. The procedure is quite standard if looked at in a natural interpretation in the category of sets:

Types

$$[\![1]\!] = \{\langle\rangle\}$$
$$[\![A \times B]\!] = [\![A]\!] \times [\![B]\!]$$
$$[\![A \Rightarrow B]\!] = [\![A]\!] \Rightarrow [\![B]\!]$$

Terms

$$[\![\langle\rangle]\!]\Theta = \langle\rangle$$
$$[\![x^A]\!]\Theta = \Theta(x^A)$$
$$[\![\langle M, N \rangle]\!]\Theta = \langle[\![M]\!]\Theta, [\![N]\!]\Theta\rangle$$
$$[\![\lambda x^A.M]\!]\Theta = \lambda a \in [\![A]\!].[\![M]\!]\Theta[x^A \mapsto a]$$
$$[\![MN]\!]\Theta = [\![M]\!]\Theta([\![N]\!]\Theta)$$

where Θ is a *variable assignment*, i.e. $\Theta(x^A) \in [\![A]\!]$. $\Theta[x^A \mapsto a]$ denotes the variable assignment which is defined as Θ except for $\Theta(x^A) = a$.

From the point of view of categories, the definition is overly generous with variables. Variable assignments can be restricted to variables which actually occur in a term without affecting the interpretation of the specific term. Let us annotate terms with a parameter list $\langle p \rangle = \langle x_1^{A_1}, \ldots, x_n^{A_n} \rangle$: $\langle p \rangle M$ states that the variables $x_1^{A_1}, \ldots, x_n^{A_n}$ may occur in M (but do not necessarily do so). We interpret the annotated terms, or *derived operators*, as functions $[\![\langle p \rangle M]\!] : [\![A_1]\!] \times \ldots \times [\![A_n]\!] \to [\![B]\!]$($a = a_1, \ldots, a_n$ is a list of arguments of suitable type):

$$[\![\langle p \rangle \langle\rangle]\!](a) = \langle\rangle$$
$$[\![\langle p \rangle x_i]\!](a) = a_i$$
$$[\![\langle p \rangle \langle M, N \rangle]\!](a) = \langle[\![\langle p \rangle M]\!](a), [\![\langle p \rangle N]\!](a)\rangle$$
$$[\![\langle p \rangle \lambda x^A.M]\!](a) = \lambda a \in [\![A]\!].[\![\langle p, x^A \rangle M]\!](a, a)$$
$$[\![\langle p \rangle (MN)]\!](a) = [\![\langle p \rangle M]\!](a)([\![\langle p \rangle N]\!](a)).$$

(Note that the definition works only because we are not too rigid, assuming only that variables *may* occur.)

Still, we do not comply with the type discipline of Cartesian closure, having used n-ary Cartesian products instead of binary products. Hence we modify parameter lists to consist either of a variable $\langle x^A \rangle$ or a pair of parameter lists $\langle \langle p \rangle, \langle q \rangle \rangle$ where, of course, $\langle p \rangle$ and $\langle q \rangle$ should have no common variables. Then annotated terms $\langle p \rangle M$ are interpreted to define functions $[\![\langle p \rangle M]\!] : [\![\langle p \rangle]\!] \to [\![B]\!]$ by induction on the structure of parameters as well as of terms (note the changed format, the interpretation of annotated terms is given directly in terms of functions):

$$[\![\langle x^A \rangle]\!] = [\![A]\!]$$
$$[\![\langle \langle p \rangle, \langle q \rangle \rangle]\!] = [\![\langle p \rangle]\!] \times [\![\langle q \rangle]\!]$$
$$[\![\langle p \rangle \langle \rangle]\!] = \langle \rangle_{[\![\langle p \rangle]\!]}$$
$$[\![\langle x^A \rangle x]\!] = 1_{[\![A]\!]}$$
$$[\![\langle \langle p \rangle, \langle q \rangle \rangle x]\!] = \begin{cases} [\![\langle p \rangle x]\!] \circ p_{[\![\langle p \rangle]\!], [\![\langle q \rangle]\!]} & \text{if } x \text{ occurs in } \langle p \rangle \\ [\![\langle q \rangle x]\!] \circ q_{[\![\langle p \rangle]\!], [\![\langle q \rangle]\!]} & \text{if } x \text{ occurs in } \langle q \rangle \end{cases}$$
$$[\![\langle p \rangle \langle M, N \rangle]\!] = \langle [\![\langle p \rangle M]\!], [\![\langle p \rangle N]\!] \rangle$$
$$[\![\langle p \rangle \lambda \langle q \rangle . M]\!] = \lambda ([\![\langle p, x^A \rangle M]\!])$$
$$[\![\langle p \rangle (MN)]\!] = apply_{[\![B]\!], [\![C]\!]} \circ \langle [\![\langle p \rangle M]\!], [\![\langle p \rangle N]\!] \rangle .$$

(I apologize for the *ad hoc* overloading of parameter lists $\langle p \rangle$ and projections $p_{A,B}$.)

This *categorical interpretation* is well defined in every Cartesian closed category.

It seems to be fair enough to talk about annotated terms as (*functional*) *procedures* or *functional terms*. In terms of a functional programming language (such as ML), an annotated term $\langle p \rangle M$ corresponds to a function definition

$$\textbf{funct } f \langle p \rangle = M$$

where f is the function identifier, $\langle p \rangle$ the list of parameters, and M the *body* of the function. Our format would rather be

$$\textbf{funct } f = \langle p \rangle M$$

which rather neatly expresses that the function $\langle p \rangle M$ is bound to the identifier f.

7.1.4 The discussion should justify our choice of language for 'procedures', which we introduced in Section 1.3 (and successively enriched in

Sections 2.1.4, 2.2.2 and 2.3.3). For convenience, we resume the features relevant here.

Types
- 1 is a type
- $A \times B$ is a type if A and B are types
- $A \Rightarrow B$ is a type if A and B are types

Parameter lists
- $\langle x^A \rangle : A$ where x is a variable,
- $\langle \langle p \rangle, \langle q \rangle \rangle : A \times B$ if $\langle p \rangle : A$ and $\langle q \rangle : B$ such that the variables in $\langle p \rangle$ and $\langle q \rangle$ are disjoint.

Functional terms
- $\langle x^A \rangle \langle \rangle : A \to 1$
- $\langle p \rangle x^B : A \to B$ if x^B occurs in $\langle p \rangle : A$
- $\langle p \rangle \langle M, N \rangle : A \to B \times C$ if $\langle p \rangle M : A \to B$ and $\langle p \rangle N : A \to C$
- $\langle p \rangle (f \cdot M) : A \to C$ if $\langle p \rangle M : A \to B$ and $f : B \to C$
- $\langle p \rangle (\lambda \langle q \rangle M) : A \to (B \Rightarrow C)$ if $\langle \langle p \rangle, \langle q \rangle \rangle M : A \times B \to C$
- $\langle p \rangle (MN) : A \to C$ if $\langle p \rangle M : A \to (B \Rightarrow C)$ and $\langle p \rangle N : A \to B$

Axioms

$(\beta \cdot)$ $\langle p \rangle ((\langle q \rangle M) \cdot N) = \langle p \rangle M[\langle q \rangle / N]$

$(\eta \cdot)$ $\langle p \rangle (f \cdot \langle p \rangle) = f$

(π) $\langle p \rangle \langle first \cdot M, second \cdot M \rangle = \langle p \rangle M$

            where $first = \langle x^A, y^B \rangle . x^A$ and
                    $second = \langle x^A, y^B \rangle . y^B$

(τ) $f = \langle p \rangle \langle \rangle$ where $f : A \to 1$

(β) $\langle p \rangle ((\lambda \langle q \rangle M) N) = \langle p \rangle M[\langle q \rangle / N]$

(η) $\langle p \rangle (\lambda \langle q \rangle (M \langle q \rangle)) = \langle p \rangle M$

     if no variable of the parameter list $\langle q \rangle$ occurs free in M.

Moreover, the standard rules for equality hold, and equality is compatible with the structure, i.e.

$$f = f$$
$$f = g \quad\quad \text{if } f = g$$
$$f = h \quad\quad \text{if } f = g \text{ and } g = h$$
$$\langle p \rangle f \cdot M = \langle p \rangle g \cdot N \quad \text{if } f = g \text{ and } \langle p \rangle M = \langle p \rangle N$$
$$\langle p \rangle \langle M_1, N_1 \rangle = \langle p \rangle \langle M_2, N_2 \rangle \quad \text{if } \langle p \rangle M_i = \langle p \rangle N_i \text{ for } i = 1, 2$$
$$\langle p \rangle \lambda \langle q \rangle M = \langle p \rangle \lambda \langle q \rangle N \quad \text{if } \langle \langle p \rangle, \langle q \rangle \rangle M = \langle \langle p \rangle, \langle q \rangle \rangle N$$
$$\langle p \rangle M_1 N_1 = \langle p \rangle M_2 N_2 \quad \text{if } \langle p \rangle M_i = \langle p \rangle N_i \text{ for } i = 1, 2.$$

We assume that parameters can be renamed for free, provided that the binding structure is preserved (technically: α-conversion holds [Barendregt, 1984]). $M[\langle p \rangle / N]$ denotes simultaneous substitution of all variables.

Though both being defined by substitution, the distinction between *parameter passing* $\langle p\rangle f \cdot M$ and *application* $\langle p\rangle((\lambda\langle q\rangle M)N)$ is substantial if looked at in terms of semantics. The semantics of parameter passing is, of course, defined as in Section 1.3.3

$$[\![\langle p\rangle f \cdot M]\!] = [\![f]\!] \circ [\![\langle p\rangle M]\!].$$

Parameter passing (or substitution on free variables) corresponds to composition, and hence can be defined in every category, while application, as defined by

$$[\![\langle p\rangle(MN)]\!] = apply_{[\![B]\!],[\![C]\!]} \circ \langle[\![\langle p\rangle M]\!], [\![\langle p\rangle N]\!]\rangle$$

and abstraction, need Cartesian closure.

7.1.5 So what about Scott's claim of a 'theory of typed λ-calculus being just a Cartesian closed category'.

Let a Λ-*signature* Σ consist of a collection B of *base types*, and a collection of elementary operations $f : A \to B$ where A, B are types. A Λ-*equation* $f = g$ is a pair of functional terms $f, g : A \to B$, and a Λ-*theory* T consists of a Λ-signature Σ and a set E of Λ-equations which comprises all axioms, and which is closed under the deduction rules of the Λ-calculus above. (Compared to 'algebraic signatures', Λ-signatures allow one to introduce higher type operators such as a fixpoint operator $Y : [A \Rightarrow A] \to A$.)

For every Λ-theory T, define a category $\mathbf{T}^*$ with objects being the types generated from the base types, and morphisms being congruence classes of functional terms, i.e. $[f] : A \to B$ where $[f] = \{g : A \to B \mid f = g \text{ in } T\}$. Composition is defined by $[g] \circ [f] = [g \circ f]$, and identities by $1_A = [\langle x^A\rangle]$.

Proposition 7.1.1.

- $\mathbf{T}$ *is a Cartesian closed category.*

- *Every Cartesian closed category* $\mathbf{C}$ *defines a Λ-theory* $T_{\mathbf{C}}$.

Proof. [Sketch]

- Compute all the properties along the lines of Section 1.3.2.

- The objects and morphisms of $\mathbf{C}$ are the base types and elementary operations of $T_{\mathbf{C}}$. The equalities of $T_{\mathbf{C}}$ identify operations on morphisms with the corresponding operations on λ-terms, e.g.

$$1_A = \langle x^A\rangle x,$$
$$g \circ f = \langle x^A\rangle g \cdot (f \cdot x)$$
$$\lambda(f) = \langle x^A\rangle\lambda\langle y^B\rangle f \cdot \langle x, y\rangle.$$

7.1.6 The relation between the Λ-calculus and Cartesian closure is even stronger in that the inherent logics are equivalent. For a change of style we present the equational logic of Cartesian closed categories by a deduction system in the tradition of type theories.

There are three forms of *judgements*

A	is a type
$f : A \to B$	f is a *C-term* with *domain A* and *codomain B*
$f = g : A \to B$	f is equal to g, both being functions with domain A and codomain B

and the following deduction rules

for *composition*

$$\frac{A}{1_A : A \to A} \qquad \frac{f : A \to B \qquad g : B \to C}{g \circ f : A \to C}$$

$$\frac{f : A \to B}{f \circ 1_A = f : A \to B} \qquad \frac{f : A \to B}{1_B \circ f = f : A \to B}$$

$$\frac{f : A \to B \qquad g : B \to C \qquad h : C \to D}{h \circ (g \circ f) = (h \circ g) \circ f : A \to D}$$

for *products*

$$\frac{f : A \to B \qquad g : A \to C}{\langle f, g \rangle : A \to B \times C} \qquad \frac{B \qquad C}{p_{B,C} : B \times C \to B} \qquad \frac{B \qquad C}{q_{B,C} : B \times C \to C}$$

$$\frac{f : A \to B \qquad g : A \to C}{p_{B,C} \circ \langle f, g \rangle = f : A \to B} \qquad \frac{f : A \to B \qquad g : A \to C}{q_{B,C} \circ \langle f, g \rangle = g : A \to C}$$

$$\frac{h : A \to B \times C}{\langle p_{B,C} \circ h, q_{B,C} \circ h \rangle = h : A \to B \times C}$$

and for *higher types*

$$\frac{f : A \times B \to C}{\lambda(f) : A \to (B \Rightarrow C)} \qquad \frac{B \qquad C}{apply_{B,C} : (B \Rightarrow C) \times B \to C}$$

$$\frac{f : A \times B \to C}{apply_{B,C} \circ (\lambda(f) \times 1_B) = f : A \times B \to C}$$

$$\frac{h : A \to (B \Rightarrow C)}{\lambda(apply_{B,C} \circ (h \times 1_B)) = h : A \to (B \Rightarrow C)}$$

Additionally we have the standard rules for equality (reflexivity, symmetry, transitivity) and we assume that the equality is compatible with operators, e.g.

$$\frac{f = f' \qquad g = g'}{g \circ f = g' \circ f'}$$

Proposition 7.1.2. *The logic of Cartesian closure and the logic of typed functional programming are equivalent, i.e. there exist codings $f^{\#}$ and $[\![g]\!]$ of C-terms f and Λ-terms g such that*

$$[\![f]\!]^{\#} = f : A \to B$$
$$[\![f^{\#}]\!] = f : A \to B$$
$$f^{\#} = g^{\#} : A \to B \quad \text{if } f = g : A \to B$$
$$[\![f]\!] = [\![g]\!] : A \to B \quad \text{if } f = g : A \to B.$$

Proof. [Sketch] The interpretation of λ-terms in a category is given in the above. The other direction is straightforward

$$id_A{}^{\#} = \langle x^A \rangle.x$$
$$(g \circ f)^{\#} = \langle x^A \rangle (g^{\#} \cdot (f^{\#} \cdot x))$$
$$p_{A,B}{}^{\#} = \langle x^A, y^B \rangle x$$
$$q_{A,B}{}^{\#} = \langle x^A, y^B \rangle y$$
$$\langle f, g \rangle^{\#} = \langle x^A \rangle \langle f^{\#} \cdot x, g^{\#} \cdot x \rangle$$
$$apply_{B,C}{}^{\#} = \langle f^{B \Rightarrow C}, x^B \rangle (fx)$$
$$\lambda(f)^{\#} = \langle x^A \rangle \lambda \langle y^B \rangle (f^{\#} \cdot \langle x, y \rangle).$$

Now the computations are straightforward though cumbersome, e.g.

$$
\begin{aligned}
[\![(g \circ f)^{\#}]\!] \quad &= [\![\langle x^A \rangle g^{\#} \cdot (f^{\#} \cdot x)]\!] = [\![g^{\#}]\!] \circ [\![\langle x^A \rangle f^{\#} \cdot x]\!] \\
&= [\![g^{\#}]\!] \circ [\![f^{\#}]\!] \circ [\![\langle x^A \rangle x]\!] \\
&= [\![g^{\#}]\!] \circ [\![f^{\#}]\!] \circ id_A = [\![g^{\#}]\!] \circ [\![f^{\#}]\!] = g \circ f \\
&\quad \text{(by inductive assumption).} \\
[\![\langle x^A \rangle (f \cdot N)]\!]^{\#} \quad &= ([\![f]\!] \circ [\![\langle x^A \rangle N]\!])^{\#} \\
&= \langle x^A \rangle [\![f]\!]^{\#} \cdot ([\![\langle x^A \rangle N]\!]^{\#} \cdot x)) \\
&= \langle x^A \rangle (f \cdot (((\langle x^A \rangle N) \cdot x)) \\
&\quad \text{(by inductive assumption)} \\
&= \langle x^A \rangle (f \cdot N).
\end{aligned}
$$

Again as in Section 1.3.3, the only non-trivial computation is the equality

$$[\![\langle x^A \rangle (((\langle y^B \rangle M) \cdot]\!] = [\![\langle y^B \rangle M]\!] \circ [\![\langle x^A \rangle N]\!]? =? [\![\langle x^A \rangle M[y/N]]\!]$$

where now a lot more cases are to be considered. ∎

We conclude that

- a denotational semantics of Λ-calculus (and of typed Λ-calculus) is sound whatever the category of denotations might be if defined along the lines above,

- Λ-calculus is an appropriate language to prove facts about Cartesian closed categories, and on the practical side,

- functional programming with or without variables is equivalent.

The first categorical axiomatization of typed λ-calculus was pioneered by Lambek in the early 60s. The message was spread to computer science by the articles [Lambek, 1980] and [Scott, 1980] and has been worked out in more detail in [Curien, 1986], [Lambek and Scott, 1986] and [Poigné, 1986].

7.2   Type equations and recursion

7.2.1   Modern functional languages offer *recursion, pattern matching,* and *user-defined (recursive) types.* Lists are typical examples of the latter. Let $list_A$ be the type of lists of type A. The inherent *constructors* are defined by code such as

$$list_A = nil_A + + append_A(A\#list_A)$$

in the typical style of a functional programming language. Case distinction is achieved by pattern matching with regard to the constructors:

$$case \ x \ in \ nil_A \mapsto f \cdot \langle\rangle, \ append_A \cdot \langle y, z \rangle \mapsto g \cdot \langle y, z \rangle \ end$$

computes $f \cdot \langle\rangle$ if $x = nil_A$, and $g \cdot \langle a, b \rangle$ if $x = append_A \cdot \langle a, b \rangle$

Standard functions on lists are then defined by (recursive) declarations, for example

$$tail_A : list_A \rightarrow list_A$$
$$tail_A \cdot x = case \ x \ in \ nil_A \mapsto nil_A, append_A \cdot \langle a, y \rangle \mapsto y \ end$$

$$maplist_{A,B} : [A \Rightarrow B] \times list_A \rightarrow list_B$$
$$maplist_{A,B} \cdot \langle f, x \rangle = case \ x \ in \ nil_A \mapsto nil_B, append_A \cdot \langle a, y \rangle \mapsto$$
$$append_B \cdot \langle f \cdot a, maplist_{A,B} \cdot \langle f, y \rangle \rangle \ end$$

Denotational semantics gives meaning to a recursive definition by means of fixpoint operators $Y : [B \Rightarrow B] \rightarrow B$ such that $Y \cdot f = f \cdot (Y \cdot f)$ for

$f : A \to [B \Rightarrow B]$. We recode the 'maplist function' as an example: $maplist_A$ is obtained as fixpoint $Y(maplistscheme)$ of the functional

$$maplistscheme : [([A \Rightarrow B]\Pi list_A) \Rightarrow list_B] \to [([A \Rightarrow B]\Pi list_A) \Rightarrow list_B]$$
$$(maplistscheme \cdot g) \langle f, x \rangle = case \; x \; in \; insert_A \cdot a \mapsto insert_B \cdot (f \cdot a),$$
$$append_A \cdot \langle a, y \rangle \mapsto append_B \cdot \langle f \cdot a, g \cdot \langle E, E \rangle \rangle \; end$$

The combination of type equations, pattern matching and recursion with Cartesian closure is in most cases sufficient to support a denotational semantics for constructs of programming languages, e.g.

$$while : [A \Rightarrow 1 + 1] \times [A \Rightarrow A] \times A \to A$$
$$while = \langle f^{[A \Rightarrow 1+1]}, g^{[A \Rightarrow A]}, x^A \rangle \; case \; (f \cdot x) \; in \; inl \; \cdot \langle \rangle \mapsto x,$$
$$inr \; \cdot \langle \rangle \mapsto while \cdot \langle f, g, (g \cdot x) \rangle \; end$$
$$if : [A \Rightarrow 1 + 1] \times A \times B \times B \to A$$
$$if = \langle f^{[A \Rightarrow 1+1]}, x^A, y^B, z^B \rangle case(f \cdot x) \; in \; inl \cdot \langle \rangle \mapsto y, \; inr \cdot \langle \rangle \mapsto z \; end$$

7.2.2 Sums

Pattern matching is closely related to coproducts (see Section 2.2.1); constructors behave like injections, and the case-statement defines a function in terms of functions on the components. Only the uniqueness condition is missing, for good reasons as we will learn below (Section 7.2.4).

Definition 7.2.1. A *binary sum* of objects A, B consists of an object $A + B$ and morphisms $inl_{A,B} : A \to A + B$, $inr_{A,B} : B \to A + B$ such that for every pair of morphisms $f : A \to C, g : B \to C$, there exists a morphism $case(f, g) : A + B \to C$ such that the equalities

$$case(f, g) \circ inl_{A,B} = f$$
$$case(f, g) \circ inr_{A,B} = g$$
$$h \circ case(f, g) = case(h \circ f, h \circ g)$$
$$case(inl, inr) = 1_{A+B}$$

hold. A functor $+ : \mathbf{C} \times \mathbf{C} \to \mathbf{C}$ is defined by $f + g := case(inl \circ f, inr \circ g)$.

We consider two examples:

- Coproducts are, of course, examples of binary sums.

- Those readers familiar with domain theory will immediately be reminded of separated sums $X +_{\perp} Y$ of directed complete posets (cf. Section 2.2.2). Every pair $f : X \to Z, g : Y \to Z$ of continuous functions induce a continuous function $case(f, g) : X +_{\perp} Y \to Z$ defined by $case(f, g)(1, x) = f(x), case(f, g)(2, y) = g(x)$, and $case(f, g)(\perp) = \perp$.

7.2.3   Recursion

Definition 7.2.2.

- Let **C** be a category with finite products. A morphism $g : A \to B$ is said to have a *fixpoint* or *solution* of $f : A\Pi B \to B$ if $g = f \circ \langle 1_A, g \rangle$.

- A category **C** *has fixpoints* if for every morphism $f : A\Pi B \to B$ there exists a specified solution $f^\dagger : A \to B$.

Observation 7.2.3. A Cartesian closed category **C** has fixpoints if and only if it has fixpoint operators $Y_A : [A \Rightarrow A] \to A$ such that

$$apply_{A,A} \circ \langle f, Y_A \circ f \rangle = Y_A \circ f \text{ for all } f : X \to [A \Rightarrow A].$$

Proof. Define $Y_A = (apply_{A,A})^\dagger$ and, vice versa, $f^\dagger = Y_B \circ \lambda(f)$. ∎

As an example for categorical functional programming we consider the '*maplist* function' which is coded as follows (where $list_A = nil_A + (A\Pi list_A)$):

$maplist = Y([A \Rightarrow B]\Pi list_A) \Rightarrow list_B](maplistscheme)$ *where*
$maplistscheme_{A,B} : [([A \Rightarrow B]\Pi list_A) \Rightarrow list_B] \to [([A \Rightarrow B]\Pi list_A) \Rightarrow list_B]$
$maplistscheme_{A,B} = apply_{[A \Rightarrow B]\Pi list_A, list_B} \circ \langle Y \circ \ulcorner\lambda(case)\urcorner \circ \langle\rangle_{[A \Rightarrow B]\Pi list_A},$
$1_{[A \Rightarrow B]\Pi list_A}\rangle$
   where $(\ulcorner f \urcorner) = \lambda(f \circ q_{1,A})$, and
      $case : [[A \Rightarrow B]\Pi list(A) \Rightarrow list(B)]\Pi([A \Rightarrow B]\Pi list(A)) \to list(A)$
        $case = case(inl_B \circ apply_{A,B} \circ \langle\pi_2, \pi_4\rangle,$
          $append_B \circ \langle apply_{A,B} \circ \langle\pi_2, \pi_4\langle\circ apply_{[A \Rightarrow B]\Pi list_A, list_B} \circ \langle\pi_1, \langle\pi_2, \pi_3\rangle\rangle\rangle) \circ distr$
        where $distr_{A,B,C} : A\Pi(B + C) \to (A\Pi B) + (A\Pi C)$
          $distr_{A,B,C} = \lambda^{-1} (case(\lambda(inl_{A\Pi B, A\Pi C} \circ reverse_{B,A}),$
                       $\lambda(inr_{A\Pi B, A\Pi C} \circ reverse_{C,A}))) \circ reverse_{A,B+C}$
        $reverse_{A,B} = \langle q_{A,B}, p \rangle : A\Pi B \to B\Pi A$

the terms π_i are the (iteratively defined) projections to the components of

$$(\dots(A_0\Pi A_1)\Pi\dots\Pi A_n).$$

We cannot claim a gain of lucidity, but the translation into the categorical language is the one used in denotational semantics, and we have clearly the advantage that the definition is valid in whatever semantic universe we choose, provided it is Cartesian closed, and also has sums and fixpoints both of morphisms and of type equations.

The category **Dcpo** of directed complete posets is an example of a Cartesian closed category with (separated) sums and fixpoints

Proposition 7.2.4. *Let* $X = (X, \sqsubseteq)$ *be a directed complete poset. Then* $Y : [X \Rightarrow X] \rightarrow X$ *by* $Y(f) = \sqcup\{f^n(\bot) \mid n \in N\}$ *defines a fixpoint operator.*

Proof.   • Y is well defined: $\bot \sqsubseteq f(\bot) \sqsubseteq f^2(\bot) \sqsubseteq \dots$ since f is monotone. We conclude that $\{f^n(\bot) \mid n \in \mathbb{N}_0\}$ is directed.

• Y is monotone: $f \sqsubseteq g$ implies $f^n(\bot) \sqsubseteq g^n(\bot) \sqsubseteq \sqcup\{g^n(\bot) \mid n \in N\}$.

• Y is continuous: $Y(f)(\sqcup G) = \sqcup\{((\sqcup G))^n(\bot) \mid n \in \mathbb{N}_0\} = \sqcup\{\sqcup\{g_n \circ \dots \circ g_1(\bot) \mid g_i \in F\} \mid n \in \mathbb{N}_0\} = \sqcup\{g_n \circ \dots \circ g_1(\bot) \mid g_i \in F, n \in \mathbb{N}_0\}$ by definition of least upper bounds and because of continuity of the g_i's. Since G is directed there exists some g in G such that $g_i \sqsubseteq g$ for all i's, hence $g_n \circ \dots \circ g_1 \sqsubseteq g$. Thus $\sqcup\{g_n \circ \dots \circ g_1(\bot) \mid g_i \in F, n \in \mathbb{N}_0\} = \sqcup\{g \circ \dots \circ g(\bot) \mid g \in F, n \in \mathbb{N}_0\} = \sqcup\{Y(f) \mid f \in G\}$.

• Y satisfies the fixpoint property: we claim that $f(Y(f)) = Y(f)$ for $f \in X \Rightarrow X$ as demonstrated by the computation

$$f(Y(f)) = f(\sqcup\{f^n(\bot) \mid n \in \mathbb{N}_0\}) = \sqcup\{f^{n+1}(\bot) \mid n \in \mathbb{N}_0\}$$
$$= \sqcup\{f^n(\bot) \mid n \in \mathbb{N}_0\} = Y(f)$$

where we make use of the fact that f is continuous, as well as of the properties of least upper bounds. Then $apply_{X,X} \circ \langle f, Y \circ f \rangle(x) = f(x)(Y(f(x)) = Y(f(x)) = Y \circ f(x)$.

(The proof is so detailed so as to compare it to its categorical version below (Section 7.3.2).)   ■

According to Scott, the elements of a directed set should be seen as partial approximations of the least upper bounds. In terms of computations: the elements of a directed set are snapshots of intermediate stages of a possibly infinite computation of a program. Directedness is somewhat of an equivalent to determinism, meaning that different finite approximations of a function cannot be compared directly, but are nevertheless compatible in that they approximate a common upper bound. In analogy, given a program, computation steps may be executed in a different order, for instance evaluating a term, without affecting the overall result if the evaluation system is Church–Rosser. (For more motivation and concise discussion we refer to the Chapter on Domain theory, Vol. 3.) This intuition will be helpful.

7.2.4   Disjoint sums do not coexist with fixpoints in Cartesian closed categories

Whoever takes a look at denotational semantics and domain theory probably wonders at some stage about why separated and coalesced sums do not behave like disjoint sums, or whether there exist disjoint sums of domains. Here is a categorical proof that disjoint sums, fixpoints and call-by-name do not coexist (the observation is implicit in [Lawvere, 1969b] and has been worked out in [Huwig and Poigné, 1990]). The proof depends on three basic observations:

(1) A Boolean algebra is trivial if negation ('$\neg$') has a fixpoint since

$$(*) \qquad \begin{aligned} \mathrm{tt} &= Y(\neg) \vee \neg(Y(\neg)) = Y(\neg) \vee Y(\neg) = Y(\neg) = \\ &= Y(\neg) \wedge Y(\neg) = Y(\neg) \wedge \neg(Y(\neg)) = \mathrm{ff}. \end{aligned}$$

(2) The coproduct $2 := 1 \amalg 1$ is a *Boolean algebra object*, i.e. there are morphisms such as $\neg : 2 \to 2$ ('not'), $\wedge : 2\Pi2 \to 2$ ('and') which satisfy the laws of Boolean algebras.

(3) $A\Pi(B \amalg C) \cong (A\Pi B) \amalg (A\Pi C)$ in a Cartesian closed category since left adjoints preserve coproducts.

Now for the proof:

Proof. The Boolean operators on 2 are defined by 'truth tables' where we exploit (3): 'not' is defined as the unique morphism in the diagram

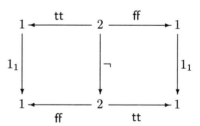

$\wedge : 2\Pi2 \to 2$ is induced by the diagrams

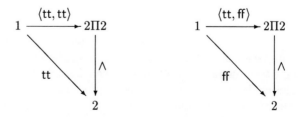

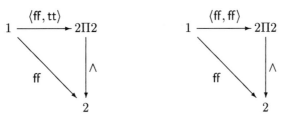

where we use that $2\Pi2 \cong 1 \amalg 1 \amalg 1 \amalg 1 \amalg 1$ is a coproduct with injections $\langle\mathsf{tt},\mathsf{tt}\rangle,\dots,\langle\mathsf{ff},\mathsf{ff}\rangle : 1 \to 2\Pi2$. One might want to check a few of the Boolean laws for exercise.

Rephrasing the chain (∗) of equations as a categorical argument proves the equality $t = f : 1 \to 2$, and $inl_{A,A} = inr_{A,A} : A \to A \amalg A$ as a consequence of (3). We use this equality, the fact that every object A has a 'global element' $\bot_A := Y_A \circ \lambda(q_{A,A}) : 1 \to A$, and the diagram

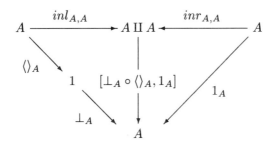

to compute $\bot_A \circ \langle\rangle_A = [\bot_A \circ \langle\rangle_A, 1_A] \circ inl_{A,A} = [\bot_A \circ \langle\rangle_A \circ 1_A] \circ inr_{A,A} = 1_A$. Moreover $\langle\rangle_A \circ \bot_A = 1_1$, because 1 is a terminal object, hence $A \cong 1$. ∎

Proposition 7.2.5. $A \cong 1$ for all objects A in a Cartesian closed category with coproducts and fixpoints.

Standard categories of domains are Cartesian closed and have fixpoint, hence coproducts cannot exist. As a variation, the categories of domains with strict functions do have coproducts being the coalesced sums, but are not Cartesian closed.

7.2.5 Recursive types

User defined recursive types such as $list_A = nil_A + {+}append_A(A\#list_A)$ are naturally interpreted to specify an isomorphism

$$\varphi_A : list_A \cong 1 + A\Pi list_A$$

of objects in a category **C**. Then list constructors are induced via the isomorphisms

$$nil_A = \varphi^{-1}{}_A \circ inl_{1,A\Pi list_A} \qquad\qquad append_A = \varphi^{-1}{}_A \circ inr_{1,A\Pi list_A}$$

as are other operators

$$tail_A = \varphi^{-1}{}_A \circ case(inl_{1,A\Pi list_A} \circ \langle\rangle_1, inr_{1,A\Pi list_A} \circ q_{A,list_A} \circ \varphi_A).$$

The general pattern is as follows. A recursive data structure X is defined by an isomorphism

$$X \cong Term(A, X)$$

where $Term(A, X)$ is a term which defines a functor $Term : \mathbf{D} \times \mathbf{C} \to \mathbf{C}$, e.g.

$$\text{LIST} : \mathbf{C} \times \mathbf{C} \to \mathbf{C}, (A, X) \mapsto 1 + (A\Pi X)$$

which may be thought of as a *specification* of a recursive data structure with *parameter* A, the type specified being a *fixpoint* or *solution* of the *type equation*.

Definition 7.2.6.

- Let $F : \mathbf{D} \times \mathbf{C} \to \mathbf{C}$ be a functor. A *solution* of F consists of a functor $S : \mathbf{D} \to \mathbf{C}$ together with a natural isomorphism $\varphi : S \cong F, \langle 1_{\mathbf{D}}, S \rangle$.

- A category $\mathbf{C}$ *has solutions of type equations* if for every functor $F : \mathbf{D} \times \mathbf{C} \to \mathbf{C}$ there exists a specified solution $F^{\dagger} : \mathbf{D} \to \mathbf{C}$.

- In the special case that $\mathbf{D} = \mathbf{1}$, we ambiguously say that an object X is a solution if $X \cong F(X)$. Note that a solution is a fixpoint in the category $\mathbf{Cat}$.

For the example of lists we for instance have the following solutions:

- The functor $\_^* : \mathbf{Set} \to \mathbf{Set}$, which maps a set A to the set A^* of finite words, is a solution to the functor $LIST : \mathbf{Set} \times \mathbf{Set} \to \mathbf{Set}$, $(A, X) \mapsto 1 + (A \times X)$ where $+$ is the disjoint union. The isomorphism is given by $\varphi_A(\epsilon) = (1, \langle\rangle), \varphi_A(aw) = (2, (a, w))$ where $a \in A$ and $w \in A^*$.

  The functor $\_^{\infty} : \mathbf{Set} \to \mathbf{Set}$, which maps a set A to the set A^{∞} of finite and infinite words, is another solution where $\varphi(\epsilon) = (1, \langle\rangle)$, $\varphi(aw) = (2, (a, w))$ where $a \in A$ and $w \in A^{\infty}$.

- Define a directed complete poset A^{ω} of finite, infinitary and infinite words, i.e. words $v, v\bot$, or w where v is a finite word, and w an infinite word over A. The order is given by $v\bot \sqsubseteq vw$. This defines a functor $\_^{\omega} : \mathbf{Dcpo} \to \mathbf{Dcpo}$ which is a solution to the functor $LIST^{\omega} : \mathbf{Dcpo} \times \mathbf{Dcpo} \to \mathbf{Dcpo}, (A, X) \mapsto 1 +_{\bot} (A\Pi X) (+_{\bot}$ denotes the separated sum).

One should not expect to solve the equations on the nose, i.e. to find an object X such that $X = F(X)$ even if such an object might exist; a semantics interprets the type operators in a fixed way, for instance $A \times B$ as a Cartesian product, and $A + B$ as $\{(0, a) \mid a \in A\} \cup \{(1, b) \mid B \in B\}$. Then $A^* \neq 1 + A \times A^*$.

The examples show that type specification $F : \mathbf{D} \times \mathbf{C} \to \mathbf{C}$ may have several solutions. Specific solutions are determined by additional criteria:

Definition 7.2.7. A solution (I, φ^I) is *minimal* if for all solutions (S, φ) there exists a unique natural transformation $\sigma : I \to S$ such that $F(1_A, \sigma_A) \circ \varphi_A^I = \varphi_A \circ \sigma_A$. A solution (T, φ^T) is *maximal* if for all solutions (S, φ) there exists a unique natural transformation $\sigma : S \to T$ such that $F(1_A, \sigma_A) \circ \varphi_A = \varphi^T{}_A \circ \sigma_A$.

For instance, the functors $\_^*$ and $\_^\infty$ are minimal resp. maximal solutions of *LIST*, and $\_^\omega$ is a minimal solution of $LIST^\omega$.

Denotational semantics typically defines the meaning of recursive type declarations in terms of minimal or *least* such *solutions*. We shall learn below about systematic ways to construct minimal as well as maximal fixpoints by means of limits and colimits.

7.2.6 There are some type equations which are quite important in denotational semantics, but which do not seem to conform with the scheme above. For instance, the right hand side of $U \cong [U \Rightarrow U]$ fails to define a functor from $\mathbf{C}$ to $\mathbf{C}$ since the object U occurs both in a covariant and in a contravariant position. Unfortunately, exactly this kind of type equations occurs invariably in a denotational semantics because of the self-referencing nature of programming languages.

However, exponentiation becomes a functor in categories of the following kind: let $\mathbf{C}$ be a Cartesian closed category. Define a category $\mathbf{C}^{\leftrightarrow}$ to have the same objects as $\mathbf{C}$ but morphism $(f, g) : A \to B$ where $f : A \to B, g : B \to A$ are morphisms of $\mathbf{C}$, and with compositions defined on the components. Then a functor $\Rightarrow : \mathbf{C}^{\leftrightarrow} \to \mathbf{C}^{\leftrightarrow}$ is defined by $A \mapsto [A \Rightarrow A]$ and $(f, g) \mapsto [g \Rightarrow f]$. Examples for such categories will be considered more closely in the chapter on 'domain theory' (Vol. 3), where $\mathbf{C} = \mathbf{Dcpo}$ and where morphisms are *projection pairs*, i.e. pairs of continuous morphisms $f : (X, \sqsubseteq) \to (Y, \sqsubseteq)$ and $g : (Y, \sqsubseteq) \to (X, \sqsubseteq)$ such that $g \circ f = 1$ and $f \circ g \sqsubseteq 1_Y$ (see also [Plotkin, 1981], the corresponding colimit construction is sketched in Section 5.3.4).

Untyped λ-calculus is the example *par excellence* for self-reference. Untyped λ-calculus strips off all type information and introduces unrestricted application MN. Hence 'functions' f can be applied to itself, e.g.

$$(\lambda x.x)(\lambda x.x).$$

This increases expressivity of the calculus substantially in that we have non-terminating computations $(\lambda x.xx)(\lambda x.xx) = (\lambda x.xx)(\lambda x.xx)$ and a fixpoint operator $Y = \lambda f.(\lambda x.(f(xx))(\lambda x.(f(xx)))$ which do not exist in the simple typed λ-calculus.

Untyped calculus is embedded into typed λ-calculus if we assume existence of a *universal object U which* comes together with operations $dec : U \rightarrow [U \Rightarrow U]$ and $code : [U \Rightarrow U] \rightarrow U$ which are the components of an isomorphism $U \cong [U \Rightarrow U]$. One translates a term M of the untyped λ-calculus to a functional term $\langle x_1^U, \ldots, x_n^U \rangle M^\#$ of the Λ-calculus with a universal element as follows ($x_1, \ldots, x_n$ is the list of free variables of M)

$$x^\# = \langle x^U \rangle x$$
$$(\lambda x_i.M)^\# = \langle x_1^U, \ldots, x_{i-1}^U, x_{i+1}^U, \ldots, x_n^U \rangle \ code \ \cdot (\lambda x_i^U.(M^\# \cdot \langle x_1^U, \ldots, x_n^U \rangle))$$
$$(MN)^\# = \langle x_1^U, \ldots, x_n^U \rangle (dec \ \cdot (M^\# \cdot \langle x_1^U, \ldots, x_n^U \rangle))(N^\# \cdot \langle x_1^U, \ldots, x_n^U \rangle).$$

The Λ-calculus with a universal type and thus the untyped λ-calculus can then be interpreted in every Cartesian close category with a universal object $U \cong [U \Rightarrow U]$.

7.2.7 Of course, we have only touched upon the subject of modelling programming languages in terms of category theory. For instance, Cartesian closure only models a specific evaluation strategy of functional programs, namely call-by-name. But other strategies, such as call-by-value, have been investigated as well, and neat axiomatizations have been found. A close relation with partial functions has been established (see for instance [Moggi, 1985]), and a variety of papers have appeared, dealing with categories of 'partial functions' (e.g. [Rosolini and Robinson, 1988] where a variety of references may be found).

Other exciting recent developments comprise the *monadic view of computations* propagated by [Moggi, 1989; Moggi, 1991], the *evaluation logic* of [Pitts, 1991], and, on the semantic side, the development of *synthetic category theory* as to be found in [Hyland, 1990], [Phoa, 1990].

7.3 Solving recursive equations

7.3.1 *F*-dynamics

For a more systematic analysis of type equations we introduce the notion of *F-dynamics* [Arbib and Manes, 1974] which will ease the technical development (and broaden the perspective). For technical convenience we restrict our attention to type equations of the form $X \cong F(X)$ where $F : \mathbf{C} \rightarrow \mathbf{C}$. (The reader may redevelop the theory for parameterized type equations as an exercise.)

Definition 7.3.1.

- Let $F : \mathbf{C} \rightarrow \mathbf{C}$ be a functor. An *F-dynamics* consists of an object A and a morphism $\delta : F(A) \rightarrow A$.

- *homomorphism* $h : (A, \delta) \to (A', \delta')$ of F-dynamics is a morphism $h : A \to A'$ such that the diagram

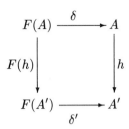

commutes. This defines a category F-**Alg** of F-dynamics.

We illustrate the definition by some examples:

- If $F : \mathbf{C} \to \mathbf{C}$ is a monotone function between partially ordered sets, F-dynamics $F(x) \le x$ are called a *prefixpoint*.

- The functor $\_\Pi A : \mathbf{C} \to \mathbf{C}$ which maps an object X to the product $X \Pi A$. Dynamics are 'transition functions' $\delta : S \Pi A \to S$.

- Let Ω be a one-sorted signature (cf. Section 3.6.1). The functor $\Omega :$ **Set** $\to$ **Set** maps a set A of 'values' to the set $\Omega(A) := \{(\omega, a_1, \dots a_n) \mid \sigma \in \Omega, a(\omega) = n, a_1, \dots, a_n \in A\}$. Ω-dynamics correspond to Ω-algebras with carrier A and operations $\omega^A(a_1, \dots, a_n) = \delta(\omega, a_1 \dots, n)$. More generally, we can define a functor $\Omega : \mathbf{C} \to \mathbf{C}$ by $\Omega(X) :=$ $\coprod_{\omega \in \Omega} X^{a(\omega)}$ (provided the products and coproducts exist) which transfers the concept of Ω-dynamics to arbitrary categories.

Proposition 7.3.2. *Let* $F : \mathbf{C} \to \mathbf{C}$ *be a functor. An initial F-dynamics is a minimal solution of* F.

Proof. By definition except that we have to check that initiality of (A, δ) implies that $\delta : F(A) \to A$ is an isomorphism. This follows from the diagram

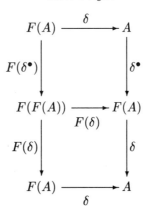

where $\delta^\bullet$ is the inverse: initiality induces $\delta \circ \delta^\bullet = 1_A$, which used in the upper quadrangle, yields $\delta^\bullet \circ \delta = 1_{F(A)}$. ∎

We reconsider the examples above:

- Given a continuous function $f : X \to Y$ of dcpo's, $\bigsqcup f^n(\bot)$ is the initial prefixpoint.

- A^* with $\delta : A^* \times A \to A^*, (w, a) \mapsto wa$ is an initial dynamics for the functor $\_ \times A : \mathbf{Set} \to \mathbf{Set}$.

- The set of Ω-terms T_Ω with $\delta : \Omega(T_\Omega) \to T_\Omega, (\omega, t_1, \ldots, t_n) \mapsto \omega(t_1, \ldots, t_n)$ is an initial dynamics with regard to $\Omega : \mathbf{Set} \to \mathbf{Set}$.

7.3.2   Construction of initial dynamics by ω-colimits

Fixpoints of continuous functions between dcpo's have been computed rather carefully because the construction of solutions for type equations will follow exactly the same pattern except that partial orders (partial order categories) are replaced by categories, and directed sets by a special kind of filtered colimits. Here is the result by brute force.

Proposition 7.3.3. *Let* $\mathbf{C}$ *be a category with filtered colimits, and let* $F : \mathbf{C} \to \mathbf{C}$ *be a functor which preserves non-empty filtered colimits. Define a filtered diagram* $F^\omega : \omega \to \mathbf{C}$ *by*

$$0 \xrightarrow{[]} F(0) \xrightarrow{F([])} F^2(0) \longrightarrow \cdots \longrightarrow F^n(0) \xrightarrow{F^n([])} F^{n+1}(0) \cdots$$

ω *being the order category* $0 \le 1 \le \ldots \le n \le \ldots$. *Then colim* F^ω *is an initial* F-*dynamics with* $[\delta^\omega] : F(colim\ F^\omega) \to colim\ F^\omega$ *being uniquely induced by the natural transformation* $\delta_n^\omega = \eta_{F^\omega(n+1)} : F(F^\omega(n)) \to colim\ F^\omega$ *where* $\eta_n : F^\omega(n) \to colim\ F^\omega$ *is the colimit cone of* F^ω.

Proof. Let $\delta : F^\omega(A) \to A$ be an F^ω-dynamics. Define a natural transformation $\sigma_n : F^\omega(n) \to A$ by

$$
\begin{array}{ccccccccc}
0 & \xrightarrow{[]} & F(0) & \xrightarrow{F([])} & F^2(0) & \longrightarrow & \cdots & \longrightarrow & F^n(0) & \xrightarrow{F^n([])} & F^{n+1}(0) & \longrightarrow \cdots \\
{\scriptstyle[]}\downarrow & & {\scriptstyle F([])}\downarrow & & {\scriptstyle F^2([])}\downarrow & & & & {\scriptstyle F^n([])}\downarrow & & {\scriptstyle F^{n+1}([])}\downarrow & \\
A & \xleftarrow{\delta} & F(A) & \xleftarrow{F(\delta)} & F^2(A) & \longleftarrow & \cdots & \longleftarrow & F^n(A) & \xleftarrow{F^n(\delta)} & F^{n+1}(A) & \longrightarrow \cdots
\end{array}
$$

The induced morphism $[\sigma] :\ colim\ F^\omega \to A$ is a homomorphism of F^ω-dynamics, i.e.

$$
\begin{array}{ccc}
F(colim\ F^\omega) & \xrightarrow{\ [\delta^\omega]\ } & colim\ F^\omega \\
{\scriptstyle F([\sigma])}\downarrow & & \downarrow{\scriptstyle[\sigma]} \\
F(A) & \xrightarrow{\ \delta\ } & A
\end{array}
$$

commutes, because of the colimit properties of $F(colim\ F^\omega)$; as argued subsequently: the diagram

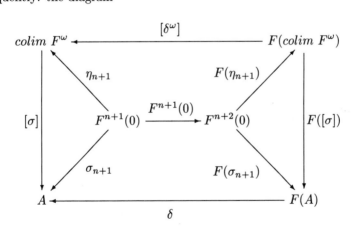

commutes because of the definitions of $\sigma, [\sigma]$ and $[\delta^\omega]$. As $F(\eta)$ defines the colimit cone of $F \circ F^\omega$, there is only one morphism from $F \circ F^\omega(n)$ to A for all n of ω. Clearly, $[\sigma]$ is the only homomorphism because of colimit properties again. (It should be helpful to correlate this proof step by step with the proof of existence of fixpoints in the category of dcpo's.) ∎

This colimit construction first appeared in Scott's work on the foundations of programming semantics [Scott, 1976] in the context of partially

ordered sets while its categorical form was first pointed out by [Smyth and
Plotkin, 1977]. [Adamek and Koubek, 1979] gives the subject a thorough
treatment extending the construction to transfinite chains.

8 Algebra categorically

We redevelop a substantial part of universal algebra in categorical terms as
a second, fundamental application of categorical definitions. Though the
theme is mathematical by nature, various applications of universal algebra
in such areas as specification theory and programming semantics should
raise enough interest for a computer scientist to have a look at this section.

8.1 Functorial semantics

8.1.1 Algebraic versus procedural logic

Lawvere was the first to point out the correspondence between categories
and theories, and between functors and models, upon which much of our
motivation in the introductory sections was based. In his pioneering thesis
[Lawvere, 1963] he observes that algebraic logic coincides with the logic of
finite products in a sense made precise below.

We translate the notions of *universal algebra* into categorical language
according to the following scheme

terms	$\mapsto$	functional terms
		(or *derived operators* in this context)
substitution	$\mapsto$	composition
equations	$\mapsto$	equations of functional terms
theories	$\mapsto$	categories with finite products
algebras	$\mapsto$	functors which 'preserve finite products'
homomorphisms	$\mapsto$	natural transformations between such functors

Functional terms are defined as in the previous section but restricted
to the components dealing with products. Explicitly, the calculus, we refer
to as Π-*calculus*, is generated by

Types
    1 is a type
    $A \times B$ is a type if A and B are types

Functional terms
    $\langle x^A \rangle \langle \rangle : A \to 1$
    $\langle p \rangle x^B : A \to B$ if x^B occurs in $\langle p \rangle : A$

$\langle p \rangle \langle M, N \rangle : A \to B \times C$ if $\langle p \rangle M : A \to B$ and $\langle p \rangle N : A \to C$
$\langle p \rangle (f \cdot M) : A \to C$ if $\langle p \rangle M : A \to B$ and $f : B \to C$

(parameter lists and equational axioms are defined as in Sections 2.1.4 resp. 7.1.4).

A Π-*signature* Σ then consists of a set B of base types, and of a set of elementary operators of the form $f : b_1 \times \ldots \times b_n \to b$. This signature is nothing but ordinary algebraic signatures as defined in Section 3.6.1 (which the reader should have a look at for notation) but in a slightly modified format.

We recall that *algebraic (equational) deduction* is given by the following deduction rules

$$\frac{}{t =_{X,b} t} \qquad \frac{t_1 =_{X,b} t_2}{t_2 =_{X,b} t_1} \qquad \frac{t_1 =_{X,b} t_2 \qquad t_2 =_{X,b} t_3}{t_1 =_{X,b} t_3}$$

$$\frac{t_1 =_{X,b} t_2}{[\Theta]t_1 =_{Y,b} [\Theta]t_2}$$

where $\Theta : X \to T_\Sigma(Y)$ is a variable assignment. As usual, an *algebraic theory* $T = (\Sigma, E)$ consists of a signature Σ and a set of equations E which are closed under deduction. **Alg**(T) denotes the category of T-algebras.

As to be expected, algebraic logic coincides with the Π-logic.

Proposition 8.1.1. *Algebraic logic and Π-logic are equivalent.*

Proof. [Sketch] Σ-terms $t \in T_\Sigma(X)$ translate to a functional term $\langle X \rangle t^{\#}$ by

- $x^{\#} = x$ where $x \in X_b$, and
- $f(t_1, \ldots, t_n)^{\#} = f \cdot \langle t_1^{\#}, \ldots, t_n^{\#} \rangle$

where $\langle X \rangle$ is a parameter list which comprises all the variables in X, typically $\langle X \rangle = \langle x_1^{b_1}, \langle \ldots, x_n^{b_n} \rangle \ldots \rangle$. Of course, several functional terms correspond to a Σ-term due to possible variations of the parameter list.

Vice versa, functional terms $\langle p \rangle M$ translate to tuples $M^{\dagger}$ of terms in $T_\Sigma(\{p\})$ defined by

- $x^{\dagger} = x$
- $(f \cdot M)^{\dagger} = f M^{\dagger}$
- $(((\langle q \rangle M) N)^{\dagger} = M^{\dagger}[\langle q \rangle / N^{\dagger}]$

- $\langle M, N \rangle^{\dagger} = \langle M^{\dagger}, N^{\dagger} \rangle.$

There is a slight imprecision in the translation in that we have to identify $f(t_1, \ldots, t_n)$ with $f\langle t_1, \langle t_2, \langle \ldots \rangle \rangle \rangle$. Equivalences of logics are then precisely stated by

- $\vdash (t^{\#})^{\dagger} =_{X,b} t$ for all terms $t \in T_{\Sigma}(X)$,

- $\vdash \langle p \rangle (M^{\dagger})^{\#} = \langle p \rangle M$ for all functional terms $\langle p \rangle M$,

- $\vdash \langle X \rangle t_1^{\#} = \langle X \rangle t_2^{\#}$ if $\vdash t_1 =_{X,b} t_2$, and

- $\vdash M^{\dagger} = M^{\dagger}$ if $\vdash \langle p \rangle M = \langle p \rangle N.$

The proof of the first two statements is a straightforward structural induction. For the other statements use induction on the deduction rules; this does not cause problems since application $(\langle p \rangle M) \cdot N$ corresponds to substitution $[\Theta]t$. ∎

Together with the restriction of Section 7.1.6 to products, we obtain an equivalence of algebraic logic with the (categorical) logic of products.

8.1.2 Algebraic theories

Every Π-theory T (or equivalently every Σ-theory) induces a category $\mathbf{T}$ with objects being the types and morphisms $[\langle p \rangle M] : A \to B$ such that $[\langle p \rangle M] = \{\langle p \rangle N \mid \langle p \rangle M = \langle p \rangle N\}$ (cf. Section 7.1.5). Vice versa, every category $\mathbf{C}$ with finite products defines a Π-theory $T_{\mathbf{C}}$. This justifies the following terminology.

Definition 8.1.2. A (small) category $\mathbf{T}$ with finite products is called an *algebraic theory*.

Those familiar with the terminology of universal algebra will note the close relation to the notion of a *clone* [Cohn, 1981], where, in our notation, the derived operators are closed under the following composition

$$\langle q \rangle t \circ \langle \langle p \rangle t_1, \ldots, \langle p \rangle t_n \rangle := \langle p \rangle t[x_i/t_i].$$

Here $\langle p \rangle t_i$ is a derived operator for each variable x_i in the parameter list $\langle q \rangle$. A clone is turned into a category by closure under tupling.

Algebraic theories (in the categorical format) abstract from specific presentations in terms of signatures and equations in that different specifications may induce the same (up to isomorphism) algebraic theory. The theory of groups is a sort of standard example to substantiate the point.

Usually groups are presented by a binary operator $x * y$, a unary operator $i(x)$ and a constant operator e, and by equations

$$x * (y * z) = (x * y) * z$$
$$x * e = x = e * x$$
$$x * i(x) = e = (x) * x.$$

Alternatively, groups can be presented by a binary operator $d(x, y)$ and a constant e, and by the equations

$$d(d(d(z, d(x, d(x, x))), d(z, d(y, d(x, x)))), x) = y$$
$$d(x, x) = e$$

(see [Manes, 1974]). There is no relation between these presentations at a first glance, but $d(x, y) = x * i(y)$ provides a bijective passage between the respective theories.

Hence 'theories' can be presented in terms of derived operations of some presentation, and of equations between the derived operators. Algebraic theories as categories consist of all derived operators and all equations between derived operators, and hence subsume all possible presentations (if considered up to isomorphism), or, rather, are *presentation independent.*

8.1.3 A functor is a model is a functor

Mathematicians often endow algebra with additional structure; topological algebras have topological spaces as carriers and continuous functions as operations, ordered algebras have partially ordered sets as carriers and monotone functions as operations, etc.

All these notions of algebras are defined according to the same scheme:

Definition 8.1.3. Let $\mathbf{V}$ be a category with finite products, a Σ-algebra $\mathcal{A}$ in $\mathbf{V}$ consists of an object $b^{\mathcal{A}}$ of $\mathbf{V}$ for every sort b, and a morphism $f^{\mathcal{A}} : b_1^{\mathcal{A}} \Pi \ldots \Pi b_n^{\mathcal{A}} \to b^{\mathcal{A}}$ of $\mathbf{V}$ for every operator $f : b_1 \ldots b_n \to b$ in Σ. With homomorphisms $h : \mathcal{A} \to \mathcal{B}$ being a family $h_b : b^{\mathcal{A}} \to b^{\mathcal{B}}$ of morphisms of $\mathbf{V}$ such that

$$
\begin{array}{ccc}
b_1^{\mathcal{A}} \Pi \ldots \Pi b_n^{\mathcal{A}} & \xrightarrow{\;f^{\mathcal{A}}\;} & b^{\mathcal{A}} \\
{\scriptstyle h_{b_1} \Pi \ldots \Pi h_{b_n}} \downarrow & & \downarrow {\scriptstyle h_b} \\
b_1^{\mathcal{B}} \Pi \ldots \Pi b_n^{\mathcal{B}} & \xrightarrow[\;f^{\mathcal{B}}\;]{} & b^{\mathcal{B}}
\end{array}
$$

this defines a category $\mathbf{V}\text{-}\mathbf{Alg}(\Sigma)$.

For $\mathbf{V} = \mathbf{Set}, \mathbf{Pos}, \mathbf{Dcpo}, \mathbf{Top}, \mathbf{Met}, \ldots$ we obtain standard Σ-algebras, *ordered Σ-algebras, directed complete ordered Σ-algebras, topological Σ-algebras, metric Σ-algebras,* ... ($\mathbf{Top}, \mathbf{Met}$ being the categories of topological resp. metric spaces).

We now relate algebras to functors. Every Σ-algebra $\mathcal{A}$ in $\mathbf{V}$ extends to a functor $\mathcal{A} : \mathbf{T}_\Sigma \to \mathbf{V}$ where $\mathbf{T}_\Sigma$ is the theory induced by the theory $\Sigma = (\Sigma, \varnothing)$:

$$\mathcal{A}(1) = 1$$
$$\mathcal{A}(A\Pi B) = \mathcal{A}(A)\Pi\mathcal{A}(B)$$
$$\mathcal{A}(\langle p \rangle \langle \rangle) = \langle \rangle_{\mathcal{A}(\langle p \rangle)}$$
$$\mathcal{A}(\langle x^A \rangle x) = 1_{\mathcal{A}(A)}$$
$$\mathcal{A}(\langle \langle p \rangle, \langle q \rangle \rangle x) = \begin{cases} \mathcal{A}(\langle p \rangle x) \circ p_{\mathcal{A}(\langle p \rangle), \mathcal{A}(\langle q \rangle))} & \text{if } x \text{ occurs in } \langle p \rangle \\ \mathcal{A}(\langle q \rangle x) \circ q_{\mathcal{A}(\langle p \rangle), \mathcal{A}(\langle q \rangle))} & \text{if } x \text{ occurs in } \langle p \rangle \end{cases}$$
$$\mathcal{A}(\langle p \rangle \langle M, N \rangle) = \langle \mathcal{A}(\langle p \rangle M), \mathcal{A}(\langle p \rangle N) \rangle$$

where $\mathcal{A}(\langle x^b \rangle) = b^{\mathcal{A}}, \mathcal{A}(\langle \langle p \rangle, \langle q \rangle \rangle) = \mathcal{A}(\langle p \rangle) \times \mathcal{A}(\langle q \rangle)$.

Proposition 8.1.4. *For every Σ-algebra $\mathcal{A}$, there exists a unique functor $\mathcal{A} : \mathbf{T}_\Sigma \to \mathbf{V}$ which preserves finite products such that $\mathcal{A}(b) = b^{\mathcal{A}}$ and $\mathcal{A}(f) = f^{\mathcal{A}}$.*

Proof. Straightforward induction along the lines of Section 1.4.5. ∎

Now let $F : \mathbf{C} \to_\Pi \mathbf{D}$ state that the functor $F : \mathbf{C} \to \mathbf{D}$ preserves finite products, and let $[\mathbf{C} \Rightarrow_\Pi \mathbf{D}]$ denote the full subcategory of the functor category $[\mathbf{C} \Rightarrow \mathbf{D}]$ with the objects being functors $F : \mathbf{C} \to \mathbf{D}$ which preserve finite products.

We demonstrate that natural transformations correspond to homomorphisms.

Lemma 8.1.5. *Let $\mathbf{C}, \mathbf{D}$ be categories with finite products, and let $F, G : \mathbf{C} \to_\Pi \mathbf{D}$. Then, for all natural transformations $\sigma : F \to G, \sigma_1 = 1_1 : 1 \to 1$ and $\sigma_{A\Pi B} = \sigma_A \Pi \sigma_B : F(A\Pi B) \to F(A\Pi B)$.*

Proof. We assume that the functors preserve specified products on the nose, i.e. $F(A\Pi B) = F(A)\Pi F(B)$ plus preservation of projections. Then the diagram

$$
\begin{array}{ccccc}
F(A) & \longleftarrow & F(A)\Pi F(B) & \longrightarrow & F(B) \\
\downarrow{\scriptstyle\sigma_A} & & \downarrow{\scriptstyle\sigma_A\Pi\sigma_B} & & \downarrow{\scriptstyle\sigma_B} \\
G(A) & \longleftarrow & G(A)\Pi G(B) & \longrightarrow & G(B)
\end{array}
$$

commutes by definition of $\sigma_A \Pi \sigma_B$ (the unlabelled morphisms being projections). If we replace $\sigma_A \Pi \sigma_B$ by $\sigma_{A\Pi B}$ the diagram still commutes since

σ is a natural transformation. By definition of products there is only one morphism to fill in the diagram, hence $\sigma_A\Pi\sigma_B = \sigma_{A\Pi B}$. ∎

Corollary 8.1.6. *Let* $\mathcal{A}, \mathcal{B} : \mathbf{T} \to_\Pi \mathbf{V}$, *and let* $\sigma : \mathcal{A} \to \mathcal{B}$ *be a natural transformation. Then the 'homomorphism' diagram*

$$
\begin{array}{ccc}
\mathcal{A}(b_1)\Pi\ldots\Pi\mathcal{A}(b_n) & \xrightarrow{\ \mathcal{A}(f)\ } & \mathcal{A}(b) \\
\downarrow{\scriptstyle \sigma_{b_1}\Pi\ldots\Pi\sigma_{b_n}} & & \downarrow{\scriptstyle \sigma_b} \\
\mathcal{B}(b_1)\Pi\ldots\Pi\mathcal{B}(b_n) & \xrightarrow[\ \mathcal{B}(f)\ ]{} & \mathcal{B}(b)
\end{array}
$$

commutes for all (derived) operators $f : b_1 \times \ldots \times b_n \to b$.

On the other hand every homomorphism $h : \mathcal{A} \to \mathcal{B}$ induces a natural transformation.

Lemma 8.1.7 (Exercise). *A natural transformation* $\sigma^h : \mathcal{A} \to \mathcal{B} : \mathbf{T} \to \mathbf{V}$ *is defined by* $\sigma^{h_1} = 1_1$, $\sigma^{h_b} = h_b$, $\sigma^{h_{A\times B}} = \sigma^{h_A}\Pi\sigma^{h_B}$.

We summarize all the observations.

Theorem 8.1.8. $\mathbf{V}\text{-}\mathbf{Alg}(\Sigma) \cong [\mathbf{T}_\Sigma \Rightarrow_\Pi \mathbf{V}]$ *for all signatures* Σ.

The result extends to algebraic theories $T = (\Sigma, E)$ in general. We say that a Σ-algebra $\mathcal{A}$ in $\mathbf{V}$ *satisfies* an equation $t_1 =_{X,b} t_2$ if $\mathcal{A}(\langle X\rangle t_1^\#) = \mathcal{A}(\langle X\rangle t_2^\#)$ where $\mathcal{A} : \mathbf{T}_\Sigma \to \mathbf{V}$ is the induced functor ($x^\# = x$, $t(f_1, \ldots, f_n)^\# = f \cdot \langle t_1^\#, \ldots, t_n^\# \rangle$ is the translation of terms to functional terms used in Section 8.1.1).

As an example, the monoid equations translate to

$$
\begin{aligned}
\langle x^M, \langle y^M, z^M\rangle\rangle x * y * z) &= \langle x^M, \langle y^M, z^M\rangle\rangle(x * y) * z \\
\langle x^M\rangle x * e &= \langle x^M\rangle x \\
\langle x^M\rangle e * x &= \langle x^M\rangle x
\end{aligned}
$$

which become diagrams

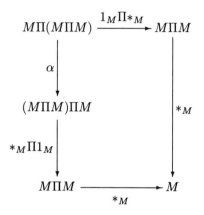

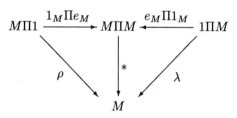

where $\alpha := \langle\langle p_{M,M\Pi M}, p_{M,M} \circ q_{M,M\Pi M}\rangle, qM, M \circ q_{M,M\Pi M}\rangle$ and $\rho = p_{M,1}$, $\lambda = q_{1,M}$ (the reader is urged to check that the diagrams define monoids in the category of sets).

We leave it to the imagination of the reader to establish the following result.

Proposition 8.1.9. $\mathbf{V\text{-}Alg}(T) \cong [\mathbf{T} \Rightarrow_\Pi \mathbf{V}]$ *for all algebraic theories* T *where* $\mathbf{V\text{-}Alg}(T)$ *is the full subcategory of* Σ-*algebras in* $\mathbf{V}$ *which satisfy the equations in* T.

8.1.4 An alternative approach

Often a different, more algebraic approach is preferred in case of $\mathbf{V} = \mathbf{Set}$. Observe that functional terms $f = \langle x_1^{b_1}, \ldots, x_n^{b_n}\rangle\langle M_1, \ldots, M_m\rangle$ correspond to variable assignments

$$\Theta_f : \{y_1^{b_1'}, \ldots, y_m^{b_m'}\} \to T_\Sigma(\{x_1^{b_1}, \ldots, x_n^{b_n}\})$$

with $\Theta_f(y_i^{b_i'}) = M_i^\dagger$ and that composition

$$\begin{aligned}
g \circ f &= \langle y_1^{b_1'}, \ldots, y_m^{b_m'}\rangle\langle N_1, \ldots, N_p\rangle \circ \langle x_1^{b_1}, \ldots, x_n^{b_n}\rangle\langle M_1, \ldots, M_n\rangle \\
&= \langle x_1^{b_1}, \ldots, x_n^{b_n}\rangle\langle N_1[\langle y_1, \ldots, y_n\rangle/\langle M_1, \ldots, M_n\rangle], \ldots, \\
&\qquad N_p[\langle y_1, \ldots, y_n\rangle/\langle M_1, \ldots, M_n\rangle]\rangle
\end{aligned}$$

corresponds to a composition

$$[\Theta_f] \circ \Theta_g : \{z_1^{b_1''}, \ldots, z_p^{b_p''}\} \to T_\Sigma(\{x_1^{b_1}, \ldots, x_n^{b_n}\}), z_j^{b_j''} \mapsto [\Theta_f]N_i$$

of variable assignments. Hence variable assignments define a dual (up to equivalence) category to the category Σ of functional terms.

More generally, every algebraic theory $T = (\Sigma, T)$ defines a category **Free**(T) with objects being B-sorted sets X and morphisms $\alpha : X \to Y$ being variable assignments $\alpha : X \to T(Y)$ where $T(Y)$ is a free T-algebra generated by Y (cf. Free Algebra Theorem 3 of Section 3.6.1). Composition $\beta \circ \alpha : X \to Z$ is defined by $\beta \circ \alpha(x) = [\beta](\alpha(x))$ where $[\beta] : T(Y) \to T(Z)$ is the unique extension. Every T-algebra $\mathcal{A}$ defines a functor $F^{\mathcal{A}} : \textbf{Free}(T)^{\text{op}} \to \textbf{Set}$ by $F^{\mathcal{A}}(X) = \{\Theta \mid \Theta : X \to \mathcal{A}$ variable assignment $\}$ and $F^{\mathcal{A}}(\alpha)(\Theta) = [\Theta] \circ \alpha$ where $[\Theta] : T(Y) \to \mathcal{A}$ is the unique extension.

Proposition 8.1.10.

- **Free**(T) *is a category with finite coproducts.*

- **Free**$(T)^{\text{op}}$ *is equivalent to* **T** *(as defined in Section 8.1.2).*

- $F^{\mathcal{A}} : \textbf{Free}(T)^{\text{op}} \to \textbf{Set}$ *preserves finite products.*

- $\sigma^h : F^{\mathcal{A}} \to F^{\mathcal{B}}, \sigma^{hx}(\Theta) = h \circ \Theta$ *defines a natural transformation for every homomorphism* $h : \mathcal{A} \to \mathcal{B}$.

Proof. • The units $\eta_X : X \to T(X)$ are the identities since $[\alpha] \circ \eta_X = \alpha$ and since $[\eta_X]$ is the identity on $T(X)$, by $[\eta_X] \circ \eta_X = \eta_X$. Composition is associative since $[\gamma] \circ [\beta] = [[\gamma] \circ [\beta]$ by the subsequent lemma. A coproduct object is given by the B-sorted disjoint union $(X + Y)_b = X_b + Y_b$ with injections $inl_{X,Y}(x) = x$ and $inr_{X,Y}(y) = y$.

• Use the translations between algebraic logic and Π-logic.

• Functorial properties hold again because of the universal property of free algebras. Finite products are preserved since $F^{\mathcal{A}}(X) \times F^{\mathcal{A}}(Y) \cong F^{\mathcal{A}}(X + Y)$, the isomorphism given by $(\Theta_1, \Theta_2) \mapsto [\Theta_1, \Theta_2] : X + Y \to \mathcal{A}$.

• $\sigma^{h_Y}(F^{\mathcal{A}}(\alpha))(\Theta) = h \circ [\Theta] \circ \alpha = [h \circ \Theta] \circ \alpha = F^{\mathcal{B}}(\sigma^{hx}(\Theta))$ ∎

The proofs use the following observation about freely generated objects.

Lemma 8.1.11. *Let* $F : \mathbf{C} \to \mathbf{D}$ *be left adjoint to* $G : \mathbf{D} \to \mathbf{C}$ *with unit* $\eta : 1_{\mathbf{C}} \to G \circ F$. *Then* $G(g) \circ [f] = [G(g) \circ f]$ *for all morphisms* $f : A \to G(B)$ *in* $\mathbf{C}$ *and* $g : B \to C$ *in* $\mathbf{D}$.

Proof. Since the diagrams

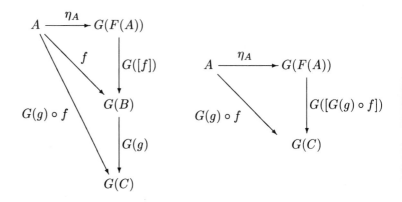

commute.   ∎

Of course, every functor $F : \mathbf{Free}(T)^{\mathrm{op}} \to \mathbf{Set}$ determines a T-algebra if, and only if, it preserves finite products and natural transformation homomorphisms. Though mathematically smooth because of exploitation of universal structure, we prefer the syntactic approach chosen above for several reasons; as a minor point, algebra is related to functional programming which may help one to appreciate algebra. Secondly, the structure naturally extends to more complex structures such as Cartesian closure. Finally, and most important, we have to extend the interpretation of operators to derived operators along the lines of Section 8.1.3 anyway in order to define satisfaction of an equation when the base category $\mathbf{V}$ is different to $\mathbf{Set}$.

8.1.5 Freely generated Σ-algebras in V

Abstractions of this kind should help to transpose results from a familiar situation to one not so well understood. As an example for such a transposition we reconsider the construction of free algebras.

Let $H : \Sigma \to \Sigma'$ be a signature morphism (cf. Section 3.6.1). Define a Σ-algebra H in Σ' with carrier objects $H(b)$ and operators $[H(f)] :$ $H(b_1) \times \ldots \times H(b_n) \to H(b)$. This induces a functor $H : \mathbf{T}_\Sigma \to_\Pi \mathbf{T}_{\Sigma'}$ (by Section 8.1.3). We recall that the *reduct* of a Σ'-algebra $\mathcal{B}$ is Σ-algebra $\mathcal{B}_H$ with carriers $\mathcal{B}_{H,b} := \mathcal{A}_{H(b)}$ and with operators $f^{\mathcal{B}}H := H(f)^{\mathcal{A}}$. Functors $H : \mathbf{T}_1 \to_\Pi \mathbf{T}_2$ correspond to interpretations of one theory in another, where operators are mapped to derived operators generalizing the idea of a morphism of algebraic theories.

Proposition 8.1.12 (Exercise). $\mathcal{B}_H : \mathbf{T}_\Sigma \to_\Pi \mathbf{V}$ *is naturally isomorphic to* $\mathcal{B} * H : \mathbf{T}_\Sigma \to \mathbf{V}$.

We use the notation $G * F$ of Section 1.5.3 for composition $G \circ F$ of functors because of the interaction of functors and natural transformations. Hence the Free Algebra Theorem 3 of Section 3.6.1 reformulates to:

Theorem 8.1.13. *For every* $\mathcal{A} : \Sigma \to_\Pi \mathbf{Set}$ *there exists a* $\mathcal{T}_H(\mathcal{A}) : \Sigma' \to_\Pi$ **Set** *and a natural transformation* $\eta_A : \mathcal{A} \to \mathcal{T}_H(\mathcal{A}) * H$ *such that for all* $\mathcal{B} : \Sigma' \to \mathbf{Set}$ *and all natural transformations* $\Theta : \mathcal{A} \to \mathcal{B} * H$ *there exists a unique natural transformation* $[\Theta] : \mathcal{T}_{H(\mathcal{A})} \to \mathcal{B}$ *such that* $([\Theta] * H) \circ \eta_A = \Theta$.

This is the definition of a left Kan extension except for the additional condition about preservation of products. We recollect the various results in Section 6.2 and state

Proposition 8.1.14. *Let* $\mathbf{V}$ *be a category with finite products, tensors and coends such that all functors* $\_\Pi B, A\Pi\_ : \mathbf{V} \to \mathbf{V}$ *preserve tensors and coends (see 6.3.7), and let* $H : \mathbf{T}_1 \to_\Pi \mathbf{T}_2$. *Then for every* $\mathcal{A} : \mathbf{T}_1 \to_\Pi \mathbf{V}$, *there exists a* $\mathcal{T}_H(\mathcal{A}) : \mathbf{T}_2 \to_\Pi \mathbf{V}$ *and natural transformation* $\eta_A : \mathcal{A} \to \mathcal{T}_H(\mathcal{A}) * H$ *such that for all* $\mathcal{B} : \mathbf{T}_2 \to_\Pi \mathbf{V}$ *and all natural transformations* $\Theta : \mathcal{A} \to \mathcal{B} * H$ *there exists a unique natural transformation* $[\Theta] : \mathcal{T}_H(\mathcal{A}) \to \mathcal{B}$ *such that* $([\Theta] * H) \circ \eta_A = \Theta$.

Proof. Section 6.3.6 plus 6.3.7. ∎

This allows one to compute free algebras over all categories $\mathbf{V}$ which are cocomplete (see Section 6.3.5). Moreover, using Section 6.5 it is possible to characterize all adjunctions between categories of algebras.

8.1.6 Deduction and completeness

Let $\mathbf{T}$ be an algebraic theory (as a category), and let $f, g : A \to B$ be morphisms in $\mathbf{T}$. Say that $f = g$ is a $\mathbf{V}$-consequence of $\mathbf{T}$, short $\mathbf{T} \models f = g$, if $\mathcal{A}(f) = \mathcal{A}(g)$ for all $\mathbf{T}$-algebras $\mathcal{A} : \mathbf{T} \to \mathbf{V}$. Since $\mathbf{T}$, being a theory, is closed under algebraic deduction (in functional form), it makes sense to say that $f = g$ is *derivable* from $\mathbf{T}$, short $\mathbf{T} \vdash f = g$, if $f = g$ in $\mathbf{T}$. Hence, algebraic logic is $\mathbf{V}$-*complete* if $\mathbf{T} \vdash f = g$ whenever $\mathbf{T} \models f = g$ for all algebraic theories T. Soundness, of course, is captured by the isomorphism $\mathbf{V}\text{-}\mathbf{Alg}(T) = [\mathbf{T} \to_\Pi \mathbf{V}]$.

Here is a proof of completeness with $\mathbf{V}$ being the category **Set** of sets.

Proposition 8.1.15. *Algebraic logic is* **Set**-*complete*.

Proof. Let $\mathbf{T}$ be the algebraic theory induced by T. T-algebras exactly correspond to product preserving functors $\mathcal{A} : \mathbf{T} \to \mathbf{Set}$. But the Hom functor $\mathbf{T}[A, \_]$ preserves finite products since $\mathbf{T}[A, B] \times \mathbf{T}[A, C] \cong \mathbf{T}[A, B \Pi C]$ by definition of products. Hence $\mathbf{T}[A, \_]$ is a $\mathbf{T}$-algebra, and $\mathbf{T}[A, f] = \mathbf{T}[A, g]$. Then $f = \mathbf{T}[A, f](1_A) = \mathbf{T}[A, g](1_A) = g$.   ∎

This proof corresponds to the standard proof of completeness via *functionally free models*, i.e. models which consist of congruence classes of terms with free variables $x_1^{b_1}, \ldots, x_n^{b_n}$ where $A = b_1 \times \ldots \times b_n$, the congruence being defined by the equalities in $\mathbf{T}$ (though there has been quite a bit of discussion about the correct notion of deduction for many-sorted algebras, cf. [Goguen and Meseguer, 1981]).

At a first sight, there is little hope to transfer this proof to other categories $\mathbf{V}$ because then $\mathbf{T}[A, B]$ should be an object of $\mathbf{V}$, and $\mathbf{T}[A, B \times C] \cong \mathbf{T}[A, B] \Pi \mathbf{T}[A, C]$ an isomorphism in $\mathbf{V}$ in order to define a product preserving functor $\mathbf{T}[A, \_] : \mathbf{T} \to \mathbf{V}$. But we can only assume that $\mathbf{T}[A, B]$ is a set.

8.2   Enriched functorial semantics

8.2.1   The disgruence between algebraic theories and $\mathbf{V}$-algebras prompts the idea of enriching the algebraic theories, namely to have Hom objects $\mathbf{T}[A, B]$ in $\mathbf{V}$ rather than Hom sets. Of course, having prepared the field in Section 4.3 there is no difficulty in defining $\mathbf{V}$-*enriched algebraic theories*. We assume that $\mathbf{V}$ is a Cartesian closed category.

Definition 8.2.1.   Call a $\mathbf{V}$-category $\mathbf{T}$ with finite $\mathbf{V}$-products a $\mathbf{V}$-*enriched algebraic theory*, and a $\mathbf{V}$-enriched functor $\mathcal{A} : \mathbf{T} \to \mathbf{V}$ which preserves the products, a $\mathbf{V}$-*enriched algebra*.

The assumption on $\mathbf{V}$ rules out examples such as **Top**, **Met** where $\_\Pi A$ does not have a right adjoint $A \Rightarrow \_$. We consider two examples with a computer science background:

8.2.2   Continuous algebras

Σ-algebras in $\omega\mathbf{Pos}$, in particular *continuous algebras*, have been used to give semantics to recursive program schemes (see for instance the book of Irène Guessarian [Guessarian, 1981]).

Definition 8.2.2.   A *continuous Σ-algebra* is a Σ-algebra in $\omega\mathbf{Pos}$ such that every set b^A has a least element $\bot$. Homomorphisms are continuous and *strict*, i.e. $h_b(\bot) = \bot$.

Continuous algebras attract attention in computer science because of the fact that natural term algebras consist of finite and *infinite* terms.

Definition 8.2.3.   Define sets $FCT_\Sigma(X)_b$ of terms of sort b as follows:

- $\bot \in FCT_\Sigma(X)_b$

- $x \in FCT_\Sigma(X)_b$ if $x \in X_b$

- $f(t_1,\ldots,t_n) \in FCT_\Sigma(X)_b$ if $f : b_1 \ldots b_n \to b \in \Sigma$ and $t_i \in FCT_\Sigma(X)_{b_i}$ for $i = 1,\ldots,n$.

On $FCT_\Sigma(X)_b$ a partial order is defined by

- $\bot \sqsubseteq t$ for all $t \in FCT_\Sigma(X)_b$

- $f(t_1,\ldots,t_n) \sqsubseteq f(t_1',\ldots,t_n')$ if $t_i \sqsubseteq t_i'$ for $i = 1,\ldots,n$.

With operation $f^{FCT_\Sigma(X)}(t_1,\ldots,t_n) = f(t_1,\ldots,t_n)$ this defines an algebra $FCT_\Sigma(X)$.

Proposition 8.2.4. *Every B-sorted function* $h : X \to \mathcal{A}$ *to a continuous algebra* $\mathcal{A}$ *induces a unique monotone and strict homomorphism* $[h] : FCT_\Sigma(X) \to \mathcal{A}$ *such that* $[h](x) = h(x)$.

Proof. Straightforward. ∎

Proposition 8.2.5.

- *Define sets* $CT_\Sigma(X)_b$ *as* ω*-completions* $CT_\Sigma(X)_b = \omega(FCT_\Sigma(X)_b)$ *(cf. Section 3.6.3).*

- *The sets* $CT_\Sigma(X)_b$ *with operations* $f^{CT_\Sigma(X)}(t_1,\ldots,t_n) = f(t_1,\ldots,t_n)$ *define a continuous* ω*-algebra.*

- *Every B-sorted function* $h : X \to \mathcal{A}$ *to a continuous algebra* $\mathcal{A}$ *induces a unique homomorphism* $[h] : CT_\Sigma(X)(X) \to \mathcal{A}$ *of continuous algebras such that* $[h](x) = h(x)$.

Proof. By 3.6.3. Completion preserves least elements. ∎

Elements of $CT_\Sigma(X)$ define finite and infinite trees in that in every chain $t_1 \sqsubseteq t_2 \sqsubseteq \ldots \sqsubseteq t_n \sqsubseteq \ldots$ basically the symbol $\bot$ is replaced by subterms to give bigger and bigger terms. If the process is infinite, the size of the terms will grow *ad infinitum*, the t_n's being finite approximations.

Such infinite trees are used to give semantics to recursive program schemes of the form

$$\Phi(x) = t(x, \Phi(x))$$

where $t(x, \Phi(x))$ is a Σ-term with free variables in the parameter list x and with *recursive calls* $\Phi(x)$. Unfolding gives an ω-chain of approximations

$$\bot \sqsubseteq t(x, \bot) \sqsubseteq t(x, t(x, \bot)) \sqsubseteq t(x, t(x, t(x, \bot))) \sqsubseteq \ldots$$

where recursive calls are replaced by $\bot$. The least upper bound of the chain in $CT_\Sigma(x)$ then defines the meaning of the recursive scheme.

8.2.3   Non-deterministic algebras

Non-deterministic programs can yield several results for a given argument according to (internal) choices made during computation. Bounded or finite non-determinism is modelled by a binary operator f or g such that $(f \text{ or } g)(x) = f(x) \text{ or } g(x)$. The *semilattice* axioms

$$
\begin{aligned}
x \text{ or } x &= x \\
x \text{ or } y &= y \text{ or } x \\
(x \text{ or } y) \text{ or } z &= x \text{ or } (y \text{ or } z)
\end{aligned}
$$

naturally specify the properties to be expected of non-determinism. Functions with several arguments may behave differently according to the chosen evaluation strategy

$$
\begin{aligned}
IO: \quad & f(x_1 \text{ or } y_1, \ldots, x_n \text{ or } y_n) &&= f(x_1, \ldots, x_n) \text{ or } f(y_1, \ldots, y_n) \\
OI: \quad & f(x_1, \ldots, x_i \text{ or } y_i, \ldots, x_n) &&= f(x_1, \ldots, x_i, \ldots, x_n) \text{ or } \\
& && \quad f(x_1, \ldots, y_i, \ldots, x_n)
\end{aligned}
$$

where OI and IO stand for 'outside in' and 'inside out', meaning basically 'call-by-name' and 'call-by-value'. The terminology is introduced in the paper [Engelfriet and Schmidt, 1977].

The IO-axioms are rather well-behaved defining IO-algebras in the category of semilattices.

Definition 8.2.6. A Σ-*IO*-*algebra* $\mathcal{A}$ consists of a semilattice $b^{\mathcal{A}}$ for each $b \in B$, and a semilattice homomorphism $f^{\mathcal{A}} : b_1^{\mathcal{A}} \times \ldots \times b_n^{\mathcal{A}} \to b^{\mathcal{A}}$ for each operator $f : b_1 \ldots b_n \to b$ in Σ. Homomorphisms are defined as usual but are also semilattice homomorphisms.

The product of semilattices is given by the Cartesian product of the underlying sets with operations being defined componentwise.

In other words, IO-algebras are Σ-algebras in **SLat** where the monoidal structure is given by products. OI-algebras are also found in **SLat** but with regard to another monoidal structure.

Definition 8.2.7. A function $f : X \times Y \to Z$ of semilattices is called *bilinear* if the function $f(x, \_) : Y \to Z, y \mapsto f(x, y)$ and $f(\_, y) : X \to Z, x \mapsto f(x, y)$ are semilattice homomorphisms. A *tensor product* of semilattices X and Y consists of a semilattice $X \otimes Y$ and a bilinear function $\gamma_{X,Y} : X \Pi Y \to X \otimes Y$ such that for every bilinear function $f : A \Pi B \to C$ there exists a unique semilattice homomorphism $[f] : X \otimes Y \to Z$ such that $[f] \circ \gamma = f$.

Explicitly, tensor products are given by the following construction: $(P_f(X \times Y), \cup)$, i.e. the set of finite, non-empty subsets of the Cartesian product $X \times Y$ with union, is a (free) semilattice (over $X \times Y$). $X \otimes Y$

is obtained by factorization by the least congruence relation generated by $\{(x, y + y')\} \sim \{(x,y),(x,y')\}$ and $\{(x + x', y)\} \sim \{(x,y),(x',y)\}$. The universal bilinear map is $\gamma : X \times Y \to X \otimes Y, (x,y) \mapsto [\{(x,y)\}]$.

The tensor product inherits a monoidal structure from the product structure since projections are bilinear. For instance

$$\alpha = \gamma_{X,Y\otimes Z} \circ [\langle[p_{X,Y}] \circ p_{X\otimes Y,Z}, \gamma_{Y,Z} \circ \langle[q_{X,Y}] \circ p_{X\otimes Y,Z}, q_{X\otimes Y,Z}\rangle :$$
$$(X \otimes Y) \otimes Z \to X \otimes (Y \otimes Z)$$

Proposition 8.2.8. $(\mathbf{SLat}, \otimes, 1)$ *is symmetric monoidal closed where*

$$\mathbf{SLat}[X \otimes Y, Z] \cong \mathbf{SLat}[X, \mathbf{SLat}[Y, Z]].$$

With these prerequisites we obtain a 'proper' definition of OI-algebras.

Definition 8.2.9. An OI-algebra is determined by a semilattice $b^{\mathcal{A}}$ for each $b \in B$ and a semilattice homomorphism $f^{\mathcal{A}} : b_1^{\mathcal{A}} \otimes \ldots \otimes b_n^{\mathcal{A}} \to b^{\mathcal{A}}$ for each operator $f : b_1 \otimes \ldots \otimes b_n \to b$ in Σ. In other words, OI-algebras are $(\mathbf{SLat}, \times, 1)$-algebras.

With regard to the tensor product, OI-algebras are an example of a whole variety of mathematical structures, the most prominent one being rings which are monoids in the category of Abelian groups, where the monoidal structure is given by tensor products with unit Z, or as a generalization, Abelian categories.

Of course, our functional view of algebra does not work in a straightforward way. But if we transpose all the definitions and replace finite products by a suitable monoidal structure the preceding may be rephrased (cf. [Pfender, 1974]).

8.2.4 Let us return to the question of completeness of algebraic logic. The enriched case provides a proof scheme for completeness because of the following observation.

Proposition 8.2.10. *The enriched Hom functor* $hom_{\mathbf{C}} : \mathbf{C} \to \mathbf{V}$ *preserves finite enriched products.*

Proof. $\mathbf{C}[A, B \times C] \cong \mathbf{C}[A, B]\Pi\mathbf{C}[A, B]$ induces

$$hom_{\mathbf{C}}(X, A\Pi B) \cong hom_{\mathbf{C}}(X, A)\Pi hom_{\mathbf{C}}(X, B)$$

since $[X \Rightarrow A\Pi B] \cong [X \Rightarrow A]\Pi[X \Rightarrow B]$ in $\mathbf{V}$. ∎

However, it is difficult to say what we mean by completeness. Let us look at a completeness result for $\mathbf{V} = \omega\mathbf{Pos}$ by [Meseguer, 1977].

Proposition 8.2.11. *Let* **T** *be an ω-complete theory. Then* $f \leq g : A \to B$ *in* **T** *iff* $\mathcal{A}(f) \leq \mathcal{A}(g)$ *for all ω-complete* **T**-*algebras* $\mathcal{A}$.

Proof. $\mathbf{T}[A, \_]$ determines an enriched algebra. Hence

$$\mathbf{T}[A, \_](f) \leq \mathbf{T}[A, \_](g)$$

and $f \leq g$ by precomposition with 1_A. ∎

A few observations may be appropriate:

- The statement refers to the internal structure of Hom objects, and hence depends on insight into the nature of a particular **V**. This is somewhat of a pity, but there is little chance of stating a 'V-completeness' result which covers the example. The best 'general' result one can expect is the following. Stipulate that $\mathbf{T} \vdash f = g$ if $f = g : I \to \mathbf{T}[A, B]$ in **V** and that $\mathbf{T} \models f = g$ if $\mathcal{A}[\_, f] \cong \mathcal{A}[\_, g]$ for all 'algebras' $\mathcal{A} : \mathbf{T} \to_{\Pi} \mathbf{V}$ where $\mathcal{A}[\_, f]$ is defined by

$$
\begin{array}{ccc}
\mathbf{T}[X, A] \cong I \otimes \mathbf{T}[X, A] & \xrightarrow{\;\; f \otimes I \;\;} & \mathbf{T}[A, B] \otimes \mathbf{T}[X, A] \\[2mm]
\mathcal{A}[X, f] \downarrow & & \downarrow \mathcal{A}[A, B] \otimes \mathcal{A}[X, A] \\[2mm]
\mathbf{V}[\mathcal{A}(X), \mathcal{A}(B)] & \xleftarrow{\quad * \quad} & \mathbf{V}[\mathcal{A}(A), \mathcal{A}(B)] \otimes \mathbf{V}[\mathcal{A}(X), \mathcal{A}(A)]
\end{array}
$$

Proposition 8.2.12. $\mathbf{T} \vdash f = g$ *if* $\mathbf{T} \models f = g$.

- Secondly, the claim of completeness would be, to say the least, unusual since no deduction system is given. However, here category theory is prescriptive in that the correct mathematical formalism is provided and that it is up to the user to find a suitable characterization.

For instance, the following is a suitable deduction system for ω-complete theories. Let $\leq$ be a second relation on morphisms besides equality and familiar deduction rules

$$
\frac{}{f \leq f} \qquad \frac{f \leq g \quad g \leq f}{f = g} \qquad \frac{f \leq g \quad g \leq h}{f \leq h} \qquad \frac{f = g}{f \leq g} \qquad \frac{f = g}{g \leq f}
$$

$$
\frac{f_n \leq f_{n+1} (n \in \omega)}{f_n \leq \bigsqcup_n f_n} \qquad \frac{f_n \leq f_{n+1} (n \in \omega)}{\bigsqcup_n f_n \leq g} \qquad \frac{f_n \leq g (n \in \omega)}{}
$$

The categorical infrastructure must be compatible with this inherent logic of ω-complete posets since it was compatible with equality. One needs new judgements $f \leq g : A \to B$, and a new operator

$$\frac{f_m \leq f_{m+1} : A \to B(n \in \omega)}{(\bigsqcup_m f_m : A \to B)}$$

to generate least upper bounds. Compatibility is than expressed by deduction rules of the form

$$\frac{f \leq f' \quad g \leq g'}{g \circ f \leq g' \circ f'} \qquad \frac{f_m \leq f_{m+1}(m \in \omega) \quad g_n \leq g_{n+1}(n \in \omega)}{(\bigsqcup_m f_m) \circ (\bigsqcup_n g_n) = \bigsqcup_k g_k \circ f_k}$$

(and similar for all other categorical operators on morphisms).

- Thirdly, the passage to enriched theories increases expressivity considerably. For example, continuous algebras can be presented by restriction to those ω-complete theories with specified morphisms $\bot_A : 1 \to A$ such that $\bot_A \circ \langle\rangle_A \leq 1_A : A \to A$. Given such a *continuous theory* the respective ω-complete algebras are continuous (see [Meseguer, 1977]).

- Finally, the familiar constructions lift to the enriched level, suitable properties of $\mathbf{V}$ assumed. For instance, the construction of right Kan extensions (or of free algebras) works in the enriched setting provided that a $\mathbf{V}$-enriched category $\mathbf{C}$ has tensors, i.e. $\mathbf{C}[X \otimes A, B] \cong \mathbf{V}[X, \mathbf{C}[A, B]]$, and coends which are both an instance of an *indexed colimit* (for details see [Kelly, 1982])

8.3 Monads

8.3.1 However abstract we have been, so far we have adhered to a rather concrete point of view, namely that algebra is about 'functions with arity'. Category theory provides an alternative, in some senses more abstract formalization of 'algebra', where term construction is the fundamental ingredient. The theory of monads or triples exploits the following observation:

Let Σ be a homogeneous (algebraic) signature. Then every Σ-algebra $\mathcal{A}$ determines a unique homomorphism $\delta^{\mathcal{A}} := [1_A] : T_{\Sigma}(A) \to A$ such that $\delta^{\mathcal{A}}(a) = a$ for all $a \in A$, where A is the carrier of the algebra $\mathcal{A}$ (due to the universal property of $T_{\Sigma}(A)$, see Section 3.6.1). These homomorphisms can be characterized diagrammatically by

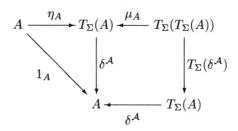

where $\eta_A : A \to T_\Sigma(A)$ is the unit of the adjunction, and where $\mu_A :=$ $[1_{T_\Sigma(A)}] : T_\Sigma(T_\Sigma(A)) \to T_\Sigma(A)$ is the uniquely induced Σ-homomorphism. We claim that there is a one-to-one correspondence between Σ-algebras $\mathcal{A}$ and mappings $\delta^{\mathcal{A}} : T_\Sigma(A) \to A$ which satisfy the diagram above (left to the reader). In fact, all that has been achieved, is a slight generalization of all the familiar definitions of algebras; we not only stipulate how to evaluate the operator σ applied to a suitable number of arguments, i.e. $\sigma^{\mathcal{A}}(a_1, \ldots, a_n)$, but evaluation homomorphically extends to all *polynomial terms*, meaning terms generated by the elements of $\mathcal{A}$, with $\delta(\sigma(a_1, \ldots, a_n)) = \sigma^{\mathcal{A}}(a_1, \ldots, a_n)$ being a specific example. Similarly, the diagram

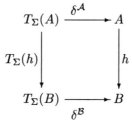

commutes for every homomorphism $h : \mathcal{A} \to \mathcal{B}$ subsuming the homomorphic squares

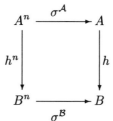

as special instances.

8.3.2   The theory of monads strips this setting down to its very bones.

Definition 8.3.1. Let **C** be a category.

- A *monad* $T = (T, \mu, \eta)$ (over **C**) consists of a functor $T : \mathbf{C} \to \mathbf{C}$ and natural transformations $\mu : T \circ T \to T$ and $\eta : 1_{\mathbf{C}} \to T$ such that the diagrams

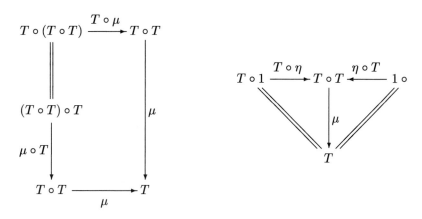

commute.

- A *T-algebra* $\mathcal{A} = (A, \delta)$ consists of an object A and a morphism $\delta : T(A) \to A$, both of **C**, such that the diagram

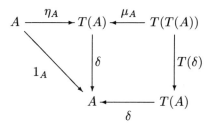

commutes.

A T-algebra *homomorphism* $h : \mathcal{A} \to \mathcal{B}$ consists of a morphism $h : A \to B$ in **C** such that the diagram

$$
\begin{array}{ccc}
T(A) & \xrightarrow{\;\delta^{\mathcal{A}}\;} & A \\
\big\downarrow{\scriptstyle T(h)} & & \big\downarrow{\scriptstyle h} \\
T(B) & \xrightarrow[\;\delta^{\mathcal{B}}\;]{} & B
\end{array}
$$

commutes. For every monad $T = (T, \mu, \eta)$, these data define a category T-**Alg**.

Obviously, Σ-algebras provide an example, but there are abundant examples, since every adjunction determines a monad, and vice versa.

Proposition 8.3.2.

- Let $F : \mathbf{C} \to \mathbf{D}$ be a left adjoint to $G : \mathbf{D} \to \mathbf{C}$ with unit $\eta : 1_{\mathbf{C}} \to G \circ F$. Then $(G \circ F, G \circ \varepsilon \circ F, \eta)$ is a monad where $\varepsilon : F \circ G \to 1_{\mathbf{D}}$ is the counit.

- Let $T = (T, \mu, \eta)$ be a monad over $\mathbf{C}$. Then the forgetful functor $U^T : T\text{-}\mathbf{Alg} \to \mathbf{C}$, $(A, \delta) \mapsto A$, has a left adjoint $F^T : \mathbf{C} \to T\text{-}\mathbf{Alg}$ which maps A to $(T(A), \mu_A)$ with unit $\eta : 1_{\mathbf{C}} \to U^T \circ F^T$.

(The first observation is a straightforward, but tedious computation (see for instance [Barr and Wells, 1985]). The second observation restates the defining diagram for T-algebras.)

Other examples are the following:

- Let $M = (M, *, e)$ be a monoid. Then the functor $M \times \_ : \mathbf{Set} \to \mathbf{Set}$, $X \mapsto M \times X$ together with natural transformations defined by $\eta_X(x) = (e, x)$ and $\mu_X(m_2 * (m_1, x)) = (m_2 * m_1, x)$ defines a monad. Algebras are M-automata.

- Let $P : \mathbf{Set} \to \mathbf{Set}$ be the covariant power set functor. Then $\mu_X(x) = \{x\}$ and $\eta_X(\{Y_i \mid i \in I\}) = \bigcup_{i \in I} Y_i$ define a monad. P-algebras are complete upper semilattices, i.e. partial orders with least upper bounds for all subsets. Homomorphisms preserve least upper bounds.

- Let $T_\Sigma(\_) : \mathbf{Set}^S \to \mathbf{Set}^S$ be the functor which maps S-sorted sets X to the S-sorted set $T_\Sigma(X)$ of terms where $\Sigma = (S, \Sigma)$. The algebras correspond to standard Σ-algebras.

It is interesting to observe that monads are monoids with regard to the monoidal category of endofunctor, where the monoidal 'tensor' is composition, and that T-algebras are just T-automata in this setting. This is one more of those observations to which I referred when arguing that, in the end, category theory is a formal game combining the structure of monoids and partial orders.

8.3.3 One may wonder whether all monads determine a category of equationally defined Σ-algebras. The answer is positive, in the case that we consider monads over the category of sets, and if we allow for infinitary operators. For instance, the contravariant power set functor specifies an infinitary operator 'least upper bound'. We refer to the literature for a proof (cf. for instance [Manes, 1974; Richter, 1979]). However, we would like to add a few remarks of a systematic nature.

While we can reconstruct a monad from its categories of algebras, starting with an adjunction, we are, in general, not able to reconstruct the adjunction from the induced monad; the forgetful functor $U : \textbf{Top} \rightarrow \textbf{Set}$ from topological spaces to underlying sets has a left adjoint mapping a set X to the discrete topological space $(X, 2^X)$. $1_{\textbf{Set}} : \textbf{Set} \rightarrow \textbf{Set}$ with identity natural transformations is the corresponding monad, the algebras of which are just sets. There is no chance of recovering the topological spaces from which we have started (cf. [MacLane, 1971]).

Hence monadic structure determines a very special class of adjunctions. At this point, we remind the reader of the arguments which have led to the definition of adjunctions, namely that certain capabilities are transported along a functor; right adjoints for instance preserve limits or more generally separation, while left adjoints preserve colimits or more generally generation. 'Monadic' adjunctions have additional such capabilities of this kind.

Proposition 8.3.3. *Let $T = (T, \mu, \eta)$ be a monad over $\textbf{C}$. Then the forgetful functor $U^T : T\text{-}\textbf{Alg} \rightarrow \textbf{C}$, which maps (A, δ) to A, creates limits. (A functor $G : \textbf{C}_1 \rightarrow \textbf{C}_2$ creates limits, whenever $(lim\ (G \circ D), B)$ is a limit in $\textbf{C}_2$, where $D : \textbf{D} \rightarrow \textbf{C}_1$ is a diagram, then there exists a unique lift (A, α) which is a limit of D in $\textbf{C}_1$. By a lift we mean that $G(A) = lim\ (G \circ B)$ and $G \circ \alpha = \beta$.)*

Proof. We leave a proof of the general case as an exercise to the reader but just consider the case of binary products. Then given T-algebras (A_1, δ_1) and (A_2, δ_2) product properties in $\textbf{C}$ induce a morphism $\langle \delta_1 \circ T(p), \delta_2 \circ T(q) \rangle : T(A_1 \Pi A_2) \rightarrow A_1 \Pi A_2$. It is easily computed that $(A_1 \Pi A_2, \langle \delta_1 \circ T(p), \delta_2 \circ T(q) \rangle)$ is a product indeed with projections being those of the underlying product. ∎

The idea should be familiar to everybody who has ever heard about the product of two monoids or similar.

Though characteristic of monads, one additional property is needed to capture monadic adjunctions. Unfortunately, this property is rather esoteric, meaning not easy to grasp. Hence we pursue the strategy again of referring to the literature (namely MacLane's book [MacLane, 1971]), but introducing a new notion which is more accessible to the less initiated, though similar in spirit. Typical categories of algebras, such as the category of monoids have the property that, whenever we have a *congruence relation* on an algebra, we can factorize the underlying set (by the underlying equivalence relation) in order to obtain the quotient algebra. More technically, there is a unique algebraic structure on the quotient set, which is induced from the original algebra. The following is one of the many ways to capture this idea in the categorical language.

Definition 8.3.4. Let $f : A \to B$ be a morphism in a category **C**. We refer to the pair of morphism $p, q : Ker(f) \to A$ in the pullback diagram

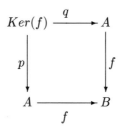

as a *kernel pair* of f.

In standard examples of categories such as those of sets, monoids or other algebraic ones, kernel pairs are a perfect description of congruence relations as the reader might check, and the coequalizer of a kernel pair is usually known as a quotient by a congruence relation. Keeping in mind that categories of algebras typically come along with a free construction, kernel pairs = congruence relations are preserved by forgetful functors (being right adjoints), hence, so to speak, congruence relations are mapped to equivalence relations. Factorization by such an equivalence should already induce the correct algebraic structure, or in categorical language:

Definition 8.3.5. A functor $U : \mathbf{D} \to \mathbf{C}$ *creates coequalizers of congruences* if, whenever we are given a pair $f, g : A \to B$ of **D** and a morphism $h : U(B) \to D$ of **C** such that $U(f), U(g) : U(A) \to U(B)$ is a kernel pair with $h : U(B) \to D$ being a coequalizer, then there exists a unique morphism $k : B \to C$ such that $U(C) = D$ and $U(k) = h$, which is a coequalizer in **D**.

Existence of a left adjoint and generation of limits and of quotients of congruence relations are the criteria which are sometimes used to speak of a forgetful functor as an *algebraic functor*, for hopefully obvious reasons. We give one of the possible categorical axiomatizations.

Definition 8.3.6. A functor $U : \mathbf{D} \to \mathbf{C}$ is called *algebraic* if it has a left adjoint and if it creates limits and coequalizers of congruences.

I must warn the reader that the attribute 'algebraic' is used in many different ways, depending on the authors. Actually, our whole discussion might be seen as parametric since there exists a variety of interpretations of 'congruence relation' in the language of category theory, but we have chosen just one of the possibilities (for instance, a functor is called *monadic* if 'congruence relation' is interpreted as *split coequalizer* [MacLane, 1971]). It is exactly this relative viewpoint that we would like to stress in this section as a systematic issue.

8.3.4 Space is not abundant enough for the discussion of even rather straightforward consequences of such restrictions of adjoint situations (or the rich literature on variations of 'algebraicity' such as *(semi-)topological functors* for instance, [MacLane, 1971; Manes, 1974; Richter, 1979; Barr and Wells, 1985]). Nevertheless, I want to give at least some indication that it might be worthwhile to consider generalities of this kind.

Proposition 8.3.7.

- *Every kernel pair is also the kernel pair of its coequalizer.*

- *Let $\mathbf{C}$ have a coequalizer of kernel pairs and intersections of kernel pairs. Then $\mathbf{C}$ has coequalizers.*

- *Let $\mathbf{C}$ be complete and well-powered. Then $\mathbf{C}$ has intersections of kernel pairs.*

  (A category is well-powered if for every object B there exists a (representative) set $Mono(B)$ of monomorphisms such that for every monomorphism $m : A \to B$ there exists a monomorphism $m' : A' \to B \in Mono(B)$ and an isomorphism $\varphi : A \to A'$ such that $m' \circ \varphi = m$.)

- *If the functor $U : \mathbf{D} \to \mathbf{C}$ is algebraic, if $\mathbf{C}$ is complete and has coequalizer of kernel pairs, and if $\mathbf{D}$ is well-powered, then $\mathbf{D}$ has all coequalizers.*

Proof. • The claim can be read off the diagram

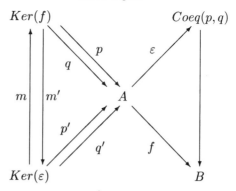

where m and m' are induced by pullback properties.

- Let $g, h : A \to B$ be a pair of morphisms. We say that a kernel pair (p, q) is greater than (g, h) if the diagrams

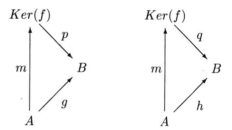

commute for some $m : A \to Ker(f)$. Since m is uniquely defined, this defines an order $(f, g) \leq (p, q)$ up to isomorphism. By intersection of kernel pairs we mean that for every non-empty class $\{(p_i, q_i) \mid i \in I\}$ of kernel pairs there exists a kernel pair (p, q) which is a lower bound, i.e. $(p, q) \leq (p_i, q_i)$ for all $i \in I$, and which is the greatest of all lower bounds. We claim that the coequalizer $\varepsilon : B \to Coeq(p, q)$ of p and q is the coequalizer of g and h if $\{(p_i, q_i) \mid i \in I\}$ consists of all kernel pairs greater than (g, h) (note that the projections $p_{B,B}, q_{B,B} : B \Pi B \to B$ are a kernel pair); clearly, $\varepsilon \circ g = \varepsilon \circ h$ since $(g, h) \leq (p, q)$. Let $k : B \to C$ be a morphism such that $k \circ g = k \circ h$. Then the kernel pair of k is greater than (g, h), and hence greater than (p, q). It follows that $k \circ p = k \circ q$, and a unique morphism $[k] : Coeq(p, q) \to C$ is induced such that $[k] \circ \varepsilon = k$.

- Observe that a kernel pair $p, q : A \to B$ defines a monomorphism $\langle p, q \rangle : A \to B \Pi B$. Given a non-empty class $\{p_i, q_i : A_i \to B \mid i \in I\}$ of kernel pairs we can assume without restriction of generality that the induced monos $\langle p_i, q_i \rangle$ are in $Mono(B)$. Thus we can assume that I is a set, and we can construct the product

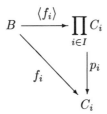

where the f_i's induce the kernel pairs. It is a straightforward exercise to prove that the kernel pair induced by $\langle f_i \rangle$ is the intersection.

- Let $f, g : A \to B$ be in $\mathbf{D}$. Consider the class of all morphisms $p, q : A \to B$ in $\mathbf{D}$ such that $(f, g) \le (p, q)$ and such that $(U(p), U(q))$ is a kernel pair. Since coequalizers of kernel pairs are created, since the kernel pairs are those of the coequalizers (part one of the proposition), and since limits are created, we can conclude that such morphisms $p, q : A \to B$ are kernel pairs of their respective coequalizers in $\mathbf{D}$. Now $\mathbf{D}$ is complete since U creates limits, and the second and the third part of the proposition can be applied. ∎

What might appear as a *tour de force* in category theory, everybody should recognize as the familiar factorization by a congruence relation, where we factorize the underlying set by an equivalence relation and where the algebraic structure is induced.

Remark 8.3.8. Just for the record, there is another, less familiar construction of quotients which gives us cause to refer to a few definitions from the standard inventory in category theory.

In the category of sets the construction works as follows: let $f, g : A \to B$ be functions, and let $Epi(f, g) = \{e : B \to C \mid e \circ f = e \circ g$ and e is surjective$\}$. We consider the diagram

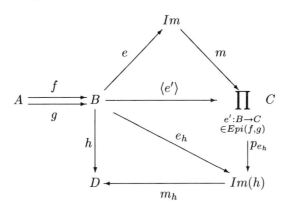

where the product is indexed by all morphisms in $Epi(f, g)$ and where $\langle e' \rangle$ is induced by product properties from the cone formed by $Epi(f, g)$. Then e, m resp. e_h and m_h are the factorizations of $\langle e' \rangle$ resp. h into a surjective followed by an injective function. The projection p_{e_h} maps to the respective component of the product, where we assume that $h \circ f = h \circ g$, and thus $e_h \circ f = e_h \circ g$. Hence the inner triangle commutes. We claim that $e : B \to Im$ is a coequalizer of f and g: since $e' \circ f = e' \circ g$ for all $e' : B \to C$ in $Epi(f, g)$ we have $\langle e' \rangle \circ f = \langle e' \rangle \circ g$ by product properties. $h : B \to D$ induces the mapping $m_h \circ p_{e_h} \circ m : Im \to D$ such that $h = m_h \circ p_{e_h} \circ m \circ e$. This is the unique such mapping since, for every other $k : Im \to D$ such that $h = k \circ e$, we compute that $k \circ e = m_h \circ p_{e_h} \circ m \circ e$. Hence $k = m_h \circ p_{e_h} \circ m$ because of surjectivity of e.

The argument is already given in terms of arrows. Abstractly, it depends on two assumptions, namely that the category is co-well-powered (the dual to being well-powered) and that *image factorization* exists, i.e. that every morphism $f : A \to B$ can be factorized into

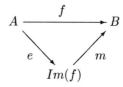

where e is an epimorphism and m is a monomorphism. We speak of an (E, M)-*category* if

- every morphism of a category $\mathbf{C}$ has an image factorization such that e and m belong to specified classes E and M of epimorphisms and of monomorphisms,

- this factorization is unique up to isomorphism, i.e. if e', m' is also a factorization, there is an isomorphism h such that

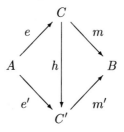

commutes, and if

- E and M are closed under composition.

These definitions and a variety of similar ones are exhaustively discussed in [Herrlich and Strecker, 1973]. It is well worth noticing that this construction typically works in categories of posets where standard factorization by means of least congruences fails. For instance, in the category **Pos** of posets an image factorization is given, where the image $Im(f)$ of a monotone function $f : (X, \leq) \rightarrow (Y, \leq)$ is given by $\{f(x) \mid |x \in X\}$ with order being the transitive closure of $f(x) \sqsubseteq f(y) :\Leftrightarrow xJ \leq y$. **Pos** is co-well-powered by a simple cardinality argument.

9 On the categorical interpretation of calculi

This section deals with the formalization of logic in terms of category theory. I cite [Lambek and Scott, 1986] instead of an overview:

"The usual development of logic in an elementary course proceeds by something like this:

(1) the propositional calculus,

(2) the predicate calculus,

(3) the theory of identity.

If one is interested in the foundations of mathematics, one then presents

(4) Peano's axioms for arithmetic,

(5) the theory of membership for set theory."

More recent developments suggest adding

(6) constructive type theories.

We shall not deal here with (4) and (5) except for those remarks about natural number objects and toposes in Sections 2.5 and 5.1.4 respectively. The other topics, though dealt with more extensively in other chapters of these volumes, are explained up to a certain elementary level, first of all, to provide yet another, quite different interpretation of categories, and secondly to provide more examples for some of the concepts introduced previously.

9.1 Category theory as proof theory

9.1.1 We reinterpret category theory as a theory of proofs where objects φ correspond to formulas, and where a morphism $f : \varphi \rightarrow \psi$ corresponds to a proof of the 'conclusion' ψ from the 'assumption' φ. Composition $g \circ f$ then states composition of proofs, and identity 1_φ that φ follows from φ.

The rules

$$(\wedge I) \quad \frac{f : \chi \to \varphi \quad g : \chi \to \psi}{\langle f, g \rangle : \chi \to \varphi \wedge \psi}$$

$$(\wedge E) \quad \frac{}{p_{\varphi,\psi} : \varphi \wedge \psi \to \varphi \quad q_{\varphi,\psi} : \varphi \wedge \psi \to \psi}$$

for products correspond to introduction and elimination rules for *conjunction*, while

$$\langle \rangle_\varphi : \varphi \to \mathsf{tt}$$

introduces *truth*, and exponentiation corresponds to implication

$$(\Rightarrow I) \frac{f : \chi \wedge \varphi \to \psi}{\lambda(f) : \chi \to [\varphi \Rightarrow \psi]} \qquad (\Rightarrow E) \frac{}{apply_{\varphi,\psi} : [\varphi \Rightarrow \psi] \wedge \varphi \to \psi}$$

Even the categorical axioms such as

$$\frac{f : \varphi \to \psi}{1_\psi \circ f = f = f \circ 1_\varphi : \varphi \to \psi} \qquad \frac{f : \varphi \to \psi, g : \psi \to \chi, h : \chi \to \delta}{h \circ (g \circ f) = (h \circ g) \circ f : \varphi \to \delta}$$

$$\frac{f : \chi \to \varphi \quad g : \chi \to \psi}{p_{\varphi,\psi} \circ \langle f, g \rangle = f : \chi \to \varphi} \qquad \frac{f : \chi \to \varphi \quad g : \chi \to \psi}{q_{\varphi,\psi} \circ \langle f, g \rangle = g : \chi \to \psi}$$

$$\frac{f : \chi \wedge \varphi \to \psi}{apply_{\varphi,\psi} \circ (\lambda(f) \times 1_\varphi) = f : \chi \wedge \varphi \to \psi}$$

make sense on the grounds that proofs may be considered as equivalent.

The logic is *intuitionistic* in that a proposition only holds if a proof of it exists, while non-existence of a proof does, of course, not imply existence of a proof of the negation.

This categorical proof theory does not exactly correspond to standard proof theory (for instance [Prawitz, 1965]) where proofs are presented by proof trees of the form

$$
\frac{(\varphi) \qquad \dfrac{\psi \qquad (\varphi)}{\dfrac{\psi \wedge \varphi}{\varphi \Rightarrow (\psi \wedge \varphi)}}}{\dfrac{\psi \wedge \varphi}{\varphi \Rightarrow (\psi \wedge \varphi)}}
$$

which would for instance correspond to the categorical proof

$$\lambda(apply \circ \langle \lambda(1_{\psi \wedge \varphi}) \circ p_{\varphi,\psi}, q_{\varphi,\psi} \rangle) : \psi \to [\varphi \Rightarrow (\psi \wedge \varphi)]$$

The reasons are manifold: on the one hand, the categorical proof does not distinguish between a list φ, ψ of assumptions and the conjunction $\varphi \wedge \psi$ (which is a defect which can be repaired if *multicategories* [Lambek, 1989] are used). On the other hand, the categorical proof is more precise in handling the data flow; for instance

$$\lambda(apply \circ \langle \lambda(p_{\psi \wedge \varphi, \varphi}), q_{\psi, \varphi} \rangle) : \psi \to [\varphi \Rightarrow (\psi \wedge \varphi)]$$

could also correspond to the given proof tree.

The categorical equalities correspond to equivalences of proof as defined in [Prawitz, 1965], e.g. the proof tree above would be equivalent to

$$\frac{\dfrac{\psi \qquad (\varphi)}{\psi \wedge \varphi}}{\varphi \Rightarrow (\psi \wedge \varphi)}$$

while

$$
\begin{aligned}
\lambda(apply \ \circ \langle \lambda(1_{\psi \wedge \varphi}) \circ p_{\varphi, \psi}, q_{\varphi, \psi} \rangle) \ &= \lambda(1_{\varphi, \psi}) \\
&= \lambda(apply \ \circ \langle \lambda(p_{\psi \wedge \varphi, \varphi}), q_{\psi, \varphi} \rangle) \\
&: \psi \to [\varphi \Rightarrow (\psi \wedge \varphi)].
\end{aligned}
$$

Joachim Lambek has probably been the first to consider *formulas-as-objects* and *proofs-as-morphisms* [Lambek, 1969; Lambek, 1972] paving the way for a categorical analysis of proof theory, an extensive account of which can be found in [Szabo, 1978].

This may (hopefully) be sufficient to indicate how a categorical version of proof theory works. Subsequently, proofs $f : \varphi \to \psi$ will be replaced by sequents $\varphi \vdash \psi$, expressing existence of a proof without actually constructing the proof, i.e. we cut down categories to preorders. However, we talk categorical language so that everybody willing to spend the effort may go through the more general argument.

9.2 Substitution as predicate transformation

9.2.1 The substitution rule (in the style of natural deduction)

$$\frac{\varphi \vdash \psi}{[\Theta]\varphi \vdash [\Theta]\varphi}$$

provides the link between individuals and formulas. (The notation $[\Theta]\varphi$ states that all free variables x in φ are replaced by the term $\Theta(x)$.) It is a major insight to interpret substitution as a *modality* or as a *predicate transformer*, which can be axiomatized in terms of equivalences of formulas.

For instance, substitution for propositional calculus is basically specified by the standard case distinction, but as logical axioms

$$[\Theta]\mathsf{tt} \dashv\vdash \mathsf{tt}$$
$$[\Theta]\mathsf{ff} \dashv\vdash \mathsf{ff}$$
$$[\Theta](\varphi \wedge \psi) \dashv\vdash [\Theta]\varphi \wedge [\Theta]\psi$$
$$[\Theta](\varphi \vee \psi) \dashv\vdash [\Theta]\varphi \vee [\Theta]\psi$$
$$[\Theta](\varphi \Rightarrow \psi) \dashv\vdash [\Theta]\varphi \Rightarrow [\Theta]\psi$$
$$[\partial \circ \Theta]\varphi \dashv\vdash [\Theta]([\partial]\varphi)$$
$$[id_X]\varphi \dashv\vdash \varphi.$$

Atomic formulas are of the form $[\Theta]\rho$ where ρ is a relational symbol.

More precisely, the substitution rule should be annotated by parameter lists

$$\text{(Sub)} \quad \frac{\varphi \vdash_{\langle p \rangle} \psi}{[\Theta]\varphi \vdash_{\langle q \rangle} [\Theta]\varphi} \quad \text{where } \Theta : \{p\} \to T_\Sigma(\{q\}) \text{ is a variable assignment}$$

to indicate the free variables which may occur in the formulas. Alternatively, propositions may be typed, replacing substitutions by functional terms to obtain

$$\mathsf{tt}_A : A$$
$$\mathsf{ff}_A : A$$
$$\varphi \wedge \psi : A \qquad \text{if } \varphi : A \text{ and } \psi : A$$
$$\varphi \vee \psi : A \qquad \text{if } \varphi : A \text{ and } \psi : A$$
$$\varphi \Rightarrow \psi : A \qquad \text{if } \varphi : A \text{ and } \psi : A$$
$$[f]\varphi : A \qquad \text{if } f : A \to B \text{ and } \varphi : B$$
$$\rho : b_1 \times \ldots \times b_n \quad \text{if } \rho : b_1 \ldots b_n \text{ is a relation symbol.}$$

The types indicate the variables as in the case of functional terms. The notation is equivalent to standard notation, e.g.

$$[\langle p \rangle \langle t_1, \ldots, t_n \rangle]\rho$$

corresponds to

$$\rho(t_1, \ldots, t_n)$$

but with the additional information that the free variables occur in $\langle p \rangle$. (We leave it to the reader to check equivalence to the standard notation.)

9.2.2 Positive Horn logic

Integrating this notation with the categorical proof rules we can axiomatize (intuitionistic) propositional logic by proof rules such as in the following conjunctive fragment

$$\text{(refl)} \quad \varphi \vdash_A \varphi \qquad\qquad \text{(trans)} \quad \frac{\varphi \vdash_A \psi \qquad \psi \vdash_A \chi}{\varphi \vdash_A \chi}$$

$$\text{(tt-E)} \quad \varphi \vdash_A \text{tt}_A$$

$$\text{($\wedge$I)} \quad \frac{\chi \vdash_A \varphi \qquad \chi \vdash_A \psi}{\chi \vdash_A \varphi \wedge \psi} \qquad\qquad \text{($\wedge$E)} \quad \varphi \wedge \psi \vdash_A \varphi \quad \varphi \wedge \psi \vdash_A \psi$$

$$\text{(sub)} \quad \frac{\varphi \vdash_B \psi}{[f]\varphi \vdash_A [f]\psi} \qquad \text{where } f : A \to B$$

$$[f]\text{tt}_B \dashv\vdash_A \text{tt}_A$$
$$[f](\varphi \wedge \psi) \dashv\vdash_A [f]\varphi \wedge [f]\psi$$
$$[1_A]\varphi \dashv\vdash_A \varphi$$
$$[f]([g]\varphi) \dashv\vdash_A [g \circ f]\varphi.$$

Positive Horn theories are specified by a signature comprising relation symbols, and a set of axioms of the form $\varphi_1 \wedge \ldots \wedge \varphi_n \vdash_{\langle p \rangle} \varphi_{n+1}$.

Here are two familiar such theories:

- Let a signature of integers consist of a base type $\underline{int}$, operator symbols $0 : 1 \to \underline{int}$, $suc, pred : \underline{int} \to \underline{int}$, $add : \underline{int} \times \underline{int} \to \underline{int}$, and relation symbols $iszero(\_) : \underline{int}$, $\_ \leq \_ : \underline{int} \times \underline{int}$. A theory of integers is then specified by the axioms:

$$\text{tt} \vdash [0] \; iszero$$
$$\text{tt} \vdash_{\underline{int}} [\langle 1_{\underline{int}}, 1_{\underline{int}} \rangle]\_ \leq \_$$
$$[\langle p_1, p_2 \rangle]\_ \leq \_ \wedge [\langle p_2, p_3 \rangle]\_ \leq \_ \vdash_{\underline{int} \times \underline{int}} [\langle p_1, p_3 \rangle]\_ \leq \_$$
$$\_ \leq \_ \vdash_{\underline{int} \times \underline{int}} [\langle p_1, suc \circ p_2 \rangle]\_ \leq \_ \quad \_ \leq \_ \vdash_{\underline{int} \times \underline{int}} [\langle pred \circ p_1, p_2 \rangle]\_ \leq \_$$
$$\_ \leq \_ \vdash_{\underline{int} \times \underline{int}} [suc \times suc]\_ \leq \_ \quad\quad \_ \leq \_ \vdash_{\underline{int} \times \underline{int}} [\langle pred \times pred \rangle]\_ \leq \_$$

- Everybody's favourite, stacks are another example: The signature is given by

Base types	$\underline{nat}, \underline{stack}$	
Operators	$0 :\to \underline{nat},$	$suc : \underline{nat} \to \underline{nat}$
	$empty :\to \underline{stack}$	$push : \underline{stack} \times \underline{nat} \to \underline{stack}$
	$pop : \underline{stack} \to \underline{stack}$	$top : \underline{stack} \to \underline{nat}$
Atomic predicates	$eq_{\underline{nat}} : \underline{nat} \times \underline{nat}$	$eq_{\underline{stack}} : \underline{stack} \times \underline{stack}$

Axioms

$$\text{tt} \vdash [\langle 0,0 \rangle] eq_{\underline{nat}} \qquad \text{tt} \vdash [\langle empty, empty \rangle] eq_{\underline{stack}}$$

$$[\langle p_1, p_1 \rangle] eq_{\underline{stack}} \wedge [\langle p_2, p_2 \rangle] eq_{\underline{nat}} \vdash [\langle pop \circ push, p_1 \rangle] eq_{\underline{stack}}$$

$$[\langle p_1, p_1 \rangle] eq_{\underline{stack}} \wedge [\langle p_2, p_2 \rangle] eq_{\underline{nat}} \vdash [\langle top \circ push, p_2 \rangle] eq_{\underline{stack}}$$

Moreover we assume axioms which state that the equality predicates are reflexive, symmetric, transitive and compatible with the operators, e.g.

$$eq_{\underline{nat}} \vdash [suc \times suc] eq_{\underline{nat}}.$$

We leave it to the reader to prove equivalence of this conjunctive fragment to one of the more familiar presentations of positive Horn logic.

9.2.3 Universal and existential quantification

Most amazingly, once more an insight due to Lawvere, the predicate transformer approach extends to *quantification*, even more, quantification turns out to define adjunctions [Lawvere, 1969a]. The standard inference rules of *existential* and *universal* quantification

$$(\exists I) \quad \frac{}{\varphi[\langle q \rangle / M] \vdash_{\langle p \rangle} \exists \langle q \rangle \varphi} \qquad (\exists E) \quad \frac{\varphi \vdash_{\langle \langle p \rangle, \langle q \rangle \rangle} \psi}{\exists \langle q \rangle \varphi \vdash_{\langle p \rangle} \psi}$$

$$(\forall I) \quad \frac{\varphi \vdash_{\langle \langle p \rangle, \langle q \rangle \rangle} \psi}{\varphi \vdash_{\langle p \rangle} \forall \langle q \rangle \psi} \qquad (\forall E) \quad \forall \langle q \rangle \varphi \vdash_{\langle p \rangle} \varphi[\langle q \rangle / M]$$

are equivalently captured by

$$\varphi \vdash_{A \times B} [p_{A,B}](\Sigma \varphi) \qquad \frac{\varphi \vdash_{A \times B} [p_{A,B}]\psi}{\Sigma \varphi \vdash_A \psi}$$

$$\frac{[p_{A,B}]\varphi \vdash_{A \times B} \psi}{\varphi \vdash_A \Pi \psi} \qquad [p_{A,B}](\Pi \varphi) \vdash_{A \times B} \varphi$$

$$[f]\Pi \varphi \dashv\vdash_A \Pi [f \times id_C] \varphi \qquad [f]\Sigma \varphi \dashv\vdash_A \Sigma [f \times id_C] \varphi$$

where

$$\Sigma \varphi : A \quad \text{if } \varphi : A \times B$$
$$\Pi \varphi : A \quad \text{if } \varphi : A \times B$$

are new modalities (the last two rules cater for preservation of structure by predicate transformation = substitution. These conditions in more general form are referred to as Beck–Chevalley conditions (see Section 9.3.1).). The reader is again invited to prove equivalence of this presentation of first order predicate logic with the standard definition.

To see that quantification defines an adjunction (here a Galois connection since we deal with preorders) observe that $[p_{A,B}] : Prop(A) \rightarrow$

$Prop(A \times B), \varphi \mapsto [p_{A,B}]\varphi$ is monotone as well as $\Sigma\varphi : Prop(A \times B) \to Prop(A)$ because of $[f]\varphi \vdash [f]\psi$ if $\varphi \vdash \psi$. Moreover $\Sigma\varphi \vdash \psi$ implies $[p_{A,B}]\Sigma\varphi[p_{A,B}]\psi$ implies $\varphi \vdash [p_{A,B}]\psi$, hence $\varphi \vdash [p_{A,B}]\psi$ iff $\Sigma\varphi \vdash \psi$. Similarly, for $\Pi\varphi$. In fact *existential quantification is left adjoint*, and *universal quantification is right adjoint to predicate transformation* (see below, or Section 3.6.2).

9.2.4 Though lurking through the arguments, the treatment still needs a more categorical formalization. *Indexed categories* are the appropriate structure to abstract predicate transformation.

Definition 9.2.1. An *indexed category* $\mathbb{C}$ consists of

- a (*base*) category $\mathbf{C}$

- a (*fibre*) category $\mathbf{C}_A$ for each object A of $\mathbf{C}$

- a functor $[f] : \mathbf{C}_B \to \mathbf{C}_A$ for each morphism $f : A \to B$ in $\mathbf{C}$ such that

  - $[1_A] : \mathbf{C}_A \to \mathbf{C}_A$ is naturally isomorphic to the identity functor,
  - $[g \circ f] \cong [g] \circ [f]$ (naturally).

An indexed category is *strict* if the natural isomorphisms are natural equalities.

We consider a few examples:

- Let $\mathbf{C} = \mathbf{Set}$ be the base category, and let $\mathbf{C}_X = P(X)$ be the power set as order category. $[f] = f^{-1} : P(Y) \to P(X), U \mapsto \{y \in Y \mid f(y) \in U\}$.

- Dijkstra's predicate transformers [Dijkstra, 1976] provide a computational example: the categories $\mathbf{C}_A$ are suitable power sets, and $[f]$ maps power sets to power sets preserving empty set, finite meets and directed joins. Syntactically, the (typed, maybe imperative) programs define the base category, and predicates allow one to express empty set, finite meets and directed joins plus some base predicates.

- Let $\mathbf{C} = \mathbf{Dcpo}$ be the category of directed complete posets, $\mathbf{C}_A = [X \Rightarrow \mathbf{2}]$ (where $\mathbf{2} = \{\perp \leq \top\}$), the set of monotone or continuous functions from X to $\mathbf{2}$. Predicate transformers are the inverse image functions resp. the composition $[f]\varphi = \varphi \circ f$ where $f : X \to Y$ and $\varphi : Y \to \mathbf{2}$. (see [Smyth, 1983]).

- Let $\mathbf{C}$ be a category with finite limits, and let slices be comma categories $\mathbf{C}_A = (\mathbf{C}, A)$. Predicate transformers $[f]\varphi = f^*\varphi$ are defined by pullbacks

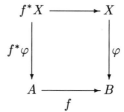

  An object $\varphi : X \to A$ of $(\mathbf{C}, A)$ may be thought of as a 'representation' of a 'subset' $Im(\varphi) := \{b \in B \mid \exists x \in X.\varphi(x) = b\}$. Then $[f]\varphi$ is a representation of $f^{-1}(Im(\varphi))$. If $\mathbf{C}$ has image factorization (cf. Section 8.3.4) one may replace $(\mathbf{C}, A)$ by the subcategory $\mathbf{Sub}(A)$ with objects being monos, approximating the categorical interpretation of first order logic as considered in [Makkai and Reyes, 1977].

- Modal operators $\Diamond$ ('possibility') and $\Box$ ('necessity') [Hughes and Cresswell, 1968] can be seen as predicate transformers; the unary modal operators (over the only base type s) generate the category $\mathbf{C}$, $\mathbf{C}_s$ consists of all modal formulas with axioms being

$$\Diamond \varphi \vdash \varphi \qquad\qquad\qquad \varphi \vdash \Box \varphi$$

$$\frac{\varphi \vdash \psi}{\varphi \vdash \Diamond \psi} \; (\varphi \text{ fully modalized}) \qquad \frac{\varphi \vdash \psi}{\Box \varphi \vdash \psi} \; (\psi \text{ fully modalized}).$$

  (This is intuitionistic system S5.)

- In order not to let the picture get too one-sided: define an indexed category $\mathbf{A}lg$ with base category being $\mathbf{Sign}$, the category of signatures and signature morphisms. Slices are the categories $\mathbf{Alg}(\Sigma)$ and the the reduct functor $\_{}_H : \mathbf{Alg}(\Sigma') \to \mathbf{Alg}(\Sigma)$ defines $[H]$ for a signature morphism $H : \Sigma \to \Sigma'$. This example does, of course, not fit into the picture of indexed categories as axiomatizations of logical theories.

9.2.5 Doctrines

As with categories, indexed categories become interesting only if additional infrastructure is at hand. We speak of *doctrines* if such infrastructure is uniformly defined. Positive Horn logic and first order predicate logic are examples for logics which determine doctrines.

Definition 9.2.2.

- *Doctrines of positive Horn logic*

  - These are indexed categories $\mathbb{D}$ such that the base category $\mathbf{D}$ and all slice categories $\mathbf{D}_A$ have finite products and if predicate transformation $[f] : \mathbf{D}_B \to \mathbf{D}_B$ preserves finite products.

- *Doctrines of intuitionistic first order logic*

  - the base category $\mathbf{D}$ has finite products,

  - all slice categories $\mathbf{D}_A$ are Cartesian closed and have finite coproducts,

  - the functors $[f] : \mathbf{D}_B \to \mathbf{D}_A$ preserve this structure,

  - $[p_{A,B}] : \mathbf{D}_A \to \mathbf{D}_{A \sqcap B}$ has a left adjoint $\Sigma_{p_{A,B}} : \mathbf{D}_{A \sqcap B} \to \mathbf{D}_A$ and a right adjoint $\Pi_{A,B} : \mathbf{D}_{A \sqcap B} \to \mathbf{D}_A$ such that for all $f : A \to B$ of $\mathbf{D}$, the canonical natural transformations

$$\Sigma_{p_{A,C}} \circ [f \sqcap 1_C] \to [f] \circ \Sigma_{p_{B,C}}$$
$$[f] \circ \Pi_{p_{B,C}} \to \Pi_{p_{A,C}} \circ [f \sqcap 1_C]$$

  are isomorphisms (Beck–Chevalley condition).

The Beck–Chevalley conditions formalizes substitution on quantified formulas.

Most of the examples below have more infrastructure than necessary in that all predicate transformers, and not only those induced by projections, have adjoints. The additional structure will turn out as relevant when modelling equality (see Section 9.3.1).

- The category **Set** of sets with power sets $P(X)$ as slices. Products in $P(X)$ are the intersections. The inverse function $[f]U := f^{-1}(U)$ has a left adjoint $\Sigma_f(U) := f(U)$ and a right adjoint $\Pi_f(U) := \bigcap \{V \mid f^{-1}(U) \subseteq V\}$ for all $f : X \to Y$. Specifically $\Sigma_{p_{A,B}}(U) = \{x \in X \mid \exists y \in Y.(x,y) \in U\}$, $\Pi_{p_{A,B}}(U) = \{x \in X \mid \forall y \in Y.(x,y) \in U\}$. As left adjoints preserve colimits and right adjoints preserve limits, one only has to check that exponentials are preserved and that the Beck–Chevalley condition holds which is straightforward.

- The category **Dcpo** of domains with slices

  - 2^X being the partially ordered set of monotone functions $\varphi : X \to 2$. Predicate transformation $[f]\varphi = \varphi \circ f$ has a left adjoint

$\Sigma_f\psi = \sqcap\{\varphi \mid \psi \leq \varphi \circ f\}$ and a right adjoint $\Pi_f\psi = \bigsqcup\{\varphi \mid \varphi \circ f \leq \psi\}$. Predicate transformation does in general not preserve exponentiation.

- $\mathbf{2}^X$ being the partially ordered set of continuous functions $\varphi : X \to \mathbf{2}$. Predicate transformation $[f]\varphi = \varphi \circ f$ only has a right adjoint as $[f]$ in general only preserves suprema and finite infima. Such morphisms are called frame morphisms while the right adjoints are continuous maps between locales [Johnstone, 1982]. Some continuous functions such as the projections also have a left adjoint (as infima are preserved).

- The category **Set** of sets with slices being comma categories $(\mathbf{Set}, X)$ and with predicate transformation defined by pullbacks (see above). $[f]$ has a left adjoint $\Sigma_f\psi := f \circ \psi$ and a right adjoint $\Pi_f\psi : \{(y,g) \mid y \in Y, g : f^{-1}(y) \to A$ s.t. $\psi(h(x)) = x$ for $x \in f^{-1}(y)\} \to Y$ projecting to the first component (where $\psi : A \to X$). One should note that $\Pi_f\psi = \{y \in Y \mid f^{-1}(y) \subseteq \psi\}$ if $\psi \subseteq X$. The case that **C** is an arbitrary category with finite limits will be discussed separately below under the name of locally Cartesian closed categories. This is a doctrine of (intuitionistic) first order logic with equality.

- Syntactical examples are generated by combining the various categories of procedures with the proof systems as defined in Sections 9.1 and 9.2. This generates 'free' doctrines just as the Λ-calculus defines free Cartesian closed categories.

9.3   Theories of equality

9.3.1   The *identity rule*

$$\Theta = \Theta', [\Theta]\varphi \vdash [\Theta']\varphi$$

(where Θ, Θ' are substitutions) states that equals can be replaced by equals which is probably the most prominent property of equality. The identity rule implies transitivity and symmetry of equality, the latter provided that equality is reflexive.

These properties are easily axiomatized categorically given the structure of an indexed category. Given a doctrine $\mathbb{D}$ of positive Horn logic, an equality predicate of type A is given by a binary object eq_A of $\mathbf{D}_{A\Pi A}$ such that

$$r_A : \mathrm{tt}_A \to [\Delta_A]eq_A$$
$$s_A : eq_A \wedge [p_{A,A}]\varphi \to [q_{A,A}]\varphi$$
$$eq_{A\Pi B} \cong [\langle p_1, p_3\rangle]eq_A \wedge [\langle p_2, p_4\rangle]eq_B$$

(the identity rule is split in order to simplify notation, the p_i's denote the projections to the respective components).

We note that certain identities hold for formulas involving equality, e.g.

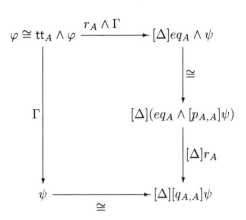

corresponds to

$$\frac{\begin{array}{c} \text{tt}_A \\ \hline t = t \end{array} \quad \begin{array}{c} \varphi \\ \Gamma \\ \psi \end{array}}{\psi} \qquad \Rightarrow \qquad \begin{array}{c} \varphi \\ \Gamma \\ \psi \end{array}$$

Lawvere was the first to observe that first order (intuitionistic) logic with equality has an elegant categorical axiomatization [Lawvere, 1969a; Lawvere, 1970] (though the equivalence to the standard axiomatization has been formally established only by [Seely, 1983]): combining first order calculus with equality allows one to extend the universal characterization of quantifiers in that

$$\sum\nolimits_{q_{A,B}}([f\Pi 1_B]eq_B \wedge [p_{A,B}]\varphi) \text{ and } \prod\nolimits_{q_{A,B}}([f\Pi 1_B]eq_B \Rightarrow [p_{A,B}]\varphi)$$

or in a more classical notation

$$\exists x \in A(f(x) = y \wedge \varphi(x)) \text{ and } \forall x \in A(f(x) = y \Rightarrow \varphi(x))$$

define a left adjoint resp. a right adjoint to the substitution functor $[f] : \mathbf{D}_B \rightarrow \mathbf{D}_A$. We obtain the structure of *doctrines of (intuitionistic) first order predicate calculus with equality*, which are doctrines $\mathbb{D}$ of first order predicate calculus with the following additional structure:

- the base category $\mathbf{D}$ has finite limits,

- $[f] : \mathbf{D}_B \rightarrow \mathbf{D}_A$ has a left adjoint $\Sigma_f : \mathbf{D}_A \rightarrow \mathbf{D}_B$ and a right adjoint $\prod_f : \mathbf{D}_A \rightarrow \mathbf{D}_B$ such that the *Beck–Chevalley condition* holds, i.e. the canonical natural transformation $\Sigma_p \circ [q] \to [f] \circ \Sigma_g$ is an isomorphism for every pullback

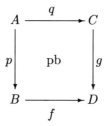

in $\mathbf{D}$ (an analogous condition for $\prod_f$ is a consequence).

For examples cf. Section 9.2.5.

Once again the Beck–Chevalley condition states that substitution distributes over quantification.

9.3.2 The equivalence of this axiomatization of the first order logic with equality to standard proof theoretic presentations is shown in [Seely, 1983]. We only indicate how equality is handled. Let

$$eq_A := \Sigma_{\Delta_A} \mathsf{tt}_A$$

which is induced by the following logical equivalences

$$x = y \dashv\vdash \exists z. z = x \wedge z = y \dashv\vdash \exists z. \Delta(z) = \langle x, y \rangle \wedge \mathsf{tt}.$$

Universal properties induce a morphism

$$r_A := \mathsf{tt}_A \eta : \mathsf{tt}_A \to [\Delta_A] \Sigma_{\Delta_A} \mathsf{tt}_A$$

which is the candidate for modelling reflexivity. For the identity rule we need two observations due to [Lawvere, 1970].

Proposition 9.3.1. *Let* $f : A \to B$, $\varphi \in \mathbf{D}_A$, $\psi \in \mathbf{D}_B$, $\alpha \in \mathbf{D}_X$. *Then*

- $\Sigma_f(\varphi \wedge [f]\psi) \cong \Sigma_f \varphi \wedge \psi$, *and*

- $\Sigma_{1_X \Pi f}(\alpha \wedge \varphi) \cong \alpha \wedge \Sigma_f \varphi$

where the functor $\wedge : \mathbf{D}_X \times \mathbf{D}_Y \to \mathbf{D}_{X \Pi Y}$ *is defined by* $\varphi \wedge \psi = [p_{X,Y}]\varphi \wedge [q_{X,Y}]\psi$.

Proof.   • The diagrams

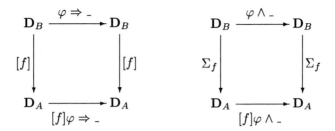

commute up to isomorphism as substitution preserves exponentiation and as the functors of the left diagram are replaced by their left adjoints.

- Since

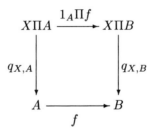

is a pullback, the Beck condition yields $\Sigma_{1_X \Pi f}[q_{X,A}]\varphi \cong [q_{X,B}](\Sigma_f \varphi)$, and thus $\alpha \wedge \Sigma_f \varphi \cong [p_{X,B}]\alpha \wedge [q_{X,B}](\Sigma_f \varphi) \cong [p_{X,B}]\alpha \wedge \Sigma_{1_X \Pi f}[q_{X,A}]\varphi \cong \Sigma_{1_X \Pi f}(([1_A \Pi f][p_{X,B}]\alpha) \wedge [q_{X,A}]\varphi) \Sigma_{1_X \Pi f}([p_{X,A}]\alpha \wedge [q_{X,A}]\varphi) \cong \Sigma_{1_X \Pi} f(\alpha \wedge \psi)$.

∎

The latter states that conjunction distributes over existential quantification if the quantified variables are 'independent'.

Then we obtain the morphism

$$s_A : eq_A \wedge [p_{A,A}]\varphi \to [q_{A,A}]\varphi$$

by

$$
\begin{aligned}
s_A := eq_A \wedge [p_{A,A}]\varphi &\cong \Sigma_{\Delta_A}([\Delta_A]([p_{A,A}]\varphi) \wedge \mathsf{tt}) \\
&\cong \Sigma_{\Delta_A}([\Delta_A]([q_{A,A}]\varphi) \wedge \mathsf{tt}) \\
&\cong eq_A \wedge [q_{A,A}]\varphi \to [q_{A,A}]\varphi
\end{aligned}
$$

while

$$eq_{A \Pi B} \cong [\langle p_1, p_3 \rangle]eq_A \wedge [\langle p_2, p_4 \rangle]eq_B$$

is a direct consequence of the second part of the proposition.

9.3.3 The notion of a *partial equivalence relation* (PER) has recently found some attention in computer science, for instance, when defining PER-models of higher type λ-calculi. A partial equivalence relation is only symmetric and transitive, but not necessarily reflexive. Elements which are reflexive are said to be total. Let us look for some motivation.

Rather naïvely, we might encounter a situation where we can write down a term which does not have a denotation, an (in-)famous example being *pop*(*empty*), the application of the pop operation to the empty stack. Then the interpration of terms becomes partial, which has some bearing on the interpretation of equality. The literature offers two notions of equality related to partiality:

Weak equality $x \doteq y$ states that 'x and y are defined and equal'
*Strong equality*J $x \equiv y$ states that 'y is defined and equal to x if x is defined, and that x is defined and equal to y if y is defined'.

These definitions are not independent since

$$x \equiv y \Leftrightarrow (x \doteq x \lor y \doteq y \Rightarrow x \doteq y)$$

and

$$x \doteq y \Leftrightarrow (Ex \land Ey \land x \equiv y)$$

where Ex is an existence predicate. We prefer weak equality here. Unfortunately the substitution rule

$$\frac{\Phi \vdash \varphi}{[\Theta]\Phi \vdash [\Theta]\varphi}$$

is not sound any more with regard to weak equality. Consider the signature

$$0 : \varepsilon \to \underline{nat}, \ suc : \underline{nat} \to \underline{nat}, \ pred : \underline{nat} \to \underline{nat}$$

with natural numbers as a model, where the 'predecessor' of 0 is undefined, the term *pred*(x) may be undefined if we assign 0 to x. The model satisfies the weak equation $pred(suc(x)) \doteq x$, but fails to satisfy the equation $pred(suc(pred(x))) \doteq pred(x)$ obtained by (syntactical) substitution. The discrepancy creeps in because satisfaction substitutes variables only by defined data, namely the elements of the model, while (syntactic) substitution replaces a variable by a term the interpretation of which is potentially undefined. By a standard strategy, the problem can be resolved by redefining the substitution rule to the effect that variables are replaced only by terms which are 'defined'.

9.3.4 [Scott, 1977] proposes a theory of identity and existence which retains the substitution rule for partial interpretation by changing the notion of a model. His models consist of 'defined' and 'undefined' elements (or just one element 'undefined' for every carrier), the latter serving as denotation for terms 'the meaning of which is not defined'. Operations are supposed to be strict, i.e. preserve undefinedness. The substitution rule is then sound with regard to weak equality, e.g. $pred(suc(x)) \doteq x$ implicitly states $x \doteq x$, and hence the existence of all values, specifically $pred(0)$, due to strictness. Hence we cannot construct the contradiction above.

A very elegant way to represent such models has been proposed in [Fourman and Scott, 1979], which brings us back to the theme of partial equivalence relation. The idea is that types (sets) A come along with a partial equivalence relation $eq : A\Pi A$ being a symmetric and transitive relation, i.e.

$$eq \vdash [\langle p_2, p_1 \rangle] eq$$
$$[\langle p_1, p_2 \rangle] eq \wedge [\langle p_2, p_3 \rangle] eq \vdash [\langle p_1, p_3 \rangle] eq$$

in our categorical framework (the reader is urged to translate this and all subsequent definitions into set theory). The extent of the type is given by

$$E_A = [\Delta_A] eq$$

which may be thought of as a predicate of *existence* or *determinedness*. Morphisms (functions) then are predicates (relations) for which the usual axioms for functions hold, but which additionally satisfy the property that determined 'elements' are mapped to determined 'elements'. Formally, we define as in [Pitts, 1981]:

9.3.5 Let $\mathbb{C}$ be an indexed category such that the base category $\mathbf{C}$ has finite products, the slice categories $\mathbf{C}_A$ have finite limits, and such that the left adjoints Σ_f exist for all predicate transformers $[f] : \mathbf{C}_B \to \mathbf{C}_A$. Moreover the predicate transformers should preserve exponentials and the Beck–Chevalley condition should hold.

Definition 9.3.2. A *partial object* $A = (\mid A \mid, eq)$ consists of an object $\mid A \mid$ of $\mathbf{C}$ and a partial equivalence relation $eq \in \mathbf{C}_A$. A *relation* on partial objects $(A_1, eq_1), \ldots, (A_n, eq_n)$ is a predicate $R \in \mathbf{C}_{A_1 \Pi \ldots \Pi A_n}$ such that

$$R(x_1, \ldots, x_n) \wedge \bigwedge_{i=1,\ldots,n} eq_i(x_i, x_i') \vdash R(x_1', \ldots, x_n')$$

(we use the standard, more intuitive notation which is easily translated to the language of indexed categories). R is strict if additionally

$$R(x_1, \ldots, x_n) \vdash \bigwedge_{i=1,\ldots,n} eq_i(x_i, x_i)$$

A strict relation F on partial objects (A, eq_A), (B, eq_B) is called *functional* if it satisfies

- F is *single-valued*, i.e.

$$F(x, y) \wedge F(x, y') \vdash eq'_B(y, y')$$

- F is *total*, i.e.

$$eq'_A(x, x) \vdash \exists y. F(x, y).$$

Definition 9.3.3. Given an indexed category as sketched above, we define a category **PER**$(\mathbb{C})$ of partial equivalence relations of $\mathbb{C}$ whose objects are partial objects (A, eq_A) and whose morphisms are functional relations modulo provable equivalence (i.e. $\varphi \dashv\vdash \psi$). Composition is given by $G \circ F = \exists y. F(x, y) \wedge G(y, z)$ and the equivalence class of eq_A determines the identity.

The definition of **PER**$(\mathbb{C})$ may look pretty heavy. However, an example might give some ideas: let us assume that a specific doctrine $\mathbb{C}_{Nat}$ of first order logic is given which is generated by the signature

$$0 : \varepsilon \rightarrow \underline{nat}, \ suc : \underline{nat} \rightarrow \underline{nat}, \ pred : \underline{nat} \rightarrow \underline{nat}$$

with an equality predicate eqnat for natural numbers. Then $eq_{\underline{nat}+}(x, y) := eq_{\underline{nat}}(x, y) \wedge isnonzero(x) \wedge isnonzero(y)$ where $isnonzero(x) = \exists n \in \underline{nat}. suc(n) = x$ defines a partial equivalence. Some computation should then prove that $pred(x) = y$ defines a functional relation from $(\underline{nat}, eq_{nat+})$ to $(\underline{nat}, eq_{nat})$. In other words, the partial predecessor function is turned into a total function on the non-zero natural numbers. Note that $0 : \varepsilon \rightarrow \underline{nat}$ is 'undefined' with regard to eq_{nat+}.

There exists a functor $\Delta : \mathbf{C} \rightarrow \mathbf{PER}(\mathbb{C})$ which is defined by $\Delta(A) = (A, \Sigma_\Delta tt_A)$ and $\Delta(f : A \rightarrow B) = \{\Sigma_{\langle i, f \rangle} tt_A\}$. One may say that Δ yields the 'extension of $\mathbf{C}$ relative to $\mathbb{C}$'. We may argue that the construction of $\mathbf{C}[\mathbf{P}]$ adds an 'axiom of extensionality', which is further substantiated by the fact that a 'constructive set theory' is obtained, provided that $\mathbb{C}$ has enough infrastructure.

Definition 9.3.4. A *tripos* is an indexed category $\mathbb{C}$ such that

- $\mathbf{C}$ has finite products,

- the categories $\mathbf{C}_A$ are biCartesian closed order categories,

- the functors $[f] : \mathbf{C}_B \rightarrow \mathbf{C}_A$ preserve exponentials and have left and right adjoints such that the Beck condition is satisfied, and

- for each object A of $\mathbf{C}$ there exists an object $P(A)$ of $\mathbf{C}$ and a predicate $\in_A$ in $\mathbf{C}_{A\Pi P(A)}$ such that there exists a morphism $\{\varphi\} : B \to P(A)$ for each $\varphi P(A\Pi B)$ such that $[1_A \Pi \{\varphi\}] \in_A \dashv\vdash \varphi$.

(This is a kind of 'logical' definition of a power object.)

Theorem 9.3.5 ([Pitts, 1981]). $\mathbf{PER}(\mathbb{C})$ *is a topos if* $\mathbb{C}$ *is a tripos.*

9.4 Type theories

9.4.1 Dependent types

The notion of *types* has been most influential to many recent developments in the theory of programming. We will see that the notion of indexed categories is instrumental in axiomatizing categorical counterparts of type theories. As a first example we consider a categorical axiomatization of a fragment of Martin-Löf's theory of dependent types [Martin-Löf, 1973; Martin-Löf, 1979], which is a fundamental construction in the field of categorical logics. The fragment consists of a judgement of the form

A	'A is a type',
$M : A$	'M is of type A',
$A = B$	'type A is equal to type B'
$M = N : A$	'M is equal to N, both of type A'.

Judgements live in an environment $\Gamma = x_1 : A_1, x_2 : A_2, \ldots, x_n : A_n$ where the variables $x_1, \ldots, x_{i-1}$ may occur free in the type A_i. The following deduction rules deal with typing

$$\overline{1[\,]}$$

$$\frac{B[\Gamma, x : A]}{\Pi x : A.B[\Gamma]} \qquad \frac{B\langle\Gamma, x : A\rangle}{\Sigma x : A.B\langle\Gamma\rangle}$$

$$\overline{\langle\rangle : 1[\,]}$$

$$\frac{M : B[\Gamma, x : A]}{(\lambda x : A.M) : \Pi x : A.B[\Gamma]} \qquad \frac{M : \Pi x : A.B[\Gamma] \quad N : A[\Gamma]}{(MN) : B[x/N][\Gamma]}$$

$$\frac{M : A[\Gamma] \quad N : B[x/M][\Gamma]}{(M, N) : \Sigma x : A.B[\Gamma]}$$

$$\frac{M : \Sigma x : A.B[\Gamma]}{first(M) : A[\Gamma]} \qquad \frac{M : \Sigma x : A.B[\Gamma]}{second(M) : B[x/first(M)][\Gamma]}$$

Propositions are here considered as *types*. This is not so unfamiliar, in set theory propositions φ determine subsets $\{x \in A \mid \varphi\}$, but *identification* of types and propositions, as proposed by [Howard, 1980] as the *formulae-as-types notion of construction*, induces a new notion: *dependent types*. In a sense, type dependency comes naturally; $B(x)[x : A]$ may be considered to state that 'B is a proposition with a free variable x of type A', and $M(x) : B(x)[x : A]$ as a notation for '$M(x)$ is a proof for the proposition $B(x)$'. Of course, we have the disturbing phenomenon that A may be a proposition, e.g. $\Pi x : C.D$, and hence that variables do not range over *individuals* only but also over *proofs*, or rather, that we cannot distinguish between proofs and individuals. Otherwise the introduction and elimination rules correspond to those for (intuitionistic) first order predicate calculus, if one forgets about the additional information of how proofs are constructed.

Type dependency is reflected by the substitution rules, since substitution of the ith variable depends on substitution for all variables of smaller index

$$\frac{A_{n+1}[\Gamma]}{A_i[\Gamma, x_{n+1} : A_{n+1}]} \qquad \frac{A_{n+1}[\Gamma]}{x_i : A_i[\Gamma, x_{n+1} : A_{n+1}]}$$

$$\frac{B[\Gamma, x : A, \Gamma'] \qquad N : A[\Gamma]}{B[x/N][\Gamma, x : A, \Gamma'[x/N]]} \qquad \frac{M : B[\Gamma, x : A, \Gamma'] \qquad N : A[\Gamma]}{M[x/N] : B[x/N][\Gamma, x : A, \Gamma'[x/N]]}$$

User-defined data can be introduced by signatures which consist of n-ary constants $C \in \Omega_n$ and by axioms of the form

$$\Gamma \vdash A(x_1, \ldots, x_n) \quad \text{or} \quad \Gamma \vdash f(x_1, \ldots, x_n) : A$$

which must be well-formed, i.e.

$\Gamma \vdash A(x_1, \ldots, x_n)$ is well-formed if $A \in \Omega_n$ and Γ!
$\Gamma \vdash f(x_1, \ldots, x_n) : A$ is well-formed if $f \in \Omega_n$ and $\Gamma \vdash A$

where the rule

$$\frac{A_{n+1}[\Gamma]}{\Gamma, x_{n+1} : A_{n+1}!}$$

introduces well-formed environments.

As an example, let $\Omega = \{nat : 0, \ 0 : 0, \ suc : 1, \ stack : 1, \ push : 2, \ pop : 1, \ empty : 0\}$ be the signature where $\_ : n$ indicates the arity. We claim that the axioms

$$\vdash nat$$
$$n : nat \vdash stack(n)$$
$$\vdash 0 : nat$$
$$n : nat \vdash suc(n) : nat$$
$$\vdash empty : stack(0)$$
$$n : nat, \ d : nat, \ s : stack(n) \vdash push(s, d) : stack(suc(n))$$
$$n : nat, \ s : stack(suc(n)) \vdash pop(s) : stack(n)$$

are well-formed.

A set theoretic argument might precede the categorical modelling: we interpret a dependent type $B(x)[x : A]$ as a family $\{b_a \in B_a \mid a \in A\}$ of sets (for convenience we identify a type A with its interpretation being a set A), and a 'dependent element' $M(x) : B(x)[x : A]$ as a family of elements $\{b_a \in B_a \mid a \in A\}$. Similarly, a dependent type $C(x, y)[x : A, y : B(x)]$ is interpreted as a family of sets $\{C_{a,b} \mid a \in A, b \in B_a\}$, and so on. These families of sets are equivalently represented by functions $first_{A,B} : B \to A$ where $B = \{(a, b) \mid a \in A, b \in B_a\}$ and $first_{C,B} : C \to B$ where $C = \{(a, b, c) \mid a \in A, b \in B_a, c \in C_{a,b}\}$ which project to the first components.

By abstraction, dependent types $B[\Gamma]$ are interpreted as morphisms $[\![B[]\!] : [\![B]\!] \to 1$ resp. $[\![B[\Gamma, x : A]\!] : [\![B]\!] \to [\![A]\!]$ in a category $\mathbf{B}$ with finite limits. Dependent elements $M : B[\Gamma]$ determine sections $[\![M : B[\Gamma]]\!] : [\![\Gamma]\!] \to [\![B]\!]$, meaning morphisms of $\mathbf{B}$ such that the diagram

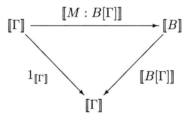

commutes where $[\![\Gamma]\!] := 1$ if $\Gamma = [\ ]$, and $[\![\Gamma]\!] := [\![A]\!]$ if $\Gamma = \Gamma', x : A$. Of course, equalities are interpreted as equalities of the corresponding morphisms.

Substitution on families of sets behaves as to be expected. In the most elementary case, a variable x of type A may be replaced by a constant term c of type A to yield the type $B(c)$. Semantically, c denotes some element $[\![c]\!]$ of A, and the type $B(c)$ the set $B_{[\![c]\!]}$. More generally, if we substitute a term $M(y)$ of type A with a free variable y of type C to obtain the dependent type $B(M(y))[y : C]$, then the corresponding family is $\{B_{[\![M]\!](c)} \mid c \in C\}$ where $[\![M]\!](c) \in A$ for every $c \in C$. Similarly, we may substitute a term $M(y, z)$ of type A with variables $y : C, z : D(y)$ in order to obtain the type $B(M(y, z))[y : C, z : D(y)]$, the semantics of which should be $\{\{B_{[\![M]\!](c,d)} \mid d \in D_c\} \mid c \in C\}$ (we hope the reader is not too confused to carry on for her/himself). Thinking in terms of functions, the substitution $\{B_{[\![M]\!](c)} \mid c \in C\}$ corresponds to pulling back $[\![B[x : A]\!]$ along $[\![M]\!] : [\![C]\!] \to [\![A]\!]$ in that the left vertical arrow in the pullback diagram

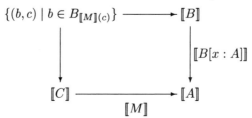

corresponds to $\{B_{[\![M]\!](c)} \mid c \in C\}$. Hence, for a categorical axiomatization, we should have all appropriate pullbacks. The following basic infrastructure has been proposed by [Hyland and Pitts, 1989]:

Definition 9.4.1. A category **D** together with a collection of morphisms $\mathcal{D}$ is called a *doctrine of dependent types* if

- **D** has finite products.

- If $f : C \to A$ of **C** and all morphisms $b : B \to A$ in $\mathcal{D}$ and if a pullback

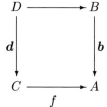

exists, then $d : D \to C$ is in $\mathcal{D}$; moreover there is a pullback square for each such f and b.

The latter property is referred to as *stability*. We use the notation $f^* : \mathcal{D}_B \to \mathcal{D}_A$ for the functors induced by specified pullbacks

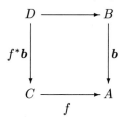

These data define a doctrine where fibres are the full subcategories $\mathcal{D}_A$ of the comma categories (B, A) with objects $b : B \to A$ in $\mathcal{D}$. The pullbacks define predicate transformation $f^* : \mathcal{D}_B \to \mathcal{D}_A$.

More infrastructure needs to be added to converge with Martin-Löf's type theory. Firstly, the type $1[\,]$ is interpreted by the identity function of the terminal object, which included in the collection $\mathcal{D}$ has the consequence that all isomorphisms of **D** belong to $\mathcal{D}$ (since every isomorphism defines a pullback of $\langle \rangle : A \to 1$ and the identity on 1). This property is referred to thus

Unit : $\mathcal{D}$ contains all isomorphisms of **D**.

Existential quantification $\Sigma x : A.B(x)$ turns a family of sets $\{B_a \mid a \in A\}$ into the set $\Sigma A.B = \{(a, b) \mid a \in A, b \in B_a\}$ and universal quantification $\Pi x : A.B(x)$ into the set $\Pi A.B = \{f : A \to \bigcup_{a \in A} B_a \mid f(a) \in B_a\}$. Note that the latter is just a way to define an A-indexed tuple where the ath component is an element of B_a. The universal character of the constructions is captured by the following observation.

Proposition 9.4.2. *There is a natural one-to-one correspondence between mappings*

- $f : X \times A \to B$ *and* $\lambda(f) : X \to \Pi A.B$ *where* $f(x, a) = (a, b)$ *for some* $b \in B_a$, *and where* $\lambda(f)(x)(a) = f(x, a)$

- $g : B \to Y \times A$ *and* $[g] : \Sigma A.B \to Y$ *where* $g(a, b) = (y, a)$ *for some* $y \in Y$ *and where* $[g](a, b) = p_{A,B}(g(a, b))$.

Proof. The first statement is only a modification of the corresponding coding for Cartesian closure, hence the notation $\lambda(f)$. For the second statement observe that $h : \Sigma A.B \to Y$ induces a mapping $\langle h, first \rangle : B \to Y \times A$ where first is the projection to the first component. ∎

Of course, we have special instances of the adjunctions for universal and existential quantification (Section 9.2.3). The following conditions are sufficient to guarantee their existence.

Sums $\mathcal{D}$ is closed under composition.

Products For all $f : A \to B$ in $\mathbf{D}$, there exists a functor $\Pi_f : \mathcal{D}_A \to \mathcal{D}_B$ which is right adjoint to $[f] : \mathcal{D}_B \to \mathcal{D}_A$ and which satisfies the Beck–Chevalley condition (see Section 9.3.2) for all pullbacks

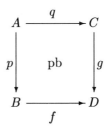

in $\mathbf{D}$ such that f and thus q is in $\mathcal{D}$.

(The reader may check the properties claimed. Specifically, $\Sigma_f(g) = f \circ g$. The Beck–Chevalley condition comes free for Σ_f.)

Assuming these properties, a categorical interpretation of Martin-Löf type theory is achieved by induction along the rules:

$$[\![1[\ ]\!]\!] = 1_1 : 1 \to 1$$

$$[\![\Pi x : A.B[\Gamma]]\!] = \Pi_{[\![A[\Gamma]]\!]}[\![B[\Gamma, x : A]]\!]$$

$$[\![\Sigma x : A.B[\Gamma]]\!] = [\![A[\Gamma]]\!] \circ [\![B[\Gamma, x : A]]\!]$$

$$[\![A_i[\Gamma, x_{n+1} : A_{n+1}]]\!] \quad = [\![A_{n+1}[\Gamma]]\!]^*[\![A_i[\Gamma]]\!] \quad \text{if } i \leq n$$
$$= 1_{[\![A_i]\!]} \qquad\qquad\qquad \text{else}$$

$$[\![\langle\rangle : 1[\ ]]\!] = 1_1 : 1_1 \to 1_1$$

$$[\![(\lambda x : A.M) : \Pi x : A.B[\Gamma]]\!] = \lambda([\![M]\!]) : 1_{[\![\Gamma]\!]} \to [\![\Pi x : A.B[\Gamma]]\!]$$

(assuming that pulling back $1_{[\![\Gamma]\!]}$ along $[\![A[\Gamma]]\!]$ gives $1_{[\![A]\!]}$)

$$[\![(MN) : B[x/N][\Gamma]]\!] = \langle [\![N]\!]^*B[\Gamma, x : A]\!], \varepsilon \circ ([\![A[\Gamma]]\!]^*[\![M]\!])\rangle : 1_{[\![\Gamma]\!]} \to$$
$$[\![B[x/N][\Gamma]]\!]$$

(note that $[\![B[x/N][\Gamma]]\!] = [\![N]\!]^*[\![B[\Gamma]]\!]$ is obtained by pullback construction and that we consider the induced morphism into the pullback object)

$$[\![(M, N) : \Sigma x : A.B[\Gamma]]\!] = \hat{M} \circ [\![N]\!] \text{ where}$$

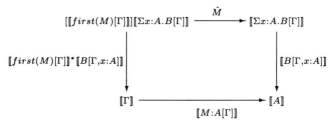

is a pullback diagram.

$$[\![first(M) : A[\Gamma]]\!] = [\![B[\Gamma, x : A]]\!] \circ [\![M]\!]$$

$$[\![second(M) : B[x/first(M)][\Gamma]]\!] = \langle 1_{[\![\Gamma]\!]}, [\![M]\!]\rangle : [\![\Gamma]\!] \to [\![B[x/first(M)]]\!]$$

(again this is a morphism into a pullback object)

$$[\![x_i : A_i[\Gamma, x_{n+1} : A_{n+1}]]\!] \quad = [\![A_{n+1}[\Gamma]]\!]^*[\![A_i[\Gamma]]\!] \quad \text{if } i \leq n$$
$$= 1_{[\![A_i]\!]} \qquad\qquad\qquad \text{else}$$

$$[\![B[x/N][\Gamma, x_1 : A_1[x/N], \dots, x_n : A_n[x/N]]]\!] = (\dots([\![N]\!]^*[\![A_1[\Gamma, x : A]]\!])^* \dots)^*[\![B[\Gamma, x : A, x_1 : A_1, \dots, x_n : A_n]]\!].$$

The definition of

$$[\![M[x/N] : B[x/N][\Gamma, x_1 : A_1[x/N], \dots, x_n : A_n[x/N]]]\!] =?$$

which should be fairly obvious by now, is left to the reader, who may as well check the soundness of the interpretation. For this it is helpful to draw

all the necessary diagrams. For completeness the reader may have a look
at [Hyland and Pitts, 1989].

Provided that $\mathcal{D}$ comprises all the morphisms of **D**, there is an alternative axiomatization under the name of *locally Cartesian closed categories*.

Definition 9.4.3. A category **C** is called *locally Cartesian closed* if all the comma categories $(\mathbf{C}, A)$ are Cartesian closed.

Proposition 9.4.4 ([**Freyd, 1972**]). *A category* **C** *is locally Cartesian closed iff the functors* $[f] : (\mathbf{C}, B) \to (\mathbf{C}, A)$ *defined by pullbacks*

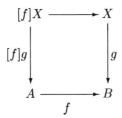

have a right adjoint and preserve exponentiation for all $f : A \to B$ *in* **C**.

Proof. For one direction one observes that the pullback is a product in
the comma category $(\mathbf{C}, B)$ and that $[f] \cong \underline{\Pi}f$. Hence $[f]$ has a right
adjoint $\Pi_{f\_} \cong f \Rightarrow_{\_} (\mathbf{C}, A) \to (\mathbf{C}, B)$. The left adjoint is defined by
$\Sigma_f h = f \circ h$ where $h : X \to A$. For the other direction observe that
$(\mathbf{C}, X)[[f]h, [f]g] \cong (\mathbf{C}, A)[h\Pi f, g]$ by pullback properties

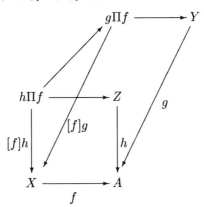

Hence $(\mathbf{C}, A)[h, \Pi_f([f]g)] \cong (\mathbf{C}, X)[[f]h, [f]g] \cong (\mathbf{C}, A)[h\Pi f, g]$ which makes
$\Pi_f([f]\_) : (\mathbf{C}, A) \to (\mathbf{C}, A)$ a right adjoint to $\underline{\Pi}f : (\mathbf{C}, A) \to (\mathbf{C}, A)$. ∎

9.4.2 Higher order type theory

Higher order type theory combines *finite products* and *power sets* as type
construction to achieve the expressivity of constructive set theory. We just

give a short overview since an extensive treatment of the subject is to be found in [Lambek and Scott, 1986]. The type constructors and terms are defined in a summary form by

$$
\begin{array}{cccc}
1 & \Omega & \tau \times \tau' & P\tau \\
\langle\rangle & a \in \alpha & \langle a, b\rangle & \{x^\tau \mid \varphi\} \\
& a = a' & &
\end{array}
$$

where α is of type $P\tau$, a, a' are of type τ, b is of type τ', and φ is a term of type Ω. Ω is a shorthand for $P1$.

Terms of type Ω are called *formulas*. Characteristic rules and axioms of deduction are the following (for the complete system cf. [Lambek and Scott, 1986]).

$$
\vdash (x \in \{x^\tau \mid \varphi\}) = \varphi
$$
$$
\vdash \forall x^1.x = \langle\rangle
$$
$$
\vdash \forall z^{\tau \times \tau'} \exists x^\tau \exists y^{\tau'}.z = \langle x, y\rangle
$$
$$
\vdash \forall u^{P\tau}.\{x^\tau \mid x \in u\} = u
$$
$$
a = b, \varphi[x/a] \vdash \varphi[x/b]
$$

$$
\frac{\Gamma, \varphi \vdash \psi \qquad \Gamma, \psi \vdash \varphi}{\Gamma \vdash \varphi = \psi}
$$

The other rules and axioms describe the usual properties of deduction and equality. Other logical symbols can be derived, e.g.

$$
\mathsf{tt} :\Leftrightarrow \langle\rangle = \langle\rangle
$$
$$
\varphi \wedge \psi :\Leftrightarrow \langle\varphi, \psi\rangle = \langle\mathsf{tt}, \mathsf{tt}\rangle
$$
$$
\varphi \Rightarrow \psi :\Leftrightarrow \varphi \wedge \psi = \varphi
$$
$$
\forall x^\tau.\varphi :\Leftrightarrow \{x^\tau \mid \varphi\} = \{x^\tau \mid \mathsf{tt}\}
$$
$$
\mathsf{ff} :\Leftrightarrow \forall x^\Omega.x
$$
$$
\varphi \vee \psi :\Leftrightarrow \forall x^\Omega.(((\varphi \Rightarrow x) \wedge (\psi \Rightarrow x)) \Rightarrow x
$$
$$
\exists x^\tau.\varphi :\Leftrightarrow \forall x^\Omega.(\forall y^\tau.\varphi \Rightarrow x) \Rightarrow x
$$
$$
\neg\varphi :\Leftrightarrow \varphi \Rightarrow \mathsf{ff}.
$$

We leave it to the reader to add all the necessary bookkeeping about variables etc. which have been omitted for a more succinct presentation.

The categorical definition related to this higher order logic is that of a topos (see Section 5.1.4 for the definition). Power types correspond to power objects $P(A)$ which represent the relation on A, i.e. $\mathbf{C}[\_, P(A)] \cong \mathbf{Rel}[\_, A]$. The interpretation of the logic in a topos is straightforward (we assume the parameter lists $\langle p\rangle$ and $\langle q\rangle$ to be of type τ and τ' respectively):

$$[\![1]\!] = 1$$
$$[\![\tau \times \tau']\!] = [\![\tau]\!]\Pi[\![\tau']\!]$$
$$[\![P\tau]\!] = P[\![\tau]\!] = [\![\tau]\!] \Rightarrow \Omega$$
$$[\![\langle p\rangle\langle\rangle]\!] = \langle\rangle_{[\![\tau]\!]} : [\![\tau]\!] \to 1$$
$$[\![\langle p\rangle a \in \alpha]\!] = apply_{[\![\tau]\!],\Omega} \circ \langle[\![\langle p\rangle\alpha]\!], [\![\langle p\rangle a]\!]\rangle : [\![\langle p\rangle]\!] \to \Omega$$
$$[\![\langle p\rangle a = a']\!] = eq_{[\![\tau]\!]} \circ \langle[\![\langle p\rangle a]\!], [\![\langle p\rangle a']\!]\rangle : [\![\langle p\rangle]\!] \to \Omega$$

where $eq_{[\![\tau']\!]}:[\![\tau']\!]\Pi[\![\tau']\!]\to\Omega$ is induced by the diagonal $\Delta:[\![\tau']\!]\rightarrowtail[\![\tau']\!]\Pi[\![\tau']\!]$

$$[\![\langle p\rangle\{\langle q\rangle \mid \varphi\}]\!] = \lambda[\![\langle\langle p\rangle, \langle q\rangle\rangle\varphi]\!] : [\![\langle p\rangle]\!] \to P[\![\langle q\rangle]\!].$$

In particular we obtain the following interpretation of logical connectives on Ω (cf. [Goldblatt, 1979], [Wraith, 1975]):

Conjunction
$\wedge : \Omega\Pi\Omega \to \Omega$ is the characteristic morphism induced by the subobject $\langle \text{tt}, \text{tt}\rangle : 1 \rightarrowtail \Omega\Pi\Omega$.

Implication
$\Rightarrow: \Omega \to \Omega$ is the characteristic morphism induced by $\leq \rightarrowtail \Omega\Pi\Omega$ which is the equalizer of $\wedge$ and $p_{\Omega,\Omega}$.

Existential quantification
$apply : (A \to \Omega)\Pi A \to \Omega$ defines a monomorphism $apply : \in_A \rightarrowtail (A \to \Omega)\Pi A$. If $f : A \rightarrowtail B$ is a monomorphism then $(1_{A \Rightarrow \Omega}\Pi f) \circ apply : \in_A \to (A \Rightarrow \Omega)\Pi B$ is a monomorphism whose characteristic morphism $(A \to \Omega)\Pi B \to \Omega$ is exponentially adjoint to $\exists_f : (A \to \Omega) \to (B \to \Omega)$.

As usual, the syntax induces a 'term model': objects are closed terms of type $P\tau$ modulo provable equality, τ being an arbitrary type. A morphism from $a : P\tau$ to $b : P\tau'$ is represented by a triple (α, β, φ), where φ is a closed formula of type $P(\tau \times \tau')$ such that

(i) $\vdash \forall x^\tau \forall y^{\tau'}.\langle x, y\rangle \in \varphi \Rightarrow (x \in \alpha \wedge y \in \beta)$
(ii) $\vdash \forall x^\tau \forall y^{\tau'}.\forall z^{\tau'}.\langle x, y\rangle \in \varphi \wedge \langle x, z\rangle \in \varphi \Rightarrow y = z.$

Morphisms are defined modulo provable equality of the three data involved.

Theorem 9.4.5 ([Lambek and Scott, 1986]). *The category so constructed is an initial object in the category of (small) toposes with morphisms being functors preserving the topos structure exactly.*

For all further results the reader should consult [Lambek and Scott, 1986].

9.4.3 Polymorphic types

We have identified a sort of polymorphism when discussing natural transformations, namely that functions can be parameterized by objects. However, there is a shortcoming in that the indexing mechanism lives somewhat outside of the respective category. We would rather have, say, a generic type of polymorphic lists with suitable generic functions, the offhand idea being to take the product over all lists, i.e. $\prod_{A \in Obj(\mathbf{C})} list(A)$, and to define generic functions such as

$$head := \prod_{A \in Obj(\mathbf{C})} head_A : \prod_{A \in Obj(\mathbf{C})} list(A) \to \prod_{A \in Obj(\mathbf{C})} A.$$

Of course, we know that such big products only exists in (pre-)order categories (see Section 6.1.2) which provide a rather boring interpretation for our 'type of all lists'. Martin Hyland has made the exciting observation that big products indexed by all objects of a category exist, if the categories and the products are internal with regard to certain categories which are even toposes [Hyland, 1988]. This observation provides the handle to give a natural interpretation of second order (and higher order) λ-calculus [Girard, 1972] which is called polymorphic λ-calculus elsewhere [Reynolds, 1974]. The idea that 'small' complete categories might exist and would provide the right setting in which to discuss models for polymorphisms seems to have been first suggested to Martin Hyland by Eugenio Moggi. Hyland then realized the connection to his earlier work on realizability [Hyland, 1982].

Let us sketch the structure of arbitrary internal products (taking up the theme of *internal categories* in Section 5.2.5, underpinning it with a substantial computer science application). We assume existence of a Cartesian closed category **E** which has finite limits.

Definition 9.4.6. Let A be an object of **E**. Then an internal category $\boldsymbol{A}$ is given by $\boldsymbol{A} = (A, A, 1_A, 1_A, 1_A, 1_A)$.

$\boldsymbol{A}$ represents the internal category with just one identity morphism for every object. Next we give an internal version of a diagonal functor $\Delta : \boldsymbol{C} \to \boldsymbol{C}^{\boldsymbol{A}}$ where $\boldsymbol{A}$ is a discrete category.

Definition 9.4.7. Let A be an object of **E**, and $\boldsymbol{C}$ be an internal category in **E**. Define an internal category $\boldsymbol{C}^{\boldsymbol{A}} = (O^A, M^A, dom^A, cod^A, *^A, e^A)$ (where $f^A = \lambda(f \circ apply)$ for $f : A \to B$ in **E**), and an internal functor $\Delta : \boldsymbol{C} \to \boldsymbol{C}^{\boldsymbol{A}}$ by $\Delta_O = \lambda(p_{O,O})$ and $\Delta_M = \lambda(p_{M,O})$.

Then the existence of *internal limits* indexed by $\boldsymbol{A}$ is guaranteed by the existence of an internal right adjoint to the constant functor $\Delta : \boldsymbol{C} \to \boldsymbol{C}^{\boldsymbol{A}}$. A more explicit picture of the adjunction is obtained by using the following characterization of [Asperti, 1990].

Proposition 9.4.8. *Every internal adjunction $\langle F, G, \Phi \rangle$ is fully determined by the following data:*

- *the functor $F : \boldsymbol{C} \to \boldsymbol{D}$,*

- an arrow $G_O : O_D \to O_C$,

- a morphism $Counit : O_D \to M_D$,

- a morphism $\varphi : D[F_-, \_] \to C[\_, G_-]$

  where $D[F_-, \_]$ and $C[\_, G_-]$ are respectively the pullbacks of

$$\langle dom, cod \rangle : M_D \to O_D \times O_D, \, F_O \times 1_{O_D} : O_C \times O_D \to O_D \times O_D$$
$$\langle dom, cod \rangle : M_C \to O_C \times O_C, \, 1_{O_C} \times G_O : O_C \times O_D \to O_C \times O_C$$

such that the equations

$$\langle q, ((Counit \circ q' \circ q) * (F_M \circ p)) \rangle \circ \varphi = 1_{D[F_-, \_]}$$
$$\varphi \circ \langle q, ((Counit \circ q' \circ q) * (F_M \circ p)) \rangle = 1_{C[\_, G_-]}$$

hold, where $g * f = * \circ \langle g, f \rangle$.

The morphism $\langle q, ((Counit \circ q' \circ q) * (F_M \circ p)) \rangle$ just states that the inverse of φ can be constructed via the counit by $counit_B \circ F(f)$ if $f : A \to G(B)$; p and q are used uniformly for projections of products and pullbacks as is pairing.

In our specific example we obtain a morphism $\forall : O^A \to O$ which 'maps an A-indexed family of objects to a (big) product $\forall_{i \in A} F(i)$'. The counit $Proj : O^A \to M^A$ 'determines an A-indexed family of mappings $Proj_i : \forall_{i \in A} F(i) \to F(i)$', and finally, the morphism Λ 'maps an A-indexed family of mappings $f(i) : B \to F(i)$ to a mapping $\Lambda(f) : B \to \forall_{i \in A} F(i)$'. The axioms then state the expected properties.

On the syntactical level, big products are reflected by adding a *second order* universal quantification $\forall \langle X \rangle A$ where X is a *type variable*, a typical example being the universal type $\forall \langle X \rangle list(X)$ of lists. We will use the notation $A[\Gamma]$ for *polymorphic types* and $\langle p \rangle M : A \to B[\Gamma]$ for *polymorphic functions* where the environment $\ldots [\Gamma]$ records which type variables may occur. Thus we propose the following syntax:

Types

$X[\Gamma]$	if X occurs in Γ
$A \times B[\Gamma]$	if $A[\Gamma]$ and $B[\Gamma]$ are types
$A \Rightarrow B[\Gamma]$	if $A[\Gamma]$ and $B[\Gamma]$ are types
$\forall \langle X \rangle A[\Gamma]$	if $A[\Gamma, X]$ is a type

Functional terms

$\langle x^A \rangle \langle \rangle : A \to 1[\Gamma]$	if $A[\Gamma]$ is a type
$\langle p \rangle x^B : A \to B[\Gamma]$	if x^B occurs in the parameter list $\langle p \rangle : A[\Gamma]$
$\langle p \rangle \langle M, N \rangle : A \to B \times C[\Gamma]$	if $\langle p \rangle M : A \to B[\Gamma]$ and

$$\langle p \rangle (f \cdot M) : A \to C[\Gamma]$$

$$\langle p \rangle N : A \to C[\Gamma]$$
if $\langle p \rangle M : A \to B[\Gamma]$ and
$$f : B \to C[\Gamma]$$

$$\langle p \rangle (\lambda \langle q \rangle M) : A \to (B \Rightarrow C)[\Gamma] \quad \text{if } \langle \langle p \rangle, \langle q \rangle \rangle M : A \times B \to C[\Gamma]$$

$$\langle p \rangle (M\,N) : A \to C[\Gamma] \quad \text{if } \langle p \rangle M : A \to (B \Rightarrow C)[\Gamma] \text{ and}$$
$$\langle p \rangle N : A \to B[\langle p \rangle]$$

$$\langle p \rangle M(C) : A \to B[X/C]\,[\Gamma] \quad \text{if } \langle p \rangle M : A \to \forall \langle X \rangle B\,[\Gamma] \text{ and } C[\Gamma]$$

Parameter lists

$$\langle x^A \rangle : A\,[\Gamma] \qquad\qquad\qquad \text{if } A[\Gamma] \text{ is a type}$$

$$\langle \langle p \rangle, \langle q \rangle \rangle : A \times B\,[\Gamma] \qquad \text{if } \langle p \rangle : A\,[\Gamma] \text{ and } \langle q \rangle : B\,[\Gamma].$$

Except for the terms dealing with universal quantification, we have just embellished the lambda calculus of Section 7.1.2 with an environment of type variables. Hence we do not bother to restate the axioms except for universal quantification where the analogy to standard abstraction and application is pretty obvious:

$$\langle p \rangle ((\Lambda \langle X \rangle M)(C)) \;=\; \langle p \rangle M[X/B] : A \to B[X/C]\,[\Gamma]$$
$$\langle p \rangle ((\Lambda \langle X \rangle M)(X)) \;=\; \langle p \rangle M : A \to B\,[\Gamma] \text{ if } X \text{ does not occur free in } M$$

(plus the usual noise about making = a congruence).

For the interpretation, we would expect to use a Cartesian closed category C which should be internal in order to allow for big products as well. Explicitly we assume to have

- a Cartesian closed (global) category **E**, and

- an internal Cartesian closed category $C = (O, M, *, e)$ with big products

  i.e. we have three internal adjunctions

$$(\langle \rangle_C, 1, \langle \rangle) : C \to 1 \qquad \text{where 1 is the internal terminal categc}$$
$$(\Delta_C, \times, \langle \_, \_ \rangle) : C \to C \times C \quad \text{where } \Delta_C \text{ is the diagonal functor}$$
$$(\times, \Rightarrow, \lambda) : C \to C \qquad\qquad \text{which is parameterized by } C$$
$$(\Delta, \forall, \Lambda) : C \to C^O \qquad\qquad \text{where } C^O \text{ is defined as above.}$$

We leave it to the reader to apply Asperti's theorem to Cartesian closure. An adjunction is parameterized if the functors $F : C \times P \to D$ and $G : P^{\mathrm{op}} \times D \to C$ come along with an isomorphism $\varphi : (F \times 1_{D^{\mathrm{op}}}) * (hom_D) \to (1_C \to G^{\mathrm{op}}) * (hom_C)$ (cf. Section 5.2.5).

Types $A[\Gamma]$ are interpreted by morphisms $[\![A[\Gamma]]\!] : O^n \to O$ where $\Gamma = (X_1, \ldots, X_n)$, and functional terms $\langle p \rangle M : A \to B\,[\Gamma]$ by morphisms

$[\![\langle p\rangle M : A \rightarrow B\,[\Gamma]\!]\!] : O^n \rightarrow M$ such that $dom \circ [\![\langle p\rangle M : A \rightarrow B\,[\Gamma]\!]\!] = [\![A[\Gamma]\!]\!]$ and $cod \circ [\![\langle p\rangle M : A \rightarrow B\,[\Gamma]\!]\!] = [\![B[\Gamma]\!]\!]$.

Types

$[\![X[\Gamma]\!]\!] = p_i : O^n \rightarrow O$ \qquad the projection to the ith component

$[\![A \times B\,[\Gamma]\!]\!] = \times \circ \langle [\![A[\Gamma]\!]\!], [\![B[\Gamma]\!]\!]\rangle$

$[\![A \Rightarrow B\,[\Gamma]\!]\!] = \Rightarrow \circ \langle [\![A[\Gamma]\!]\!], [\![B[\Gamma]\!]\!]\rangle$

$[\![\forall\langle X\rangle A\,[\Gamma]\!]\!] = \forall \circ \lambda([\![A[\Gamma, X]\!]\!])$

Functional terms

$[\![\langle x^A\rangle\langle\rangle[\Gamma]\!]\!] = p \circ \langle\rangle \circ \langle\langle\rangle_{O^n}, \langle [\![A[\Gamma]\!]\!], \langle\rangle_{O^n}\rangle\rangle$

$[\![\langle x^B\rangle x^B\,[\Gamma]\!]\!] = e \circ [\![B[\Gamma]\!]\!]$

$[\![\langle\langle p\rangle, \langle q\rangle\rangle x\,[\Gamma]\!]\!] = [\![\langle p\rangle x\,[\Gamma]\!]\!] * (p \circ \langle [\![A[\Gamma]\!]\!], [\![B[\Gamma]\!]\!]\rangle)$ \quad if x occurs in $\langle p\rangle : A[\Gamma]$

$\qquad\qquad\qquad = [\![\langle q\rangle x\,[\Gamma]\!]\!] * (q \circ \langle [\![A[\Gamma]\!]\!], [\![B[\Gamma]\!]\!]\rangle)$ \quad if x occurs in $\langle q\rangle : B[\Gamma]$

$[\![\langle p\rangle\langle M, N\rangle\,[\Gamma]\!]\!] =$
$p \circ \langle\_,\_\rangle \circ \langle\langle [\![\langle p\rangle M\,[\Gamma]\!]\!], [\![\langle p\rangle N\,[\Gamma]\!]\!]\rangle, \langle [\![A[\Gamma]\!]\!], \langle [\![A[\Gamma]\!]\!], [\![A[\Gamma]\!]\!]\rangle\rangle\rangle$

$[\![\langle p\rangle(f \cdot M)[\Gamma]\!]\!] = [\![\langle p\rangle M[\Gamma]\!]\!] * [\![f[\Gamma]\!]\!]$

$[\![\langle p\rangle(\lambda\langle q\rangle M)[\Gamma]\!]\!] = p \circ \lambda \circ \langle [\![\langle\langle p\rangle, \langle q\rangle\rangle M[\Gamma]\!]\!], \langle [\![A \times B\,[\Gamma]\!]\!], [\![C[\Gamma]\!]\!]\rangle\rangle$

$[\![\langle p\rangle(M\,N)\,[\Gamma]\!]\!]$
$= Eval * (\langle\_,\_\rangle \circ \langle\langle [\![\langle p\rangle M[\Gamma]\!]\!], [\![\langle p\rangle N[\Gamma]\!]\!]\rangle, \langle [\![A[\Gamma]\!]\!], \langle [\![B \Rightarrow C\,[\Gamma]\!]\!], [\![B[\Gamma]\!]\!]\rangle\rangle\rangle)$

$[\![\langle p\rangle\Lambda\langle X\rangle M)\,[\Gamma]\!]\!] = p \circ \Lambda \circ \langle \lambda[\![\langle p\rangle M[\Gamma, X]\!]\!], \langle [\![A[\Gamma]\!]\!], \lambda[\![B[\Gamma, X]\!]\!]\rangle\rangle$

$[\![\langle p\rangle M(C)\,[\Gamma]\!]\!] = (eval \circ \langle Proj \circ \lambda[\![B[\Gamma, X]\!]\!], [\![C[\Gamma]\!]\!] * [\![\langle p\rangle M[\Gamma]\!]\!]\rangle).$

A diagrammatic argument for $[\![\langle p\rangle\langle M, N\rangle\,[\Gamma]\!]\!]$ might enhance understanding of the definitions:

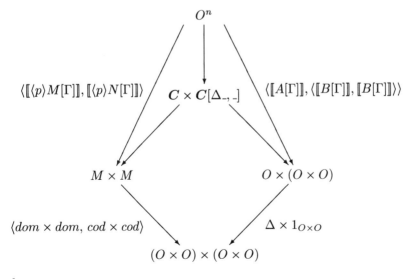

and

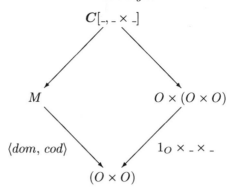

together with

$$C \times C[\Delta_-, \_] \xrightarrow{\quad \langle \_, \_\rangle \quad} C[\_, \_ \times \_].$$

This being another exercise in 'limit programming', we refrain from discussing models which for sure will be done elsewhere in these volumes. Let us just mention that such models are discussed in the book [Asperti and Longo, 1991] where all this can be found as well.

10 A sort of conclusion

I have neither touched upon every subject in computer science where category theory has been used nor are my references in any sense complete. In fact, many more applications should have been discussed in order to demonstrate the use so far made of category theory in computer science. However, these volumes will comprise many examples for *Category Theory and Computer Science* as do the conference proceedings of the same title [Pitt *et al.*, 1985; Pitt *et al.*, 1987; Pitt *et al.*, 1989; Pitt *et al.*, 1991]. I apologize to everybody for not referring to particular work or for misinterpretation. As I stated in the beginning, these notes are biased by my own experiences and preferences.

Finally, I want to express deep gratitude to all friends and colleagues who, at various stages, took part in my efforts to appreciate category theory, specifically Gisbert Dittrich, Hagen Huwig and Wolfgang Merzenich with whom I studied MacLane's book for the first time, Samson Abramsky and Steve Vickers for many stimulating discussions at Imperial College, and to Bill Lawvere for his papers which I still struggle to understand, though they are probably the best reading in category theory, of course excluding MacLane's masterwork. Last but not least, I would very much like to acknowledge the masterly typesetting of Ruy de Queiroz, without which the text would have never surfaced in its present form. Most diagrams were typeset with Paul Taylor's TeX macro package.

11 Literature

11.1 Textbooks

[Arbib and Manes, 1975] M. A. Arbib and E. G. Manes. *Arrows, Structures and Functors – The Categorical Imperative*. Academic Press, New York-San Francisco-London, 1975.

[Asperti and Longo, 1991] A. Asperti and G. Longo. *Categories, Types and Structures – An Introduction to Category Theory for the Working Computer Scientist*. MIT Press, 1991.

[Barr and Wells, 1985] M. Barr and C. Wells. *Toposes, Triples and Theories*. Springer Verlag, 1985.

[Barr and Wells, 1990] M. Barr and C. Wells. *Category Theory for Computer Science*. Prentice-Hall, 1990.

[Goldblatt, 1979] R. Goldblatt. *The Categorical Analysis of Logic*. North Holland, 1979.

[Herrlich and Strecker, 1973] H. Herrlich and G. E. Strecker. *Category Theory*. Allyn and Bacon, 1973.

[Herrlich and Strecker, 1991] H. Herrlich and G. E. Strecker. *Category Theory at Work*. Heldermann, Berlin, 1991.

[Johnstone and Paré (eds), 1978] P. T. Johnstone and R. Paré (eds). *Indexed Categories and Their Application*, volume 661 of *Lecture Notes in Mathematics*. Springer-Verlag, 1978.

[Johnstone, 1977] P. T. Johnstone. *Topos Theory*. Cambridge University Press, 1977.

[Johnstone, 1982] P. T. Johnstone. *Stone Spaces*. Cambridge University Press, 1982.

[Kelly, 1982] G. M. Kelly. *Basic Concepts of Enriched Category Theory*, volume 64 of *Lecture Notes of the London Mathematical Society*. Cambridge University Press, 1982.

[Lambek and Scott, 1986] J. Lambek and P. Scott. *Introduction to Higher Order Categorical Logic*, volume 7 of *Cambridge Studies in Advanced Mathematics*. Cambridge University Press, 1986.

[MacLane, 1971] S. MacLane. *Category Theory – For the Working Mathematician*. Springer, 1971.

[Makkai and Reyes, 1977] M. Makkai and G. E. Reyes. *First Order Categorical Logic*, volume 611 of *Lecture Notes in Mathematics*. Springer Verlag, 1977.

[Manes, 1974] E. Manes. *Algebraic Theories*. Springer Verlag, 1974.

[Mitchell, 1965] B. Mitchell. *Theories of Categories*. Academic Press, 1965.

[Richter, 1979] G. Richter. *Kategorielle Algebra*. Studien zur Algebra und ihren Anwendungen. Akademie Verlag, Berlin, 1979.

[Schubert, 1970] Schubert. *Categories*. Springer, 1970.

[Szabo, 1978] M. Szabo. *The Algebra of Proofs*. North Holland, 1978.

11.2   References

[Abramsky, 1987] S. Abramsky. Domain Theory in Logical Form. In *Symposium on Logic in Computer Science*. Computer Science Press of the IEEE, 1987. Also in: *Annals of Pure and Applied Logic* 51, 1991.

[Adamek and Koubek, 1979] J. Adamek and V. Koubek. Least Fixed Point of a Functor. *JCSS*, 19, 1979.

[Arbib and Manes, 1974] M. A. Arbib and E. G. Manes. Machines in a Category: an Expository Introduction. *SIAM Rev.*, 16, 1974.

[Asperti and Martini, 1989] A. Asperti and S. Martini. Categorical Models of Polymorphism, 1989. To appear in *Information and Computation*.

[Asperti, 1990] A. Asperti. *Categorical Topics in Computer Science*. PhD thesis, Dipartimento di Informatica, Università di Pisa, 1990. Report TD 7/90.

[Bachmayr et al., 1986] L. Bachmayr, N. Dershowitz, and J. Hsiang. Proof Orderings for Equational Proofs. In *Proc. LICS '86*, 1986.

[Barendregt, 1984] H. Barendregt. *The Lambda Calculus*. North Holland, 1984.

[Casley et al., 1989] R. Casley, R. F. Crew, J. Meseguer, and V. Pratt. Temporal Structures, 1989. In [Pitt et al., 1989].

[Chevalley, 1956] C. Chevalley. *Fundamental Concepts of Algebra*. Academic Press, 1956.

[Cohn, 1981] P. M. Cohn. *Universal algebra*, volume 6 of *Mathematics and its applications*. D. Reidel, Dordrecht, 1981.

[Cosineau *et al.*, 1987] G. Cosineau, P.-L. Curien, and M. Mauny. The Categorical Abstract Machine. *Science of Computer Programming*, 8:203–211, 1987.

[Curien and Ghelli, 1990] P.-L. Curien and G. Ghelli. Coherence of Subsumption. In *Proc. CAAP '90*, volume 431 of *Lecture Notes in Computer Science*. Springer, 1990. Extended version to appear in *Mathematical Structures in Computer Science*.

[Curien, 1986] P.-L. Curien. *Categorical Combinators, Sequential Algorithms and Functional Programming*. Pitman, London, 1986.

[Curien, 1990] P.-L. Curien. Substitution up to isomorphism. Technical Report LIENS-90-9, Ecole Normale Supérieure, Paris, 1990.

[Dijkstra, 1976] E. Dijkstra. *A Discipline of Programming*. Prentice Hall, 1976.

[Ehrig and Mahr, 1985] H. Ehrig and B. Mahr. *Fundamentals of Algebraic Specification 1*, volume 6 of *EATCS Monographs on Theoretical Computer Science*. Springer Verlag, 1985.

[Ehrig and Mahr, 1990] H. Ehrig and B. Mahr. *Fundamentals of Algebraic Specification 2*, volume 21 of *EATCS Monographs on Theoretical Computer Science*. Springer Verlag, 1990.

[Ehrig *et al.*, 1974] H. Ehrig, K. D. Kiermeier, H.-J. Kreowski, and W. Khnel. *Universal Theory of Automata*. Teubner, 1974.

[Ehrig, 1978] H. Ehrig. Introduction to the algebraic theory of graph grammars. In *International Workshop on Graph Grammars and Their Application to Computer Science and Biology*, volume 73 of *Lecture Notes in Computer Science*, Bad Honnef, 1978. Springer.

[Engelfriet and Schmidt, 1977] J. Engelfriet and E. M. Schmidt. IO and OI. *JCSS*, 15(3):328–353, 1977. And 16(1):67–99, 1978.

[Fourman and Scott, 1979] M. Fourman and D. S. Scott. Sheaves and logic. In *Applications of Sheaves (Proc. Durham)*, volume 753 of *Lecture Notes in Mathematics*. Springer Verlag, 1979.

[Freyd, 1972] P. Freyd. Aspects of topoi. *Bull. Austral. Math. Soc.*, 7, 1972.

[Gaifmann and Pratt, 1987] H. Gaifmann and V. Pratt. Partial order models of concurrency. In *Proc. IEEE Symp. on Logic in Computer Science*, Ithaca, New York, 1987.

[Gaifmann, 1989] H. Gaifmann. Modelling concurrency by partial orders and nonlinear transition systems. In *Proc. REX School/Workshop on*

Linear Time, Branching Time and Partial Order in Logics and Models for Concurrency. Springer, 1989.

[Girard, 1972] J.-Y. Girard. *Interpretation fonctionelle et elimination des coupure dans l'arithmetic d'ordre superieur.* These de Doctorat d'Etat, Université de Paris, 1972. See also J.-Y. Girard's 'The System F of Variable Types: Fifteen Years After', in *TCS*, 1987.

[Goguen and Meseguer, 1981] J. A. Goguen and J. Meseguer. Completeness of many-sorted equational logic. *ACM SIGPLAN Notices*, 16.7, 1981.

[Goguen and Meseguer, 1982] J. Goguen and J. Meseguer. Universal Realization, Persistent Interconnection and Implementation of Abstract Modules. In *ICALP '82*, volume 140 of *Lecture Notes in Computer Science.* Springer, 1982.

[Guessarian, 1981] I. Guessarian. *Algebraic Semantics*, volume 99 of *Lecture Notes in Computer Science.* Springer Verlag, 1981.

[Hennessy and Plotkin, 1979] M. Hennessy and G. Plotkin. Full Abstraction for a Simple Parallel Programming Language. In *Proc. MFCS*, volume 74 of *Lecture Notes in Computer Science*, 1979.

[Howard, 1980] W. A. Howard. The Formulae-as-Types Notion of Construction. In J. P. Seldin and J. R. Hindley, editors, *To H. B. Curry: Essays on Combinatory Logic, Lambda-Calculus and Formalism.* Academic Press, 1980. Manuscript already circulated in 1969.

[Hughes and Cresswell, 1968] G. E. Hughes and M. J. Cresswell. *An Introduction to Modal Logic.* Methuen, 1968.

[Huwig and Poigné, 1990] H. Huwig and A. Poigné. On Inconsistencies Caused by Fixpoints in A Cartesian Closed Category. *TCS*, 1990.

[Hyland and Pitts, 1989] J. M. E. Hyland and A. M. Pitts. The Theory of Constructions: Categorical Semantics and Topos-theoretic Models. *Contemporary Mathematics*, 92, 1989.

[Hyland, 1982] J. M. E. Hyland. The effective topos. In A. S. Troelstra and D. van Dalen, editors, *The L.E.J. Brouwer Centenary Symposium.* North-Holland, 1982.

[Hyland, 1988] J. M. E. Hyland. A small complete category. *Annals of Pure and Applied Logic*, 40, 1988.

[Hyland, 1990] J. M. E. Hyland. First steps in synthetic domain theory. In A. Carboni, M. C. Pedicchio, and G. Rosolini, editors, *Category Theory '90.* Springer, 1990. To appear.

[Lambek, 1969] J. Lambek. *Deductive Systems and Categories II*, volume 86 of *Lecture Notes in Mathematics*. Springer, 1969.

[Lambek, 1972] J. Lambek. *Deductive Systems and Categories III*, volume 274 of *Lecture Notes in Mathematics*. Springer, 1972.

[Lambek, 1980] J. Lambek. From Lambda Calculus to Cartesian Closed Categories. In J. P. Seldin and J. R. Hindley, editors, *To H.B. Curry: Essays on Combinatory Logic, Lambda-Calculus and Formalism*. Academic Press, 1980.

[Lambek, 1989] J. Lambek. Multicategories revisited. *Contemporary Mathematics*, 92, 1989.

[Lawvere, 1963] F. W. Lawvere. Functorial Semantics of Algebraic Theories. In *Proc. Nat. Acad. Sci. USA*, 1963.

[Lawvere, 1969a] F. W. Lawvere. Adjointness in Foundations. *Dialectica*, 23, 1969.

[Lawvere, 1969b] F. W. Lawvere. Diagonal Arguments and Cartesian Closed Categories. In *Category Theory, Homology Theory and their Applications*, volume 92 of *Lecture Notes in Mathematics*, 1969.

[Lawvere, 1970] F. W. Lawvere. Equality in Hyperdoctrines and Comprehension Schema as an Adjoint Functor. *Proc. Amer. Math. Soc.*, Applications of Categorical Algebra, 1970.

[Lawvere, 1973] F. W. Lawvere. Metric Spaces, Generalized Logic, and Closed Categories. In *Rendiconti del Seminario Matematico e Fisico di Milano, XLIII*. Tipografia Fusi, Pavia, 1973.

[Martin-Löf, 1973] P. Martin-Löf. An Intuitionistic Theory of Types. In *Logic Colloquium '73*. North Holland, 1973. (Published 1975).

[Martin-Löf, 1979] P. Martin-Löf. Constructive Mathematics and Computer Programming. In L. J. Cohen *et al.*, editor, *Sixth International Congress for Logic, Methodology and Philosophy of Science*. North-Holland, 1979. (Published 1982).

[Meseguer, 1977] J. Meseguer. On Order-Complete Universal Algebra and Enriched Functorial Semantics. In *Proc. FCT '77*, volume 56 of *Lecture Notes in Computer Science*. Springer, 1977.

[Milner, 1980] R. Milner. *A Calculus of Communicating Systems*, volume 92 of *Lecture Notes in Computer Science*. Springer, 1980.

[Moggi, 1985] E. Moggi. Partial Morphisms and the λ_p-Calculus, 1985. In [Pitt *et al.*, 1985].

[Moggi, 1989] E. Moggi. Computational Lambda Calculus and Monads. In *LICS '89*. IEEE, Comp. Soc. Press, 1989.

[Moggi, 1991] E. Moggi. Notions of computation and monads. *Information and Computation*, 93(1), 1991.

[Pfender, 1974] M. Pfender. Universal Algebra in s-monoidal categories. Algebra Berichte Nr. 20, Universität München, 1974.

[Phoa, 1990] W. Phoa. Effective Domains and Intrinsic Structure. In *Proc. of IEEE 5th LICS*, 1990.

[Pitt et al., 1985] D. Pitt, S. Abramsky, A. Poigné, and D. Rydeheard, editors. *Category Theory and Computer Programming: Tutorial and Workshop (Guildford)*, volume 240 of *Lecture Notes in Computer Science*. Springer, 1985.

[Pitt et al., 1987] D. Pitt, A. Poigné, and D. Rydeheard, editors. *Category Theory and Computer Programming: Tutorial and Workshop (Edinburgh)*, volume 283 of *Lecture Notes in Computer Science*. Springer, 1987.

[Pitt et al., 1989] D. Pitt, D. Rydeheard, P. Dybjer, A. M. Pitts, and A. Poigné, editors. *Category Theory and Computer Programming: Tutorial and Workshop (Manchester)*, volume 389 of *Lecture Notes in Computer Science*. Springer, 1989.

[Pitt et al., 1991] D. Pitt, P.-L. Curien, S. Abramsky, A. M. Pitts, A. Poigné, and D. Rydeheard, editors. *Category Theory and Computer Programming: Tutorial and Workshop (Paris)*, volume 389 of *Lecture Notes in Computer Science*. Springer, 1991.

[Pitts, 1981] A. M. Pitts. *The Theory of Triposes*. PhD thesis, Department of Mathematics, Cambridge University, 1981.

[Pitts, 1987] A. M. Pitts. Polymorphism is Set-theoretic, Constructively, 1987. In [Pitt et al., 1987].

[Pitts, 1991] A. M. Pitts. Evaluation Logic. In G. Birthwistle, editor, *IV Higher Order Workshop, Banff 90*. Springer, 1991.

[Plotkin, 1976] G. Plotkin. A powerdomain construction. *SIAM J. Comp.*, 5, 1976.

[Plotkin, 1981] G. Plotkin. Postgraduate Lecture Notes in Advanced Domain Theory (incorporating the "Pisa Notes"). Technical report, Dept of Computer Science, University of Edinburgh, 1981.

[Poigné, 1985] A. Poigné. Note on Distributive Laws and Power Domains, 1985. In [Pitt et al., 1985].

[Poigné, 1986] A. Poigné. On Specifications, Theories, and Models with Higher Types. *Information and Control*, 1986.

[Poigné, 1989] A. Poigné. Bisimulation as Cointersection. *EATCS Bulletin*, 39, 1989.

[Power, 1989] A. J. Power. An Abstract Formulation for Rewrite Systems, 1989. In [Pitt *et al.*, 1989].

[Pratt, 1991] V. Pratt. Modelling Concurrency with Geometry. In *Proc. POPL '91*, 1991.

[Prawitz, 1965] D. Prawitz. *Natural Deduction: A Proof-Theoretical Study.* Almqvist and Wiksell, Stockholm, 1965.

[Reynolds, 1974] J. Reynolds. Toward a Theory of Type Structures. In *Colloque sur la Programmation*, volume 19 of *Lecture Notes in Computer Science*. Springer, 1974.

[Rosolini and Robinson, 1988] G. Rosolini and E. P. Robinson. Categories of Partial Maps. *Information and Computation*, 79, 1988.

[Rydeheard and Stell, 1987] D. Rydeheard and J. Stell. Foundations of Equational Deduction: A Categorical Treatment of Equational Proofs and Unification Algorithms, 1987. In [Pitt *et al.*, 1987].

[Scott, 1976] D. S. Scott. Data Types as Lattices. *SIAM J. Comput.*, 5, 1976.

[Scott, 1977] D. S. Scott. Identity and Existence in Intuitionistic Logic. In M. P. Fourman, C. J. Mulvey, and D. S. Scott, editors, *Applications of Sheaves*, volume 753 of *Lecture Notes in Mathematics*. Springer, 1977.

[Scott, 1980] D. S. Scott. Relating Theories of Lambda Calculus. In J. P. Seldin and J. R. Hindley, editors, *To H. B. Curry: Essays on Combinatory Logic, Lambda-Calculus and Formalism*. Academic Press, 1980.

[Scott, 1982] D. S. Scott. Domains for Denotational Semantics. In *Proc. ICALP '82*, volume 140 of *Lecture Notes in Computer Science*. Springer, 1982.

[Seely, 1983] R. A. G. Seely. Hyperdoctrines, Natural Deduction and the Beck Condition. *Zeitschr. f. math. Logik und Grundlagen d. Math.*, 29, 1983.

[Seely, 1984] R. A. G. Seely. Locally Cartesian Closed Categories and Type Theory. *Math. Proc. Camb. Phil. Soc.*, 95, 1984.

[Smyth and Plotkin, 1977] M. B. Smyth and G. D. Plotkin. The category-Theoretic Solution of Recursive Domain Equations. In *Proc. 18th FOCS*, 1977. Full paper published in *SIAM Journal on Control*, 1983.

[Smyth, 1978] M. B. Smyth. Powerdomains. *JCSS*, 16, 1978.

[Smyth, 1983] M. B. Smyth. Powerdomains and Predicate Transformers: A Topological View. In *Proc. ICALP '83*, volume 154 of *Lecture Notes in Computer Science*. Springer, 1983.

[Winskel, 1987] G. Winskel. Event Structures. In *Applications and Relationships to Other Models of Concurrency, Advances in Petri Nets 1986, Part II*, volume 255 of *Lecture Notes in Computer Science*. Springer, 1987.

[Wraith, 1970] G. C. Wraith. Algebraic Theories. Lecture Notes Series No. 22, Dept. of Mathematics, Aarhus, 1970.

[Wraith, 1975] G. C. Wraith. Lectures on Elementary Topoi. In F. W. Lawvere, C. Maurer, and G. C. Wraith, editors, *Model Theory and Topoi*, volume 445 of *Lecture Notes in Mathematics*. Springer, 1975.

Topology

M.B. Smyth

Contents

1   Observable properties

Consider a device which outputs a binary sequence. We suppose that the
device continues for ever, and also that there is no bound on the time
taken to compute each digit. Thus the device may produce an infinite
sequence, but may also 'diverge', that is, produce some finite sequence and
then compute for ever without further output. The set of possible outputs
is $\Sigma^\infty = \Sigma^\star \cup \Sigma^\omega$, where $\Sigma = \{0,1\}$ and $\Sigma^\star, \Sigma^\omega$ are the sets, respectively,
of finite and of infinite words (sequences) over Σ. More generally, we may
admit that Σ is any countable set. As a slight variant, we may suppose that
the device has the (additional) capacity to halt. In that case, representing
the halt by an end-marker, the set of possible outputs may be taken to be
$\Sigma^\infty \cup \Sigma^\star e$.

An observer inspects the output sequence as it proceeds, noting various
properties of it. His judgements must be based only on the finite segment
which has been output so far (the device is a black box). It is evident that
there are some properties which, holding in certain instances, are such that
those instances will never be detected by the observer: for example, that
there are infinitely many occurrences of 0 in the sequence. We ask: what
properties are observable, supposing that they are present? In considering
this question we shall, at least for the time being, understand 'property'
extensionally, that is, as simply a subset of Σ^∞. Technically, then, the
condition of (finite) observability of a property A may be stated as follows:
for any $\sigma \in \Sigma^\infty$, if $\sigma \in A$ then there is some finite initial segment $\sigma \upharpoonright n$
of σ such that every extension of $\sigma \upharpoonright n$ belongs to A. Informally, the
condition is just that, if A holds of a sequence σ, then knowledge of some
initial finite segment of σ (some finite approximation of σ, we might say)
suffices to establish this. In view of the generalization to other types of
data, it is more useful to grasp the informal version than the technical
one. Notice that there is no requirement that the *absence* of the property
be detectable. The requirement that both presence and absence of the

property be detectable would, indeed, be rather a severe one in the present context: as a little reflection will show, only the trivial properties True and False would count as observable (given a property P, such that both P and $\neg P$ are observable, consider whether the null sequence satisfies P).

Two important closure properties of the class of observable properties can immediately be noted

1. *Finite Conjunction.* Suppose that $A_1, \ldots, A_n$ are observable properties, and that $A_1 \cap A_2 \cap \ldots \cap A_n$ holds of a sequence σ. Then $\sigma \in A_1 \cap \ldots \cap A_n$ is known (secured) by stage $max(k_1, \ldots, k_n)$ where $k_1, \ldots, k_n$ are the stages (of the output) at which $\sigma \in A_1, \ldots, \sigma \in A_n$ are known. Thus: $A_1, \ldots, A_n$ observable $\Rightarrow A_1 \cap \ldots \cap A_n$ observable.

2. *Arbitrary Disjunction.* Suppose that $\mathbf{A}$ is a collection of observable properties. If $\sigma \in \cup\mathbf{A}$, then $\sigma \in A$ for some $A \in \mathbf{A}$. Since A is observable, $\sigma \in A$ is known at some finite stage n; but then $\sigma \in \cup\mathbf{A}$ is also secured at stage n. Thus $\cup\mathbf{A}$ is observable.

This merely says that the class of observable properties is a *topology*.

Definition 1.0.1. A *topology* on a set S is a collection of subsets of S that is closed under finite intersection and arbitrary union. A set S together with a topology $\mathcal{T}$ on S is a *topological space* $(S, \mathcal{T})$; the elements of $\mathcal{T}$ are the *open sets* of the space.

Notice that the empty set $(= \cup\varnothing)$ and S itself $(= \bigcap\varnothing)$ are open in any topology over S. Indeed, any topology on S, considered as a poset ordered by the subset relation is a complete lattice (because it has arbitrary joins: Exercise 1.0.5(3)) having least element $\varnothing$ and greatest element S.
Note: A summary of basic information on lattices and related matters is provided in the Appendix to the chapter.
Notation Given a topological space X we denote the topology of X considered as a complete lattice by ΩX.

Another complete lattice which presents itself here is the collection of all possible topologies on a set S, again ordered by subset; this is seen to be a complete lattice since it has arbitrary *meets* (Exercise 1.0.5(4)). It has the greatest, or *finest* element $\mathcal{P}S$ (the so-called *discrete* topology on S) and the least, or *coarsest* element $\{\varnothing, S\}$ (the *indiscrete* topology).

Typically, a topology $\mathcal{T}$ on a set S is specified in terms of some convenient subset of $\mathcal{T}$. The most useful concepts here are given by:

Definition 1.0.2. A *base* of a topology $\mathcal{T}$ is a subset $\mathcal{B}$ of $\mathcal{T}$ such that every open set is the union of elements of $\mathcal{B}$. A space having a countable base is said to be *second countable*. If $\mathcal{A} \subseteq \mathcal{P}S$, and $\mathcal{T}$ is the least topology containing $\mathcal{A}$, we say that $\mathcal{A}$ is a *subbase* of $\mathcal{T}$.

Every base of $\mathcal{T}$ is also a subbase of $\mathcal{T}$; on the other hand, $\mathcal{A}$ is a subbase of $\mathcal{T}$ iff the collection of finite intersections of elements of $\mathcal{A}$ is a base of $\mathcal{T}$ (Exercise 1.0.5(5)). Second countability is especially significant for us, as a necessary condition for a space to be computationally meaningful. (There is a weaker notion of *first* countability. A space X is *first countable* if, for every point x of X, the filter of open sets which contain x has a countable base.)

Example 1.0.3. A (countable) base of the topology on Σ^∞ considered above is given by the sets of the form $\uparrow \sigma$ for finite σ. Here we presuppose the ordering of $\Sigma^\infty : \sigma \leq \sigma'$ iff σ' extends σ (σ is a prefix of σ'); and $\uparrow x$ denotes as usual $\{y \mid x \leq y\}$.

An interesting refinement of this topology is obtained if we take as open also the sets of the form $\Sigma^\infty - \uparrow \sigma$ (σ finite). That is, we now take $\{\uparrow \sigma \mid \sigma$ finite$\} \cup \{\Sigma^\infty - \uparrow \sigma \mid \sigma$ finite$\}$ as subbase. This topology and its possible computational significance will be considered in detail later (Section 7.3). For the moment, the reader might care to verify that, in the refined topology, singleton sets $\{\sigma\}$ (σ finite) are open iff the alphabet Σ is finite.

Needless to say, we shall in due course consider various ways of constructing new topological spaces from old ones. For the purposes of our initial list of examples (Section 2) it will be convenient to have available just the following:

Definition 1.0.4. Let (S, $\mathcal{T}$) be a topological space, and $S' \subseteq$ S. Then the *subspace* topology on S' induced by $\mathcal{T}$ is the relativization of $\mathcal{T}$ to S'; that is, the sets open in the subspace are precisely those of the form $S' \cap O$ where $O \in \mathcal{T}$.

It is trivial to verify that the sets $S' \cap O$ constitute a topology on S'. In terms of 'observable properties', we in effect keep $\mathcal{T}$ unchanged in going to the subspace (the *observations* are unchanged; the set over which they are made is reduced).

For further general discussion of 'finite observability', see the Postscript to Section 5, and Section 6.4.

Acknowledgements

Discussion of topological themes with colleagues at Imperial, notably Samson Abramsky and Steve Vickers, has been a valuable stimulus. I have benefited also from comments and suggestions concerning previous drafts of the chapter by various people, in particular, Henk Barendregt, Wilson de Oliveira and Prakash Panangaden. I would also like to thank Mrs Jane Spurr for the typing of this manuscript.

General remarks

The emphasis of this chapter is very much on computability and semantics, and on 'topology as logic', rather than on geometry. Geometrical considerations are virtually confined to the Exercises of Section 5, dealing with the new area of topology applied to digital image processing. I believe this topic to be much more important and promising than its relegation to these exercises might suggest.

The chapter deals exclusively with general, or analytic (as opposed to algebraic) topology. This is probably a fair reflection of the subject at the present time. However,for some pointers to the possible relevance of algebraic topology one could consult, for example, [Beck, 1980] and [Smale, 1987]. Also, it seems likely that concepts of algebraic topology will play a part in the development of the work on digital images mentioned previously.

Exercises 1.0.5.

1. Let S be a two point set. Display the four topologies which can be defined on S.

2. Let Σ^∞ have the topology with base $\{\uparrow \sigma \mid \sigma \text{ finite }\}$. Consider various properties which a sequence σ may have, such as: σ does not begin with 000; 000 occurs in σ; 000 only occurs when immediately followed by 111; 0 occurs infinitely often in σ; σ is eventually always 0. (The last property may be considered vacuously true in case σ is finite. Note that temporal language is often convenient in specifying properties.) For each property P, consider informally what it would mean to (not necessarily finitely) 'observe' the presence of P. Formally, determine the topological status of P: is P open? If it is not open, can it be expressed simply in terms of open sets, say as the complement of an open set, or as the intersection of a sequence of open sets? Do any of these answers change if it is known that σ is infinite (that is, if we restrict to the subspace of infinite sequences)?

3. Recall the standard result that any poset having arbitrary joins has also arbitrary meets, and is therefore a complete lattice. Verify that in any topology $\mathcal{T}$, ordered by set inclusion, the join of an arbitrary collection $\mathcal{C}$ (of open sets) is just the set union $\cup \mathcal{C}$. Show also that, in case $\mathcal{T}$ is (for example) the usual topology of Σ^∞, the meet of a collection $\mathcal{C}$ need not be $\cap \mathcal{C}$.

4. Show that the ordered set of all the topologies on a set S is a complete lattice, by verifying that any collection $\mathcal{K}$ of topologies has $\cap \mathcal{K}$ as meet. Show that the join of two topologies is not in general their set union (unless S has ≤ 2 elements).

5. Let $\mathcal{A}$ be a subbase of open sets of a space X. Show that a set $S \subseteq X$ is open iff S can be expressed as a union of finite intersections of members of $\mathcal{A}$.

6. Let $\mathcal{B}$ be an arbitrary base of a second-countable space X. Show that there is a countable subset of $\mathcal{B}$ which is also a base of X. [Hint: it suffices to show that each member of the assumed countable base of X can be expressed as the union of countably many members of $\mathcal{B}$.]

2   Examples of topological spaces

2.1   Sierpinski space

This is the two-element space in which exactly one of the two singleton subsets is taken as open. We represent it as $\{\top, \bot\}$, where $\{\top\}$, but not $\{\bot\}$, is open. It may be regarded as the space of 'truth values' for observations. Here, we think of $\top$ as representing the positive outcome of an observation (or test), while $\bot$ represents failure of the test. A positive outcome is itself detectable at some finite stage, while failure is not (in general); hence the asymmetry of the topology.

2.2   Scott Topology

By a *dcpo* we understand a poset $(D, \sqsubseteq)$ in which every directed subset S has a least upper bound (sup, or join) $\bigsqcup_\uparrow S$. If in addition $\bigsqcup \varnothing$ exists (that is, D has the least element, $\bot$), D is a *cpo*. A subset O of dcpo $(D, \sqsubseteq)$ is taken as open in the *Scott topology* of D provided

(i) $x \in O \Rightarrow \uparrow x \subseteq O$ (O is an upper set)

(ii) $\bigsqcup_\uparrow S \in O \Rightarrow S \cap O \neq \varnothing$ (equivalently, in view of (i): S is eventually in O).

The idea of this topology is as follows: (i) if information x suffices to indicate that test O has succeeded, then any greater information is sufficient *a fortiori*; (ii) if the limit of a 'generalized sequence' (directed set) of better and better approximations passes a test O, then some of the approximants already pass.

Requirement(ii) is connected with the idea that open sets correspond to *finitary* tests. For an adequate rendering of ideas of finiteness, however, more structure needs to be imposed on the dcpo's. In particular, a dcpo

D is *algebraic* if every element of D is the sup of a directed set of finite elements of D, where $a \in D$ is *finite* iff:

$$a \sqsubseteq \sqcup_\uparrow S \Rightarrow a \sqsubseteq x \text{ for some } x \in S.$$

We sometimes refer to the set of finite elements of an algebraic dcpo D as the *basis* of D, and denote it by B_D; if the basis of D is countable (which is always the case in computational contexts), D is said to be *ω-algebraic*. The Scott topology of an algebraic dcpo can be simply characterized: it has as a base the collection of sets $\uparrow a$, where a is finite. Our previous discussion of Σ^∞ (or $\Sigma^\infty \cup \Sigma^\star e$) provides an illustration: the data type in question, under the prefix ordering is an ω-algebraic cpo, and its 'topology of observable properties' is the Scott topology.

A less severe restriction than algebraicity is that of *continuity* of a dcpo. This is axiomatized in terms of a *relative* notion of finiteness; see Exercise 2.7.1(2).

For the purpose of giving some of the examples, we have in this chapter to assume that the reader has a basic knowledge of cpo's and 'domain theory', or can readily consult expositions thereof.

2.3 Spaces of maximal elements. Cantor space

That spaces come equipped with non-trivial 'information orderings' is, of course, characteristic of (Scott's) domain theory. We shall have much more to say about the topological significance of these orderings in due course. At this stage we may remark that many of the spaces traditionally studied in mathematics may be recovered as (sub)spaces of *maximal* elements of domains; roughly speaking, the traditional spaces may be obtained by throwing away the partial, or incompletely specified, elements of domains.

If D is a cpo with Scott topology, let us denote by $\text{Max}(D)$ the set of maximal elements of D taken with the subspace topology. Consider in particular $\text{Max}(\Sigma^\infty)$. As a set this is Σ^ω, and the topology has as a base the collection of sets $\uparrow \sigma \cap \Sigma^\omega (\sigma \text{ finite})$; the latter point follows from the trivial observation that if $\mathcal{B}$ is a base for the topology of a space X, and $Y \subseteq X$, then the relativization of $\mathcal{B}$ to Y is a base for the subspace topology of Y. Taking Σ as $\{0, 1\}$ we obtain in this way *Cantor space*; taking Σ as **N** we obtain *Baire space*.

See Exercise 2.7.1(5) for the construction of Euclidean space as $\text{Max}(D)$ for suitable D.

2.4 Alexandroff topology

This is a technically convenient, if not always computationally appropriate, topology for a poset P: one takes as open sets simply the upper sets of

P. That is, the Alexandroff topology is obtained by using only condition (i) in the definition of the Scott topology, and is thus (for a dcpo) finer than the Scott topology. We could say that the Alexandroff topology is computationally appropriate when only finite elements are in question. For example, on a finite dcpo the two topologies obviously coincide; see also Exercise 2.7.1(6).

Our next two examples illustrate an extremely important method of construction of topological spaces. Indeed, it can be argued that this method - in a sufficiently general form — provides all the spaces which we ever need to consider in computer science. Schematically, the construction is as follows. We start with a lattice, L; in the intended applications, L will usually represent a 'logic' of observable propositions (or properties). Meet and join in L will represent conjunction and disjunction, respectively, and the lattice will normally be distributive. The order in L represents logical implication. Next, we choose some set X of filters of L. [Recall: $F \subseteq L$ is a (proper) *filter* provided that

    (a) $a \in F, a \leq b \Rightarrow b \in F$

    (b) $a \in F, b \in F \Rightarrow a \wedge b \in F$

    (c) $F \neq L$ (equivalently, $0 \notin F$)].

The idea is that a filter represents a deductively closed, consistent theory (or specification) in the logic. The choice of X is determined by, for example, whether we want to consider only theories that are as complete (specifications that are as precise) as possible, or to allow some degree of latitude (incompleteness, partiality), etc. Then we topologize X by taking as basic open subsets all those of the form

$$X_a = \{F \mid a \in F\}, a \in L.$$

We could say that X_a collects together the specifications in which (at least) the property a is satisfied.

It should be pointed out that, in many cases, the lattice L will have additional structure, which may permit more flexible choice of the space X of filters.

2.5 Stone spaces

Let $(L, \wedge, \vee, \neg)$ be a Boolean algebra, and choose for X the space of ultrafilters of L [Recall: an ultrafilter is simply a maximal filter of a Boolean algebra — which we can also characterize as a filter F such that, for every $a \in L$, either $a \in F$ or $\neg a \in F$. It is the classical notion of a complete (and consistent) theory.] The spaces which can be constructed in this way are known as *Stone spaces*.

Suppose in particular that L is the Boolean algebra of propositions over a countable set of variables $p_0, p_1, p_2, \ldots$. From elementary logic we know that the ultrafilters, or complete theories, are in $(1, 1)$-correspondence with the valuations, that is, the maps from $p_0, p_1, \ldots$ into $\{0, 1\}$, which we may regard simply as binary sequences. A little reflection will now convince the reader that the space of complete propositional theories is 'essentially the same' as Cantor space: under the $(1, 1)$-correspondence just mentioned, open sets also correspond. The standard term for the equivalence of spaces involved here is *homeomorphism*, which we shall define formally a little later.

Thus a Stone space could more accurately be defined as a space that is homeomorphic to the space of ultrafilters of some Boolean algebra. It will prove convenient to have a slight reformulation of this. Given a Boolean algebra L, with its space X of ultrafilters, let $\mathcal{B}$ be the base $\{X_a \mid a \in L\}$; $\mathcal{B}$ is itself a Boolean algebra of subsets of X. Now if $\mathcal{F}$ is an ultrafilter (= prime filter) of $\mathcal{B}$, it is evident that $\bigcap \mathcal{F}$ is a singleton subset of X (namely $\{x\}$, where $x = \{a \mid X_a \in \mathcal{F}\}$). On the other hand, if Z is any space having a base $\mathcal{B}$ such that $\mathcal{B}$ is a Boolean algebra of sets, each singleton $\{x\}$ of X is the meet of the prime filter $\mathcal{F}_x = \{A \in \mathcal{B} \mid x \in A\}$ of $\mathcal{B}$. We arrive at the following characterization: X is a Stone space if and only if X has a base $\mathcal{B}$ such that $\mathcal{B}$ is a Boolean algebra of sets and the meet of any prime filter of $\mathcal{B}$ is a singleton.

2.6 Spectral spaces

It is often inappropriate to assume that classical negation is available. So suppose now only that $(L, \wedge, \vee)$ is a distributive lattice, and let X be the space of prime filters of L. [A filter F is *prime* if, whenever $a \vee b \in F$, either $a \in F$ or $b \in F$.] The idea of this choice of filters can be that we want to exclude specifications which are overtly ambiguous. The spaces (homeomorphic to those) which can be constructed in this way are the *spectral spaces*.

Although primality (of filters) is in general a less stringent requirement than maximality, the two notions are equivalent in case L is a Boolean algebra (Exercise 2.7.1(10)). Thus (2.5) is a special case of (2.6): every Stone space is spectral.

Where shall we find some spectral spaces that are not Stone spaces? Simple but instructive examples are provided by *finite posets*. Indeed, let P be a finite poset with Alexandroff (equivalently, Scott) topology. Let L be the collection of upper sets of P; as a sublattice of $\mathcal{P}(\mathrm{P})$, L is distributive. (L is of course the topology of P, although we do not lay stress on this, in view of the generalization which follows.) It is rather evident that a filter F of L is prime iff F is the collection of supersets ($\in L$) of some $a \in P$, and that this gives us a $(1, 1)$-correspondence — indeed a homeomorphism — between P and the space of prime filters of L. Thus P is spectral.

For a satisfying generalization of these finite spaces, consider algebraic cpo's (*cpo* ≡ dcpo with least element). If D is an algebraic cpo, taken with Scott topology, an open set $O \subseteq D$ is said to be *compact* if O is the join of *finitely* many basic open sets $\uparrow a$ (a finite in D). (This definition is only a temporary convenience; later we will study compactness in proper generality.) Suppose now that D satisfies the condition that, for any $a, b \in B_D$, $\uparrow a \cap \uparrow b$ is compact. By distributivity it follows that the meet of any two compact open sets is compact, from which we can conclude that the collection L of compact open sets is a sublattice of $\mathcal{P}(D)$. Now by essentially the same argument as in the finite poset case, we get a homeomorphism between D and the prime filter space of L, showing that D is spectral.

Notice in particular that D is spectral if it satisfies one of the most commonly assumed conditions on domains, namely that of *bounded completeness* (every bounded subset of D has a lub): for in that case we have $\uparrow a \cap \uparrow b = \uparrow (a \sqcup b)$. But weaker conditions than this on the existence of minimal upper bounds are sufficient for spectrality. For the reader having some familiarity with domain theory, we can point out that the spectral cpo's coincide with the '2/3-SFP' domains of Plotkin, [Plotkin, 1981].

The alert reader may have noticed that our spectral cpo's do not provide a strict generalization of the finite posets, since a cpo has to have a least element. We can take care of this, if desired, by amending the requirement that a least element exists in D to: the set M of minimal elements of D is finite, and $\uparrow M = D$.

The class of spectral (d)cpo's contains all the varieties of spaces that are most frequently used in the domain-theoretic approaches to semantics. On the other hand, as we shall see later, the spaces which are obtained when semantics is studied in terms of *metrics* are typically Stone spaces. We can fairly say that spectral spaces are ubiquitous in computer science.

2.7   The reals

We will assign $\mathbb{R}$ its standard topology, with basic open sets those of the form $(p, q) = \{x \mid p < x < q\}$, where p, q may be taken as rational, with $p < q$. Does this topology fit with our view of open sets as observable properties? To answer this question we have to consider some system of representation of the reals, such as by infinite decimal expansions (a better, more flexible, system would be: rational sequences with prescribed rate of convergence). Denote by $\bar{r}$ the real number represented by expansion r. Then we can see that a set $O \subseteq \mathbf{B}$ is open if, and only if, for any r such that $\bar{r} \in O$, there is a finite initial segment r_0 of r such that every infinite extension of r_0 represents an element of O. If this is not immediately apparent, it should become clear on viewing the topology in metric terms (as we shall do formally in a later section): a set O is open iff, for any x

$\in O$, every point within some positive distance ε of x is in O. Thus the equation

$$open = finitely\ observable$$

holds up well in this case.

Despite the pre-eminent role of $\mathbb{R}$ in mathematics generally, its status in the theory of computing is somewhat uncertain. Exact computation over the reals (as opposed to some bounded-precision approximation of them) has been considered as a realistic option in [Wiedmer, 1980; Boehm *et al.*, 1986; Vuillemin, 1987].

We may perhaps have to incorporate $\mathbb{R}$ into semantic domains if we are to handle probabilistic programs adequately [Saheb-Djahromi, 1980; Kozen, 1981; Jones and Plotkin, 1989]. But none of this is entirely convincing (in establishing the role of $\mathbb{R}$).

An explanation may be suggested as follows. The discrete character of digital computation means that — even admitting infinite data and processes as ideal elements — the topology of a data space is necessarily an *algebraic* lattice (as is the case for all spectral spaces). But this excludes $\mathbb{R}$, since its topology is not algebraic (Exercise 2.7.1(3)).

Can one envisage any (more computational) alternatives to $\mathbb{R}$? The question has hardly been investigated. But one candidate may be the *p*-adic numbers (Vickers, 1988).

Exercises 2.7.1.

1. Let D be a dcpo with Scott topology, and $x \in D$. Show that the set $D - \downarrow x$ (note : $\downarrow x = \{y \mid y \sqsubseteq x\}$) is open.

2. Let D be a dcpo, and $x, y, \in D$. We say that x is *way below* (or *finite relative to*) y if, for any directed set S such that $y \sqsubseteq \sqcup S$, we have $x \sqsubseteq s$ for some $s \in S$. Notation: $x << y$. Write $\Uparrow x$ for the set $\{y \mid x << y\}$ and similarly for $\Downarrow x$. The dcpo D is said to be *continuous* if, for every $x \in D$, $\Downarrow x$ is directed and $x = \sqcup \Downarrow x$. A lattice which, as an ordered set, is a continuous (d)cpo is a *continuous lattice* (it is necessarily a complete lattice).
   Show that the unit interval $[0,1]$ with its usual order is not an algebraic lattice, but is a continuous lattice, and identify its Scott topology.
   Show that, in any continuous dcpo, the sets of the form $\Uparrow x$ form a base of the Scott topology.

3. Show that the topology of an algebraic dcpo, considered as a complete lattice, is necessarily an algebraic lattice. Show that the Scott topology of the dcpo $(I, \leq)$, where I is the unit interval, is a continuous lattice, but is not an algebraic lattice. Show that the same is true of the *usual* (Euclidean) topology of I.

4. Let D be a dcpo with Scott topology, and let $\mathrm{Max}(D)$ be the subspace of the maximal elements of D. Show that, if x, y are any two distinct points of $\mathrm{Max}(D)$, there is an open set which contains x but not y. (This is the 'T_1 separation' property of $\mathrm{Max}(D)$: Section 4.)

   For a rather harder exercise, suppose now that D is a Scott domain (or more generally, a bounded-complete continuous dcpo). Show that, if $x, y \in \mathrm{Max}(D)$ are distinct, there are disjoint open sets U, V such that $x \in U, y \in V$ ('T_2', or Hausdorff, separation). See also Examples 4.1.4.

5. Let D be the poset whose elements are pairs of real numbers (x, y) such that $x \le y$, ordered by:

$$(x, y) \sqsubseteq (x', y') \quad \Leftrightarrow \quad x \le x' \le y' \le y.$$

   The idea of this ordering is that the smaller of two intervals $[x, y]$ in the order of set inclusion is the greater in information content, when viewed as an approximation. Show that D is a continuous dcpo, and that $\mathrm{Max}(D)$ can be identified with the Euclidean space $\mathbb{R}$. (The example is essentially Scott's *interval lattice*: [Scott, 1970])

6. Check that the Scott and Alexandroff topologies coincide for a finite dcpo (= finite poset). For which infinite dcpo's is this true?

7. Show that the Alexandroff topology of a poset P satisfies the following:

   (a) for every point $x \in P$, there is a smallest open set which contains x, say O_x;

   (b) the map $x \mapsto O_x$ is injective, that is $x \ne y \Rightarrow O_x \ne O_y$.

   Prove also the converse: any topological space P satisfying conditions (a) and (b) is the 'Alexandroff space' derived from a suitable partial order on P.

8. Point out that the definition of the Alexandroff topology of an ordered set $(P, \le)$ works just as well if $\le$ is only a pre-order. Show that every finite topological space arises from a suitable pre-order in this way. Try to formulate a result to the effect that finite topological spaces are "essentially the same" as finite pre-ordered sets.

9. Let D be an algebraic dcpo taken with its Scott topology, and B_D the basis of D. Show that the subspace topology of B_D is the Alexandroff topology derived from the ordering of B_D (as a sub-poset of D).

10. Let F be a filter of the Boolean algebra B. Show that the following are equivalent:

(a) F is an ultra-filter;

(b) F is prime;

(c) for each $b \in B$, either $b \in F$ or $\neg b \in F$.

11. Verify the following observations: the Stone space $St(B)$ of a finite Boolean algebra B has the discrete topology, and may be identified with the set of atoms of B. Up to isomorphism, B may be recovered from $St(B)$ as its power set. If B, B' are finite Boolean algebras, the Boolean homomorphisms from B to B' are in $(1, 1)$-correspondence with the mappings from $St(B')$ to $St(B)$. The category of finite Boolean algebras (with Boolean homomorphisms) is dual to the category of finite sets (with arbitrary maps).

Extend the preceding observations to the finite distributive lattices and the corresponding spectral spaces, leading to a duality with the category of finite posets (with monotonic maps).

3 Alternative formulations of topology

The definition of a topological space which we have given in Definition 1.0.1 is the most standard one. However, many different formulations exist for this fundamental notion. We shall consider here a few of the more useful of these. Apart from the change in perspective which they provide, some of these formulations suggest interesting generalizations and extensions of the basic idea of topology.

3.1 Closed sets

To begin with a rather trivial variant, we can use closed sets in place of open sets. A subset of a topological space is *closed set* iff it is the complement of an open set. Thus — dually to the case of open sets — finite joins and arbitrary *meets* of closed sets are again closed. We can treat closed sets rather than open sets as fundamental, defining a *topology* on a set S to be a collection of subsets of S that is closed under finite join and arbitrary meet.

Have closed sets any direct computational significance, which might justify such a change of viewpoint?

In line with our previous analysis, we can say that a closed set is the extension of a property which holds iff a certain (finitely) observable event does *not* occur. This is surely what is commonly known as a *safety property*.

3.2 Neighbourhoods

A subset S of a space X is a *neighbourhood* of $x \in X$ if, for some open set $O, x \in O \subseteq S$. In terms of our informal understanding of topological notions, the idea that S is a neighbourhood of x iff $x \in S$, can be established on the basis of a finite amount of information about x. The following is then evident informally, as well as being trivial to prove formally:

Proposition 3.2.1. *A set is open iff it is a neighbourhood of every one of its points.* ∎

In the light of this, it is not too surprising that *neighbourhood* can be used, in lieu of *open* or *closed*, as the primitive notion of topology: [Kelley, 1955] for the details, especially the axiomatization of *neighbourhood system*. *Open* can be defined as in 3.2.1, and thence all other topological notions. It is interesting to observe, however, that in the case of *closed* and related notions, we also have available a direct, 'positive' style of definition (not deriving from open sets via classical complementation). We begin with:

Definition 3.2.2. If S is a subset of the space X, then $x \in X$ is an *adherent point* of S if every neighbourhood of x meets S. Notation: $\mathrm{Adh}(S)$ — or, more suggestively, $\mathrm{Cl}(S)$ — for the set of adherent points of S.

Note that we have $S \subseteq \mathrm{Cl}(S)$ trivially.

Proposition 3.2.3. *Let S be any subset of the space X.*

Then

1. $\mathrm{Cl}(S)$ is closed;

2. S closed $\Rightarrow S = \mathrm{Cl}(S)$.

Proof.   1. Given $x \notin \mathrm{Cl}(S)$, let O be an open neighbourhood of x disjoint from S. Then O is also disjoint from $\mathrm{Cl}(S)$; for any $y \in \mathrm{Cl}(S) \cap O$, then y would have a neighbourhood (namely O) disjoint from S, contradicting the definition of Cl. Thus we have shown that $X - \mathrm{Cl}(S)$ is a neighbourhood of each of its points.

2. Suppose S is closed. If $y \notin S$, the neighbourhood $X - S$ of y is disjoint from S, so that $y \notin \mathrm{Cl}(S)$. This shows that $\mathrm{Cl}(S) \subseteq S$.   ∎
   Thus we see that, starting with *neighbourhood* a *closed* set can be defined as one which contains all its adherent points.   ∎

Proposition 3.2.3 can be strengthened to give the following characterization of Cl:

Proposition 3.2.4. *Cl(S) is the smallest closed set containing S.*

Proof. We know that Cl(S) is closed and contains S. Suppose that Q is closed and contains S. Since $S \subseteq Q$, Cl(S) $\subseteq$ Cl(Q) $= Q$. ∎

It is convenient now to introduce the following important definition, even though it involves a slight digression from the main theme of this section:

Definition 3.2.5. A subset S of the topological space X is *dense* in X if Cl(S) $= X$. Equivalently, S is dense if every non-empty open set meets S.

3.3 Examples

1. Q is dense in $\mathbb{R}$.

2. If D is an algebraic dcpo with Scott topology, the set B_D of finite elements of D is dense.

3. Let P be a poset with a greatest element $\top$. If P is taken with Alexandroff topology, or any coarser topology (such as the Scott topology, if appropriate), then $\{\top\}$ is dense in P

4. Let X be the space Σ^ω (Section 1, or Section 2.3). Then $S \subseteq X$ is dense iff

   $$\forall \sigma \in \Sigma^*.\exists y \in \Sigma^\infty.\sigma y \in S$$

   (equivalently: $\forall \sigma \in \Sigma^*. \uparrow \sigma \cap S \neq \varnothing$).

With regard to (4), note that, in verifying the density of a set S in a space X with a given base for its topology, it is of course sufficient to show that S meets every non-empty *basic* open set.

Example (4) is significant in connection with *liveness properties*. At first one might think that a liveness property ('something good happens') is the exact dual of a safety property ('something bad doesn't happen'), and so should be explained simply as an open set. However, the situation is not quite as simple as this. In the first place, the 'good thing' involved in a liveness property may be required to happen repeatedly. For example, the liveness property of starvation freedom requires that a process makes progress *infinitely often*. Perhaps a liveness property has to be the intersection of a sequence of open sets.

This may or may not be so; but a second aspect of liveness is, according to Alpern and Schneider [Alpern and Schneider, 1985] even more significant. This is that, with respect to a liveness property, a partial execution is always remediable. That is, any partial execution can be extended in a way that satisfies the property. Formally, this is just the condition (on S) stated in Example (4). It is argued by Alpern and Schneider that this condition is necessary and sufficient for S to be a liveness property; thus, in topological terms, their proposal is

$$\text{liveness} = \text{density}.$$

Another computationally interesting aspect of density is that, in favourable cases, an uncountable space may be presented in terms of a countable dense subset; cf. Examples (1), (2). The case of metric spaces will be discussed below.

It is by no means always true that a dense subset of a space contains complete information about the space: Example (3) provides a rather extreme illustration.

A countable dense subset is the main alternative to a countable base as a vehicle for effectively presenting a space. A space which has a countable dense subset is said to be *separable*. The condition is weaker than that of having a countable base.

Proposition 3.3.1. *Any second-countable space is separable.*

Proof. Suppose that the space X has a countable base B. Let S be a countable set such that each non-empty element of B has an element in S. Then S is dense as well as countable, since every non-empty open set meets S. ∎

3.4   Closure operators

We have presented Proposition 3.2.3 as showing that the closed sets can be defined directly, or positively, in terms of neighbourhoods. However, it can equally well be read as an indication that the 'closure operator' Cl (that is, Adh) could itself be taken as primitive, the closed sets then being introduced as its fixed points. This is in fact a very common approach to the axiomatization of topology.

Let $M : P(S) \to P(S)$ be an operator over a set S, and consider the following 'closure axioms'.

1		$A \subseteq M(A)$	(M is increasing)
2		$A \subseteq B \Rightarrow M(A) \subseteq M(B)$	(M is monotonic)
3		$M(A) = M(M(A))$	(M is idempotent)
4	(i)	$M(\varnothing) = \varnothing$	(M is finitely additive)
	(ii)	$M(A) \cup M(B) = M(A \cup B)$	

The axioms are not independent. Indeed, 4(ii) implies 2: if 4(ii) is satisfied, we have $A \subseteq B \Rightarrow A \cup B = B \Rightarrow M(A) \cup M(B) = M(B) \Rightarrow M(A) \subseteq M(B)$.

Definition 3.4.1. The operator M is said to be a *closure operator* if it satisfies 1, 2, and 3; a *Cech closure* if it satisfies 1 and 4; and a *Kuratowski* (or *topological*) closure if it satisfies 1, 3, and 4 (that is, if it satisfies all the closure axioms).

Proposition 3.4.2. *In any topological space, the operator Cl (Definition 3.2.2) is a topological closure.*

Proof. Closure axioms 1 and 4(i) are evident, while 3 follows from Proposition 3.2.3. For 4(ii), use the characterization 3.2.4: $M(A) \cup M(B)$, being the union of two closed sets, is closed, and contains $A \cup B$; hence $M(A \cup B) \subseteq M(A) \cup M(B)$. ∎

We have shown that, if one starts with the closed sets of a topology, 3.2.4 yields a topological closure, from which the original closed sets are recovered as its fixed points. To complete the proof that topologies are equivalent to topological closures, we have to consider the reverse order of carrying out these constructions:

Proposition 3.4.3. *Let M be a topological closure on S, and let F be the set of fixed points of M. Then F is the collection of closed sets of a topology on S, and the closure Cl derived from this topology coincides with M.*

Proof. That F is closed under finite joins is immediate from axiom 4. Let C be any subset of F. The set $\bigcap C$ is contained in each member of C, and therefore, using the monotonicity of M:

$$M(\bigcap C) \subseteq \bigcap \{M(C) \mid C \in C\} = \bigcap C.$$

Thus $\bigcap C \in F$, showing that F is the collection of closed sets of a topology. Now, for any $A \subseteq S$, $\mathrm{Cl}(A)$ is the least closed set, that is, the least fixed point of M, containing A. But since $M(A) = M(M(A))$, and $A \subseteq F \in F \Rightarrow M(A) \subseteq M(F) = F$, $M(A)$ is also the least fixed point of M which contains A. Thus $\mathrm{Cl} = M$. ∎

A corresponding, but simpler result holds for closure operators in general (Exercise 3.5.3(4)). As that result indicates, closure operators that are not necessarily topological are of very common occurrence: wherever we have a distinguished collection of subsets of a set closed under arbitrary meets, (for example, the subalgebras of an algebra), there we have a natural closure operation.

Another rich source of closure operations is provided by Galois connections:

Definition 3.4.4. Let P, Q be posets. A *Galois connection* between P, Q is a pair of maps $l \colon P \to Q, r \colon Q \to P$ such that

1. l, r are order reversing: $p \leq p' \Rightarrow l(p') \leq l(p)$, and similarly for r;

2. $p \leq r(l(p)), q \leq l(r(q))$.

The following observation is standard [Birkhoff, 1967]. We omit the easy proof.

Proposition 3.4.5. Let $l \colon P \to Q, r \colon Q \to P$ be a Galois connection. Then the map $r \circ l \colon P \to P$ is monotonic and idempotent. ∎

Since $r \circ l$ is increasing by definition, we have in the case that P is of the form $(\mathcal{P}(S), \subseteq)$ that $r \circ l$ is a closure operation on P.

Example 3.4.6. Let L be a set (thought of as a language), W another set, and $\models \subseteq W \times L$ a binary relation (read as *satisfaction*). Given $E \subseteq L$, define $\mathrm{Mod}(E)$, the set of *models* of E, to be $\{s \in W \mid \forall\, U \in E.\ s \models U\}$. Similarly, given $S \subseteq W$, the *theory* of S is $Th(S) = \{U \in L \mid \forall s \in S.\ s \models U\}$. Then the maps Mod, Th give a Galois connection between $\mathcal{P}(L)$ and $\mathcal{P}(W)$, as is easily checked. Thus we have a closure operator Mod∘Th on $\mathcal{P}(W)$.

This example has been taken as the starting point for applications of topology to program semantics by Parikh [Parikh, 1983]. One takes W, in particular, to be a set of programs, and L a set of partial correctness (Hoare logic) assertions for W. So far we have said nothing to indicate that the closure Mod∘Th is topological. But it needs only some very weak assumptions on the 'logic' of L to ensure that this is the case:

Definition 3.4.7. L has *falsehood* if there is an element ff of L such that $\mathrm{Mod}(\mathrm{ff}) = \varnothing$. L has *disjunction* if, for every U, U' in L, there exists V such that $\mathrm{Mod}(V) = \mathrm{Mod}(U) \cup \mathrm{Mod}(U')$.

Proposition 3.4.8. If L has falsity and disjunction then the closure operator $J = \mathrm{Mod} \circ Th$ is topological.

Proof. Assume that L has falsity and disjunction. Since $Th(\varnothing) = L$, we have $J(\varnothing) \subseteq \mathrm{Mod}(\mathrm{ff}) = \varnothing$.

It remains to show that $J(A \cup B) \subseteq J(A) \cup J(B)$. Suppose then that $w \notin J(A) \cup J(B)$. This means that we have $U \in Th(A), U' \in Th(B)$ such that $w \not\models U$ and $w \not\models U'$. Now we have $A \cup B \subseteq \mathrm{Mod}(U) \cup \mathrm{Mod}(U') = \mathrm{Mod}(V)$

(for suitable V), so that $V \in \mathrm{Th}(A \cup B)$. Since $w \notin \mathrm{Mod}(V)$, it follows that $w \notin \mathrm{Mod}(\mathrm{Th}(A \cup B)) = J(A \cup B)$. ∎

For more on Parikh's topology, see Exercises 3.5.3(5), (6) (and Exercise 7.7.8(9)).

Related to open sets in the same way that closure operators are to closed sets, we have *interior operators*. Given a topology on a set S, we can define the *interior* $\mathrm{Int}(A)$ of $A \subseteq S$ to be the largest open set contained in A; on the other hand, given an operator satisfying the appropriate 'interior axioms', we can extract a topology by declaring the open sets to be its fixed points. The appropriate axioms are the expected duals of the closure axioms given previously:

1. $(A) \subseteq A$

2. $A \subseteq B \Rightarrow I(A) \subseteq I(B)$

3. $I(A) = I(I(A))$

4. (a) $I(S) = S$

   (b) $I(A \cap B) = I(A) \cap I(B)$.

The reader will have observed that the closure and interior axioms strongly resemble the customary axioms for modal (propositional) logic. Reading closure as possibility and interior as necessity, axiom 4(b) in particular suggests the normality (K) condition of modal logic, axiom (1) the T axiom, and the axiom set as a whole the $S4$ system.

To make more of this, we can consider modal algebra as the bridge between topology and modal logic. We take a *modal algebra* to be (at least) a poset together with one or more unary, monotonic operations. Varieties of modal algebras are introduced by axioms of the type 1–4 , where we of course read '$\subseteq$' as the partial order, and assume some (semi)lattice structure in order to interpret 4(b) . The topological approach to modal semantics of McKinsey and Tarski [McKinsey and Tarski, 1944] or Rasiowa and Sikorski [Rasiowa and Sikorski, 1963] proceeeds by abstracting a modal algebra from a modal theory (as its Lindenbaum algebra), and representing the modal algebra as the algebra of subsets of a suitable topological space. We note that, as it stands, this approach is only available for modal logics at least as strong as $S4$ (since the counterparts of the $S4$ axioms are necessarily satisfied by the subsets of a space).

In the other direction, one may think of abstracting from topological spaces to modal algebras. This is the path taken by Nöbeling [Nöbeling, 1954]; it is one of the two main approaches to the formulation of topology-without-points (or 'pointless topology'). Nöbeling shows that much basic analytic topology can be made to go through without mentioning points, and on the basis of closure axioms substantially weaker than those listed above. There is currently a revival of interest in pointless topology, due largely to its constructive flavour, and more particularly, to its relevance for computer science (see, for example, [Johnstone, 1982; Smyth, 1983b; Vickers, 1989]); the current work, however, follows the alternative approach via 'frames' (complete Heyting algebras), and Nöbeling's work remains rather neglected.

Can the association between modality and topology be explained in terms of our basic scheme of ideas? The connection is most readily understood if one thinks of the epistemic modalities. Indeed, interior and closure are surely epistemic operators: $x \in I(A)$ means that it can be *known* (by a finite investigation) that x is in A, while $x \in \mathrm{Cl}(A)$ holds iff x cannot be (finitely) distinguished from A. Knowledge operators are typically introduced, in computing contexts (and also in economics: [Aumann, 1976]), in terms of relations R_i of indistinguishability (one for each 'agent' i) between states of a system. For each R_i, there is then the closure M_i defined by:

$$M_i(A) = \{y \mid \exists x \in A . x R_i y\}.$$

If, as is often done, R_i is assumed to be an equivalence relation, the result is topologically rather trivial. However, it may be plausibly argued that, in practice, indistinguishability is not a transitive relation (Poincaré [Poincaré, 1905]; and much work on the sorites paradox, logic of vagueness, etc.).

Suppose now that we assume only that R is reflexive and symmetric; then the resulting closure operator is easily seen to be a Cech closure, which is Kuratowski iff R is an equivalence. Concerning the potential of this, perhaps more realistic, view of indistinguishability, we have two observations. First, in his text [Čech, 1966], Cech has demonstrated, on a larger scale than Nöbeling, that much of general topology can be formulated on the basis of Cech closure operators. Secondly, and more specifically, the work of Poston [Poston, 1971] must be considered. Poston defines a *fuzzy space* to be a set equipped with a reflexive symmetric relation, shows that fuzzy spaces have a rich 'topological' theory, and argues that much basic mathematical (and even physical) theory can be reworked in a finitist sense in this framework. Poston's remarkable work does not (yet) seem to have been followed up.

3.5   Convergence

The topic of convergence may be presented either in terms of *nets* (generalizing *sequences*) of points, or in terms of *filters*. Since we wish to reduce the emphasis on points, we take the latter approach:

Definition 3.5.1. Let $\mathcal{F}$ be a filter of subsets of a topological space X. We say that $\mathcal{F}$ *converges* to $x \in X$, written $\mathcal{F} \to x$, if $\mathcal{F}$ refines the neighbourhood filter $\mathcal{N}(x)$ of neighbourhoods of x; that is, if $\mathcal{N}(x) \subseteq \mathcal{F}$.

Viewing filters as specifications, we can say that $\mathcal{F} \to x$ provided that x does not satisfy more properties than required, or contain too much information, *vis-à-vis* the specification $\mathcal{F}$. One might well ask whether a more stringent conception of the relation between $\mathcal{F}$ and x, under which x contains *just the right* information, is not possible and desirable: see [Smyth, 1987]. Here, however, we shall adhere to the standard definition, which is 3.5.1.

Proposition 3.5.2. *The relation of convergence enjoys the following three properties:*

1. $\mathcal{PF}(x) \to x$, *where* $\mathcal{PF}(x)$ *is the principal filter at* x, *that is, the filter of all supersets of* x.

2. *If* $\mathcal{F} \to x$, *then* $\mathcal{F}' \to x$ *for any filter* $\mathcal{F}'$ *which refines* $\mathcal{F}$.

3. *If* $\mathcal{F} \not\to x$, *then there is a filter* $\mathcal{F}'$ *which refines* $\mathcal{F}$ *such that for no filter* $\mathcal{F}''$ *refining* $\mathcal{F}'$ *do we have* $\mathcal{F}'' \to x$.

Proof. 1., 2.: Immediate.

3. Suppose that $\mathcal{F}$ does not converge to $x \in X$. Then there is a neighbourhood N of x such that $N \notin \mathcal{F}$. In that case, the complement $X - N$ of N meets every member of $\mathcal{F}$, and we can consider the filter

$$\mathcal{F}' = \{A \cap (X - N) \mid A \in \mathcal{F}\}$$

Then $\mathcal{F}'$ refines $\mathcal{F}$, but any filter $\mathcal{F}''$ which refines $\mathcal{F}'$ has $X - N$ as a member, and so cannot converge to x. ∎

In the other direction, let S be given, together with a relation $\to$ between filters over $\mathcal{P}(S)$ and S satisfying the three conditions of 3.5.2. Then we obtain a topology on S by taking as open any set O such that, for any $x \in O$, and any filter $\mathcal{F}, \mathcal{F} \to x \Rightarrow O \in \mathcal{F}$ (equivalently: N is a neighbourhood of a point x if N is a member of every filter $\mathcal{F}$ such that $\mathcal{F} \to x$).

Exercises 3.5.3.

1. Let S be any subset of a space X. Show that $S \cup (X - \mathrm{Cl}(S))$ is dense in X. Deduce that S is the meet of a dense set and a closed set. (According to Alpern and Schneider [Alpern and Schneider, 1985],

who consider the case that X is a space Σ^ω of state sequences, this shows that every property may be specified as the conjuncion of a liveness property and a safety property.)

2. In any topological space, a set which is the intersection of a countable family of open sets is said to be a G_δ set. Show that a (finite or) countable intersection of G_δ sets, and likewise a finite union of G_δ sets, is again a G_δ set. Show that in Cantor space (and likewise in $\mathbb{R}$), every closed set is G_δ. (It was suggested in [Smyth, 1983a] that a specification is a sequence of finitary properties, understood as a conjunction, so that a specifiable set is necessarily G_δ. We have expressed the same idea at various places in the present work by saying that a specification is a theory, each of whose propositions represents an open set.)

3. Denote the temporal operator 'eventually' by $\Diamond$. Recalling Exercise 1.0.5(2), assume that temporal propositions may be treated as properties of infinite sequences of 'states', that is, as subsets of a space Σ^ω (it will suffice to take $\Sigma = \{0,1\}$). Give a reasonable semantic definition of $\Diamond$, and point out that any set S of the form $\Diamond P$, where P is non-empty, is dense. Show also that there is a dense set S, such that for no non-empty P is it true that $\Diamond P \subseteq S$. [Hint: consider a property S which ties what happens eventually to what happens initially.]

4. Define an *intersection system* over a set X to be a collection of subsets of X that is closed under arbitrary intersections. Show, in analogy with Proposition 3.4.3 (and the discussion preceding it) that intersection systems over X are, in effect, the same as closure operators over X.

5. Let L be a Boolean algebra, let W be the set of ultrafilters of L, and define the satisfaction relation $\vDash\, \subseteq W \times L$ by : $W \vDash b \Leftrightarrow b \in W$. Show that the Stone topology of W (Section 2.5) coincides with the topology of W induced by the closure operator Mod $\circ$ Th (Example 3.4.6)
Let W be the unit interval $[0,1]$, and L the set of finite unions of open rational intervals $(a,b), 0 \le a \le b \le 1$, $a,b \in \mathbb{Q}$. Define the satisfaction relation by: $w \vDash V$ iff $w \in V$. What topology is induced on W by the closure Mod $\circ$ Th?

6. (a) If X is a topological space, we say that a collection $\mathcal{K}$ of closed sets of X is a *base of closed sets* for X provided that every closed set can be expressed as the intersection of members of $\mathcal{K}$. Recalling Example 3.4.6, show that if L has falsity and disjunction,

then the collection $\{\text{Mod}(A) \mid A \in L\}$ is a base of closed sets for Parikh's topology on W.

(b) It would be more in keeping with our general approach to regard a language L of basic propositions as determining *open* sets rather than closed sets. Formulate sufficient (or, as a harder exercise, necessary and sufficient) conditions on L so that $\{\text{Mod}(A) \mid A \in L\}$ is a base of open sets for a topology on L. Comment on the preceding exercise (5) in the light of your findings.

4 Separation, continuity and sobriety

4.1 Separation conditions

Definition 4.1.1. Let T be a topology on a set S. The *specialization (pre)order*, $\leq_T$, induced by T on S is defined by:

$$x \leq_T y \equiv \forall O \in T.x \in O \to y \in O. \tag{1}$$

We leave to the reader the easy verification that the relation defined by (1) is indeed a preorder. The significance of this preorder for us is that it can be seen as an *information ordering* of S: $x \leq y$ means that every property of x expressible in the topology is also a property of y, so that y has (or encapsulates) all the information that x has.

In practice we will almost always want to identify any two points which have the *same* information content; thus, spaces of computational interest are typically such that their specialization preorder is a partial order. In a moment we will introduce this condition formally as the 'T_0 separation property'. In mathematical practice a much stronger condition is typically imposed: namely that the specialization order is trivial or, in other words, is the identity relation ('T_1 separation': see below).

Definition 4.1.2. A space X is said to be T_0 if, for any two distinct points x, y of X, there is an open set which contains one, but not both, of the points. The space is T_1 if, for any pair of distinct points x, y, there is an open set which contains x but not y. Finally, the space is T_2 (or *Hausdorff*, or just *separated*) if, whenever x, y are distinct points, there are disjoint open sets containing x, y respectively.

Proposition 4.1.3.

1. *A space X is T_0 iff the specialization order of X is a partial order.*

2. For any space X, the following are equivalent:

    (a) X is T_1

    (b) the specialization order $\leq_X$ is the identity

    (c) every singleton subset of X is closed.

Proof.   1. Trivial.

  2. We consider just the equivalence of (a) and (c). The condition that X is T_1 can be formulated as follows: for any $y \in X$, and $x \neq y$, $X - \{y\}$ is a neighbourhood of x. But this just says that $X - \{y\}$ is open, that is, $\{y\}$ is closed.

■

Examples 4.1.4. All the spaces considered in Section 2 are T_0. Any space of the form $\text{Max}(D)$ (Section 2.3) is T_1 — Exercise 2.7.1(4). For an example of a space $\text{Max}(D)$ that is T_1 but not T_2, see Exercise 4.4.6(1). Stone spaces, $\mathbb{R}$, and spaces $\text{Max}(D)$ for reasonably well-behaved D are T_2 (for clarification of this last point, see Exercise 2.7.1(4)). Spaces that are T_1 but not T_2 tend to be rather artificial. Relatively straightforward examples result from the following considerations: by the axioms for closed sets, a space X is T_1 iff every *finite* subset of X is closed. But, for any set X, the finite subsets of X together with X itself evidently constitute the closed sets of a topology on X, which is thus the *least* T_1 topology on X; it is called the *cofinite* topology on X (since open sets are cofinite). Now it is evident that, if X is infinite, the cofinite topology is not T_2.

A useful consequence of the Hausdorff property is the unicity of limits:

Proposition 4.1.5. *A space X is Hausdorff iff it is the case that every filter over X converges to at most one point.*

Proof. ONLY IF: let x, y be any two distinct points of the Hausdorff space X, and let O, O' be disjoint neighbourhoods of x, y respectively. Then a filter $\mathcal{F}$ cannot converge to both x and y, since then we would have $O \in \mathcal{F}, O' \in \mathcal{F}$ and so $\varnothing = O \cap O' \in \mathcal{F}$.

IF: suppose that X is not Hausdorff. Let x, y be points of X such that $x \neq y$, while every neighbourhood of x meets every neighbourhood of y. Let $\mathcal{F}$ be the collection of all such meets; that is; $A \in \mathcal{F}$ iff $A = B \cap C$ where B, C are neighbourhoods of x, y respectively. Then $\mathcal{F}$ is a filter (as is easily checked) and $\mathcal{F}$ converges to each of x, y. ■

Further 'separation' conditions arise by requiring that (certain) disjoint pairs of *subsets* of a space be separable by open sets. Considering closed sets in this context, we obtain:

Definition 4.1.6. A space X is said to be *regular* if, for every $x \in X$ and closed $Q \subseteq X$ such that $x \notin Q$, there are disjoint open sets, one containing x and the other containing Q. (Equivalently, a space is *regular* if every neighbourhood of a point contains a closed neighbourhood of the point.) A regular T_1 space is said to be T_3. The space is *normal* if any two disjoint closed sets can be separated by open sets; a normal T_1 space is T_4.

Evidently, we have $T_4 \Rightarrow T_3 \Rightarrow T_2 \Rightarrow T_1 \Rightarrow T_0$.

As already mentioned, some of the standard separation conditions (which include certain conditions not defined here, such as $T_{3\frac{1}{2}}$ and T_5) are of limited significance in the computer science context. Notice, for example, that a regular T_0 space is necessarily T_1, and therefore T_3; since there is little reason to consider non-T_0 spaces, there is not much point in distinguishing explicitly between regular and T_3 spaces (in fact, regular spaces are sometimes required to be T_1 by definition, so that regular $\equiv T_3$). A more substantial observation is that, for second-countable spaces (which include all spaces of computational significance) the separation conditions from T_3 upwards are all equivalent, collapsing in fact to metrizability (see Section 6). Moreover, naturally occurring T_1 spaces that are not T_3 seem to be fairly rare (although in mathematics there is the important example of the Zariski topology, used in algebraic geometry). The hierarchy of separation conditions can thus be seen to play rather a small role in the theory of computation. We would be inclined to place more weight on the class of *sober* spaces (defined below) and, within that, of metrizable spaces.

4.2 Continuous functions

Definition 4.2.1. Suppose that X, Y are topological spaces, and that $f : X \to Y$. Then f is *continuous* at a point $x \in X$ if, for any neighbourhood N of $f(x)$, there is a neighbourhood M of x such that $f(M) \subseteq N$. If f is continuous at every point of X, we say simply that f is *continuous*. Equivalently, f is *continuous* if, for every set O that is open in Y, $f^{-1}(O)$ is open in X.

Functions considered in topology are almost always assumed to be continuous. From our point of view, continuity is important since it can be seen as a necessary condition for computability. For f to be computable, it must be the case that, to obtain finite information about the 'result' $f(x)$ — that is, to come to know that $f(x)$ possesses a particular finitely observable property — it suffices to have finite information about the 'datum' x. Hence the requirement of continuity at x. Alternatively, we may argue for the second formulation of continuity as follows. A computable function should compose with a finitary test to give a finitary test. That is, if O is a (finitely) observable property over Y, which we may consider as a (finitary) test, or partial map to $\{tt\}$, then by composing with the computable function $f : X \to Y$, we have that $f^{-1}(O)$ is observable over X: namely, to 'test' $x \in f^{-1}(O)$, one computes $f(x)$ and tests $f(x) \in O$. Hence the formulation of continuity as preservation of open sets by the inverse function.

Since, for any function $f : X \to Y$, the inverse function $f^{-1} \colon \mathcal{P}Y \to \mathcal{P}X$ preserves complement (indeed, all Boolean operations), we see at once that continuity of f can equally well be formulated as preservation of *closed* sets by f^{-1}. Many further useful formulations can be given. Certainly, each of the alternative formulations of topology considered in the preceding section gives rise to a natural formulation of the continuity notion. The next two propositions provide a sample of these ideas. It will be convenient now to use $\overline{S}$ as an abbreviation for $\mathrm{Cl}(S)$.

Proposition 4.2.2.

1. *A map $f : X \to Y$ is continuous at $x \in X$ iff, for any $S \subseteq X, x \in \overline{S} \Rightarrow f(x) \in \overline{f(S)}$.*

2. *f is continuous iff, for any $S, f(\overline{S}) \subseteq \overline{f(S)}$ (equivalently, iff $\overline{f^{-1}(T)} \subseteq f^{-1}(\overline{T})$ for any $T \subseteq Y$).*

Proof.   1. Suppose that f is continuous at x, and that $f(x) \notin \overline{f(S)}$. Then the inverse image of the neighbourhood $Y - \overline{f(S)}$ of $f(x)$ is a neighbourhood of x, thus $x \notin \overline{S}$. This proves the necessity of the condition. For sufficiency, suppose that f is not continuous at x. Then there is a neighbourhood T of $f(x)$ such that $f^{-1}(T)$ is not a neighbourhood of x. This means that we have $S \subseteq X$ disjoint from $f^{-1}(T)$ with $x \in \overline{S}$ (S can be taken as $X - f^{-1}(T)$); but then $f(S)$ is disjoint from T and so $f(x) \notin \overline{f(S)}$. Thus the condition fails.

2. An immediate consequence of (1), noting that $f(\overline{S}) \subseteq \overline{f(\overline{S})}$ is equivalent to: $\forall x \in \overline{S}.f(x) \in \overline{f(S)}$. ∎

For the next proposition, note that if $\mathcal{F}$ is a filter over X, and $f : X \to Y$, then the collection $\mathcal{B} = \{f(A) \mid A \in \mathcal{F}\}$ is a filter-base over Y (that is, $\forall B, B' \in \mathcal{B} \exists C \in \mathcal{B}.C \subseteq B \cap B'$). By abuse we will denote by $f(\mathcal{F})$ the filter generated by $\mathcal{B}$ (that is, the collection $\{C | \exists B \in \mathcal{B}.B \subseteq C\}$).

Proposition 4.2.3. *The function $f : X \to Y$ is continuous at x iff, for every filter $\mathcal{F}$ which converges to $x, f(\mathcal{F})$ converges to $f(x)$.*

Proof. Continuity at x is evidently equivalent to the assertion that $f(\mathcal{N}(x))$ refines the neighbourhood filter at $f(x)$ — that is, $f(\mathcal{N}(x)) \to f(x)$. But this implies $f(\mathcal{F}) \to f(x)$ for any filter $\mathcal{F}$ which refines $\mathcal{N}(x)$. ∎

The next few propositons concern mainly the interaction between continuity and the specialization order.

Proposition 4.2.4. *Suppose that* $f : X \to Y$ *is continuous. Then* f *is monotonic with respect to the specialization orders of* X, Y.

Proof. Suppose $x \leq_X y$, that is, $x \in \overline{\{y\}}$. By Proposition 4.2.2, continuity of f yields $f(x) \in \overline{\{f(y)\}}$, that is, $f(x) \leq_Y f(y)$. ∎

'Continuity implies monotonicity' is of course a familiar lemma of elementary cpo theory. Proposition 4.2.4 is indeed a generalization of this lemma, via the Scott topology of a cpo, as will be apparent in a moment. Let us distinguish by the prefix 'order' - the continuity notion usual in cpo theory: a map $f : D \to E$, where D, E are dcpo's, is *order-continuous* if f preserves lubs of all directed subsets of D. We will establish the standard result that order-continuity is equivalent with Scott-continuity; with a view to later developments, however, our approach will be somewhat more elaborate than is usual.

Proposition 4.2.5. *Let* $(D, \sqsubseteq)$ *be a dcpo. The Scott topology of* D *is the finest topology on* D *enjoying the following properties:*

(1) *for all* $x \in D, \overline{\{x\}} = \downarrow x$ *(that is, the specialization order coincides with* $\sqsubseteq$*);*

(2) S *directed,* $\bigsqcup S \in O, O$ *open* $\Rightarrow S \cap O \neq \emptyset$ *(equivalently:* S *directed,* $S \subseteq Q, Q$ *closed* $\Rightarrow \bigsqcup S \in Q$*).*

Proof. For any x the set $\downarrow x$, as a lower set closed under directed sups, is Scott-closed. Since any Scott-closed set which contains x must contain $\downarrow x, \downarrow x$ is the Scott-closure of $\{x\}$. Thus the Scott topology satisfies (1); and it satisfies (2) by definition. Further, it is the finest such topology; indeed it is by definition the finest topology satisfying (2) together with a condition weaker than (1), namely

(1)′ *every closed set* Q *is a lower set* $(Q = \downarrow Q)$. ∎

Definition 4.2.6. *A topology for a dcpo* D *will be called* order-consistent *if it satisfies the conditions (1), (2) of Proposition 4.2.5.*

We shall argue in due course that every computationally meaningful topology is an order-consistent topology for an appropriate dcpo. For the moment, we notice that the defining conditions for order-consistency can be expressed more concisely:

Proposition 4.2.7. *A topology for a dcpo* $(D, \sqsubseteq)$ *is order-consistent iff, for every directed* $S \subseteq D$, $\overline{S} = \downarrow \bigsqcup S$.

Proof. IF: immediate.

ONLY IF: suppose that (1), (2) of 4.2.5 hold, and that S is directed. By (2), any closed set which contains S contains $\bigsqcup S$. But by (1), $\downarrow \bigsqcup S$ is the smallest closed set containing $\bigsqcup S$; it is therefore the closure of $\bigsqcup S$. ∎

Proposition 4.2.8. *Let D, E be dcpo's, and $f : D \to E$.*

     1. *If f is continuous with respect to (arbitrary) order-consistent topologies on D, E, then f is order-continuous.*

     2. *If f is order-continuous, then it is Scott-continuous.*

Proof.   1. Suppose that f is continuous w.r.t. (given) order-consistent topologies. Since the specialization order coincides with $\sqsubseteq$ on each of D, E, we have by Proposition 4.2.4 that f is monotonic. By proposition 4.2.8 (and 4.2.2), for any directed $S \subseteq D, f(\downarrow \bigsqcup S) \subseteq \downarrow \bigsqcup f(S)$, and hence $f(\bigsqcup S) \sqsubseteq \bigsqcup f(S)$. In conjunction with monotonicity, this gives order-continuity.

     2. If f is order-continuous, we easily check that the inverse image of a Scott-closed subset of Y is Scott-closed in X.   ∎

Corollary 4.2.9. *A map between dcpo's is order-continuous iff it is Scott-continuous.*

Rather than give numerous examples of continuous functions at this point, we merely note that continuity is evidently preserved by composition, and that identity maps are continuous, so we have a category:

Definition 4.2.10. The category **Top** of *spaces* has topological spaces as objects and continuous functions as morphisms. A *homeomorphism* is an isomorphism in **Top**; that is, spaces X, Y are homeomorphic iff there are continuous maps f: $X \to Y, g : Y \to X$ such that $g \circ f = Id_X$ and $f \circ g = Id_Y$

4.3   Predicate transformers and sobriety

A continuous function from X to Y gives rise to a map in the 'opposite' direction, from the topology (lattice of open sets) of Y to that of X. This apparently trite observation leads to an extremely rich theory — indeed to an entire approach to the subject — so we shall take the trouble to state it more formally. Recall that ΩX is the topology of a space X, regarded as a complete lattice. Although ΩX is complete as a lattice, joins and meets do not have equal status. For arbitrary joins, but only finite meets, in ΩX coincide, in general, with the corresponding set operations (union and intersection), and it is only these that we can expect to be well-behaved.

Notably, we have distributivity of *finite* meets over arbitrary joins, which we can express by:

$$a \wedge \bigvee S = \bigvee \{a \wedge s | s \in S\} \tag{1}$$

Definition 4.3.1. A complete lattice satisfying the infinite distributive law (1) is known as a *frame* (or *locale*).

If $f : X \to Y$ is any map, then f^{-1}, considered as a map from $\mathcal{P}Y$ to $\mathcal{P}X$, preserves all unions and intersections (as well as complements); so if f is continuous, the inverse restricted to ΩY, which we shall denote by Ωf, preserves joins and finite meets (since these coincide with the set-theoretic operations. In view of this, we take as frame *(homo)morphisms*, the maps which preserve joins and finite meets. Thus we can rephrase the opening sentence of this paragraph:

a continuous function $f : X \to Y$ gives rise to a frame homo-morphism $\Omega f : \Omega Y \to \Omega X$.

Now, in the computing context, a map of type $\Omega Y \to \Omega X$, and in particular a frame morphism, is known as a *predicate transformer*. At this point it will be objected that in the classical sources on predicate transformers, especially [Dijkstra, 1976], there is no mention of topology, open sets, or frames; so the remark certainly needs a little amplification. Our claim is, roughly, that the weakest precondition operator, wp, is a special case of the operator Ω (see below for the definition of wp); and that Ω really *is* the appropriate generalization when one wants to work with domains of computation that are more complex (in terms of information ordering) than those considered in, say [Dijkstra, 1976]. To give substance to this claim, we may start with the idea of a program as represented by a 'state transformer': a transformation of type $\Sigma \to \Sigma$ (Σ a set of states). There are two complications. First, the program may be 'non-deterministic'. To represent this adequately, the state transformer would have to be a many valued function (or relation). We have not yet considered many-valued functions (we will do so briefly at the end of this section), so for the moment we exclude this possibility: the program is to give at most one result for a given initial state. Secondly, the program may fail to terminate for some initial states. The standard way to accommodate this is to expand Σ to the 'flat domain' $\Sigma_\perp$, non-termination now being represented by $\perp$. It is true that Dijkstra does not do this explicitly, but the reasons for doing so are fairly compelling, particularly when we *do* admit non-determinism: for we need somehow to distinguish between the program which, say, always terminates in state σ, and a program of which some executions terminate in σ while others (the remainder) fail to terminate (see, for example, de Roever [de Roever, 1976]). Thus our program is represented by $f : \Sigma \rightarrow \Sigma_\perp$. As topologies we naturally have for Σ the discrete topology, and for $\Sigma_\perp$ the Scott topology $\mathcal{P}(\Sigma) \cup \{\Sigma_\perp\}$; thus we get a frame morphism $\Omega f : \Omega(\Sigma_\perp) \rightarrow \Omega\Sigma$. Recall [Dijkstra, 1976] that, if P is any 'predicate' over Σ, the predicate wp(f, P) is true of $x \in \Sigma$ just in case the program f, when executed with the initial state as x, is guaranteed to terminate in a state satisfying P. By simply comparing definitions, we see that Dijkstra's wp$(f, -)$ is the restriction of Ωf to $\mathcal{P}(\Sigma)$; equivalently, Ωf is wp$(f, -)$ trivially extended by the case $\Omega f(\Sigma_\perp) = \Sigma$. Dijkstra's 'healthiness conditions' also have their counterpart: they correspond to the condition that a frame map be a *homomorphism*. (The healthiness conditions do not read quite the same as the homomorphism condition. The differences are due to our restriction to deterministic programs, and to the replacement of Σ by $\Sigma_\perp$ as the result space.) That Ω (or rather, a non-deterministic modification of Ω) is the appropriate generalization of wp was argued for by the present author [Smyth, 1983b] on mathematical grounds, and on the ground that it provides a basis for more adequate programming logics. The latter claim has since been substantiated, with a wealth of detail, by Abramsky with his 'Logic of Domains' [Abramsky, 1991].

To what extent, or under what circumstances, is a continuous map characterizable by (or determined by) a frame morphism? The question is certainly a natural one, given that we are interested in characterizing programming constructs by predicate transformers (cf. [Dijkstra, 1976]). For *unicity* (of a continuous map yielding a given frame morphism) the answer is very simple: we need only to assume that the spaces involved are T_0.

Proposition 4.3.2. *Suppose that Y is T_0, and that f, g are distinct continuous maps from X to Y. Then $\Omega f \neq \Omega g$.*

Proof. Let $x \in X$ be such that $f(x) \neq g(x)$. Then $\Omega f(O), \Omega g(O)$ are distinct, since exactly one of them contains x. ∎

The question of existence is much more subtle. We can easily see that there may be no continuous map yielding a given frame morphism, even when the spaces are T_0. For example, let X be the one point space, and Y be $\omega = 0 \leq 1 \leq 2 \leq \ldots$ with the Alexandroff topology. Define $\phi : \Omega Y \rightarrow \Omega X$ by:$\phi(O) = X$ for O non-empty, $\phi(\varnothing) = \varnothing$. Then ϕ is obviously a homomorphism. However there is no corresponding point map: if $f : X \rightarrow Y$ is given by $f(\cdot) = k$, then $\Omega f(\uparrow (k+1)) = \varnothing$, and so $\Omega f \neq \phi$. One has the impression that there is a point "missing" from Y; we would like to add the infinite point ω to it (making it a cpo!) and take $f(\cdot) = \omega$. This example can be developed so as to suggest a general solution to the problem. Notice that ΩX, for X the one point space, is the lattice $2(= 0 \leq 1)$. For a space Y to be satisfactory, it is required that for every frame homomorphism $p : \Omega Y \rightarrow 2$ there be a point $y \in Y$ such that p is $\Omega(\lambda x.y)$. Now the open neighbourhoods of the point y (if it exists) are precisely the elements of ΩY which are mapped to 1 by p. On the other hand, we see that a map $\phi : \Omega Y \rightarrow 2$ (indeed, a map from any frame into 2) is a frame morphism iff $\phi^{-1}(1)$ is a completely prime filter, where a filter F in a complete lattice L is said to be *completely prime* if, for any $a \in F$ and $S \subseteq L, a \leq \bigvee S \Rightarrow \mathcal{F} \cap S \neq \varnothing$. (Main step in the argument: suppose ϕ is a frame morphism, $\phi(a) = 1$, and $a \leq \bigvee S$. Then $\phi(\bigvee S) = 1$; hence $\phi(s) = 1$ for some $s \in S$, since if $\phi(s) = 0$ for all $s \in S$ we would have $\phi(\bigvee S) = 0$ by preservation of joins.) Thus we have the following necessary and sufficient condition for the existence of a continuous map from X to Y corresponding to each frame morphism from ΩY to ΩX, in the case that X is the one point space: every completely prime filter in ΩY is the (open) neighbourhood filter of some point of Y.

We pause here for some remarks on neighbourhood filters in general (*open* neighbourhoods being understood here). An easy check shows that, in any space Y, the neighbourhood filter of any point is completely prime (over ΩY). In terms of the discussion of filters and theories, or specifications, in Section 2.4, we can think of a completely prime filter as a specification that is as precise as possible, in the sense that it admits no ambiguity: if a disjunction $\bigvee S$ is in the specification, then some disjunct s is also in it. A T_0 space is a space in which distinct points cannot have the same neighbourhood filter (or specification!). We see that the T_0 condition, for a space Y, is also equivalent to: every completely prime filter over ΩY is the neighbourhood filter of at most one point.

At this point the following definition is appropriate:

Definition 4.3.3. A space is *sober* if every completely prime filter of open sets is the neighbourhood filter of a unique point. (See also Exercise 4.4.6(4).)

Thus a sober space is one which is determined (up to homeomorphism) by its frame — namely as the space of completely prime filters over the frame. We shall see in a moment that the space $Pt(L)$ of completely prime filters over a frame L is always sober, whether or not L is the frame of a space.

Lemma 4.3.4. Let $\phi : L \rightarrow M$ be a frame morphism, and $\mathcal{F}$ a completely prime filter over M. Then $\phi^{-1}(\mathcal{F}) = \{a \mid \phi(a) \in \mathcal{F}\}$ is a completely prime filter (over L).

Proof. By direct verification, or by noting that $\phi^{-1}(\mathcal{F})$ is the inverse image of 1 under the composite $L \xrightarrow{\phi} M \xrightarrow{\pi} 2$, where π is the homomorphism corresponding to $\mathcal{F}$. ∎

In the following (Proposition 4.3.5, Lemma 4.3.6), L is an arbitrary frame, and $c : L \to \Omega Pt(L)$ is defined by $c(a) = \{x \in Pt(L) | a \in x\}$. Thus, for any $a \in L$, a point $x \in Pt(L)$ has $c(a)$ as a neighbourhood iff $a \in x$.

Proposition 4.3.5.

  1. c is a surjective frame map.

  2. $Pt(L)$ is sober.

Proof. 1. That c preserves finite meets is trivial. For join, we have: $x \in c(\bigvee S)$ iff $\bigvee S \in x$ iff $s \in x$ for some $s \in S$ (since x is completely prime) iff $x \in c(s)$ for some $S \in s$. It then follows that c is surjective, since every open set of $Pt(L)$ is a join of basic open sets $c(a)$.

  2. Suppose $x \in Pt(L), \mathcal{F} \subseteq \Omega Pt(L)$. As an immediate consequence of definitions, we have: $\mathcal{F}$ is the set of neighbourhoods of x of the form $c(a)$ iff $x = c^{-1}(\mathcal{F})$. Moreover, since c is surjective, we can omit the qualification 'of the form $c(a)$' from the preceding statement. Now assume that $\mathcal{F}$ is a completely prime filter. By the preceding lemma, $c^{-1}(\mathcal{F})$ is completely prime. Thus $\mathcal{F}$ is the neighbourhood filter of a unique point, namely $c^{-1}(\mathcal{F})$. ∎

Recalling at last the question with which we began the current discussion, namely (in effect) 'When can a frame morphism stand duty for a continuous map?', a simple answer is at hand (utilizing Lemma 4.3.4): when the map is between sober spaces. However we shall now see that, with only a little further effort, a much more comprehensive and satisfying answer can be given.

Lemma 4.3.6. *Let X be a space and $\phi : L \to \Omega X$ a frame morphism. Then there is a unique continuous map $\overline{\phi} : X \to Pt(L)$ such that $\Omega\overline{\phi} \circ c = \phi$.*

Proof. The requirement which $\overline{\phi}$ has to satisfy can be stated as:

$$\overline{\phi}(x) \in c(a) \Leftrightarrow x \in \phi(a).$$

By the definition of c, this can be restated:

$$a \in \overline{\phi}(x) \leftrightarrow x \in \phi(a) \qquad (1)$$

This determines $\overline{\phi}$ uniquely as $x \mapsto \phi^{-1}(\mathcal{N}_0(x))$, where $\mathcal{N}_0(x)$ is the filter of open neighbourhoods of x (in X). Notice that (1) also shows that $\overline{\phi}$ is continuous, since $\{y|a \in y\}$, where $a \in \overline{\phi}(x)$, is a typical basic neighbourhood of $\overline{\phi}(x)$. ∎

It has been implicit in our work up to this point that the 'operator' Ω is actually a functor; and the import of Lemma 4.3.6 is that Pt also defines a functor, (right-)adjoint to Ω. In order to make this precise, we must of course specify the range of Ω:

Definition 4.3.7. The category of frames (and frame morphisms) is denoted **Frm**; the category **Loc** of *locales* is the opposite of **Frm**.

The point is that **Loc** has its arrows in the 'same direction' as those of **Top**, thus making the comparison of the categories more convenient. In particular, we can (and do) consider Ω as an ordinary (covariant) functor from **Top** to **Loc**, whereas considered as a functor from **Top** to **Frm** it would be contravariant. That being so, we deduce in the usual way from Lemma 4.3.6:

Theorem 4.3.8. *The assignment $L \mapsto Pt(L)$ is the object part of a functor Pt: **Loc** → **Top**, right adjoint to Ω.* ∎

A sober space is, in effect, one which can be recovered from its frame (or locale). That is, a space X is sober iff $X \cong Pt \circ \Omega(X)$. The corresponding notion for locales is given by:

Definition 4.3.9. A locale L is *spatial* if $L \cong \Omega \circ Pt(L)$.

Both sobriety and spatiality can be considered as notions of *completeness* — though in rather different senses. Sobriety is quite analogous to completeness of metric and uniform spaces — or of partial orders. On this last point, the example of the space ω (discussion following Proposition 4.3.2) is typical.

Definition 4.3.10. The *sobrification* of a space X is $Pt \circ \Omega(X)$.

Adopting the above, we have: ideal completion of a poset (as in cpo theory) is a special case of sobrification (Exercise 4.4.6(6)). Moreover, it can be shown that sobrification and the completion of uniform spaces are both instances of a still more general process of completion Smyth (1987a). In contrast with this, as it were, mathematical notion of completeness, spatiality of a locale is analogous to the completeness of a logic. This can be seen as follows. A *satisfaction* relation (cf. Example 3.4.6) for a locale L is a binary relation $\vDash \subseteq W \times L$ satisfying:

$$w \vDash \bigwedge S \text{ (S finite)} \iff w \vDash s \text{ for each } s \in S;$$

$$w \vDash \bigvee S \qquad\qquad \Leftrightarrow w \vDash s \text{ for some } s \in S.$$

This gives us a relation of 'semantic entailment' for $L : a \vdash b$ iff, for every satisfaction relation $(W, \vDash)$ for L and $w \in W$, if $w \vDash a$, then $w \vDash b$. Then it is not difficult to show that the semantic entailment of L coincides with the ordering of L iff L is spatial. (In outline: L is spatial iff the ordering of L agrees with the inclusion over $Pt(L)$. But $Pt(L)$ serves as a 'canonical model' for L: observe that $Th(w)$, where $w \in W$ and $(W, \vDash)$ is any satisfaction relation for L, is always a completely prime filter.) For an extended treatment of 'locales as logics' one may consult [Vickers, 1989].

Dualities between spaces and lattices (regarded as 'logics') — Stone dualities, broadly conceived — are assuming increasing importance in the theory of computation. Most of them may be regarded as arising by suitable restriction of the adjunction 4.3.8 . These 'dualities' are, of course 'equivalences' if one works with **Loc** rather than **Frm**. The equivalence between sober spaces and spatial locales is just the broadest of these.

Proposition 4.3.11. *The adjunction of Theorem 4.3.8 restricts to an equivalence of categories between the (full) subcategory* **Sob** $\subseteq$ **Top** *of sober spaces and the subcategory* **SLoc** $\subseteq$ **Loc** *of spatial locales.* ∎

Returning now to the connection between sobriety and cpo's, we have the following striking result:

Proposition 4.3.12. *Every sober space is, with respect to its specialization order, a dcpo with order-consistent topology.*

Proof. In any complete lattice, the union of a directed collection of completely prime filters is evidently a completely prime filter. Now if S is a directed (w.r.t. the specialization order) subset of the sober space X, then $\{\mathcal{N}_0(x) \mid x \in S\}$ is a directed collection of (open) neighbourhood filters. Thus $\mathcal{F} = \bigcup_{x \in S} \mathcal{N}_0(x)$ is completely prime; the point at which $\mathcal{F}$ is the neighbourhood filter is clearly the lub of S.

For order-consistency, we have only to check condition (2) of Proposition 4.2.5 ((1) being satisfied by definition). But we have just seen that, for any directed set S in X, O is a neighbourhood of $\bigvee S$ if it is a neighbourhood of some $s \in S$. ∎

The converse is false: in fact, by an example of Johnstone [Johnstone, 1981a] a cpo may fail to be sober even in its Scott topology.

It has been argued by Melton and Schmidt [Melton and Schmidt, 1986] that, in certain computational contexts, the Scott topology is too restrictive, and that dcpo's with order-consistent topology provide a more satisfactory framework. We argue that computationally reasonable spaces are sober. Proposition 4.3.12 shows that our view does not contradict that of Melton and Schmidt, but refines it: 'good' spaces are sober, and hence are dcpo's with order-consistent topology. (In searching for the convenient category for computing, we may of course need to refine further, that is, to select some proper subcategory of **Sob**.)

Before leaving the topic of sobriety, we should consider how it relates to the separation properties. We already know that sober implies T_0.

Lemma 4.3.13. *Let $\mathcal{F}$ be a completely prime filter of open sets of a space X. Then $\mathcal{F}$ is convergent. (Strictly speaking, $\mathcal{F}$ is not necessarily a filter, but only a filter-base, over X. But the definition of convergence is the same.)*

Proof. Suppose the conclusion is false. Then every point of X has an open neighbourhood which is not a member of $\mathcal{F}$. But then the union of all these neighbourhoods, namely X, is not a member of $\mathcal{F}$, which is absurd. ∎

Proposition 4.3.14. *Every Hausdorff space is sober.*

Proof. Let $\mathcal{F}$ be a completely prime filter of open sets of the Hausdorff space X. By Lemma 4.3.13, we have $\mathcal{N}_0(x) \subseteq \mathcal{F}$ for some $x \in X$. But the reverse inclusion also holds. For suppose, if possible, that $O \in \mathcal{F}$ while $x \notin O$. By the Hausdorff property, we have for every $y \in O$ an open neighbourhood O_y of y disjoint from some open neighbourhood of x. But this is impossible, since by complete primality some O_y is a member of $\mathcal{F}$. ∎

This leaves the relation with T_1 separation to be settled. Of course, sobriety does not imply the T_1 property. The reverse implication also fails: an infinite set equipped with the cofinite topology (Examples 4.1.4) is T_1, but is not sober since the collection of all non-empty open sets is a completely prime filter which is not the neighbourhood filter of any point. Thus sobriety and the T_1 property are incomparable.

4.4 Many-valued functions

Many-valued functions are significant for us as the denotations of non-deterministic programs. Following Dijkstra, we seek to characterize them by predicate transformers. The definition of the weakest precondition operator wp given above (following Definition 4.3.1) takes account of the possibility of non-determinism; it reflects what is often called the 'total correctness' aspect of the logic of programs, as opposed to the 'partial correctness' aspect which we will touch on later. Assuming that predicates are to be identifed with open sets, we are led to the following notion of continuity for many-valued functions (multifunctions). Notation: if $h : X \to Y$

is a multifuction, then $h^+(S)$ is $\{x \mid h(x) \subseteq S\}$ for $S \subseteq Y$, while h^- is the relational inverse, so that $h^-(S)$ is $\{x \mid h(x) \cap S \neq \varnothing\}$.

Definition 4.4.1. A multifunction $h : X \to Y$ is *upper semicontinuous* (usc) if $h^+(O)$ is open in X whenever O is open in Y. The operator taking a usc multifunction $h : X \to Y$ to the 'inverse' mapping of frames is denoted Ω^+; that is, we have $\Omega^+ h : \Omega Y \to \Omega X$ where $\Omega^+ h(O) = h^+(O)$.

As before, we can put: $\mathrm{wp}(f, P) = \Omega^+ f(P)$. Any predicate transformer ϕ which is of the form $\Omega^+ f$ for some (usc) function $f : X \to Y$ clearly satisfies the conditions:

1. $P \subseteq Q \Rightarrow \phi(P) \subseteq \phi(Q)$

2. $\phi(P \cap Q) = \phi(P) \cap \phi(Q)$

3. $\phi(\varnothing) = \varnothing$

4. $\phi(\textit{true }) = \textit{ true}$ (ie $\phi(Y) = X$),

of which (1) is technically redundant (it is a consequence of (2)). Let us compare what we have so far with Dijkstra's analysis of wp ([Dijkstra, 1976] or [Gries, 1981]). To make the comparison we need to take $X = \Sigma, Y = \Sigma_\perp$ (Σ the set of states). Since Dijkstra does not work with $\perp$, his wp is in effect the restriction of ours to $P(\Sigma)$. This explains why, for Dijkstra, $\mathrm{wp}(f, -)$ does not need to satisfy condition (4). But there is an obvious $(1,1)$ correlation between the predicate transformers over $P(\Sigma)$ which satisfy (1)–(3) and the predicate transformers over $\Omega(\Sigma_\perp)$ which satisfy (1)–(4).

The notion of upper semi-continuity is not needed in Dijkstra's treatment, since Σ has the discrete topology: *every* map with domain Σ is upper semi-continuous. Later we will see some non-vacuous applications of the usc condition.

Now we have to take into account the very important condition of *bounded non-determinacy* ([Dijkstra, 1976], Chapter 9). The activity of a computing mechanism for a given initial state s can be represented by a tree having its root labelled s and with nodes along a path labelled by successive states of the execution, where branching notes represent (non-deterministic) choice points. Since no realizable mechanism can, in one step, make a choice between infinitely many possiblities, the tree may be assumed finitary. By König's Lemma we can infer that, if every execution sequence terminates, the set of possible output states (labels of the tree) is finite. Thus, reasonable state transformers $f : \Sigma \to \Sigma_\perp$ are such that either $\perp \in f(s)$ or $f(s)$ is finite. Can this be expressed as a condition on $\Omega^+(f)$?

In fact, it is easy to see that it can: it is precisely the condition that $\Omega^+(f)$ be Scott-continuous, that is, that it preserve directed joins of open sets. To see this, assume first that f is of bounded non-determinacy. Suppose that $s \in \phi(\bigcup \mathcal{L})$, where $\phi = \Omega^+ f$ and $\mathcal{L}$ is a directed collection of predicates (open subsets of $\Sigma_\perp$) such that $\bigcup \mathcal{L} \neq \Sigma_\perp$. Then $s \in \phi(B)$ for some finite $B \subseteq \bigcup \mathcal{L}$ (by bounded non-determinacy), and so $s \in \phi(P)$ for some $P \in \mathcal{L}$. The case that $\bigcup \mathcal{L} = \Sigma_\perp$ is trivial, since in that case some member of $\mathcal{L}$ contains $\perp$, that is, $\Sigma_\perp \in \mathcal{L}$. Again, if f is not of bounded non-determinacy, so that $f(s)$ is an infinite subset of Σ for suitable s, we see that ϕ is not Scott-continuous by considering a directed collection of proper subsets of $f(s)$ whose join is $f(s)$. We thus have condition

5. ϕ is Scott-continuous,

a weakening of the corresponding condition, considered above, for deterministic transformers (preservation of arbitrary joins).

Besides wp there is, according to Dijkstra, another operator which is needed for the full characterization of non-deterministic systems: the weakest *liberal* precondition, wlp. The idea is that an initial state s satisfies $\text{wlp}(f, P)$ provided that the system f is guaranteed not to terminate in a state *not* satisfying P. It seems best to view wlp as being concerned with safety properties, that is, with *closed* sets, and the following notion of continuity is appropriate.

Definition 4.4.2. A multifunction $h : X \to Y$ is *lower semicontinuous* (lsc) if $h^+(Q)$ is closed in X whenever Q is closed in Y. (See also Exercise 4.4.6(9).)

Thus a lower semicontinuous map h gives rise to a 'predicate transformer' $\psi (= h^+)$ taking *closed* sets to closed sets, and we can define wlp by: $\text{wlp}(h, Q) = \psi(Q)$. It is an elementary exercise to establish the equivalence of this formulation with that of Dijkstra, in the case that we are working with discrete state sets. Of course, the condition of lower semicontinuity is itself vacuous in the discrete case, but we will see a non-vacuous instance of it shortly. In the discrete case one may remark, as does Dijkstra, that wlp is weaker than wp: if $P \subseteq \Sigma$, then $\text{wp}(f, P) \subseteq \text{wlp}(f, P \cup \{\perp\})$ (here, of course, we have had to take the second argument of wlp as the closure of P in $\Sigma_\perp$, whereas in Dijkstra's formulation we simply have $\text{wlp}(f, P)$). In the general case, however, wp and wlp (as we have defined them) are incomparable. For a view of wlp as having to do with 'partial correctness', in contrast with the 'total correctness' aspect of wp, see Exercise 4.4.6(8).

The three conditions on many-valued functions which we have obtained by trying to put Dijkstra's ideas in a general setting, namely upper and lower semi-continuity and (a suitable formulation of) bounded non-determinacy are fairly standard in topology. But the topic is rarely discussed in textbooks (as opposed to the journal literature): exceptions are [Berge, 1959; Kuratowski, 1961]. Notice that bounded non-determinacy appears in [Ber59] as the condition that images of points are compact sets, and is incorporated into the definition of upper semicontinuity. The formulation in terms of compactness is not yet available to us, but we will see (in Exercise 4.4.6(13) and remarks following Definition 7.1.1) that it is equivalent to condition(5). We shall continue here to regard condition (5) as a third continuity condition, additional to upper and lower semicontinuity.

As our example of many-valued functions, we will consider the merging of two streams of data. The following will prove useful:

Definition 4.4.3. Let X_1, Y_2 be topological spaces, and $\leqslant_1, \leqslant_2$ their specialization orders. A multifunction $h : X_1 \to X_2$ is *upper monotonic* if, for any x, y in X_1 with $x \leqslant_1 y$ and any $z \in h(y)$, there exists $w \in h(x)$ such that $w \leqslant_2 z$.

Clearly, upper monotonicity reduces to ordinary monotonicity in the case of single-valued functions.

Proposition 4.4.4. Let $h : X_1 \to X_2$ be an upper semi-continuous multifunction. Then h is upper monotonic.

Proof. Assume $x \leqslant_1 y$. Suppose, if possible, that there is a point $z \in h(y)$ such that, for every $w \in h(x), w \not\leqslant z$. This means that, for each $w \in h(x)$, we have an open neighbourhood O_w of w such that $z \notin O_w$. Take $O = \bigcup_{w \in h(x)} O_w$. Then $h^+(O)$ contains x but not y. This, however, is impossible since $h^+(O)$ is open. ∎

Lower monotonicity may be defined in an exactly analogous way to upper monotonicity. Strangely enough, however, lower semi-continuity does *not* imply lower monotonicity (Exercise 4.4.6(12)).

Example 4.4.5. The merging of two streams of data is an operation which, notoriously, gives rise to difficulties in presenting the semantics of data flow networks. One is assuming here that the merge is to be represented as a process (node of the network) whose possible output streams are all the 'fair' merges of its two input streams. Thus, if Σ is the alphabet of atomic data, we may view the merge as (denoting) the multifunciton $FM : \Sigma^\infty \times \Sigma^\infty \to \Sigma^\infty$, where $z \in FM(x,y)$ iff there exist sequences $(x^i), (y^j)$ of finite words such that $x = x^0 x^1 \ldots, y = y^0 y^1 \ldots$, and $z = x^0 y^0 x^1 y^1 \ldots$. We consider the continuity properties of FM.

First, we can show that FM is lower semi-continuous.

To this end we note that, although FM is not lower monotonic (Exercise 4.4.6(12)), the following restricted case is easily verified: for any $u, v \in (\Sigma^\infty)^2$ with u finite and $u \sqsubseteq v$ and any $w \in FM(u)$, there exists $z \in FM(v)$ such that $w \sqsubseteq z$. Suppose now that Q is a (Scott-)closed subset of

Σ^∞, and that $(<x_i, y_i>)_i$ is an increasing sequence of finite elements of $FM^+(Q)$; we have to show that $\downarrow \bigsqcup_i <x_i, y_i> \subseteq FM^+(Q)$ (to conclude that $FM^+(Q)$ is Scott-closed). Thus, suppose $x \sqsubseteq \bigsqcup x_i, y \sqsubseteq \bigsqcup y_i$. Let z be any element of $FM(x,y)$ and let $x^0 x^1 \ldots, y^0 y^1 \ldots$ be decompositions of x, y into finite strings such that $z = x^0 y^0 x^1 y^1 \ldots$. If $u \sqsubseteq < x, y >$ is finite, we have $u \sqsubseteq < x_i, y_i >$ for some i, so that (by 'restricted lower monotonicity' and the fact that Q is $\downarrow$-closed) $FM(u) \subseteq Q$. It follows that $x^0 y^0 \ldots x^i y^i \in Q$ for all i, so that (since Q is Scott-closed) $z \in Q$. This shows that $< x, y > \in FM^+(Q)$.

On the other hand, FM is not upper semi-continuous. In fact it is not upper monotonic: $001 \in FM(00, 1)$, but 001 is not an extension of any element of $FM(0,1)(= \{01, 10\})$.

Moreover, FM fails on bounded non-determinacy (which we take to be expressed by condition (5) above). To see this, take $\Sigma = \{0, 1\}$, and let X be the set of all infinite binary sequences in which there are occurrences of both 0 and 1. For each positive integer n, let X_n be the set of infinite binary sequences x such that both 0 and 1 occur in the first $n+1$ characters of x. Clearly, $X_1 \subseteq X_2 \subseteq \ldots$, each X_n is open, and $X = \bigcup_n X_n$. Now $FM(0^\omega 1^\omega)$ is, trivially, a subset of X, while it is not a subset of any X_n. Thus $FM^+(X) \neq \bigcup_n FM^+(X_n)$.

Thus the fair merge fails rather badly on our various continuity conditions. The situation is distinctly better if one restricts attention to *infinite* sequences, that is, if we consider $FM^\omega : \Sigma^\omega \times \Sigma^\omega \to \Sigma^\omega$ (the definition being just as for FM). Since FM is lsc, its restriction FM^ω is also (Exercise 4.4.6(10)). FM^ω differs from FM in being usc; the proof is left to the reader. But bounded non-determinacy still fails (same argument as for FM).

Exercises 4.4.6.

1. Give an example of an algebraic dcpo D such that $\mathrm{Max}(D)$ is not T_2. (Suggestion: construct two increasing chains $(a_i), (b_i)$, and another sequence of elements (c_i) such that, for each i, c_i is an upper bound of a_i, b_i, while elements c_i are incomparable with each other.)

2. A space X is said to be $T_{\frac{1}{2}}$ if every singleton subset of X is either open or closed. Verify that $T_1 \Rightarrow T_{\frac{1}{2}} \Rightarrow T_0$. Show that the finite $T_{\frac{1}{2}}$ spaces may be identified with the finite posets in which no chain $a \leq b \leq \ldots$ contains more than two distinct points.

3. Let $2 = \{\bot, \top\}$ be the Sierpinski space, with its (specialization) order $\bot \leq \top$. For any space X, show that the poset of continuous maps from X to 2 (ordered pointwise) is isomorphic with the lattice of open sets of X.// Note: we have already seen, in the discussion following Proposition 4.3.2, that 2 is usefully regarded also as a frame. In the

terminology of [Johnstone, 1982], Chapter VI, 2 is a 'schizophrenic' object of the categories **Top** and **Frm**, and plays a dual rôle with respect to them in generating, respectively, the spatial frames and the sober spaces.)

4. (Alternative definition of *sober*.) Given an arbitrary subset S of a space X, let $\mathcal{M}(\mathcal{S})$ be the collection of all those open sets which meet S. Verify that $\mathcal{C} = \mathcal{M}(\mathcal{S})$ enjoys the following properties:

   (a) $U \in \mathcal{C}, \mathcal{U} \subseteq \mathcal{V} \Rightarrow \mathcal{V} \in \mathcal{C}$

   (b) $U \in \mathcal{C}, \mathcal{B}$ is an open cover of $U \Rightarrow \mathcal{B} \cap \mathcal{C} \neq \varnothing$.

   Next, let $\mathcal{C}$ be any collection of open sets satisfying (a), (b). Let Q be the set of points $\{x \mid \mathcal{N}_0(x) \subseteq \mathcal{C}\}$, where $\mathcal{N}_0(x)$ is the set of open neighbourhoods of x. Show that Q is closed, and that $\mathcal{C} = \mathcal{M}(\mathcal{Q})$. Deduce that there is a (1, 1)-correspondence between the closed sets of X and the collections of open sets satisfying (a), (b).
   A non-empty closed set Q is said to be *irreducible* if Q cannot be expressed as the union of two closed sets, each of which is a proper subset of Q. Show that Q is irreducible if and only if $\mathcal{M}(\mathcal{Q})$ is a (completely prime) filter. Deduce that X is sober if and only if we have:

   (S) for every irreducible closed set Q of X, there is a unique point x such that $Q = \text{Cl}(\{x\})$.

   Condition (S) is very often taken as the definition of sobriety.

5. Reformulate the definition(4.3.10) of the *sobrification* of a space X in terms of the irreducible closed subsets of X.

6. Let P be a poset. Show that an ideal of P is the same thing as an irreducible Alexandroff-closed subset of P. The *ideal completion* $\overline{P}$ of P is the collection of ideals of P, ordered by subset inclusion. Verify that $\overline{P}$ is a dcpo. Show that $(\overline{P}, \text{Scott})$ (i.e. $\overline{P}$ taken with its Scott topology) is the sobrification of $(P, \text{Alexandroff})$.

7. Formulate a simple necessary and sufficient condition for a multifunction $f : \Sigma_\perp \to \Sigma_\perp$ (Σ discrete) to be lsc.

8. Let Σ be a discrete state set, and $f : \Sigma \to \Sigma_\perp$ a multifunction such that $f(x)$ is non-empty for each x. A (Hoare-style) *partial correctness assertion* for f takes the form $\{P\}f\{Q\}$, where $P, Q \subseteq \Sigma$; the meaning of this assertion is that, if $x \in P$, then $f(x) \subseteq Q \cup \{\perp\}$ ('if f terminates, then it does so in a state satisfying Q').
   Rewrite the partial correctness assertion in terms of wlp. Show that the set of true partial correctness assertions for f characterizes f.

Try to generalize these ideas to the non-discrete case (note that in the general case the set of true "partial correctness" assertions will not characterize f, but at best the closed-sets predicate transformer corresponding to f, or $\text{wlp}(f, -)$).

9. Show that a multifunction $h : X \to Y$ is lsc if and only if, for every open set U of Y, $h^-(U)$ is open in X.

10. (a) Suppose that $f : X \to Y$ is a continuous map, and that $f(A) \subseteq B$, where A, B are subspaces of X, Y. Show that the restriction of f to A, considered as a map from A to B, is continuous. Show that the corresponding proposition is true also in the cases that f is a usc or a lsc multifunction.

    (b) Point out that a sufficient condition for continuity of a function $f : X \to Y$ is that $F^{-1}(A)$ is open for every member A of a *base* (or even subbase) of open sets of Y. Do corresponding results obtain in the case that f is a usc or a lsc multifunction?

11. Let X, Y be posets with Alexandroff topology. Show that a multifunction from X to Y is lower (upper) monotonic iff it is lower (upper) semicontinuous.

12. Show that FM is not lower monotonic. Devise also a simpler example to show that a lower semicontinuous function need not be lower monotonic.

13. Given a subset S of a topological space X, denote by $\text{Th}(S)$ the collection of open supersets of S. (cf. Example 3.4.6 the 'satisfaction' relation we have here is simply that of membership between points and open sets.) Note that $\text{Th}(S)$ is a filter of ΩX. Show that a usc map $f : X \to Y$ is of bounded non-determinacy (that is, $\Omega^+(f)$ satisfies condition (5) of Section 4.4) if and only if, for every $x \in X$, $\text{Th}(f(x))$ is open in the Scott topology of ΩX. (A Scott-open filter of open sets might be regarded as a 'finitary' theory or specification. The condition that $\text{Th}(S)$ is Scott-open may thus be taken as stating that S is 'finitarily specifiable' — although we may prefer to say that $\text{Mod} \circ \text{Th}(S)$, that is $\bigcap \text{Th}(S)$, that is $\uparrow S$, is the set that is specified by $\text{Th}(S)$.)

5 Constructions: new spaces from old

In this section we consider some of the simplest and most frequently used constructions of spaces: subspace and quotient, sum and product, direct and inverse limits. Two extremely important constructs are omitted from the present section as being too complex: function space and hyperspace (or 'power domain'). These will receive some attention in later sections.

The six which we consider in this section fall into three dual pairs. **Subspace, Quotient.** The subspace has already been defined in 1.3. Here we remark that it can be characterized in the following manner:

Proposition 5.0.1. *Let S be a subset of a space X, and $i : S \to X$ the inclusion map. Then the subspace topology on S is the coarsest topology on S for which i is continuous.* ∎

This proposition is only an instance of a much more general situation:

Proposition 5.0.2. *Let S be a set, $(Y_i)_{i \in I}$ a family of spaces and f_i a map from S to Y_i (for each $i \in I$). Let $\mathcal{T}$ be the topology on S having as subbase the collection of sets of the form $f_i^{-1}(O)$ ($i \in I$, O open in Y_i). Then $\mathcal{T}$ is the coarsest topology on S for which all the maps f_i are continuous.*

Proof. For a given topology on S, the maps f_i are continuous iff the sets $f_i^{-1}(O)$ are open. But $\mathcal{T}$ is by definition the least topology in which all these sets are open. ∎

Definition 5.0.3. Under the conditions of Proposition 5.0.2, we say that $\mathcal{T}$ is the *initial topology* on S for the family $(f_i)_{i \in I}$.

Thus the subspace topology is the initial topology for the inclusion map.

The notion of initial *topology* is only an instance of a notion of initial *structure* which makes sense in many categories besides **Top**. This should be plain from the following characterization:

Proposition 5.0.4. *Let $S, Y_i, f_i, \mathcal{T}$ be as in Proposition 5.0.2. Then $\mathcal{T}$ is the unique topology on S such that (i) the maps f_i are continuous, and (ii) for any space Z and map $G : Z \to S$, g is continuous if all the maps $f_i \circ g$ are continuous.*

Proof. For (i) to be satisfied, the topology on S must be at least as fine as $\mathcal{T}$. However, for (ii) to be satisfied, the topology must be at least as coarse as $\mathcal{T}$, as we see by taking Z to be the space $(S, \mathcal{T})$. Thus no topology other than $\mathcal{T}$ satisfies (i) and (ii). But we already know that $\mathcal{T}$ does satisfy (i) and it satisfies (ii) since: the maps $f_i \circ g$ are continuous iff the sets $(f_i \circ g)^{-1}(O)$ (O open in Y_i) are open in Z iff the sets $g^{-1}(f_i^{-1}(O))$ are open in Z iff g is continuous (cf. Exercise 4.4.6(10b)). ∎

Corollary 5.0.5. *Let S be a subset of a space X. Then the subspace* topology is the unique topology on S such that (i) the inclusion is continuous, and (ii) for any space Z a map $g : Z \to S$ is continuous if it is continuous as a map into X. ∎

Dually to the preceding, we have:

Proposition 5.0.6. *Let S be a set, $(X_i)_{i \in I}$ a family of spaces, and f_i a map from X_i to S (for each $i \in I$). Let $\mathcal{T}$ be the collection of subsets O of S such that $f_i^{-1}(O)$ is open in X_i for each $i \in I$. Then $\mathcal{T}$ is a topology, which can be characterized in either of the following ways:*

(a) *$\mathcal{T}$ is the finest topology on S for which all the maps f_i are continuous;*

(b) *$\mathcal{T}$ is the unique topology on S such that (i) the maps f_i are continuous, and (ii) for any space Z and map $g : S \to Z$, g is continuous if all the maps $g \circ f_i$ are continuous.* ∎

Under these conditions, $\mathcal{T}$ is said to be the *final topology* on S for the family $(f_i)_{i \in I}$.

Notation 5.0.7. If R is an equivalence relation on a set S, we denote by S/R the quotient set of S by R, and by $[x]_R$ (or just $[x]$) the equivalence class $\{y \mid xRy\}$ containing a given $x \in S$.

Definition 5.0.8. Let R be an equivalence relation on a space X. The *quotient topology* on X/R is the final topology on X/R with respect to the canonical map $x \mapsto [x]$ of X onto X/R. The quotient set X/R taken with the quotient topology is the *quotient space* of X by R.

In concrete terms, a set O is open in the quotient space X/R iff $\phi^{-1}(O)$ is open in X where ϕ is the canonical map. Equivalently, the open sets of X/R are obtained as the images of the saturated open sets of X, where a set $S \subseteq X$ is *saturated* (w.r.t. R) if $x \in S \& xRy \Rightarrow y \in S$.

If $f : X \to S$ is a surjective map of the space X onto the set S, we have an evident homeomorphism between S endowed with its final topology for f and the quotient space $X/\equiv$, where $x \equiv y$ means $f(x) = f(y)$. We will sometimes speak, loosely, of S (with the final topology) as the *quotient* of X by f, under these circumstances.

Some further preliminary observations are incorporated into the following examples:

Examples 5.0.9.

1. Any quotient of a discrete space (that is, a space in which every subset is open) is discrete.

2. Let X be Cantor space. Let us regard two elements of X as equivalent if they begin with the same number of 0's; that is, $\sigma \equiv \tau$ iff, for some

n, σ and τ both have $0^n 1$ as an initial segment, or else $\sigma = \tau = 0^\omega$. The quotient we obtain may be described as follows. The points may be identified with the natural numbers together with ∞; thus the underlying set is $\mathbb{N} \cup \{\infty\}$. As for the topology, each singleton $\{n\}$ is open, since $\uparrow 0^n 1$ is open in X, and so every subset of $\mathbb{N}$ is open; while a set containing ∞ is open iff it contains also all numbers $\geq k$, for some k. This space, say Y, is usually known as the *one-point compactification* of (the discrete space) $\mathbb{N}$: see Section 7. Now suppose we map Y onto the Sierpinski space 2 by taking $g(\infty) = \bot, g(n) = \top (n \in \mathbb{N})$. It should be clear that the final topology induced on the set $\{\top, \bot\}$ by g is the Sierpinski topology; 2 is a quotient of Y. Since a quotient of a quotient is evidently a quotient, we deduce also that the Sierpinski space is a quotient of Cantor space. Thus a quotient of a Hausdorff space, unlike the case with subspaces, may very well be non-Hausdorff. Notice however that, in view of the preceding Example (1), we must use an *infinite* Hausdorff space in order to get 2 as a quotient.

3. Let X be Baire space. Define $f : X \to \mathcal{P}\mathbb{N}$ by $f(\sigma) = \{n | n + 1 \in$ range $(\sigma)\}$. We claim that the sets $\uparrow A (= \{S \subseteq \mathbb{N} | A \subseteq N\})$, A finite, constitute a base for the final topology induced on $\mathbb{N}$. Indeed each $\uparrow A$ is open in the final topology, since $f^{-1}(\uparrow A) = \{\sigma \mid n + 1 \in$ range (σ) for each $n \in A\}$, which is clearly open in X. At the same time, the image of any basic open set $\uparrow \alpha$ (α finite) of X under f is $\uparrow A$ for some finite $A \subseteq \mathbb{N}$. Thus the topology having the $\uparrow A$ as basic open sets is both coarser and finer than the final topology. Now the $\uparrow A$ are of course a base for the Scott topology of $\mathcal{P} \mathbb{N}$(as a complete lattice). Thus we have represented the Scott universal domain $\mathcal{P}\omega$ $(= \mathcal{P}\mathbb{N})$ as a quotient of Baire space.

The domain $\mathcal{P}\omega$ is universal inasmuch as all the relevant domains (in this case, the countably-based continuous lattices) are representable as *retracts* of it.

Retracts are particularly interesting in the present context since they are, in effect, simultaneously subspaces and quotients.

Definition 5.0.10. A space Y is said to be a *retract* of the space X if there are continuous maps $r : X \to Y$ and $e : Y \to X$ such that $r \circ e = Id_Y$. Evidently, r is surjective and e is injective. Further:

Proposition 5.0.11. *Suppose that Y is a retract of X via (r, e). Then the topology of Y is both the initial topology for e and the final topology for r.*

Proof. Let $g : Z \to Y$ be any map with codomain Y. Then $g = r \circ (e \circ g)$. Hence, g is continuous if $e \circ g$ is. By Proposition 5.0.4, the topology of Y is initial for e. The proof that Y is the quotient of X by r is similar. ∎

A *retraction* of a space X is a continuous map $r : X \to X$ such that $r \circ r = r$. Given a retraction $r : X \to X$, let $i : Im(r) \to X$ be the inclusion of the image of r in X. We see that $Im(r)$ is a retract of X via (r, i). Conversely, if Y is a retract of X via (r, e), then the composite $e \circ r$ is a retraction of X, and $Y \cong Im(e \circ r)$. Thus a calculus of retractions over a space X (such as $\mathcal{P}\omega$) gives us a means of defining and manipulating a range of 'data types' — namely those which can be represented as retracts of X (see [Scott, 1976; Plotkin, 1978]).

Despite the elegance and power of this approach, it seems that we have to go beyond retractions to get an adequate general account of the representation of data types. We would normally consider that the data type $\mathcal{P}\omega$ — a type of *sets* — is, from a practical point of view itself in need of representation by some more concrete type, say a type of strings or sequences. Thus in example (3) (5.0.8) we have represented $\mathcal{P}\omega$ as a quotient of Baire space, though it is evidently no retract of Baire space. One might question how natural this 'representation' is. But the point remains that, in the practice of computing, a representation of an 'abstract' type by a type whose elements we can directly construct and manipulate would usually be regarded as involving a quotient (albeit that we may have algebraic rather than topological structure in mind), but not necessarily a retract: there does not have to be a computable map giving a canonical representative for each abstract element. This conforms to the pattern of introduction/representation of sets commonly used in constructive mathematics: to introduce a set S, explain how to construct (a concrete representative of) an element of S, and define an equivalence relation over these concrete elements.

We should consider briefly how the representation of (computable) *functions* works under the view of representations as quotients. We have:

Definition 5.0.12. Suppose that Y is a quotient of X by ϕ. Then a function $g : Y \to Y$ is *represented* (w.r.t. ϕ) by $h : X \to X$ provided that the following diagram commutes:

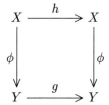

(We have here given a simplified definition, ignoring such possibilities as partial representations, functions of several arguments, etc). Representations as quotients may be unsatisfactory due to failure to represent enough functions:

Example 5.0.13. Representation of the reals. Consider first the ordinary decimal representation. For simplicity consider just the unit interval $I = [0, 1]$, so that we have a surjection $\phi : \Sigma^\omega \to I : \sigma \mapsto \cdot\sigma$, where $\Sigma = \{0, 1, \ldots, 9\}$. It is easy to show that, if Σ^ω has the usual topology, ϕ induces (as a quotient) the standard topology of I. But ϕ fails in terms of representation of functions: as is well-known, trivially computable functions such as multiplication by 3 are not representable. (Informally: $1/30$ has 03^ω as its only representative. Multiplication by 3 cannot be computed in terms of the decimal representation, since no matter how much of the input 03^ω we have, we can never safely begin the output — that is, choose safely between $1\ldots$ and $09\ldots$)

Much better is a representation using *signed digits* [Avizienis, 1961; Wiedmer, 1980]. In particular, let $\Sigma = \{-1, 0, +1\}$, and define $\Psi : \Sigma^\omega \to [-1, +1]$ by: $\Psi(a_1 a_2 \ldots) = \Sigma_{n=1}^{\infty} a_n 2^{-n}$. As before, we obtain the standard topology of the real interval $J = [-1, +1]$ as quotient. But the position is quite different with respect to representation of functions: it is not difficult to show that every continuous function over J is representable.

Representation by Baire space has been taken as the basis of a general theory of Type 2 computability by K. Weihrauch and his associates: see Weihrauch [Wei87] for a comprehensive account. The analogy is with Ershov's theory of numbered sets, based on representation by $\mathcal{B}N$. Baire space representation is a very powerful technique. Indeed, we can claim that every space of potential computational interest can be handled with it, due to the following fact: every second countable T_0 space is a quotient of a subspace of Baire space (Exercise 5.1.1(9)). At the same time Baire space is itself said to be especially convenient computationally. On the convenience of Baire space we harbour some doubts: the space is not locally compact, in particular see Section 7. We certainly consider that there are many representations, other than by Baire space, which are of great interest (for example, signed digit representation of the reals); and that much remains to be done in this area.

Sum, Product. These can be disposed of fairly quickly. The *sum* $X + Y$ of topological spaces X, Y is the disjoint union of X, Y (as sets), endowed with the final topology with respect to the canonical injections $i_1 : X \to X + Y, i_2 : Y \to X + Y$. Thus, a set O is open in $X + Y$ iff O is of the form $i_1(U) \cup i_2(V)$, where U, V are open in X, Y respectively. The sum of an arbitrary family of spaces, $\Sigma_{i \in I} X_i$, is defined similarly. It follows from the characterization of the final topology given in Proposition 5.0.6(b) that the sum of topological spaces is the coproduct in the category **Top**.

Example 5.0.14. The sum of two (indeed, of an arbitrary family of) discrete spaces is discrete. Let now $\mathbb{N}$ be the space of natural numbers with discrete topology. Define the map $\phi : 1 + \mathbb{N} \to \mathbb{N}$ by $\phi(\cdot) = 0, \phi(n) =$ succ $(n)(= n+1)$. (Strictly speaking, we should write $\phi(i_2(n)) = $ succ (n), etc.) Then ϕ is bijective. Since all maps between discrete spaces are continuous we have, rather trivially, a homeomorphism

$$\mathbb{N} \cong 1 + \mathbb{N}$$

Example 5.0.15. More generally, we can see that a space Z satisfies the 'domain equation'

$$X \cong 1 + X$$

iff Z is of the form $\mathbb{N} + Y$, which amounts to saying that Z contains a sequence of isolated points.

Dually, the *product* $\Pi_{i \in I} X_i$ of a family $(X_i)_{i \in I}$ of spaces is the Cartesian product endowed with initial topology with respect to the projections $\pi_i : \Pi X_i \to X_i$. Thus, the topology of the product space has as subbasic open sets the sets of the form $\pi_i^{-1}(U)$ where $i \in I$ and U is open in X_i. Using Proposition 5.0.4, we verify that the product space is indeed the product in the category **Top**.

Example 5.0.16. The Cantor space C can be represented as the product of countably many copies of the discrete two-point space $2 = \{0, 1\}$. Let $c : 2 \times C \to C$ be defined by $c(a, \sigma) = a.\sigma$; that is, if $\sigma \in C$ is the sequence $(b_0, b_1, \ldots)$, then $c(a, \sigma)$ is $(a, b_0, b_1, \ldots)$. Then c is bijective, and gives a homeomorphism

$$C \cong 2 \times C.$$

Similar remarks apply to Baire space as the product off countably many copies of $\mathbb{N}$.

Direct and Inverse Limits. The *direct limit* of a directed system of spaces is the set-theoretic direct limit endowed with the final topology. In practice we need to consider only *sequences* of spaces rather than general directed systems. Suppose then, that

$$\Delta = X_0 \xrightarrow{f_0} X_1 \xrightarrow{f_1} \ldots$$

is a sequence of spaces and continuous maps. If $m \leq n$, write f_{mn} for the composition $f_{n-1} \circ f_{n-2} \circ \ldots \circ f_m$.

Define the relation $\sim$ on $\Sigma_i X_i$ by

$$x \sim y \Leftrightarrow \exists p \geq m, n.f_{mp}(x) = f_{np}(y) \text{ (where } x \in X_m, y \in X_n).$$

It is easily checked that $\sim$ is an equivalence relation. Then the disjoint limit of $\Delta, \varinjlim \Delta$, is the quotient of the direct sum ΣX_i by $\sim$; a subset O of $\varinjlim \Delta$ is open if and only if, for each of the canonical maps $e_i : X_i \rightarrow \varinjlim \Delta, e_i^{-1}(O)$ is open in X_i.

Most typically, Δ is just a sequence of subspace embeddings: $X_0 \subseteq X_1 \subseteq \ldots$. In this case the direct limit can be described rather simply, as the union $\cup X_i$ endowed with the least topology for which each X_i is a subspace of it: a subset O of $\varinjlim \Delta$ is open if and only if, for each $i, O \cap X_i$ is open in X_i.

Although direct limits (especially of embedding sequences) are used heavily in mathematical practice, they do not prove to be so useful in the computer science context. This may perhaps be attributed to the fact that the completeness properties in which we are interested tend not to be preserved by direct limits.

Example 5.0.17. For $n = 0, 1, 2, \ldots$, let X_n be the ordinal number n (that is, the ordered set $0 \leq 1 \leq \ldots \leq n-1$) taken with the Alexandroff topology, and $f_n : X_n \rightarrow X_{n+1}$ the subspace embedding. Then the direct limit is the ordinal ω. Here, each X_n is (trivially) a Scott domain, but $\varinjlim \Delta$ is not even a sober space.

The reader is perhaps familiar with the observation that, in the context of the 'D_∞' approach to solving domain equations, the limit of an embedding-projection sequence is both a direct limit (w.r.t. the embeddings) and an inverse limit (w.r.t. the projections). The 'direct limit' involved in that formulation is actually the colimit in a restricted category of spaces, for example the category of Scott domains. The direct limit construction we have given above, on the other hand, is the colimit in the category **Top**.

Turning now to inverse limits, let

$$\Delta = X_0 \xleftarrow{g_0} X_1 \xleftarrow{g_1} \ldots$$

be an (inverse) sequence of $\Delta, \varprojlim \Delta$, is the set $\{(x_n) \in \Pi X_n \mid \forall n \geq 0.g_n(x_{n+1}) = x_n\}$ considered as a subspace of ΠX_n. If $\mathcal{B}_n$ is a base of open sets of $X_n (n \geq 0)$, we may consider, for each $A \in \mathcal{B}_n$, the set $L_A = \{(x_i) \in \varprojlim \Delta \mid x_n \in A\}$; then the collection of sets $L_A, A \in \cup_n \mathcal{B}_n$, is a subbase of open sets of $\varprojlim \Delta$. Of course, the set $L_A(A \in \mathcal{B}_n)$ is just $\pi_n^{-1}(A)$, where $\pi_n : \varprojlim \Delta \rightarrow X_n$ is the projection, and the topology of $\varprojlim \Delta$ is the initial topology for the p_n. From this it follows (by Proposition 5.0.4) that inverse limits are limits in the category **Top**.

In contrast with the situation regarding colimits, limits (of inverse sequences) in most of the "useful" subcategories of **Top** coincide with the inverse limits just described. Moreover, many important classes of spaces may usefully be characterized as the closures under inverse limits of certain classes (categories) of particularly simple spaces, for example finite spaces. This is illustrated by the following:

Proposition 5.0.18. *Let X be a second-countable space. Then X is a Stone space if and only if X is homeomorphic with the inverse limit of a sequence of finite, discrete spaces.*

Proof outline

1. Suppose that X is the Stone space corresponding to the (countable) Boolean algebra B. We can express B as the union of a sequence $B_0 \subseteq B_1 \subseteq \ldots$ of finite Boolean algebras. For each n, let X_n be the set of minimal non-zero elements of B_n, taken with discrete topology. Notice that the clopen subsets (that is, subsetes that are both closed and open) of X corresponding to the members of X_n form a partition of X. Define $g_n : X_{n+1} \to X_n$ by: $g_n(a) = b \in X_n$ such that $a \geq b$. It is easy to see that the limit of the resulting inverse sequence of spaces is homeomorphic with X.

2. Suppose that X is the inverse limit of a sequence (x_n) of finite discrete spaces, the canonical projections being $\pi_n : X \to X_n$. A base of the topology of X is given by the finite unions of sets of the form $\pi_n^{-1}(a)(a \in X_n, n \in \mathbb{N})$; this base, say B, is evidently a Boolean algebra of clopen subsets of X. Applying König's lemma to the finitary tree of finite initial sections (prefixes) of elements of X, we see that the meet of any filter of (non-empty) members of B is non-empty. At the same time it is clear that the meet of a *prime* filter in B cannot have more than one element. By the discussion in Section 2.5, X is a Stone space. ∎

A more elegant proof of Proposition 5.0.16 might be couched in terms of the *duality* between the category of Stone spaces (with continuous maps) and the category of Boolean algebras (with Boolean homomorphisms). This and other 'Stone-like' dualities are implicit in several of our discussions in this Chapter, but we prefer not to develop the machinery needed to present them fully. For an account of the Stone space/Boolean algebra duality one may see, for example, [Halmos, 1963], and for a more comprehensive account of related dualities, [Johnstone, 1982].

For the reader who has some acquaintance with the 'information system' approach to Domain Theory ([Scott, 1982; Larsen and Winskel, 1984]), let us just point out that these dualities can show how to extend this approach to many classes of spaces besides Scott domains. Consider in particular the second-countable Stone spaces. We can consider the (countable) Boolean algebras as information systems for them. It is easy to see that these Boolean algebras form an ω-cpo (poset with least element and lubs of ω-chains) under the subalgebra relation $\trianglelefteq$: the least element is the two-point Boolean algebra, and lubs of chains are given by set union. Call this ω-cpo BA. (It may be objected that BA is not a set. One reply to this is that it suffices that BA be a class — a 'large cpo' (Larsen and Winskel, 1984); alternatively, we may agree to consider only Boolean algebras whose members are drawn from a fixed set of 'tokens'.) Type constructors are to be represented as (ω-)continuous functions over BA, so that domain equations can be solved by means of ordinary least fixed points. For example, the homeomorphism $C \cong 2 \times C$ of 5.0.14 suggests that we consider the equation

$$B = \quad F + B \qquad\qquad (1)$$

where F is the four-point Boolean algebra $\{0, 1, b, \neg b\}$ corresponding to the two-point discrete space, and we take the sum (coproduct) of Boolean algebras since this is dual to the product of spaces. The reader is invited to give a concrete definition of $+$, to check that it is ω-continuous over BA, and to show that the least solution of (1) is, in effect, the Boolean algebra of classical propositional calculus in countably many variables (more abstractly: the countable atomless Boolean algebra); as we saw in Section 2.5, this is indeed the algebra, or information system, for Cantor space.

The preceding is little more than a hint of the required theory, of course. An information system account along these lines was proposed (for various classes of locally compact spaces) in [Smyth, 1985], but the detailed account has not yet been provided.

Meanwhile, as an example of a substantial application of domain equations over Stone spaces, see [Abramsky, 1988].

5.1   Postscript: effectiveness and representation

We pause here to review the brief account of the computational significance of topology provided in Section 1 above. That account approximately followed our previous account in [Smyth, 1983b], but with a significant omission: the previous account contained a cursory discussion of *effectiveness*, but this has disappeared from the version above. The main part of our review will be to consider whether, and if so how, effectiveness should be taken into account; our presentation of this extremely complex topic will be particularly sketchy.

In the previous account, [Smyth, 1983b], a variety of phrases was used to express the intuition which is (supposedly) captured by the use of open sets: computable property, semi-decidable property; uniform procedure yielding (positive) results in a finite time; tests of processes. It was pointed out that, strictly speaking, these ideas do not correspond to (classical) topology, but rather to some kind of 'effective' topology, in which one studies effectively open sets rather than arbitrary open sets. In practice, however, good enough results in semantics can usually be obtained without dealing explicitly with effectiveness (beyond, perhaps, requiring that spaces be second countable). It was on these pragmatic grounds that we left on one side the rather undeveloped subject of effective topology.

In place of the variegated terminology of computable properties, uniform procedures, etc., used to convey the informal notion of an open set in [Smyth, 1983b], we have in this chapter used the convenient phrase 'finitely observable property' suggested by Abramsky in [Abramsky, 1991]. For Abramsky, it is true, this was more than a convenience. He considered that by introducing observability we could achieve a principled (not merely pragmatic) justification for working with ordinary topology rather than effective topology: specifically, that we could thereby remain 'ontologically neutral', avoiding commitment to an 'effective universe'. This, however, is rather questionable. A (finite) observer, in the intended sense, must surely be an effective observer: his (or its) testing must be algorithmic. It follows that 'open' cannot be strictly identified with 'finitely observable'. To make the point concretely, let H be the Turing machine halting set. The relevant topology on the natural numbers is just the discrete topology. This means that both H and $\mathbb{N} - H$ are (trivially) open. The identification of openness with observability would force us to conclude that it is finitely observable whether *or not* an arbitrary Turing machine halts; which is absurd.

What has gone wrong here? Let us review the argument for closure of the collection of observable properties under arbitrary disjunction. Succinctly (as in [Abramsky, 1988]), this goes: to observe that a disjunction holds, we need only observe that one of its disjuncts holds. But we must ask: what is the experiment, or test, corresponding to the disjunction? To pick a disjunct (at random?) and test for it? This would not be a test of the kind we intend. We require a test which is guaranteed to yield *true* when (and only when) the property holds. It seems that, if talk of observation and experiment is to be taken seriously, the test for the disjunction has to be: run all the tests in (pseudo-) parallel, until a positive result is obtained from one of them. This is possible only if the disjuncts are presented to us in an effective listing (enumeration).

If the slightly more detailed argument for the disjunction axiom presented above (Section 1) is inspected closely, it will be seen to contain a subtle (gross?) equivocation between 'observable' and 'securable'. The sense of 'securable' here can be taken to be, briefly: holds by virtue of a finite amount of information about the subject (of the investigation). Note: *holds* in the sense of classical truth, not in the sense of being known, observed, etc. With this classical notion of securability, which separates the question of finiteness of information from whether any possible observer can determine the information, the argument for arbitrary disjunctions clearly goes through. Thus the non-halting set $\mathbb{N} - H$ is (trivially) securable: any positive instance $n \in \mathbb{N} - H$ holds on the basis of finite information, namely just the information embodied in n.

Do these considerations provide the basis for a principled justification of the use of classical topology? Perhaps they do; but this does not imply that it is not important to consider also effective topology. It is desirable, in studying computational processes, to work with basic properties that are observable or testable (as well as being determined by finite information); and this will lead us to effective open sets. One may surmise that when comprehensive 'logics of domains' (along the lines of [Abramsky, 1991]) come to be developed, a useful (partial) criterion of expressive adequacy will be that every effective open set is represented by a formula of a suitably restricted form. (Notice that [Abramsky, 1991] is not comprehensieve in the relevant sense, since only compact open sets are considered.)

We should remark that the association of topology with finiteness of information is hardly new, since the requirement of continuity of functions is often motivated in these terms. This indeed is the basis for the usefulness of topology in recursion theory, as emphasized particularly by Normann [Normann, 1980].

As an illustration of the use of topological methods in recursion theory, which draws also on computer science, we may cite [Barendregt and Longo, 1983]. Here in particular, the Scott topology and the Cantor topology of $\mathcal{P}\omega$ are used in studying the various reducibility notions (enumeration, Turing, truth-table, ...) or recursion theory (Rogers, 1967). (For the 'Cantor topology' of $\mathcal{P}\omega$, cf. Exercise 7.7.8(8) below.)

We conclude with a few remarks on effective topology. These remarks are little more than pointers to some of the relevant literature.

A starting point for effective topology could be the theory of numberings, as developed especially by Ershov ([Ershov, 1973]). A *numbering* of a set X is simply a surjective map $\nu : \mathbb{N} \to X$. One of the basic concepts of the theory is the following: given a numbering $\nu : \mathbb{N} \to X$, the *Ershov topology* on X is the topology having as base the collection

$$\Gamma = \{A \subseteq X \mid \nu^{-1}(A) \text{ is recursively enumerable (r.e.)}\}.$$

The members of Γ may be considered to be the *effective* Ershov-open sets (it is easy to see that Γ is closed under countable effective unions, suitably defined). In this rendering of the idea that open $\equiv$ semi-decidable, the criterion of semi-decidability is applied to the set of indices of the elements of a given set.

How does this compare with topologies explained above in terms of finite observability? Consider $\mathcal{P}\omega$. Any Scott-open subset of $\mathcal{P}\omega$ can be expressed as ↑ F, where F is a (countable) collection of *finite* sets. The obvious way to strengthen this to an effective notion is to ask that F be r.e., in terms of a suitable indexing of the finite sets. 'Suitable' indexings of $\mathcal{P}_{\text{fin}}\omega$ are undoubtedly the *canonical* indexings cf. [Rogers, 1967] (we shall shortly explain this choice of indexings in terms of effective domain theory). The effective *elements* of $\mathcal{P}\omega$, on the other hand, are simply the r.e. sets; let $\rho : \mathbb{N} \to RE$ be a standard numbering of these (derived in the usual way from an acceptable indexing of the partial recursive functions). We thus get two notions of effective open subset of RE: one derived from the Scott topology of $\mathcal{P}\omega$, and one from the numbering ρ via Ershov's definition. The content of the classical *Rice–Shapiro theorem* is precisely that these two notions coincide. (The Rice-Shapiro theorem is perhaps more usually formulated in terms of partial recursive fucntions rather than r.e. sets; see for example [Cutland, 1980]. The version in terms of r.e. sets might better be called the Rice–McNaughton–Myhill–Shepherdson theorem; see [Myhill and Shepherdson, 1955].)

The topological formulation of the Rice–Shapiro theorem shows us how to put the theorem in a general setting. Indeed, let D be an *effectively given* Scott domain [Egli and Constable, 1976; Smyth, 1977]. This means in particular that we have an indexing (or numbering) $E : \mathbb{N} \to B_D$ of the basis of D such that the order $\sqsubseteq$ of B_D and the consistency of pairs of elements of B_D are recursive as predicates in the indices. Notice that, of the various types of indexings of $\mathcal{P}_{\text{fin}}\omega$ considered by Rogers, only the canonical indexings enable $\mathcal{P}\omega$ to be presented as an effectively given domain. An element x of D is *computable* if x can be given as the lub of an increasing sequence in B_D which is recursive in indices. We can thus obtain a numbering $\rho : \mathbb{N} \to C_D$ of the computable elements from an acceptable numbering of the partial recursive functions. Exactly as in the special case of $\mathcal{P}\omega$, we can now define two notions of effective open subset of C_D, based on e and the Scott topology and on ρ and the Ershov topology, respectively. A generalized version of the Rice–Shapiro theorem states that these coincide: succinctly,

$$\text{Scott} = \text{Ershov}$$

(Generalized Rice–Shapiro theorems have been rediscovered several times: see for example [Plotkin, 1981; Giannini and Longo, 1983; Weihrauch, 1987]. A further generalization, to *continuous* domains, is given in [Spreen, 1989].)

Unfortunately, this approach will not work for other spaces in which we are interested. If we try to use the above Ershov-style definition of 'effective open' in connection with the Baire space B, for example, the Rice–Shapiro theorem itself tells us that the only effective open (completely r.e.) subset of B is $\varnothing$. One attempt at a solution could be to consider spaces as embedded into suitable Scott domains. In fact, every second-countable space X embeds in a rather trivial way into $\mathcal{P}\omega$: if $B_0, B_1, \ldots$ enumerates a base of open sets of X, the map

$$M : x \mapsto \{n \mid x \in B_n\}$$

gives X as a subspace of $\mathcal{P}\omega$.

The solution which we consider in a moment will incorporate this idea. It will also incorporate a more radical departure from the Ershov approach, which we now try to motivate. The Ershov view of open sets as semi-decidable may be said to be an *intensional* view: a subset S of RE is Ershov (effective-)open if the set of *programs* that produce members of S is semi-decidable. By contrast, the version based on Scott topology is extensional: a member x of the effective open set S need be given only as a stream of outputs for the decision to accept x to be made correctly (the program which generates x may be considered as a black box).

The intensional approach presupposes an effectively enumerable domain of computable elements: it commits us to an 'effective universe'. The extensional approach requires only an enumeration of a base (or basis), and is comfortable with domains having continuum many elements.

For an account offering the advantages of the extensional approach, we may turn to the theory of representation by Baire space [Kreitz and Weihrauch, 1985; Weihrauch, 1987]. A representation of a set X is taken to be, in general a *partial* surjective map $\delta : \mathbb{B} \to X$. It can be shown [Weihrauch, 1987] that the basic theory of representations can be developed in a way that closely parallels the theory of numberings. Topology plays a much larger role in the theory of representations, however. Under a representation $\delta : \mathbb{B} \to X$, X is considered with the quotient topology. From the previously noted facts that $\mathcal{P}\omega$ is representable (Examples 5.0.8) and that every second countable T_0 space embeds into $\mathcal{P}\omega$, we see that *any second countable T_0 space is representable*. (Significantly, however, it is not true that every representable space is second countable; we will return to this point shortly.) Given two representations δ, δ' of X, X' respectively, a map g from X to X' induced by a map $F : \mathbb{B} \to \mathbb{B}$ (cf. Definition 5.0.11) is continuous if f is continuous: we may say that g is *computable* if it is induced in this way by computable f. But some care is needed with this definition of computability, due to anomalies such as that of Example 5.0.13: a 'bad' choice of representation of $\mathbb{R}$ may result in a function as simple as $\lambda x.3*x$ being considered non- computable. An interesting general solution to this problem is provided by Weihrauch's theory. Analogously to the theory of numberings, a preorder is defined over the representations of a second countable space X by: $\delta \le \delta'$ iff there is a continuous function $f : \mathbb{B} \to \mathbb{B}$ such that $\delta = \delta' \circ f$. (This may be thought of as saying that δ can be *translated* into δ'.) A representation is *admissible* if it is maximal in this order. Then we have the result: provided that the representations δ, δ' of X, X' are admissible, any continuous map $G : X \to X$ is induced by some continuous $F : \mathbb{B} \to \mathbb{B}$. The anomaly involving $\lambda x.3 * x$ clearly cannot arise if an admissible representation of $\mathbb{R}$ is used.

It should be stressed that in this approach the theory of effectively representable functions (in which we require that the representing maps over $\mathbb{B}$ be computable rather than just continuous) is, for the most part, a minor variant of the purely topological theory.

Weihrauch's theory is presented as a theory of type 2 computability, and certain difficulties arise if one tries to extend it to higher type (total) functionals. The functionals in question are the Kleene–Kreisel countable functionals, of which the most convenient exposition is [Normann, 1980]. Up to a point, the theory of representation will work quite well: indeed, the technique of *associates* used in defining the successive types $\mathbb{N}^{(1)} = \mathbb{N} \to \mathbb{N}, \mathbb{N}^{(2)} = \mathbb{N}^{(1)} \to \mathbb{N}, \ldots$, of countable functionals amounts to constructing representations of these types by $\mathbb{B}$. Unfortunately, for $k \geq 2$, the topology of $\mathbb{N}^{(k)}$ (considered as a quotient of $\mathbb{B}$) is not very convenient. In particular, it is not second countable (or even first countable; see [Hyland, 1979] or [Normann, 1980]). This means that the main results of the (Weihrauch's) theory of representations do not hold for it, and generally it does not conform to our idea of a computationally reasonable topology. Hyland [Hyland, 1979] has shown that these problems disappear if one presents the types (of countable functionals), not as topological spaces, but as spaces of a more general kind: filter spaces, or limit spaces (L-spaces). To define filter spaces, one gives a *weak* axiomatization of the notion of convergence of filters: in effect, the notion is required to satisfy just the first two properties of convergence given in Proposition 3.5.2. Limit spaces are defined similarly via an axiomatization of convergence of sequences. Unfortunately it is outside the scope of this chapter to pursue these topics.

A truly *constructive* topology goes beyond the preceding inasmuch as it requires not only that the objects of study should be constructive or computable in some sense, but that the methods of reasoning employed in the theory should be constructively valid. It is difficult to get very far with conventional point-set topology without continual recourse to powerful set-theoretic principles. Point free approaches offer more promise, and indeed this is one of the principal motivations for frame/locale theory [Johnstone, 1983]. A noteworthy source of ideas for a thoroughgoing constructive (and point-free) treatment is the unjustly neglected [Martin-Löf, 1970]. A more systematic presentation, which however does not emphasize the point-free aspect is [Grayson, 1981; Grayson, 1982].

Exercises 5.1.1. The bulk of the exercises for this section are concerned, not entirely appositely, with basic concepts in the recently developed area of topological (as opposed to graph-theoretic) *digital topology* — that is to say, the topology of digital images (references later). The relegation of this topic to exercises by no means reflects its importance. Notice that these exercises represent almost our only concession to topology as geometry in the chapter!

1. A map $f : X \to Y$ is *open* if the image of any open set of X under f is open in Y. Show that the projection maps from a product space onto its components are open. Show also that the quotient mapping from Baire space onto $\mathcal{P}\omega(= \mathcal{P}\mathbb{N})$ (Example 5.0.8(3)) is open.

   Show that if $F : X \to Y$ is continuous, open, and bijective (that is, (1,1) and onto), then f is a homeomorphism.

2. Let D, E be algebraic dcpo's taken with Scott topology. Show that the Scott topology of the dcpo $D \times E$ agrees with the product topology. (A very much harder exercise is to show that this does *not* hold for dcpo's in general. See [Barendregt, 1984], Exercise 1.3.12.)

3. A space X is said to be *connected* if X cannot be expressed as the union of two disjoint non-empty open subsets. Equivalently, X is *connected* if, whenever $\{A, B\}$ is a cover of X by non-empty open sets, $A \cap B$ is non-empty. (Note: replacing 'non-empty' here by 'inhabited', we get the *positive* formulation of connectedness which is suitable for constructive topology. See [Grayson, 1981]).

   Let $(P, \leq)$ be a poset, which we view both as a directed graph (an arc goes from x to y iff $x \leq y$) and as a space (Alexandroff topology). Verify that P is connected in the graph-theoretic sense iff P is a connected space.

4. A subset S of a space X is said to be *connected* if, endowed with the subspace topology, S is a connected space. Moreover, S is a *component* of X if it is a maximal connected subset of X (that is, it is connected but is not a proper subset of any other connected subset). Show that, if S, T are connected subsets of a space X such that $S \cap T$ is non-empty, then $S \cup T$ is connected. Deduce that the components of any space X from a partition of X.

For the remaining exercises, we assume that the set $\mathbb{Z}$ of integers is equipped with the *connected* topology which has the sets $\{2n+1\}$ and $\{2n-1, 2n, 2n+1\}(n \in \mathbb{Z})$ as basic open sets. This may be depicted as follows:

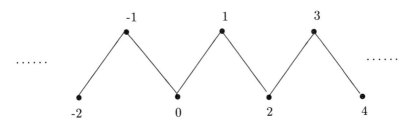

The intended topology is the Alexandroff topology of the depicted poset. This space, or poset, may be thought of as representing the subdivision of the real line $\mathbb{R}$ into 'cells' (open simplexes) $(n, n+1)$ together with their vertices $n, n+1$; in fact, the point $2n+1$ of $\mathbb{Z}$ represents $(n, n+1), 2n$ represents the vertex n, and $k \sqsubseteq l$ if k is a vertex of the cell l. The main interest of $\mathbb{Z}$ lies in the product spaces $\mathbb{Z}^2, \mathbb{Z}^3$. In particular, $\mathbb{Z}^2$ may be thought of as representing pixels (considered as unit squares of $\mathbb{R}^2$) together with their edges and vertices; the reader should try to supply details for this. In this way a more satisfactory basis can be provided for the study of connectedness, closed curves, etc., as required for digital image processing, than by the use of purely graph-theoretic ideas: see in particular [Kovalevsky, 1989; Khalimsky *et al.*, 1990]. Note: the spaces $\mathbb{Z}, \mathbb{Z}^2, \ldots$, have been termed collectively *Khalimsky space*, and attributed to [Khalimsky, 1977].

5. Show that the map $x \mapsto 2x$ (x an integer), $x \mapsto 2n+1 (x \in (n, n+1))$, gives $\mathbb{Z}$ as a quotient of $\mathbb{R}$. Show that $\mathbb{Z}^2$ is similarly a quotient of $\mathbb{R}^2$.

6. Verify that $\mathbb{Z}$ is a $T_{\frac{1}{2}}$-space, and that $\mathbb{Z}^2$ is not a $T_{\frac{1}{2}}$-space. Give an example of a finite $T_{\frac{1}{2}}$-space that is not a subspace of $\mathbb{Z}$.

7. A *connected ordered topological space* [Khalimsky *et al.*, 1990], or COTS, is a connected space X satisfying: for any subset T of X with exactly three elements, there is a point p of T such that the other two points of T lie in different compoents of $X - \{p\}$. Verify that $\mathbb{Z}, \mathbb{R}$, and all intervals of $\mathbb{Z}$ and $\mathbb{R}$, are COTS. Show that there is just one COTS that is not T_0, namely the two point indiscrete space. Show that every finite COTS with at least three points is $T_{\frac{1}{2}}$. (In fact this remains true if we delete 'finite': [Khalimsky *et al.*, 1990].) Show further that every finite COTS with at least three elements is homeomorphic with a subspace (an interval) of $\mathbb{Z}$.

8. A subset s of a space X is said to be a *path* (respectively, *arc*) if S is the continuous (resp., homeomorphic) image of a COTS in X. (These definitions [KKM90] are non-standard. Customarily one works just

with mappings from the unit interval [0, 1] into X.) Give a direct description of a finite arc in X in terms of the specialization order of X. (Note: one may as well restrict all spaces to be T_0 to simplify the discussion.) Show that any finite arc with end-points x, y is minimal among the connected sets containing x and y. Prove the converse of this statement under the assumption that X is finite (the converse is actually true without restriction on X: [Khalimsky *et al.*, 1990]).

A space X is said to be *pathwise (arcwise) connected* if any two points of X are contained in a path (arc) in X. Prove that, if X is finite, the following are equivalent:

(a) X is arcwise connected

(b) X is pathwise connected

(c) X is connected.

9. Strictly speaking, our remarks on $\mathbb{B}$ and $\mathcal{P}\omega$ (Examples 5.0.8 and Section 5.1) have shown only that every second countable T_0 space can be represented as a subspace of a quotient of $\mathbb{B}$, whereas what is required for Weihrauch's theory is that each such space be a *quotient of a subspace* of $\mathbb{B}$. Show that the argument can be amended to give the desired result. More ambitiously, recall Exercise (1) and prove the following general result: for any topological spaces X, Y, if Y is a subspace of a quotient of X, where the quotient map is open, then Y is (homeomorphic with) a quotient of the appropriate subspace of X. (It may be helpful to consult [Bourbaki, 1966], Sections 3.6 and 5.2.)

6 Metric Spaces

Metric spaces constitute the main alternative to cpo's as 'domains' for denotational semantics. As for cpo's, we have for metric spaces a convenient fixed point theorem with the aid of which we can interpret recursive definitions. The metric fixed point theorem indeed has the advantage of giving us *unique* fixed points. Moreover, unlike cpo's and Scott domains, metric spaces are familiar from 'ordinary' mathematics.

What, then, determines the choice between cpo's and metric spaces? Topologically, the choice is a stark one: only the trivial one-point space can be viewed in both ways. A familiar observation is that the metric space view is appropriate for the (Hausdorff) space of the *maximal*, or total, elements of a computational domain, while the cpo/Scott topology approach is needed if we want to take into account also the partially defined elements. This observation has much to commend it. Yet it is not the whole truth: as we shall see, one and the same 'domain' can sometimes be appropriately viewed both as a cpo and (endowed with a different topology) as a metric space. A relevant question here is whether the metric is thought of as having any intrinsic significance. Sometimes it may seem that the metric has been introduced solely as a technical device with which to obtain fixed points, and that neither the metric nor the topology which it induces have any meaning beyond this. Despite this, we shall argue (or at least: suggest) that the metrics we are interested in can generally be explained in a way that is fully consonant with the interpretation of topology in terms of observable properties. This explanation has, in particular, the consquence that it is very natural to consider *quasi*-metrics; the appearance of this topic will be one of the more unusual features of our presentation, although one which we will be able to discuss only very briefly in the present context.

It should be admitted that our presentation of metrics in this section, as regards their 'meanings' and the applications envisaged, is somewhat narrow in focus. Metrics surely occur in other branches of computing science than denotational semantics, in some cases even with the 'meaning' of a geometrical distance! As a notable example, we will just mention M Barnsley's [Barnsley, 1988] work on the generation of fractal images. One may observe that the main metrical tools employed by Barnsley — fixed points and the Hausdorff distance — are also central in the semantical applications.

6.1 Basic definitions

Let X be a set, $d : X \times X \to \mathbb{R}^{0+}$ a map into the non-negative reals. We consider a number of conditions, or axioms, concerning (X, d). Notice that each of the axioms 2, 4 has a weaker and a stronger version.

1. $d(x, x) = 0$
2. (a) $d(x, z) \leq d(x, y) + d(y, z)$
   (b) $d(x, z) \leq max\{d(x, y), d(y, z)\}$
3. $d(x, y) = d(y, x)$
4. (a) $(d(x, y) = d(y, x) = 0) \Rightarrow x = y$
   (b) $(d(x, y) = 0) \Rightarrow x = y$.

Each condition is to be understood as universally quantified with respect to x, y, z. Axiom 1 may be termed *reflexivity*. Axiom 2 is the *triangle inequality* (2(b) being the *strong* triangle inequality). Axiom 3 is *symmetry*. Axiom 4 might be called *identity of indiscernibles*. In the presence of symmetry, the two versions of 4 are of course equivalent.

The most familiar combination of these axioms is $1 + 2(a) + 3 + 4(b)$, which says that (X, d) is a *metric space*. Metrics occurring in semantics typically satisfy the strong inequality 2(b); in that case we speak of an *ultra-metric* space (the metric is also said to be *non-Archimedean*). Rather less familiar is the combination $1 + 2(a) + 4(a)$: this defines a *quasi-metric* space. (The dropping of the symmetry axiom is not as bizarre as it may seem. Think of the 'distance' from x to y as a measure of the effort involved in going from x to y. Then think of hilly terrain ... .) We should point out that the notion of a quasi-metric space is often defined in a more restrictive way than has been adopted here, namely by the combination $1 + 2(a) + 4(b)$. Finally, if the axiom 4 is dropped from any of the formulations considered so far, the resulting weakening is expressed by means of the prefix 'pseudo-': thus, a *pseudo-metric* space is defined by $1 + 2(a) + 3$, a *pseudo-quasi-metric* by $1 + 2(a)$, etc.

Given (X, d) satisfying at least axiom 1, the (standard) topology *induced* on X is defined by taking a set $U \subseteq X$ to be open iff, for any $x, x \in U \Rightarrow B_\varepsilon(x) \subseteq U$ for some $\varepsilon > 0$; here $B_\varepsilon(x) = \{y \mid d(x, y) < \varepsilon\}$. Notice that if (X, d) is at least a pseudo-quasi-metric space, any 'ε-ball' $B_\varepsilon(x)$ is necessarily an open set, since if $y \in B_\varepsilon(x)$ then $B_\delta(y) \subseteq B_\varepsilon(x)$ for any $\delta < \varepsilon - d(x, y)$ (in detail: $z \in B_\delta(y) \Rightarrow d(x, z) \leq d(x, y) + d(y, z) < d(x, y) + \delta < \varepsilon$). In that case, the (open) ε-balls evidently constitute a base of open sets for the topology.

In order to spare the use of cumbersome and unfamiliar terminology, we will in the sequel generally state our definitions and results in a more restricted form than necessary. Specifically, we will usually state them for *metric* spaces even when they apply without modification to pseudo-quasi-metric spaces. Comment will be reserved for those situations in which, say, the dropping of the symmetry axiom *does* make a significant difference.

Resuming the series of definitions, we now consider maps between metric spaces. Let $(X, d), (X', d')$ be metric spaces, $f : X \rightarrow X'$ a map. We say that f is an *isometry* if, for all $x, y \in X, d'(f(x), f(y)) = d(x, y)$; f is *non-expansive* if, for all $x, y, d'(f(x), f(y)) \leq d(x, y)$; while f is *uniformly continuous* provided

$$\forall \varepsilon > 0 \, \exists \delta > 0 \, \forall x, y \in X. \ d(x, y) \leq \delta \Rightarrow d'(f(x), f(y)) \leq \varepsilon.$$

Isometry is the strictest notion of a structure-preserving map between metric spaces. Clearly we have: f is isometric $\Rightarrow$ f non-expansive $\Rightarrow$ f uniformly continuous. The idea of uniform continuity is that if a pair of points is close in X their images are to be close in X', uniformly across the space X. The reader will have no difficulty in showing that, if f is uniformly continuous, then it is continuous with respect to the induced (metric) topologies of X, X'. Each of these classes of maps is closed under composition, and gives us a category with objects the metric spaces.

Two metrics d, d' on the same set X are *uniformly equivalent* if the identity map on X, Id_X, is uniformly continuous both as a map from (X, d) to (X, d'), and as a map from (X, d') to (X, d); stated differently, if Id_X is an isomorphism in the category of uniformly continuous maps. For example, if (X, d) is any metric space, then d is uniformly equivalent to each of the metrics d_1, d_2 defined by:

$$\begin{aligned} d_1(x, y) &= 2.d(x, y) \\ d_2(x, y) &= min(d(x, y), 1). \end{aligned}$$

We say that (X, d) is *bounded* if $\exists k \forall x, y.d(x, y) \leq k$. The metric d_2 just defined is obviously bounded; we conclude that every metric space is uniformly equivalent to a bounded space.

Next we consider approximation and limits. The basic notion is that of a Cauchy sequence. A sequence $(x_n)_{n \in \mathbb{N}}$ of points of a metric space (X, d) is *Cauchy* provided that:

$$\forall \varepsilon > 0 \; \exists k \; \forall m, n \geq k.d(x_m, x_n) \leq \varepsilon. \tag{1}$$

In words: however small an $\varepsilon > 0$ is given, eventually all the terms of the sequence are within ε of each other. A point x is a *limit* of a sequence (x_n) if, eventually, the terms of the sequence are (arbitrarily) close to x; that is, $x = Lim_{n \to \infty} x_n$ provided that

$$\forall \varepsilon > 0 \exists k \; \forall n \geq k.d(x_n, x) \leq \varepsilon \tag{2}$$

Less formally: $d(x_n, x) \to 0$ as $n \to \infty$.

An easy argument shows that, if x, x' are limits of a sequence (x_n), then $d(x, x') = 0$. Thus, if we are strictly in a *metric* space, limits are unique when they exist. The preceding definitions may be applied without change in the context of pseudo-metric space; in that case, limits (even of Cauchy sequences) may be non-unique, but all will be distant zero from each other. *Quasi*-metrics require rather more discussion here. At least seven non-equivalent definitions of *Cauchy sequence* have been proposed for quasi-metric spaces: see [Reilly *et al.*, 1982]. In this section, we shall adopt definitions which involve the least possible departure from what happens in metric spaces. Thus we take (1), unchanged, as the criterion of a Cauchy sequence. (In the context of the other proposed definitions of Cauchyness in quasi-metric spaces, sequences satisfying (1) may be termed *bi*-Cauchy: see Exercise (1)). For a point x to be a *limit* of (x_n), we require that the terms are (eventually) close to x 'both ways'; that is, we say that $x = Lim_{n \to \infty} x_n$ if

$$\forall \varepsilon > 0 \; \exists k \; \forall n \geq k.d(x_n, x) \leq \varepsilon \; \& \; d(x, x_n) \leq \varepsilon. \tag{2'}$$

Alternatively, $d(x_n, x) \to 0$ and $d(x, x_n) \to 0$ as $n \to \infty$. Again, such limits might be termed more specifically *bi*-limits. The preceding remarks about unicity of limits clearly extend to these bilimits.

Next, we consider a useful process for 'symmetrizing' a quasi-metric. Let (X, d) be any quasi-metric space. Define the *conjugate* of d, written d^{-1}, by:

$$d^{-1}(x, y) = d(y, x).$$

One easily checks that d^{-1} is a quasi metric on X. Then we obtain a (symmetric) metric space (X, d^*) by taking

$$d^*(x, y) = max(d(x, y), d^{-1}(x, y)).$$

Notice that a sequence (x_n) in a quasi-metric space (X, d) is bi-Cauchy iff it is an ordinary Cauchy sequence in (X, d^*); and x is a bilimit of (x_n) iff x is an ordinary limit of (x_n) w.r.t. (X, d^*).

Finally, we remark that any pseudo-metric gives rise in a rather trivial way to a *metric*, by identifying points distant 0 from each other. That is, if (X, d) is a pseudo-metric space, we have the derived metric space (X', d') where $X' = X/\equiv$ is the set of equivalence classes of points of X under the equivalence given by

$$x \equiv y \quad \leftrightarrow \quad d(x, y) = 0,$$

and $d'([x], [y]) = d(x, y)$.

6.2 Examples

1(a) $\mathbb{R}$ with the standard metric $d(x, y) = | x - y |$. This, of course, induces the usual (Euclidean) topology.

(b) $\mathbb{R}$ with the metric d' given by:

$$d'(x, y) = | \int_x^y e^{-|t|} dt | .$$

This induces the same topology as the standard metric, but the two metrics are not uniformly equivalent. With d', it is harder to 'see' the difference between points x, y of large (say, positive) value than with d; indeed $d'(x, y) \to 0$ as $x, y \to \infty$. The metric d' is bounded, and extends straightforwardly to the extended real line $\mathbb{R} \cup \{-\infty, +\infty\}$ (by simply allowing the bounds of the definite integral to be infinite). Question: what are the neighbourhoods of $+\infty$ in the topology induced by d' on the extended line?

2(a) The *discrete* metric for any set S is defined by

$$d(x, y) = \begin{cases} 0 & \text{if} \quad x = y \\ 1 & \text{if} \quad x \neq y. \end{cases}$$

This induces the discrete topology on S.

(b) As a generalization of the discrete metric, we have the $(0, 1)$-valued quasi-metric for a poset $(P, \leq)$:

$$d(x, y) = \begin{cases} 0 & \text{if} \qquad x \leq y \\ 1 & \text{otherwise.} \end{cases}$$

This quasi-metric induces the Alexandroff topology on P. If posets P, Q are endowed with the $(0, 1)$ quasi-metric, and $f : P \to Q$, then the following statements are equivalent: f is non-expansive; f is uniformly continuous; f is (Alexandroff-) continuous; f is monotonic.

Clearly a *pre*-order can be represented in the same way by a *pseudo*-quasi-metric.

3(a) Given two (finite or infinite) words in an alphabet Σ, we define the distance between them to be 2^{-n}, where n is the first position at which they differ. More formally, we consider the metric defined on Σ^∞ by

$$d(x, y) = inf\{2^{-n} \mid x \restriction n = y \restriction n\},$$

where $x \restriction n$ denotes the n-truncation of x (that is, the result of deleting all terms of x after the first n). Note that we have chosen to derive $d(x, y)$ via the initial segment(s) of agreement between x and y; and that 'inf' rather than 'min' is required simply to allow for the possiblity that $x \restriction n = y \restriction n$ for all n, that is, $x = y$ (in which case, of course, we require $d(x, y) = 0$). This is the most commonly considered metric on Σ^∞; as is easily seen, it is in fact an *ultra*-metric.

The topology it induces may be described by taking as a base of open sets all the sets of the forms $\{\sigma\}$ and $\uparrow \sigma$, where σ is a finite word. Evidently, it strictly refines the Scott topology, and coincides with the refinement previously considered (Example 1.0.3) in case Σ is finite.

(b) If we want to generate the Scott topology, we can use a *quasi*-metric. Indeed, we have only to modify the previously defined metric to:

$$d'(x, y) = inf\{2^{-n} \mid x \restriction n \leq y \restriction n\},$$

where $\leq$ is the usual prefix ordering of strings. Outline proof that d' induces the Scott topology: every basic Scott-open set $\uparrow \sigma$ (σ finite) is of the form $B_\varepsilon(\sigma)$, where we choose $\varepsilon < 2^{-|\sigma|}$ ($|\sigma| = $ length of σ). At the same time, each $B_\varepsilon(\sigma)$ is easily seen to be the join of the sets $\uparrow \sigma'$ (σ finite) such that $\sigma' \in B_\varepsilon(\sigma)$.

4. This example and the next are relevant to the modelling of *processes*. In this first version we regard a process, naïvely, as just a *set* of possible sequences of actions (the *traces* of the process). Thus, denoting

the set of (atomic) actions by Σ, we seek to define a convenient metric on $\mathcal{P}(\Sigma^\infty)$. The idea is that the distance from S to T is the greatest possible distance from an element s of S to T, where the 'distance' from s to T is the least (ordinary) distance from s to any element t of T. This gives us at best a quasi-metric, so we have to symmetrize. More precisely:

$$d_L(S,T) \quad = \quad \underset{s \in S}{sup} \quad \underset{t \in T}{inf} \quad d(s,t);$$

$$d_H(S,T) = d_L^*(S,T)(= max\{d_L(S,T), d_L(T,S)\}).$$

What we have defined here is simply an instance of the *Hausdorff* (pseudo-)metric, to be presented in generality later (Section 6.5). We confine ourselves here to the following two remarks:

(i) The singleton map $x \mapsto \{x\}$ is an isometry of (Σ^∞, d) (Example 3) into $(\mathcal{P}(\Sigma^\infty), d_H)$.

(ii) Strictly speaking, d_H is at best a pseudo-metric. For example, if $S = a^*$, while $T = a^* \cup \{a^\omega\}$, we have $d_H(S,T) = 0$. Also, $d_L(S, \varnothing) = \infty$ (for $S \neq \varnothing$), if we try to take the above definition literally (see Exercise 6.6.2(2)). In order to obtain a metric, we usually define d_H only over the collection $\mathcal{P}_H(\Sigma^\infty)$ of closed, non-empty subsets of Σ^∞.

For a slightly different description of d_H in the present example, see Exercise 6.6.2(3).

5. **Synchronization trees.** We now take note of the distinction [Milner, 1980] which may be drawn between, for example, the two processes depicted as follows:

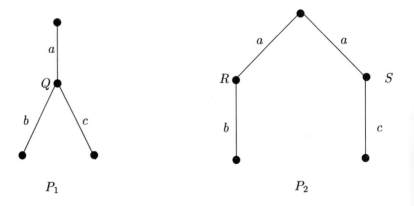

$$P_1 \hspace{6cm} P_2$$

In P_1 we have a process which can engage in the action a (or: accept the experiment a), after which it has two capabilities, namely to engage in b and to engage in c. With P_2, however, after engaging in a we are either at position R, in which only the action b is possible, or at S, in which only c is possible. The two processes have the same trace-set, so we are making a finer distinction than in Example (4).

Our 'processes' will now be *finite* arc-labelled trees, the labels being drawn from alphabet Σ (of 'actions'). Given trees P, Q, and $a \in \Sigma$, the notation $P \xrightarrow{a} Q$ means that there is an arc, labelled a, from the root of P to a node q such that Q is the subtree (of P) with root q. Thus, with P_1 as above, we can write $P_1 \xrightarrow{a} Q, Q \xrightarrow{b} NIL, Q \xrightarrow{c} NIL$, where NIL is the one-point tree (the process which refuses all actions). We consider a sequence $R_0, R_1, \ldots$ of finer and finer equivalence relations over processes [Milner, 1980]. The relation R_n is supposed to represent indistinguishability by experiments of depth $\leq n$. The definition is inductive:

$$
\begin{aligned}
P R_0 Q \;\; &\equiv \;\; \text{True} \\
P R_{n+1} Q \;\; &\equiv \;\; \forall a.[P \xrightarrow{a} P' \Rightarrow \exists Q'.(Q \xrightarrow{a} Q' \,\&\, P' R_n Q')] \\
& \qquad \&\forall a.[Q \xrightarrow{a} Q' \Rightarrow \exists P'.(P \xrightarrow{a} P' \,\&\, P' R_n Q')].
\end{aligned}
$$

For instance, with P_1, P_2 as above, we have $P_1 R_1 P_2$ but not $P_1 R_2 P_2$.

It is now natural to introduce a distance function, the idea being that the harder two trees, or processes, are to distinguish (by experiment), the closer they are:

$$
d(P, Q) = \; \inf \; \{2^{-n} \mid P R_n Q\}.
$$

As usual, we have in the first instance a pseudo-metric here; thus, with P, Q as

respectively, we have trivially that $d(P, Q) = 0$. For accuracy, then, we may introduce the metric space (*FProc*, d_p) of (finite) *processes* as the metric space derived from the just-defined pseudo-metric over *Trees*.

As we shall see in the next subsection (Example 6.3.11), infinite processes are obtained on *completing* the space *FProc*.

6. **A space of types.** Let D be an algebraic cpo, B_D the set of finite elements of D, and $\mathcal{I}(D)$ the set of (Scott-)closed subsets of D. A *rank function* for D is a map $r : B_D \rightarrow \mathbb{N}$. Observe that closed sets I, J are equal iff I, J contain the same *finite* elements. Given a rank function r, we can consider the sequence $R_0 \supseteq R_1 \supseteq \ldots$ of relations over $\mathcal{I}(D)$ where $I R_n J$ says that I, J contain the same finite elements of rank $< n$. As in previous examples, we now have a distance defined by:

$$d(I, J) = \inf \{ 2^{-n} \mid I R_n J \}.$$

This makes $(\mathcal{I}(D), d)$ an (ultra-) metric space.

Such metrics are considered by [MacQueen *et al.*, 1986]. In providing semantics for a polymorphic type system, they define a certain domain V of *values*, satisfying a domain equation of the form

$$V \cong T + \mathbb{N} + [V \rightarrow V] + \ldots$$

(so that V contains the truth-values, the natural numbers, and its own function space). Types are interpreted as closed subsets (or *ideals*) of V, a suitable rank function is defined, and $\mathcal{I}(V)$ is then considered as a metric space (as above).

In the next subsection we will briefly indicate why it is necessary to equip $\mathcal{I}(V)$ with a metric, rather than (as would be expected from domain-theoretic precedents) just a partial order.

6.3   Completeness

In view of previous remarks about approximation and Cauchy sequences, the following definition will not be a surprise:

Definition 6.3.1. A metric space X is said to be *complete* if every Cauchy sequence in X has a limit.

Example 6.3.2. A discrete space X is trivially complete, since a Cauchy sequence in X must be eventually constant. The space of finite synchronization trees (Example (5) above) is *not* complete: e.g. the sequence $a.NIL, a.a.NIL, \ldots$ (that is,

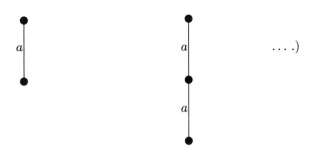

is Cauchy, but has no (finite) limit.

The definition of completeness is the same in the case of a *quasi*-metric space X, though we shall sometimes use the term *bi*complete in this situation (since other definitions of completeness are possible with quasi-metric spaces). Thus, a partial order with the $(0, 1)$ quasi-metric is necessarily bicomplete, for the same reason that a discrete metric space is complete. More generally, a quasi-metric space (X, d) is bicomplete if and only if (X, d^*) is a complete metric space.

It is not difficult to show directly that the remaining examples of metric spaces given in Section 6.2 are complete. Alternatively, one may use the criterion to be given in a moment (Proposition 6.3.4).

Proposition 6.3.3. *Let (X, d) be a metric space, $S \subseteq X, y \in X$. We have:*

1. *$y \in \text{Cl}(S)$ if and only if there is a Cauchy sequence (y_n) of points of S such that $y = Lim_{n \to \infty} y_n$*

2. *S is closed if and only if the limit, if it exists, of any Cauchy sequence of points of S is also an element of S.*

Proof. 1. Suppose that $y = Lim_{n \to \infty} y_n, (y_n)$ Cauchy, $y_n \in S$. Then every open ε-ball $B_\varepsilon(y)$ contains terms of (y_n). Thus, every neighbourhood of y meets S, which is to say: $y \in \text{Cl}(S)$. For the converse, suppose $y \in \text{Cl}(S)$. Every ball $B_\varepsilon(y)$ meets S. For each $n \geq 0$, choose $y_n \in S \cap B_{2^{-n}}(y)$. Clearly, $d(y_n, y) \to 0$ as $n \to \infty$.

2. Immediate consequence of 1. ∎

It is convenient to formulate the next proposition in terms of *closed* ε-balls $\overline{B_\varepsilon}(x)$, where we define

$$\overline{B_\varepsilon}(x) = \{y \mid d(x, y) \leq \varepsilon\}.$$

It is easy to check that, in any (pseudo-)metric space, closed ε-balls are indeed closed in the metric topology. This observation depends on the symmetry of the distance function (Exercise 6.6.2(5)); we will not attempt to provide versions of the next two propositions for quasi-metric spaces.

Proposition 6.3.4. *Let (X, d) be a metric (or pseudo-metric) space. Then (X, d) is complete if and only if the following condition is satisfied:*

(N) *if $B_0 \supseteq B_1 \supseteq \ldots$, where $B_n = \overline{B}_{\varepsilon(n)}(x_n)$, is a nested sequence of closed balls, and $\varepsilon(n) \to 0$ as $n \to \infty$, then $\bigcap_n B_n$ is non-empty.*

Proof.   1. Suppose X is complete. Let (B_n) be a nested sequence satisfying the hypotheses of (N). For each n, choose $x_n \in B_n$. Then (x_n) is clearly a Cauchy sequence. Moreover, for each $k \geq 0, x_n \in B_k$ for all $n \geq k$, so that by Proposition 6.3.3 (2) $Lim_{n \to \infty} x_n \in B_k$. Thus $Lim x_n \in \bigcap_k B_k$.

2. Suppose that condition (N) is satisfied. Let (x_n) be a Cauchy sequence. The statement that (x_n) is Cauchy is equivalent to: for every n we have $k(n)$ such that $\overline{B}_{2^{-n}}(x_{k(n)})$ contains all terms x_j with $j \geq k(n)$. If we put $B_n = \overline{B}_{2^{-n+1}}(x_{k(n)})$, then the sequence (B_n) clearly satisfies the hypotheses of (N); and if $x \in \bigcap_n B_n$, then $Lim_{n \to \infty} d(x_n, x) = 0$.   ∎

Example 6.3.5. Returning to $\mathcal{I}(D)$ (Example (6), preceding subsection), we note that the distance between any two distinct points is of the form 2^{-n}, and that the closed 2^{-n}-balls are simply the equivalence classes for the relation R_n. (This situation is typical for ultrametric spaces, as we shall see in the next subsection.) Thus, for any closed ball B, there is a positive integer $k(B)$ and a set $F(B) \subseteq B_D$ such that $I \in B$ iff $F(B)$ is precisely the set of finite elements of I of rank $< k(B)$. If $B_0 \supseteq B_1 \supseteq \ldots$ is any nested sequence of closed balls, then $F(B_0) \subseteq F(B_1) \subseteq \ldots$; we can conclude that $Q = \bigcap_n B_n$ is non-empty, since in particular $\mathrm{Cl}(\bigcup_n F(B_n)) \in Q$. By the preceding proposition, $\mathcal{I}(D)$ is complete.

Consider again part (1) of the proof of Proposition 6.3.4. We see that this part of the proof still goes through if we suppose only that X has a dense subset S, such that every Cauchy sequence in S has a limit (in X): we merely have to restrict the choice of x_n to $B_n \cap S$. Combining this with part (2) we obtain:

Proposition 6.3.6. *Let (X, d) be a metric space which has a dense subset S such that every Cauchy sequence in S has a limit in X. Then X is complete.*   ∎

As already mentioned, one of the principal benefits of completeness is the availability of a (Banach's) fixed point theorem. To formulate this we need:

Definition 6.3.7. A map $f : (X, d) \to (Y, d')$ is a *contraction* if there is a real number $c < 1$ such that, for all $p, q, \in X, d'(f(p), f(q)) \le c.d(p, q)$.

Then we have:

Theorem 6.3.8. *Let $f : X \to X$ be a contraction of the complete metric space (X, d). Then f has a unique fixed point.*

Proof. Let $c < 1$ be such that $d(f(x), f(y)) \le c.d(x, y)$.

1. Existence. Let x_0 be any point of X, and define the sequence (x_n) by: $x_n = f^n(x_0)$. This sequence is Cauchy, since if $k < l$ we have

$$
\begin{aligned}
d(x_k, x_l) &\le d(x_k, x_{k+1}) + \ldots + d(x_{l-1}, x_l) \\
&\le d(x_0, x_1).c^k(1 + c + c^2 + \ldots)
\end{aligned}
$$

which can be made as small as we please by choosing k large enough. Let $x = Lim(x_n)$. For any n we have $d(x_{n+1}, f(x)) \le c.d(x_n, x)$. Since $d(x_n, x) \to 0$ as $n \to \infty$, it follows that $d(x_{n+1}, f(x)) \to 0$ as $n \to \infty$, and so $d(x, f(x)) = 0$.

2. Uniqueness. Suppose that x and y are fixed points. Then $d(x, y) = d(f(x), f(y)) \le c.d(x, y)$. Hence $d(x, y) = 0$. ∎

Theorem 6.3.8 holds even if X is only a (bi)complete *quasi*-metric space: the proof goes through with minor amendments, as the reader will easily verify. If X is pseudo-metric, the theorem holds with a suitably weakened conclusion. (Exercise 6.6.2(6)).

Examples 6.3.9. (A) Buffers. We model a simple buffer with the naïve process semantics of 6.2 Example (4). We assume that the buffer has only the actions *in*, *out*, and that the capacity of the buffer is 2. Thus, in its initial empty state (in state *buffer0*), the buffer can engage only in *in*, going to state *buffer1*; in state *buffer1* it can engage in *out*, returning to *buffer0*, or in *in*, going to state *buffer 2*, etc. This behaviour is specified, in an obvious notation, by the equations *in.buffer*

$$
\begin{aligned}
buffer0 &= in.buffer1 \\
buffer1 &= out.buffer0 + in.buffer2 \\
buffer2 &= out.buffer1.
\end{aligned}
$$

We may proceed by noting that we can eliminate *buffer0* and *buffer2* from these equations, obtaining a single equation of the form *buffer1* = $f(buffer1)$; noting that f represents a contraction map, we conclude that there is a unique (closed, non-empty) set satisfying the specification of *buffer1*; *buffer0* is then obtained from the first equation. More elegantly, we can argue that the whole specification can be considered as an equation $z = h(z)$, with $h : X^3 \to X^3$ a contraction (X, of course, is $\mathcal{P}_H(\{in, out\}^\infty)$). Product metrics will be considered in Section 6.5.

Rather than work directly with the definition of the Hausdorff metric, in this example and the next, the alternative approach via 'depth n indistinguishability' (Exercise (3)) may be found preferable. Certainly, the latter approach renders trivial the verification that the relevant semantic maps (which typically involve prefixing) are contractive; it also meshes well with the account of the 'meaning of metrics' to be included in the next subsection.

(B) Context-free grammars. Suppose we have a context-free grammar G satisfying the Greibach condition, say

$$S \to aSc \mid b.$$

The alphabet of terminals is $\Sigma = \{a, b, c\}$, and the language generated, $L(G)$, is of course $\{a^n bc^n \mid n \in \mathbb{N}\}$. It is well-known (and easy to see directly) that $L(G)$ is a fixed point of a certain operator $T(G) : \mathcal{P}(\Sigma^*) \to \mathcal{P}(\Sigma^*)$ associated with G; in the present example, $T(G)$ is given by $T(G)(X) = \{axc \mid x \in X\} \cup \{b\}$. What is interesting is that, to define $L(G)$ following this 'denotational' approach, we can use *either* of our standard fixed point techniques. In the first place, we have that $T(G)$ is continuous as a cpo map, and that $L(G)$ is its least fixed point (the Greibach condition is not necessary for this to work). But, secondly, we can define a metric on $P(\Sigma^*)$ in our usual manner (see especially Example 6.2.1(b)) via relations R_n, where $X R_n Y$ holds iff X and Y contain the same words of length $< n$; that is,

$$d(X, Y) = inf\{2^{-n} \mid X R_n Y\}.$$

This may be proved complete by the method of Proposition 6.3.4 (taking the closed sets to be equivalence classes under the R_n, as in Example 6.3.5).

Alternatively, the reader may prefer to show directly that, if (X_i) is a Cauchy sequence, its limit is given as $\{x \mid x \in X_i$ for all but finitely many $i\}$. In any case, $T(G)$ is easily seen to be contracting (because of the Greibach condition), and so $L(G)$ is its *unique* fixed point.

(C) The need for a metric on the 'space of types', $\mathcal{I}(V)$. First we note how function types are construed in this model. If I, J are types (closed subsets of V), then $[I \to J]$ is construed as the set of maps from V to V which map I into J, regarded as itself a subset of V via the canonical embedding of $[V \to V]$ into V. Next, we observe that it is desired to accommodate recursive types, for example, a type satisfying

$$I \cong [I \to T].$$

The problem now is that the function type constructor which we have adopted is not monotonic (still less continuous) in the first argument; indeed it is anti-monotone (with respect to the inclusion ordering). Thus the usual methods of cpo theory cannot be applied. However it turns out that, if the rank function is chosen suitably, the map $\lambda I.[I \to T]$ is a contraction (as are many other useful semantic maps), and so Theorem 6.3.8 gives us a unique fixed point. For the details, see [MacQueen *et al.*, 1986].

Our last topic for this subsection is that of *completion*. Rather as a poset may be 'completed' (via ideals — or sequences) to a dcpo, so any metric space has a completion, that is, a complete metric space into which it is isometrically embedded as a dense subset.

Let (X, d) be a metric space. We may define a distance function over the Cauchy sequences of X by:

$$d'((x_n), (y_n)) = Lim_{n \to \infty} d(x_n, y_n).$$

It is easily checked that d' is well-defined, and that it constitutes a pseudo-metric. Let us denote the metric space corresponding to this pseudo-metric by $(\hat{X}, \hat{d})$. Thus, an element of $\hat{X}$ is an equivalence class of Cauchy sequences under the equivalence defined by:

$$(x_n) \equiv (y_n) \Leftrightarrow Lim_{n \to \infty} d(x_n, y_n) = 0.$$

Theorem 6.3.10. *Let (X, d) be a metric space. Then $(\hat{X}, \hat{d})$ is complete and X is isometric to a dense subset of $(\hat{X}, \hat{d})$; moreover, $(\hat{X}, \hat{d})$ is determined up to isometry by these conditions.*

Proof. X is isometrically embedded as a dense subset of $\hat{X}$ by the map $i : x \mapsto (x, x, \ldots)$. Any Cauchy sequence (x_n) in X (more precisely: Cauchy sequence $(i(x_n))_n$ in $i(X)$) has the limit $[(x_n)]$ in $\hat{X}$; hence by Proposition 6.3.4, $\hat{X}$ is complete.

Now suppose that (Y, d') is a complete metric space which has (an isometric copy of) X as a dense subset. By Proposition 6.3.3, every point of Y is the limit of a Cauchy sequence in X. Suppose that $x, y \in Y, x = Lim x_n, y = Lim y_n, x_n \in X, y_n \in X$. Given $\varepsilon > 0$, we can choose k large enough so that $d(x_k, x) \leq \varepsilon, d(y_k, y) \leq \varepsilon$. Using the triangle inequality, we find that $d(x, y) \leq d(x_k, y_k) + 2\varepsilon$ and $d(x_k, y_k) \leq d(x, y) + 2\varepsilon$. We deduce that $Lim_{n \to \infty} d(x_n, y_n) = d(x, y)$. Thus (Y, d') is isometric with $(\hat{X}, \hat{d})$. ∎

The most familiar example of this construction is the completion of the rationals to the reals. Somewhat less familiar may be the following:

Example 6.3.11 (Processes and non-well-founded sets). In Section 6.2, Example (5), we presented a metric space *FProc* of finite processes.

What is unsatisfactory about this space, from our present point of view, is that it is incomplete. If a is a particular atomic action, for example, the sequence of processes $a, a^2, a^3, \ldots$ (in an obvious notation) is Cauchy, but has no limit. Intuitively, its limit should be the infinite process a^ω.

In view of this, we may propose the completion $\widehat{FProc}$ as a more satisfactory space of processes, which will allow for processes with infinite behaviour (and for specifying processes as fixed points). This is, in effect, the approach of de Bakker and Zucker [de Bakker and Zucker, 1982]. Here we are interested in a slightly different line of development of the example. Let us restrict attention to the case that there is just one atomic action (say a), so that, in depicting trees, the arc labels can be ignored. Then finite trees have a natural interpretation as *hereditarily finite sets*. Specifically, a tree T represents the set of the immediate subtrees of T (and so recursively): thus, NIL represents $\varnothing$, and the tree depicted as:

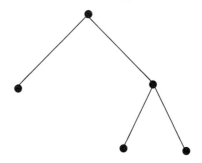

represents $\{\varnothing, \{\varnothing\}\}$ (there is some redundancy in the representation). It is easily seen that two trees are metrically equivalent (that is, the distance between them is zero) if and only if they represent the same set. Denote this space (with just one atomic action) by F.

It has been suggested by Abramsky [Abramsky, 1988] that the completion $\hat{F}$ can reasonably be considered as a space of *non-well-founded sets*. For example, the set whose only element is itself, characterized as the solution of the equation

$$x = \{x\},$$

is constructed as $Lim_n a^n$ (the 'infinite process' mentioned previously). Abramsky provides several alternative characterizations of the space $\hat{F}$, and shows that many useful constructions can be carried out within it. This space of 'finitary' non-well-founded sets seems to offer a pleasant alternative to the purely set-theoretic approach of, for example, [Aczel, 1988; Barwise and Etchemendy, 1988].

6.4 Topology and metric

Our topic now is that of the relation between topological and metrical properties. We consider some elementary topological properties of metric spaces, with emphasis on the non-Archimedean case. In the other direction, we are interested in the conditions on a topological space which suffice to ensure that a compatible (ultra-)metric may be defined on it. In addition, we proffer some speculations on the 'meaning of metrics', in line with our basic thesis on the computational meaning of topology.

Definition 6.4.1. A topological space (X, T) is *metrizable* if a metric can be defined on X such that T is the metric topology.

Similarly, we can define *ultra-metrizable, quasi-metrizable*, etc. With regard to the standard (and rather well-developed) theory of metrizability we shall be extremely selective. In fact, we will confine ourselves to the statement of the following classical result.

Theorem 6.4.2 (Urysohn). *Every second-countable T_3 space is metrizable.* ∎

For the proof of Urysohn's metrization theorem, one may consult any textbook of general topology. Also to be found in any textbook is the demonstration that metric spaces enjoy strong separation properties: in particular, that every metric space is T_4. Assuming that every computationally reasonable space is second countable, this situation helps to explain (as already mentioned in Section 4) why the separation conditions do not loom large in our work.

Our treatment of *ultra*-metrizability will be somewhat less perfunctory than the preceding. Before taking that up, we pause for some reflections on the possible computational significance of (ultra-)metrics in general. If this more speculative material is not to the reader's taste, it may safely be ignored.

Let us recall the basis on which our claims for the computational significance of topology rest. This is the idea that an open set represents a finitely verifiable (or observable) property. In particular (Section 4), the specialization order of space (X, T) is construed as an information ordering: $x \leq_T y$ means that every finitely observable property of x holds also for y. We now propose to refine these ideas by means of the notion of a *boundedly* observable property: a property whose instances can be verified within a number of steps fixed in advance. Thus, whereas (the universe being Σ^ω) '111 occurs in the sequence' is a finitely observable property, '111 occurs in the first fifty terms of the sequence' is a boundedly observable property. It is rather clear in this case that the boundedly observable properties constitute a base for the intended topology (every finitely observable property is a disjunction of boundedly observable properties).

Supposing that the notion of a boundedly observable property can be suitably axiomatized, what can be done with it formally? Let us understand by a *depth k property* a boundedly observable property whose (positive) instances can be verified within k steps. Here, the 'steps' are best thought of as steps in the elaboration of an object, or of refinement of observation of the object. Then we can refine the information (or specialization) order into a sequence of 'bounded' orders: say $x \leq^k y$ iff every depth k property of x holds also for y. The sequence $(\leq^k)_k$ is a sequence of finer and finer pre-orders, from which we can define a distance function exactly as was done previously on the basis of a sequence of equivalence relations (Example 6.2(6)). Since we start with pre-orders rather than equivalence relations, the distance obtained will in general be only a quasi-metric. But the result is that a distance $d(x,y)$ is 'small' iff a 'deep' property (test, or experiment) is needed to distinguish x from y. Our claim is, briefly, that this natural extension of the basic idea of an observable property suffices to account for the presence of metrical notions in the theory of computing.

As for the axiomatization, it seems reasonable to assume that, for each k, the depth k properties constitute a finite lattice, say $\Delta(k)$. Finite since, with a bound on the depth (or precision) of testing, only finitely many distinct outcomes of a test can be envisaged; and a lattice since, if properties p, q require for their verification at most k stages of the elaboration of an object, so presumably do $p \wedge q, p \vee q$. The finiteness of $\Delta(k)$ will be taken account of in due course, although we will not need the lattice aspect. For the immediate purpose of defining a quasi-metric, on X, only very weak assumptions on $\Delta(k) \subseteq \mathcal{P}X$ are needed, viz.: $\Delta(0) \subseteq \Delta(1) \subseteq \ldots$, and $\Delta = \bigcup_k \Delta(k)$ is a subbase of a T_0 topology (that is, for $x, y \in X, (\forall A \in \Delta.x \in A \Leftrightarrow y \in A) \Rightarrow x = y)$.

The theory being hinted at here is still in process of development. Rather than attempt a systematic expositon therefore, it seems best to confine our discussion of it to a series of brief observations on points of special note:

1. One might ask whether the *negation* of a depth k property p is also of depth k (so that p is in effect decidable in k steps). This is not necessarily so, however, if partial objects are under consideration. An object may be insufficiently defined for p to be either affirmed or denied of it.

2. The analysis of metrics in terms of bounded depth properties can at best account for *non-Archimedean* quasi-metrics. Therefore, the analysis is too restricted, even if the non-Archimedean situation is the most usual one in semantics. We may admit that this is true, for the analysis as presented above. However, that presentation was only a simplified version. For a fuller analysis, as in [Smyth,], we

work with a sequence of *relations* $<_k$ (of 'strong inclusion') between properties, rather than just sets $\Delta(k)$ of properties. The meaning of $A <_k B$, roughly, is that for any $x \in A$ we can tell within k steps that $x \in B$. We can define $\Delta(k)$ in terms of $<_k$ by: $A \in \Delta(k) \Leftrightarrow A <_k A$ (but $\Delta(k)$ may well have $\varnothing$ and X as its only elements). The analysis in terms of strong inclusions is able to account for quasi-metrics (or quasi-uniformities) quite generally.

3. We should mention, as an alternative to our topological approach, the logical view of Rowlands-Hughes [Rowlands-Hughes, 1987]. Here, the idea is to construe a 'distance' $d(x, y)$ as the logical value of the assertion $x \sqsubseteq y$: the degree to which x approximates y, one might say (with degree 0 meaning simple truth, or perfect approximation). With this approach, it is possible to present the theory of quasi-metrics as simply the real-valued theory of partial orders, and topics such as completeness and continuity can be presented as a natural extension of the usual domain-theoretic treatment. Connections between the logical and the topological approach exist, but are not yet fully understood.

4. It may be objected that we have, at best, accounted for distances in-the-small. Indeed, there is a certain arbitrariness about the derivation of specific numbers (as distances) from the relations $\leq_k$; technically, it could be said that our procedure determines the metric only up to uniform equivalence. In fact, in previous expositions [Smyth, ; Smyth, 1987], we have claimed only to be developing an approach to quasi-uniformities, not properly to quasi-metrics. (Some remarks on uniformities may be found at the end of this section). We must admit that it remains to be seen whether a convincingly quantitative theory can be developed on the basis proposed here.

5. The question of *uniform* continuity of functions merits more attention than it usually receives in the metric approach to semantics and computability. Consider a function taking bit streams to bit streams. Informally, uniform continuity means that there is a modulus m such that, in attempting to obtain k bits of output, no more than $m(k)$ bits of input need be processed. It can be argued that the modulus serves as a bound on the complexity of computation of the function, or at least as a necessary condition for the existence of such a bound (cf. [Ko and Friedman, 1982]). It should be noted that our formulations of uniform continuity are specifically intended to allow for the case that *partial* (finite) streams are admitted, so that we have to do with quasi-metric spaces (as in Example 6.2(3b)). In this more general setting we have to be careful not to read the condition on the modulus m as saying that $m(k)$ bits of input *suffice* to give k bits of

output (for that would imply that the stream function is total over infinite streams). The idea is that, even if $m(k)$ input bits do not suffice, nothing is gained by processing more than $m(k)$ bits. A 'thesis' to the effect that properly computable functions are uniformly continuous has been suggested in [Smyth, 1989].

Returning now to the more technical development, we briefly take up the question of quasi-metrizability. Quasi-metrization is in general much easier than metrization. Indeed, every second-countable T_0 space can be quasi-metrized, by an almost trivial construction; moreover, quasi-metrics of a very restricted type suffice for the task.

Definition 6.4.3. A metric space (X, d) is *totally bounded* if for every $\varepsilon > 0$ there is a finite set $E \subseteq X$ such that for every $x \in X$ there exists $e \in E$ with $d(x, e) \leq \varepsilon$ (if the space is only quasi-metric, we require here $d^*(x, e) \leq \varepsilon$).

Theorem 6.4.4. *Let $(X, \mathcal{T})$ be a second countable T_0 space. Let $\Delta(0) \subseteq \Delta(1) \subseteq \ldots$ be an increasing sequence of finite subsets of $\mathcal{T}$ such that $\Delta = \cup_k \Delta(k)$ is a subbase of $\mathcal{T}$. Define the relations $\leq^k$ by:*

$$x \leq^k y \Leftrightarrow (\forall A \in \Delta(k).x \in A \Rightarrow y \in A),$$

and the distance function d by:

$$d(x, y) = inf\{2^{-n} \mid x \leq^n y\}.$$

Then d is a totally bounded, non-Archimedean quasi-metric which induces the topology $\mathcal{T}$ on X.

Outline of proof. That d is a non-Archimedean quasi-metric is clear. Suppose that $\varepsilon > 0, x \in X$. Then the ball $B_\varepsilon(x)$ is $\cup\{A \in \Delta(n) \mid x \in A\}$, where n is the least k such that $2^{-k} < \varepsilon$. Since $\Delta(n)$ is finite, $B_\varepsilon(x)$ is open in $\mathcal{T}$. Suppose on the other hand that $x \in U \in \mathcal{T}$. We can choose n and $A \in \Delta(n)$ so that $x \in A \subseteq U$. Then if $\varepsilon \leq 2^{-n}$ we have $B_\varepsilon(x) \subseteq A \subseteq U$. Thus U is a join of open balls; the metric topology coincides with $\mathcal{T}$.

Finally, d is totally bounded. Let $\varepsilon = 2^{-n}$ be given. Define a partition P of X by putting x, y in the same class of P just in case x, y belong to exactly the same members of $\Delta(n)$. Select one element from each equivalence class (of P), giving a (finite) set E. Then each $x \in X$ satisfies $d^*(x, e) \leq \varepsilon$ for some $e \in E$. ∎

From a computational point of view, total boundedness of a (quasi-) metric is a natural requirement: informally, it means that, for any given level of precision of measurement (or depth of testing), only finitely many points can be distinguished. The 'natural requirement' that is *not* satisfied, in general, by the quasi-metric introduced in Theorem 6.4.4 is that of completeness. To ask for completeness in addition to total boundedness would, in fact, impose a severe resriction on the induced T_0 topology; we will return to this point at the end of the subsection.

We now look more specifically at ultrametric spaces. These have many remarkable features not shared with metric spaces in general. We begin with some elementary facts about the system of closed balls of an ultrametric space:

Proposition 6.4.5. *Let (X, d) be an ultrametric space.*

1. *If $x, y \in X, \varepsilon > 0$ and $y \in \overline{B}_\varepsilon(x)$, then $\overline{B}_\varepsilon(x) = \overline{B}_\varepsilon(y)$.*

2. *If two balls meet, then one is contained in the other. In more detail: if $x, y, \in X$ and $\varepsilon \geq \delta > 0$, then either $\overline{B}_\varepsilon(x) \cap \overline{B}_\delta(y) = \varnothing$ or $\overline{B}_\varepsilon(x) \supseteq \overline{B}_\delta(y)$.*

3. *Every closed ball is also open in the metric topology.*

Proof. 1. Assume $y \in \overline{B}_\varepsilon(x)$. If $z \in \overline{B}_\varepsilon(y)$, we have $d(x, z) \leq max\{d(x, y), \varepsilon\}$ and so $z \in \overline{B}_\varepsilon(x)$. Thus $\overline{B}_\varepsilon(y) \subseteq \overline{B}_\varepsilon(x)$. Since it is also true that $x \in \overline{B}_\varepsilon(y)$, we have $\overline{B}_\varepsilon(x) \subseteq \overline{B}_\varepsilon(y)$ by symmetry.

2. If $\overline{B}_\varepsilon(x) \cap \overline{B}_\delta(y) \neq \varnothing$, then $y \in \overline{B}_\varepsilon(x)$ and so, by (1), $\overline{B}_\varepsilon(y) = \overline{B}_\varepsilon(x)$; hence $\overline{B}_\delta(y) \subseteq \overline{B}_\varepsilon(x)$.

3. For any element y of the ball $\overline{B}_\varepsilon(x), \overline{B}_\varepsilon(y) \subseteq \overline{B}_\varepsilon(x)$ by (1); so $\overline{B}_\varepsilon(x)$ is open. ∎

The word *clopen* abbreviates 'closed and open'; thus the closed balls of an ultrametric space are clopen sets w.r.t. the metric topology (it is also true that the open balls are clopen: Exercise 6.6.2(7)). A topological space is *zero-dimensional* if the topology has a base consisting of clopen sets. A closed ball in a metric space is said to have *radius ε* if it can be expressed as $\overline{B}_\varepsilon(x)$ (a closed ball may have many different radii).

The following proposition is now an immediate consequence of Proposition 6.4.5:

Proposition 6.4.6. *Let X be an ultrametric space.*

1. *The topology of X is zero-dimensional.*

2. *If $\varepsilon > 0$, the closed balls of radius ε form a partition of X. (Denote the partition by $P_\varepsilon(X)$.)*

3. *If $\varepsilon \geq \delta \geq 0$, then $P_\delta(X)$ refines $P_\varepsilon(X)$.* ∎

We are now ready for the following result, characterizing ultrametrizable topologies:

Theorem 6.4.7. *Let X be a topological space.*

1. *X is ultrametrizable $\Leftrightarrow$ X is homeomorphic to a subspace of a countable product of discrete spaces.*

2. *(The topology of) X is induced by a complete ultrametric $\Leftrightarrow$ X is homeomorphic to the inverse limit of a sequence of discrete spaces.*

3. *X is induced by a complete, totally bounded ultrametric $\Leftrightarrow$ X is homeomorphic to the inverse limit of a sequence of finite discrete spaces.*

Proof. ($\Rightarrow$) Assume given an ultrametric which induces X. For each $n \in \mathbb{N}$, we will consider the partition $P_{2^{-n}}(X)$ (which we write more simply as P_n) as a discrete topological space. For each n, the map $f_n : X \to P_n : x \mapsto \overline{B}_{2^{-n}}(x)$ is continuous (since closed balls are open sets of X). The induced map $f : X \to \Pi_n P_n$ is injective (distinct points of X belong to distinct elements of the partition P_n, for sufficiently great n). Moreover, regarded as a map from X onto $f(X)$, f is open: each basic open set $B = \overline{B}_{2^{-n}}(x)$ is mapped to $\{p \in \Pi_n P_n \mid p(n) = B\} \cap f(X)$, which is open in the subspace $f(X)$, since $\{B\}$ is open in P_n. Hence (cf. Exercise 5.1.1(1)) f is a homeomorphism of X onto $f(X)$.

Observe now that each $p \in f(X)$ is such that $p(0) \supseteq p(1) \supseteq \dots$. This means that, if we define the maps $j_n : P_{n+1} \to P_n$ by letting $j_n(B)(B \in P_{n+1})$ be the (unique) member of P_n which contains B, then $f(X)$ is in fact a subspace of the inverse limit $L = Lim_{\leftarrow}(P_n, j_n)_n$. Every element of L is a sequence $B_0 \supseteq B_1 \supseteq \dots$ of closed balls of X such that B_n has radius 2^{-n} (all n); the elements of $f(X)$ are precisely the sequences which have non-empty meet. Hence, by Proposition 6.3.4, completeness of the ultrametric implies $f(X) = L$.

The third part of the Theorem (left to right implication) now follows on observing that, if the ultrametric is totally bounded, each P_n is finite.

($\Leftarrow$) (We proceed now a little more rapidly.) Let $V_0, V_1, \dots$ be a sequence of discrete spaces. The formula

$$d(x, y) = inf\{2^{-n+1} \mid x(i) = y(i) \text{ for all } i < n\}$$

defines an ultrametric on ΠV_n which is easily seen to induce the product topology. Thus ΠV_n (and, *a fortiori*, any subspace of it) is ultrametrizable.

Next, suppose that connecting maps $j_n : V_{n+1} \to V_n$ are given. We consider the restriction of d to $L = Lim_{\leftarrow}(V_n, j_n)$.

If $d(x, y) \leq 2^{-n}$, then $x(n) = y(n)$; thus we have for each n an injection $e_n : P_n \to V_n$, where P_n is the partition of L into closed balls of radius 2^{-n}. If $B_n \in P_n, B_{n+1} \in P_{n+1}$ are such that $B_{n+1} \subseteq B_n$, then $j_n(e_{n+1}(B_{n+1})) =$

$e_n(B_n)$ and if $B_0 \supseteq B_1 \supseteq \ldots$ is any nested sequence with $B_n \in P_n$, we have $(e_n(B_n))_n \in \bigcap_n B_n$. By Proposition 6.3.4, L is complete.

Finally, if in addition the spaces V_n are finite, then the partitions P_n are finite, and so L is totally bounded. ∎

Recalling Proposition 5.0.16, we infer the following:

Corollary 6.4.8. *A topological space $(X, \mathcal{T})$ is a second-countable Stone space if and only if $\mathcal{T}$ is induced by a complete, totally bounded ultrametric.*

We note without proof that Theorem 6.4.7(3) and its corollary can be extended to the quasi-metric case, giving for example the following series of equivalences: complete, totally bounded non-Archimedean quasi-metric space ≡ inverse limit of finite T_0 spaces ≡ second-countable spectral space (the second part of the pair of equivalences is well-known; for the first, see [Smyth, 1991]).

If we want to remove the 'non-Archimedean' restriction here, we cannot expect still to have a characterization as an inverse limit of finite spaces, but we have for example (using terminology to be defined in Section 7): a space is given by a complete, totally bounded quasi-metric if and only if it is second-countable stably compact. Unfortunately it is outside the scope of this chapter to discuss results of this type in detail, even though it can be argued (Smy[91]) that they are of considerable significance for the theory of computation.

6.5 Constructions

We now consider some of the principal ways of constructing new metric spaces from old. As this section is already disproportionately long, the presentation will be fairly terse.

A few words about the characterization of the various constructs are appropriate here. Most of the constructs can be seen to be categorical, provided one has chosen for morphisms the *non-expansive maps*. As in the case of topological spaces (Section 5) we also have useful intermediate characterizations in terms of *initial* and *final* structures. Let us order the distance functions on a set X in the following way:

$$d_1 \leq d_2 \Leftrightarrow \forall x, y \in X. d_1(x, y) \leq d_2(x, y).$$

Given, then, a family of maps $f_i : S \to Y_i, i \in I$, where S is a set and each Y_i is a metric (or quasi-metric, etc.) space, the *initial* structure for S will be the least distance function on S such that all the maps f_i are non-expansive; likewise for *final* structure. We will not discuss here the existence of initial and final structures in general (for categories of metric spaces); rather, we will use the idea, when convenient, to characterize particular constructs.

6.5.1   Subspace, quotient

The obvious subspace construction has been taken for granted already in the preceding. If $S \subseteq X, (X,d)$ a metric space, the *subspace* metric for S is just the restriction of d to S. It is the initial structure for the injection of S into X.

The quotient construct is rather more interesting. Strangely, however, it is rarely discussed in the literature. Given a surjection $h : (X,d) \to S$, we require the greatest distance function d' on S which renders h non-expansive. It can be constructed as follows. Define an *h-path* from y to y' (where $y, y' \in S$) to be a finite sequence $x_0, x_1, \ldots x_n$ of points of X such that $h(x_0) = y, h(x_n) = y'$. Define the *length* of the h-path $\pi = x_0, \ldots x_n$ by:

$$l(\pi) = \Sigma \{d(x_i, x_{i+1}) \mid h(x_i) \neq h(x_{i+1})\}.$$

That is, a segment (x_i, x_{i+1}) contributes its 'length' to the length of the h-path provided that x_i, x_{i+1} maps to distinct points of S (and otherwise contributes 0). Finally, we can define d' by:

$$d'(y, y') = inf\{l(\pi) \mid \pi \text{ is an } h\text{-path from } y \text{ to } y'\}.$$

For an example of the quotient construct, and some observations regarding it, see Exercises 6.6.2(9–12).

6.5.2   Sum, product

Given a collection of metric spaces $(X_i, d_i)(i \in I)$, what is the appropriate distance function on the disjoint sum $X = \Sigma X_i$? It is not difficult to decide that, if $x, y, \in X$ are in the same component X_i, then $d(x,y) = d_i(x,y)$. However, if x, y lie in different components, the requirement that the inclusion maps (of the X_i into X) be non-expansive, places no restriction on $d(x,y)$, and so $d(x,y)$ should be assigned the greatest value in $\mathbb{R}^{0+}$ — which does not exist. Faced with this situation, we have the following alternatives:

1. To admit ∞ into our range of values, so that distance functions are taken to be maps to $[0, \infty]$;

2. To choose a fixed bound, say 1, and define the sum only within the subcategory of bounded metric spaces having the bound 1;

3. To abandon the universal characterization of the sum, and fix $d(x,y)$ in some arbitrary way for x, y lying in different components.

Since the sum construct will not be required in any subsequent work, we here make no decision between these alternatives.

We now consider products. Suppose first that $(X_i, d_i)_{i \in I}$ is a *finite* collection of metric spaces. Define a metric d on the product ΠX_i by

$$d((x_i), (y_i)) = max\{d(x_i, y_i) \mid i \in I\}. \tag{1}$$

Then d is clearly the least distance function for which the projections are non-expansive, and the product space so defined is categorical. Moreover, the metric topology of this product is the topological product (of the components X_i considered as topological spaces).

Infinite products are more problematic. If we simply replace *max* by sup in (1), we have of course the problem that the supremum may not exist. Even when the supremum is guaranteed to exist, there is the further difficulty that the metric topology may not agree with the product topology. For example, consider X^ω where X is $\{0, 1\}$ taken with the discrete metric. With the product topology, this is of course Cantor space. However, if the product metric is defined as in (1), we obtain the discrete topology, since for any $x \in X^\omega$ the open ball $B_{\frac{1}{2}}(x)$ is $\{x\}$. For these reasons, one usually defines the product metric on a sequence of bounded metric spaces $(X_i, d_i), i = 0, 1, \ldots$, all having the same bound, as follows (other cases, especially that of uncountable products, are not considered):

$$d((x_i), (y_i)) = sup\{2^{-i}.d(x_i, y_i) \mid i \in \mathbb{N}\}.$$

Infinite products can be of use in computer science where we have a specification that takes the form of an infinite set of equations.

Example 6.5.1. With the notation of Example 6.3.9(A), an *unbounded* buffer could be specified by an infinite sequence of equations

$$\begin{aligned} buffer0 &= in.buffer1 \\ buffer1 &= out.buffer0 + in.buffer2 \\ buffer2 &= out.buffer1 + in.buffer3 \\ &\vdots \end{aligned}$$

The whole sequence can be considered as a single equation $z = h(z)$, with $h : X^\omega \to X^\omega$ a contraction.

The infinite product also finds a more technical use in connection with the solution of metric space 'domain equations' [America and Rutten, 1989].

6.5.3 Function space

A function space construct is well-behaved, categorically speaking, if it yields for each pair of objects X, Y an *exponential* object Y^X; the category as a whole (which is assumed to have finite products) is then *Cartesian closed.* For an account of these ideas we refer to the chapter on Basic Category Theory. For the categories with which we are concerned here, the condition reduces in effect to the existence of *evaluation* and *currying* maps associated with each function space $[X \to Y]$. These maps are defined as usual by:

$$\mathbf{eval} \; : [X \to Y] \times X \to Y :< f, x > \mapsto f(x).$$

$$\mathbf{curry} \; : (W \times X \to Y) \to (W \to [X \to Y]) : \; \mathbf{curry} \; (h)(w)(x) = h(w, x).$$

Here, **eval** is required to be a morphism of the category, while **curry** maps any morphism $h : W \times X \to Y$ to a morphism $\mathbf{curry}(h) : W \to [X \to Y]$ but is not itself assumed to be a morphism.

If X, Y are metric spaces with the distance functions d_X, d_Y, we will take as $[X \to Y]$ the collection of all non-expansive maps from X to Y with distance $d_{X \to Y}$ defined by:

$$d_{X \to Y}(f, g) = sup_{x \in X} d_Y(f(x), g(x))$$

We omit the trivial verification that this makes $[X \to Y]$ a metric space. As usual in this subsection, we take as morphisms the non-expansive maps. It is readily checked that, under these assumptions, **curry** is well-defined: that is, it takes non-expansive maps to non-expansive maps. On the other hand, in order to have **eval** as a non-expansive map we need to assume that the codomain Y is *ultra*metric. Making that assumption we check (with $f, g \in [X \to Y], x, x' \in X$):

$$d_Y(f(x), g(x')) \leq \max \{d_Y(f(x), g(x)), d_Y(g(x), g(x'))\}$$
$$\leq \max \{d_{X \to Y}(f, g), d_X(x, x')\}$$

which is by definition the distance from $< f, x >$ to $< g, x' >$ in $[X \to Y] \times X$. On noting that $[X \to Y]$ is ultrametric if X, Y are, we obtain:

Proposition 6.5.2. *The category of ultrametric spaces (with non-expansive maps) is Cartesian closed.* ∎

We are more particularly interested in complete, totally bounded spaces; the position here is quite favourable:

Proposition 6.5.3. *Suppose that X, Y are metric spaces. Then*

1. Y complete $\Rightarrow [X \to Y]$ complete

2. X, Y totally bounded $\Rightarrow [X \to Y]$ totally bounded.

Proof.   1. Limits are calculated pointwise. Indeed, a trivial check shows
that if (f_i) is a Cauchy sequence in $[X \to Y]$ then $\lambda x.Lim_{i \to \infty} f_i(x)$
is non-expansive and is the limit of (f_i).

2. Assume X, Y totally bounded. Let $\varepsilon > 0$ be given. Choose $x_1, \dots, x_m$
$\in X$ so that every point of X is within a distance $\varepsilon/3$ of some x_i, and
$y_i, \dots, y_n \in Y$ similarly.

Given a map π from $\{x_1, \dots, x_m\}$ to $\{y_1, \dots, y_n\}$, we say that $f : X \to$
Y is of *class* π if $d_Y(f(x_i), \pi(x_i)) \leq \frac{\varepsilon}{3}$ for each i. Obviously, every $f : X \to$
Y is of class π for some π. From each non-empty class π, arbitrarily choose
a (non-expansive) map f_m. We claim that each member f of $[X \to Y]$ is
within a distance ε of some f_π (specifically, any f_π such that f is of class
π).

Indeed for each $x \in X$ we have:

$$
\begin{aligned}
d_Y(f(x), f_\pi(x)) \quad \leq \quad & d_Y(f(x), f(x_i)) + \\
& d_Y(f(x_i), f_\pi(x_i)) + d_Y(f_\pi(x_i), f_\pi(x)) \\
& \text{where } x_i \text{ is chosen to be within } \tfrac{\varepsilon}{3} \text{ of } x \\
\leq \quad & d_X(x, x_i) + \tfrac{\varepsilon}{3} + d_X(x_i, x) \\
\leq \quad & \varepsilon.
\end{aligned}
$$

∎

This enables us to conclude:

Theorem 6.5.4. *The category of complete totally bounded ultrametric
spaces is Cartesian closed.* ∎

Since nothing in the preceding discussion depended on symmetry of the
distance functions, a similar conclusion (with completeness understood as
bicompleteness) holds for the non-Archimedean quasi-metric spaces.

6.5.4 Hyperspace

The Hausdorff metric over the closed, non-empty subsets of a metric space
X has, in effect, already been defined in 6.2, Example (4). To recapitulate,
we put

$$
d_L(S, T) \quad = \quad \sup_{s \in S} \ \inf_{t \in T} \ d_X(s, t)
$$

and then obtain the Hausdorff distance by symmetrizing:

$$
d_H(S, T) = d_L^*(S, T).
$$

This evidently defines a pseudo-metric. To show that it is actually a metric,
suppose $d_H(S, T) = 0$. This means that, for every $s \in S$, $\inf_{t \in T} d(s, t) = 0$.

The latter equation implies that every neighbourhood of s meets T, and so (since T is closed) $s \in T$. This shows that $S \subseteq T$; by symmetry, $T \subseteq S$.

A more restricted 'hyperspace' construct than this is often considered, in which one takes the Hausdorff distance only over the collection $\mathcal{P}_C(X)$ of totally bounded, closed, non-empty subsets of X (the subscript C may be read as *compact*: see the next section). The rationale for considering $(\mathcal{P}_C(X), d_H)$ is somewhat as follows. Note first that a non-empty subset S of X is totally bounded if and only if, for every $\varepsilon > 0$, there is a finite subset E of X with $d_H(E, S) \leq \varepsilon$. Thus the non-empty finite subsets of X constitute a dense subset of $\mathcal{P}_C(X)$. On the other hand, let (E_i) be a Cauchy sequence of non-empty finite subsets of X such that $d_H(E_i, E_{i+1}) \leq 2^{-i}$ for all i. Let S be the set of limit points of sequences (e_i) such that $e_i \in E_i$ and $d(e_i, e_{i+1}) \leq 2^{-i}$. Then $d_H(E_i, \mathrm{Cl}(S)) \to 0$ as $i \to \infty$; hence $Cl(S)$ (is totally bounded and) is the limit of E_i in $P_C(X)$. Thus (Proposition 6.3.6, Theorem 6.3.10) $P_C(X)$ is the completion of the collection of finite subsets of X w.r.t. d_H. Roughly speaking, $P_c(X)$ is the least complete space of subsets of X which includes the singleton sets and is closed under finite joins; it is thus the appropriate metric construct with which to discuss bounded non-determinism.

6.6 A note on uniformities

Whereas a topology gives us, in the neighbourhoods of a point, a local notion of 'closeness', a metric gives us degrees (or relations) of closeness that apply globally; it is this global notion of closeness that enables us to define Cauchy sequences and completeness, uniform continuity of functions, etc. In many of the examples in this section, we have proceeded by first defining a family of relations of closeness (or degrees of indistinguishability), and then assigning numbers as distances in a slightly arbitrary fashion. If the main benefit we seek is a global (or uniform), rather than a local, notion of closeness, it might well be asked whether we could not work directly with the family of relations, bypassing the assignment of numbers. This is precisely what the notion of a quasi-uniformity accomplishes:

Definition 6.6.1. A *quasi-uniformity* on a set X is a filter $\mathcal{U}$ of relations (i.e. subsets of $X \times X$), called 'entourages', satisfying:"

1. $\forall U \in \mathcal{U} \forall x \in X. x U x$

2. $\forall U \in \mathcal{U} \exists V \in \mathcal{U}. V \circ V \subseteq U.$

We are not going to prove any theorems about quasi-uniformities here, but the following remarks are appropriate:

1. In metric terms, a typical entourage may be thought of as
   $U_\varepsilon = \{< x, y > |\ d(x, y) \le \varepsilon\} (\varepsilon > 0)$ — or, more generally, as a super-
   set of some U_ε.

2. In condition (2) one may think of V as $U_{\varepsilon/2}$, if U is given as U_ε; the
   condition corresponds to the triangle inequality for metrics.

3. The base of the quasi-uniformity $\mathcal{U}$ is a collection $\mathcal{B}$ of relations on
   X such that $\mathcal{U}$ is the collections on supersets of members of $\mathcal{B}$. The
   family of relations U_ε obtained from a quasi-metric d (as in (1)) forms
   a base of a quasi-uniformity. Moreover, the families of relations we
   have used throughout the section as auxiliaries in defining (quasi)-
   metrics are quasi-uniform bases.

4. A quasi-uniformity $\mathcal{U}$ satisfying a suitable symmetry condition ($U \in$
   $\mathcal{U} \Rightarrow U^{-1} \in \mathcal{U}$) is a *uniformity*, which can be considered as an abstraction from a metric.

In computer science, bases of uniformities occur very commonly in the form
of sequences of finer and finer equivalence relations; more generally, bases of
quasi-uniformities appear as (decreasing) sequences of transitive relations.
As already indicated, it is for many purposes sufficient to work with these
families of relations, rather than introduce explicit distances. However, the
case of fixed point definitions is somewhat problematic: we do not have,
for uniform spaces, a simple canonical result like the Banach theorem.

Exercises 6.6.2.

1. The following is one of the more natural 'Cauchy' conditions for a
   sequence (x_n) in a quasi-metric space (X, d):

   $$\forall \varepsilon > 0\ \exists k.\ k \le l \le m\ \Rightarrow\ d(x_l, x_m) \le \varepsilon. \qquad (1)$$

   A sequence satisfying (1) might be called *forward* Cauchy (there is
   no standard terminology). Verify that a sequence is (bi-) Cauchy
   in (X, d) iff it is forward Cauchy with respect to both d and d^{-1}.
   Determine what forward and bi-Cauchy sequences amount to in the
   case of a poset with the $(0, 1)$ quasi-metric.

2. In connection with the definition of Hausdorff distance, carefully explain why, if the empty set were admitted, we would have to put (in
   general) $d_H(S, \varnothing) = \infty$. Suppose that we extend the definition of a
   metric space by allowing the distance function to be a map into the
   extended non-negative reals (i.e. admitting ∞ as a distance), interpreting the triangle inequality appropriately. Show that an extended
   Hausdorff metric can be defined over the arbitrary closed subsets of

a metric space (that is, there is no need to exclude the empty set and unbounded sets).

3. Assume the truncation operation extended to *sets* of sequences, that is, put

$$S \restriction n = \{x \restriction n \mid x \in S\},$$

where S is a set of sequences (words) in some alphabet Σ. Show that the Hausdorff distance for sets of sequences (Example (4) above) can alternatively be defined by:

$$d_H(S, T) = \inf\{2^{-n} \mid S \restriction n = T \restriction n\}.$$

4. Let D be an ω-algebraic cpo, and $r : B_D \to \mathbb{N}$ a rank function (cf. Example (6) above) such that $r^{-1}(n)$ is a finite set for each $n \in \mathbb{N}$. Define a distance function over D by

$$d(x, y) = \inf\{2^{-n} \mid e \sqsubseteq x \Rightarrow e \sqsubseteq y \text{ for every } e \text{ of rank } \leq n\}.$$

Show that d is a totally bounded quasi-metric which induces the Scott topology of D.

5. Show that the 'closed ε-balls' of a quasi-metric space are not in general topologically closed.

6. Formulate a version of Theorem 6.3.8 (fixed point theorem) for pseudo-metric spaces.

7. Show that, in an ultrametric space, open ε-balls are topologically closed (as well as open).

8. Returning to Exercise (4), show that if D is actually a Scott domain, then (D, d) is a (bi-)complete. (It follows from general results, not proved in this chapter, that (D, d) is complete iff D is spectral in its Scott topology, which amounts to saying that D is '2/3-SFP' in the terminology of [Plotkin, 1981].)

9. Let C be the (Cantor) space of infinite binary sequences, with metric defined as in Example (3(a)) above. By considering the elements of C as representing reals in the usual way ($\sigma \mapsto \cdot\sigma$), we have a surjection $q : C \to I(= [0, 1])$. Show that the quotient metric induced by q is the usual Euclidean metric of the unit interval. (Hint: consider 'grids' on I composed of dyadic rationals $m/2^n$.)

10. Show by examples that (a) the quotient of a metric space may be only *pseudo*-metric, and (b) the topology induced by the quotient (pseudo-)metric need not coincide with the quotient topology.

11. Let G be a directed graph, its arcs labelled with non-negative real numbers. Show that, if $d(x, y)$ is defined as the shortest path length

from x to y (more precisely: $d(x,y) = inf$ {length (π) | π is a finite path from x to y}), then (Nodes $(G), d$) is a pseudo-quasi-metric space. Noting that any metric space X may trivially be considered as an arc-labelled graph, show that the quotient of X by a given equivalence relation may be constructed by taking the shortest path length (pseudo-)metric in the graph obtained by identifying the equivalent nodes.

12. Graph-theoretic terminology (Question 11) makes it easy to generalize the quotient metric construction to 'final structures' for (pseudo-quasi-)metrics. Let S be a set, $(X_i)_{i \in I}$ a family of metric spaces, and d_i a map from X_i to S (for each $i \in I$). Build a graph G having S as its set of nodes, by inserting arcs as follows: for each $i \in I$, and $x, y \in X_i$, insert an arc from $f_i(x)$ to $f_i(y)$ labelled with $d_i(x,y)$. Show that the shortest path length function over $S \times S$ is the greatest distance (metric) definable over S such that the maps f_i are non-expansive. [Note: the distance ∞ has to be admitted in this construction, unless restrictions are imposed on S.]

7 Compactness

7.1 Compactness and finiteness

Besides being central to topology, the notion of compactness is of great computational interest. This is because of the finite (or, better, *finitary*) aspect of compact spaces and sets. In this subsection we will pursue the finitary theme via a series of alternative formulations of the basic notion.

Definition 7.1.1. A topological space X is *compact* if, for every collection $\mathcal{C}$ of open sets which covers X, some finite subset of $\mathcal{C}$ already covers X; briefly, if every open cover of X has a finite subcover. A subset S of X is *compact* if it is compact as a subspace of X; equivalently, S is *compact* if every cover of S by open subsets of X has a finite subcover.

Noting that, if $\mathcal{C}$ is any cover of a set S, the joins of the finite subsets of $\mathcal{C}$ form a *directed* cover of S, we see that the definition can also be given as follows: $S \subseteq X$ is *compact* if, whenever $\mathcal{C}$ is a directed open cover of S, some member of $\mathcal{C}$ already contains S. Thus, we can also say that a space X (likewise, any open subset of X) is compact iff it is a finite (or compact!) element of the lattice $\Omega(X)$, in the usual sense of domain theory or lattice theory.

More generally, a set S is compact iff $Th(S)$ is Scott-open (the condition studied in Exercise 4.4.6(13)). Thus, W is compact iff S (or, better, $\uparrow S$)

is 'finitarily specifiable'.

As trivial examples of compact spaces we have: all finite spaces; and any space having a least element in its specialization order (in particular, any cpo with Scott topology). Non-trivial examples are rather harder to come by. Indeed we can only obtain them by appealing to results such as Theorems 7.1.3 and 7.5.3 below. Since all such results are non-constructive, it can be argued (as by Bishop and Bridges [Bishop and Bridges, 1985]) that there are no non-trivial compact spaces.

A formulation of compactness dual to that of Definition 7.1.1 often proves useful. A collection $\mathcal{F}$ of sets is said to have the *finite intersection property* if each finite subset of $\mathcal{F}$ has non-empty meet. We have then the following criterion for compactness in terms of closed sets.

Proposition 7.1.2. *A space X is compact if and only if the meet of each collection of closed subsets of X having the finite intersection property is non-empty.*

Proof. If $\mathcal{C}$ is a collection of subsets of X, then, by de Morgan's formulas, $X - \bigcup \{S \mid S \in \mathcal{C}\} = \bigcap \{X - S \mid S \in \mathcal{C}\}$; hence $\mathcal{C}$ fails to cover X iff the meet of the complements of the members of $\mathcal{C}$ is non-empty. Now X is compact iff each collection of open sets, such that no finite subcollection covers X, fails to be a cover, and this is true iff each collection of closed sets having the finite intersection property has non-empty meet. ∎

Theorem 7.1.3. *A metric space is compact if and only if it is complete and totally bounded.*

Proof. ONLY IF: suppose the metric space (X, d) is compact. For every $\varepsilon > 0$, the cover of X by all (open) ε-balls has a finite subcover. Hence X is totally bounded. A nested sequence of closed balls clearly has the finite intersection property. Hence, by Propositions 6.3.3 and 7.1.2, X is complete.

IF: let $\mathcal{C}$ be a collection of open sets of the complete, totally bounded metric space (X, d). Suppose that each finite subset of $\mathcal{C}$ fails to cover X. Choose a point x_0 such that no finite subset of $\mathcal{C}$ covers $\overline{B}_1(x_0)$; this is possible since, if every 1-ball had a finite cover by members of $\mathcal{C}$, we could by total boundedness obtain a finite cover of X. By the same token we may choose successively $x_1, x_2, \dots$ such that $d(x_n, x_{n+1}) \leq 2^{-n}$ and no finite subset of $\mathcal{C}$ covers $\overline{B}_{2^{-n}}(x_n)(n = 0, 1, \dots)$. Let x be the limit of the Cauchy sequence (x_n). Then x does not belong to any member of $\mathcal{C}$; for, if $x \in A \in \mathcal{C}$, we would have $\overline{B}_{2^{-n}}(x_n) \subseteq A \in \mathcal{C}$ for some (sufficiently great) n. Thus $\mathcal{C}$ does not cover X. ∎

Example 7.1.4.

1. (Heine–Borel Theorem) A closed, bounded interval of the real line is compact.

2. By the corollary to Theorem 6.4.7, every second-countable Stone space is compact.

The proof that complete + totally bounded $\Rightarrow$ compact is not acceptable constructively. The response of Bishop and Bridges [Bishop and Bridges, 1985] is to dispense with compactness as usually defined and work explicitly with complete totally bounded metric spaces. This approach is not so convenient for us, as we do not wish to be confined to metric (therefore Hausdorff) spaces; Theorem 7.1.3 clearly fails (even in the ONLY IF direction) for *quasi*-metric spaces.

For a characterization that works in general sober spaces, we recall the discussion of Scott-open filters from Exercise 4.4.6(13). If S is any compact subset of a space X, the collection $\mathcal{N}(S)$ of open supersets of S is evidently Scott-open in $\Omega(X)$. If S is an upper set in the specialization order of $X(S =\uparrow S)$, then $S = \bigcap\mathcal{N}(S)$. Hence any compact upper set is the meet of a Scott-open filter[1] of open sets. Our aim now is to establish a converse of this (given that X is sober), as a result of which we can identify the compact (upper) sets as the sets which are 'finitarily specifiable'.

The proof makes use of a certain correspondence between the points of a space X and the *prime* elements of the lattice $\Omega(X)$:

Definition 7.1.5. An element a of a lattice L is *prime* if $a \neq 1$ and, for any $b, c \in L, b \wedge c \leq a$ implies that $b \leq a$ or $c \leq a$.

Observe that, if x is any point of a space X, then $int(x - \{x\})$ (which is the largest open set U such that $x \notin U$) is a prime element of $\Omega(X)$, since, if $B \cup C$ is not a neighbourhood of x, at least one of B, C is not a neighbourhood. On the other hand, if a is any prime element of a locale L, we can verify that the set $F(a) = \{b \in L \mid b \not\leq a\}(= L- \downarrow a)$ is a completely prime filter over L, and that a is the largest element of L that does not belong to $F(a)$. The reader will have no difficulty in completing the proof of the following:

Proposition 7.1.6. Let L be a locale. Then the map $F :$ Prime $(L) \to Pt(l) : a \mapsto L- \downarrow a$ is a bijection between the primes of L and the completely prime filters over L, with inverse given by $F^{-1}(\mathcal{F}) = \bigvee\{a \mid a \notin \mathcal{F}\}$

Of course this gives us, for any sober space X, a bijection between X and Prime $(\Omega(X))$.

---

[1]In this context only, we allow filters that are not proper. The empty (compact) set is the meet of the improper filter $\Omega(X)$.

Suppose now that $\mathcal{F}$ is a Scott-open, but not necessarily completely prime, filter over a locale L. Then we can no longer conclude that $\bigvee(L - \mathcal{F}) \notin \mathcal{F}$, nor that this join is prime (in L). However, the existence of suitable primes is assured by:

Lemma 7.1.7. *Suppose that $\mathcal{F}$ is a Scott-open filter of the locale L, and $a \notin \mathcal{F}$. Then there is a prime b such that $a \le b$ and $b \notin \mathcal{F}$.*

Proof. Let $Z = {\uparrow}a - \mathcal{F}(= \{c \mid a \le c \& c \notin \mathcal{F}\})$. If C is any totally ordered chain in Z, then $\bigcup C \in Z$ (since if $\bigcup C \in \mathcal{F}$, then, by the Scott-open property, some element of C is in $\mathcal{F}$). Hence by Zorn's Lemma, Z has a maximal element, say b. To show that b is prime, suppose $c \wedge d \le b$. The elements $b \vee c, b \vee d$ are not both in the filter $\mathcal{F}$, since their meet is b. Say $b \vee c \notin \mathcal{F}$. Then, since b is maximal with this property, $b \vee c = b$; thus $c \le b$. ∎

The desired result is now easily proved:

Theorem 7.1.8. *Let X be a sober space. Then the assignment $\mathcal{F} \mapsto \bigcap \mathcal{F}$ defines an order-reversing bijection between the Scott-open filters of $\Omega(X)$ and the compact upper sets of X.*

Proof. Assume that $\mathcal{F}$ is an open filter of $\Omega(X)$. Suppose that A is an open superset of $\bigcap \mathcal{F}$. We claim that $A \in \mathcal{F}$. For, if not, we would have by Lemma 7.1.7 a prime B of $\Omega(X)$ such that $A \subseteq B$ and $B \notin \mathcal{F}$; thence, by Proposition 7.1.6, a completely prime filter $\mathcal{F}'$ such that $\mathcal{F} \subseteq \mathcal{F}'$ and $B \notin \mathcal{F}'$ (and so $A \notin \mathcal{F}'$); and thence (since X is sober) a point x such that $x \in \bigcap \mathcal{F}$ while $x \notin A$. This also shows that $\bigcap \mathcal{F}$ is compact, since if $\mathcal{C}$ is a directed open cover of $\bigcap \mathcal{F}$, we have $\bigcup \mathcal{C} \in \mathcal{F}$, hence $A \in \mathcal{F}$ for some $A \in \mathcal{C}$, and hence $\bigcap \mathcal{F} \subseteq A$. In sum: $\bigcap \mathcal{F}$ is compact, and $\mathcal{N}(\bigcap \mathcal{F}) = \mathcal{F}$. We have previously observed that, if S is a compact upper set, then $\mathcal{N}(S)$ is Scott-open and $\bigcap \mathcal{N}(S) = S$. We thus have a bijection; that it is order-reversing is evident. ∎

In practice, the spaces with which we deal in the computing context typically have an ample supply of compact open sets, and the 'finitary theories' can be taken to be, rather than the Scott-open filters of general open sets, simply the filters of compact open sets. This is the topic of the next subsection.

7.2 Spectral spaces

We have defined (Section 2.6) a space to be *spectral* if it is homeomorphic to the space of prime filters of a distributive lattice. The discussion of compactness enables us to give a more direct characterization of the spectral spaces. The task is largely accomplished, indeed, once we have identified the compact (upper) sets of a spectral space. One way to do this is to construct the appropriate locale (by 'ideal completion') from the given distributive lattice, and then apply Theorem 7.1.8. However, it is also possible to bypass the construction of locales, repeating the development in the latter part of Section 7.1 in a simplified form for the special case of spectral spaces. In view of the central importance of spectral spaces for computer science, we have thought it worthwhile to adopt the latter course here.

Denote by $Sp(L)$ the spectral space derived from a distributive lattice L. The points of $Sp(L)$ are the prime filters of L. We intend to show that the compact upper sets of $Sp(L)$ can be identified with the (arbitrary) filters of L — and thus, one may perhaps say, with the 'theories', or specifications, over L.

In place of Lemma 7.1.7 it will be appropriate to use the closely related, but more standard Prime Ideal Theorem, which may be formulated as follows:

Theorem 7.2.1. *Suppose that F is a filter of the distributive lattice L and I is an ideal disjoint from F. Then there exists a prime ideal M of L such that M extends I and is disjoint from F.* ∎

The proof via Zorn's Lemma of this theorem follows the pattern of the proof of 7.1.7.

A prime ideal of L gives us a point of $Sp(L)$ simply by taking the complement. In fact, we have:

Proposition 7.2.2. *Let L be a lattice. Then complementation (that is, the operation $I \mapsto L - I$) gives a bijection between the prime ideals of L and the prime filters of L.*

Proof. Easy verification. ∎

Given $a \in L$, L a distributive lattice, we denote by $P(a)$ the set of prime filters F of L such that $a \in F$. The collection $\{P(a) \mid a \in L\}$ is of course our standard base for the topology of $Sp(L)$. We now have the following key result:

Proposition 7.2.3. *Let F be a filter of the distributive lattice L. Then $K = \bigcap \{P(a) \mid a \in F\}$ is a compact upper subset of $Sp(L)$ such that $a \in F$ iff $K \subseteq P(a)$.*

Proof. To see that K is compact, let $\mathcal{C}$ be a directed open cover of K. The set $I = \{a \in L \mid \exists U \in \mathcal{C}.P(a) \subseteq U\}$ is clearly an ideal of L. Suppose, if possible, that I is disjoint from F. Then by Theorem 7.2.1 there is a prime ideal M which extends I and is disjoint from F. This means that $L - M$

is a prime filter which extends F and is therefore an element of K, while $L - M$ is not an element of $P(a)$ for any $a \in I$. But this is impossible, since $\{P(a) \mid a \in I\}$ covers K. Hence some element of I is in the filter F and this implies that some member of $\mathcal{C}$ contains K; thus K is compact. The same argument also shows that, if $K \subseteq P(a)$, then $a \in F$ (here we take for I the ideal $\downarrow a$). This completes the proof of the non-trivial parts of the proposition. ∎

Now suppose that K is a compact upper set of $Sp(L)$, and let F be the filter $\{a \in L \mid K \subseteq P(a)\}$. If U is an open superset of K, then by compactness of K we have $K \subseteq P(a) \subseteq U$ for some basic open set $P(a)$. Since K, being an upper set, is the meet of its open supersets, we have $K = \bigcap\{P(a) \mid a \in F\}$. Altogether we have demonstrated:

Theorem 7.2.4. *Let L be a distributive lattice. Then the assignment $F \mapsto \bigcap\{P(a) \mid a \in F\}$ defines an order reversing bijection between the filters of L and the compact upper sets of $Sp(L)$.* ∎

Applying this theorem to principal filters (of the form $\uparrow a$), we obtain:

Corollary 7.2.5. *For each $a \in L$, the open set $P(a)$ is compact. Moreover, $a \le b$ iff $P(a) \subseteq P(b)$.* ∎

Since P is thus an order isomorphism (between L and $P(L)$), we can see that, if $\mathcal{F}$ is any completely prime filter(-base) of basic open sets of $Sp(L)$ (cf. the Appendix), then $F = \{a \in L \mid P(a) \in \mathcal{F}\} (= P^{-1}(\mathcal{F}))$ is a prime filter of L, and moreover F is the unique point of $Sp(L)$ such that $\mathcal{F}$ is a base of neighbourhoods of F. This shows that spectral spaces are sober. Noting that any compact open subset of $Sp(L)$ is of the form $P(a)$ for some $a \in L$, we have one half of:

Theorem 7.2.6. *A space X is spectral if and only if X is sober and the collection $K\Omega(X)$ of compact open subsets of X satisfies*

1. *$K\Omega(X)$ is a base of the topology*

2. *$K\Omega(X)$ is a sublattice of $\Omega(X)$.*

Proof. To complete the proof, assume that X is a sober space satisfying (1) and (2). Since X is sober, X is homeomorphic with the space of completely prime filter-bases of $K\Omega(X)$ (that is, of filters of $K\Omega(X)$ which, considered as filter-bases of $\Omega(X)$, are completely prime). But, due to the compactness of the open sets in question, this is simply the space of prime filters of $K\Omega(X)$. Hence X is spectral. ∎

Comparing the development of Theorems 7.2.1 – 7.2.4 with 7.1.6 – 7.1.8, the simplification we have achieved by restricting to the spectral case is perhaps not very great. The passage from the lattice theoretic formulations to the corresponding spatial ones inevitably involves rather complex argumentation which is, moreover, non-constructive (in its dependence on maximality principles such as Zorn's Lemma). In the light of this, one may wonder whether it would not be better to remain entirely on the lattice-theoretic side of the divide: speaking, for example, of specifications as filters, and making no mention of compact upper sets. Ultimately this may indeed be the best way to proceed. However, if we are to remain in touch with the present state of the subject (topological aspects of computation), we have to take account of both sides, and expend a certain amount of energy in establishing equivalences between them.

7.3 Positive and negative information: patch topology

The spaces in which we are typically most interested have partial as well as total or completed elements. Typically, the topology of such a space can be considered as being based on the detection of positive features in finite (partial) elements; in this context the criterion of a 'positive' feature is that, once it holds of an approximate or partial element a, it holds for all extensions of a.

From time to time we have alluded to the possibility of also taking negative properties into account in defining the topology. Recalling in particular the simple example of Section 1, let us view Σ^∞ as representing a class of rudimentary processes, where a process is conceived to be just a sequence of possible actions. If we now suppose that a process must either accept or *refuse* a proffered action (and does not have the possibility of diverging, that is of proceeding indefinitely without either acceptance or refusal), then we may argue that the appropriate topology for the space of processes is that in which the sets $\Sigma^\infty - \uparrow \sigma$ (σ finite) as well as the sets $\uparrow \sigma$ are taken as open. For example, the set $\Sigma^\infty - \uparrow ab$ corresponds to the observation that a given process either refuses a, or accepts a and then refuses b. Notice that, if Σ is infinite, the topology so obtained is different from the metric topology of Section 6 (Example 6.2(3a)): singleton sets $\{\sigma\}$, σ finite, are open in the metric topology, but not in the 'positive and negative' topology. The latter topology is the one in which we find that 'NIL is not observable'; in order to observe that a process is NIL (or $\wedge$) we would have to observe that it refuses every action in the set Σ, which is impossible if Σ is infinite.

A most important feature of the topology of positive and negative information is that it is *compact*; more than that, it defines a *Stone Space*. (By contrast, it is an easy exercise to show that Σ^∞ with the metric topology is not compact if Σ is infinite.) With a view to formulating this result in a general setting, let us try to describe in abstract terms the steps by which we arrived at the 'positive and negative' topology for Σ^∞. We begin with the 'positive' topology — namely, for Σ^∞, the Scott topology $\mathcal{T}$. This topology is spectral and, as such, has a base $\mathcal{C}$ of compact open sets — these being, in the present example, the finite unions of sets of the form $\uparrow \sigma$ (σ finite). Next, we introduce the collection $\mathcal{C}^*$ of the complements of the members of $\mathcal{C}$. Then the topology $\mathcal{T}'$ of positive and negative information derived from $\mathcal{T}$ is that which has $\mathcal{C} \cup \mathcal{C}^*$ as subbase. Equivalent descriptions of $\mathcal{T}'$: it has a base consisting of the sets $U \cap V (U \in \mathcal{C}, V \in \mathcal{C}^*)$; and it has as a base the sub-Boolean-algebra of $\mathcal{P}\Sigma^*$ generated by $\mathcal{C}$.

As this description suggests, the construction makes sense for any spectral space. By definition, every Stone space is spectral; what the construction will show is that, in a certain sense, every spectral space can be considered as a Stone space.

We begin by slightly reformulating the definition of *spectral* (and *Stone*). The following terminology will be used: if $\mathcal{B}$ is a base for the topology $\mathcal{T}$ of a space X, then the $\mathcal{B}$-*neighbourhood filter*, $\mathcal{N}_\mathcal{B}(x)$, of a point $x \in X$ is the set $\{U \in \mathcal{B} \mid x \in U\}$. Thus $\mathcal{N}_0(x)$ (in the notation of Section 4) can be written also as $\mathcal{N}_\mathcal{T}(x)$. If $\mathcal{B}$ happens to be a sublattice of $\mathcal{T}$ (equivalently, of $\mathcal{P}X$), then $\mathcal{N}_\mathcal{B}(x)$ is evidently a prime filter of $\mathcal{B}$ for any $x \in X$. By slightly rearranging the definitions, then, we obtain:

Lemma 7.3.1. *A space X is spectral iff its topology has a base $\mathcal{B}$ such that*

1. *$\mathcal{B}$ is a sublattice of $\mathcal{P}X$*

2. *every prime filter of $\mathcal{B}$ is the $\mathcal{B}$-neighbourhood filter of a unique point of X.*

In particular, X is a Stone space iff it has a base $\mathcal{B}$ which is a sub-Boolean-algebra of $\mathcal{P}X$, and such that every ultrafilter of $\mathcal{B}$ is $\mathcal{N}_\mathcal{B}(x)$ for a unique $x \in X$ ∎

From the results of the preceding subsection, we know that if a base $\mathcal{B}$ satisfies the conditions of Lemma 7.3.1, then $\mathcal{B}$ is the lattice of compact open subsets of X.

Definition 7.3.2. Let $(X, \mathcal{T})$ be a spectral space, and $\mathcal{C}$ the lattice of compact open sets of X. Put $\mathcal{C}^* = \{X - U \mid U \in \mathcal{C}\}$. The *patch topology*, $\mathcal{T}'$ of X is the topology having $\mathcal{C} \cup \mathcal{C}^*$ as subbase.

The notation of Definition 7.3.2 will be presupposed in the proof of the following theorem. In addition, we will denote by $\mathcal{B}$ the sub-Boolean-algebra of $\mathcal{P}X$ generated by $\mathcal{C}$. It is clear that the members of $\mathcal{B}$ are precisely the finite unions of sets of the form $U \cap V$ where $U \in \mathcal{C}, V \in \mathcal{C}^*$. The patch topology has $\mathcal{B}$ as base.

Theorem 7.3.3. *Let* $(X, \mathcal{T})$ *be a spectral space, with patch topology* $\mathcal{T}'$. *Then* $(X, \mathcal{T}')$ *is a Stone space.*

Proof. Let $\mathcal{F}$ be a prime filter of $\mathcal{B}$. Then $\mathcal{F} \cap \mathcal{C}$ is clearly a prime filter of $\mathcal{C}$. Since X is spectral, there is a unique point x such that $\mathcal{F} \cap \mathcal{C} = \mathcal{N}_{\mathcal{C}}(x)$; we will show that this point x also satisfies $\mathcal{F} = \mathcal{N}_{\mathcal{B}}(x)$.

1. Every $A \in \mathcal{F}$ is a neighbourhood of x. Indeed, since $\mathcal{F}$ is prime, we have $A \supseteq U \cap V \in \mathcal{F}$, where $u \in \mathcal{C}, V \in \mathcal{C}^*$. Now $x \in U$, since $u \in \mathcal{F} \cap \mathcal{C}$; but $x \in V$ also, since if not we would have $x \in X - V$, which implies that $X - V \in \mathcal{F} \cap \mathcal{C} \subseteq \mathcal{F}$, and hence that $\varnothing = V \cap (X - V) \in \mathcal{F}$.

2. Every $\mathcal{B}$-neighbourhood of x is a member of $\mathcal{F}$. Clearly it suffices to show this for any neighbourhood of x of the form $U \cap V (u \in \mathcal{C}, V \in \mathcal{C}^*)$. Now if $x \in U \cap V, U \in \mathcal{F} \cap \mathcal{C} \subseteq \mathcal{F}$; but also $V \in \mathcal{F}$, since if not we would have (since $\mathcal{F}$ is an ultrafilter of $\mathcal{B}$), $X - V \in \mathcal{F} \cap \mathcal{C}$, implying that $x \notin V$.

Further if $y \neq x$, then $\mathcal{N}_{\mathcal{B}}(y) \neq \mathcal{N}_{\mathcal{B}}(x)$ since $\mathcal{N}_{\mathcal{C}}(y) \neq \mathcal{N}_{\mathcal{C}}(x)$ (and $\mathcal{C} \subseteq \mathcal{B}$). Thus we have shown that x is the unique point satisfying $\mathcal{F} = \mathcal{N}_{\mathcal{B}}(x)$. ■

The patch topology $\mathcal{T}'$ of a spectral space $(X, \mathcal{T})$ can equivalently (to Definition 7.3.2) be described by saying that $\mathcal{T}'$ extends $\mathcal{T}$ by counting the compact $\mathcal{T}$-open subsets of X as $\mathcal{T}'$-closed. Now, in a spectral space, every compact upper set may be expressed as the intersection of compact open sets; hence we may equivalently describe the extension by saying that it takes the compact upper sets as $\mathcal{B}'$-closed. The point of this remark is to indicate that, although we are applying the patch topology only in the context of spectral spaces, it can in fact be given a reasonable definition for arbitrary spaces: the *patch* of a space $(X, \mathcal{T})$ is the least extension of $\mathcal{T}$ in which the compact (w.r.t. $\mathcal{T}$) upper (w.r.t $\leq_{\mathcal{T}}$) sets are closed.

'Positive and negative information' is often introduced in terms of *Lawson* topology rather than patch topology. The Lawson topology is best regarded as a topology for a dcpo finer than the Scott topology.

Definition 7.3.4. Let D be a dcpo. The *Lawson topology* of D is the least extension of the Scott topology for which each set of the form $\uparrow x$ is closed.

Since every set $\uparrow x$ is trivially compact, we see that the Lawson topology of a dcpo D is coarser than the patch of the Scott topology of D. But it is also easy to see that, when D is sufficiently well-behaved (for example, when it is a Scott domain), the Lawson topology coincides with the patch of the Scott topology (Exercise 7.7.8(8)). For an extended treatment of the Lawson topology, and some discussion of its connection with patch topology, see [Gierz et al., 1980].

The patch topology (under the name of Cantor topology) has been used by Plotkin (1976) in the description of his power domain construction. For a recent example of the use of patch (or Lawson) topology in connection with the semantics of logic programming, see [Batarekh and Subrahmanian, 1989].

We claimed above that the patch construction would show that ' ... every spectral space can be considered as a Stone space.' Now this claim is in need of some qualification, since it is rather evident that, in taking the patch, we may lose some information about the original topology: a topology cannot, in general, be recovered from its patch. For example, if (P, T) is a finite poset with Alexandroff topology, then the patch, T', is simply the discrete topology on P. In fact, what is 'lost' (in general) in going from a spectral topology T on a set X to its patch T' is precisely the order, $\leq_T$. The original topology can be recovered from the 'ordered Stone space' $(X, T', \leq_T)$: the open sets of T are the open upper (w.r.t. $\leq_T$) sets of T'.

For the full statement of this, involving the equivalence between spectral spaces and (suitably defined) *ordered Stone spaces*, we refer to [Priestley, 1972] or [Johnstone, 1982]. Here we remark that ordered Stone spaces are a species of compact *ordered spaces*, which have been widely studied since their introduction by [Nachbin, 1965]. This provides the possibility of an alternative approach to spectral spaces, and thus to Scott domains.

7.4   Hyperspaces

In this subsection we present a purely topological hyperspace construction. We make no attempt to achieve maximum generality with this presentation; our goal, rather, is a version which is easy to compare with the previous metrical construction (Section 6.5).

Definition 7.4.1. Let X be a topological space, and S be a collection of subsets of X. The *finite* (or *Vietoris*) topology on S has as a subbase the collection of all sets of the following two types, with O ranging over $\Omega(X)$:

$$U_O = \{S \mid S \subseteq O\}$$
$$L_O = \{S \mid S \cap O \neq \varnothing\}.$$

We may remark that the topologies obtained by using the U_O alone, or the L_O alone, as subbase are of great interest also; along with the Vietoris topology, they are closely related to the standard power domain constructions of computer science. In the present context, however, we will consider only the Vietoris construct; moreover, we restrict attention to the case that X is Hausdorff (indeed, metric).

We are going to show that the finite topology coincides with the Hausdorff metric topology over the *compact* subsets of a metric space. The following notation will be used: if S is a subset of a metric space (X, d), and $\varepsilon > 0$, then $\varepsilon(S)$ is $\{y \mid \exists x \in S.d(x, y) \leq \varepsilon\}$. Thus $\varepsilon(x)$ (strictly, $\varepsilon(\{x\})$), for $x \in X$, is the closed ball $\overline{B}_\varepsilon(x)$.

Lemma 7.4.2. *Let K be a compact subset of the metric space X, and O an open subset of X such that $K \subseteq O$. Then we can find $\varepsilon > 0$ such that $\varepsilon(K) \subseteq O$.*

Proof. For each $x \in K$, let $\delta(x) > 0$ be such that $B_{\delta(x)}(x) \subseteq O$. By compactness, there is a finite subset A of K such that $\{B_{\delta(x)/2}(x) \mid x \in A\}$ covers K. It suffices to take $\varepsilon = \min\{\delta(x)/2 \mid x \in A\}$. ∎

For the proof of the next result, note that if $d_H(K, L) < \varepsilon$, where d_H is Hausdorff distance, then $L \subseteq \varepsilon(K)$ and $K \subseteq \varepsilon(L)$. We denote by $K(X)$ the collection of non-empty compact subsets of a (Hausdorff) space X; thus, if X is a complete metric space, $K(X)$ is the same as $\mathcal{P}_C(X)$ of Section 6.

Theorem 7.4.3. *Let (X, d) be a metric space. Then the finite topology and the Hausdorff metric topology coincide on $K(X)$.*

Proof. 1. Let U_O be a (sub)basic open set of the first type, in the finite topology. Let $K \in U_O$. By Lemma 7.4.2 we have $\varepsilon > 0$ with $\varepsilon(K) \subseteq O$. Thus, if $d_H(K, L) < \varepsilon$, then $L \subseteq O$, which is to say $L \in U_O$. This shows that U_O is open in the Hausdorff metric topology. Again, consider a subbasic open set L_O. If K meets O, then there exist $x \in K$ and $\varepsilon > 0$ such that $\varepsilon(x) \subseteq O$. Then, if $L \subseteq X$ is such that $K \subseteq \varepsilon(L)$, L also meets O. Thus, as before, $d_H(K, L) < \varepsilon \Rightarrow L \in L_O$, showing that L_O is open w.r.t. d_H.

2. Suppose that $N \subseteq K(X)$ is a Hausdorff metric neighbourhood of the compact set K; thus, we have $\varepsilon > 0$ such that if $d_H(K, L) < \varepsilon$ then $L \in N$. Let $\{B_1, \ldots, B_n\}$ be a (finite) cover of K by open ε-balls $B_i = B_\varepsilon(x_i)$, where $x_i \in K$. Put $B = \cup_i B_i$. Then it should be evident that, for any $L \in U_B \cap L_{B_1} \cap \ldots \cap L_{B_n}, d_H(K, L) \leq \varepsilon$. Thus N is a neighbourhood of K with respect to the finite topology. ∎

7.5   Tychonoff's theorem

This asserts that compactness is preserved under (arbitrary) products of
spaces, and thus provides us with a large supply of compact spaces. Sometimes regarded as the most important single result in general topology, it is
false (or at least highly questionable) from a constructive point of view. We
will provide a simple version of the proof for the case of second-countable
spaces and countable products (the only case which concerns us), and comment briefly on the (non-)constructive aspect.

We begin by remarking that, in stating the criterion for compactness of
a space X, one can evidently work with any particular base for the topology
of X, in place of $\Omega(X)$:

Proposition 7.5.1. *Let $\mathcal{B}$ be a base of open sets for the space X. Then
X is compact if and only if every cover of X by members of $\mathcal{B}$ has a finite
subcover.* ∎

For the proof of the Tychonoff theorem, we need to know that the same
holds for an arbitrary *sub*base of the topology. In contrast with Proposition
7.5.1, this is a highly non-trivial result:

**Proposition 7.5.2 (Alexander's subbase lemma, second-countable
case).** *Let $\mathcal{S}$ be a countable subbase of open sets of the (second-countable)
space X. Then X is compact if and only if every cover of X by members
of $\mathcal{S}$ (every subbasic cover, we shall say) has a finite subcover.*

Proof. ONLY IF: nothing to prove.
IF: assume that every subbasic cover has a finite subcover. Let $\mathcal{B}$ be the
open base of X comprising the finite meets of members of $\mathcal{S}$. Let $\mathcal{C} =
\{\bigcap E_1, \bigcap E_2, \ldots\}$, where each E_i is a finite subset of $\mathcal{S}$, be a $\mathcal{B}$-cover of
X; we need to show that $\mathcal{C}$ has a finite subcover. To this end let T be
the full finitary tree with cross-sections E_i. More precisely, the notes of T
can be taken to be the finite sequences $< A_1, \ldots, A_n >$ where $A_i \in E_i$.
The subbasic sets occurring on any infinite path through T form a cover
of X (simply because $\bigcap E_i \subseteq A$, when $A \in E_i$). Hence, by the assumption
on subbasic covers and (the contrapositive of) König's Lemma [see the
Appendix], there is an integer n such that the subbasic sets at (up to)
each node at depth n in T form a cover of X; that is, every collection
$\{A_1, \ldots, A_n\}$, where $A_i \in E_i$, is a cover. By elementary manipulation of
sets, $\{\bigcap E_1, \ldots, \bigcap E_n\}$ is a cover. ∎

Classically, one proves the Alexander Lemma without restriction of cardinality, by appeal to one of the set-theoretic maximality principles (equivalent to the Axiom of Choice): see for example [Kelley, 1955].

Theorem 7.5.3 (Tychonoff's Theorem, countable case.). *The product of countably many second-countable compact spaces is compact.*

Proof. Let $X = \Pi_{i \in \mathbb{N}} X_i$, each X_i second-countable compact. Let $\mathcal{B}_i$ be a countable base for $X_i, i = 0, 1, \ldots$, and let $\mathcal{S}$ be the (countable) subbase for X consisting of all sets of the form $\pi_i^{-1}(B), B \in \mathcal{B}_i$. Let $\mathcal{A} \subseteq \mathcal{B}$ be such that no finite subset of $\mathcal{A}$ covers X. By Definition 7.3.2 the proof is complete if we can show that a fails to cover X. For each index i, let a_i be $\{B \in \mathcal{B}_i \mid \pi_i^{-1}(B) \in \mathcal{A}\}$. Then no finite subset of $\mathcal{A}_i$ covers X_i (since the preimage by π_i of a cover of X_i is evidently a cover of X), hence by compactness we have a point x_i such that $x_i \in X_i - B$ for each $B \in \mathcal{B}_i$. The point $< x_i >_{i \in \mathbb{N}} \in X$ then belongs to no member of $\mathcal{A}$; $\mathcal{A}$ fails to be a cover. ∎

As to the (non-)constructive character of these results, our countable version of Alexander's Lemma can actually be considered as intuitionistically acceptble, since the Fan Theorem can be used for the contrapositive of König's lemma [Appendix]. But the proof given for the Tychonoff Theorem itself is not acceptable: the negative formulation of compactness used in it is not (constructively) equivalent to the (positive) definition, and moreover is not strong enough to guarantee the existence of the points x_i.

We conclude these unsystematic remarks by noting that [Johnstone, 1981b; Johnstone, 1982] has obtained a constructive version of Tychonoff's theorem by recasting it in purely localic form.

7.6   Locally compact spaces

The requirement of compactness (for a space to be 'computationally reasonable') is in some respects too stringent, in others too weak.

Too stringent, inasmuch as it excludes some of the most familiar spaces, such as $\mathbb{N}$ or $\mathbb{R}$. But can $\mathbb{N}, \mathbb{R}$ and the like be considered as 'data types'? Perhaps they can; but it is relevant to point out that, in practice, the domain over which one is working, in a given context, is a restricted (bounded, or compact) range. (Another possiblilty is to embed $\mathbb{N}$ or $\mathbb{R}$ in a *larger* compact space and work over that; see below.)

On the other hand compactness is a very *weak* requirement, given that we admit non-Hausdorff spaces into consideration. For example, any space having a least element in its specialization order is trivially compact regardless of how badly behaved it may be locally.

In view of such considerations as these, the following is appropriate:

Definition 7.6.1. A space X is *locally compact* if, for every point x and neighbourhood N of x, N contains a compact neighbourhood of x.

In the case that X is Hausdorff, the criterion can be given more simply:

Proposition 7.6.2. *Let X be a Hausdorff space. Then X is locally compact if and only if every point of X has a compact neighbourhood.*

Proof. Suppose that $x \in X$ has a compact neighbourhood C. As a subspace of X, C is a compact Hausdorff space, and is therefore regular (Exercise 7.7.8(1d)). Moreover, since C is a neighbourhood of x (in X), any subset of C is a neighbourhood of x in C iff it is a neighbourhood of x in X. Let N be any neighbourhood of x (in X). Then $N \cap C$ is a neighbourhood of x in C which, by regularity of C, contains a closed neighbourhood Q of x. By Exercise 7.7.8(1a), Q is compact. This argument shows that, if every point has a compact neighbourhood, then X is locally compact. ∎

Corollary 7.6.3. *Every compact Hausdorff space is locally compact.*

It is remarkable, in view of the great importance of the notion, that several non-equivalent definitions of local compactness are current. Quite often, the criterion we have given in Proposition 7.6.2 for the Hausdorff case (every point has a compact neighbourhood) is taken as the general definition; but other versions are seen as well. This situation may be explained by the fact that, usually, only Hausdorff spaces are studied with any seriousness in general topology. Indeed, it is only when applied to non-Hausdorff spaces that conflict arises between the various definitions. The version we have given (Definition 7.6.1) has the right form to be the 'local' version of compactness. In addition, there is ample evidence (some of which we will touch on below) from recent studies of frame theory and continuous lattices that this is the right definition for general spaces.

Any discrete space (such as $\mathbb{N}$) is trivially locally compact, since singletons are both compact and open in such a space. In fact, any space having a base of compact open sets is locally compact: this gives us, in particular, the spectral spaces and the algebraic posets with Scott topology. The real line is locally compact: here we can use closed intervals as the required compact neighbourhoods. Any continuous poset (with Scott topology) is locally compact: sets of the form $\uparrow x$ suffice as the required compact neighbourhoods. Further examples, as well as non-examples, may be found by applying the following elementary propositions.

Proposition 7.6.4. *An open subset of a locally compact space is locally compact (as a subspace).*

Proof. Suppose that X is locally compact, and $U \subseteq X$ open. A neighbourhood N in the subspace U of $x \in U$ is also a neighbourhood in X. Thus we have (in X) a compact neighbourhood C of x, where $C \subseteq N$. Again using the principle that a subset of U is open in U iff it is open in X, we have that C is compact in U and is a neighbourhood of x in U. ∎

Proposition 7.6.5. *The image of a locally compact space by a continuous, open map is locally compact.*

Proof. Assume that X is locally compact, and that $f : X \to Y$ is an open, continuous surjection. Let $y = f(x)$ be an arbitrary point of Y, and V a neighbourhood of y. Since f is continuous, $f^{-1}(V)$ is a neighbourhood of x. Let C be a compact neighbourhood of x contained in $f^{-1}(V)$. Then $f(C)$ is compact (by Exercise 7.7.8(1b)), is a neighbourhood of y (since f is open), and is contained in V. ∎

Proposition 7.6.6. *Let* $X_0, X_1, \ldots$ *be locally compact spaces, with all but finitely many being also compact. Then the product* ΠX_n *is locally compact.*

Proof. [Sketch] Suppose that $x \in \Pi X_n$ and that $U = \bigcap \pi_i^{-1}(U_i) \mid i \in F$, where F is finite and U_i is open in X_i, is a basic open neighbourhood of x. Let G be a finite set of indices such that X_i is compact for all $i \notin G$. For $i \in F$, choose C_i a compact neighbourhood of $x_i (= \pi_i(x))$ contained in U_i; for $i \in G - F$, let C_i be any compact neighbourhood of x_i; and for $i \notin F \cup G$, put $C_i = X_i$. Then ΠC_i is a compact neighbourhood of x contained in U. ∎

What is interesting is that the converse of this proposition is also true. We can deduce at once from Proposition 7.6.5 that, if a product is locally compact, then every coordinate space is locally compact, since projection maps are (continuous and) open. It remains to show that, if infinitely many coordinate spaces are non-compact, then the product is not locally compact. In fact, in this situation, local compactness of the product fails spectacularly. For a space to be locally compact, every point must have an ample supply of compact neighbourhoods. Now, if infinitely many coordinate spaces are non-compact, the product has no compact neighbourhoods at all. To be precise, we have:

Proposition 7.6.7. *Let* $X = \Pi_{i \in I} X_i$ *be a product space having at least one compact subset with non-empty interior. Then all but finitely many of the spaces* X_i *are compact.*

Proof. Assume that C is a compact neighbourhood of $x \in X$. Then C contains a neighbourhood of x of the form $\bigcap \{ \pi_i^{-1}(U_i) \mid i \in F \}$ where F is finite and each U_i is open in X_i. If $i \in I - F$, then $\pi_i(C) = X_i$ and so X_i, being the continuous image of a compact set, is compact. Thus, all but finitely many coordinate spaces are compact. ∎

Example of Proposition 7.6.7.
Baire Space is not locally compact.

We now briefly consider the important topic of *compactification*, focusing on the simple case of one-point compactification. Informally, this is the

process of adding a 'point-at-infinity' to a space. A familiar instance is the
one-point compactification of the Euclidean plane, viewed in terms of the
projection of a sphere onto the plane. We imagine the sphere, S, as resting
on the origin and having its north pole, p, at $(0,0,1)$. Via projection from p,
the plane is seen to be homeomorphic with $S - \{p\}$. Thus S is a compact
space which can be considered as the extension of the plane by a single
point. Viewing S as an extension of $\mathbb{R}^2$ in this way, how is the topology of
S determined by that of $\mathbb{R}^2$? Evidently, an open set of S not containing p
is simply an open set of $\mathbb{R}^2$. But a set containing p is open in S if and only
if its complement is compact in $\mathbb{R}^2$. (Equivalently, a neighbourhood of p is
any set whose complement is a bounded subset of the plane).

The general construction is as follows. Let X be a topological space.
We adjoin a new element, written ∞, to X, giving $X^* = X \cup \{\infty\}$. Take
as the open sets of X^*, (i) the open sets of X (type 1 open sets), and (ii)
the complements of the closed compact subsets of X (type 2 open sets).
Of course, we must show that this defines a topology on X^*. Now, a finite
meet of type 2 sets is of type 2, since a finite join of closed compact sets
is closed and compact; while a finite meet of 'open' sets, at least one of
which is of type 1, is evidently of type 1. An arbitrary join of type 1 sets
is of type 1. If U is a join of 'open' sets, at least one of which is of type
2, then $X^* - U$, as a meet of closed sets, is closed (in X) and, as a closed
subset of a compact set, is compact; thus U is of type 2. So we have a
topology on X^*; moreover, X is a subspace of X^*. The most significant
property of X^* is that it is compact. Indeed, if $\mathcal{C}$ is an open cover of X^*,
we obtain a finite subcover by taking a member U of $\mathcal{C}$ which contains ∞
together with a finite subcover of the compact set $X - U$. The construct
is known as the *one-point* (or *Alexandroff*) compactification of X. It is
particularly well-behaved in the case that X is locally compact Hausdorff.
We have:

Proposition 7.6.8. *The one point compactification X^* of X is Hausdorff
if and only if X is locally compact and Hausdorff.*

Proof. If X^* is Hausdorff, then X is locally compact by Corollary 7.6.3 and
Proposition 7.6.4. Suppose conversely that X is locally compact Hausdorff.
To conclude that X^* is Hausdorff, it will suffice to show that, if $x \in X$,
then there are disjoint neighbourhoods of x and ∞. In fact, since X is
locally compact, there is a compact neighbourhood K of x in X, which is
closed since X is Hausdorff (Exercise 7.7.8(1c)); $X^* - K$ will then serve as
the required neighbourhood of ∞. ∎

Consider again a non-compact 'data type' such as $\mathbb{N}$. In practice, the
data must be represented by strings in some finite alphabet. For simplicity,
let us assume unary representation. Now, we can argue that, if arbitrarily long finite unary strings are accepted by a program, then the infinite

string 1^ω must be acceptable as well; for at what stage in the reading of 1^ω could the type-checking mechanism issue an error message? Thus the type should be considered to be $\mathbb{N}^*$ rather than $\mathbb{N}$. In this representation of natural numbers, we have implicitly assumed that finite strings have end-markers (numbers are total, not partial, data elements), yet computation may proceed before reading of the input data has been completed. This is perhaps somewhat unusual. Suppose instead we require that the end-marker of the input be reached before computation of the output can take place. In that case we read 1^ω for ever without obtaining any usable information, and so this input is best thought of as $\bot$. On this view, the type is really $\mathbb{N}_\bot$: a different compactification of $\mathbb{N}$.

We now consider very briefly the general topic of compactification. The standard treatment of this is concerned purely with the Hausdorff case. A *Hausdorff compactification* of a space X is a pair (c, Y) (also written (c, cX), or just cX), where Y is a compact Hausdorff space and c is a homeomorphism of X onto a dense subspace of Y. The Hausdorff compactifications of X are ordered by writing $cX \geq c'X$ iff there is a continuous map h of cX onto $c'X$ such that $h \circ c = c'$. By a trivial verification, $\geq$ is indeed a pre-order. Moreover, two compactifications of X which are equivalent under this pre-order are homeomorphic; thus we can regard the relation as a partial order, up to homeomorphism. It can be shown that, if X has a Hausdorff compactification at all (namely, if X is a Tychonoff, or $T_{3\frac{1}{2}}$, space — terms which we shall not define here), then it has a greatest such compactification: the Stone–Cech compactification, βX. Rather easier to show is that, if X is locally compact Hausdorff, then it has a least compactification (Exercise 7.7.8(10)).

One way to get a view of the compactifications of a space X is to see them as spaces of filters of open sets of X. For just a hint of this, consider the case $X = \mathbb{N}$, and suppose we are only interested in what we shall call *Stone* compactifications: that is, compactifications which happen to be Stone spaces. We can construct these from bases of the topology which are Boolean algebras. In the present case, since the topology is discrete, these bases are the sub-Boolean-algebras of $\mathcal{P}\mathbb{N}$ which contain all the singletons $\{x\} \in \mathcal{P}\mathbb{N}$. The smallest such base is clearly the Boolean algebra of finite and cofinite subsets of $\mathbb{N}$. The space of ultrafilters (or prime filters) of this algebra coincides with the one-point compactification of $\mathbb{N}$. The largest 'Boolean base' is $\mathcal{P}\mathbb{N}$ itself; the space of ultrafilters ofthis is $\beta \mathbb{N}$. From a computational point of view the Stone–Cech compactification βX is absurdly large (not even first-countable), and impossible to grasp in any effective sense. The analogy between compactness and finiteness needs to be treated with caution!

The preceding has been generalized to T_0 spaces in [Smyth, 1992]. Instead of compact Hausdorff spaces, we use *stably compact spaces*. These are the compact, locally compact sober spaces such that meets of pairs of compact subsets are compact. (They have been mentioned above, in connection with quasi-uniformities, and they have several other interesting characterizations as well: for example, they are the retracts of the spectral spaces.) Of course, for a Hausdorff space, stable local compactness reduces to compactness. Given a T_0 space X we consider 'strong inclusion' relations $<<$ on the lattice $\Omega(X)$ (in effect, these relations are *quasi-proximities* for X, see for example [Fletcher and Lindgren, 1982]). Each structure $(\Omega(X), <<)$ gives rise to a space of filters, which is a 'stable compactification' of X; and all such compactifications arise in this way. The standard theory of Hausdorff compactification can be extracted from this as a special case.

With a view to a localic treatment, one may ask whether local compactness of a space can be characterized in terms of the lattice of open sets alone. The notion of relative compactness is useful here. For open sets U, V of a space X, we say that U is *(relatively) compact in V* if every cover of V by open sets admits a finite subcover of U. This relation is simply Scott's 'way-below' relation for the complete lattice ΩX and is usually denoted $<<$ (note, however, that it is not necessarily a strong inclusion in the sense of the preceding paragraph). If K is a compact set contained in an open set V, then the interior of K is evidently compact in V. This suggests the following modification of the notion of local compactness:

Definition 7.6.9. A space X is *quasi-locally compact*, or *core compact*, if, for every $x \in X$ and open set V containing x, there is an open neighbourhood U of x which is compact in V.

In view of the immediately preceding remark, we have: X locally compact $\Rightarrow$ X core compact. Now, core compactness is especially interesting because of the following restatement of the defining condition for it: every open set V is the join of the open sets compact in V. In other words, a space X is core compact if and only if ΩX is a continuous lattice.

Are core compactness and local compactness equivalent? The answer is negative, as is shown by the examples of Isbell [Isbell, 1975]. However, when X is given to be sober, we can assert: X core compact $\Rightarrow$ X locally compact (this may be regarded as an exercise). Thus, for sober spaces, the two statements are equivalent to: $X \cong Pt(L)$ for L a distributive continuous lattice.

7.7 Function spaces

We have considered the function space construct in the context of metric spaces in Section 6.5; we now consider it in the general topological context. Two major questions arise. First, for which pairs of spaces X, Y does a satisfactory function space $[X \rightarrow Y]$ exist? (It certainly does not exist in every case.) Secondly (and more vaguely), can we find a good *Cartesian closed* category of 'computationally reasonable' (at least second-countable sober) spaces? The first question has been well explored, and we will sketch the received answer to it (with one or two embellishments). As to the second question, the only detailed information we have on it lies in the area of domain theory, and for that reason we shall do no more than touch on it here.

The question with which we are mainly concerned is, a little more precisely, that of finding the 'right' topology on the set $[X \to Y]$ of continuous functions from X to Y. A topology T on $[X \to Y]$ is said to be *splitting* if (for any W) the currying map

$$\text{curry}: \quad [W \times X \to Y] \to [W \to [X \to Y]]$$

is well-defined; that is, provided that whenever $g: W \times X \to Y$ is continuous, $\hat{g} = \text{curry}(g)$ is also continuous. Conversely, T is said to be *conjoining* if the continuity of $\hat{g}$ always implies that of g; this is equivalent to saying that the evaluation, or application, map

$$\text{eval}: \quad [X \to Y] \times X \to Y :< f, x > \mapsto f(x)$$

is continuous.

The elementary facts about splitting and conjoining topologies may be summarized as follows ([Arens and Dugundji, 1951; Dugundji, 1966]).

Proposition 7.7.1.

1. *Any topology on $[X \to Y]$ finer (larger) than a conjoining topology is also conjoining.*

2. *Any topology coarser (smaller) than a splitting topology is also splitting.*

3. *Any conjoining topology is finer than any splitting topology.*

Proof. 1. If **eval**: $[X \to Y] \times X \to Y$ is continuous for a given topology T on $[X \to Y]$ it is evidently continuous for any finer topology $T' \supseteq T$.

2. Similar to 1.

3. Suppose that T is conjoining and that T' is splitting. Thus, **eval**: $[X \to Y]_T \times X \to Y$ is continuous and, moreover, curry (eval: $[X \to Y]_T \to [X \to Y]_{T'}$ is continuous. But the latter map is just the identity on $[X \to Y]$, and its continuity amounts to the fact that T is finer than T'. ∎

Corollary 7.7.2. *A function space $[X \to Y]$ has at most one topology that is both conjoining and splitting; such a topology, if it exists, is both the coarsest conjoining topology and the finest splitting topology.*

For which spaces does the conjoining, splitting topology exist, and how may it be described explicitly? Our general approach would suggest that we consider finitely observable properties of functions. Unfortunately, this is not so straightforward. One suggestion for a 'finite observation' of a function might be: generate an input and see whether the output satisfies a specified (finitary) property. This would give us the *point-open* topology of $[X \to Y]$, namely the topology which has as a subbase the collection of sets

$$[x, V] = \{f \in [X \to Y] \mid f(x) \in V\},$$

for $x \in X$, V open in Y. Now it is easy to show that the point-open topology is always splitting. And we are familiar from domain theory with the result that, if X, Y are domains of a suitable kind, then the point-open topology is conjoining (see Scott (1972)) for the case of continuous lattices). However, this certainly does not hold generally, for example if X is a typical Hausdorff space (see Exercise 7.7.8(14)). One approach to this problem could be to argue that the result fails only because X does not have (enough) *partial* elements, and that what is needed is to embed X into a domain for which the point-open topology 'works'; this approach is pursued in Exercise 15. Another method of obtaining the correct general solution from the point-open topology for domains will be considered below (Isbell topology). But for the moment let us pursue the standard approach, which is to introduce the compact-open topology without further ado, and show that it 'works' provided we restrict to locally compact spaces.

Definition 7.7.3. Let X, Y be topological spaces. The *compact-open* topology of $[X \to Y]$ has as a subbase the collection of sets

$$[K, V] = \{f \in [X \to Y] \mid f(K) \subseteq V\},$$

for K a compact subset of X, V open in Y.

Proposition 7.7.4. *Let X be locally compact, and Y be any space. The compact-open topology for $[X \to Y]$ is conjoining.*

Proof. Suppose that $x \in X, f \in [X \to Y]$, and V is a neighbourhood of $f(x)$ in Y. Since f is continuous, $f^{-1}(V)$ is a neighbourhood of x. Since X is locally compact, $f^{-1}(V)$ contains a compact neighbourhood of x, say K. Now $f \in [K, V]$. Thus, if $[X \to Y]$ is taken with the compact-open topology, $U = [K, V] \times K$ is a neighbourhood of $< f, x >$ in $[X \to Y] \times X$ such that $\mathbf{eval}(U) \subseteq V$; this shows that $\mathbf{eval}$ is continuous. ∎

Notice that the *compactness* of K was not really used in this proof. All that was needed was the fact that the compact neighbourhoods of a point constitute a neighbourhood base. Compactness is used more essentially in connection with the splitting property:

Proposition 7.7.5. *The compact-open topology is always splitting.*

Proof. Suppose that $g : W \times X \to Y$ is continuous. We consider $\hat{g} = \mathrm{curry}(g) : W \to [X \to Y]$, where $[X \to Y]$ is taken with the compact-open topology. Assume that $w \in W$, and let $[K, V]$ be a subbasic neighbourhood of $\hat{g}(w)$ in $[X \to Y]$. Thus, for each $x \in K$, we have $g(w, x) \in V$. By continuity of g we have (for each $x \in K$) neighbourhoods O_x, U_x of w, x, respectively, such that $g(O_x \times U_x) \subseteq V$. By compactness of K, finitely many of the U_x cover K. Let O be the meet of the O_x corresponding to these U_x. Then $g(O \times K) \subseteq V$, which means that $\hat{g}(O) \subseteq [K, V]$, while O is a neighbourhood of w. This proves that $\hat{g}$ is continuous. ∎

From the preceding results we get:

Theorem 7.7.6. *Let X be locally compact, and Y any space. Then $[X \to Y]$ has a (unique) conjoining, splitting topology, namely the compact-open topology. In categorical terms: each locally compact space is exponentiable in* **Top**.

Are there any exponentiable spaces other than the locally compact ones? Recall (Definition 7.6.9) that a space X is core compact (or quasi-locally compact) iff $\Omega(X)$ is a continuous lattice. Mention of frames as well-behaved 'domains' may suggest to us a new approach to defining function–space topologies. Consider the collection $\Phi = [\Omega(Y) \to \Omega(X)]$ of frame maps corresponding to a function space $[X \to Y]$. In the case that the frames are continuous lattices, would it not be reasonable to endow them with their Scott topology (frame maps are of course Scott-continuous), and take the point-open topology of Φ? Pursuing this idea, we arrive at the Isbell topology:

Definition 7.7.7. *Let X, Y be topological spaces. The Isbell topology of $[X \to Y]$ has as a subbasis the collection of sets of the form*

$$N(H, V) = \{f \in [X \to Y] \mid f^{-1}(V) \in H\},$$

for H a Scott-open subset of $\Omega(X)$ and Y an open set of Y.

Similarly to Propositions 7.7.4, and 5 one may prove: if X is core compact, the Isbell topology for $[X \to Y]$ is conjoining and splitting. This yields the conclusion that the core compact spaces are exponentiable in **Top**. The converse follows from the results of Day and Kelly [Day and Kelly, 1970], as pointed out by Lawson [Lawson, 1987]: in fact, if X is not core compact, $[X \to 2]$ admits no topology that is both conjoining and splitting. There is a corresponding result for locales, due to [Hyland, 1981; Johnstone, 1982]: a locale is exponentiable in **Loc** if and only if it is locally compact (i.e., is a continuous lattice).

The consideration of core compact spaces and Isbell topology is, to be sure, a little excessive from our point of view. When the spaces are sober, Isbell topology reduces to compact-open topology: indeed, if X is locally compact (even if not sober), we see that the Isbell topology and the compact-open topology for $[X \to Y]$ coincide, simply because both are conjoining and splitting.

Sobriety is not the only constraint which concerns us, of course. For the semantic modelling of languages which admit function types, we would like to work with a category of spaces which is Cartesian closed. The preceding considerations by no means give us this: the function space of locally compact spaces need not be locally compact, as the example of Baire space $(= [\mathbb{N} \to \mathbb{N}])$ shows. To get a Cartesian closed category of spaces, topologists have proposed the *compactly generated*, or *Kelley*, spaces (). The compactly generated spaces are problematic for us, in two respects. In the first place they rule out non-Hausdorff spaces, such as Scott domains. This deficiency could possibly be overcome. The second problem is that our requirement that spaces be second-countable is quickly violated as soon as we start to iterate the function space construction (cf. the example of $C^{(2)}$ mentioned in Section 5; the spaces of countable functionals are compactly generated).

Cartesian closed categories (ccc's) of second-countable sober spaces are rather scarce. Essentially the only examples we have are various categories of domains. We will conclude with some tentative conjectures on largest ccc's together with references to the domain theory literature. It is assumed that we are concerned only with *full* subcategories of **Top** (all continuous maps count as morphisms).

From the preceding analysis and discussion it seems plausible that any (maximal) ccc will be a subcategory of the locally compact spaces. Let us consider for the moment just the locally compact sober spaces having a (countable) base of *compact* open sets (that is, having a topology which is an ω-algebraic lattice). For these one might venture the conjecture that the answer will be the same as for the ω-algebraic 'domains' (dcpo's) [Smyth, 1983a; Jung, 1989]: that is, the largest ccc is ω-**Bif**, the category of ω-bifinite dcpo's. (A dcpo is ω-*bifinite* if it is the limit of a Scott embedding projection sequence of finite posets. For a general introduction to these domains, see [Gunter, 1987]. The term 'bifinite', as opposed to 'profinite', is due to [Taylor, 1986]. Note that ω-**BIF** is ω-**FB** in Jung's notation.) Going on then to the general locally compact sober spaces (and thus, if the previous suggestion is correct, to the second countable sober spaces), it is natural to suggest the category $\mathbf{r}\omega$-**BIF** of *retracts* of ω-bifinite domains as the largest ccc. In this form the suggestion would include the conjecture that Jung's (countably based) 'FS domains' [Jung, 1990] are in fact in $\mathbf{r}\omega$-**BIF**, since Jung proved that ω-**FS** is the largest ccc of ω-continuous cpo's. Clearly our overall conjecture, or suggestion, has a number of doubtful components.

The categories of domains to which we are apparently led by asking for Cartesian *sub*categories of the sober spaces may seem rather too small and restrictive. In that case, as mentioned already in a related context (Section 5), one might consider the option of *expanding* the category, to filter spaces or the like. But this option cannot be explored here.

Exercises 7.7.8.

1. Prove the following elementary facts about compactness (or look them up in a textbook):

   (a) Any closed subset of a compact space is compact.

   (b) The image of a compact set under a continuous function is compact.

   (c) Any compact subset of a Hausdorff space is closed. (Hint: proceed by separating each point not in the compact set A from points of A, in accordance with the Hausdorff property.)

   (d) Every compact Hausdorff space is regular.

2. Show that if the space $(X, \mathcal{T})$ is compact, and $\mathcal{T}' \subseteq \mathcal{T}$, then $(X, \mathcal{T}')$ is compact.

3. (a) Show that the meet of an arbitrary collection of closed, compact subsets of a topological space is closed and compact.

   (b) Give an example of a poset P and two compact upper subsets C, C' of P such that $C \cap C'$ is not compact (assume Alexandroff topology).

4. Give a necessary and sufficient condition for a poset to be compact in its Alexandroff topology.

5. Rephrase Questions 3(b) and 4 (and their solutions) in terms of dcpo's with Scott topology rather than posets with Alexandroff topology.

6. Give an example of a non-algebraic cpo in which every open subset is compact. Deduce that a cpo which is spectral (in its Scott topology) need not be algebraic.

7. Characterize the ω-algebraic cpo's which are spectral (in fact, they are the '2/3-SFP' domains of [Plotkin, 1981]).

8. Show that, in any spectral algebraic dcpo, the Lawson topology coincides with the patch of the Scott topology. Show that, using the extended definition of the patch, the same holds for algebraic (and even continuous) dcpo's in general. Show that the patch (or Lawson) space of $(\mathcal{P}\omega, \text{Scott})$ is homeomorphic with Cantor space.

9. Recall Section 3.4.6 and Exercises 3.5.3(5) and (6). Assume that L has conjunction and truth, so that $\{\text{Mod}(A) \mid A \in L\}$ can be taken as the base of open sets of a topology $\mathcal{T}$ on W. Say that L is *compact* if every cover of a basic open set (that is, a set $\{\text{mod}(A)\}$ by basic open sets has finite subcover). Show that if L is compact:

   (a) $\mathcal{T}$ is spectral

   (b) assuming L has falsity and disjunction, Parikh's topology $\mathcal{T}_p$ on W is the topology of negative information corresponding to $\mathcal{T}$.

   Note that if L has also (classical) negation, there is no difference between $\mathcal{T}$ and $\mathcal{T}_p$.

10. Let T be a finitary tree. Denote the set of nodes at depth n in T by X_n (with $X_o = \{\text{root}_T\}$), and endow X_n with the discrete topology. Consider the set P of infinite paths of T as a subset of ΠX_n, in the obvious way. By expressing P as the meet of a certain collection of closed subsets of ΠX_n having the finite intersection property, show that Tychonoff's Theorem for the product ΠX_n yields König's Lemma for T.

11. Verify that, if X is a subspace of Y, any compact subset of X is also compact as a subset of Y. Prove that the one-point compactification of a locally compact Hausdorff space X is the least Hausdorff compactification of X.

12. (Rudimentary knowledge of continuous lattices advisable.) Let X be a sober, core-compact space. For an open subset U of X, let $\uparrow U = \{V \mid U << V\}$ (i.e. the collection of open sets V such that

U is non-empty), then $\cap\!\uparrow\!U$ is compact. Deduce that X is locally compact.

13. Verify that the compact-open topology of $[\mathbb{N} \to \mathbb{N}]$ agrees with the topology of $\mathbb{B}$ as the product of countably many copies of the discrete space $\mathbb{N}$.

14. Give an example of spaces X, Y, with X compact Hausdorff, such that the point-open topology of $[X \to Y]$ is not conjoining.

15. (A possible way to derive the compact-open topology from domain-theoretic considerations.) Given a space X, let $P_U(C)$ be the collection of compact saturated subsets of X, ordered by reverse inclusion. (Note: 'saturated set' = upper set in the specialization order. The ordering of $P_U(X)$ is: $A \sqsubseteq B$ iff $B \subseteq A$.) Prove that $P_U(X)$ is a dcpo if X is sober, and is a continuous dcpo if X is locally compact. Assume now that X, Y are locally compact spaces, and take $P_U(X), P_U(Y)$ with Scott topology. Show that the assignment $x \mapsto \uparrow x$ embeds X as a subspace of $P_U(X)$. Show also that each continuous map $f : X \to Y$ has a natural 'extension' to $P_U(f) : P_U(X) \to P_U(Y)$, where $P_U(f)(A) = \uparrow f(A)$, and that in this way $[X \to Y]$ can be regarded as a subset of $[P_U(X) \to P_U(Y)]$. Finally, show that the subspace topology of $[X \to Y]$ induced by the point-open topology of $[P_U(X) \to P_U(Y)]$ is the compact-open topology of $[X \to Y]$.

16. Show that a second-countable locally compact space has, in a suitable sense, a countable base of compact saturated sets. Deduce that if X is locally compact and X, Y are both second-countable, then $[X \to Y]$ (with compact-open topology) is second-countable.

8 Appendix

In this Appendix, we collect together some of the general (mainly order-theoretic) definitions and facts presupposed in the text. A brief discussion of König's lemma and the Fan Theorem is also provided. For more details on the order-theoretic topics, a convenient reference is [Davey and Priestley, 1990].

A *pre-order* is a reflexive, transitive binary relation. A *partial order* is an anti-symmetric pre-order. A *partially ordered set*, or *poset*, is then a set P together with a partial order $\leq$ defined on P. A *lattice*, as we are using the term in this chapter, is a poset $(L, \leq)$ in which (i) there is a least element 0 and a greatest element 1, and (ii) each pair $x, y \in L$ has the least upper bound (lub) $x \vee y$ and the greatest lower bound (glb) $x \wedge y$. It is equivalent to say that each *finite* subset of L has both a lub and a glb. 'Lattice' is often defined in such a way that only condition (ii) has to be satisfied; in that case the lattices as we have them are *bounded* lattices.

A lattice L is *distributive* if, for all $a, b, c \in L, a \wedge (b \vee c) = (a \wedge b) \vee (a \wedge c)$. By an easy exercise, this condition is equivalent to its dual: for all $a, b, c \in L$,

$$a \vee (b \wedge c) = (a \vee b) \wedge (a \vee c).$$

A *Boolean* lattice (or Boolean algebra) is a distributive lattice in which every element b has a (necessarily unique) *complement* b'; this means that $b \wedge b' = 0$ and $b \vee b' = 1$. Strictly, it would be preferable to define a Boolean algebra as an algebraic structure $(B; \wedge, \vee, 0, 1')$ where the five operations satisfy suitable axioms, but we shall not trouble with this. What we choose to regard as the operations of the structure is reflected in the notion of homomorphism adopted. In this chapter, a *lattice homomorphism* is required to preserve $\vee, \wedge, 0$ and 1. For a *Boolean* homomorphism (of Boolean algebras) we could then add: preservation of $'$. However, it is easy to show that a lattice homomorphism of Boolean algebras automatically preserves complements, so that Boolean homomorphisms reduce to lattice homomorphisms.

A *complete* lattice is a lattice in which every subset has both a lub and a glb. It is easy to show that each half of the condition actually suffices for the definition: precisely, if in a poset P every subset has a lub, then it is also true that every subset has a glb, and vice versa. A weaker notion of 'completeness' is also of great interest: a *dcpo*, or directed-complete poset (= *pre-cpo* of [DP90]) is a poset in which every directed subset has a lub. (A subset $D \subseteq (P, \leq)$ is *directed* provided (i) D is non-empty, and (ii) $x, y \in D \Rightarrow \exists z \in D.x \leq z \& y \leq z$.) A *cpo* is then a dcpo having a least element. A complete lattice is *a fortiori* a cpo; an *algebraic* lattice is a complete lattice which, as a (d)cpo, is algebraic (see Section 2.2).

Now we consider filters and ideals. Let $(P, \leq)$ be a poset. An *ideal* of P is a directed set $I \subseteq P$ such that $I =\downarrow I$; in brief, a directed down-set of P. The *ideal completion* $\overline{P}$ of P is the collection of all ideals of P, ordered by inclusion. The ideal completion of P is necessarily an algebraic dcpo; conversely, any algebraic dcpo D is isomorphic (as a poset) with $\overline{B}_D$, where B_D is the ordered set of finite elements of D. The notion of a *filter* is dual to that of an ideal. Thus, a filter of a lattice L (the only context in which we have occasion to consider filters) is a non-empty upward-closed subset F of L such that, if $a, b \in F$, then $a \wedge b \in F$. The filter is *proper* if $F \neq L$ (that is, $0 \notin F$). In this chapter, we generally assume that filters (and ideals) of lattices are proper. The filter F is *prime* if, whenever $a \leq \vee A$ (for $a \in F$ and A a finite subset of L), it follows that some member of A belongs to F. An equivalent definition: F is *prime* provided (i) F is proper, and (ii) $a \leq b \vee c \& a \in F \Rightarrow b \in F$ or $c \in F$. (The properness of F

comes from considering $0 \leq \bigvee \varnothing$.) Prime ideals dually.

Suppose now that the lattice L is complete. Then we have the notion of a *completely prime* filter of L; the definition is exactly as for *prime filter*, except that the word 'finite' is deleted (that is, A can be an arbitrary subset of L).

Corresponding to each of these types of filter, there is the notion of a *base* of a filter of that type: we say that $B \subseteq L$ is an X *filter base* if $\uparrow B$ is an X filter. It is easy to see that (unrestricted) filter bases, also known as filtered sets, are dual to directed sets. Exercise: state the definition of a completely prime filter base as a direct condition on the subset $B \subseteq L$, rather than by way of $\uparrow B$. Lastly, maximal filters. A *maximal filter*, or *ultra-filter*, of a lattice L is a filter F of L which is maximal in the ordered (by inclusion) set of proper filters of L; thus, F is proper, and no filter properly contains F other than L itself. We are concerned with ultra-filters only in the case that L is Boolean, and here we have the following standard result:

Let F be a subset of the Boolean lattice (or algebra) L. Then the following are equivalent:

1. F is an ultra-filter

2. F is a prime filter

3. for each $b \in L$, exactly one of b, b' belongs to F.

Notation 8.0.1. Note that L is often itself a lattice of sets (typically $\mathcal{P}(X)$ for a set X, or $\Omega(X)$ for a topological space X), and in that case we use script (calligraphic) notation for filters and filter bases of L : $\mathcal{F}, \mathcal{B}$, etc. We may also speak of filters *over* X, meaning the same as filters *of* $\mathcal{P}(X)$.

We turn now to covers and partitions. Let $\mathcal{C}$ be a collection of subsets of a set X, and $A \subseteq X$. Then $\mathcal{C}$ is said to be a *cover* of A if $A \subseteq \bigcup \mathcal{C}$. In that case, a *subcover* of $\mathcal{C}$ is any subset of $\mathcal{C}$ which also covers A; further, a cover $\mathcal{B}$ of A is said to *refine* $\mathcal{C}$ if, for every $B \in \mathcal{B}$, there exists $C \in \mathcal{C}$ such that $B \subseteq C$. In case X is a topological space, we have the notion of an *open* cover (of a set $A \subseteq X$): this simply means that every member of the cover is an open set. A *partition* of a non-empty set A is a cover of A by mutually disjoint, non-empty subsets of A. The partitions of A are in an evident (1,1)-correspondence with the equivalence relations over A.

We come at last to König's lemma and the Fan Theorem. König's lemma is the statement that in any finitary tree with infinitely many nodes, there is at least one infinite path. (A *finitary* tree is a tree in which each node has only finitely many immediate successors.) The proof of the lemma inevitably involves uses of negation which are constructively unacceptable. Thus, one argues: any node having infinitely many descendants must have an immediate successor with the same property, for if *not*, then ...; here the argument involves (at least) an illicit step of the form $\neg\neg P \vdash P$. However, no such objection applies to the contrapositive of the lemma ('If every path of a finitary tree T is finite, then T is finite'), which indeed is a consequence of the Fan Theorem of intuitionism, suitably formulated. For this, we take a *fan* (finitary spread) to be a finitary tree in which each node has a successor (so that all paths are infinite), and the elements of the spread to be the paths. Let the variable α range over the paths of such a tree, or fan, T. Let R be a set of nodes of T (R will play the rôle of the set of end-points of terminated paths, for the contrapositive of König's lemma). Then one version of the Fan Theorem (termed by Dummett the 'General Fan Theorem', [Dummett, 1977]) states:

$$\forall\alpha\exists n.\alpha(n) \in R \Rightarrow \exists k\forall\alpha\exists n \leq k.\alpha(n) \in R.$$

To apply this to a tree T which has terminating paths, we extend T to a fan in some trivial way. Then, if all paths of T terminate, the General Fan Theorem gives us a bound on the lengths of such paths, from which we conclude that T is finite.

These remarks on König's lemma and its intuitionistic counterpart were prompted by the analysis of the Subbase lemma (Section 7.5). Indeed it is apparent that the General Fan Theorem is exactly what is needed for our constructive proof of Proposition 7.5.2.

References

[Abramsky, 1988] S. Abramsky. A cook's tour of the finitary non-well-founded sets. *Abstract published in EATCS Bulletin*, 36:233–234, 1988. (Unpublished notes).

[Abramsky, 1991] S. Abramsky. Domain theory in logical form. *Annals of pure and applied logic*, 51:1–77, 1991.

[Aczel, 1988] P. Aczel. *Non-well-founded sets*. CSLI Lecture notes 14. Stanford University, Stanford, 1988.

[Alpern and Schneider, 1985] B. Alpern and F. B. Schneider. Defining liveness. *Information Processing Letters*, 21:181–185, 1985.

[America and Rutten, 1989] P. America and J. Rutten. Solving reflexive domain equations in a category of complete metric spaces. *Journal of Comupter Systems and Science*, 39:343–375, 1989.

[Arens and Dugundji, 1951] R. Arens and J Dugundji. Topologies for function spaces. *Pacific Journal of Mathematics*, 1:5–32, 1951.

[Aumann, 1976] R. Aumann. Agreeing to disagree. *The Annals of Statistics*, 4:1236–9, 1976.

[Avizienis, 1961] A. Avizienis. Signed-digit number representations for fast parallel arithmetic. *IRE transactions on Electronic Computers*, 10:389–400, 1961.

[Barendregt and Longo, 1983] H. Barendregt and G. Longo. Recursion theoretic operators and morphisms on numbered sets. *Fundamenta Mathematica*, CXIX:49–62, 1983.

[Barendregt, 1984] H. Barendregt. *The Lambda Calculus: Its Syntax and Semantics*. North-Holland, revised edition, 1984.

[Barnsley, 1988] M. Barnsley. *Fractals Everywhere*. Academic Press, Boston, 1988.

[Barwise and Etchemendy, 1988] J. Barwise and J. Etchemendy. *The Liar: an Essay in Truth and Circularity*. Oxford University Press, 1988.

[Batarekh and Subrahmanian, 1989] A. Batarekh and V. Subrahmanian. The query topology in logic programming. In B. Monien and R. Cori, editors, *STACS '89*, pages 375–387. Springer-Verlag, 1989. LNCS 349.

[Beck, 1980] J. M. Beck. On the relationship between algebra and analysis. *Journal of Pure and Applied Algebra*, 19:43–60, 1980.

[Berge, 1959] C. Berge. *Espaces Topologiques: Functions Multivoques*. Dunod, Paris, 1959.

[Birkhoff, 1967] G. Birkhoff. *Lattice theory*. AMS Colloquium Publication, third edition, 1967.

[Bishop and Bridges, 1985] E. Bishop and D. Bridges. *Constructive Analysis*. Springer-Verlag, Berlin, 1985.

[Boehm et al., 1986] H. J. Boehm, R. Cartwright, M. J. O'Donnell, and M. Riggle. Exact real arithmetic: a case study in higher order programming. In *Proc. ACM conference on Lisp and Functional Programming*, pages 162–173, 1986.

[Bourbaki, 1966] N. Bourbaki. *Elements of Mathematics: General Topology*, volume I and II. Hermann/Addison-Wesley, 1966. Translation of 3rd edition of *Topologie Générale*.

[Cutland, 1980] N. Cutland. *Computability: an Introduction to Recursive Function Theory*. Cambridge University Press, 1980.

[Davey and Priestley, 1990] B. Davey and H. Priestley. *Introduction to Lattices and Order*. Cambridge University Press, 1990.

[Day and Kelly, 1970] B. Day and G. Kelly. On topological quotient maps preserved by pullbacks or products. *Proc. Cambridge Philosophical Society*, 67:553–558, 1970.

[de Bakker and Zucker, 1982] J. de Bakker and J. Zucker. Processes and the denotational semantics of concurrency. *Information and Control*, 54:70–120, 1982.

[de Roever, 1976] W. de Roever. Dijkstra's predicate transformers,nondeterminsim, recursion and termination. In A. Mazurkiewicz, editor, *MFCS 1976*, pages 472–481. Lecture Notes in Computer Science 45. Springer, 1976.

[Dijkstra, 1976] E. W. Dijkstra. *A Discipline of Programming*. Prentice-Hall, Englewood Cliffs, New Jersey, 1976.

[Dugundji, 1966] J. Dugundji. *Topology*. Allyn and Bacon, 1966.

[Dummett, 1977] M. Dummett. *Elements of Intuitionism*. Clarendon Press, Oxford, 1977.

[Egli and Constable, 1976] H. Egli and R. Constable. Computability concepts for programming language semantics. *Theoretical Computer Science*, 2:133–145, 1976.

[Ershov, 1973] Ju. L. Ershov. Theorie der enumierungen I. *Zeitschr. Math. Logik*, 19(4):289–388, 1973.

[Fletcher and Lindgren, 1982] P. Fletcher and W. Lindgren. *Quasi-uniform spaces*, volume 77 of *Lecture Notes in Pure and Applied Mathematics*. Marcel-Dekker, 1982.

[Giannini and Longo, 1983] P. Giannini and G. Longo. Effectively given domains and lamda calculus semantics. Technical report, DSipt. Infomatica, Corso Italia 40, 56100 Pisa, 1983.

[Gierz et al., 1980] G. Gierz, K. H. Hofmann, K. Keimel, J. D. Lawson, M. Mislove, and D. S. Scott. *A Compendium of Continuous Lattices*. Springer-Verlag, Berlin, 1980.

[Grayson, 1981] R. J. Grayson. Concepts of general topology in constructive mathematics and in sheaves I. *Annals of Mathematical Logic*, 20:1–41, 1981.

[Grayson, 1982] R. J. Grayson. Concepts of general topology in constructive mathematics and in sheaves II. *Annals of Mathematical Logic*, 23:55–98, 1982.

[Gries, 1981] D. Gries. *The Science of Programming*. Springer-Verlag, New York, 1981.

[Gunter, 1987] C. Gunter. Universal profinite domains. *Information and Computation*, 72:1–30, 1987.

[Halmos, 1963] P. R. Halmos. *Lectures on Boolean Algebras.* van Nostrand, 1963.

[Hyland, 1979] J. M. Hyland. Filter spaces and the continuous functionals. *Annals of Mathematical Logic*, 16:101–143, 1979.

[Hyland, 1981] J. M. Hyland. Function spaces in the category of locales. *Continuous Lattices, Lecture Notes in Math.*, 871:264–281, 1981.

[Isbell, 1975] J. Isbell. Function spaces and adjoints. *Symposia Math.*, 36:317–339, 1975.

[Johnstone, 1981a] P. Johnstone. *Scott is not always sober.* Springer-Verlag, 1981.

[Johnstone, 1981b] P. Johnstone. Tychonoff's theorem without the axiom of choice. *Fund. Math.*, 113:21–35, 1981.

[Johnstone, 1982] P. T. Johnstone. *Stone Spaces*, volume 3 of *Cambridge Studies in Advanced Mathematics.* Cambridge University Press, Cambridge, 1982.

[Johnstone, 1983] P. Johnstone. The point of pointless topology. *Bull. Am. Math. Soc.*, 8:41–53, 1983.

[Jones and Plotkin, 1989] C. Jones and G. D. Plotkin. A probabilistic powerdomain of evaluations. In *Proceedings of the IEEE Fourth Annual Sumposium on Logic in Computer Science*, pages 186–195. Computer Society Press, 1989.

[Jung, 1989] A. Jung. *Cartesian closed categories of domaisn.* PhD thesis, Technische Hochschule Darmstadt, 1989.

[Jung, 1990] A. Jung. The classification of continuous domains. In *Symposium on Logic in Computer Science.* Computer Society Press (IEEE), 1990.

[Kelley, 1955] J. L. Kelley. *General Topology.* Van Nostrand, Princeton, 1955.

[Khalimsky *et al.*, 1990] E. Khalimsky, R. Kopperman, and P. Meyer. Computer graphics and connected topologies on finite ordered sets. *Topology and its applications*, 35, 1990.

[Khalimsky, 1977] E. Khalimsky. *Ordered Topological Spaces.* Naukova Dumka Press, Kiev, 1977.

[Ko and Friedman, 1982] K. I. Ko and H. Friedman. Computational complexity of real functions. *Theoretical Computer Science*, 20, 1982.

[Kovalevsky, 1989] V. Kovalevsky. Finite topology as applied to image analysis. *Compu. vision graphics image porcess*, 46:141–161, 1989.

[Kozen, 1981] D. Kozen. Semantics of probabilistic programs. *Journal of Computer and System Sciences*, 22:328–350, 1981.

[Kreitz and Weihrauch, 1985] C. Kreitz and K. Weihrauch. Theory of representations. *Theoretical Computer Science*, 38:35–53, 1985.

[Kuratowski, 1961] C. Kuratowski. *Topologie.* Warsaw, 1961.

[Larsen and Winskel, 1984] K. G. Larsen and G. Winskel. Using information systems to solve recursive domain equations effectively. In G. Kahn, D. B. MacQueen, and G. Plotkin, editors, *Semantics of Data Types*, pages 109–130. Springer-Verlag, Berlin, 1984. Lecture Notes in Computer Science Vol. 173.

[Lawson, 1987] J. D. Lawson. The versatile continuous order. In M. Main, A. Melton, M. Mislove, and D. Schmidt, editors, *Third Workshop on Mathematical Foundations of Programming Languages Semantics*, pages 134–160. Springer-verlag, Berlin, 1987. Lecture Notes in Computer Science Vol. 298.

[MacQueen *et al.*, 1986] D. MacQueen, Gordon D. Plotkin, and R. Sethi. An ideal model for recursive polymorphic types. *Information and Control*, 71:95–130, 1986.

[Martin-Löf, 1970] P. Martin-Löf. *Notes on Constructive Mathematics.* Almqvist and Wicksell, Stockholm, 1970.

[McKinsey and Tarski, 1944] J. C. McKinsey and A. Tarski. The algebra of topology. *Annals of Mathematics*, 44:141–191, 1944.

[Melton and Schmidt, 1986] A. Melton and D. Schmidt. A topological framework for cpo's lacking bottom elements. In *Lecture notes in computer science 239*, pages 196–204. Springer-Verlag, 1986.

[Milner, 1980] R. Milner. *A Calculus for Communicating Systems*, volume 92 of *Lecture Notes in Computer Science*. Springer-Verlag, Berlin, 1980.

[Myhill and Shepherdson, 1955] J. Myhill and J. C. Shepherdson. Effective operations on partial recursive functions. *Zeitschr. f. math. Logik und Grundlagen d. Math*, 1:310–317, 1955.

[Nachbin, 1965] L. Nachbin. *Topology and Order*, volume 4 of *New York Math. Studies*. Van Nostrand, 1965. Princeton, N.J.

[Nöbeling, 1954] G. Nöbeling. *Grundlagen der Analytischen Topologie.* Springer-Verlag, Berlin, 1954.

[Normann, 1980] D. Normann. *Recursion on the countable functionals.* Lecture Notes in Mathematics 811. Springer, 1980.

[Parikh, 1983] R. Parikh. Some applications of topology to program semantics. *Math. Systems Theory*, 16:111–131, 1983.

[Plotkin, 1978] G. D. Plotkin. t^ω as a universal domain. *J. Computer and System Sciences*, 17:209–236, 1978.

[Plotkin, 1981] G. D. Plotkin. Post-graduate lecture notes in advanced domain theory (incorporating the 'Pisa Notes'). Dept. of Computer Science, University of Edinburgh, 1981.

[Poincaré, 1905] H. Poincaré. *La Valeur de la Science.* Flammarion, Paris, 1905.

[Poston, 1971] T. Poston. *Fuzzy Geometry.* PhD thesis, Warwick University, 1971.

[Priestley, 1972] H. A. Priestley. Ordered topological spaces and the representation of distributive lattices. *Proceedings of the London Mathematical Society (3)*, 24:507–30, 1972.

[Rasiowa and Sikorski, 1963] H. Rasiowa and R. Sikorski. *The Mathematics of Metamathematics.* PWN - Polish Scientific Publishers, Warszawa, 1963. Monografie Matematyczne tom 41.

[Reilly *et al.*, 1982] I. L. Reilly, P. V. Subrahmanyam, and M. K. Vamanamurthy. Cauchy sequences in quasi-pseudo-metric spaces. *Monatshefte für Mathematik*, 93:120–127, 1982.

[Rogers, 1967] H. Rogers, Jr. *Theory of Recursive Functions and Effective Computability.* McGraw-Hill Book Company, 1967.

[Rowlands-Hughes, 1987] D. Rowlands-Hughes. Domains versus metric spaces. Master's thesis, Department of Computing, Imperial College, 1987.

[Saheb-Djahromi, 1980] N. Saheb-Djahromi. CPO's of measures for nondeterminism. *Theoretical Computer Science*, 12(1):19–37, 1980.

[Scott, 1970] D. S. Scott. Outline of a mathematical theory of computation. In *4th Annual Princeton Conference on Information Sciences and Systems*, pages 169–176, 1970.

[Scott, 1976] D. S. Scott. Data types as lattices. *SIAM Journal on Computing*, 5:522–587, 1976.

760 *References*

[Scott, 1982] D. S. Scott. Domains for denotational semantics. In M. Nielson and E. M. Schmidt, editors, *Automata, Languages and Programming: Proceedings 1982*. Springer-Verlag, Berlin, 1982. Lecture Notes in Computer Science 140.

[Smale, 1987] S. Smale. On the topology of algorithms I. *Journal of Complexity*, 3:81–99, 1987.

[Smyth,] M. B. Smyth. Completeness of quasi-uniform spaces in terms of filters. Department of Computing, Imperial College, 1987. To appear: *J. Lon. Math. Soc.*

[Smyth, 1977] M. B. Smyth. Effectively given domains. *Theoretical Computer Science*, 5:257–274, 1977.

[Smyth, 1983a] M. B. Smyth. The largest cartesian closed category of domains. *Theoretical Computer Science*, 27:109–119, 1983.

[Smyth, 1983b] M. B. Smyth. Powerdomains and predicate transformers: a topological view. In J. Diaz, editor, *Automata, Languages and Programming*, pages 662–675, Berlin, 1983. Springer-Verlag. Lecture Notes in Computer Science Vol. 154.

[Smyth, 1985] M. B. Smyth. Finite approximation of spaces. In D. Pitt, S. Abramsky, A. Poigné, and D. Rydeheard, editors, *Category Theory and Computer Programming*, volume 240 of *Lecture Notes in Computer Science*, pages 225–241. Springer-Verlag, Berlin, 1985.

[Smyth, 1987] M. B. Smyth. Quasi-uniformities: reconciling domains with metric spaces. In *Third Workshop on Mathematical Foundations of Programming Language Semantics*. Springer-Verlag, Berlin, 1987. Lecture Notes in Computer Science.

[Smyth, 1989] M. B. Smyth. Uniform continuity and computability: a thesis. Lecture at British Theor. Comp. Sci. Symp., Egham. Abstract in EATCS Bulletin, 1989.

[Smyth, 1991] M. B. Smyth. Totally bounded spaces and compact ordered spaces as domains of computations. In *Oxford Topology Symposium'89*. Oxford University Press, 1991.

[Smyth, 1992] M. B. Smyth. Stable compactification I. *Journal of the London Mathematical Society*, 1992. to appear.

[Spreen, 1989] D. Spreen. *A characterization of effective topological spaces*. Proc. 1989 Oberwolfach Meeting on Recursion Theory: Lecture Notes in Mathematics. Springer, 1989.

[Taylor, 1986] P. Taylor. *Recursive Domains, Indexed category Theory and Polymorphism.* PhD thesis, Cambridge University, 1986.

[Čech, 1966] E. Čech. *Topological Spaces.* Interscience Publishers, London/New York, 1966. revised edition.

[Vickers, 1989] S. J. Vickers. *Topology Via Logic.* Cambridge Tracts in Theoretical Computer Science. Cambridge University Press, 1989.

[Vuillemin, 1987] J. Vuillemin. Exact real computer arithmetic with continued fractions. Technical Report Rapports de recherche 760, INRIA, 1987.

[Weihrauch, 1987] K. Weihrauch. *Computability.* Springer-Verlag, Berlin, 1987.

[Wiedmer, 1980] E. Wiedmer. Computing with infinite objects. *Theoretical Computer Science*, 10:133–155, 1980.

Model Theory and Computer Science: An Appetizer

J. A. Makowsky

Contents

For M.N.Y., who emerged into my life
while I was preparing this chapter.

1 Introduction

The purpose of this chapter is to give an account of those aspects of model theory which we think are relevant to theoretical computer science. We start with a general outline of the evolution of model theory, which will serve as an exposition of the major themes. In the subsequent section we shall elaborate some of these themes and put them into the context of theoretical computer science.

We assume that the reader is familiar with the basics of First Order Logic, Computability Theory, Complexity Theory and Basic Algebra. Whenever possible we shall refer to textbooks and monographs rather than the original papers. Only material not treated in standard texts will be quoted in the original (or by referring to a subsequent paper which contains the result in the most readable form).

This chapter is not meant to be an exhaustive scholarly survey of model theoretic methods in theoretical computer science. It is more of a personal guided tour into a well mapped but still largely unexplored landscape. It has definite autobiographical traits. No author can completely escape that. It proposes to some extent a unifying view which ultimately should lead to the disappearance of the personal touch. However, for that to happen more research and reinterpretation of classical results is needed. Logic and model theory are relatively old disciplines which enjoy renewed interest. They can serve as one explanatory paradigm for foundational problems in theoretical computer science. But the gap between the traditional logicians and mathematicians, and the working computer scientists is first of all cultural in the sense of R. Wilder's [Wilder, 1981]. His studies deserve special attention especially when one has in mind the evolution and development of programming languages, operating systems, user interfaces and other paradigms of computing, but also in addressing foundational questions, cf. [Makowsky, 1988].

Wilder's studies clearly show several phenomena: that the evolution of concepts to widely accepted norms of practice takes much longer and needs more than just the availability of such concepts; that the evolution of concepts is not due to individuals but is embedded in one (or several competing) cultural systems which are themselves embedded in host cultural systems; that nevertheless the fame and prestige of the protagonists of science and scientific progress do play an important, possibly also counterproductive rôle; that cultural stress and cultural lag play a crucial rôle in the evolution of concepts; that periods of turmoil are followed by periods of consolidation after which concepts stabilize; that diffusion between different fields usually will lead to new concepts and accelerated growth of science; that environmental stresses created by the host culture and its subcultures will elicit observable response from the scientific culture in question; and, finally, revolutions may occur in the metaphysics, symbolism and methodology of computing science, but not in the core of computing itself. Wilder has developed in [Wilder, 1981] a general theory of 'Laws' governing the evolution of mathematics, from which I have adapted the above statements. It remains a vast research project to assimilate Wilder's theory into our context, but it is an indispensable project if we want to adjust our expectation of progress in computing science to realistic hopes. Wilder's work also sheds some light into the real problems underlying the so called 'software crisis': the cultural lag of programming practice behind computing science and the absence of various cultural stresses may account for the abundance of programming paradigms without the evolution of rigorous standards of conceptual specifications.

I can only hope that I may contribute my small share to the slow process of bridging that gap and further the logical foundations of computer science.

Acknowledgements

I am indebted to many colleagues who encouraged me at several stages to pursue my research of model theoretic methods in computer science. Among them I would like to mention Erwin Engeler, Eli Shamir, Shimon

Even, Vaughan Pratt, Catriel Beeri, Saharon Shelah, David Harel and Yuri
Gurevich. I am also indebted to the graduate students of my department
whose work contributed to my understanding. Among them are Ariel Calò,
Yaniv Bargury, Yachin Pnueli and Avy Sharel. I would like to thank S.
Ben-David who allowed me to include his unpublished results in section
5 and Y. Pnueli, who helped in the writing of Section 7. Finally I would
like to thank D. Gabbay for inviting me to write this chapter and for his
insisting that I give him this version for publication.

2   The set theoretic modelling of syntax and semantics

Model theory is the mathematical (set theoretical) study of the interplay
between Syntax and Semantics. Historically it has its roots in the various
attempts at reducing first mathematics to logic (Frege, Hilbert), then logic
to number theory (Skolem, Gödel) and finally, at modelling logic within
set theory (Tarski, Vaught). The first two reductions were motivated by the
fundamental questions of the foundations of mathematics, whereas the latter accepts Bourbaki's view that set theory is the foundational framework
of mathematics. It is this latter approach which forms the background of
model theory proper. Let us elaborate this further: We take some Naïve
Set Theory for granted and attempt to model all objects of mathematical
study within this Set Theory. Without having to bother too much about
the choice of set theory we can model the natural numbers, finite strings,
finite graphs within set theory. We accept the axiom of choice as a fact of
life. With this we can model also most of the concepts of classical algebra
(field theory, ring theory, group theory, but not necessarily cohomology
theory) within set theory. The natural numbers, fields, graphs are mathematical structures which serve as the prime examples for models of logical
theories. We usually think of models of a logical theory rather than of a
single model, and the models form usually a proper class (the class of all
groups, rings, etc.). If we restrict ourselves to finite mathematical structures we can additionally consider recursive sets of models or sets of models
of lower complexity classes (Logarithmic Space or Polynomial Time recognizable classes of models). Next we observe that logical theories are just
sets of formulas and that formulas can be viewed again as either strings
over some alphabet or as some kind of labelled trees. Most people think
of formulas as inherently finite objects, but infinite formulas (then better
viewed as trees) are easily conceivable. So formulas and sets of formulas
can also be modelled in our set theory. If we think of finite formulas as
strings it makes sense to bring in also concepts of recursion theory and
complexity theory.

The basic relationship between sets of formulas and models is the satisfaction relation. We view it here as a ternary relation $M(\Sigma, \mathcal{A}, z)$, where Σ is a set of formulas, $\mathcal{A}$ is a structure, i.e. a generalized algebra over some vocabulary (similarity type) and z is an assignment function mapping free variables of the formulas Σ into elements of the universe of the structure $\mathcal{A}$. If $M(\Sigma, \mathcal{A}, z)$ holds for every z we simply write $\mathcal{A} \models \Sigma$ and say that $\mathcal{A}$ is a model of Σ. The characteristic function of the satisfaction relation is often called meaning function. The meaning function can also be modelled in set theory.

2.1 First order structures

It is customary to model algebraic structures as sets equipped with functions and relations. This view has its origins in algebra as understood in the nineteenth century. A structure consists of a set, the *universe*, equipped with some relations, functions and constants, which model the *primitives*.

A *group* then is a set equipped with a binary function, called *multiplication*, a unary function, called the *inverse*, and a constant called the *unit element*. An ordered group is additionally equipped with a binary relation, called the *order relation*. In similar ways we can define *fields*, *rings* or the structure of *arithmetic on the natural numbers*. In computer science other *data structures* are defined similarly, such as *words, stacks, lists, trees, graphs, Turing machines etc.*. A word of length n over the alphabet $\{0, 1\}$ can be viewed as a set of n elements with a binary relation which linearly orders that set and a unary relation, which indicates which places in the word are occupied by the letter 1. A graph is just a set with a binary relation. In each case it is required that the functions, relations and constants satisfy some interrelating properties which make it into a group (field, word, graph, Turing machine etc.).

Sometimes, it is more practical to model structures with several underlying sets, as in the case of vector spaces. These sets form several universes and are called *sorts*. We then speak of *many-sorted structures*. A Turing machine consists of two sets: a set of states and a set of letters; a binary relation between states and letters; a unary relation, the set of final states; and a constant, the initial state. Many-sorted structures allow us to model also concepts which involve sets of sets, such as topologies, families of subgroups or whatever comes to ones mind. This last statement is not just a sloppy way of saying something vague. It really expresses a belief, or rather experience, that everything which can be modelled in set theoretic terms with finitely many basic concepts can be modelled by such structures.

In modern terms a structure is a tuple of sets of specified characteristics. The primitive concepts have *names* and these names form again a set, called the *vocabulary*. A structure then is an *interpretation* of a vocabulary. More precisely, a (first order) vocabulary τ is a set of sort symbols, function symbols, relation symbols and constant symbols. The function, relation and constant symbols have an *arity* which specifies the number and sorts of the arguments and values. The arity is mostly assumed to be *finite*. In this way we can naturally associate with a vocabulary τ the proper class of all τ-structures, which we denote by $STR(\tau)$.

2.2   The choice of the vocabulary

The notion of a τ-structure evolved naturally in mathematics, more precisely in algebra. Groups and fields are usually described as sets with operations, ordered fields are sets with operations and relations. The choice of the basic operations is in no way trivial. Should we add the inverse operation as basic or not ? In the case of arithmetic we have the successor relation, addition and multiplication. The first order theory of arithmetic is undecidable, but if we leave out multiplication, it becomes decidable. This is a dramatic change. Subtraction is definable by a first order formula, so leaving it out or adding it, does not affect decidability. But it does affect the set of substructures.

In the case of graphs the modelling issue is more subtle. It is customary to describe a graph as a set with an incidence relation. Thus there is quantification over vertices but not over edges. If we choose to allow quantification over edges we change the notion of structure. To what extent this matters has been studied by Courcelle in a series of papers [Courcelle, 1990a; Courcelle, 1990b]. Finite graphs can also be described by their incidence matrix, which does not fit the notion of a τ-structure in a natural way. However, we can consider the incidence matrix itself as a τ-structure in many ways.

Logic and model theory take the notion of τ-structures for granted. How to choose the particular vocabulary depends on many extra-logical issues. Discussing some of these issues is a discipline in itself called Data Modelling. The issues discussed there come from data processing and data bases.

First order logic allows quantification only for elements of the underlying universe. This looks like a severe restriction, as in mathematics we quantify very often also over subsets and more complex objects. However, this restriction only affects the modelling issue. In set theory all objects are sets, and second order arithmetic can be formalized using first order τ-structures, where the universe consists of points and sets with a unary predicate distinguishing between them. It is in this sense that the notion of τ-structures is as universal as the set theoretic modelling of mathematical situations.

More surprisingly, τ-structures can also capture situations of modal and temporal propositional logic. A propositional variable may be true in some moments of time and false in others. So let the universe of our discourse be time and propositions be unary predicates [Burgess, 1984]. This is almost obvious. In the case of modal logic it needed Kripke's ingenuity to make use of this idea [Bull and Segerberg, 1984]. The universe now is a set, the set of all possible worlds or situations, propositions are again unary predicates, but the relationship between possible worlds is described by an accessibility relation. From here, it is natural to continue and consider several accessibility relations (to model for example the distinction between the legally and the morally possible). In the theory of program verification this was used to model the behaviour of abstract programs (Dynamic Logic [Harel, 1984]). In AI this approach was extended further to model reasoning about knowledge [Emerson, 1990]. The interested reader will find more also in Section 3.8 and [Gabbay *et al.*, 1992]. For reasons of space we shall not treat these issues much further in this chapter.

The point we want to emphasize here is that the framework of τ-structures is flexible enough to model everything which can be modelled in mathematics, more precisely in set theory. The choice of vocabulary is sometimes difficult and guided by various issues, including user friendliness, explicitness and technicalities of the field of application.

2.3 Logics

The most prominent logic is First Order Logic. Although we argued that τ-structures are sufficiently general to model all situations which can be treated mathematically, First Order Logic has a limited expressive power. This means that a description in First Order Logic of a situation will allow what are called *non-standard models*. In other words, it will have models of that description which do not capture all the intended features.

Other logics we shall consider are Second Order Logic (allowing quantification over subsets and relations without making them into objects of the model), Monadic Second Order Logic (allowing quantification only over subsets), infinitary logics (allowing infinite conjunctions and disjunctions) and logics with generalized quantifiers. The latter will be discussed in detail in Section 7.

A logic itself again can be modelled within set theory. It consists of a family of τ-formulas $Fm(\tau)$ with associated meaning functions M_τ subject to several conditions. The most fundamental among them is the *Isomorphism Condition* which asserts that isomorphic τ-structures cannot be distinguished by τ-formulas. The other conditions assert that the most basic operations such as conjunction, disjunction, negation, relativization and quantification over elements are well defined. Such logics are called regular logics. If negation is omitted we call the logics semi-regular. The model theory of such logics has been extensively studied, cf. [Barwise and Feferman, 1985].

For applications in computer science the relevant logics have two additional features: The set of τ-formulas Fm_τ is recursive for finite τ and the meaning functions M_τ are *absolute* for set theory, i.e. they do not depend on the particular model of Zermelo-Fränkel set theory that we are working in. If we additionally require that the tautologies of such a logic are recursively enumerable, we call such a logic a *Leibniz Logic*. It now follows from work of Lindström and Barwise that every Leibniz Logic is in some precise sense equivalent to First Order Logic, cf. [Barwise and Feferman, 1985]. In other words, a proper extension of First Order Logic is either not regular or not absolute or its tautologies are not recursively enumerable. If we restrict ourselves to finite structures the latter is unavoidable even for First Order Logic, but then the satisfiable formulas are recursively enumerable. Semi-regular logics on finite structures where the satisfiable formulas are recursively enumerable have many applications to computer science and are studied in Section 7.

3 Model theory and computer science

As we discuss here applications of model theory to computer science let us clarify what we intend both by model theory and theoretical computer science.

3.1 Computer science

Concerning Computer Science we take a pragmatic approach. Any mathematically modelled situation which captures any issue arising in the dealings with computers is a possible topic for computer science. This includes hardware, software, data modelling, interfaces and more. Some of the more classical fields of theoretical computer science have already matured into well established subdisciplines. Among them we find computability theory, algorithmics, complexity theory, database theory, data and program specification, program verification and testing etc. However, we feel that a certain confusion in the definitions of these fields is obfuscating the issues involved. It very much depends whether our point of view is *method-oriented* or *application-oriented*. Computability and complexity theory deal with the clarification of our notion of what is computable. This represents a clear case of a well defined method-oriented subdiscipline of computer science and the foundations of mathematics. Database theory on the other hand is a field which grew from an application-oriented approach. From a method-oriented point of view, database theory tends to fall apart into subfields, such as *finite model theory, operating systems, file systems, user interfaces* and *algorithmics*, where each of these transcend the boundaries of the database applications. Scientifically speaking, the *ad hoc* collection of methods bound together by a vaguely defined common application is unsatisfactory. It is justified only for didactic purposes such as training application-oriented engineers. But such training is detrimental to a deeper understanding of the craft and the science and leads to chaotic duplicity (and multiplicity) of research and research subcultures each disguised in its own terminology and provincialisms.

In this paper we try to exhibit a method and a scientific framework, *model theory*, and discuss typical problems whose discussion in this framework is beneficial to our understanding.

3.2 The birth of model theory

Model theory deals with the mathematical study of the satisfaction relation or its characteristic function, the meaning function. For a specific syntactic system which we call logic, the meaning function singles out the pairs of first order structures and formulas which we interpret as asserting that the given formula holds in the given structure. Any mathematically proven statement about the meaning function is a model theoretic theorem.

The first result of mathematical logic which could be called model theoretic was the famous Löwenheim–Skolem Theorem:

Theorem 3.2.1 (Löwenheim–Skolem Theorem). *Let* Σ *be a set of formulas of first order logic such that there is an infinite* $\mathcal{A}$ *with* $\mathcal{A} \models \Sigma$. *Then there are models* $\mathcal{B}$ *of arbitrary infinite cardinalities* $\kappa > |\tau| + \aleph_0$ *such that* $\mathcal{B} \models \Sigma$.

The most basic model theoretic theorem is the compactness theorem for first order logic. We say that a set Σ of formulas is satisfiable if there is a structure $\mathcal{A}$ such that $\mathcal{A} \models \Sigma$. The compactness theorem now states that:

Theorem 3.2.2 (Compactness Theorem). *A set* Σ *of first order formulas is satisfiable iff every finite subset of* Σ *is satisfiable.*

It follows from Gödel's completeness theorem for countable Σ and was proven for arbitrary Σ by Mal'cev. A model theoretic proof of the completeness theorem was given independently by Hasenjäger, Henkin and Hintikka in 1949. This proof, most widely known as Henkin's method, was instrumental in shaping the further developments of logic and model theory.

The completeness theorem usually refers to some specific *deduction method* and states that a τ-formula ϕ is derivable from a set of τ-formulas Σ

iff ϕ is a semantical consequence of Σ. The notion of semantical consequence is model theoretic. It says that for every τ-structure $\mathcal{A}$ and every assignment z such that $M(\Sigma, \mathcal{A}, z) = 1$ we also have $M(\phi, \mathcal{A}, z) = 1$. A purely model theoretic statement which captures the essence of the completeness theorem without reference to the particular deduction is the following:

Theorem 3.2.3. *For every recursive enumerable set Σ of τ-formulas the set τ-formulas ϕ which are semantical consequences of Σ are recursive enumerable.*

3.3   Definability questions

The next ten years of evolving model theory were marked by explorations of the compactness theorem and the Löwenheim–Skolem Theorem. The first of these explorations concerns definability questions, both negative and positive results.

On the negative side we have that many important mathematical concepts cannot be captured by first order formulas. Among them are the concept of well-orderings, connectivity of binary relations and Cauchy completeness of linear orders. This was first perceived as a blow to the foundation of mathematics, as it led to the 'non-standard' models of Natural Numbers, Real Numbers and Set Theory. However, A. Robinson realized that those non-standard models had their own usefulness for developing genuine first order mathematics. For theoretical computer science, the non-standard models of number theory and set theory only recently started to play a rôle. We shall not discuss their use in this paper, but refer the reader to [Andréka *et al.*, 1982; Makowsky and Sain, 1989; Pasztor, 1990; Gergely and Úry, 1991].

On the positive side we have Beth's theorem on implicit definitions and its various generalizations. Those theorems were mostly proven first by syntactic methods, but the model theoretic proofs found later make those theorems independent of the particular formalism of first order logic. Let Σ be a set of first order formulas over some vocabulary τ, and let P be an n-ary relation symbol not in τ. We say that a formula $\phi(P)$ over $\tau \cup \{P\}$ defines P *implicitly* using Σ, if in each model $\mathcal{A}$ of Σ there is *at most one* interpretation of P. We say that the predicate implicitly defined by ϕ using Σ has an *explicit* definition for which there is a formula $\theta(x_1, x_2, \ldots, x_n)$ over τ such that

$$\Sigma \cup \phi(P) \models \forall x_1, x_2, \ldots, x_n(\theta(x_1, x_2, \ldots, x_n) \leftrightarrow P(x_1, x_2, \ldots, x_n)).$$

Now Beth's theorem can be stated as follows:

Theorem 3.3.1 (Beth). *Let Σ be a set of first order formulas and let $\phi(P)$ be an implicit definition of P using Σ. Then there is an explicit* *definition of P using* Σ.

Beth's theorem is trivially true for second order logic, and false for first order logic when restricted to finite structures. In the latter case, implicit definitions allow us to define classes of structures recognizable in **NP ∩ co-NP**, whereas first order formulas define classes recognizable in **L**. We shall discuss the consequences of this observation in Section 7. Beth's theorem is mainly appealing as a closure property of a logic. There are surprisingly few genuine applications of Beth's theorem and its relatives. One of them, in the axiomatic treatment of specification theory, is relevant to theoretical computer science (cf. [Maibaum, to appear]). More recently Kolaitis has studied implicit definability on finite structures and related it to issues in complexity theory [Kolaitis, 1990].

3.4   Preservation theorems

Another line of exploration of the compactness theorem was initiated by Tarski. He observed that universal first order formulas are preserved under substructures. In other words, if Σ is a set of first order formulas in prenex normal form with universal quantifiers only and $\mathcal{A} \models \Sigma$ and $\mathcal{B} \subseteq \mathcal{A}$ is a substructure of $\mathcal{A}$ then $\mathcal{B} \models \Sigma$. The same is true for any Σ_1 equivalent to Σ. By an ingenious application of the compactness theorem he proved the converse of this observation:

Theorem 3.4.1 (Substructure Theorem).   *A set* Σ *of first order formulas is preserved under substructures iff* Σ *is equivalent to some set of universal formulas.*

The Substructure Theorem set a pattern for further investigations whose results are called preservation theorems. It led to similar syntactic characterizations for formulas preserved under unions of chains, homomorphisms, products, intersections and other algebraic operations. There are also some surprising interrelationships between a generalization of Beth's theorem and preservation theorems for a wide class of operations between structures, cf. [Makowsky, 1985]. Some of these preservation theorems have variations and interpretations which are of importance in database theory [Makowsky, 1984] and the foundations of logic programming, [Makowsky, 1987]. Questions related to such preservation theorems also occur naturally in the compositional approach to model checking for various temporal logics [Emerson, 1990]. The latter is a subdiscipline of program verification. It still remains an open avenue of research to find the preservation theorems which will be useful for model checking, in particular those preservation theorems which will reflect the compositionality of programs.

3.5   Disappointing ultraproducts

With these early investigations centring around the compactness theorem and the preservation theorems, an alternative proof of the compactness theorem was discovered using ultraproducts. The method of ultraproducts also led to alternative proofs of preservation theorems and dominated research in model theory throughout the 60s (cf. [Chang and Keisler, 1990]), but it had almost no impact on theoretical computer science. Although Kochen and Kripke [Kochen and Kripke, 1982] used bounded ultraproducts to give a model theoretic proof of the Paris–Harrington Theorem, Kanamori and McAloon [Kanamori and McAloon, 1987] gave a model theoretic proof of this theorem without bounded ultraproducts.   In the language of theoretical computer science this theorem can be stated as follows:

Theorem 3.5.1 (Paris–Harrington). *There are programs (number theoretic functions) which*

 (i) *always terminate (are total) but*

 (ii) *such a termination proof does not exist within the formalization of Peano arithmetic.*

A very picturesque version of this theorem is due to Kirby and Paris [Kirby and Paris, 1982]. The function described there is a winning strategy for the fight of Hercules against the Hydra, where the Hydra grows n new heads after the nth blow it receives. The underlying theme of this theorem is *fast growing functions.* We return to this topic in Section 5.

3.6   Complete theories and elimination of quantifiers

Another line of early investigations was the study of complete theories. A set Σ of formulas (over a fixed vocabulary τ) is complete if for every formula ϕ either $\Sigma \models \phi$ or $\Sigma \models \neg\phi$. The original interest for complete theories stems from questions of decidability. A set of formulas Σ is decidable if its set of consequences is recursive.

Theorem 3.6.1. *If Σ is recursive and complete then Σ is decidable.*

Proofs of completeness were often obtained using the method of elimination of quantifiers. Tarski used these ideas to show that there is a decision procedure for Elementary Geometry, which he identifies with the first order theory of real closed fields. This theorem led recently to interesting applications in robotics.   But the method of elimination of quantifiers has not yet received the attention it deserves among researchers in automated theorem proving. The state of the art in automated theorem proving for elementary geometry is best discussed in [Chou, 1988; Schwartz *et al.*, 1987].

Another way of proving completeness of first order theories is based on a simple but ingenious observation due to Vaught, which shows the power of model theoretic reasoning. Let Σ be a complete theory. If Σ has a model $\mathcal{A}$ which is finite, then it is unique up to isomorphism. If $\mathcal{A}$ is infinite, then by the Löwenheim–Skolem Theorem, Σ has models of arbitrary infinite cardinalities. Now, if all models of Σ of infinite cardinality κ are isomorphic, we say that Σ is κ-categorical. Note that if $\mathcal{A}$ and $\mathcal{B}$ are isomorphic then they satisfy the same first order sentences.

Theorem 3.6.2 (Vaught). *If Σ is κ-categorical for some infinite κ and Σ has no finite models, then Σ is complete.*

Proof. Assume, for contradiction, that there is ϕ such that neither $\Sigma \models \phi$ nor $\Sigma \models \neg\phi$. As Σ has no finite models, using the Löwenheim–Skolem Theorem we can find models $\mathcal{A}$ and $\mathcal{B}$ such that $\mathcal{A} \models \Sigma \cup \{\phi\}$ and $\mathcal{B} \models \Sigma \cup \{\neg\phi\}$, both of cardinality κ. But then $\mathcal{A}$ is isomorphic to $\mathcal{B}$, which contradicts the fact that $\mathcal{A} \models \phi$ and $\mathcal{B} \models \neg\phi$.

Classical mathematical results establish categoricity of a few natural first order theories. Hausdorff and Cantor showed that any two countable dense linear orderings are isomorphic, and a similar argument shows the same for countable atomless Boolean algebras. Steinitz showed that any two uncountable algebraic closed fields of characteristic zero of the same cardinality are isomorphic. So Vaught's theorem quickly establishes that these theories are complete and therefore decidable.

3.7 Spectrum problems

The study of categoricity of first order theories was the driving force behind the deepest results of model theory. Ryll-Nardzewski, Svenonius and Engeler independently characterized ω-categorical theories, and Morley proved the following generalization of Steinitz's theorem:

Theorem 3.7.1 (Morley). *If Σ is categorical for some uncountable κ then Σ is categorical for every uncountable κ.*

If Σ is not categorical, then it is natural to look at the following: let Σ be a set of formulas and denote by $I(\Sigma, \kappa)$ the number of non-isomorphic models of cardinality κ. $I(\Sigma, \kappa)$ is called the spectrum of Σ. The study of $I(\Sigma, \kappa)$ for infinite κ was initiated by Morley and Vaught (cf. [Chang and Keisler, 1990]). A complete analysis of the infinite case dominated the research efforts in model theory and culminated in Shelah's theorem [Shelah, 1990]:

Theorem 3.7.2 (Shelah's Spectrum Theorem). *For uncountable κ $I(\Sigma, \kappa)$ is non-decreasing in κ and, in fact either*

(i) $I(\Sigma, \kappa) = 2^\kappa$ *or*

(ii) $I(\Sigma, \omega_\alpha) < BETH_{\omega_\alpha}(card(\alpha))$.

The infinite spectrum and its ramifications are the core of a highly sophisticated development in model theory called stability theory. Although it is of extreme mathematical depth and beauty I can so far see no fruitful interplay between stability theory and computer science.

Instead of $I(\Sigma, \kappa)$ for finite κ, we shall look at the finite cardinal spectrum $Spec(\Sigma)$ of finite sets of formulas Σ. $Spec(\Sigma)$ is the set of natural numbers n such that there is a finite model of Σ of cardinality n. The study of $Spec(\Sigma)$ was initiated by Scholz. For the historic remarks cf. [Fagin, 1990]). In contrast to stability theory, the study of the finite cardinal spectrum $Spec(\Sigma)$ led to very interesting interactions between model theory and complexity theory, through the pioneering work of Büchi, Fagin and Immerman (cf. [Büchi, 1960; Fagin, 1974; Immerman, 1987].

Büchi studied the interplay between Monadic Second Order Logic and automata theory. He looked at words over a finite alphabet as finite linearly ordered structures with unary predicates. Recall that a set of words is regular if it is recognizable by a finite automaton. His theorem states:

Theorem 3.7.3 (Büchi 1960). *A set of words is regular iff it is definable by an existential formula of monadic second order logic.*

Trakhtenbrot independently found a similar Theorem [Trakhtenbrot, 1961]. Fagin studied the finite spectrum and was led to the following theorem:

Theorem 3.7.4 (Fagin). *A set of finite structures is in* **NP** *iff it is definable by an existential (full) second order sentence.*

Let ϕ be a first order formula over a vocabulary τ. We note that $Spec(\phi)$ can be viewed as the set of finite models of Φ over the empty vocabulary, where Φ is obtained from ϕ by existentially quantifying all the predicate symbols of τ. So Fagin's theorem generalizes both the spectrum problem as well as Büchi's theorem.

Immerman characterized similarly sets of ordered finite structures in **L**, **NL**, **P**. We shall discuss the interplay between model theory and complexity theory in Section 7.

3.8 Beyond first order logic

In this introduction we have already come across features which go beyond
first order logic. We have tacitly introduced quantification over relations in
Büchi's theorem, and we have mentioned the semantic restriction to finite
structures. These mark the two independent directions that generalizations
might take: more sentences vs. more complex models.

The model theoretic study of richer logics over τ-structures in the usual
sense was initiated in the late 50s independently by A. Tarski and his students, and E. Engeler for infinite first order formulas, and by A. Mostowski
for generalized quantifiers. The book [Barwise and Feferman, 1985] contains an excellent bibliography and historic account. From a naive model
theoretic point of view it is natural to ask whether for those generalized
logics the compactness theorem and the Löwenheim–Skolem theorem are
still true. For infinite formulas compactness fails trivially. It was also
observed that in all the examples of generalized quantifiers studied one of
the two usually failed. In 1966 P. Lindström published a paper which was
hardly noticed till 1970. In it the following fundamental result was stated
and proved:

Theorem 3.8.1 (Lindström).   *Let $\mathcal{L}$ be a regular logic over τ-structures
which both satisfies the compactness theorem and the Löwenheim–Skolem
theorem. Then $\mathcal{L}$ is, up to semantic equivalence, first order logic.*

A logic is *regular* if it is closed under Boolean operations, quantification and relativization and does not distinguish between isomorphic τ-
structures. This theorem was followed by intense investigations of model
theories of particular logics and the evolution of a framework for 'abstract
model theory'. The fruits of these investigations were collected in the mon-
umental volume [Barwise and Feferman, 1985].

In 1965 S. Kripke initiated the model theoretic study of logics different from classical first order or propositional logic, such as intuitionistic
logics, modal logics and temporal logics.   His main idea was to look
at, say propositional modal or temporal logic, as a special case of first
order logic. A Kripke-structure is a first order structure with a binary relation for accessibility to possible states (worlds in the case of modal logic,
points in time in the case of temporal logic). Propositions then are unary
predicates in Kripke-structures. The modal and temporal operators (neces-
sarily/possibly, always/sometimes) now become first order definable. The
axioms of modal or temporal logic shape the accessibility relation. In this
way Kripke was able to state precisely the semantics of modal logic and
prove, for the first time, completeness theorems. To illustrate this let us
state here the case of the modal system T, which captures the unproblem-
atic aspects of 'necessity'. The formula $\Box\phi$ is read as 'necessarily ϕ'. The
system T contains all substitutions of propositional tautologies, the axioms

$\Box(\phi \to \psi) \to (\Box\phi \to \Box\psi)$ and $\Box\phi \to \phi$, and the two deduction rules *modus ponens* and from ϕ infer $\Box\phi$.

Theorem 3.8.2 (Kripke). *A modal formula ϕ is provable in T iff ϕ is true in all Kripke-structures with a reflexive accessibility relation.*

We speak of temporal logic when the accessibility relation is a partial order, in the most natural case a discrete linear order. The formula $\Box\phi$ is now read as 'always ϕ'. It is natural to ask whether the introduction of one temporal operator (or for that matter, modal operator) suffices, or whether there are many hitherto undiscovered temporal operators. Obviously we have operators corresponding to 'next', 'previously', 'always in the future', 'always in the past', 'ϕ until ψ' and 'ϕ since ψ'. We note that all these operators are first order definable over linearly ordered Kripke-structures. H. Kamp now proved the following remarkable theorem.

Theorem 3.8.3 (Kamp). *Let $TO(p_1, \ldots, p_n)$ be an n-ary temporal operator which is first order definable over discrete, complete linear orderings. Then $TO(p_1, \ldots, p_n)$ is definable from the operators 'next', 'previously', 'until' and 'since'.*

The theorems of Kripke and Kamp are two prime examples of model theoretic theorems in non-standard logic. The underlying techniques, however, are applicable in a much wider context and have not yet been systematically developed. Good surveys are [Burgess, 1984; Bull and Segerberg, 1984; Ryan and Sadler, 1992].

Both types of generalizations of first order logic, more formulas and richer semantic structures, found rich applications in theoretical computer science. Engeler was the first to observe that infinitary logic can serve as a framework to formulate the input/output behaviour of programs. His approach was considered awkward. V. Pratt and D. Kozen used a Kripke-like semantics for their approach to axiomatize the input/output behaviour of programs, which was finally called 'Dynamic Logic'. This was received enthusiastically. However, it was soon observed that the two approaches were equivalent. Burstall suggested modal and Pnueli temporal logic for the axiomatic description of program behaviour. Kripke-structures are also abundant in foundational research in AI, especially in the theory of knowledge.

3.9 The hidden method

One model theoretic tool of central importance does not usually appear in the statement of theorems, but mostly in their proofs. This is the 'back-and-forth' characterization of n-equivalent structures, i.e. structures satisfying the same sentences of quantifier rank n. This characterization originated in the early work of R. Fraïssé and was popularized in an influential paper by A. Ehrenfeucht. Ehrenfeucht also generalized the method to monadic second order logic, and further generalizations for infinitary logic and logics with generalized quantifiers and predicate transformers were developed subsequently, cf. [Barwise and Feferman, 1985]. We shall devote section 8 to an extensive discussion of this method, which we call Ehrenfeucht–Fraïssé games. Here we only list some of its applications.

Originally, Ehrenfeucht–Fraïssé games were used to prove that certain concepts are not definable by first order formulas even if restricted to finite structures. Among such concepts we find the connectivity and planarity of graphs. The deepest and most surprising application of Ehrenfeucht-Fraïssé games occurs in the proof of Lindström's theorem. A close analysis of this proof also shows that Beth's theorem can be proven using this method, as well as various preservation theorems. Ehrenfeucht's generalization of the method to monadic second order logic can be used to give a model theoretic proof of Büchi's theorem. It was used in [Ferrante and Rackoff, 1979] to establish lower and upper bounds for the complexity of decidable theories such as Presburger Arithmetic and the theory of two successors functions. And finally, it can be also used to prove the 0–1 law for first order logic over finite structures, due independently to R. Fagin and Glebskiï, and Kogan, Ligon'kiï and Talanov.

3.10 0–1 Laws

To state the 0–1 Theorem, let τ be a vocabulary without function symbols and let ϕ be a first order τ-formula. We think of a structure of size n as having the universe $\{0, 1, \ldots, n-1\}$. Let $S_\tau(n)$ be the number of τ-structures of size n. Recall that $I(\phi, n)$ is the number of different structures of size n satisfying ϕ. Let $P(n, \phi) = I(\phi, n)/S_\tau(n)$.

Theorem 3.10.1 (0–1 Law of First Order Logic). *For every τ without function symbols and every first order τ-sentence ϕ the limit*

$$\lim_{n \to \infty} P(n, \phi)$$

is well defined and is either 0 or 1.

Definition 3.10.2 (Almost true formulas). A First Order Formula ϕ is almost true if $\lim_{n \to \infty} P(n, \phi) = 1$.

In contrast to First Order Validity over finite structures, which is undecidable (cf. Trakhtenbrot's theorem, 7.1.1), the set of first order sentences true in almost all structures is decidable. In fact Grandjean proved [Grandjean, 1983]:

Theorem 3.10.3 (Grandjean). *Assume that τ has no function symbols. The problem of deciding, whether a first order τ-formula ϕ is almost true, is **P**–Space complete.*

0–1 Laws were investigated also for extensions of First Order Logic. For a further discussion of similar theorems the reader should consult [Compton, 1987; Fagin, 1990] and the literature quoted therein. Striking applications of 0–1 Laws in Computer Science are still missing. They may emerge in the context of Average Case Complexity Theory [Gurevich, 1991], Graph Algorithms [Gurevich and Shelah, 1987] and the like.

4   Preservation theorems

Preservation theorems of First Order Logic characterize syntactic classes of formulas in terms of their semantic properties. In Section 3.4 we have given the simplest example, the substructure theorem. Its proof is a simple but ingenious use of the compactness theorem. This specific application of compactness was termed the *Method of Diagrams* and may be found in every introduction to model theory. Alternative proofs of preservation theorems were given using ultra-products [Chang and Keisler, 1990].

The classical preservation theorems characterize formulas which are preserved under

- Unions of chains

- Homomorphisms

- Products and reduced products

- Relativization

- Intersection of models

- Monotone predicates

and many more. All these preservation theorems were inspired by analysing the constructions used by the *working algebraist*. They always exploit the fact that infinite structures are part of our discourse. In fact, most preservation theorems fail, if restricted to finite models. One exception was found by Gurevich and Shelah cf. [Gurevich, 1990].

4.1   Horn formulas

Both, in Relational Database Theory and Logic Programming, first order formulas form the syntactic background of the field. In both fields it was observed that certain syntactically defined classes of formulas play a special rôle. For a detailed discussion of first order logic's rôle in database theory one may consult [Vardi, 1988; Kannelakis, 1990] and the corresponding chapter in this handbook [Makowsky, to appear]. The most prominent such class of formulas are called *Universal Horn formulas*. They also play a certain rôle in the Specification of Abstract Data Types.

Definition 4.1.1 (Horn formulas).

(i) A quantifier-free Horn formula is a formula of the form

$$P_1 \wedge \ldots \wedge P_k \rightarrow P_0$$

where all the $P_i, i \leq k$ are atomic formulas.

(ii) A Universal Horn formula is a formula of the form $\forall x_1, \ldots, x_m \Phi$ in which Φ is a quantifier-free Horn formula.

The classical theorem of model theory gives the following characterization of Universal Horn formulas.

Theorem 4.1.2 (Mal'cev). *Let K be a class of τ-structures which are exactly the models of a set of first order τ-formulas Σ. Then K is closed under substructures and products iff Σ is equivalent to a set of Universal Horn sentence.*

It is now tempting to use this characterization of Universal Horn formulas in order to explain their special properties in terms of Databases and Logic Programming. Fagin has done this in [Fagin, 1982] for the case of databases. Mahr and Makowsky have done this for the case of Specification of Abstract Datatypes [Mahr and Makowsky, 1984] extracting the model theoretic content of [Goguen *et al.*, 1977]. The latter was based on ideas from Category Theory and Universal Algebra. A more general discussion may be found in [Makowsky, 1984]. Clearly, neither the closure under substructures nor under products has any explanatory power *per se* in these contexts. It would be more satisfactory, if the formation of products and the closure under substructures could be replaced by some activity stemming from handling databases. This was achieved with moderately satisfactory results in [Makowsky, 1981; Makowsky and Vardi, 1986].

The predominant rôle that Horn formulas play in Logic Programming can be explained syntactically by the similarity of Horn formulas to deterministic rules or instructions. Semantically, the situation is similar to Abstract Data Types in as much as one thinks of a unique minimal interpretation. An exact model theoretic analysis of Horn formulas in Logic Programming was proposed in [Makowsky, 1987]. Its relevance for Negation by Failure was discussed in Shepherdson's [Shepherdson, 1984; Shepherdson, 1985;

Shepherdson, 1988]. The exact formulation of this analysis is unfortunately
not possible in this survey. An excellent exposition of special properties
of Horn formulas is [Hodges, 1992]. Formulas preserved under relativization play a vital rôle in relational database theory, especially in connection with *safe* queries, cf. [Ullman., 1982; Topor and Sonenberg, 1988;
Makowsky and Vardi, 1986]. Horn formulas preserved under intersections
were analyzed in [Makowsky, 1987]. Finally, formulas with monotone predicates can be characterized as formulas with positive occurrence of the
predicate and play an important rôle in the theory of computable fixed
points and related topics [Moschovakis, 1974]. The use of preservation
theorems in Database Theory will be discussed in [Makowsky, to appear].

5 Fast growing functions

5.1 Non-provability results in second order arithmetic

It is questionable whether the model theoretic proof of the Paris–
Harrington theorem (3.5.1) really captures the essence of the matter completely. The original proof has a proof theoretic flavour and for various
generalizations of this theorem no purely model theoretic proofs are known.
A prominent example is Friedman's theorem:

Theorem 5.1.1 (H. Friedman). *There are programs (number theoretic
functions) which*

  (i) *always terminate (are total) but*

  (ii) *such a termination proof does not exist within the formalization of
       various fragments of Second Order Arithmetic.*

A technical and philosophical discussion of such theorems may be found
in [Harrington *et al.*, 1985]. A presentation of this theorem and related
results accessible to computer scientists may be found in [Gallier, 1991].

5.2 Non-provability in complexity theory

The following variation of the above theorems is due to S. Ben-David [Ben-
David and Halevi, 1991]:

Theorem 5.2.1 (S. Ben–David). *There is a language (a set of words
recognizable by a Turing machine) L_{PH} such that*

  (i) *L_{PH} is in* **Co–NP**;

  (ii) *L_{PH} is not context free, but*

(iii) it is not provable in Peano Arithmetic that L_{PH} is not regular.

Recently, S. Ben-David has analysed these results further and related them to discuss the prospect of $\mathbf{P} \neq \mathbf{NP}$ not being provable in some formalized system such as Peano Arithmetic or fragments of Second Order Arithmetic [Ben-David and Halevi, 1991]. The key notion here are functions *extremely close to polynomials* where extremely close depends on the growth rate of functions not provably total in the formal system in question. His theorem states the following:

Theorem 5.2.2 (S. Ben–David).   *If $\mathbf{P} \neq \mathbf{NP}$ is not provable in some fragment of Second Order Arithmetic S then every problem P in* **NP** *can be solved by an algorithm with run time upper bound* **S**-*extremely close to a polynomial.*

5.3   Model theory of fast growing functions

The underlying theme of these theorems are fast growing functions. We review now some basic facts from the proof theory of Arithmetic and follow closely the exposition in [Ben-David and Halevi, 1991]. We refer the reader to [Smorynski, 1980; Smorynski, 1983] for an elaborated and truly enjoyable discussion of this topic. The basic idea goes back to Kreisel [Kreisel, 1952]. For every recursive formal theory which is sound for Arithmetic there exist total recursive functions such that the theory cannot prove their totality. Such functions can be characterized by their rate of growth.

Wainer [Wainer, 1970] supplies a useful measuring rod for the rate of growth of recursive functions (from natural numbers to natural numbers) — the *Wainer hierarchy*. The Wainer hierarchy is an extension to infinite countable ordinal indices of the more familiar Ackermann hierarchy of functions, and is defined as follows:

$$F_1(n) = 2n$$
$$F_{k+1}(n) = F_k^{(n)}(n)$$
$$F_\delta(n) = \sup_{\alpha < \delta} F_\alpha(n) \text{ for } \delta \text{ a countable limit ordinal}$$

where $F^{(n)}$ denotes the n'th iterate of F. The famous Ackermann Function is F_ω — quite low in this hierarchy.

Definition 5.3.1.

(i) We say that a function f *dominates* a function g if for all large enough n, $g(n) < f(n)$.
Note that for every $\alpha < \beta$, F_β dominates F_α. It is also worthwhile

to recall that the Ackermann Function, F_ω, dominates all primitive recursive functions.

(ii) ϵ_0 is the first ordinal α satisfying $\omega^\alpha = \alpha$. (The exponentiation here is ordinal exponentiation and ϵ_0 turns out to be a countable ordinal - the limit of the sequence $\omega, \omega^\omega, \omega^{\omega^\omega}, \ldots$).

(iii) A function f is *provably recursive* in a formal theory T if there is an algorithm A that computes f for which T proves (using some fixed recursive coding of algorithms) that A halts on every input.

Theorem 5.3.2 (Wainer).

(i) *For every ordinal $\alpha < \epsilon_0$, the α'th function in the Wainer hierarchy, F_α, is provably recursive in PA.*

(ii) *If a total recursive function $f : N \to N$ is provably recursive in PA then it is dominated by some F_α in $\{F_\alpha : \alpha < \epsilon_0\}$.*

Furthermore, Fortune *et al.*[Fortune *et al.*, 1983] prove that the set of functions that are provably recursive in PA equal that set of functions that can be computed in (deterministic) time dominated by some F_α in $\{F_\alpha : \alpha < \epsilon_0\}$. Let $\mathcal{N}$ be the standard model of Peano Arithmetic. Let PA_1 be the first order theory consisting of the first order Peano Axioms together with all the Π_1 formulas true in $\mathcal{N}$. A formula is Π_1 if it is of the form $\forall x \phi$ where all the quantifiers in ϕ are bounded. The model theoretic statement which shows the existence of functions not provably recursive in PA_1 is now the following theorem [Ben-David and Dvir, 1991].

Theorem 5.3.3 (S. Ben-David and M.Dvir). *Let $f : N \to N$ be a recursive function which is not dominated by any F_α for $\alpha < \epsilon_0$. Then every non–standard model $\mathcal{B}$ elementarily equivalent to $\mathcal{N}$ has a submodel $\mathcal{B}_0$ such that*

(i) *$\mathcal{B}_0$ is an initial segment of $\mathcal{B}$ and*

(ii) *$\mathcal{B}_0 \models PA_1+$ 'f is not total'.*

This theorem is a substantial improvement on [Kanamori and McAloon, 1987] where a similar result is proven for some special combinatorial functions.

6   Elimination of quantifiers

In this section we discuss a model theoretic method useful in Computer Aided Mathematics. We first discuss in some detail the knowledge needed for turning mathematical practice into recordable and automatically reproducible experience. After all, the computer is a kind of a phonograph or rather epistograph, which will replay recorded procedures. If there is nothing to record, no programming technique can overcome the void.

6.1 Computer aided theorem proving in classical mathematics

What we call here *classical mathematics* comprises Number Theory, Differential and Integral Calculus and Elementary Geometry. In each of these fields we state precisely a class of problems and discuss the prospects of mechanization and implementation of theorem proving. The point we wish to make here is that success or failure very much depend on a thorough understanding of particular features of the class of problems, and that, superficially speaking, *minor variations* may change the situation radically. Since very often advocates of the imminent breakthrough in Artificial Intelligence argue that mathematical Expert Systems like MACSYMA, REDUCE, CALEY, MAPLE, MATHEMATICA, etc. prove their point, we would, on the contrary, argue, that, upon closer analysis, these systems rather prove the point of Natural Ingenuity. More precisely, they show how the collective experience of generations of mathematicians, having reached a deep understanding of the subject, allows the mechanization of their knowledge, and that the invocation of Artificial Intelligence, at least in these cases, only adds confusion to an otherwise clearly understandable situation.

Number Theory was called by Gauss the Queen of Mathematics. It has many facets and recently number theoretic results have increasingly found applications in Computer Science. But the true purpose of Number Theory was always to test the depth of our mathematical understanding. The oldest branch of Number Theory is the theory of Diophantine Equations. Challenging examples are the Catalan equation

$$x^y - z^u = 1$$

or the Fermat equation
$$x^n + y^n = z^n$$

where the variables range over the integers and the problem is to characterize the set of its solution quantitatively (empty, finite, infinite) or qualitatively (by another such equation holding between the values of the variables which form a solution). More precisely, a Diophantine Equation is of the form

$$Term = 0$$

where terms are built from variables and natural numbers by addition, subtraction, division and exponentiation. Concerning the above two equations it is conjectured that Catalan's equation has only one solution ($x = u = 3, y = z = 2$) and that Fermat's equation has none for $3 \le n$. The solutions for $n \le 2$ have been completely characterized. Both conjectures are still open, but in the last fifteen years Tijdeman, based on the results of A. Baker [Baker, 1975], has proved that Catalan's equation has only finitely many solutions and Faltings proved that for every $3 \le n$ Fermat's equation has only finitely many solutions (cf. also [Baker, 1984]). In 1900 D.Hilbert asked whether there is one general algorithm which decides whether a given Diophantine equation has a solution (Hilbert's Tenth Problem). The cumulative efforts of Davis, Putnam, Julia Robinson and Matijasevič finally showed in 1970 that *no such algorithm exists* [Matijasevič, 1970]. On the other hand the state of the art seems to suggest that if we restrict the problem to Diophantine equations in two variables, then all the successful methods known so far exhaust all the cases, which have been distinguished, and therefore such an algorithm might exist [Baker, 1984].

A much simpler class of problems is the class of Arithmetical Identities. An Arithmetical Identity is an equation of the form

$$Term_1 = Term_2$$

such as

$$(a + b)^2 = a^2 + 2ab + b^2.$$

Here the terms are formed from variables and natural numbers by addition and multiplication only and the identity holds if it holds for all values that the variables can assume. This problem has been mechanized completely. Its solution basically states that the methods you have (supposedly) learned in High School are complete. An excellent discussion of the arithmetical identities may be found in [Henkin, 1977].

Integration in Finite Terms is the problem of whether there is a function term $f(x)$ such that

$$f(x) = \int g(x)$$

for a given function term $g(x)$. If there is such a function term the problem also requires one to find it. Here the function terms are built from one variable x and the natural numbers by addition, subtraction, multiplication, division, the formation of exponentials and logarithms to basis e and the taking of nth roots. The cumulative knowledge and expertise from Leibniz and Newton till 1969 have led to Risch's theorem [Risch, 1969], which completely solves the problem. [Rosenlicht, 1972] gives an elementary exposition. The solution is much more complicated than the solution of the corresponding problem for Algebraic Equations, i.e. whether zeros of polynomials with integer coefficients can represented by terms in the coefficients built from the natural numbers by addition, subtraction, multiplication, division and the formation of nth roots. The latter problem was solved completely by Galois and the solution of the problem of Integration in Finite Terms is a grandiose completion of Galois' methods. Attempts at generalizing Galois theory to arbitrary structures were initiated by E.Engeler [Engeler, 1981]. This is a very promising and challenging program, which combines model theory and algorithmics, but which remains beyond the scope of this chapter.

Elementary Geometry is the geometry of *Ruler and Compass* in the Euclidean Plane and its generalization to the n-dimensional Euclidean Space. In contrast to the above problems, whose solutions are all based on the cumulative effort of generations of mathematicians, Elementary Geometry can be mechanized (and even implemented) based on rather little mathematical knowledge. The method which works here, however, is not based on a generally applicable method, like logical deduction, but on a method whose scope is limited and which has been completely characterized: the method of *Elimination of Quantifiers*. This section is exclusively dedicated to this method. A good textbook introduction may be found in [Kreisel and Krivine, 1967].

6.2   Tarski's theorem

Tarski studied the first order theories of fields, in particular algebraically closed fields and real closed fields. Let $\tau^{field} = \{0, 1, -1, +, *\}$ where $0, 1, -1$ are constant symbols and $+, *$ are binary function symbols. A field is a structure over the vocabulary τ^{field} which satisfies the field axioms, where free variables are quantified universally:

Commutativity: $x + y = y + x$, $x * y = y * x$;

Associativity: $(x + y) + z = x + (y + z)$, $(x * y) * z = x * (y * z)$;

Neutral elements: $x + 0 = x$, $x * 1 = x$, $1 \neq 0$;

Distributivity: $x * (y + z) = (x * y) + (x * z)$;

Inverse elements: $x + (-1 * x) = 0$, $x \neq 0 \rightarrow \exists y (x * y = 1)$.

A field is of characteristic $n > 0$ if n is least such that:

$$\underbrace{1 + \ldots + 1}_{n} = 0,$$

if such an n exists, and characteristic 0 if not. A field is algebraically closed if for every polynomial (τ_{field}-term) $t(x) = y_n * x^n + \ldots + y_1 * x + y_0$ with one free variable x and coefficients y_i we have

$$\exists x\, t(x) = 0 \vee \forall y, z\, t(y) = t(z).$$

Note that n is the degree of the polynomial provided that $y_n \neq 0$. We denote the axioms of algebraically closed fields of characteristic 0 by ACF_0.

The field of complex numbers is an algebraically closed field of characteristic 0. The class of algebraically closed fields was structurally analysed by Steinitz ([Steinitz, 1910]), and his analysis served as a paradigm for algebra and model theory. Classical algebraic geometry is the study of algebraic varieties over the complex numbers. An algebraic variety is a set of n-tuples of complex numbers all of which satisfy some set of polynomial equations. In logic we replace equalities by arbitrary first order formulas, including quantifiers and inequalities. Tarski now proved the following:

Theorem 6.2.1 (Tarski, Elimination of quantifiers for algebraically closed fields). Let $\phi(x_1, \ldots, x_n)$ be a τ_{field}-formula. Then there is a quantifier-free τ_{field}-formula $\psi(x_1, \ldots, x_n)$ with the same free variables such that $ACF_0 \models \phi \leftrightarrow \psi$. Furthermore, ψ is computable from ϕ in double exponential time and simple exponential space.

The complexity result is due to J. Heintz [Heintz, 1983] and there are several refinements. For an up to date discussion, cf. [Ierardi, 1989]. The proof uses the field axioms to bring terms into polynomial normal form and then reduces the problem to the case where ϕ is an existential formula with no disjunctions. The last step exploits the fact that every polynomial has a zero. This proof does not invite generalizations. However, Shoenfield [Shoenfield, 1967] found the model theoretic contents of Tarski's theorem.

Definition 6.2.2. Let Σ be a set of first order formulas over some vocabulary τ.

(i) Σ *admits elimination of quantifiers* if for every τ-formula, $\phi(x_1, \ldots, x_n)$ there is a quantifier-free τ-formula $\psi(x_1, \ldots, x_n)$ with the same free variables such that $\Sigma \models \phi \leftrightarrow \psi$.

(ii) Σ satisfies the isomorphism condition, if for every two models $\mathcal{A}$ and $\mathcal{B}$ of Σ and every isomorphism g of a substructure of $\mathcal{A}$ and a substructure of $\mathcal{B}$, there is an extension of g which is an isomorphism of a submodel of $\mathcal{A}$ and a submodel of $\mathcal{B}$.

(iii) Σ satisfies the submodel condition, if for every model $\mathcal{B}$ of Σ and every submodel $\mathcal{A}$ of $\mathcal{B}$, and every sentence

$$\phi = \exists x_1, \ldots, x_n B(x_1, \ldots, x_n)$$

with B quantifier-free, we have that $\mathcal{B} \models \phi$ iff $\mathcal{A} \models \phi$.

Theorem 6.2.3 (Shoenfield). *If Σ satisfies both the isomorphism condition and the submodel condition, then Σ admits elimination of quantifiers.*

To prove Tarski's theorem one has to verify the isomorphism and the sub-model condition, which again uses particulars of the theory of algebraically closed fields.

Tarski proved the same theorem also for real closed fields. A τ_{field}-structure is a real closed field if it satisfies the field axioms and the following two axioms:

Square roots exist: $\forall x(\exists y(y*y = x) \vee \exists y(y*y = -1*x))$, and $\neg\exists y(y*y = -1)$.

Polynomials of odd degree have roots:
$\forall y_0, \ldots, y_{2n+1} \exists x t(x, y_0, \ldots, y_{2n+1})$ for every term $t(x, y_0, \ldots, y_{2n+1})$ of the form $y_{2n+1} * x^{2n+1} + y_{2n} * x^{2n} + \ldots + y_1 * x + y_0$.

There is an x such that $t(x) = 0$. We denote the axioms of real closed fields by RCF. Real closed fields are always of characteristic 0. Moreover, one can define an order relation $x < y$ by the formula $\exists z(y + (-1*x) = z*z)$, which makes a real closed field into an ordered field.

Theorem 6.2.4 (Quantifier elimination for real closed fields). *RCF admits quantifier elimination. Furthermore, given ϕ the equivalent quantifier-free formula ψ can be computed in double exponential time and simple exponential space.*

Corollary 6.2.5. *The first order theories ACF_0 and RCF are decidable in double exponential time and simple exponential space.*

The complexity results are due to Collins and Ben-Or, and Kozen and Reif, cf. [Schwartz *et al.*, 1987].

The question arises whether other first order theories of fields are decidable as well. Tarski conjectured and Ziegler [Ziegler, 1982] proved the following.

Theorem 6.2.6 (Ziegler). *Let T be a finite set of formulas over τ_{field} consistent with either ACF_0 or RCF. Then every subset T' of T is undecidable. In particular, the first order theory of fields is undecidable.*

Corollary 6.2.7. *Let T be a finite set of formulas over τ_{field} consistent with either ACF_0 or RCF. Then T does not admit elimination of quantifiers.*

In other words, elimination of quantifiers is a rather rare phenomenon.

6.3 Elementary geometry

The most important application of the method of elimination of quantifiers is in Elementary Geometry. Elementary Geometry is, by Descartes' reduction, identified with the normed vector spaces $\mathbf{R}^n$. Statements about configurations of points in $\mathbf{R}^n$ can be expressed in first order logic over the vocabulary τ_{field} using the axioms of RCF. Tarski's theorems, therefore, show that any such statement is decidable, using the method of elimination of quantifiers. Although the general method is doubly exponential in time and simply exponential space, special cases have been singled out with lower complexity. Applications of this method to Robotics are surveyed and discussed in [Schwartz *et al.*, 1987], applications to automated theorem proving in [Chou, 1988].

6.4   Other theories with elimination of quantifiers

We briefly list in this section other cases where the method of elimination of quantifiers was applied successfully.

- Dense orders, linear orders, well-orderings, monadic theory of linear order. [Rosenstein, 1982]

- Applications to temporal logic, such as Kamp's theorem, [Gabbay *et al.*, 1980; Flum, 1991].

- Boolean algebras, Abelian groups and modules and similar theories also with generalized quantifiers, [Baudisch *et al.*, 1985].

7   Computable logics over finite structures

The presentation of the material in this section has been developed by the author. It was never published but has been used in lectures since 1984. I am indebted to Y. Pnueli who once worked out the notes of my lectures on the subject.

   In this section we shall frequently speak of complexity classes $\mathbf{C}$ such as $\mathbf{L}$ (LogSpace), $\mathbf{NL}$ (Non-deterministic LogSpace), $\mathbf{P}$ (Polynomial Time), $\mathbf{NP}$ (Non-deterministic Polynomial Time), ..., recursive, recursively enumerable. We do not give an abstract definition, and it is enough to have these and similar examples in mind.

7.1   Computable logics

When we restrict ourselves to finite τ-structures first order logic is not any more characterizable in a way similar to Lindström's theorem. We have to allow for extensions of first order logic. We still require that logics should be regular. We shall require that a formula is satisfiable iff it has a finite model.

However, we cannot require that the set of valid sentences is recursively enumerable because of the following classical result [Trakhtenbrok, 1950]:

Theorem 7.1.1 (Trakhtenbrot). *There is a finite vocabulary τ such that the set of first order τ-sentences which is true in all finite τ-structures is not recursively enumerable.*

On the other hand we have the following observation.

Definition 7.1.2.

(i) We say that a finite structure of cardinality n is *natural*, if its universe consists of the set $\{0, 1, \ldots, n-1\}$.

(ii) We say that a finite structure of cardinality n is *naturally ordered*, if it is natural and its vocabulary contains a binary relation symbol whose interpretation is the customary linear order on $\{0, 1, \ldots, n-1\}$.

Theorem 7.1.3. *Let τ be a finite vocabulary and ϕ be a first order τ-sentence. Then the set of finite natural τ-structures $\mathcal{A}$ such that $\mathcal{A} \models \phi$ is recursively enumerable. In fact, it is even in* **L.**

These considerations lead us to the following development.

Definition 7.1.4. Let **C** be a complexity class. A semi-regular logic L (see section 2.3) is **C**-*computable* if all of the following hold:

(i) For every finite vocabulary τ the set of τ-formulas is computable in **C.**

(ii) The meaning function is invariant under isomorphisms. In other words, for every τ, every two τ-isomorphic τ-structures $\mathcal{A}, \mathcal{B}$ and every τ-formula ϕ we have $\mathcal{A} \models \phi$ iff $\mathcal{B} \models \phi$.

(iii) Let τ be a finite vocabulary and ϕ be a τ-sentence. Then ϕ has a model-checker of complexity less than **C**, i.e. the set of finite natural τ-structures $\mathcal{A}$ such that $\mathcal{A} \models \phi$ is recognizable in **C.**

Definition 7.1.5. Let $\mathcal{L}$ be a semi-regular computable logic and **C** be a complexity class. We say that $\mathcal{L}$ *captures* **C** if for every class of naturally ordered τ-strucures K we have that K is in **C** iff K is definable by a τ-formula of $\mathcal{L}$.

Now the analogue of Lindström's theorem over finite structures would be the identification of semi-regular logics $\mathcal{L}$ which capture complexity classes **C** for suitable **C**.

For the case where **C** equals the recursively enumerable (recursive) sets of natural structures such a characterization is easy. Let $\mathcal{L}_{r.e.}$ be like first order logic but with infinite disjunctions over r.e. sets of formulas and such that infinite disjunctions appear only under an even number of negations. Similarly, let $\mathcal{L}_{rec}$ be like first order logic but with infinite disjunctions over recursive sets of formulas, iteratively applied a finite number of times without restrictions. Clearly $\mathcal{L}_{r.e.}$ and $\mathcal{L}_{rec}$ are semi-regular logics satisfying the above requirements. On the other hand any finite structure can be described up to isomorphism by a first order sentence, therefore any r.e. set of finite structures can be described by a simple r.e. disjunction of such first order sentences. In other words:

Theorem 7.1.6.

(i) $\mathcal{L}_{r.e.}$ *is a r.e.-computable semi-regular logic which captures the r.e. sets of structures.*

(ii) $\mathcal{L}_{rec}$ *is a recursive semi-regular logic which captures the recursive sets of structures.*

A different characterization of r.e.-computable logics was given in the fundamental paper of Chandra and Harel [Chandra and Harel, 1980], in terms of computable queries and Pascal-like programs.

As already mentioned in Section 3.7, Fagin noted that

Theorem 7.1.7 (Fagin). *Let K be a class of natural structures. Then K is in* **NP** *iff K is definable by an existential formula of Second Order Logic.*

In our terminology the set of existential formulas of Second Order Logic form a semi-regular logic which captures **NP**. This is a special case of a more general theorem due to J. Lynch, generalizing a result by Stockmeyer [Lynch, 1982], noting that **NP** is just one level in the polynomial hierarchy **PH**. Recall, cf. [Garey and Johnson, 1979], that **PH** is the union of the complexity classes $\Sigma_n{}^P$ and $\Pi_n{}^P$, where $\Sigma_1{}^P$ is **NP** and $\Pi_1{}^P$ is **Co − NP**. Furthermore, a Σ_n-formula (Π_n-formula) of Second Order Logic is a formula in prenex normal form starting with existential (universal) second order quantifiers and having $n - 1$ alternations of second order quantifiers.

Theorem 7.1.8 (Lynch–Stockmeyer). *Let K be a class of natural structures.*

(i) *K is in $\Sigma_n{}^P$ ($\Pi_n{}^P$) iff K is definable by Σ_n-formula (Π_n-formula) of Second Order Logic.*

(ii) *K is in* **PH** *iff K is definable by a formula of Second Order Logic.*

In our terminology, Second Order Logic captures **PH**.

7.2 Computable quantifiers

We now define computable quantifiers. Our presentation follows the definition of the Lindström Quantifiers in [Ebbinghaus, 1985] and combines the theory of generalized quantifiers with the idea of computable queries of [Chandra and Harel, 1980].

Let $\mathbf{C}$ be a complexity class. Let σ be a vocabulary without function or constant symbols, K be a set of natural σ-structures in $\mathbf{C}$. Further, let $\mathcal{L}$ be some regular $\mathbf{C}$-computable logic. We define $\mathcal{L}_{qK}$, the extension of $\mathcal{L}$ with a $\mathbf{C}$-computable quantifier, as follows :

Definition 7.2.1 (Syntax).

(i) All the rules for forming formulas in $\mathcal{L}$ are rules for forming formulas in $\mathcal{L}_{qK}$.

(ii) Let $\phi_1, \phi_2, ..., \phi_n$ be formulas in $\mathcal{L}_{qK}$ satisfying the following condition: for each relation symbol R_i of arity a_i in σ, there is a formula ϕ_i with a_i free variables which do not appear in any other of the ϕ's. Then the following is a formula in $\mathcal{L}_{qK}$:

$$Qx_1x_2...x_m(\phi_1, \phi_2, ..., \phi_n)$$

where m is the sum of the arities a_i. The variables $x_1, ..., x_m$ are bound. We refer to such a formula as a formula of form (*).

(iii) The formula (*) above is considered as the syntax of a quantifier of *size* 1. A quantifier of *size* k is then defined as the formula we get by replacing each of the variables x_i with a vector of variables of size k (each element of each such vector can appear only in a single formula among the ϕ's and ψ's). The syntax of $\mathcal{L}_{qK}$ is then extended to include quantifiers of any size.

Definition 7.2.2 (Semantics). The meaning function $M_{\mathcal{L}_{qK}}(\theta, \mathcal{A}, z)$ is defined by extending the meaning function $M_{\mathcal{L}}$ of $\mathcal{L}$ as follows: let θ be a τ-formula of the form (*), let $\mathcal{A}$ be a τ-structure and let z be an assignment of the free variables in θ to elements in A, the universe of $\mathcal{A}$. Now $\theta, \mathcal{A}$ and z define a σ-structure $\mathcal{A}_\theta$ as follows:

(i) Its universe is the universe A of $\mathcal{A}$ (in the case where the quantifier is of some size $k > 1$, each element in the defined structure is a vector of size k of elements from A).

(ii) Let $R = R_i^{a_i}$ be the ith relation in σ (its arity is a_i). We associate with it the formula ϕ_i ,which has a_i free variables not appearing in any other of the formulas of (*) : $x_{j+1}, ..., x_{j+a_i}$, and define

$R(b_1, b_2, ..., b_{a_i}) = M_{\mathcal{L}_{qK}}(\phi_i, \mathcal{A}, z')$. Where z' is a substitution as follows : for each free variable x_{j+p} from $x_{j+1}, ..., x_{j+a_i}$ we substitute the element b_p. For each other free variable in ϕ_i (which is also free in θ) we substitute the corresponding element from z.

Now we put $M_{\mathcal{L}_{qK}}(\theta, \mathcal{A}, z) = 1$ iff $\mathcal{A}_\theta \in K$.

Example 7.2.3. Let σ consist of one binary predicate and let $3COL$ be the class of σ-structures, which are 3-colourable graphs. A formula of the form (*) is true in a τ-structure $\mathcal{A}$ if ϕ_1 defines a 3-colourable graph on the universe of $\mathcal{A}$. The case of the same quantifier but this time choosing size 2 binds four variables. A formula of the form (*) is true in a structure $\mathcal{B}$ if the 4-ary relation defined by ϕ_1 defines a 3-colourable graph on the pairs of elements of the universe of $\mathcal{B}$.

We have the following general theorem:

Theorem 7.2.4. Let **C** be a complexity class containing at least **L**. If K is in **C** and in co-**C** then $\mathcal{L}_{qK}$ is a **C**-computable logic, i.e. every formula in the extended logic $\mathcal{L}_{qK}$ has a model checker of complexity **C**.

This theorem requires some tedious checking, but is not surprising. However, if **C** does not contain **L**, serious problems arise. A similar theorem can be formulated for semi-regular logics by only requiring that K be in **C** and restricting the quantifier to positive occurrences.

7.3 Computable predicate transformers

In some cases it is convenient to work with predicate transformers instead of generalized quantifiers. A predicate transformer is a mapping from relations to relations. A guiding example is the transitive closure of a binary relation. Other examples are computable queries in Database theory.

We are interested in extensions of computable logics $\mathcal{L}$ by predicate transformers. This generalizes the construction introduced in [Immerman, 1987].

To set up our framework we need some machinery.

Definition 7.3.1. Let σ be a vocabulary without function symbols or constant symbols that contains at least one relation symbol. Let $\sigma' = \sigma \cup \{R\}$ where R is a relation symbol of arity r not in σ.

(i) We say that a σ'-structure $\mathcal{A}'$ is an *expansion* of a σ-structure $\mathcal{A}$, if they have the same universe and all the relations of σ are identical.

(ii) Let J be a set of natural structures over σ'. J is a *predicate transformer* if for every σ-structure $\mathcal{A}$ there is at most one expansion $\mathcal{A}' \in J$.

(iii) Let **C** be a complexity class and J be a predicate transformer. We now say that J is **C**-*computable* if, given a σ-structure $\mathcal{A}$ and an r-tuple of elements from the universe, $(b_1, b_2, ..., b_r)$, the problem of whether $R(b_1, b_2, ..., b_r)$ is true in $\mathcal{A}'$ can be decided in **C**. (Similarly we can define when J is in co-**C**).

We define $\mathcal{L}_{pJ}$, the extension of $\mathcal{L}$ with a **C**-computable predicate transformer by the addition of one more formation rule as follows:

Definition 7.3.2 (Syntax).

(i) Let $\phi_1, \phi_2, ..., \phi_n, \psi_1$ be formulas in $\mathcal{L}_{pJ}$ such that for each relation symbol R_i of arity a_i in σ, there is a formula ϕ_i with a_i free variables which do not appear in any other of the ϕ's.
Then the following is a formula in $\mathcal{L}_{pJ}$:

$$P x_1, x_2, ..., x_m y_1, y_2, ..., y_r (\phi_1, \phi_2, ..., \phi_n)$$

where the variables $x_1, x_2, ..., x_m$ are bound, m is the sum of the a_i's and the variables $y_1, y_2, ..., y_r$ are free. We refer to such a formula as a formula of form (**).

(ii) As in the case of quantifiers we can consider the formula above as a predicate of *size* 1 and in a similar way define predicates of any size.

Also the meaning function for $\mathcal{L}_{pJ}$ is very similar to the case of quantifiers. We only give the modifications needed.

Definition 7.3.3 (Semantics). The meaning function $M_{\mathcal{L}_{pJ}}(\theta, \mathcal{A}, z)$ is defined as the extension of the meaning function of $\mathcal{L}$ in the following way. Let θ be a τ-formula of the form (**), let $\mathcal{A}$ be a τ-structure. Let z be an assignment of the free variables in θ to elements in A the universe of $\mathcal{A}$ and z' be the restriction of z to the variables different from the y's.
Then $M_{\mathcal{L}_{pJ}}(\theta, \mathcal{A}, z) = 1$ iff the σ-structure defined by $\theta, \mathcal{A}$ and z' can be extended to a unique σ'-structure $\mathcal{B} \in J$ for which $R(z(y_1), z(y_2), ..., z(y_r))$ is true.

Example 7.3.4.

(i) Let σ consist of one binary relation symbol R and $\sigma' = \sigma \cup \{S\}$ with S another binary relation symbol. TC (DTC) consists of the class of σ'-structures such that the interpretation of S is the (deterministic) transitive closure of the interpretation of R. Both can be used to define predicate transformers of arbitrary size.

(ii) Similarly one can define ATC for the alternate transitive closure of a relation over a unary predicate, cf. [Immerman, 1987].

Theorem 7.3.5. *Let $\mathcal{L}$ be a C-computable logic. If J is in C and in co-C then $\mathcal{L}_{pJ}$ is a C-computable logic.*

A similar theorem can be formulated for semi-regular logics by only requiring that K be in C and restricting the predicate transformer to positive occurrences.

7.4 $\mathcal{L}$-Reducibility

Immerman and Dahlhaus [Dahlhaus, 1982; Dahlhaus, 1983; Immerman, 1987] independently introduced the notion of classes K of σ-structures complete for a complexity class C by first order reductions. We present here a slight generalization:

Definition 7.4.1. Let K_1 be a class of σ_1-structures and K_2 be a class of σ_2-structures.

(i) K_2 is $k - \mathcal{L}$-*reducible* to K_1 for a natural number k if K_2 is definable by a formula of the form

$$Q^k{}_{K_1}(\phi_1, \ldots, \phi_n)$$

where all the ϕ's are $\mathcal{L}(\sigma_2)$ formulas and $Q^k{}_{K_1}$ is the quantifier defined by K_1 of size k.

(ii) K_2 is $\mathcal{L}$-reducible to K_1 if it is $k - \mathcal{L}$-reducible for some natural number k.

(iii) If $\mathcal{L}$ is First Order Logic we speak of k-first order reducible. (Immerman in [Immerman, 1987] considers the case where the ϕ's are even quantifier-free).

An analogous definition can be given for predicate transformers instead of quantifiers.

Definition 7.4.2. Let K be a class of natural σ-structures. Let $C \subseteq D$ be complexity classes and let $\mathcal{L}$ be a C-computable logic. We say that K *is D-complete for $\mathcal{L}$-reductions* if

(i) K is in D and

(ii) every class K_1 of σ_1-structures in **D** is $\mathcal{L}$-reducible to K.

The traditional definition of **NP**-completeness says that K is **NP**-complete iff K is **NP**-complete for **P**-reductions. Our definition is a model theoretic counterpart of the more usual computational definition of K being **D**-complete for **C**-reductions. The exact relationship between the two definitions is given by the following proposition and is proven using Theorem 7.2.4 or Theorem 7.3.5

Proposition 7.4.3. *Let K be a class of natural σ-structures. Let* **C** $\subseteq$ **D** *be complexity classes and let $\mathcal{L}$ be a* **C**-computable logic.

(i) *If K is* **D**-complete for $\mathcal{L}$-reductions then K is **D**-complete for **C**-reductions.

(ii) *There are K's which are* **D**-complete for **C**-reductions, but which are not **D**-complete for $\mathcal{L}$-reductions.

The following are some examples of such complete classes K, cf. [Immerman, 1987; Immerman, 1988; Stewart, 1991a; Stewart, 1991b].

Proposition 7.4.4.

(i) **Immerman.** *DTC is* **L**-complete, *TC is* **NL**-complete and *ATC is* **P**-complete for first order reductions.

(ii) **Stewart.** *The class 3COL of three-colourable graphs is* **NP**-complete for first order reductions.

Note that DTC, TC and ATC give rise to computable quantifiers with constants or, better, to predicate transformers, whereas $3COL$ can best be used as a computable quantifier.

7.5 Logics capturing complexity classes

The framework developed so far allows us to state a very general theorem about logics capturing complexity classes. Theorems of this type were first stated in this form by Immerman [Immerman, 1987], but were preceded and motivated by Fagin's work [Fagin, 1974]. We first need a definition.

Definition 7.5.1. Let K be a class of natural structures which is **C**-complete for $\mathcal{L}$ reductions. We denote by $\mathcal{L}(K)$ the logic obtained from $\mathcal{L}$ by adding the quantifiers (or predicate transformers) associated with K

for all sizes $k \in \mathbf{N}$. If $\mathcal{L}$ is First Order Logic we write $FOL(K)$ instead of $\mathcal{L}(K)$.

Theorem 7.5.2. *Let* $\mathbf{C}$ *be a complexity class which contains* $\mathbf{L}$. *Let* $\mathcal{L}$ *be a semi-regular* $\mathbf{C}$*-computable logic. Let* K *be a class of natural structures* $\mathbf{C}$*-complete for* $\mathcal{L}$*-reductions. Then* $\mathcal{L}(K)$ *captures* $\mathbf{C}$.

Corollary 7.5.3 (Immerman).

  (i) $FOL(DTC)$ *captures Logarithmic Space* $\mathbf{L}$.

 (ii) $FOL(TC)$ *captures Non-deterministic Logarithmic Space* $\mathbf{NL}$.

(iii) $FOL(ATC)$ *captures Polynomial Time* $\mathbf{P}$.

Recently, I. Stewart [Stewart, 1991b; Stewart, 1991a] has taken a similar, but less general approach to show that

Corollary 7.5.4 (Stewart). $FOL(3COL)$ *captures Non-deterministic Polynomial Time* $\mathbf{NP}$.

These results make the problems $\mathbf{L} \neq (?)\mathbf{NL}$, $\mathbf{P} \neq (?)\mathbf{NP}$ or even $\mathbf{L} \neq (?)\mathbf{NP}$ into problems of definability by computable quantifiers. We shall see in the next section how to convert them into problems of model theory.

8 Ehrenfeucht–Fraïssé games

In this section we introduce a very powerful tool with many applications in the analysis of the expressive power of First Order Logic. The tool is commonly known under the name *Ehrenfeucht–Fraïssé Games* $\mathcal{E}_n^\tau(\mathcal{A}, \mathcal{B})$.

8.1 The games

Informally the *Ehrenfeucht–Fraïssé* Games are described as follows:

- $\mathcal{E}_n^\tau(\mathcal{A}, \mathcal{B})$ is played by two players I and II, called *spoiler* and *duplicator* respectively. The game is played for n moves, where n is a natural number. The 'board' of the game consists of two τ-structures $\mathcal{A}$ and $\mathcal{B}$ with universe A and B respectively. Here we make a restriction on τ in as much as τ may not contain any function symbols.

- The moves of the game look as follows: In the kth move, player I chooses one of the structures $\mathcal{A}$ (or $\mathcal{B}$) and an element $a_k \in A$ ($b_k \in B$) and player II replies by choosing an element $b_k \in B$ ($a_k \in A$).

- After the n moves they have chosen elements $(a_1, \ldots, a_n)$ and $(b_1, \ldots, b_n)$. Player II has won, if the map $f(a_i) = b_i, i \leq n$ is a partial isomorphism between the τ-structures $\mathcal{A}$ and $\mathcal{B}$.

- We look also at the infinite game $\mathcal{E}_\infty^\tau(\mathcal{A}, \mathcal{B})$ which is defined similarly, but with the moves numbered n for every $n \in \mathbf{N}$.

The interesting case here is when player II has a *winning strategy*. To make this notion precise we need some notational effort. Intuitively, a winning strategy is a catalogue of moves which lists all possible moves of the opponent with at least one answer which will guarantee ultimately that player II wins. With this informal definition in mind, we proceed now further:

- Let $\mathcal{A}$ and $\mathcal{B}$ be two τ-structures. We say that $\mathcal{A}$ and $\mathcal{B}$ are n-equivalent (∞-equivalent) and write $\mathcal{A} \equiv_n \mathcal{B}$, if player II has a winning strategy in the game $\mathcal{E}_n^\tau(\mathcal{A}, \mathcal{B})$ ($\mathcal{E}_\infty^\tau(\mathcal{A}, \mathcal{B})$).

- The relation $\mathcal{A} \equiv_n \mathcal{B}$ between τ-structures is indeed an equivalence relation, i.e. we have:

(i) $\mathcal{A} \equiv_n \mathcal{A}$;

(ii) $\mathcal{A} \equiv_n \mathcal{B}$ iff $\mathcal{B} \equiv_n \mathcal{A}$; and

(iii) If $\mathcal{A} \equiv_n \mathcal{B}$ and $\mathcal{B} \equiv_n \mathcal{C}$ then $\mathcal{A} \equiv_n \mathcal{C}$.

Examples 8.1.1.

(i) If $\mathcal{A} \simeq \mathcal{B}$ then $\mathcal{A} \equiv_n \mathcal{B}$ for every $n \in \mathbf{N}$.

(ii) Let τ consist of one binary relation symbol and let $\mathcal{G}_n$ be the τ-structure which, viewed as a graph, is the complete graph on n elements, and let $\mathcal{H}_n$ be the τ-structure which, viewed as a graph, is the complete bipartite graph on n elements. Analyse for small l, m and n whether $\mathcal{G}_l \equiv_m \mathcal{H}_n$.

After playful definitions we are motivated enough to give a definition of n-equivalence of τ-structures in terms of partial functions. The winning strategies for $\mathcal{E}_n^\tau(\mathcal{A}, \mathcal{B})$ will be described by a finite family of non-empty sets of partial isomorphisms $\{I_m\}_{m=0}^n$ in the following way.

Definition 8.1.2. Let $\mathcal{A}$ and $\mathcal{B}$ be two τ-structures. We say that $\mathcal{A}$ is *finitely isomorphic* to $\mathcal{B}$ if and only if there is a family $\{I_n\}_{n=0}^{\infty}$ of nonempty sets of partial isomorphisms from A to B such that:

Constants: for every $f \in I_n$ and for every constant symbol $c \in \tau$, $I_A(c) \in Dom(f)$.

    Forth: for every $f \in I_n$ and for every $a \in A$ there is a $g \in I_{n-1}$ extending f such that $a \in Dom(g)$.

    Back: for every $f \in I_n$ and for every $b \in B$ there is a $g \in I_{n-1}$ extending f such that $b \in Im(g)$.

The following are two special cases of this definition which are of particular interest:

Definition 8.1.3.

(i) If $I_n = I$ constant for every $n \in N$, we say that $\mathcal{A}$ is *partially isomorphic* to $\mathcal{B}$.

(ii) If there is only a finite family $\{I_m\}_{m=0}^{n}$ of partial isomorphisms with the back and forth properties, we say that $\mathcal{A}$ is *n-isomorphic* to $\mathcal{B}$.

Proposition 8.1.4. *Let $\mathcal{A}$, $\mathcal{B}$ be two τ-structures.*

(i) *If $\mathcal{A}$ and $\mathcal{B}$ are partially isomorphic then $\mathcal{A}$ and $\mathcal{B}$ are finitely isomorphic.*

(ii) **(Cantor)** *If $\mathcal{A}$ and $\mathcal{B}$ are both countable and partially isomorphic then $\mathcal{A}$ and $\mathcal{B}$ are isomorphic.*

(iii) *If $\mathcal{A}$, $\mathcal{B}$ are finite, then $\mathcal{A}$ is finitely isomorphic to $\mathcal{B}$ if and only if $\mathcal{A}$ is isomorphic to $\mathcal{B}$.*

The relationship between the game and the families of partial isomorphisms is given in the following:

Proposition 8.1.5. *Let $\mathcal{A}$ and $\mathcal{B}$ be two τ-structures.*

(i) *$\mathcal{A}$ is n-equivalent to $\mathcal{B}$ iff $\mathcal{A}$ is n-isomorphic to $\mathcal{B}$;*

(ii) *$\mathcal{A}$ is n-equivalent to $\mathcal{B}$ for every $n \in \mathbf{N}$ iff $\mathcal{A}$ is finitely isomorphic to $\mathcal{B}$;*

(iii) *$\mathcal{A}$ is ∞-equivalent to $\mathcal{B}$ iff $\mathcal{A}$ is partially isomorphic to $\mathcal{B}$.*

As an exercise, prove Proposition 8.1.4 both, in the formalism of games and in the formalism of families of partial isomorphisms.

Proposition 8.1.6.

(i) Let $\tau = \emptyset$: τ-structures then consist of their universe only. Let $\mathcal{A}$ and $\mathcal{B}$ be two τ-structures. $\mathcal{A}$ and $\mathcal{B}$ are n-equivalent iff either both have at most n elements and have the same number of elements, or both have at least n elements.
(*Find the corresponding statement for linear orders*).

(ii) Let τ have no function symbols. There are for every n two τ-structures $\mathcal{A}$ and $\mathcal{B}$ such that $\mathcal{A}$ and $\mathcal{B}$ are n-equivalent, $\mathcal{A}$ has exactly n elements and $\mathcal{B}$ is infinite.

(iii) Let τ have exactly one binary relation symbol. There are for every n two τ-structures $\mathcal{A}$ and $\mathcal{B}$ such that $\mathcal{A}$ and $\mathcal{B}$ are n-equivalent, and $\mathcal{A}$ is a connected (planar, Hamiltonian) graph and $\mathcal{B}$ is not connected (planar, Hamiltonian).

8.2 Completeness of the game

We are now ready to formulate the connection between First Order Logic and the winning strategies for the game $\mathcal{E}_n^\tau(\mathcal{A}, \mathcal{B})$.

Notation 8.2.1. We denote by $FOL_{k,n}(\tau)$ the set of τ-formulas with all variables among $v_1, \ldots, v_n$ and all free variables among $v_{k+1}, \ldots, v_n$. $FOL_{n,n}$ are the formulas without free variables of quantifier depth n.

Theorem 8.2.2 (Ehrenfeucht–Fraïssé). Let τ be without function symbols and finite. Let $\mathcal{A}$ and $\mathcal{B}$ be two τ-structures.
Let $a_1, \ldots, a_{n-k}$ and $b_1, \ldots, b_{n-k}$ be elements of $\mathcal{A}$ and $\mathcal{B}$ respectively.

(i) Player II has a winning strategy for k more moves in the game $\mathcal{E}_n^\tau(\mathcal{A}, \mathcal{B})$ starting in the position described by $a_1, \ldots, a_{n-k}$ and $b_1, \ldots, b_{n-k}$ iff $\mathcal{A}$ and $\mathcal{B}$ satisfy the same formulas of $FOL_{k,n}(\tau)$, where the variable v_{k+m} takes the value a_m or b_m respectively.

(ii) $\mathcal{A}$ and $\mathcal{B}$ are n-equivalent iff they satisfy the same formulas of quantifier depth n.

For many applications the following corollary is most useful:

Corollary 8.2.3. A class of finite τ-structures K, τ finite and without function symbols, is definable by a formula ϕ of quantifier depth n iff K is closed under n-equivalence.

Ehrenfeucht–Fraïssé-games can be used to obtain non–definability results. To prove them we combine Theorem 8.2.2 and Proposition 8.1.6.

Theorem 8.2.4. *The following classes of finite structures are not first order definable:*

  (i) *the class of connected graphs;*

 (ii) *the class of planar graphs;*

(iii) *the class of Hamiltonian graphs;*

 (iv) *the class of finite linear orders with an even number of elements.*

For a detailed discussion, cf. [Gaifman, 1982; Ajtai and Fagin, 1990].

8.3 Second order logic and its sublogics

Ehrenfeucht–Fraïssé–games can be generalized and adapted for various extensions of first order logics. In the case of Second Order Logic (Monadic Second Order Logic), one simple adds a new type of moves where the players choose relations (unary relations). Additionally the winning condition has to be modified correspondingly. For the transitive closure logics **DTC** = $FOL(DTC)$, **TC** = $FOL(TC)$ and **ATC** = $FOL(ATC)$ introduced in [Immerman, 1987] and Section 7, such games were explicitly studied in [Caló, 1989; Caló and Makowsky, to appear 1992]. The existence of similar games as introduced in this paper, already follows from successive papers culminating in [Makowsky and Mundici, 1985]. There, they are defined for logics with generalized quantifiers rather than predicate transformers. Recently, a logic equivalent to **TC** based on generalized quantifiers only was exhibited in [Shen, 1991]. However, the explicit use of such games for the case of predicate transformers stemming from transitive closure operators was first introduced in [Caló, 1989]. It has been motivated by, but is distinctly different from, the games introduced in [Makowsky and Ziegler, 1980]. It should also be noted that our explicit definition of these games is more straightforward than their derivation in the framework of abstract model theory and generalized quantifiers as described in [Barwise and Feferman, 1985] and [Makowsky and Mundici, 1985]. In [de Rougemont, 1987] similarly motivated games are introduced for Fixed Point Logic which is, on ordered structures, of the same expressive power as **ATC**. However, the game introduced in [de Rougemont, 1987] is only shown to be sound, which suffices for the non-definability result discussed there. Recently Grädel used similar games to investigate the fine structure of **TC**, cf. [Grädel, to appear].

8.4 More non-definability results

We discuss here briefly the case of **TC**, the other cases being similar, but not trivially so. The games naturally induce a sequence of equivalence relations $\equiv_n^{TC}$ between structures which we call n-isomorphic for **TC**. In [Calò, 1989] Calò proves soundness and completeness of these equivalence relations in the sense that two structures are n-isomorphic for **TC** iff they satisfy the same formulas of quantifier depth n. As all these logics contain predicate transformers of arbitrary arity k, the logic **TC** can be viewed as a sequence of logics $\mathbf{TC}^k$ with the arity of the predicate transformer not exceeding k. $\mathbf{TC}^1$ is the logic with the transitive closure applied only to binary relations of 1-tuples. Our games also allow us to characterize definability in $\mathbf{TC}^k$.

In [Immerman, 1987] it is shown that a class of finite ordered structures is definable in these logics iff its recognizability is in certain complexity classes. In this sense we speak of logics capturing complexity classes and we have for ordered structures, by abuse of notation, **L** = **DTC**, **TC** = **NL** and **ATC** = **P**. **NP** was shown in [Fagin, 1974] to capture existential Second Order Logic. As an application of our work we state a necessary and sufficient condition for separating the complexity classes **L**, **NL**, **P** and **NP** respectively which is of pure model theoretic character. In the case of **NL** $\neq$ **NP** this condition can be stated as follows:

Let *HALFCLIQUE* be the set of ordered graphs which contain a clique of half its size. Let *HAM* be the set of ordered graphs which contain a Hamiltonian path. Note that *HALFCLIQUE* and *HAM* are **NP**-complete, cf. [Garey and Johnson, 1979].

Theorem 8.4.1. **NL** $\neq$ **NP** *iff there is a sequence of pairs of ordered graphs* G_n, H_n *such that*

(i) $G_n \equiv_n^{TC} H_n$ *and*

(ii) $G_n \notin HALFCLIQUE$ *but* $H_n \in HALFCLIQUE$.

The same holds for HALFCLIQUE replaced by HAM or any other **NP**– *complete problem.*

The construction of such families of graphs may be very hard and possibly requires probabilistic methods similar to the ones used in [Ajtai and Fagin, 1990]. The following result nevertheless sheds some light on the problem.

Theorem 8.4.2. *HALFCLIQUE and HAM are not definable in Monadic Second Order Logic (with arbitrary alternation of quantifiers) and hence not definable in* $\mathbf{DTC}^1, \mathbf{TC}^1, \mathbf{ATC}^1$.

Proof. We present here a new and quick but surprising proof of the above result, cf. [Makowsky, 1991]. Recall that Büchi's theorem states that a language L is regular (=recognizable by a finite automaton) iff the set L of words considered as first order structures with a linear order is definable

by a formula in Monadic Second Order Logic. Note that the proof of
Büchi's theorem in [Ladner, 1977] also uses Ehrenfeucht–Fraïssé games.
    The proof has several stages:

(i) First we note that the language $\{a^n b^n : n \in \mathbf{N}\}$ is not regular, cf.
    [Hopcroft and Ullman, 1980]. This is usually proved by the Pumping
    Lemma, which is in some sense an automata theoretic counterpart of
    the Ehrenfeucht–Fraïssé games.

(ii) Next we use Büchi's theorem and conclude that the set of words cor-
    responding to $a^n b^n$ is not definable in Monadic Second Order Logic.

(iii) Note that a complete bipartite graph is Hamiltonian iff both sets have
    the same cardinality.

(iv) Now assume, for contradiction, that HAM were definable by a τ-
    formula ϕ of Monadic Second Order Logic. Let $w \in \{a, b\}^*$ be a
    word. We define a binary relation E_w on w by $(i, j) \in E_w$ iff $i \in P_a$
    and $j \in P_b$. E_w makes w into a complete bipartite graph. E_w is
    definable by a first order formula θ over τ_{words}. Substituting R in ϕ
    by θ would give us a formula in Monadic Second Order Logic which
    assures that this graph is Hamiltonian, and hence would define the
    language $\{a^n b^n\}$, a contradiction.

(v) For $HALFCLIQUE$ we proceed similarly. We first note that the lan-
    guage $HALF(a)$ where at least half the letters are a's, is not regular,
    again by the Pumping Lemma. Hence $HALF(a)$ is not definable in
    Monadic Second Order Logic.

(vi) We now define E_w by $(i, j) \in E_w$ iff $i \in P_a$ or $j \in P_a$. E_w makes
    w into a graph where the a's form a clique. Using this in the above
    argument completes the proof.

Note however, that the specific sequence of pairs of ordered graphs G_n, H_n,
which we construct to show this, can be separated by a $\mathbf{TC^2}$-formula, i.e.
a formula with the predicate transformer TC of size 2.

Corollary 8.4.3.   *$HALFCLIQUE$ and HAM are not definable in*
$\mathbf{DTC^1}, \mathbf{TC^1}, \mathbf{ATC^1}$, *the First Order Logics augmented by the predicate*
transformers DTC, TC, ATC of size 1.

    This gives us a quick example for interesting families of pairs of ordered
finite structures which are n-isomorphic in $\mathbf{TC^1}$.

Corollary 8.4.4. *There are functions $f, g : \mathbf{N} \mapsto \mathbf{N}$ such that for ev-*
ery $n \in \mathbf{N}$, $f(n) \neq g(n)$ and the words $a^{f(n)} b^{f(n)}$ and $a^{f(n)} b^{g(n)}$ are n-
isomorphic (equivalent) in $\mathbf{TC^1}$.

In [de Rougemont, 1987] it is only proved that *HAM* is not definable in existential Monadic Second Order Logic.

8.5 The games and pumping lemmas

The various games described in this section are used to show non-definability results. Corollary 8.4.4 is a model theoretic version of the well known Pumping-Lemma for regular languages. Theorem 8.4.1 is a model theoretic version of a Pumping-Lemma for **L**. Similar Pumping-Lemmas can be formulated for other complexity classes **C**, provided **C** is captured by some logic, for which there are suitable games. It remains a challenging open problem, how to exploit this observation. Ressayre has taken a first step in this direction [Ressayre, 1988; Ressayre, to appear].

9 Conclusions

We, that is if the reader is still with me, have travelled through the landscape of Model Theory coming from the land of Theoretical Computer Science. There are many places we have not visited, and even where we did, we did not explore them enough. We have seen some of the land's history and cultural system and hinted at its connections with Database Theory and Logic Programming. We have explored its deeper connections to Computer Aided Geometry. But we have spent most of our time exploring the model theoretic aspects of the question $\mathbf{P} \neq \mathbf{NP}$ and its provability in formal systems of arithmetic.

The chapter is called an 'Appetizer'. It should tempt more than one reader to travel more. If your appetite has been whetted, let me invite you to a treat:

Castagnaccio: This is a popular sweet in Tuscany of which I have learnt during my own travels in the land of Model Theory (and Tuscany) from A. Marcja. We then worked together on the model theory of modal logic and wrote the paper [Makowsky and Marcja, 1977]. Some time ago you could still find street vendors in winter selling castagnaccio which was baked on an open fire. The ingredients are simple: chestnut flour, olive oil, sugar, raisins and pinenuts. The original recipe does not appear in Artusi's classical cooking book, but it does appear in the very enjoyable and nostalgic book by E. Servi-Machlin [Servi-Machlin, 1981]. Incidentally, she is the sister of the Italian logician M. Servi who did some work on the model theory of categories with finite products [Servi, 1971]. As it is difficult to find chestnut flour outside of Tuscany, let me give a modified version of the dish.

Take one can of canned, unsweetend chestnut purée. Add a third of the can of olive oil and half the can of sugar and mix in a food processor to a homogeneous paste. Add a third of the can of regular flour and keep mixing till smooth.

Spread the paste on a flat cooking sheet, not more than a small finger thick. It is advisable to rub the sheet with margarine or olive oil before spreading the paste. Now stick the pinenuts and raisins into the paste to your liking. Bake medium hot for about 40 minutes (till the paste hardens) but be careful not to burn it.

Let cool and break into pieces. Best served with coffee (espresso) and Grappa (di Brunello).

References

[Ajtai and Fagin, 1990] M. Ajtai and R. Fagin. Reachability is harder for directed than for undirected finite graphs. *Journal of Symbolic Logic*, 55(1):113–150, 1990.

[Andréka *et al.*, 1982] H. Andréka, I. Nemeti, and I. Sain. A complete logic for reasoning about programs via non-standard model theory, parts I and II. *Theoretical Computer Science*, 17:193–212 and 259–278, 1982.

[Baker, 1975] A. Baker. *Transcendental Number Theory*. Cambridge University Press, 1975.

[Baker, 1984] A. Baker. *A Concise Introduction to the Theory of Numbers*. Cambridge University Press, 1984.

[Barwise and Feferman, 1985] J. Barwise and S. Feferman, editors. *Model-Theoretic Logics*. Perspectives in Mathematical Logic. Springer Verlag, 1985.

[Baudisch *et al.*, 1985] A. Baudisch, D. Seese, P. Tuschik, and M. Weese. Decidability and quantifier-elimination. In *Model-Theoretic Logics*, chapter 7. Springer Verlag, 1985.

[Ben-David and Dvir, 1991] S. Ben-David and M. Dvir. Non-standard models for independent arithmetical statements. Technical report, Technion-Israel Institute of Technology, Haifa, Israel, 1991.

[Ben-David and Halevi, 1991] S. Ben-David and S. Halevi. On the independence of P versus NP. Technical report, Technion-Israel Institute of Technology, Haifa, Israel, 1991.

[Büchi, 1960] J.R. Büchi. Weak second-order arithmetic and finite automata. *Zeitschrift für mathematische Logik und Grundlagen der Mathematik*, 6:66–92, 1960.

[Bull and Segerberg, 1984] R.A. Bull and K. Segerberg. Basic modal logic. In D. Gabbay and F. Günthner, editors, *Handbook of Philosophical Logic, Volume 2*, chapter 1. D. Reidel Publishing Company, 1984.

[Burgess, 1984] J. Burgess. Basic tense logic. In D. Gabbay and F. Günthner, editors, *Handbook of Philosophical Logic, Volume 2*, chapter 2. D. Reidel Publishing Company, 1984.

[Calò and Makowsky, to appear 1992] A. Calò and J.A. Makowsky. The Ehrenfeucht–Fraïssé games for transitive closure. In *Proceedings of the Symposium on Logical Foundations of Computer Science, LFCS '92: Logic at Tver*, LNCS, to appear, 1992.

[Calò, 1989] A. Calò. The expressive power of transitive closure. Master's thesis, Faculty of Computer Science, Technion–Israel Institute of Technology, Haifa, Israel, 1989.

[Chandra and Harel, 1980] A. K. Chandra and D. Harel. Computable queries for relational data bases. *Journal of Computer and System Sciences*, 21(2):156–178, October 1980.

[Chang and Keisler, 1990] C.C. Chang and H.J. Keisler. *Model Theory*. Studies in Logic, vol 73. North-Holland, 3 edition, 1990.

[Chou, 1988] Shang-Ching Chou. *Mechanical Geometry Theorem Proving*. Mathematics and its Applications. D. Reidel Publishing Company, 1988.

[Compton, 1987] K.J. Compton. A logical approach to asymptotic combinatorics I: First-order properties. *Advances in Mathematics*, 65:65–96, 1987.

[Courcelle, 1990a] B. Courcelle. Graph rewriting: An algebraic approach. In J. van Leeuwen, editor, *Handbook of Theoretical Computer Science, Volume 2*, chapter 5. Elsevier Science Publishers, 1990.

[Courcelle, 1990b] B. Courcelle. The monadic second-order theory of graphs I: Recognizable sets of finite graphs. *Information and Computation*, 85:12–75, 1990.

[Dahlhaus, 1982] E. Dahlhaus. *Combinatorial and Logical Properties of Reductions to some Complete Problems in NP and NL*. PhD thesis, Technische Universität Berlin, Germany, 1982.

[Dahlhaus, 1983] E. Dahlhaus. Reductions to NP-complete problems by interpretations. In E. Börger *et al.*, editor, *Logic and Machines: Decision Problems and Complexity*, Lecture Notes in Computer Science, 171. Springer Verlag, 1983.

[de Rougemont, 1987] M. de Rougemont. Second-order and inductive definability on finite structures. *Zeitschrift für mathematische Logik und Grundlagen der Mathematik*, 33:47–63, 1987.

[Ebbinghaus, 1985] H.D. Ebbinghaus. Extended logics: The general framework. In *Model-Theoretic Logics*, Perspectives in Mathematical Logic, chapter 2. Springer Verlag, 1985.

[Emerson, 1990] E.A. Emerson. Temporal and modal logic. In J. van Leeuwen, editor, *Handbook of Theoretical Computer Science, Volume 2*, chapter 16. Elsevier Science Publishers, 1990.

[Engeler, 1981] E. Engeler. Generalized Galois theory and its application to complexity. *Theoretical Computer Science*, 13:271–293, 1981.

[Fagin, 1974] R. Fagin. Generalized first-order spectra and polynomial time recognizable sets. In R. Karp, editor, *Complexity of Computation*, American Mathematical Society Proc 7, pages 27–41. Society for Industrial and Applied Mathematics, 1974.

[Fagin, 1982] R. Fagin. Horn clauses and database dependencies. *Journal of ACM*, 29(4):952–985, 1982.

[Fagin, 1990] R. Fagin. Finite model theory — a personal perspective. In S. Abiteboul and P.C. Kannelakis, editors, *ICDT'90*, Lecture Notes in Computer Science, 470, pages 3–24. Springer Verlag, 1990.

[Ferrante and Rackoff, 1979] J. Ferrante and C.W. Rackoff. *The Computational Complexity of Logical Theories*. Lecture Notes in Mathematics. Springer Verlag, 1979.

[Flum, 1991] J. Flum. On bounded theories. to appear in CSL'91, 1991.

[Fortune *et al.*, 1983] S. Fortune, D. Leivant, and M. O'Donnell. The expressiveness of simple and second-order type structures. *Journal of ACM*, 30(1):151–185, 1983.

[Gabbay *et al.*, 1980] D. Gabbay, A. Pnueli, S. Shelah, and J. Stavi. On temporal analysis of fairness. In *Proceedings of the 7th ACM-POPL*, pages 163–173, 1980.

[Gabbay *et al.*, 1992] D. M. Gabbay, C. J. Hogger, and J. A. Robinson, editors. *Handbook of Logic in Artificial Intelligence and Logic Programming*, volume 1–6. Oxford University Press, 1992.

[Gaifman, 1982] H. Gaifman. On local and non-local properties. In J. Stern, editor, *Logic Colloquium '81*, pages 105–135. North-Holland publishing Company, 1982.

[Gallier, 1991] J.H. Gallier. What is so special about Kruskal's theorem and the ordinal γ_0 ? A survey of some results in proof theory. *Annals of Pure and Applied Logic*, 53:199–260, 1991.

[Garey and Johnson, 1979] M.G. Garey and D.S. Johnson, editors. *Computers and Intractability*. W. H. Freeman and Company, 1979.

[Gergely and Úry, 1991] T. Gergely and L. Úry. First-order programming theories. *EATCS Monographs on Theoretical Computer Science*, 24, 1991.

[Goguen et al., 1977] J. Goguen, J.W. Thatcher, E.G. Wagner, and J.B. Wright. Initial algebra semantics and continuous algebras. *Journal of ACM*, 24:68–95, 1977.

[Grädel, to appear] E. Grädel. On transitive closure logic. In *Proceedings of the 5th workshop in computer science logic*. Springer Verlag, to appear. CSL '91, Bern 1991.

[Grandjean, 1983] E. Grandjean. Complexity of the first-order theory of almost all structures. *Information and Control*, 52:180–204, 1983.

[Gurevich and Shelah, 1987] Y. Gurevich and S. Shelah. Expected computation time for Hamiltonian path problem. *SIAM Journal for Computing*, 16(3):486–502, 1987.

[Gurevich, 1990] Y. Gurevich. On finite model theory. In S.R. Buss and P.J. Scott, editors, *Perspectives in Computer Science*. Birkhäuser Verlag, 1990.

[Gurevich, 1991] Y. Gurevich. Average case completeness. *Journal of Computer and System Sciences*, 42:346–398, 1991.

[Harel, 1984] D. Harel. Dynamic logic. In D. Gabbay and F. Günthner, editors, *Handbook of Philosophical Logic, Volume 2*, chapter 10. D. Reidel Publishing Company, 1984.

[Harrington et al., 1985] L.A. Harrington, M.D. Morley, A. Ščedrov, and S.G. Simpson. *Harvey Friedman's Research in the Foundations of Mathematics*, volume 117 of *Studies in Logic and the Foundations of Mathematics*. North-Holland, 1985.

[Heintz, 1983] J. Heintz. Definability and fast quantifier elimination in algebraically closed fields. *Theoretical Computer Science*, 24:239–277, 1983.

[Henkin, 1977] L. Henkin. The logic of equality. *American Mathematical Monthly*, 84:597–612, 1977.

[Hodges, 1992] W. Hodges. Logical features of horn clauses. In D. Gabbay et al., editor, *Handbook of Logic in Artificial Intelligence and Logic Programming*, volume 1. Oxford University Press, 1992.

[Hopcroft and Ullman, 1980] J. E. Hopcroft and J. D. Ullman. *Introduction to Automata Theory, Languages and Computation*. Series in Computer Science. Addison-Wesley, 1980.

[Ierardi, 1989] D. Ierardi. Quantifier elimination in the theory of an algebraically closed field. In *STOCS'89*, pages 138–147. ACM, 1989.

[Immerman, 1987] N. Immerman. Languages that capture complexity classes. *SIAM Journal on Computing*, 16(4):760–778, August 1987. Also appeared as a preliminary report in Proceedings of the 15th Annual ACM Symposium on the Theory of Computing.

[Immerman, 1988] N. Immerman. Nondeterministic space is closed under complementation. *SIAM Journal of Computing*, 17(5):935 – 938, 1988.

[Kanamori and McAloon, 1987] A. Kanamori and K. McAloon. On Gödel incompleteness and finite combinatorics. *Annals of Pure and Applied Logic*, 33:23–41, 1987.

[Kannelakis, 1990] P.C. Kannelakis. Elements of relational database theory. In J. van Leeuwen, editor, *Handbook of Theoretical Computer Science, Volume 2*, chapter 17. Elsevier Science Publishers, 1990.

[Kirby and Paris, 1982] L. Kirby and J. Paris. Accessible independence results for Peano arithmetic. *Bulletin of the London Mathematical Society*, 14:285–293, 1982.

[Kochen and Kripke, 1982] S. Kochen and S. Kripke. Non-standard models of Peano arithmetic. In V. Strassen, E. Engeler, and H. Läuchli, editors, *Logic and Algorithmic, An international Symposium held in honour of E. Specker*, pages 275–296. L'enseignement mathématique, 1982.

[Kolaitis, 1990] P. G. Kolaitis. Implicit definability on finite structures and unambiguous computations. In *FOCS'90*, pages 168–180. IEEE, 1990.

[Kreisel and Krivine, 1967] G. Kreisel and J.L. Krivine. *Elements of Mathematical Logic: Model Theory*. North-Holland, 1967.

[Kreisel, 1952] G. Kreisel. On the concepts of completeness and interpretation of formal systems. *Fundamenta Mathematicae*, 39:103–127, 1952.

[Ladner, 1977] R.E. Ladner. Application of model theoretic games to discrete linera orders and finite automata. *Information and Control*, 33:281–303, 1977.

[Lynch, 1982] J.F. Lynch. Complexity classes and theories of finite models. *Mathematical Systems Theory*, 15:127–144, 1982.

[Mahr and Makowsky, 1984] B. Mahr and J.A. Makowsky. Characterizing specification languages which admit initial semantics. *Theoretical Computer Science*, 31:49–60, 1984.

[Maibaum, to appear] T. Maibaum. Axiomatizing specification theory. In *Handbook of Logic in Computer Science, Volume 5*. Oxford University Press, to appear.

[Makowsky and Marcja, 1977] J.A. Makowsky and A. Marcja. Completeness theorems for modal model theory with the Montague–Chang semantics, I. *Zeitschrift für mathematische Logik und Grundlagen der Mathematik*, 23:97–104, 1977.

[Makowsky and Mundici, 1985] J.A. Makowsky and D. Mundici. *Abstract equivalence relations*, chapter 19. Perspectives in Mathematical Logic. Springer Verlag, 1985.

[Makowsky and Sain, 1989] J. A. Makowsky and I. Sain. Weak second order characterizations of various program verification systems. *Theoretical Computer Science*, 66:299–321, 1989.

[Makowsky and Vardi, 1986] J.A. Makowsky and M. Vardi. On the expressive power of data dependencies. *Acta Informatica*, 23(3):231–244, 1986.

[Makowsky and Ziegler, 1980] J. A. Makowsky and M. Ziegler. Topological model theory with an interior operator: Consistency properties and back and forth arguments. *Archiv für mathematische Logik und Grundlagenforschung*, 20:37–54, 1980.

[Makowsky, 1981] J. A. Makowsky. *Characterizing database dependencies*, volume 115 of *Lecture Notes in Computer Science*, pages 86–97. Springer Verlag, 1981.

[Makowsky, 1984] J. A. Makowsky. Model theoretic issues in theoretical computer science, part I: Relational databases and abstract data types. In G. Lolli al., editor, *Logic Colloquium '82, Studies in Logic*, pages 303–343. North-Holland, 1984.

[Makowsky, 1985] J.A. Makowsky. *Compactness, embeddings and definability*, chapter 18. Perspectives in Mathematical Logic. Springer Verlag, 1985.

[Makowsky, 1987] J. A. Makowsky. Why Horn formulas matter for computer science: Initial structures and generic examples. *Why Horn formulas matter for computer science: Initial structures and generic examples*, 34(2/3):266–292, 1987.

[Makowsky, 1988] J.A. Makowsky. Mental images and the architecture of concepts. In R. Herken, editor, *The Universal Turing Machine. A Half–Century Survey*. Oxford University Press, 1988.

[Makowsky, 1991] J.A. Makowsky. The impact of model theory on theoretical computer science. In *Proceedings of the 9th International Congress of Logic, Methodology and Philosophy of Science*, 1991. to appear.

[Makowsky, to appear] J.A. Makowsky. Database theory. In *Handbook of Logic in Computer Science, Volume 5*. Oxford University Press, to appear.

[Matijasevič, 1970] Ju. V. Matijasevič. Enumerable sets are diophantine. *Sov. Math. Dokl.*, 11:354–357, 1970.

[Moschovakis, 1974] Y. Moschovakis. *Elementary Induction on Abstract Structures*, volume 77. North-Holland, 1974.

[Pasztor, 1990] A. Pasztor. Recursive programs and denotational semantics in absolut logics of programs. *Theoretical Computer Science*, 70:127–150, 1990.

[Ressayre, 1988] J.P. Ressayre. Formal languages defined by the underlying structure of their words. *Journal of Symbolic Logic*, 53(4):1009–1026, 1988.

[Ressayre, to appear] J.P. Ressayre. Formal languages defined by the underlying structure of their words, part II. *Journal of Symbolic Logic*, to appear.

[Risch, 1969] R.H. Risch. The problem of integration in finite terms. *Transactions of the American Mathematical Society*, 139:167–189, 1969.

[Rosenlicht, 1972] M. Rosenlicht. Integration in finite terms. *American Mathematical Monthly*, 79(9):963–972, 1972.

[Rosenstein, 1982] J.G. Rosenstein. *Linear Orderings*. Academic Press, 1982.

[Ryan and Sadler, 1992] M. Ryan and M. Sadler. Valuation systems and consequence relations. In D. M. Gabbay *et al.*, editor, *Handbook of Logic in Computer Science, Volume 1*. Oxford University Press, 1992.

[Schwartz et al., 1987] J.T. Schwartz, M. Sharir, and J. Hopcroft, editors. *Planning, Geometry, and Complexity of Robot Motion.* Ablex Series in Artificial Intelligence. Ablex Publishing Corporation, 1987.

[Servi-Machlin, 1981] E. Servi-Machlin. *The Classical Cuisine of the Italian Jews.* Everest House, 1981.

[Servi, 1971] M. Servi. Una questione die teoria dei modelli nelle categorie con prodotti finiti. *Matematiche*, 26:307–324, 1971.

[Shelah, 1990] S. Shelah. *Classification Theory and the number of non-isomorphic models.* Studies in Logic and the Foundations of Mathematics. North-Holland, 2 edition, 1990.

[Shen, 1991] Enshao Shen. Manuscript. unpublished, Department of Mathematics, University of Freiburg, Germany, 1991.

[Shepherdson, 1984] J.C. Shepherdson. Negation as failure: A comparison of clark's completed data base and Reiter's closed world assumption. *Journal of Logic Programming*, 1:51–81, 1984.

[Shepherdson, 1985] J.C. Shepherdson. Negation as failure II. *Journal of Logic Programming*, 3:185–202, 1985.

[Shepherdson, 1988] J.C. Shepherdson. Negation in logic programming. In J. Minker, editor, *Foundations of deductive data bases and logic programming*, chapter 1. Morgan Kaufmann Publishers, 1988.

[Shoenfield, 1967] J. Shoenfield. *Mathematical Logic.* Series in Logic. Addison-Wesley, 1967.

[Smorynski, 1980] C. Smorynski. Some rapidly growing functions. *Mathematical Intelligencer*, 2:149–154, 1980.

[Smorynski, 1983] C. Smorynski. Big news from Archimedes to Friedman. *Notices of the AMS*, 30:251–256, 1983.

[Steinitz, 1910] E. Steinitz. Algebraische Theorie der Körper. *Journal für reine und angewandte Mathematik*, 137:167–309, 1910.

[Stewart, 1991a] I. A. Stewart. Comparing the expressibility of languages formed using np complete operators. *Journal of Logic and Computation*, 1(3):305–330, 1991.

[Stewart, 1991b] I.A. Stewart. On completeness for NP via projection translations. In *CSL'91*, 1991. to appear.

[Topor and Sonenberg, 1988] R.W. Topor and E.A. Sonenberg. On domain independent databases. In J. Minker, editor, *Foundations of deductive data bases and logic programming*, chapter 6. Morgan Kaufmann Publishers, 1988.

[Trakhtenbrok, 1950] B. Trakhtenbrok. On the impossibility of an algorithm for the decision problem in finite domains. *Dok. Akad. Nank. SSSR*, 70:569–572, 1950. in Russian.

[Trakhtenbrot, 1961] B. Trakhtenbrot. Finite automata and the logic of monadic predicates. *Dokl. Akad. Nank. USSR*, 140:326–329, 1961. In Russian.

[Ullman., 1982] J.D. Ullman. *Principles of Database Systems*. Principles of Computer Science Series. Computer Science Press, 2 edition, 1982.

[Vardi, 1988] M. Vardi. Fundamentals of dependency theory. In E. Börger, editor, *Trends in Theoretical Computer Science*, chapter 5. Computer Science Press, 1988.

[Wainer, 1970] S.S. Wainer. A classification of ordinal recursive functions. *Archiv für Mathematische Logik und Grundlagen der Mathematik*, 13:136–153, 1970.

[Wilder, 1981] R.L. Wilder. *Mathematics as a Cultural System*. Pergamon Press, 1981.

[Ziegler, 1982] M. Ziegler. Einige unentscheidbare Körpertheorien. In V. Strassen, E. Engeler, and H. Läuchli, editors, *Logic and Algorithmic, An international Symposium held in honour of E. Specker*, pages 381–392. L'enseignement mathématique, 1982.

Author index

Index